HYDROCARBON LIQUID TRANSMISSION PIPELINE AND STORAGE SYSTEMS – DESIGN AND OPERATION

M. Mohitpour
M.S. Yoon
J.H. Russell

ASME PRESS

Library of Congress Cataloging-in-Publication Data

Mohitpour, Mo
Hydrocarbon liquid transmission pipeline and storage systems : design and operation / M. Mohitpour, M.S. Yoon, J.H. Russell.
p. cm.
Includes bibliographical references and index.
ISBN 978-0-7918-6000-7 (alk. paper)
1. Petroleum pipelines–Design and construction. 2. Liquefied natural gas pipelines–Design and construction. 3. Pipelines–Design and construction. I. Yoon, Mike II. Russell, J. H. (James Hooper), 1947- III. Title.
TN879.53.M64 2012
665.5'44–dc23

2012016731

IN MEMORIAM

This book is dedicated to its primary author, Dr. Mo Mohitpour, who passed away on August 21, 2012. Mo devoted much of his career to the development of the education of aspiring engineers in the field of pipeline engineering. As well as providing technical direction to many pipeline corporations worldwide, Mo dedicated his time to education. In his later years, Mo devoted many hours to continuing education courses at the University of Calgary and throughout the world. He provided leadership in the development of six highly regarded books published through ASME Press on pipeline design and operation. Mo will always be remembered for his dedication to the profession and to his encouragement of others.

Mo was educated at the University of London where he achieved a Ph.D. in Mechanical Engineering. He maintained language skills in English, Spanish, Portuguese and his native Farsi. Mo was a Fellow of the Institute of Mechanical Engineers (UK), Fellow of the Engineering Institute of Canada, Fellow of the American Institute of Mechanical Engineers as well as a PE in Texas and a P.Eng. In Alberta and British Columbia, Canada. His credentials transcended boundaries.

In addition to being a consummate professional, Mo was known for being jovial and welcoming to all with whom he developed relationships. Mo will truly be missed for his dedication to educating the next generation of professionals in the field of pipeline engineering. His colleagues and many friends will remember him forever.

Again, in his most gracious way, Mo provided the dedications on the following page before he passed away.

DEDICATIONS

This book is dedicated to the Pipeline Industry whose prime objective has been, and continues to be, transporting hydrocarbon energy efficiently but with utmost safety and reliability. It is dedicated to the experts and professionals in the industry whose breadth of expertise and continued effort has led to advancements that have been the cornerstone of integrity and safety in pipeline energy transportation.

To our wives Carol, Julie, and Fern whose patience, unyielding support, and love allowed us to complete this book. They made us to wonder in our dreams to achieve our aspiration. THANK YOU.

TABLE OF CONTENTS

PREFACE

This book is a sequel to and augments the series of ASME-initiated pipeline books and monograms, documents published since year 2000. The following include a partial list of such publications:

- "Pipeline Design and Construction — A Practical Approach," 3rd Edition 2007, M. Mohitpour, H. Golshan and A. Murray
- "Pipeline Operation and Maintenance — A Practical Approach," 2nd Edition 2010, M. Mohitpour, T. Van Hardeveld, B. Peterson and J. Szabo
- "Energy Supply and Pipeline Transportation — Challenges and Opportunities," 2008, M. Mohitpour
- "Pipeline Pumping and Compression Systems — A Practical Approach," 2008, M. Mohitpour, K.K. Botros and T. Van Hardeveld
- "Pipeline Integrity Assurance — A Practical Approach," 2010, M. Mohitpour, A. Murray, I. Colquhoun and M. McManus
- "Pipeline Transportation of Carbon Dioxide containing Impurities," M. Mohitpour, P. Seevam, K.K. Botros, B. Rothwell and C. Ennis, 2011
- "Pipeline System Automation and Control," M. Yoon, C. Warren and S. Adam, 2007
- "Pipeline Geo-Environmental Design and Geohazard Management", edited by M. Rizkalla, 2008
- "Pipeline Geomatics," edited by S. Adam and K. Davis, 2009

This book brings together the entire spectrum of hydraulics, design, and operating requirements for pipeline transportation and storage of hydrocarbon liquids, the essence of our energy supply. It is a professional reference, training tool, or comprehensive text for specialized university courses. The contents cover a broad range of subjects essential in knowing the elements making up hydrocarbon liquid pipeline and storage systems and how to most reliably design and operate such facilities with the least environmental impact and energy transportation disruption.

Chapters of the book have been written based on the collective experience of the authors and research of appertaining published materials available from the pipeline industry journals and documents published by individual professionals, experts, operators, educators, and scientific research works. Each chapter has been written with the intent that it would stand alone as far as possible without referencing other chapters. In this way, professionals can source their search topic of interest more conveniently without recourse to other parts of the book. However, where appropriate, referencing has been alluded to.

In this book, mostly metric units have been used. However in some chapters both imperial and metric units are referred to. This was justified because the industry continues to use the unit systems interchangeably. A conversion table is provided within this section.

The authors have exercised care to ensure correctness of the content, acknowledgement of other publications, copyright permissions, and referencing documents and names. It is not intended that specific techniques, examples, or applications be applied or copied for turnkey use. Readers are very much encouraged to check and assess all details before use and application. The authors and ASME welcome notification of corrections, omissions, and attributions. These will be attended to in the next edition and contributors acknowledged as such.

Mo Mohitpour, White Rock, British Columbia, Canada
Mike S. Yoon, Calgary, Alberta, Canada
Jim H. Russell, Edmonton, Alberta, Canada
2012

ACKNOWLEDGMENTS

Writing this book was a tremendous enjoyment for the authors. The concept of writing this book rekindled the authors' long time association of over 30 years (going back to 1982–1983 with Canuck Engineering, in Calgary, Alberta, Canada) and, most of all, reconnected us with leaders and colleagues throughout the pipeline industry. The encouragement and enthusiasm from the industry and the continued support of our publisher; ASME Press was indeed the backbone of our interest and commitment to bring to fruition this document.

The authors' connection with the pipeline industry goes as far back as mid 1970s when as junior engineers entering the industry we were mentored by professionals in the industry, gained knowledge by hands-on work, field assignments, and direct involvement with leading edge pipeline technology projects globally. For example, while the use of X100 pipe is development of the past decade, its deployment and application goes as far back as 1975 when API X110 was contemplated for transportation of large quantities (6650 MMSCFD) of high-pressure dense phase natural gas to 3500 psig. Exxon-Mobil and Nippon Steel Corporation developed API X120 and now-a-days the use of high strength steel (API X80 and over) is common place in our industry (for both oil and gas pipeline use).

It is the unyielding support of our mentors and associates in the industry that led us to initiate and contribute to the series of pipeline books published by ASME Press since 2000.

Thanks are due to our colleagues in the industry who, in many ways, encouraged and contributed to the review of preparatory manuscripts, updating, corrections, additions, and the supply of materials for this book. The authors wish to express sincere thanks to and acknowledge the valuable contribution of all of the following colleagues for their intensive reviews of various chapters, verification of content, and suggestions;

Dr. Alan Murray, P.Eng, Principia Consultant
John Kazakoff, P.Eng, Silver Fox Engineering Consultants
Hal Oliver, PE
John A. Jacobson PE, CB&I, Texas
Ed Seiders, PE, Willbros Engineering, Tulsa
Mike McManus, P.Eng, Enbridge
Andres Mendizbal, President, OCP Ecuador
Wagner Carrera, Operation Supervisor, OCP Ecuador
Dr. John M. Shaw, P.Eng University of Alberta
Ms. Nafiseh Dodgostar, University of Alberta
Jakob Buchert, Sr. Engineer, Energy Solutions International
Dick Spiers, Sr. Consultant, Energy Solutions International
Scott Lauchlan, Solution Team Leader, Telvent North America
Dave Jardine, President and Chairman, Telvent North America
Mike Doxey, Executive Vice President, HMT Inc
Jim Enarson, Consultant
Mike Fillipof, Actenum Corp

Special thanks are due to our longtime colleague Dr. Alan Murray for his contributory reviews.

We also acknowledge the leadership, management, and continued unyielding support of the ASME Press and staff who have made this book possible. Acknowledgments are due to Mr. Phillip DiVietro, Ms. Mary Grace Stefanchik, and Ms. Tara Collins Smith for their tremendous encouragement, dedication, commitments, and, above all, their patience toward the completion of this book. And where would we be without ASME's marketing efforts? Thank you so much Annette Missouri and John Yelavich, and the ASME Press staff for all your support over the years and for the future sale of this book. THANK YOU.

Mo Mohitpour, Ph.D., PE, P.Eng, F. I. Mech. Eng., FASME, FEIC
mo.mohitpour@shaw.ca
Mike Yoon, Ph.D., MASME
mike.yoon@hotmail.com
Jim Russell, P.Eng, MASME
j_frussell@shaw.ca

2012

ACCREDITATION

American Society of Mechanical Engineers and the authors would like to hereby accredit all organizations and individuals for the use and/or granting their kind permission to reprint or reproduce illustrations, photos, and other materials in this book. Where contacts were not possible, such organization or individuals are referenced and accredited in each chapter, as appropriate, and are included herein.

American Petroleum Institute (API)
ASME (American Society of Mechanical Engineers) 2008
Buckeye Equipment
CEPA (Canadian Energy Pipeline Association)
CB&I
Corrpro Canada
Daniel Meters
Enbridge
Endress+ Hauser
Energy Solutions International
Flowserve
FMC Technologies
GPSA
HMT Inc. www.hmttank.com
Hydraulics Institute
ITA
Kobe Steel Ltd. (KOBELCO)
Micro Motion
Pembina Pipeline Corporation
OCP Ecuador, Ecuador
OTEC, Singapore
Resource Protection International, Dr. Niall Ramsden
Smith Meters
Telvent North America, a Schneider Electric company
TransCanada Corporation

FOREWORDS

Foreword From TransCanada

The use of pipe for petroleum transportation was conceptualized in 1863 by Dmitri Mendelev. The development of hydrocarbon liquid pipeline transportation over long distances goes back to the late1800s when oil was exploited in large quantities first in the USA and in the Persian Gulf area and Baku, Azerbaijan. Pipelines have since proven to be the safest, the most reliable and economical means for transporting such oil and petroleum products from sources of supply to market areas.

The series of pipeline books commenced with the authors' training courses delivered for the industry and those held at the University of Calgary, Alberta, Canada, since mid-1980s.

We at TransCanada supported the authors and ASME Press with the publication of the series of books in pipeline development that commenced in 2000 with the publication of "Pipeline Design and Construction — A Practical Approach," Mohitpour, Golshan, and Murray. The book is now in its 3rd edition and the authors and colleagues have now seven such publications by ASME Press, New York.

This book "Hydrocarbon Liquid Transmission Pipeline and Storage Systems — Design and Operation" is a culmination of the series. This book is a comprehensive resource that marks a significant contribution for the pipeline industry. The book brings together the entire spectrum of liquid pipeline transportation including pumping, storage, measurement, automation, design, and operation from supply to delivery points.

TransCanada is very pleased to provide our support for the series of pipeline books published by ASME Press. This is a significant achievement by the authors to bring together the knowledge and expertise and condenses this important information in a single reference guide.

Andrew Jenkins, P.Eng
Vice President TransCanada

TransCanada
In business to deliver

Foreword From Enbridge Inc.

The business of transporting a wide variety of hydrocarbon liquids by pipeline is multifaceted, technically intensive and continually evolving. Liquid pipelines are absolutely vital to the support and growth of our society, and they represent the safest and most efficient means of transport, particularly over long distances, for crude oils, refined products, and a host of other energy commodities.

To undertake the writing of a comprehensive book on liquid pipelines is a monumental task. ***Hydrocarbon Liquid Pipelines and Storage Systems — Design and Operation*** is equal to that task. The authors have explained the history and purpose of pipelines; the origin and characteristics of the fluids they transport; the technical design philosophy and features of pipelines; the purpose and design of ancillary tanks and equipment; pipeline operations; batching and safety considerations, including the detection of leaks from the system. This book will well serve the need for a single source of learning for new entrants to the business and industry veterans alike, and I can envision the book becoming a vital teaching tool in pipeline company engineering departments, University programs, pipeline regulators' offices and anywhere that a deeper understanding of how pipelines really work is sought.

It has been my privilege to have known and worked with Messrs. Mohitpour, Yoon and Russell. They represent many decades of theoretical and practical experience in the pipeline industry and have condensed their broad and deep knowledge into a logically organized book. It is my honor to recommend it to you.

Stephen J. Wuori
President, Liquids Pipelines
Enbridge Inc.

ENBRIDGE

Foreword From Willbros Engineering

I feel honored to have been asked to prepare the foreword for "Hydrocarbon Liquid Transmission Pipeline and Storage Systems — Design and Operation." This is the latest in a noteworthy series of technical books which have added immensely to the available literature on what is perhaps the most important bulk transportation technology in the world today. Indeed, today's pipelines move almost anything imaginable from crude oils to refined products, from natural gas to carbon dioxide, from coal slurry to drinking water in a safe, efficient, reliable, and quiet way.

In 2000, when this series was originally begun, there was little organized information available for the practicing pipeline professional, and what was available was mostly out dated and in sore need of bringing up to speed with today's needs. I, like most other pipeline engineers, had amassed a collection of technical papers, vendors' catalogs, engineering handbooks, magazine articles and such that I had found to contain the information that was needed in my work. Shortly after the ASME Pipeline Systems Division was organized in 2000, we began hearing from academic institutions that wanted to include some aspect of pipeline engineering in their curricula and asking where they could find suitable texts. The answer at the time was that there were very few.

"Pipeline Design and Construction — a Practical Approach" by Mohitpour, Golshan and Murray was first published in 2000, with a second, updated edition published in 2003, followed by a third edition in 2009. It is a true pipeline engineer's text, which includes not only detailed technical explanations for the theories and equations that are needed to design safe, efficient and reliable pipeline systems, but also has many practical examples for analyzing, planning and constructing those systems. It has undoubtedly resulted in better pipeline transportation systems around the world. This text has been followed on a regular basis by others, each of which delves into more detail on specialized aspects of pipeline technology: pipeline construction and maintenance, pipeline integrity management, pumping and compression systems, pipeline transportation of carbon dioxide, pipeline automation and control, and pipeline geomatics. The result is that today we have a well-documented, well-indexed collection of reference materials for the pipeline specialist, of which "Hydrocarbon Liquid Transmission Pipeline and Storage Systems — Design and Operation" is an invaluable addition.

Ed Seiders
Senior Technical Advisor
WILLBROS Engineering
Tulsa, Oklahoma

METRIC CONVERSION OF SOME COMMON UNITS

To Convert From Customary Unit	To Define Unit As	Symbol	Multiply By
barrel per hour	liters per second	L/s	0.044 163
barrel per day	cubic meters per day	m^3/d	0.158 987
MMBOD	cubic meters per day	m^3/d	0.158 987 106
Btu/second	kilowatt	kW	1.055 056
Btu/hour	watt	W	0.293 071
Btu/lbm	kilojoule per kilogram	kJ/kg	2.326
Btu/lbm-°F	kilojoule per kilogram-Kelvin	kJ (kg K)	4.1868
Btu/lbm-mole-°R	joule per mole-Kelvin	J/(mol K)	4.1868
Btu/°R	kilojoule per Kelvin	kJ/K	1.8991
Btu/ft^2-hr.	joule per sq. meter-second	J/(m^2 s)	3.154 591
Btu/ft-hr-°F	joule per meter-second-Kelvin	J/(m/s K)	1.730 735
Btu/ft^2-hr-°F	joule per square meter-second Kelvin	J/(m^2 s K)	5.678 263
Centipoise	milliPascal-second	cP	1
Centistoke	square millimeter per second	cSt	1
Foot	meter	m	0.3048
foot-pound force (ft. lbf)	joule	J	1.355 818
$foot^2$	square meter	m^2	0.092 903
$foot^3$	cubic meter	m^3	0.028 316 85
$foot^3$/minute	liter per second	L/s	0.471 947
$foot^3$/hour	cubic meter per day	m^3/d	0.679 604
MMSCFD	cubic meter per second	m^3/s	0.327 774
gallon/minute (GPM)	liter per second	L/s	0.063 090

$inch^2$	square centimeter	cm^2	6.451 600
$inch^3$	cubic centimeter	cm^3	16.387 064
kilowatt-hour (kWh)	megajoule	MJ	3.6
mile per hour	kilometer per hour	km/h	1.609 344
pound	kilogram	Kg	0.453 592 37
pound force	newton	N	4.448
pound mass/$foot^3$ (lbm/ft^3)	kilogram per cubic meter	kg/m^3	16.018 463
pound mass/gallon	kilogram per liter	kg/L	0.119 826
pound mass/hour	kilogram per hour	kg/h	0.453 592
psi	kiloPascal	kPa	6.894 757
psi/foot	kiloPascal per meter	kPa/m	22.620 59
psi/mile	Pascal per meter	Pa/m	4.284 203
Watt-hour	kilojoule	kJ	3.6
$yard^2$	square meter	m^2	0.836 127
$yard^3$	cubic meter	m^3	0.764 555
acre	square meter	m^2	4046.856
atmosphere (std)	kilopascal	kPa	101.325
barrel (42 US gal)	cubic meter	m^3	0.158 987
Btu (International Table)	kilojoule	kJ	1.055 056
calorie (Thermochemical)	joule	J	4.184
degree F	degree Celsius	°C	5/9 × (°F-32)
degree R	degree Kelvin	K	5/9
foot	meter	m	0.3048
gallon (US liquid)	liter	L	3.785 412
horsepower (US)	kilowatt	kW	0.7457
inch (US)	millimeter	mm	25.4
inch of mercury (60°F)	kilopascal	kPa	3.376 85
inch of water (60°F)	kilopascal	kPa	0.248 843
mil	micrometer	μm	25.4
mile (US Statute)	kilometer	km	1.609 344
ounce (US fluid)	milliliter	mL	29.573 53
poise	Pascal-second	Pa.s	0.1
SSU	Saybolt-Universal-Seconds	cSt	See graph on next page
stokes	square centimeter per second	cm^2/s	1
ton, long (2240 lbm)	ton	t	1.016 047
ton, short (2000 lbm)	ton	t	0.907 184 74

ton of refrigeration	kilowatt	kW	3.516 853
yard (US)	meter	m	0.9144

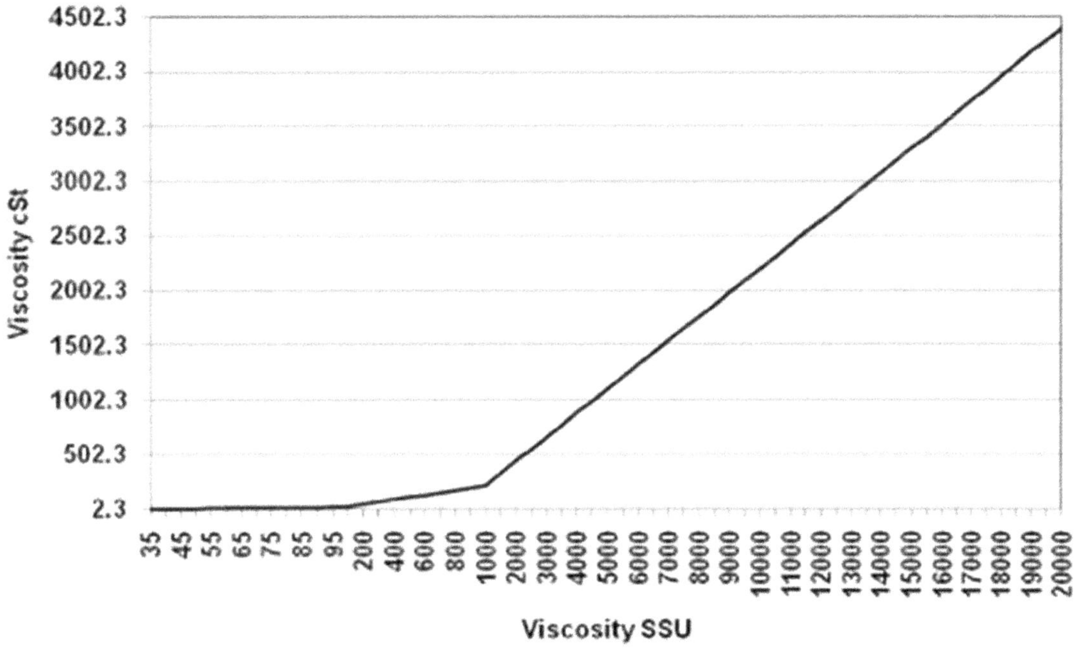

Relationship between SSU and cSt

Chapter 1

INTRODUCTION TO HYDROCARBON LIQUID PIPELINES

1.1 LIQUID HYDROCARBON TRANSPORTATION SYSTEM SCOPE

The liquid pipeline transportation systems that apply to hydrocarbon liquids include, crude oil, natural gas liquids, refined petroleum products, liquid petroleum gases (LPG), gas to liquids (GTL), anhydrous ammonia, alcohols, and carbon dioxide. The systems generally consist of the following:

- pipelines (mainline and laterals);
- measurement facilities;
- pumping systems;
- pressure reduction facilities;
- pipeline appurtenances (scraper traps, flow restrictive devices (valves), drag reducer injection facilities, etc.);
- Control systems for the above.

The limits of the pipeline transportation systems are well defined by applicable codes, an example of which is shown in ANSI/ASME B31.4 — Pipeline Transportation Systems for Liquid Hydrocarbons and Other Liquids (see Figure 1-1).

Also included within the scope of the system are the following:

- primary and associated auxiliary liquid petroleum and liquid anhydrous ammonia piping at pipeline terminals (marine, rail, and truck), tank farms, pump stations, pressure reducing stations, and metering stations, including scraper traps, strainers, and meter prover loops;
- storage and working tanks, including pipe-type storage fabricated from pipe and fittings, and piping interconnecting these facilities;
- liquid petroleum and liquid anhydrous ammonia piping located on property which has been set aside for such piping within petroleum refinery, natural gasoline, gas processing, ammonia, and bulk plants;
- those aspects of operation and maintenance of Liquid Pipeline Systems relating to the safety and protection of the general public, operating company personnel, environment, property, and the piping systems

1.2 HYDROCARBON LIQUID PIPELINES

Hydrocarbon liquid pipelines are utilized for transporting liquid hydrocarbons including crude oil or natural gas liquids from producing fields to refineries, where

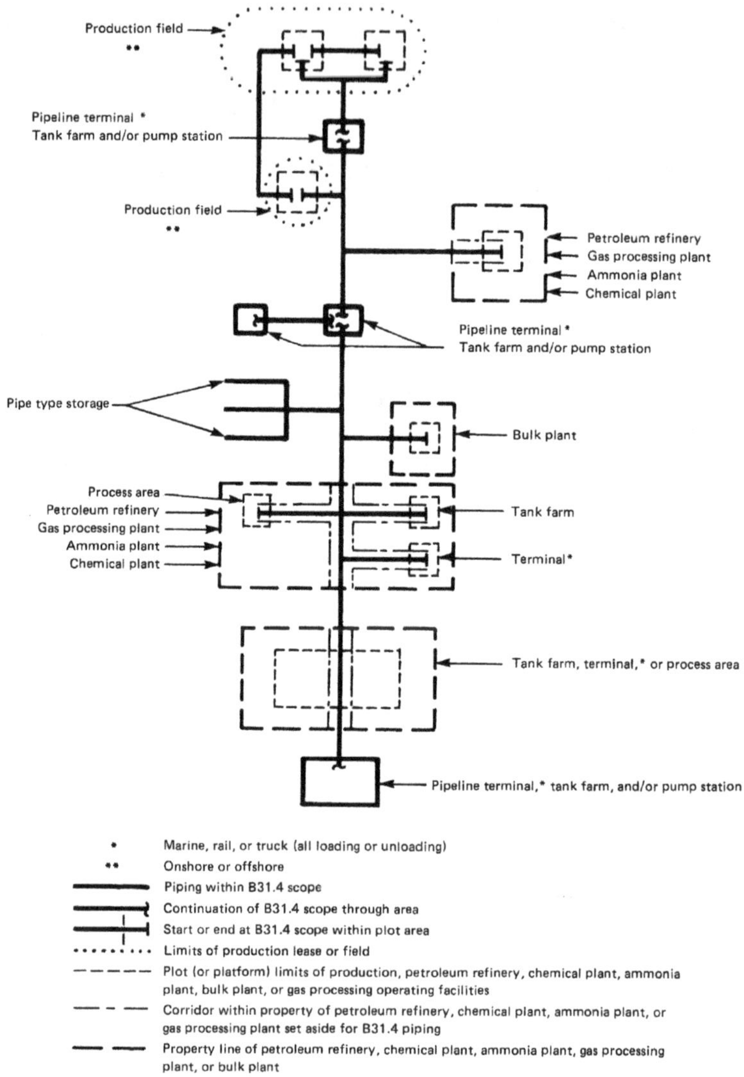

Figure 1-1. Liquid Hydrocarbons Pipeline System Definitions (ANSI/ASME B31.4)

they are turned into gasoline, diesel and other petroleum products. Some liquids pipelines are also used to transport these finished petroleum products from refineries to terminals and distribution centers. Petroleum products include fuels such as gasoline, aviation fuel, diesel, and heating oil, as well as hundreds of products such as solvents and lubricants, as well as raw materials for manufacturing petrochemicals.

Liquid hydrocarbon pipelines can be used to move different batches or grades of liquids or varieties of crude oil. In a closed system, each batch of product is transported along the pipeline at the same velocity and may or may not be separated by mechanical devices. Where the two batches do come in contact with each other, there is a small amount of mixing that occurs — these small interface volumes, known as transmix, are segregated and eventually reprocessed at a refinery.

Output from one barrel of oil after processing (in % terms) is shown in Figure 1-2.

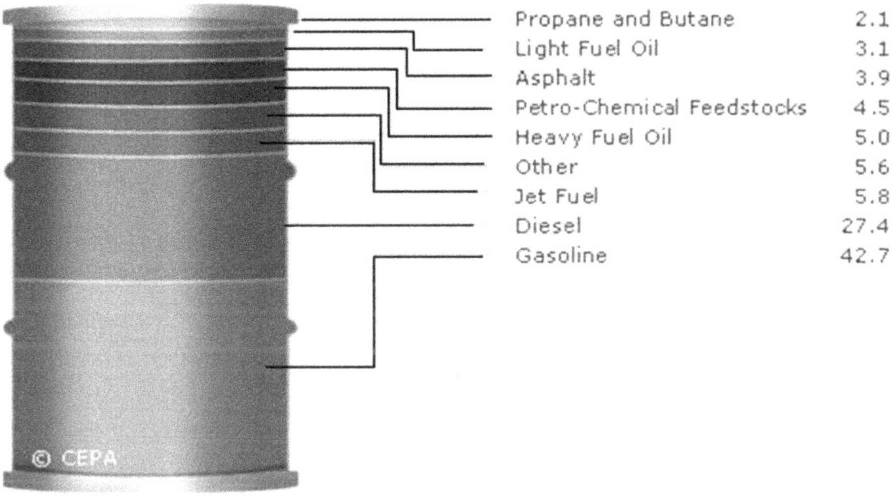

Figure 1-2. Output from A Barrel of Oil (%) (courtesy of CEPA, 2011) [1]

In actual volume, from refining processes, one barrel of crude oil (42 US gallons) the following amount of petroleum products are GENERALLY produced (in US Gallons):

- Gasoline — 19.4
- Distillate fuel — 9.7
- Kerosene type fuel — 4.3
- Coke — 2.0
- Residual fuel oil — 1.9
- Liquefied refinery gases — 1.9
- Still gas — 1.8
- Asphalt or road oil — 1.4
- Petrochemical feed stocks — 1.1
- Lubricant oils — 0.5
- Kerosene — 0.2
- Other — 0.4

It may be noted that the total is over by 2.4 US Gallons due to processing gain.

1.3 LIQUID PIPELINE TRANSPORTATION SYSTEMS

Liquid petroleum systems include production wells and gathering lines for crude oil production, processing plants and refineries, transmission pipelines, pump stations, valve and metering stations, and aboveground/belowground storage facilities as shown in Figure 1-3.

Liquid transmission and distribution system components can consist of the following:

- Field production facilities (onshore and offshore);
- Gathering pipeline facilities;
- Treatment/refining/processing facilities;
- Pipeline terminals (tank farms and/or pump stations);
- Receipt meter/custody transfer stations;

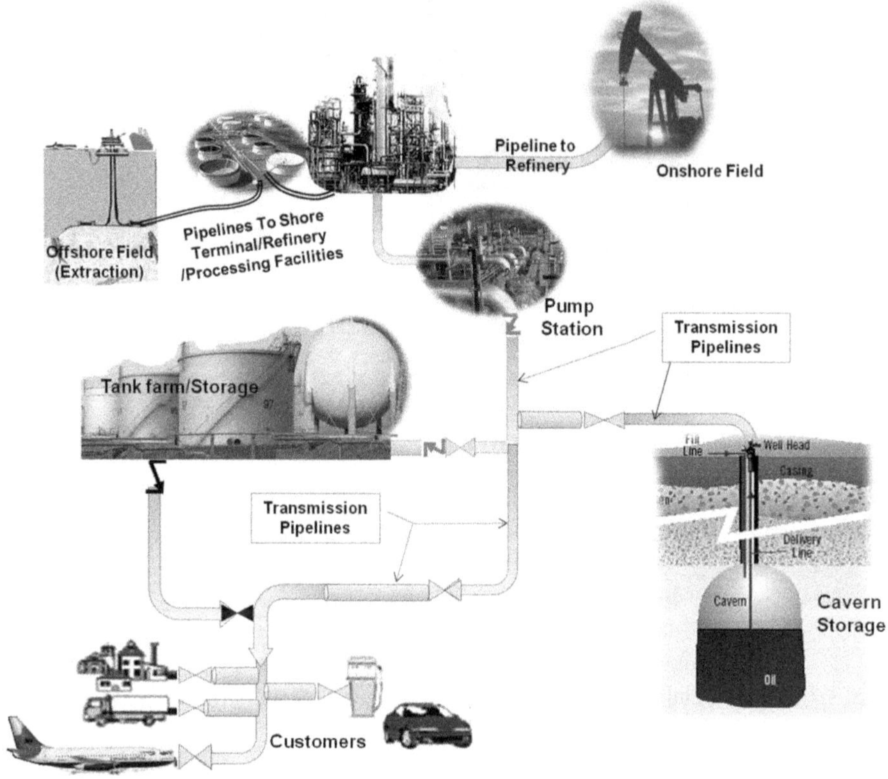

Figure 1-3. Components of oil production, transmission, and distribution systems

- Lateral lines;
- Mainline transmission pipelines;
- Mainline control valves to regulate pressure or flow;
- Emergency flow restrictive devices (valves, check valves);
- Manifolds (piping/valving arrangements);
- Mainline pumping facilities (including prime movers, station controls, etc.);
- Pipeline inspection devices (scraper traps, pigging stations);
- Delivery/custody transfer meter stations;
- Product quality control and sampling;
- Storage facilities (tankage/caverns);
- Protection/corrosion control systems;
- Supervisory Control and Data Acquisition (SCADA) systems;
- Leak detection systems;
- Communication systems.

Crude oil must undergo refining before it can be used as a product. Once oil is pumped from the ground, it travels through gathering pipelines to tank batteries. A typical tank battery contains a separator to separate oil, gas, and water.

After the crude oil is separated from other production fluids such as natural gas and water, it is usually stored in field production tankage or underground caverns. The crude oil is then transported to central pipeline terminal facilities. The oil is then moved through large-diameter, long-distance transmission pipelines to refineries, other storage tanks, tanker ships, or railcars. The pressure in the main lines is initiated and maintained by pumps to overcome friction, changes in elevation, or other pressure-decreasing factors.

Drag reducing agents (DRAs) are sometimes used to improve throughput by decreasing the effects of friction. Pump stations are located at the beginning of the line and are spaced along the pipeline at regular intervals to adequately propel the oil along.

1.4 TYPES OF TRANSMISSION PIPELINES

There are two types of liquid transmission pipelines; single product (transporting a petroleum product in a single stream (e.g., crude oil) or multi-products pipelines (transporting generally multiples of hydrocarbon liquids in a batch form).

Single product pipelines (e.g., crude lines) are broken down into two types:

- Gathering lines (generally NPS 6 or below) that transport crude oil or other hydrocarbon liquids to concentration points such as pipeline terminals; and,
- Trunk or transmission lines transporting liquid hydrocarbon products usually from pipeline terminals and tank farms, refineries/processing plants to market area.

From regulatory points of view, pipelines are categorized as follows:

- Offshore (i.e., marine);
- Onshore (i.e., land based);
- Gathering or flow lines (from wells);
- Transfer lines;
- Piping and manifolds;
- Breakout piping (tanks);
- Distribution pipelines/pipes.

It may be noted that above category liquid pipeline includes "breakout tanks" as in US DOT CFR Section§ 195.1: "Breakout tank means a tank that is used to (a) relieve surges in a hazardous liquid pipeline system or (b) receive and store hazardous liquid transported by a pipeline for reinjection and continued transportation by pipeline" [2].

Also, it may be noted that distribution pipelines can be composed of several interconnected pipelines with small diameters, used to deliver products to the final consumer. Pipelines at terminals for distributing products to tanks and storage facilities are included in this group.

1.5 LIQUID PETROLEUM PIPELINE NETWORKS

Worldwide, there are vast networks of pipelines that transport liquid hydrocarbons from the sources of supply to terminals, refineries, processing plants, and market area. Global spread of pipeline transmission networks is indicated in Figure 1-4.

It is significant to note that liquid hydrocarbon transmission pipelines mostly serve shorter hauls in comparison to natural gas transmission pipelines which are generally for long haul and cross border/international transportation.

Liquid hydrocarbon production (crude oil and natural gas liquids) can be transported to refining facilities by a number of methods due to its low vapor pressure characteristics. The determination of the most economical method of getting the liquids from production areas that are often in remote areas to refineries can be based on a choice of either transmission pipelines, marine tankers, rail cars, or some combination of these transportation modes. In many cases, refineries are located in areas that can be

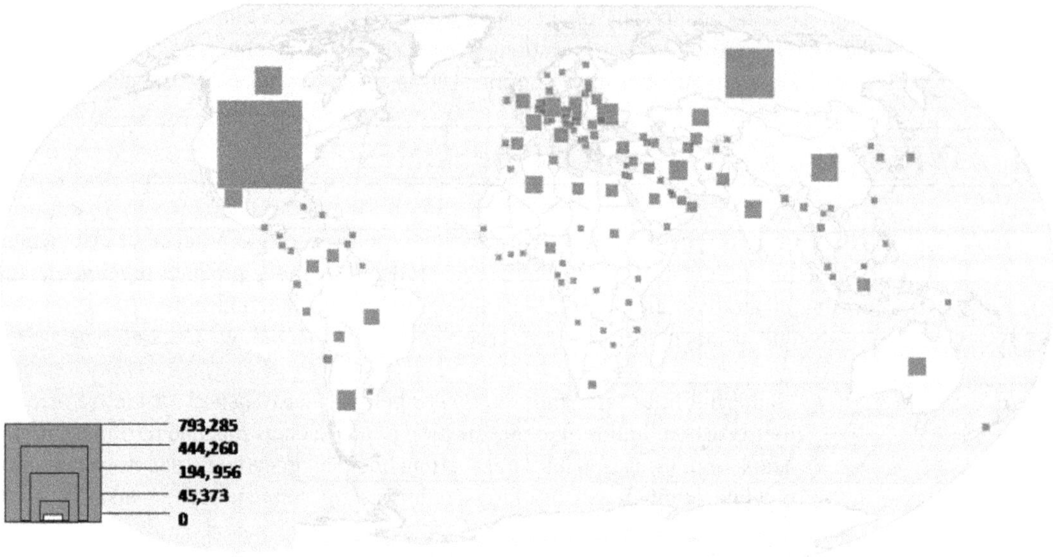

Figure 1-4. World Regional Pipeline Transportation Spread (KM)

accessed by marine tankers and, if the production facilities are located near tidewater, tankers are often the transportation system of choice. Much of the world's production of crude oil is transported to market by a tanker.

Natural gas, unless it is liquefied, cannot be economically transported by a tanker and therefore is typically moved to market directly through long haul transmission systems. Natural gas pricing recently has trended well below the energy-equivalent for crude oil (Figure 1-5). This and the fact that natural gas has an inherently reduced emission of CO_2 have resulted in the increased demand for natural gas in power generation and other industrial uses.

In Figure 1.5, it should be noted that the comparison of oil and gas prices are on an energy-equivalent basis, natural gas prices are increased by a factor of 5.8 since one barrel of oil = 5,800,000 BTUs, and natural gas is priced in dollars per million BTUs.

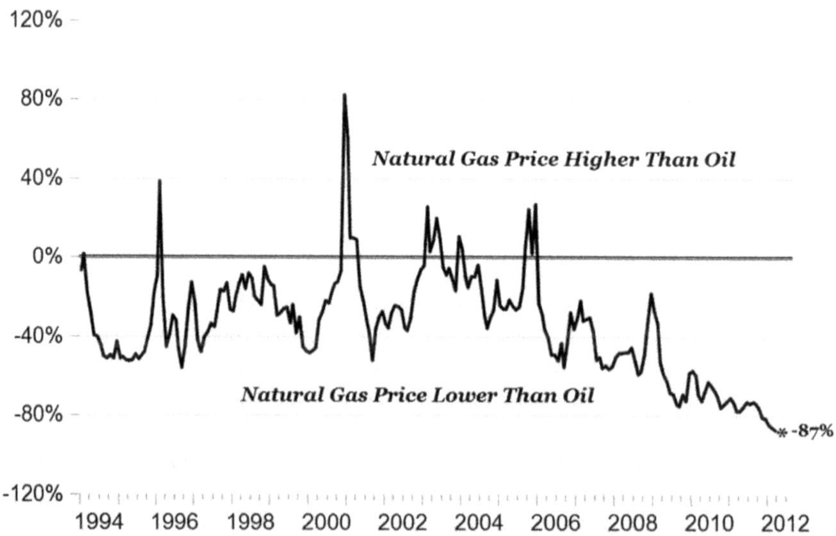

Figure 1-5. Natural gas versus oil-energy equivalent value [3]

Therefore, when natural gas is selling for about $2.30 per million BTUs, its energy equivalent price for a barrel of oil would be $13.34 or 87% below the price of West Texas Intermediate (WTI) oil at $104.71. When measured on an energy-equivalent basis, natural gas has never been cheaper in comparison to oil than it is today [3].

Long haul overland liquid hydrocarbon pipeline transportation provides superior safety performance and efficiency in comparison to truck or rail transportation. They are a secure form of liquid hydrocarbon transportation within a nation or between friendly neighbors.

World on-shore operating liquid hydrocarbon pipelines are listed in Table 1-1 [4], CIA 2009. World breakdown of liquid hydrocarbon pipelines are indicated in Figure 1-6.

TABLE 1-1. World Onshore Hydrocarbon Liquid Operating Pipelines [5]

COUNTRY	Oil	Condensate-gas	LPG - LNG	Refined Products	Water/Oily Water/othe	TOTAL	COUNTRY	Oil	Condensate-gas	LPG - LNG	Refined Products	Water/Oily Water/other	TOTAL
Afghanistan							Kyrgyzstan	16					16
Albania	207					207	Laos				540		540
Algeria	7611	2600	3447	144		13802	Latvia	0			415		415
Angola	87					87	Liberia	4					4
Argentina	6166		41	3631	13	9851	Libya	6960	776				7736
Australia	3258		240		1	3499	Lithuania				114		114
Austria	663			157		820	Luxembourg				27		27
Azerbaijan	1424	1				1425	Macedonia	120					120
Bahrain	29					29	Malaysia	30	3	155	114	19	321
Barbadous	62			4		66	Mexico	7499		2152	7264	37	16952
Belarus	1528			1730		3258	Morocco	439					439
Belgium	154			535		689	Mozambique				278		278
Bolivia	2510		51	1627		4188	Netherlands	365			716		1081
Bosnia and Herzegovina	9					9	New Zealand	288	331	172	198		989
Brazil	3729	62	352	4684		8827	Nicaragua	54					54
Brunei	18	33				51	Nigeria	3441	26	97	4090		7654
Bulgaria	346			156		502	Norway 2006 (inc offshore)	2444	31			457	2932
Burma	551					551	Oman	3558	107		263		3928
Cameroon	886		5			891	Pakistan	2013			787		2800
Canada 2010	23564		75000			98564	Panama	128					128
Chad	265					265	Papua New Guinea	195					195
Chile	895		517	768		2180	Peru	1566		679	15		2260
China	23470			13706		37176	Philippines, 2006	107			112		219
Colombia	6334			3309		9643	Poland	1384	35		777		2196
Congo, Democratic Republic of the	39			756		795	Portugal	11			188		199
Congo, Republic of the	273					273	Qatar	382	177	90			649
Costa Rica				662		662	Romania	2424					2424
Côte d'Ivoire	92	86				178	Russia	77630	122	127	13658	38	91575
Croatia	532					532	Saudi Arabia	4232	212	1183	1151		6778
Cuba	230					230	Senegal				8		8
Czech Republic	547			94		641	Serbia and Montenegro	323					323
Denmark	107					107	Slovakia	416					416
Dominican Republic 2006	99					99	Slovenia	5					5
Ecuador	1812			1262		3074	South Africa	980	11		1382		2373
Egypt	4314	333	959	895	3	6504	Spain	560			3441		4001
France	2939			5084		8023	Sudan	4070			1613		5683
Gabon	893					893	Suriname	50					50
Georgia	1258					1258	Switzerland	94			7		101
Germany	3687			4875		8562	Syria	1997					1997
Ghana	5			312		317	Tajikistan	38					38
Gibraltar	0					0	Tanzania	888			8		896
Greece	75					75	Thailand			85	1099		1184
Guatemala	480					480	Trinidad and Tobago	334					334
Hungary	984			361		1345	Tunisia	1323			453		1776
India	7448	2	2152	10486		20088	Turkey	3636					3636
Indonesia	5984	885		617	50	7542	Turkmenistan	1457					1457
Iran	7123	19	570	7937		15649	Ukraine	4514			4211		8725
Iraq	5104		918	1637		7659	United Arab Emirates	1437	458	220	212	90	2417
Israel	442			261		703	United Kingdom 2006	595	8	59	4907		5569
Italy	1241					1241	United States 2010				244620		244620
Japan	171				53	224	Uruguay	155					155
Jordan	49					49	Uzbekistan	868					868
Kazakhstan	11201	658		1095	1465	14419	Venezuela	7674			1620		9294
Kenya	4			928		932	Vietnam		38		206		244
Korea, North	154					154	Yemen	1367		22			1389
Korea, South				864		864	Zambia	771					771
Kuwait	540			57		597	Zimbabwe	0			270		270
							Total Length Km	148842	2335	5041	294764	641	451623
Total Length Km	141593	4679	84252	68634	1591	300749	**Grand Total Length Km**	**290435**	**7014**	**89293**	**363398**	**2232**	**752372**

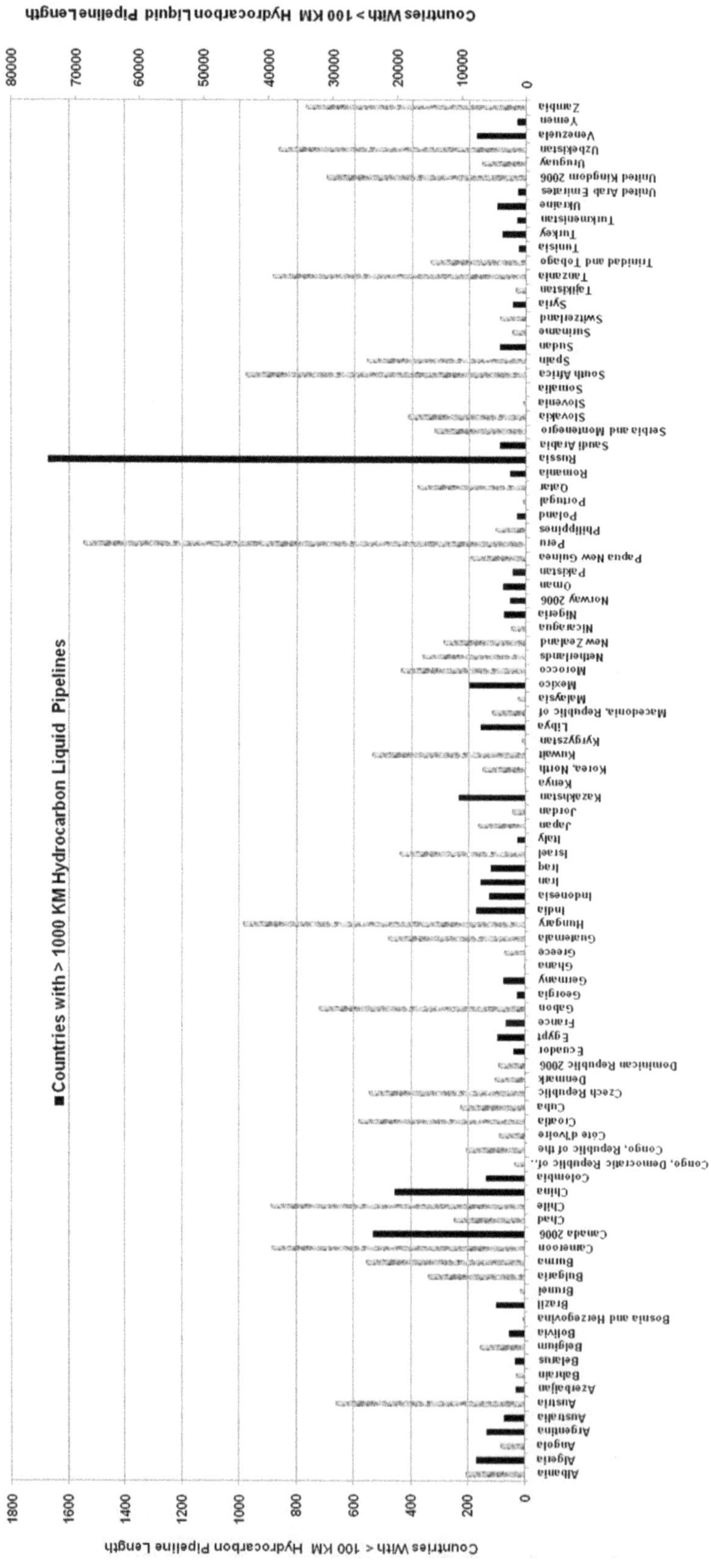

Figure 1-6. Breakdown of onshore hydrocarbon liquid pipelines (including heavy oil) market share (CIA 2009, [4])

TABLE 1-2. World major daily oil imports and exports, 2008 in thousand barrels [6]

Country	Crude Imports	Product Imports	Crude Exports	Product Exports
US	9756	3116	139	1829
Canada	693	288	1931	566
Mexico	40	500	1459	150
Southern and Central America	610	864	2479	1136
Europe	10859	2892	281	1741
Former Soviet Union	1	149	6235	1948
Middle East	220	202	17925	2203
North Africa	329	170	2618	642
West Africa	18	281	4466	121
East and Southern Africa	509	189	340	17
Australasia	489	400	274	37
China	3581	812	75	313
India	2557	458	^	718
Japan	4069	856	-	356
Singapore	1092	1593	40	1542
Other Asia Pacific	4627	2390	931	1460
Unidentified *	-	17	257	394
Total World	**39451**	**15175**	**39451**	**15175**

The breakdown of oil including product imports and exports by country is provided in Table 1-2. In 2008, in United States alone, liquid pipelines transported over 38 million barrels per day (B/D). This was roughly equivalent to twice actual US consumption of oil. The following indicates the share of domestic shipments, in barrel barrel-miles in the United States [7]:

- Pipelines — 66%,
- Water carriers — 28%,
- Trucks — 4%,
- Rail — 2%.

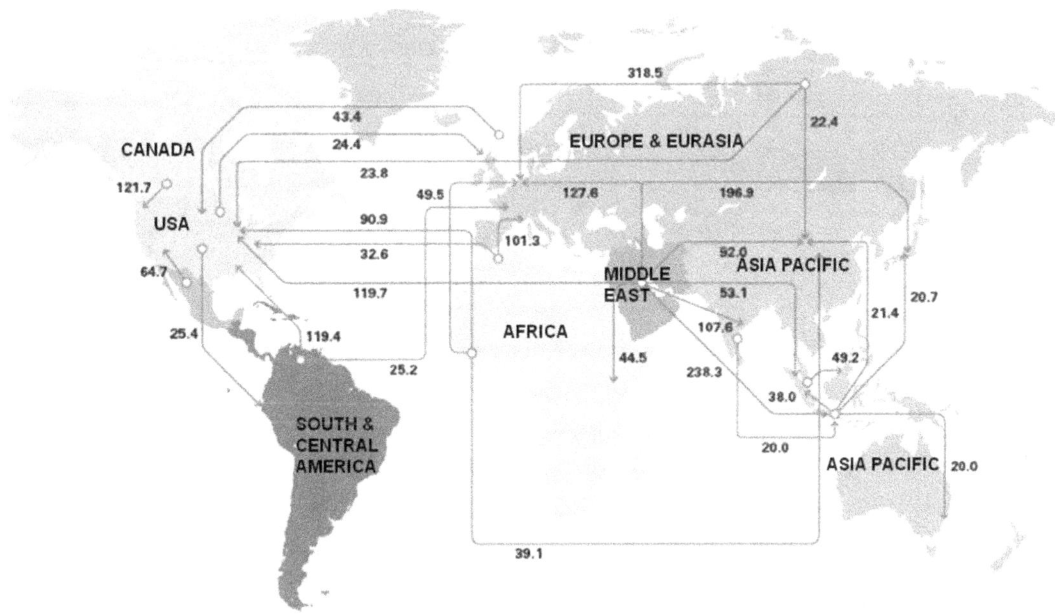

Figure 1-7. World major oil trade movements 2008 (million tons), source [6]

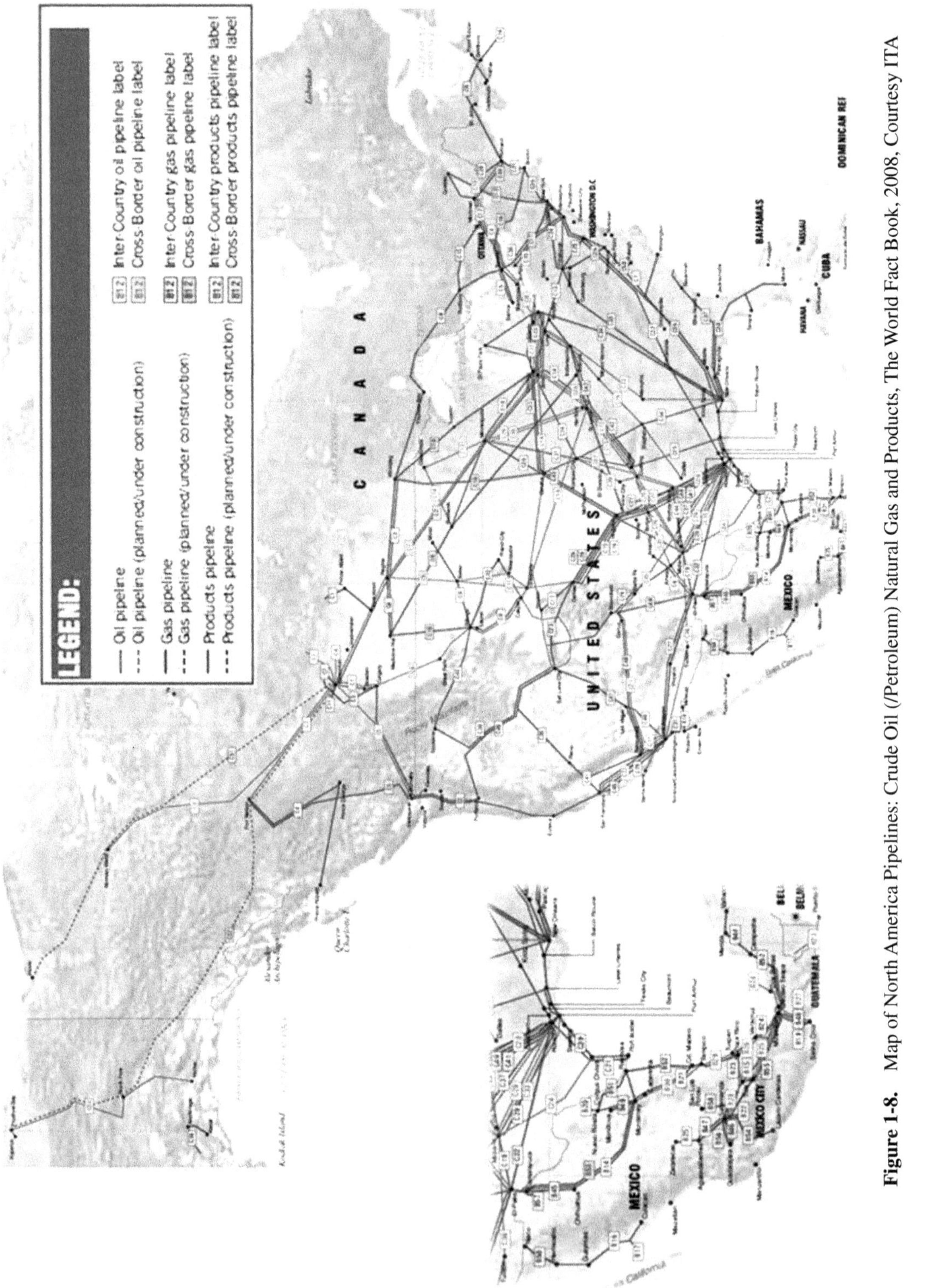

Figure 1-8. Map of North America Pipelines: Crude Oil (/Petroleum) Natural Gas and Products, The World Fact Book, 2008, Courtesy ITA

It may be noted that, in the US, pipelines carry 17% of US freight at 2% of the cost.

Globally, from Table 1-2, it may be noted that a net movement of about 2000 billion barrels of oil is realized annually. Figure 1-7 shows the corresponding world major oil trade movement provided in Table 1-2. The North American pipeline network is shown in Figure 1-8.

The previous Figure 1-8 illustrates North American pipelines including those for oil, gas, and products such as gasoline, propane, and ethylene. Similar regional maps can be accessed by referring to website http://www.theodora.com/pipelines/north_america_oil_gas_and_products_pipelines.html.

1.6 SINGLE VERSUS MULTIPLE PRODUCTS PIPELINE

As indicated previously, liquid pipelines are either designed to carry single products (i.e., transporting a petroleum product in a single stream) such as low vapor pressure fluids (e.g., crude oil, bitumen, water, etc.), refined products, high vapor pressure fluids (e.g., condensate, propane, etc.) or a number of products (low and/or high vapor pressure) in a batch form. Others include slurry (mixed product pipelines) which are in essence single products pipeline but are designed to transport a heavier fluid/solids using a carrier fluid, such as water, carbon dioxide, air, etc.

1.6.1 Refined Petroleum Products

Flammable and combustible liquids are the most common products transported by liquid transmission pipelines. These include refined products such as gasoline, aviation gas and jet fuel, and distillates, such as home heating and diesel fuels. It may be noted that refined gasoline is not the same as "natural gasoline" or "casing head gasoline", which is the liquid hydrocarbon typically referred to as "natural gas condensate" that is recovered from "wet" natural gas production.

Liquid hydrocarbon transportation in a batch form allows multiple products to be shipped in the same pipeline (Figure 1-9). Sequential movement of liquids in a batch form is commonly exercised by refineries and pipeline companies to transport multitudes of products through a single pipeline. This form of transportation includes batch transportation of low as well as high vapor pressure fluids.

Liquid pipeline companies can transport and deliver a multiple of petroleum liquid products to many customers. For example, in year 2000, Colonial Pipeline delivered 90 different products for 85 customers to 270 terminals and into more than 1000 storage tanks, Jacobs, 2002. Typical products commonly transported as single or batched products together with their properties are shown in Table 1-3 [4]. The industry expects

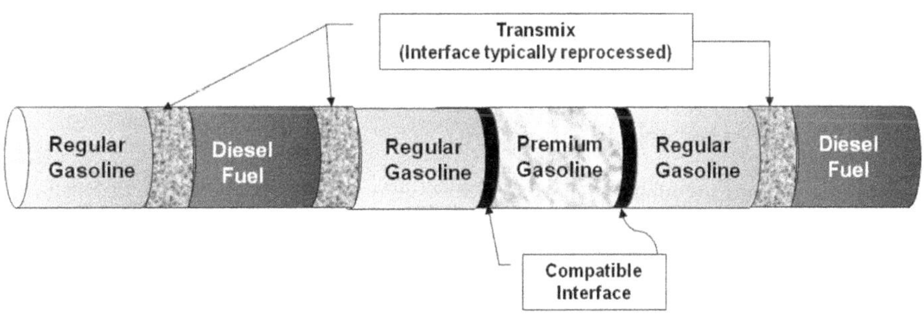

Figure 1-9. Typical sequence of petroleum products flow through a pipeline

TABLE 1-3. Typical petroleum products and their properties (at 101.325 kPa and 15°C) transported by pipelines [4]

Commodity	Viscosity (CS)	Temperature (°C)	Density (kg/m³)*
Diesel	6.86	5	847
	5.10	15	820
Gasoline (leaded)	0.68	5	711.3
	0.61	15	–
Gasoline (unleaded)	0.7	5	699
	0.63	15	690 (assumed)
Jet A Fuel	8	29	774
	1.5	0	–
Jet B Fuel	1.9	15	–
	1.5	35	–
Kerosene	3	15	–
	2.2	35	–
Condensate (sweet)	0.599	15	708.8
	0.548	25	688.8
Condensate (raw)	0.199	15	572.3
	0.171	30	547.5
Propane (@1000 kPa)	0.218	20	500.6
	0.199	30	483.5
	0.166	52	446.9
Butane (at 470–520 Kpa)	0.237	25	560.8
	0.235	44	535.12
Very Heavy Oil	0.212	52	529
	21.1	83	
	37.8	3.75	
	40	3.28	
Heavy Crude	50	2.44	
	21.1	37	
Medium Crude	37.8	19	
	21.1	16.2	
Light Crude	37.8	9.41	
	21.1	10.2	
	37.8	6.25	

that all products transported meet the physical constraints of the system and generally are free of water or other unacceptable impurities. Refer to **Chapter 5** for Batching in Products pipelines. Properties of some petroleum products (condensate and diluents) and crude oils (Heavy, light, medium, sweet/sour) generally available and transported through pipeline in United States and Canada are shown in Table 1-4.

There are other crude oil and products characteristics that are not included in Tables 1-3 and 1-4. Such characteristics/properties are included Chapters 2, 3, 5, 6, and 8 where their applications are demonstrated.

1.7 LIQUID PIPELINE DEVELOPMENT HISTORY/CHRONOLOGY

1.7.1 Historical Overview

Pipeline history for the petroleum industry is about two centuries old over the span of which the pipeline industry has grown into a worldwide transportation system handling well over 2 billion barrels of oil per annum. Liquid hydrocarbon pipelines are able to transport multiple grades and types of product without significant mixing. The pipelines transport crude oil economically and safely from the oil producing fields to the refineries and refined products from the refineries to centers of population.

TABLE 1-4. Typical US and Canadian crude and products characteristics

Product Designation	Crude Type (Long Name)	Total Sulphur (% by wt.)	Reid Vapour Pressure (kPa)	Pour Point (deg. C)	Density (kg/m3)	Viscosity (cSt) at Specified Temperature (° C)				
						10	20	30	40	45
ACB	Amoco Condensate	0.07	92.330	<-30	677.37	0.55	0.50	0.45		
BG	Bonnie Glen (SW)	0.45	59.250	<-30	818.00	5.63	4.12	3.29	2.58	2.34
BGO	Bonnie Glen Sour	1.31	41.340	-21	875.90	17.87	12.14	8.67	6.44	5.76
BR	Bow River	2.43	28.940	<-30	913.20	111.00	63.60	40.30	26.00	21.91
CAM	Amoco (Strachan) Condensate	0.11	72.804	<-30	711.81	0.69	0.62	0.57	0.51	0.49
CBG	Bonnie Glen Condensate	0.16	83.620	<-30	703.20	0.63	0.57	0.52	0.48	0.45
CCA	Caroline Condensate	0.96	57.244	<-30	761.70	0.71	0.64	0.58	0.53	0.51
CFD	Federated Condensate	0.17	95.082	<-30	701.25	0.68	0.62	0.57	0.53	
CFT	Ft. Saskatchewan Condensate	0.08	90.087	<-30	674.05	0.52	0.47	0.44	0.42	
CGB	Gibsons Condensate	0.07	90.029	<-30	711.78	0.65	0.56	0.54	0.51	
CL	Cold Lake	3.66	44.100	<-27	926.20	274.04	143.98	85.34	53.11	42.41
CPC	Petro-Can Condensate	0.03	12.747	<-30	751.11	0.87	0.75	0.68	0.61	0.58
CPM	Pembina Condensate	0.11	68.670	<-30	756.76	1.11	0.95	0.84	0.76	0.68
CPR	Peace River Condensate	0.29	79.350	<-30	739.65	0.92	0.81	0.73	0.66	0.61
CRB*	Rainbow Condensate	0.08	84.517	n/a	738.11	n/a	n/a	n/a	n/a	n/a
CRM	Rimbey Condensate	0.07	84.001	<-30	691.16	0.54	0.49	0.46	0.43	0.43
CRW	Condensate Blend (Diluent)	0.18	77.600	<-29	738.50	0.96	0.85	0.76	0.70	0.68
F	Fosterton	2.97	33.760	<-30	917.10	163.99	85.05	51.76	31.51	25.85
FD	Federated	0.27	52.360	<-30	826.00	5.27	3.90	3.01	2.48	2.26
FHM*	Foreign Sweet-Mokena	0.36	44.100	n/a	828.60	8.75	5.09	4.08	3.29	2.99
FOM	Foreign Sour-Mokena	1.47	35.000	-15	871.90	27.34	15.97	11.20	8.30	7.28
HO	Husky	0.85	55.120	-3	836.10	9.68	5.77	4.06	3.18	2.81
HSB	Husky Synthetic Blend	0.09	44.100	<-30	853.90	8.03	5.94	4.59	3.62	3.25
ISH	ISH Energy	0.85	39.960	<-30	858.20	10.92	8.24	6.25	4.79	4.37
KAL	Koch Alta	1.12	57.880	-12	858.20	13.34	8.44	6.01	4.63	4.22
LB	Lube Blend	0.40	55.700	-8.52	827.50	6.26	4.58	3.63	2.95	2.72
LES	Leduc Sour	1.10	49.610	<-30	851.50	12.19	8.69	6.59	5.00	4.57
LLB	Lloydminster Hardisty	3.42	70.970	<-30	927.10	264.34	141.85	82.46	50.96	41.48
LLC	Lloydminster Chauvin	2.78	36.520	<-30	926.40	245.58	130.91	77.15	47.39	38.26
LLE	Lloydminster Echo	3.41	59.250	<-30	922.90	260.26	133.63	78.52	48.12	38.87
LLG	Gibsons Blend	3.16	38.580	<-30	926.80	279.59	146.46	85.85	52.64	42.63
LLK	Lloydminster Kerrobert	2.82	31.690	<-30	926.30	261.36	140.86	80.47	49.00	39.86
LLW	Lloydminster Wainwright	2.84	35.830	<-30	927.30	253.98	136.45	78.85	48.25	39.13
LSB	Light Sour Blend	1.27	57.933	-7.59	847.90	10.91	6.80	4.99	3.92	3.57
M	Midale	2.35	34.067	-12.22	894.87	44.74	26.47	16.22	11.38	9.85
MBL	Hardisty Light	1.75	37.900	<-30	883.70	41.76	25.88	17.45	12.22	10.40
MCG	Waskada Light Sour ICG	0.78	68.900	-12	842.07	15.96	6.77	4.94	3.74	3.42
MGS	Gibsons Sour	1.66	63.390	<-30	849.80	12.89	8.86	6.96	5.33	4.77
MJT	Moose Jaw Tops	1.69	33.760	-21	848.50	5.92	4.49	3.59	2.94	2.60
MLN	Joarcam	0.18	41.340	-15	838.90	26.82	11.15	7.53	5.29	4.73
MM	Manitoba Medium	2.15	20.670	<-30	905.16	62.47	37.25	23.56	15.88	13.19
MML	Manitoba Light	1.36	77.860	<-30	851.48	10.17	7.14	5.14	4.23	3.79
MPR	Peace River	0.36	63.390	-12	822.40	6.03	4.47	3.39	2.69	2.48
MSM	Westspur Midale	2.49	46.850	-9	880.48	34.13	17.73	12.03	8.44	7.00
MSN	Manitoba Sweet	0.42	21.360	-9	848.62	8.57	5.68	4.94	3.92	3.52
MSY	Smiley Light	0.27	35.830	<-30	854.50	20.58	9.22	6.52	5.05	4.51
MWL	Wizard Lake	0.27	59.250	<-30	807.23	2.80	2.28	1.97	1.67	1.65
NSB	Newgrade Synthetic Blend	0.28	48.230	<-30	844.90	8.84	6.65	5.26	4.07	3.78
NW	Norman Wells	0.34	100.590	<-30	824.30	4.90	3.90	3.23	2.69	2.46
OSA	Suncor - A	0.27	35.830	<-30	864.90	8.58	6.26	4.80	3.71	3.39
OSB	Suncor - B	3.55	24.120	<-30	927.90	103.86	60.67	37.02	25.58	21.62
OSE	Suncor - E	3.85	26.180	<-30	921.00	16.61	11.04	7.71	5.57	4.99
OSF	Suncor - F	2.05	7.580	<-30	899.30	12.70	8.68	6.26	4.66	4.13
OSV	Suncor - V	3.01	17.910	<-30	921.70	79.99	43.30	25.79	16.16	13.55
P	Pembina	0.36	59.250	-12	832.70	7.91	5.30	4.14	3.22	2.88
PH	Peace River Heavy	4.69	56.500	<-30	919.81	206.66	114.16	69.81	46.69	35.43
RA	Rainbow	0.36	77.860	-12	825.30	5.78	4.27	3.39	2.74	2.52
RW	Redwater	0.56	64.770	<-30	857.50	12.09	8.20	6.22	4.82	4.27
SC	Smiley Coleville	2.91	34.450	<-30	929.40	270.28	146.21	84.09	50.50	41.65
SES	South East Saskatchewan	1.35	62.010	-9	849.40	11.98	7.13	5.27	4.04	3.61
SO	Mixed Blend Sour	1.68	43.067	<-29	875.10	22.40	14.92	10.03	7.49	6.64
SPR	Sour Peace River	1.19	62.010	-21	838.00	7.72	5.33	4.23	3.31	3.01
SSS	Sarnia Special	0.85	53.740	-9	893.40	215.28	96.97	58.94	35.78	29.34
SW	Mixed Blend Sweet	0.44	57.233	-16.85	833.00	6.58	4.71	3.73	3.02	2.78
SYN	Syncrude	0.11	43.410	<-30	868.00	7.55	5.47	4.20	3.33	3.03
UCH*	U.S. Condensate - Hartsdale	0.21	47.900	n/a	771.80	1.93	1.67	1.43	1.26	1.17
UCM	U.S. Condensate - Mokena	0.29	46.400	-26.11	792.60	3.35	2.65	2.19	1.87	1.74
UHC	U.S. High Sweet - Clearbrook	0.35	43.500	-20.56	824.10	8.41	4.98	3.84	3.08	2.80
UHG	U.S. High Sweet - Griffith	0.42	48.200	<-29	834.70	7.73	5.77	4.53	3.68	3.39
UHL	U.S. High Sweet - Lewiston	0.35	71.833	5.37	794.67	n/a	n/a	n/a	1.87	1.75
UHM	U.S. High Sweet - Mokena	0.42	30.700	-7.59	832.70	8.81	6.26	4.81	3.83	3.47
UMC	U.S. Midale - Clearbrook	2.40	46.900	-9.44	891.30	34.26	17.50	10.99	7.99	7.01
UOC	U.S. Sour - Clearbrook	1.86	43.533	10	872.30	94.34	18.73	9.55	6.48	5.57
WH	Wabasca Heavy	4.05	35.830	<-30	928.72	255.95	133.77	77.67	48.26	39.89
ZA	Zama Delivery	0.33	99.220	<-30	824.20	4.99	3.81	3.26	2.67	2.53

The history of liquid pipelines can be traced to early 19th century. In the United States liquid pipelines can be traced to the late 1800s in Ohio, Pennsylvania, and New Jersey [8].

The primary driver for the use of pipelines has always been economic. When oil was discovered and production commenced, crude oil volumes were small and a distributed transportation system, such as horse-drawn wagons, trucks, and railroads, was the most efficient means of transporting the oil to refineries where it would be converted into products desired by oil consumers [9].

Late 1800s	Pioneering of pipeline transportation by Vladimir Shukhov and the Branobel company
1857	The birth of Canadian petroleum industry (James Miller Williams discovery of an oil well in Enniskillen Township (Oil Springs), Ontario
1859	Colonel Drake Strikes Oil, Titusville, Pennsylvania
1859	World's first dependable internal combustion engine by French engineer J.J. Etienne Lenoir
1860	Samuel Duncan Karnes conceptualize transportation of oil by pipelines, Parkesburg WV, USA [10]
1860	First oil pipeline in the USA (by Oil Transport Association), 2-in. (51 mm) wrought iron pipeline over a 6-mile (9.7 km) track from an oil field in Pennsylvania to a railroad station in Oil Creek,
1862	Use of iron pipe in oil wells
1862	The first oil refinery (using atmospheric distillation to produce kerosene)
1863	The Teamsters & Pipeline Gathering (first transportation of oil by whiskey barrels)
1865	The first oil pipeline in the United States, following the discovery of oil in Pennsylvania. Van Syckel 5 miles oil line (wrought iron, lap welded) [11, 12].
1870	Creation of Standard Oil by John D. Rockefeller
1879	Construction of the first Crude trunk line, Tidewater
1880s	The rise of Russian oil industry (discovery of oil near the Russian sea town of Baku and building of first pipeline east of Baku to)
1880-1905	Oil discoveries in Ohio, Oklahoma and Kansas) Oil discovery in California at the turn of the century) USA
1890	Anti Trust Act the USA
1891	The first high-pressure, long-distance pipeline was constructed between Chicago and fields in Indiana, a distance of almost 200 km
1903	US inventor, Henry Ford, builds the first Model-A automobile
1905	Cross country crude oil pipelines (Texas, Oklahoma and Kansas to the refineries in the Eastern USA)
1906	1st US pipeline regulation (Hepburn Act making interstate pipelines common carriers thus offering services at equal cost to all shippers) under Interstate Commerce Act
1911	Construction of the first pipeline from Masjed Soleiman to Abadan, Iran
1912	Dissolution of Standard Oil into seven regional oil companies.
1914	Discovery of Oil and wet natural gas In Alberta (Turner Valley), Canada
1917	World War I
1925	First introduction of crude oil to Iranian markets
1938	First offshore producing oil well is in the Gulf of Mexico
1939	World War II
1940	Beginning of oil pipelines in Brazil
1945	Growth of product pipelines to eastern sea world, USA
1948	Conception of first oil transmission pipeline in Canada
1950	Beginning of construction of 1800 km long Interprovincial crude oil

pipeline (Redwater oil field in Alberta, Canada to Superior Wisconsin, USA)

1951 Incorporation of the TransMountain Oil Pipe Line Company

1953 Incorporation of Great Canadian Oil Sands Limited

1954 Formation of Stanolind (Indiana Standard pipeline company), the largest liquid pipeline carrier in North America

1959 Formation of National Energy Board (NEB) of Canada

1960s First Brazilian long distance oil pipeline (TORGUÁ and REDUC)

1968 Discovery of Oil in Purdue Bay, Alaska

1969 National Environmental Policy Act (NEPA)

1970 Peak production of oil in the United States (9.4 million barrels of production per day in the Lower 48 states

1970s–1990s The Advent of Specialty Pipes

1973 Trans-Alaska Pipeline Authorization Act (by President Nixon)

1976 Creation of Call Before You Dig "DigAlert"

1977 Creation of FERC (US Federal Energy Regulatory Commission under Department of Energy Organization Act)

1977 Trans-Alaska Pipeline Operation Start

1979 The Hazardous Liquid Pipeline Safety Act, USA

1985 Construction of 30-inch pipeline to link South-Central coast oil with Freeport, the Celeron Pipeline Company of California.

1992 US Energy Policy Act

1999 Formation of BP Amoco Pipelines (North America)

2003 1st Use of X120 steel pipe

2006 BTC (Baku-Tbilisi-Ceyhan) Pipeline became operational, July 13th.

1.7.2 Codes, Standards and Regulations (Addressing Liquid Pipeline Design, Construction and Operation)

1.7.3 Codes

The primary liquid pipeline design codes that are widely used are as follows:

- **ASME B31.4,** Pipeline Transportation Systems for Liquid Hydrocarbons and Other Liquids
- **BS 8010,** Code of practice for pipelines. Pipelines on land, parts 1 and 2 (published by BSI in 2004, later withdrawn to support BS EN 14161 now ISO 16623)
- **EN 14161,** European code for design of petroleum and gas transport systems
- **ISO 13623,** Petroleum and natural gas industries - Pipeline transportation systems (an International code)
- **DNV OS-F101,** Rules for Design, Construction and Inspection of Marine Pipelines and Pipeline Risers (The design section of this code is mainly based on ASME codes)
- **API RP-1111** (Design, Construction, Operation, and Maintenance of Offshore Hydrocarbon Pipelines)
- **AS 2885 parts 0, 1, 2, 3, 4 and AS/NZS 2885.5** "Pipelines—Gas and Liquid Petroleum"
- **CSA Z 662,** Oil and gas pipeline systems

Other widely used codes for pipeline and facilities are listed in Table 1-5.

TABLE 1-5. Major applicable hydrocarbon liquid pipeline codes

Satndard/Code	Desciiption	Type
API RP 1107	Pipeline Maintenance Welding Practices	Recommended Practice
API 6D	Specification for Pipeline Valves (Gate, Plug, and Check Valves)	Specification
ANSI/API 1104	Welding of Pipelines and Related Facilities	Standard
API 5L & Supplements	Specification for Line Pipe,"	
API 1102	Steel Pipelines Crossing Railroads and Highways,"	Standard
API 1109	Marking Liquid Petroleum Pipeline Facilities	Standard
API 1113	Developing a Pipeline Supervisory Control Center	Guideline
API 1117	Recommended Practice for Movement in inservice Pipelines	Recommended Practice
Api 1130	Computational Pipeline Monitoring for Liquids Pipelines	
API 1149	Pipeline Variable Uncertainties and Their Effects on Leak Detectability	
API 1155	Evaluation Methodology for Software-Based Leak Detection Systems	
API 1157	Hydrostatic Test Water Treatment and Disposal	
API 1160	Managing System Integrity for Hazardous Liquid Pipelines	
API 1162 RP	Public Awareness Programs for Pipeline Operators	Recommended Practice
API 1163	In-Line Inspection Systems Qualification Standard	Standard
API 1165	Recommended Practice for Pipeline SCADA Displays	Recommended Practice
API 1168	Pipeline Control Room Management	Recommended Practice
ANSI/API RP 1110,	Pressure Testing of Liquid Petroleum Pipelines,"	Recommended Practice
API RP 2200	Repairing Crude Oil, Liquid Petroleum Gas, and Product Pipelines	Recommended Practice
API 570	Piping Inspection Code: Inspection, Repair, Alteration, and Rerating of In-service Piping Systems	
ASME A13.1	Standard for the Identification of Piping Systems	Standard
ASME B31.1	Power Piping	
ASME B31G	Manual for Determining the Remaining Strength of Corroded Pipelines:	A Supplement to ASME B31 Code for Pressure Piping
ASME B31.11	Slurry Transportation Piping Systems	
ASME B 31.3	Process Piping	
ASME B 31.5	Refrigeration and Heat Transfer Components	
IEEE	National Electrical Safety Code	Institute of Electrical and Electronics Engineers
IPS (Institute of petroleum)	Area Classification Code for Petroleum Installations	Code of Safe Practice
NFPA Code	National Electrical Code Handbook	National Fire Protection Association
Building Codes	Uniform Building Codes	International Conference of Building Officials,
Fire Codes	Uniform Fire Codes	International Conference of Building Officials,
NACE RP 0169	Control of External Corrosion on Metallic Piping Systems	Recommended Practice
NACE RP 0102	In-Line Inspection of Pipelines	Recommended Practice
SNiP 2.05.06	Construction regulations and rules. Transmission pipelines	
DIN 2470	Steel gas pipelines for permissible working pressures exceeding 16 bar; requirements for pipeline components	Standard
NACE RP 0102	Requirements for pipeline systems , Parts 1 to 6	

1.7.4 Regulations [13,14]

Generally, governmental authorities, such as federal, state, and municipal agencies, have the right to issue regulations defining minimum requirements for the pipeline and related facilities.These regulations are legally binding for the design, construction, and operation of pipeline system facilities which come under the jurisdiction of the relevant authority.

The American Society of Mechanical Engineers (ASME) Code for Pressure Piping ASME B31.4, *Pipeline Transportation Systems for Liquid Hydrocarbons and Other Liquids* is generally the industry's norm for design and operation of hazardous liquid pipelines. The main design principle of this code is that the pipeline is assessed as a pressure vessel, by limiting the pipe hoop stress to a specific fraction of minimum specified yield strength (SMYS). In US, the code is incor-

TABLE 1-6. US and Canadian Regulatory Definition

PIPELINE TYPE	US DOT PIPELINE & HAZARDOUS MATERIALS SAFETY ADMINISTRATION	US MINERALS MANAGEMENT SERVICE	US EPA	NATIONAL ENERGY BOARD OF CANADA
ON-SHORE TRANSMISSION LINE OR PIPELINE SYSTEM	The regulations applicable to hazardous liquid pipelines, which include both crude and petroleum product lines are found at 49 CFR Parts 195 and 199. The statutory provisions are found at 49 USC § 60101 *et seq.* In 49 CFR Part 195.2, the definition of pipeline is: "*Pipeline* or *pipeline system* means all parts of a pipeline facility through which a hazardous liquid or carbon dioxide moves in transportation, including, but not limited to, line pipe, valves, and other appurtenances connected to line pipe, pumping units, fabricated assemblies associated with pumping units, metering and delivery stations and fabricated assemblies therein, and breakout tanks." PHMSA does not use the terms transmission or transportation line with respect to hazardous liquid pipelines.See http://ops.dot.gov/regs/regsindex.htm for more information. The general rule is that PHMSA regulates all pipeline transportation "in or affecting interstate or foreign commerce, including pipeline facilities on the Outer Continental Shelf." § 195.1(a). There are many exceptions to applicability. § 195.1(b)	Not regulated by MMS	Pipelines are transportation-related and pursuant to Executive Order 12777, the responsibilities are split between DOI and DOT. To clarify EPA's role, an MOU was signed by DOI, DOT and EPA in late 1993/early 1994. Appendix B of 40 CFR Part 112 contains this MOU. If a discharge occurs from a pipeline, then the US EPA would typically get involved because the discharge may be a violation of the Clean Water Act as amended by the Oil Pollution Act of 1990 (OPA 90). Also, EPA's definition of "facility" at 40 CFR Part 112.2 includes pipelines.	The NEB regulates inter-provincial and international pipelines. Pipeline is defined in the *National Energy Board Act* section 2 as follows: "pipeline" means a line that is used or to be used for the transmission of oil, gas or any other commodity and that connects a province with any other province or provinces or extends beyond the limits of a province or the offshore area as defined in section 123, and includes all branches, extensions, tanks, reservoirs, storage facilities, pumps, racks, compressors, loading facilities, inter-station systems of communication by telephone, telegraph or radio and real and personal property, or immovable and movable, and works connected to them, but does not include a sewer or water pipeline that is used or proposed to be used solely for municipal purposes. **Regulations** • Onshore Pipeline Regulations, 1999 - SOR/99-294 • National Energy Board Processing Plant Regulations - SOR/2003-39 • National Energy Board Pipeline Crossing Regulations, Part I - SOR/88-528 • National Energy Board Pipeline Crossing Regulations, Part II - SOR/88-529 **Guidance notes** • Guidance Notes for the Onshore Pipeline Regulations, 1999 - Amendment 1 (20 January 2003) • Guidance Notes for the National Energy Board Processing Plant Regulations (28 July 2003)

TABLE 1-6. Continued

PIPELINE TYPE	US DOT PIPELINE & HAZARDOUS MATERIALS SAFETY ADMINSTRATION	US MINERALS MANAGEMENT SERVICE	US EPA	NATIONAL ENERGY BOARD OF CANADA
OFF-SHORE TRANSMISSION LINE OR PIPELINE SYSTEM	As noted above, PHMSA regulates all pipeline transportation "in or affecting interstate or foreign commerce, including pipeline facilities on the Outer Continental Shelf." §195.1(a), although there are many exceptions to applicability with respect to offshore lines. §195.1(b) "*Offshore* means beyond the line of ordinary low water along that portion of the coast of the United States that is in direct contact with the open seas and beyond the line marking the seaward limit of inland waters." §195.2	30 Code of Federal Regulations (CFR) 250.1001: "DOI pipelines include: (1) Producer-operated pipelines extending upstream (generally seaward) from each point on the OCS at which operating responsibility transfers from a producing operator to a transporting operator; (2) Producer-operated pipelines extending upstream (generally seaward) of the last valve (including associated safety equipment) on the last production facility on the OCS that do not connect to a transporter-operated pipeline on the OCS before crossing into State waters; (3) Producer-operated pipelines connecting production facilities on the OCS; (4) Transporter-operated pipelines that DOI and DOT have agreed are to be regulated as DOI pipelines; and (5) All OCS pipelines not subject to regulation under 49 CFR parts 192 and 195." And: "Pipelines are the piping, risers, and appurtenances installed for the purpose of transporting oil, gas, sulfur, and produced water. (Piping confined to a production platform or structure is covered in Subpart H, Production Safety Systems, and is excluded from this subpart.)"	US EPA does not regulate offshore pipelines.	East coast: the NEB jointly regulates with the Canada-Nova Scotia Offshore Petroleum Board Sable Offshore Energy Inc. pipeline. West coast: there are no offshore pipelines.

TABLE 1-6. Continued

PIPELINE TYPE	US DOT PIPELINE & HAZARDOUS MATERIALS SAFETY ADMINISTRATION	US MINERALS MANAGEMENT SERVICE	US EPA	NATIONAL ENERGY BOARD OF CANADA
GATHERING OR FLOW LINE	"*Gathering line* means a pipeline 219.1 mm (8 5/8 in) or less in nominal outside diameter that transports petroleum from a production facility." § 195.2. PHMSA does not regulate "onshore gathering lines in rural areas except gathering lines in the inlets of the Gulf of Mexico subject to §195.413." §195.1(b)(4) PHMSA does not regulate "flow lines" which are considered part of onshore production. §195.1(b)(7). "Production facility" is defined in § 195.1; "flow line" is not. The agency plans to initiate rulemaking on the regulation of liquid gathering lines in 2006. The Regulatory Identification Number for the rulemaking on the OMB Unified Agenda is 2137-AD98.	MMS does not have a specific definition for gathering or flow lines. However, MMS does have flow line regulations covering these types of lines at 30 CFR 250.803; and pipelines at 30 CFR 250 Subpart J.	Gathering lines or flowlines are not explicitly defined, but production facility is defined in 40 CFR Part 112.2 as follows: "*Production facility* means all structures (including but not limited to wells, platforms, or storage facilities), piping (including but not limited to flowlines or gathering lines), or equipment (including but not limited to workover equipment, separation equipment, or auxiliary non-transportation-related equipment) used in the production, extraction, recovery, lifting, stabilization, separation or treating of oil, or associated storage or measurement, and located in a single geographical oil or gas field operated by a single operator." See: http://ecfr.gpoaccess.gov/cgi/t/text/text-idx?c=ecfr&sid=5416e5c98f51011b0ed79559353b49dc&rgn=div8&view=text&node=40:21.0.1.1.7.1.6.2&idno=40	See Pipeline definition above
PUMP STATION	PHMSA has no definition, but the definition of pipeline includes "pumping units" and "fabricated assemblies associated with pumping units" (195.2) § 195.262 provides special regulations for pumping equipment.	Not regulated by MMS	Same as facility piping above	See Pipeline definition above

TABLE 1-6. Continued

PIPELINE TYPE	US DOT PIPELINE & HAZARDOUS MATERIALS SAFETY ADMINISTRATION	US MINERALS MANAGEMENT SERVICE	US EPA	NATIONAL ENERGY BOARD OF CANADA
FACILITY PIPING	§195.2: "*In-plant pipeline system* means piping that is located on the grounds of a plant and used to transfer hazardous liquid or carbon dioxide between plant facilities or between plant facilities and a pipeline or other mode of transportation, not including any device and associated pipeline that are necessary to control pressure in the pipeline under § 195.406(b)." PHMSA does not regulate "transportation of a hazardous liquid or carbon dioxide through onshore production,(including flow lines), refining, or manufacturing facilities, or storage or in-plant pipeline systems associated with such facilities" § 195.1(b)(7). Note: A 2003 Federal Register document added a different § 195(b)(7) without renumbering. This will be corrected in a future Code of Federal Regulations.	MMS does not have a specific definition defining facility piping. However, MMS does regulate all facility piping that is part of MMS regulated facilities. Some of the regulations can be found at 30 CFR 250.802.	Authority is in 40 CFR Parts 112.8(d), *Facility transfer operations, pumping, and facility process (all other onshore facilities)* and 112.9(d), *Facility transfer operations, oil production facility.* **The** definition of "facility" at 40 CFR Part 112.2 includes piping. "Facility means any mobile or fixed, onshore or offshore building, structure, installation, equipment, pipe, or pipeline (other than a vessel or a public vessel) used in oil well drilling operations, oil production, oil refining, oil storage, oil gathering, oil processing, oil transfer, oil distribution, and waste treatment, or in which oil is used, as described in Appendix A to this part. The boundaries of a facility depend on several site-specific factors, including, but not limited to, the ownership or operation of buildings, structures, and equipment on the same site and the types of activity at the site."	See Pipeline definition above
TRANSFER LINE	PHMSA does not use this term.	MMS does not use this term.	See Pipeline definition above	See Pipeline definition above

TABLE 1-6. Continued

PIPELINE TYPE	US DOT PIPELINE & HAZARDOUS MATERIALS SAFETY ADMIN.	US MINERALS MANAGEMENT SERVICE	US EPA	NATIONAL ENERGY BOARD OF CANADA
BREAKOUT TANK	The definition of pipeline includes "breakout tanks" as in § 195.1: "Breakout tank means a tank used to (a) relieve surges in a hazardous liquid pipeline system or (b) receive and store hazardous liquid transported by a pipeline for reinjection and continued transportation by pipeline."	Any atmospheric or pressure vessel located on an MMS regulated faci ity is regulated by the MMS.	Regarding EPA authority to regulate, see jurisdiction memo at: http://www.epa.gov/oilspill/pdfs/dotepa.pdf. This memorandum clarifies jurisdictional issues over breakout tanks and bulk storage containers, and establishes mutual goals for the EPA and DOT. Examples of complex facilities (those facilities containing both transportation-related and non-transportation-related components) are provided, with diagrams showing jurisdictional delineations. EPA's definition at 40 CFR Part 112.2 states: "*Breakout tank* means a container used to relieve surges in an oil pipeline system or to receive and store oil transported by a pipeline for reinjection and continued transportation by pipeline. "	See Pipe ine definition above
OTHER PIPELINE TYPE	No other pipeline types are regulated by PHMSA.	No other pipeline types are regulated by MMS.	EPA regulates piping at non-petroleum oil facilities, such as animal fat and vegetable oil facilities defined at 40 CFR Part 112.2: "Non-petroleum oil means oil of any kind that is not petroleum-based, including but not limited to: Fats, oils, and greases of animal, fish, or marine mammal origin; and vegetable oils, including oils from seeds, nuts, fruits, and kernels."	No other pipeline type regulated
AGENCY CONTACT	Barbara Betsock, Attorney Advisor Office of Pipeline Safety Pipeline and Hazardous Materials Safety Administration 400 Seventh Street SW Rm 8417 Washington. D.C. 20590 Phone: 202 366 4361; Fax: 202 366 7041 Barbara.Betsock@dot.gov http://phmsa.dot.gov	Theresa Bell, Petroleum Engineer Office of Facilities, Safety and Enforcement U.S. Minerals Management Service 770 Paseo Camarillo, Camarillo, CA 93010 Phone: (805) 389-7554; Fax: (805) 389-7592 theresa.bell@mms.gov www.mms.gov	J. Troy Swackhammer, Chemical Engineer Office of Emergency Management, Regulation and Policy Development Division, U.S. EPA 1200 Pennsylvania Ave, NW, Washington D.C. 20460, Mail Code 5104A Phone: 202-564-1966; Fax: 202-564-2625 swackhammer.j-troy@epa.gov www.epa.gov/oilspill	Robert LeMay, Emergency Management Specialist Operations Business Unit National Energy Board of Canada 444 Seventh Avenue SW Calgary. Alberta, Canada T2P 0X8 Phone: 403-299-3187; Fax: 403-292-5875 rlemay@neb-one.gc.ca www.neb-one.gc.ca

porated by reference in US DOT Office of Pipeline Safety Code of Federal Regulation **49 C.F.R. PART 195—TRANSPORTATION OF HAZARDOUS LIQUIDS BY PIPELINE.**

This regulation is overseen and enforced by the US Pipeline and Hazardous Materials Safety Administration (PHMSA) under US Hazardous Liquid Pipeline Safety Act of 1979.

AS 2885-1:1997—Pipelines, gas and liquid petroleum, design and construction relates to design, construction, testing, operations, and maintenance of gas and petroleum pipelines that operate at pressures in excess of 1050 kPag.

In Canada National Energy Board (NEB) regulates the inter-provincial and international pipelines, i.e., pipelines that cross Canadian border to USA. Table 1-6 provides the regulatory definition for US and Canadian pipelines (PS/BC):

1.8 MAJOR PIPELINE FACILITIES LAYOUT

1.8.1 Pump Station

Pump stations are required to boost the pressure of the liquid hydrocarbon for transportation through the transmission pipeline. At a minimum, pump stations include pumps and prime movers to the pumps. Centrifugal pumps are most commonly used in liquid transmission pipelines due to their ability to economically handle large volumes of product (see Chapter 4). Prime movers are primarily electric motors due to their relatively low cost and inherent reliability but in areas where electric power is not available, gas turbine or diesel internal combustion engines may be utilized.

Pump station piping consists of piping manifolds and valves separate from the main pipeline. These facilities allow the connection of the pumping units to the station piping and control functions. Often, multiple pump units are installed in series, parallel or combined configurations to meet operational requirements.

Pump stations require infrastructure that addresses waste handling such as sump systems and pump out facilities. Also, the handling and injection of additives, such as drag reducers, often occurs at pump stations. Pump stations are typically equipped with sufficient emergency power generation to support monitoring and control systems to accomplish an immediate safe shutdown.

Pump stations can in certain instances have co-located facilities that support pipeline operation or facilitate shutdowns or maintenance on pipeline segments. This may include breakout tankage for temporary storage of product or for use in managing line pressures and controlling product surges. Finally, pump stations are in some instances, co-located at pipeline terminal facilities.

A typical pump station layout is shown in Figure 1-10.

1.8.2 Metering/Measurement

Measurement is a reference standard used for the quantitative comparison of properties such as fluid flow (mass, volume, energy) and static storage tank gauging (refer to Chapter 7).

There are two types of flow measuring devices:

- Direct measurement devices (energy extractive). These devices use the fluid properties as direct measuring parameters to determine the fluid flow rate. Such parameters are mass, density, viscosity, temperature, pressure, etc. Examples are Positive Displacement Meters and Mass Flow Meters.

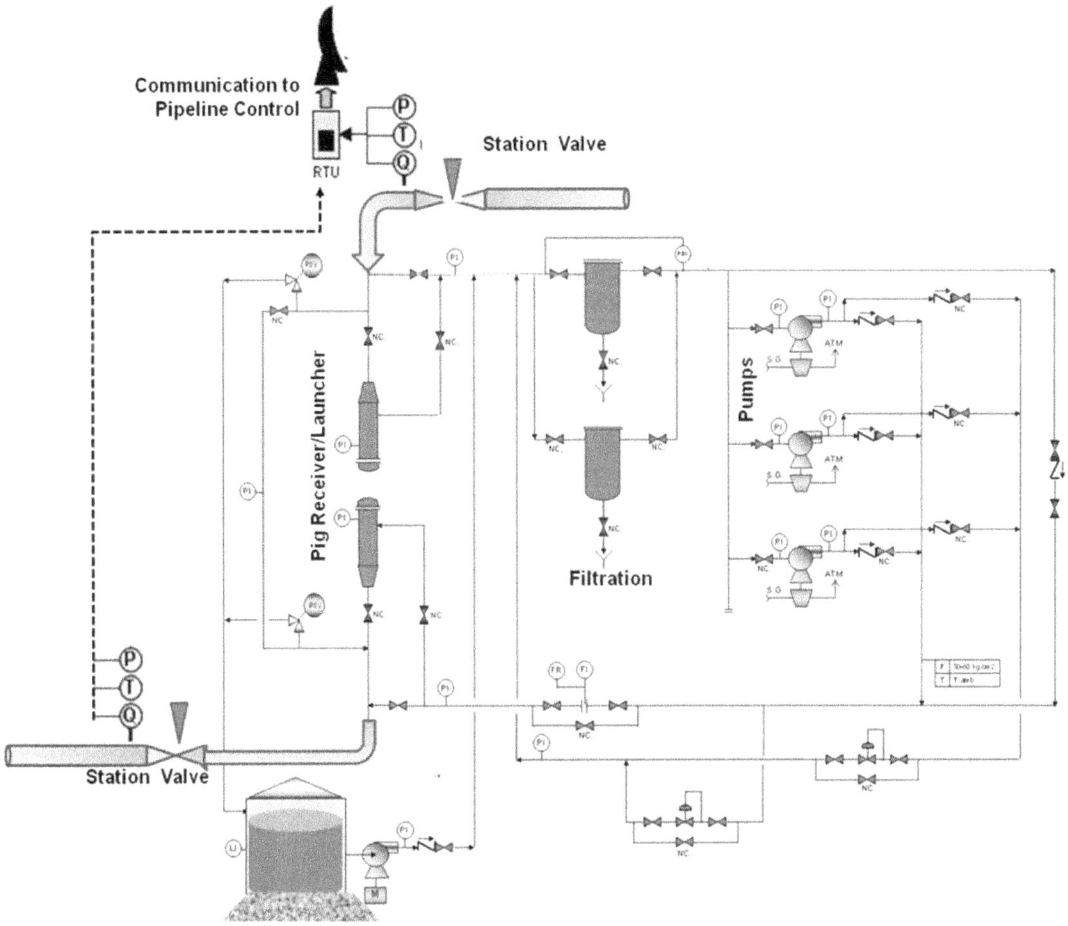

Figure 1-10. Typical pump station mechanical/piping layout

- Inferential Measurement Devices (energy additive). Inferential measuring devices use parameters other than the fluid properties, such as electronic pulse counts, meter factors, system factors, linear and rotational velocities form components of the measuring devices to infer the fluid flow rate. Examples are: turbine, orifice plate, vortex shedding, venturi and ultrasonic and Coriolis mass flow meters.

Generally, hydrocarbon liquid measurement systems for pipeline applications may include any or all of the following:

- Metering
  - Primary devices
  - Secondary devices
  - Tertiary devices
- Control
  - Pressure regulation
  - Flow control
  - Overpressure protection

- Verification
  - Proving
  - Sampling
  - Quality
- Others
  - Filtration / Separation
  - Heating
- Temperature, Static and Differential Pressure Measuring Devices/Recording as required
  - Chart Recorders,
  - Manometers,
  - DP Cells, etc.
  - Densitometers
  - Viscometers

Achieving consistent high accuracy measurements is purpose prime requirement for the pipeline industry. Custody transfer measurement facilities can achieve an accuracy of better than + 0.25%. Figure 2-17 show the measurement accuracy profiles for custody transfer/royalty metering [15].

Figure 1-11 refers to two general types of metering; Custody Transfer and Royalty metering as explained below:

a) Custody Transfer. A measurement or transfer of a deliverable at the point of change of responsibility/ownership, providing quantity and quality information used for the physical and fiscal documentation of a change in ownership and/or

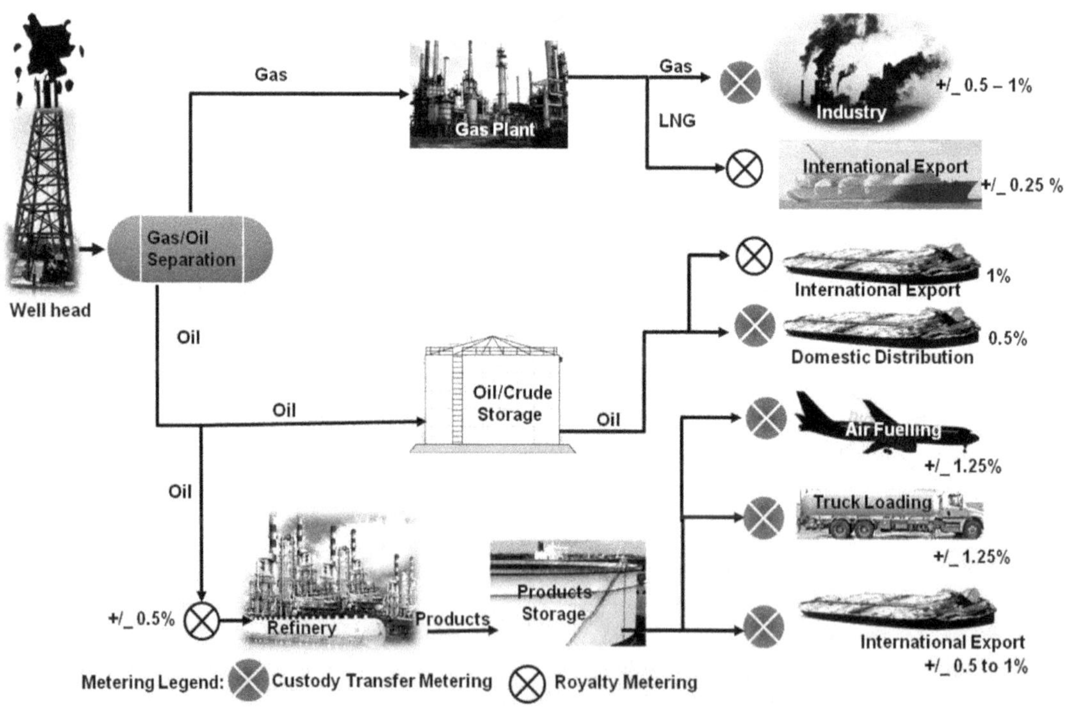

Figure 1-11. Typical spread of measurement uncertainties (adapted from [15])

responsibility for the commodities between two parties possessing a contractual agreement and bound by the terms and conditions of such a contract.

b) Royalty Transfer. A specialized form of measurements and is the basis for paying a fee or percentage of the revenues generated by the sales (royalty) to owner(s).

Measurement is the basis of trade between oil producers, royalty owners, pipeline transportation companies, refiners, marketers, regulators, governmental authorities and the general public.

Therefore, measurement of hydrocarbon fluids applies to steady-state mass flow conditions for fluids that, for all practical purposes, are considered to be clean, single phase, homogeneous and Newtonian (if possible) under the operating conditions of the measurement facility/metering station.

Chapter 7 describes the type of metering used for pipeline applications.

To some extent, the piping components for liquid meter stations (Figure 1-12) vary depending on the metering purpose; receipt, custody transfer, transmission, or delivery. But in each case, they require components to isolate and control the flow (valves), components to protect the meter from contaminants (filtration), flow conditioning elements, and the meters themselves. Additionally the station may require piping arrangements for a meter prover and appertaining valving and supports. A meter prover (Figure 1-13) is generally required to calibrate the meter for custody transfer purposes.

Although primarily utilized to measure the volume, quality, and consistency of product for billing purposes and delivery receipts, storage tank monitoring and product metering can be used with line pressure monitors to verify that pipeline integrity has not been compromised. Any discrepancy could indicate some sort of system leak.

Meter provers are designed to prove the accuracy of meters under operating conditions. They use detector switches and known volumetric section of pipe. Two typical configurations are:

- unidirectional
- bidirectional

For design, operation and maintenance of liquid metering system and measurement techniques refer to [4]. Further details are provided in Chapter 7.

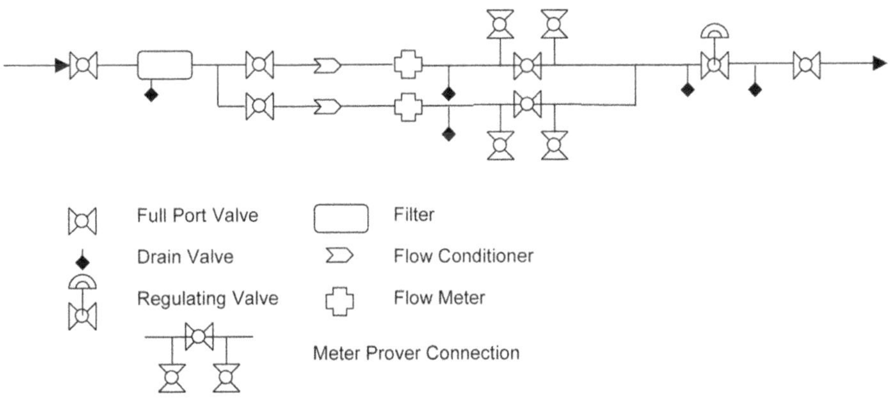

Figure 1-12. Typical liquid hydrocarbon meter station piping schematics

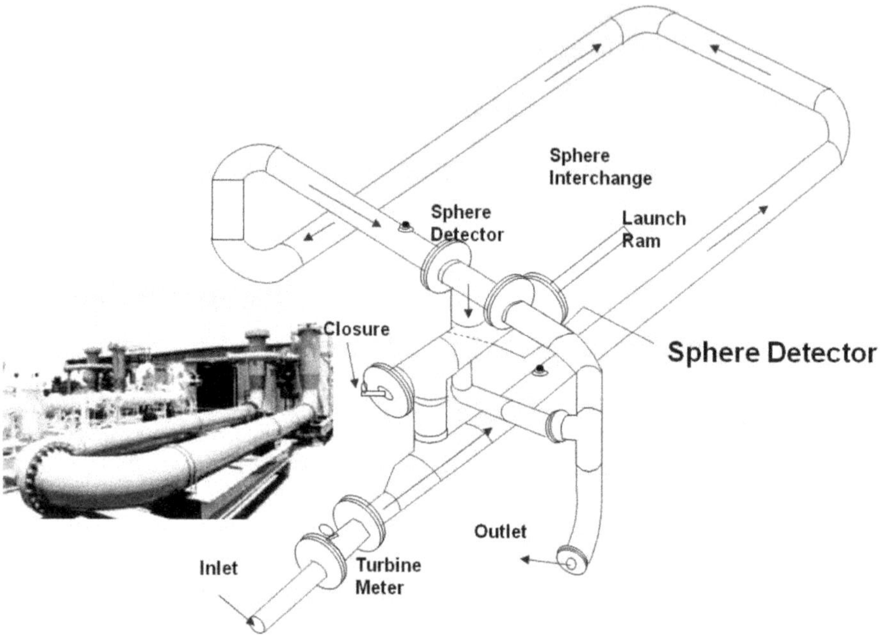

Figure 1.13. Typical liquid meter proving components and piping arrangement

1.8.3 Valve and Manifolds

1.8.3.1 Valves

Valves in pipelines are generally referred to as Emergency Flow Restricting Devices (EFRD) or sectionalizing valves on gas and liquids pipelines. This wording is well defined in API 1130 (2001).

Valves have been utilized in pipeline transmission systems ever since pipelines have been used for transportation of liquid hydrocarbons. Pipeline companies rely on their liquid network of pipelines to consistently deliver a required amount of products to customer companies/clients at delivery points. Valves and valve operators are an integral part of this system.

The actual type of valve selected is dictated by parameters such as pressure drop, leakage (seat and system), type of fluid, actuation requirements, cost and maintenance requirements. Figure 1-14 schematically presents the most commonly used valves in transmission pipelines [4] and provides an outline and design features of such valves utilized in hydrocarbon pipeline transmission systems.

There are four types of valves used by as mainline transmission valves. These are:

- Gate valves
- Ball valves
- Plug valves and
- Check valves

There are also valves that are used for control, limit and/or relief of pressure and flow. These mostly used in pump/compressor/metering stations, they are also used on the mainline where required:

- Globe valve
- Relief valve
- Control valve

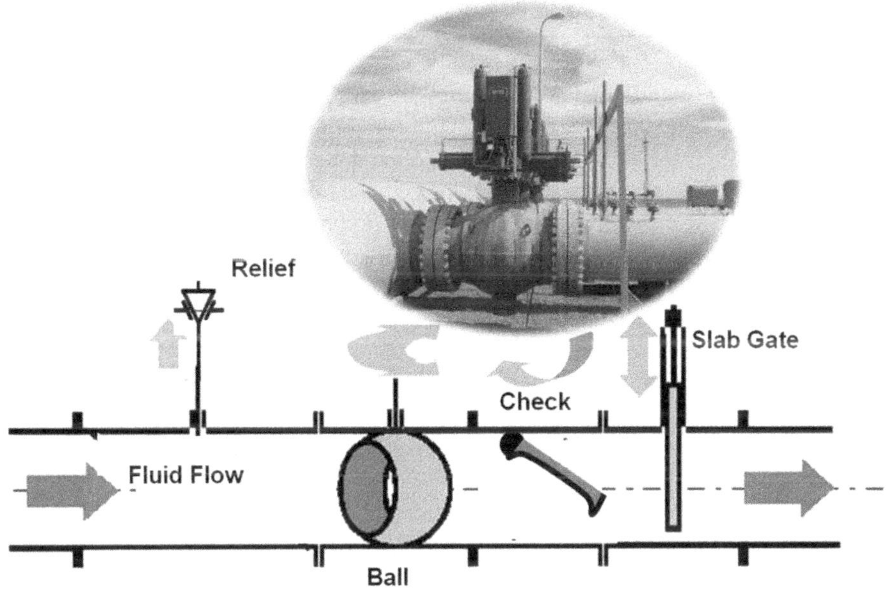

Figure 1-14. Schematic of most commonly used valves in transmission pipelines and a ball valve with an operator in an oil pipeline [4]

1.8.3.2 Manifolds

Manifolds are piping with valving arrangements (Figure 1-15) to direct or divert flow of hydrocarbon liquids to and from pumps and pump stations, terminals and tank farms, and field gathering systems.

The manifold design allows for both isolation of a facility and/or diversion of flow. Piping must be designed to avoid mechanical and flow induced vibration and allow flexibility of operation as well as pressure and thermally induced stress conditions.

Figure 1-15. Typical pump station piping manifold

1.9 GENERAL PIPELINE OPERATIONS

Liquid pipeline operations are influenced by the design of the system either carrying a dedicated single product or a combination of products in a sequential batch form. Operation of a batched pipeline which allows multiple products (high and low vapor pressure) to be shipped through the same pipeline is however more complex. Batching is generally achieved by injecting liquids sequentially into the pipeline and thus pipeline operation is dependent on the following factors:

- Number of products batched
- Nominations/contracts
- Batch injection methods
- Batch sequencing (schedule) and cycle
- Batch size, contamination level/quality
- Storage/capacity available
- Degree of automation

Liquid hydrocarbon pipeline operation commences with the pipeline operator receiving a request for shipment from various sources/customers (Shippers) for single or different products and delivery of product(s) to single or multiple locations. This is referred to as a Nomination. Typically, this is achieved through the following steps by the pipeline operator:

- Scheduling department accepts nominations and develops a pumping schedule;
- Control Center which remotely operates the systems commences the pumping operation;
- Product is pumped into the pipeline system. Flow velocity usually 5-13 km h (3-8 mph);
- Different products can be batched;
- Product is transported to the intended destination, measured and tested;
- Customers (Shippers) are billed for product movements through system.

The pipeline operator constantly tracks the location of the shipment (through an SCADA system, Figure 1-16) as it moves from the initial receipt point to its delivery point. The SCADA system is comprised of field devices (instrumentation, data gathering units) and communication systems. The field Instrumentation includes

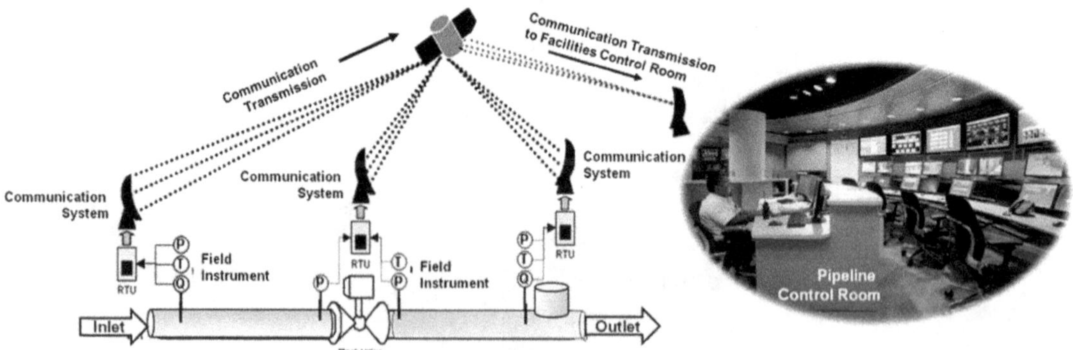

Figure 1-16. Typical pipeline supervisory data acquisition and control system

flow, pressure and temperature gauges/transmitters, and other devices to measure the relevant data required. These instruments are installed along the pipeline on some specific locations, such as injection or delivery stations, pump stations, and block valve stations.

The information measured by field instruments is gathered in local Remote Terminal Units (RTUs). It is usual to find RTUs installed at every station along the pipeline. The RTUs transfer the field data to a central control center in real time using communication systems, such as satellite channels, microwave links, land line or cellular phone connections. In the control center, all the data related to field measurement is consolidated in one central database. Such a data is also utilized for accounting purposes.

The amount charged (with respect to pipeline transportation costs) to a Shipper is usually dependent upon the product transported, the volume, and the transportation distance (i.e., the distance between the receipt and delivery points).

Control room operators monitor hazardous liquid pipelines through various computer equipment that record and display operational information about the pipeline system. Critical functions, such as pressure, flow rates, and valve positions, allow controllers to make informed decisions about what actions to take. The operators are the first line of defense for assessment of any potential hazard to the pipeline. These operators can take appropriate action to lessen any impact that may threaten the pipeline, the public, and the environment.

In the US, the functions and qualification of the operators/controllers in control room who are assigned to monitor and control all or part of a pipeline facility through a SCADA system are defined by regulations [16]. The following definitions apply:

- The "operator of a pipeline facility" refers to the company that operates the pipeline. Thus, the regulation applies to the pipeline operating company. The controller is not the operator.
- Controller is defined as a qualified individual who remotely monitors and controls the safety-related operations of a pipeline facility via a SCADA system from a control room and who has operational authority and accountability for the remote operational functions of the pipeline facility on behalf of the operating company.
- Control room is defined as an operations center staffed by personnel charged with the responsibility for remotely monitoring and controlling a pipeline facility.
- Supervisory Control and Data Acquisition (SCADA) system is defined as a computer-based system or systems used by a controller in a control room that collects and displays information about a pipeline facility and may have the ability to send commands back to the pipeline facility

Control room management is regulated by two sets of codes through US DOT, Pipeline and Hazardous Material safety Administration [17, 18]. While the final rule was issued in Dec 3, 2009, it has been effective through amendments to DOT, CFR part 195-93 since Feb 2010. It may be noted that LNG excluded entirely from regulations.

REFERENCES

[1] CAPP (Canadian Association of Petroleum Producers), 2011, "Crude Oil - Forecast, Markets & Pipelines" http://www.capp.ca/getdoc.aspx?DocId=190838.

[2] PS/BC (Pacific States/British Columbia Oil Spill Task Force), 2006, "Pipeline Type: Regulatory Definition," June, http://www.oilspilltaskforce.org/docs/project_reports/PipelineDefinitions.pdf.

[3] Perry, M., 2012, "Chart Of The Day: Natural Gas vs. Oil," *Daily Markets*, March 13, http://www.dailymarkets.com/economy/2012/03/13/chart-of-the-day-natural-gas-vs-oil/.

[4] Mohitpour, M., Van Hardeveld, T., Peterson, W., and Szabo, J., 2010, *Pipeline Operation and Maintenance – A Practical Approach*, 2nd Edition, ASME Press, New York.

[5] CIA 2010, "The World Fact Book - Field Listing Pipelines," https://www.cia.gov/library/publications/the-world-factbook/fields/2117.html?countryName=&countryCode=&regionCode=%C5%A1.

[6] BP, 2009, "Statistical Review of World Energy," June http://www.bp.com/multipleimagesection.do?categoryId=9023755&contentId=7044552.

[7] Smith, G., 2008, "Land Based Liquid Pipelines" *API Tanker Conference*, Houston, TX, June 23, http://www.api.org/meetings/topics/marine/upload/Greg_Smith.pdf.

[8] Rabinow, R. A., 2004, "The Liquid Pipeline Industry in the United States: Where It's Been Where It's Going," Report for the US Association of Oil Pipe Lines, April, http://www.aopl.org/pdf/rabinow.pdf.

[9] Alder, K., 2002, "Oil Refining Changes Transportation, History, and Ways of Life Chemical Week," Sept. 18, http://www.findarticles.com/cf_0/m3066/36_164/92232563/print.jhtml.

[10] Miller, E. C., 1974, *Pennsylvania's Oil Industry*, The Pennsylvania Historical Association, 69 pp.

[11] Giddens, P. H., 1938, *The Birth of the Oil Industry*, The Macmillan Company, 216 pp.

[12] Tait, S. W., Jr., 1946, *The Wildcatter*, Princeton University Press, 218 pp.

[13] NEB (National Energy Board of Canada), 2011.

[14] Reed, S., 2009, "The History of Oil Pipeline Regulation" Report for Association of Oil Pipelines," Sept. 17, http://www.ferc.gov/help/pub-ref-rm/history-oil-pipeline-regulation.pdf.

[15] Basrawi, Y.F., 1999, "Crude and Hydrocarbon Measurement Technologies", Society of Petroleum Engineers (SPE) Annual Technical Conference and Exhibition, 3–6 October, Houston, Texas.

[16] Byrd, W. R., (Bill), 2011, "New Control Room Management Regulations Require Structured Management Approach," *Oil & Gas Journal*, **238**(8), Aug., http://pipelineandgasjournal.com/new-control-room-management-regulations-require-structured-management-approach?page=show.

[17] PHMSA, 2010, "Control Room Management," http://publicutilities.utah.gov/phmsapart192controlroommanagement.pdf.

[18] Klinger, P., 2011, "U.S. Department of Transportation Takes Steps to Improve Pipeline Control Room Management," US DOT Directive PHMSA 10-11, June 15 http://www.phmsa.dot.gov/staticfiles/PHMSA/DownloadableFiles/Press%20Releases/phmsa10011.pdf.

[19] BP, 2002, "Our History," http://www.bppipelines.com/history.html.

Chapter 2

HYDROCARBON LIQUID PROPERTIES

This chapter outlines and describes properties and parameters important to the design and operational issues related to pipelines transporting hydrocarbon liquids. It describes various liquid property terms and provides either data for use or equations for predicting/calculating such properties.

2.1 HYDROCARBON LIQUIDS

Petroleum products are mixtures of hydrocarbons (of varying density and viscosity), or molecular compounds of hydrogen and carbon. The products range from natural gases to crude oils. The differences in petroleum products are due to varying properties of hydrogen and carbon making up the petroleum molecule. Natural gas contains a high ratio of hydrogen to carbon (H/C) molecules at the light end. On the other hand, bitumen contains much lower H/C ratio at the heavy end.

Crude oils, differ in color from almost clear to amber, green, brown, or black (Figure 2-1). Crude oil is classified as light crude (high API gravity), intermediate crude, heavy crude, and extra heavy crude (oil) or bitumen (lowest API gravity usually 8 to 10), refer to Chapter 6 for details. Crude oil can also be sweet or sour, according to the sulfur (S) content as follows:

- Sweet: $S < 0.5\%$ by weight,
- Intermediate: $0.5\% < S < 1.0\%$ (greater than 0.5% but less than 1.0%)
- Sour or high: $S > 1.0\%$.

In extraction from an oil reservoir, the crude oil will contain some amount of saltwater and particulate matter (sediment or mud) plus associated gas from the reservoir formation. Crudes (depending on the field) will have varying water content. Large quantities may be present if oil extraction is enhanced using water injection technology, see Chapter 6. Petroleum products from wellheads will generally require treatment and upgrading for pipeline transportation. Pipeline transportation specifications limit the following products specifically to an acceptable level to meet product quality and operational safety standards:

- Sediment & Water (S&W)
- H_2S
- Other impurities.

Liquid petroleum products can be generalized in a number of ways; here we will consider a break down by density. There are three categories:

Figure 2-1. Color of crude oils

1. Light density hydrocarbon liquids (including pure liquids; ethylene and propylene, mixture of light components such as ethane, propane, normal butane, and iso-butane). These may contain small amounts of other hydrocarbon liquids; e.g., ethane stream (>90% of ethane, and small amounts of propane, carbon dioxide, etc.);
2. medium density/mixed light products (including Natural Gas Liquids (NGL), natural gas condensate, natural gasoline, and Liquid Petroleum Gas (LPG);
3. heavy hydrocarbon products (include conventional, heavy crude, waxy crude and bitumen).

The petroleum product properties are reflected in the pipeline system designs and operations.

In the mixed light/medium density hydrocarbon liquids, NGL is a light hydrocarbon mixture extracted from natural gas and includes propane, butane, pentanes$^+$ and also may include traces amount of ethane. NGLs generally are classified according to their vapor pressure as:

- Condensate (composed of pentanes, hexane, heptanes, and a small amount of heavier hydrocarbons);
- Natural gasoline (composed of pentanes$^+$ plus and some amounts of butanes); and
- Liquefied petroleum gas (LPG-composed of propane, normal and iso-butane).

The vapor pressure of condensate is low, natural gasoline intermediate, and LPG high. Natural gasoline has an intermediate vapor pressure between condensate and LPG.

Condensate is typically recovered from field separation facilities (has a gravity of about 80°API) and has a low vapor pressure but the highest density among the three types of NGLs.

While the vapor pressure of condensate is lower than that of natural gasoline, the density of condensate is similar to but tends to be higher than that of natural gasoline, GPSA [1].

LPG (with typical gravity of around 120°API) is liquefied under pressure that is higher than its vapor pressure. LPG can be extracted from NGL and is often used as fuel and chemical feedstock.

The medium density products may include light or medium crudes, refined products such as gasoline and diesel, naphtha, condensate, etc. The changes in the density and viscosity of these products are relatively insensitive to temperature and pressure.

Heavy hydrocarbon products include conventional heavy crude, waxy crude, and bitumen.

2.2 HYDROCARBON LIQUIDS PHASE BEHAVIOR

To understand the properties of hydrocarbon liquids, the basic principles of phase behavior of a hydrocarbon system must be realized. Phase behavior of hydrocarbon liquids directly affects liquid pipeline system design and operation.

Depending on the pressure and temperature, the fluid density or volume changes, Hydrocarbons can exist as a liquid, gas or a mixture of both (i.e., two-phase flow). The relationship between fluid density or specific volume, pressure and temperature can be very complex and has to be expressed in an equation of state (EOS). To facilitate an easy understanding of phase behavior, the relationship is usually presented graphically. The graphical representation of the relationship is called a phase diagram. In the context of types of hydrocarbon liquids, it is necessary to identify and define the following key points on a typical phase diagram (Figure 2-2).

- The **bubble point** is the point at which the first drop of a liquid mixture begins to vaporize. Line **AC** defines the loci of the bubble points
- The **dew point** is the point at which the first drop of a gaseous mixture begins to condense. Line **BC** defines the loci of the dew points
- The **critical point C** is the state of pressure and temperature at which all intensive properties of the gas and liquid phases are equal. At the critical point, the corresponding pressure and temperature are called the critical pressure P_c and critical temperature T_c of the mixture

It may be noted that the dense phase is defined to be the region between the *critical temperature* and the *cricondentherm* if the pressure is above the *cricondenbar*. In practice, there is no clear line (i.e., *critical temperature*) dividing dense phase from liquid phase or other single line (i.e., *cricondentherm*) dividing the dense phase from the gas phase. It should also be noted that the shape of the phase diagram will alter depending upon the hydrocarbon constituents present in the fluid.

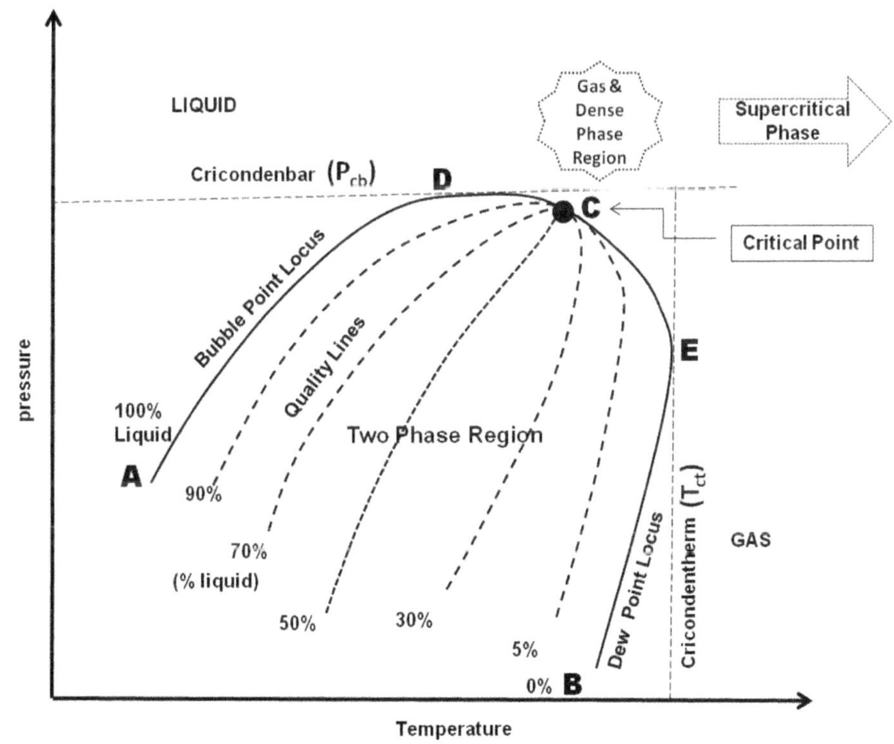

Figure 2-2. Typical phase diagram: definitions of terms [2]

Generally, with hydrocarbon liquids, the two-phase region is demarcated by the dew point curve at the bottom and a bubble point curve at the top of the phase diagram, as indicated in Figure 2-2. The loci of critical points, or the critical loci lie on a line of higher pressure and lower temperature. For the pressure-enthalpy relationship diagram, refer to Figure 2-11 that is detailed further in this chapter.

Additionally, the limits where the two phases of gas and liquid mixtures can also coexist must be defined. These are the Cricondentherm and the Cricondenbar. Figure 2-2 will be useful in describing what follows:

- **Cricondentherm (T_{ct})** — is the maximum temperature at which two phases (liquid and vapor) can coexist. The Cricondentherm is thus the maximum temperature above which liquid cannot be formed regardless of pressure (point E). The corresponding pressure is termed the Cricondentherm pressure $\boldsymbol{P}_{cb}$.
- **Cricondenbar (P_{cb})** — the maximum pressure at which two phases (liquid and vapor) can coexist. It is thus the maximum pressure above which no gas can be formed regardless of temperature (point D). The corresponding temperature is called the Cricondenbar temperature $\boldsymbol{T}_{cb}$.
- **Quality lines** — the dashed lines indicated in Figure 2-2 within the phase diagram are defined as quality lines. They describe the pressure and temperature conditions for constant percentage volumes of liquids. It may be noted that the quality lines converge at the critical point C.

It may be noted that heavier hydrocarbon liquids such as crude oils remain mostly in liquid form for transportation while light hydrocarbons such as ethane can be transported in a dense phase. The objective of this section is to review the basic principles of phase behaviors of a hydrocarbon system and their particular applications to liquid pipeline system design and operation.

In a phase diagram, a dense phase region lies above the critical point and to the right. The liquids in a dense phase have physical properties somewhere between that of the liquid and gas phases. They have the density of a liquid and viscosity of a gas. If the pressure on a liquid increases at constant temperature, there is no phase change as the liquid begins to enter the dense phase region. For the pressure and temperature ranges commonly used for pipeline applications, the dense phase can be encountered in high vapor pressure products such as ethane and ethylene and gases such as CO_2 and natural gas at very high pressures. The dense phase fluids except natural gas can be treated as liquid in liquid hydraulic calculation [3].

2.2.1 Phase Diagram Determination

An Equation of State (EOS) is generally utilized to determine the phase behavior of a hydrocarbon liquid, in particular, its pressure-temperature relationship which determines the thermodynamic state of the liquid as it is transported through a pipeline.

An equation of state describes the thermodynamic state of matter under a given set of physical conditions and is expressed in terms of temperature, pressure, density, or volume. Thus, it is useful in describing the relationships between thermodynamic properties (such as temperature, pressure, enthalpy, density or volume.) of fluids and mixtures of fluids. The functional form of an EOS can be expressed as:

$$f\left(P,V,T,a_k,k=1,n_p\right)=0 \qquad (2-1)$$

where a_k = EOS parameters

There are five universally accepted methods for predicting fluid properties for gas and liquid pipelines. These are the:

- Generalized natural gas correlations (Sarem)
- Benedict-Webb-Rubin-Starling (BWRS) EOS
- Soave modification to the original Redlich-Kwong (SRK) EOS
- Peng-Robinson (Peng) EOS
- The large acentric factor correction to Peng Robinson

Liquids are much less compressible than gasses. Even when a liquid is described with an equation similar to a gas equation, the constants in the equation will result in much less dramatic changes in volume with a change in temperature. Also, at constant volume, a temperature change will result in a much larger pressure change than would be the case for gases.

A common equation of state used for both liquids and solids is [4, 5]:

$$V_m = C_1 + C_2T + C_3T^2 - C_4p - C_5pT \qquad (2-2)$$

where
V_m = molar volume
T = temperature
p = pressure
C_1, C_2, C_3, C_4, C_5 = empirical constants

where the empirical constants are all positive and specific to each substance.

For constant pressure processes, this equation is often shortened to

$$V_m = V_{mo}\left(1 + AT + BT^2\right) \qquad (2-3)$$

where
V_m = molar volume
V_{mo} = molar volume at 0°C
T = temperature
A, B = empirical constants

Note: A and B are positive constants.

The equation of state created by Peng and Robinson has been found to be useful for both liquids and real gasses, particularly for phase equilibrium calculations.

$$p = \left[R \times T / \left(V_m - b\right)\right] - \left[a(T) / \left[V_m\left(V_m + b\right) + b\left(V_m - b\right)\right]\right] \qquad (2-4)$$

where
p = pressure
a = empirical constant
V_m = molar volume
R = ideal gas constant
b = empirical constant
T = temperature

However, for liquid pipeline applications for light hydrocarbons (such as ethane or propane) where the compositions of a fluid are known, Benedict, Webb, Rubin and

Starling (BWRS) is utilized by the pipeline industry as it allows a more rigorous analysis of the fluid properties [6].

$$P = \rho RT + \left(B_0 RT - A_0 - \frac{C_0}{T^2} + \frac{D_0}{T^3} - \frac{E_0}{T^4} \right) \rho^2 + \left(bRT - a - \frac{d}{T} \right) \rho^3$$
$$+ \alpha \left(a + \frac{d}{T} \right) \rho^6 + \frac{c\rho^3}{T^2} \left(\beta + \gamma \rho^2 \right) \exp\left(-\gamma \rho^2 \right) \qquad (2-5)$$

ρ = the molar density.

The BWRS equation method is a parametric equation of state. Values of the various parameters for up to 15 substances (including methane, ethane, ethylene, propane, propylene, isobutene, n-butane, isopentane, n-pentane, hexane, heptane, octane, carbon dioxide, hydrogen sulfide, carbon dioxide, and some pure components, hydrogen, nitrogen, are detailed elsewhere [7].

However, for heavier hydrocarbon liquids, a bulk equation of state is used for pipeline applications. It is expressed in terms of bulk modulus and thermal expansion coefficient for heavier hydrocarbons see section 2.3.3 on "Compressibility, Bulk Modulus and Thermal Expansion."

It is based on the assumption that the change rate of the liquid density is constant with respect to a change in pressure or temperature. The volume or density change rate with respect to the applied pressure at a constant temperature is called isothermal bulk modulus, and that with respect to the temperature at a constant pressure, the isobaric thermal expansion coefficient.

From the definitions of bulk modulus (see later in this chapter) and thermal expansion, a bulk equation of state can be expressed as:

$$\rho(P,T) = \rho(P_b, T_b) * \text{Exp}\left((P - P_b)/K \right) * \text{Exp}\left(-\alpha * (T - T_b) \right) \qquad (2-6)$$

where

$\rho(P,T)$ = density or specific gravity at pressure P and T
$\rho(P_b,T_b)$ = density or specific gravity at P_b and T_b
K = bulk modulus of the liquid
α = thermal expansion coefficient
P = flowing pressure
P_b = reference or base pressure
T = flowing temperature
T_b = reference or base temperature

Bulk modulus K and thermal expansion coefficient α depend on pressure (P) and temperature (T). The magnitude of change is small for heavier hydrocarbon liquids and can thus be treated as a constant. However, they are relatively large for lighter hydrocarbon liquids such as propane and ethane.

The following equation is often used for volume correction, particularly for custody transfer to a base condition [8]:

$$\rho_{(P,T)} = \rho_b \times C_T \times C_p \qquad (2-7)$$

where

ρ_b = density at base pressure and temperature (gm/cm^3 or 0.001 kg/m^3)
C_T = $e^{[-\alpha^{\delta}T\,(1 + 0.8\,\alpha^{\delta}T)]}$

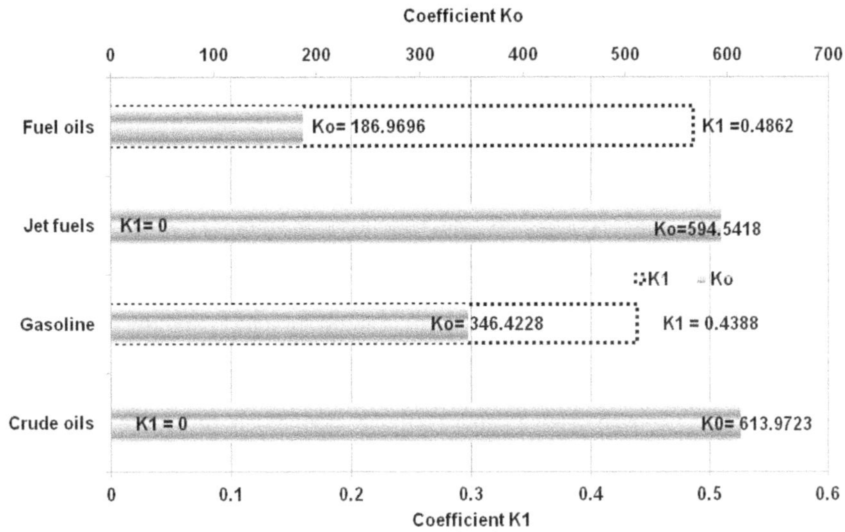

Figure 2-3. Value of coefficients K_o, K_1 for typical hydrocarbon liquids

$^{\delta}T$ = difference between the flowing temperature and base temperature
α = coefficient of thermal expansion at base temperature
= $(K_o + K_1 * \rho_b)/\rho_b^2$
K_o, K_1 = product dependent constants, see Figure 2-3 above.
C_P = $1/(1 - C_f * P)$
P = difference between the flowing pressure and base pressure (normally the base pressure is zero)
C_f = $\text{Exp}\ [-1.62080 + 0.00021592 * T_f + (0.87096/\rho_b^2) + (0.0042092 * T_f/\rho_b^2)] * 10^{-6}$

This equation is valid for those petroleum products whose density is greater than 635 kg/m^3 or API gravity is up to 90°API.

Since the density of light hydrocarbon liquids are highly sensitive to pressure and temperature, the equation of state is complex. For custody transfer of high vapor pressure liquids, whose density ranges from 350 kg/m^3 to 635 kg/m^3 or greater than 91°API, API bulletin 11.2.2 can be used [8].

2.3 PROPERTIES OF PETROLEUM LIQUIDS

The following properties of petroleum liquids have to be known for pipeline system design and determining operational limitations [9].

- Mass, or Volume
- Density, compressibility or bulk modulus, and thermal expansion
- Specific gravity and API gravity
- Viscosity (Viscosity (cP), or kinematic viscosity (cSt))
- Blending/diluting Ratio of hydrocarbon liquids (if applicable)
- Vapor pressure
- Heat capacity and thermal conductivity
- Pour point/Cloud Point
- Flash point (safety issues only)

2.3.1 Mass, Volume, and Density

Mass: is the amount of matter contained in a body, and is a measure of the inertial property of that body, i.e., its resistance to change of motion (Inertia). Inertial mass and gravitational mass are identical.

Mass is sometimes interchangeably used in place of weight; however, mass is different from weight. Weight is a vector quantity and is a measure of the attraction of the earth due to gravity which changes depending upon distance to the center of the earth. Equal masses at the same location in a gravitational field have equal weights. However, a mass in outer space may have nearly zero weight.

In common speech, mass and weight are generally referred to in units of kilograms (kg) or pounds (lb) but technically they are referred to respectively as kilogram-mass (kgm) or pound-mass (lbm), and kilogram-force (kgf) or pound force (lbf).

Mass is independent of temperature and pressure.

Volume: is the space occupied by a particular mass. Unlike mass, it is dependent upon temperature and pressures. The volume of a liquid increases slightly with increase in temperature but pressure has very little effect on volume especially when compared to gases. Bulk modulus relates pressures and temperatures for a particular volume of a liquid, see below.

Density: Liquid density is defined as mass per unit volume. Since mass does not change with temperature or pressure but volume does change, density thus changes with pressure and temperature. Therefore, like volume, density also depends upon temperature and pressure. Liquid density varies with temperature; decreasing with an increase in liquid temperature and vice versa. Liquid density increases with increase in pressure while volume decreases. The density unit is kg/m^3 in SI units and lbm/ft^3 in imperial units.

2.3.2 Density and Thermal Expansion

As noted above, liquid density decreases with increase in temperature while volume increases. The decreasing ratio with increasing temperature is referred to as **thermal expansion coefficient**. Liquid density increases with increase in pressure while volume decreases. The increasing ratio with increasing pressure is referred to **bulk modulus**, see Bulk Modulus.

2.3.3 Compressibility, Bulk Modulus, and Thermal Expansion

2.3.3.1 Compressibility: is the extent to which a fluid can be compressed. A change in pressure applied to a fluid changes the volume of the fluid (Figure 2-4).

The compressibility expressed as

$$K = -\left(1/{}^{\delta}PP\right)\left({}^{\delta}V/V\right) \qquad (2-8)$$

where

K = bulk modulus elasticity
${}^{\delta}P$ = differential change in pressure
${}^{\delta}V$ = differential change in volume
V = initial volume

Or Bulk Modulus of Elasticity can be alternatively expressed as

$$K = d\rho/\left(d\rho/\rho\right) \qquad (2-9)$$

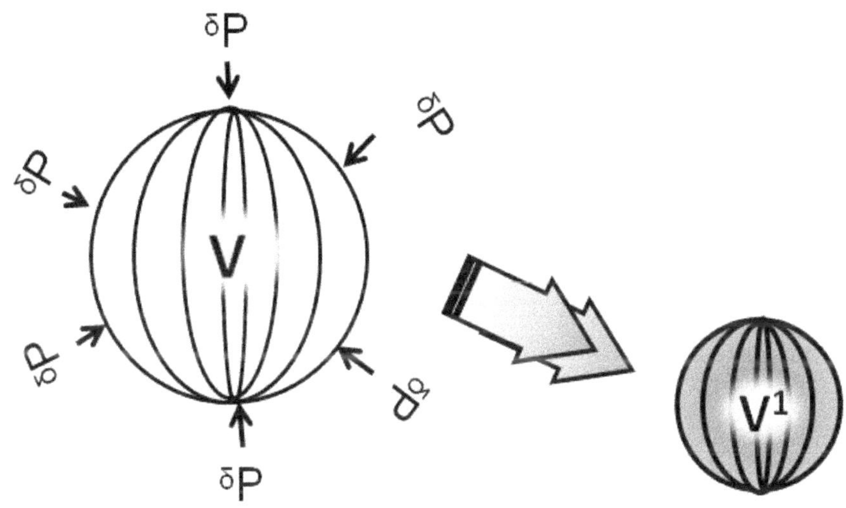

Figure 2-4. A unit liquid volume under uniform pressure

where
dρ = differential change in density
ρ = initial density

An increase in the pressure will decrease the volume. A decrease in the volume will increase the density

- The SI unit of the bulk modulus elasticity is N/m^2 (Pa or kPa)
- The imperial unit is lb_f/in^2 (psi) = 6.895 10^3 N/m^2 (Pa) or 6.895 kPa

A large Bulk Modulus (K) indicates a relatively incompressible fluid.

The value obtained for the bulk modulus in Eq. (2.8) is negative because the volume shrinks due to the increased pressure.

2.3.3.2 Bulk Modulus K: as shown above bulk modulus is the inverse of compressibility and is more frequently used than compressibility for liquid pipeline applications. Bulk modulus therefore defines the compressibility of a liquid. The higher the bulk modulus, the stiffer the liquid. Even though the liquid compressibility is generally small for heavier hydrocarbon liquids, it is the main cause of pressure surge in pipeline systems. Refer to Chapters 3 and 5 for a detailed discussion of surge phenomena.

TABLE 2-1. Comparison of bulk modulus of some liquids

	SI Units	Imperial Units
Bulk Modulus— *K*	(10^9 Pa, N/m^2)	(10^5 psi, $lb_f/N/in^2$)
Mercury	28.5	41.4
Crude oil	1.66	2.41
Oil (range)	1.4	2.03
Bitumen-condensate	1.53	2.22
Gasoline	1.07–1.49	1.55–2.16
Motor oil (SAE 30)	1.5	2.2
Seawater	2.34	3.39
Water	2.15	3.12

The bulk modulus (K) of a liquid is defined as the pressure required producing a unit change in its volume, expressed as

$$K = -^{\delta}P(V/\mathrm{d}V) = {}^{\delta}P\left(\rho/{}^{\delta}\rho\right) \qquad (2-10)$$

where dV is the change in volume corresponding to a change in pressure $^{\delta}P$, Refer to Figure 2-1.

2.3.3.3 Thermal Expansion: is the property of liquids to expand as their temperature rises and is defined by the coefficient of thermal expansion of the liquid (α): Thermal expansion of a unit volume of fluid can be defined as:

$$\alpha = -(1/{}^{\delta}T)\left({}^{\delta}\rho/\rho\right) \qquad (2-11)$$

where

α = coefficient of thermal expansion
ρ = density, $^{\delta}\rho$ = change in density
$^{\delta}T$ = temperature change

Thermal expansion coefficient is a function of fluid pressure and temperature. It does not change very significantly for heavy hydrocarbon liquids over the range of temperatures that are in common use in pipelines. However, it changes significantly for light hydrocarbon liquids.

The thermal expansion coefficient can be estimated from the temperature correction term of the API equation.

Figure 2-5 shows typical bulk modulus and thermal expansion coefficients of various crude oils and lighter products. The values of bulk modulus and thermal expansion

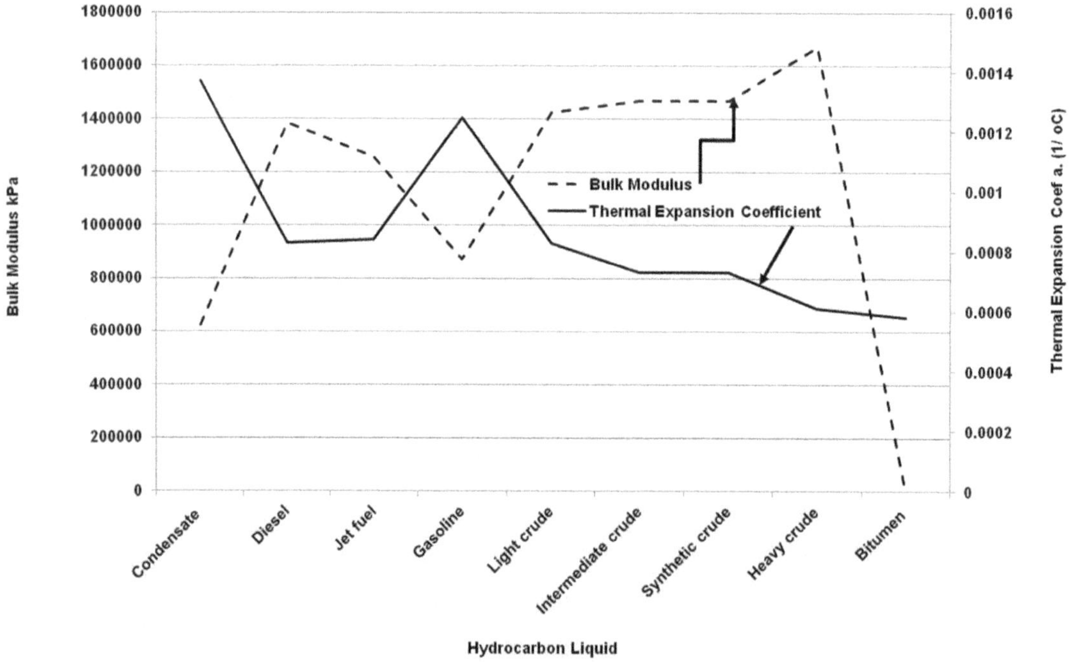

Figure 2-5. Bulk modulus and thermal coefficient of expansion for typical hydrocarbon liquids transported through pipelines

coefficients are approximate and are the API equation at 15°C and the atmospheric pressure.

2.3.3.4 Calculating Bulk Modulus for Various Fluids

The literature does not often provide values of the bulk modulus for various fluids. The following relationships are thus provided for determination of fluid densities at different pressure and temperatures and bulk modulus K.

Liquid density (ρ) at various pressures (P) and temperatures (T) can be expressed by the following relationship:

$$\rho = \rho_b \left[\left(1 + \frac{P - P_b}{K} \right) - \alpha (T - T_b) \right] \qquad (2-12)$$

where

ρ_b = density at base condition
P_b = pressure at base condition
T_b = temperature at base condition
α = liquid temperature coefficient of density
K = bulk modulus

For isobaric conditions (i.e., at constant pressure) $\frac{P - P_b}{K} = 0$, and Eq. (2-12) can be rewritten as:

$$-\alpha = \frac{\frac{\rho - \rho_b}{\rho_b}}{T - T_b} = \frac{^{\delta}\rho}{\rho_b \, ^{\delta}T} \qquad (2-13)$$

where

$^{\delta}\rho = \rho - \rho_b$ = change in density

If the liquid temperature coefficient of density α is known, it is possible to compute liquid densities at different pressures.

For isothermal conditions (i.e., at constant temperature), $T - T_b = 0$, so Eq. (12-12) can be rewritten as follows:

$$\rho = \rho_b \left[1 + \frac{P - P_b}{K} \right] \qquad (2-14)$$

or

$$K = \frac{^{\delta}P}{^{\delta}p} \rho_b \qquad (2-15)$$

where

$^{\delta}P = P - P_b$ change in pressure

Example: Calculate the bulk modulus and liquid coefficient of density for liquid CO_2 if the pressure drop across a pipeline segment is 3100 kPa. The inlet pressure is 13100 kPa. The density at base pressure and temperature is 968.5 kg/m^3.

Solution: Given the density at inlet (13100 kPa) is 1073.5 kg/m^3 and the density at the outlet (1000 kPa) is 1064 kg/m^3. Then, $\Delta\rho$ = 1073.5 to 1064.0 = 9.5 kg/m^3.

Therefore,

$$K = \frac{\Delta P \rho_b}{\Delta \rho} = \frac{3100 \times 968.5}{9.5} = 316{,}036 \text{ kPa} \quad (2-16)$$

Assume the following:

T_1 = –20°C corresponding ρ = 1073.5 kg/m^3
T_2 = +20°C corresponding ρ = 882.0 kg/m^3
ρ_o = (Base Density) at 15 °C = 968.5 kg/m^3

$$-\alpha = \frac{\Delta \rho}{\rho_b\, \Delta T} = \frac{1073.5 - 882}{968.5 \times (40)} = -0.005 \text{ kg/m}^3\ ^\circ\text{C}$$

2.3.3.5 Other Techniques for Calculating Bulk Modulus

Some measurement standards such as API 1101 provided formulae for calculation of hydrocarbon liquids bulk modulus based on specific or API gravity [10]. For example, API 1101 refers to the following expression:

$$K = 10\,\hat{}\left(5.722708 - 0.00819 \times {}^\circ\text{API} - 0.00219 \times T\right) \quad (2-17)$$

where
K = Bulk Modulus in psig, T = Temperature °F

Another example is the use of Caragoe equation as shown below:

$$\text{Bulk Modulus } K, \text{(PSI)} = 100000 \times \exp[1.9947 - 0.00013427 \times T - 0.79392/\text{SG\^{}2} - 0.002326 \times T/\text{SG\^{}2}] \quad (2-18)$$

The bulk modulus of a heavier hydrocarbon liquid can be estimated by either using the pressure correction term of the API equation given above or the Arco correlation as follows:

$$K = 2.619 * 10^6 + 9.203 * P - 1.417 * 10^5 * T^{1/2} + 73.05 * T^{3/2} - 341.0 * ({}^\circ\text{API})^{3/2} \quad (2-19)$$

where
P = pressure in psig,
T = temperature in °R and
°API = API gravity of the liquid

In general, the bulk modulus for heavier hydrocarbon liquids, e.g., crudes is relatively constant with respect to pressure and becomes smaller as the liquid temperature increases and larger as the temperature decreases.

The bulk modulus for lighter hydrocarbon liquids, e.g., propane varies strongly with pressure and temperature.

2.4 SPECIFIC GRAVITY AND API GRAVITY

Specific gravity (also known as relative density) of a liquid is the ratio of its density to the density of water at the same pressure and temperature. It is a measure of how heavy a liquid is compared with water. It is dimensionless and has no units. Since the densities of water and the comparing liquid change differently with pressure and temperature, specific gravity changes with pressure and temperature too.

However, accurate determination of the density, relative density (specific gravity), or API gravity of petroleum and its products is necessary for the conversion of measured volumes to volumes or masses, or both, at the standard reference temperatures during custody transfer and/or for facilities design.

There are several methods in use expressing specific gravity (SG) of hydrocarbon liquids. One method is the ratio of the specific weight of the liquid at 60°F to the specific weight of water at 60°F. Another method makes use of the Degree API (°API) and is the method used often by the petroleum industry.

The following provides the formula used to define the API gravity of hydrocarbon liquids in relation to specific gravity (SG).

$$\text{Degrees API Gravity} = (141.5 / \text{Specific Gravity} @ 60°\text{F}) - 131.5 \qquad (2-20)$$

Conversely, the specific gravity of hydrocarbon liquids can be derived from the API gravity value as

$$\text{Specific Gravity at } 60°\text{F} = 141.5/(\text{API Gravity} @ 60°\text{F} + 131.5) \qquad (2-21)$$

For example, oil with a specific gravity of 1.0 (i.e., with the same density as pure water at 60°F) would have an API gravity of:

$$[141.5/1.0] - 131.5 = 10.0 \text{ °API}.$$

There are also methods that provide adjustments for temperature. ASTM [11] describes the methodology for temperatures corrections. Alternatively the following correction factors can be used to allow for temperature effects (for crude oils relative to 15°C (59°F). They are divided into 3 ranges:

All temperatures are in expressed in °C.

- *For temperatures less than 3.98°C:*

$$\text{Correction factor} = -0.000032692*\text{C to } 0.000740644$$

- *For temperatures less than 50.0°C and greater than or equal to 3.98°C:*

$$\text{Correction factor} = -0.0008031922 \text{ to } 0.0000473773*\text{T} + 0.000007231263*\text{T}*\text{T} - 0.00000003078278*\text{T}*\text{T}*\text{T}$$

- *For temperatures greater than or equal to 50.0°C:*

$$\text{Correction factor} = -0.005431719 + 0.0001963596*\text{T} + 0.000002661056*\text{T}*\text{T}$$

Therefore, SG corrected = SG (at 15°C , 60°F)+/– correction factor. (– for temperatures below 3.98°C and above 50°C, + for temperatures between 3.98°C and 50°C).

A third method for expressing the specific gravity of hydrocarbon liquids is the use of Degrees Baume. It is named after the French chemist Antoine *Baumé* (1728 to 1804).

For liquids lighter than water:

$$\text{Degree Baume} = \left[140/\text{SG}\right] - 130 \qquad (2-22)$$

For liquids heavier than water (e.g., heavy bitumen):

$$\text{Degree Baume} = \left[145(1 - 1/\text{SG})\right] \qquad (2-23)$$

It may be noted that an older version of the scale for liquids heavier than water, at a reference temperature of 15.5°C (59.9°F), uses 144.32 rather than 145.

The relationship between API Gravity, Specific Gravity and Density (at 60°F) is summarized in Figure 2-6.

Densities and API gravities for some hydrocarbon liquids typically transported through pipelines are shown in Table 2-2.

2.4.1 Specific Gravities of Blended Products

When two or more petroleum products are blended, the specific gravity of the resultant liquid (provided that the gravities are measured at the same pressure and temperature) can be calculated using the following weighted average method.

$$\text{SG}_b = \sum(V_i\,\text{SG}_i)/\sum(V_i) = \sum(Q_i\,\text{SG}_i)/\sum(Q_i) \qquad (2-24)$$

where
SG_b = specific gravity of the blended liquid
V_i = volume of each product
Q_i = flow rate of each product
SG_i = specific gravity of each product

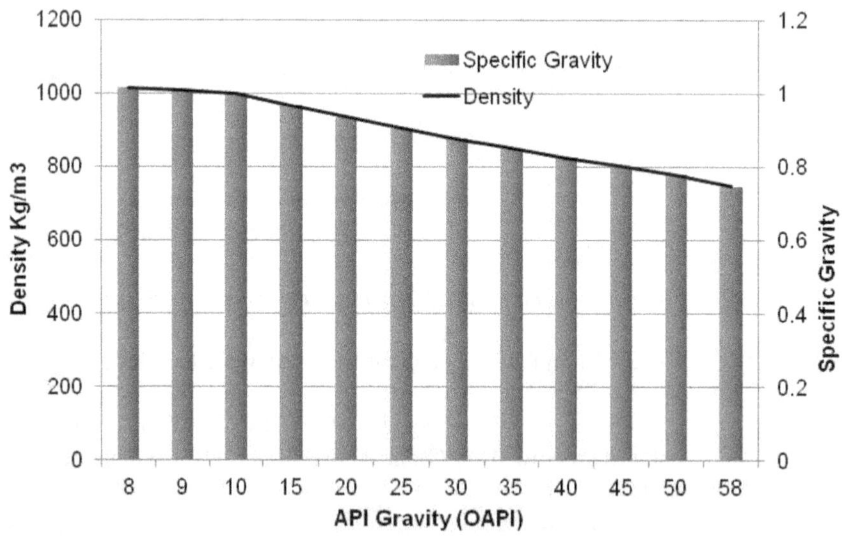

Figure 2-6. API gravity, specific gravity, and density (at 60°F)

TABLE 2-2. Values of density and API gravity for typical hydrocarbon liquid transported through pipelines

Hydrocarbon Liquids	Typical Density (kg/m^3)	°API
Condensate	669.0	80.0
Diesel	832.0–850.0	35.0–38.6
Jet fuel	775.0–840.0	51.0–37.0
Gasoline	713.0–767.0	52.0–67.0
Light crude	<870	>31.1
Intermediate crude	870.8–920.0	22.3–31.0
Synthetic crude	865.4–870.8	31.1–32.0
Heavy crude	920.6–1000	10.0–22.2
Bitumen	1000.7–1029.1	6.0–9.9

The above method cannot be directly applied when the gravities are expressed in °API. The °API values must be first converted to specific gravities before applying Eq. (2-24).

2.5 VISCOSITY, NEWTONIAN VERSUS NON-NEWTONIAN

Viscosity is a relative measure of resistance to flow. It can also be defined as a measure of friction between adjacent layers of a flowing fluid. Consider in pipe flow that the flow velocity is zero at a thin layer adjacent to pipe wall, and each subsequent layer above this has a different velocity compared with the layer below. This difference in the velocity of the liquid layers results in a velocity gradient caused by viscosity.

When a fluid is flowing, a frictional force exists within the fluid that opposes the flow. This frictional force, caused by shear stress, acts between the two adjacent layers of fluid. Similarly, the velocity with which an individual layer moves relative to neighbouring layers is known as shear rate. Shear stress is a function of pressure, and shear rate is a function of geometry and the average velocity of a fluid. The relationship between shear stress and shear rate defines the flow behavior of the fluid.

A fluid's rheology depends on its shear stress shear rate relationship. The shear stress (τ) between adjacent layers of a flowing fluid is proportional to the velocity gradient ($\Delta u/\Delta y$). The proportional constant is called as the absolute or dynamic viscosity (μ).

For a two-dimensional flow, the shear stress is

$$\tau = \mu\left(\Delta u/\Delta y\right) \tag{2-25}$$

If a fluid shows constant μ, it is said to be Newtonian; otherwise, it is non-Newtonian.

The viscosity of a fluid is dependent on temperature, shear rate ($\acute{\varepsilon}$) and time. Liquids that have a constant shear rate ($\acute{\varepsilon}$) with respect to shear stress (σ) at any given temperature are termed Newtonian fluids (e.g., water, crude oil), and the viscosity is a function of temperature only, increasing with decreasing temperatures.

Therefore, a linear relationship between shear stress and shear rate on a Cartesian plot, which passes through the origin, indicates that a fluid exhibits Newtonian characteristics.

Non-Newtonian fluids such as bitumen have viscosities which are not only a function of temperature, but also of shear rate, and, in some cases, time (i.e., shrinkage) [12, 13]. There are a number of different fluids that can exhibit non-Newtonian behavior. These

can include dilatants (e.g., starch, quicksand), pseudoplastic fluids (e.g., lime solution), and Bingham plastics [14].
Generally, non-Newtonian fluids are grouped in classes as:

1. Time-dependent non-Newtonian fluids.
2. Time-independent non-Newtonian fluids.
3. Viscoelastic non-Newtonian fluids.

a-Time dependent non-Newtonian fluids: Depending on how viscosity changes with time the flow behavior is characterized as:

- Thixotropic (time thinning, i.e., viscosity decreases with time), for example, yoghurt, paint materials which become less viscous over time when shaken, agitated, or otherwise stressed.
- Rheopetic (time thickening, i.e., viscosity increases with time), for example, gypsum paste, honey which become more viscous over time when shaken, agitated, or otherwise stressed.

Thixotropic describes materials that are gel-like at rest but fluid-like when agitated. Thixotropic fluids are quite common in the chemical as well as in the food industry. Rheopetic fluids are very rare.

It may noted that some fluids (like bitumen) show time thinning behavior due to breakdown of structure. This phenomenon is sometimes known as rheomaiaxis.

b-Time-independent non-Newtonian fluids: The viscosity of a time independent non-Newtonian fluid is dependent not only on temperature but also on shear rate.

Depending on how viscosity changes with shear rate the flow behavior is characterized as follows:

- shear thinning—the viscosity decreases with increased shear rate. Shear thinning liquids are very commonly, but misleadingly, described as thixotropic.
- shear thickening—the viscosity increases with increased shear rate.
- plastic—exhibits a so-called yield value, i.e., a certain shear stress must be applied before flow occurs.

Shear thinning fluids are also called pseudoplastic and shear thickening fluids are also called dilatant.

The time-independent non-Newtonian fluids can be characterized by the flow curves of shear stress versus shear rate as shown in Figure 2-7, which are as follows:

a. Bingham plastic fluid. A Bingham plastic is a material that behaves as a solid at low stresses but flows as a viscous fluid at high stresses.
b. Plastics are complex, non-Newtonian fluids in which the shear force is not proportional to the shear rate. most drilling muds are plastic fluids.
c. Pseudoplastics have the capability of changing apparent viscosity with a change in shear rate. Apparent viscosity is the measure of viscosity of fluid at a given shear rate at a fixed temperature.
d. Pseudoplastic fluids gain viscosity when subjected to a decrease in shear rate, Pseudo-plastic fluids (also known shear thinning), exhibit a so-called yield value, i.e., a certain shear stress must be applied before flow occurs.
e. Dilatant fluids (shear thickening).

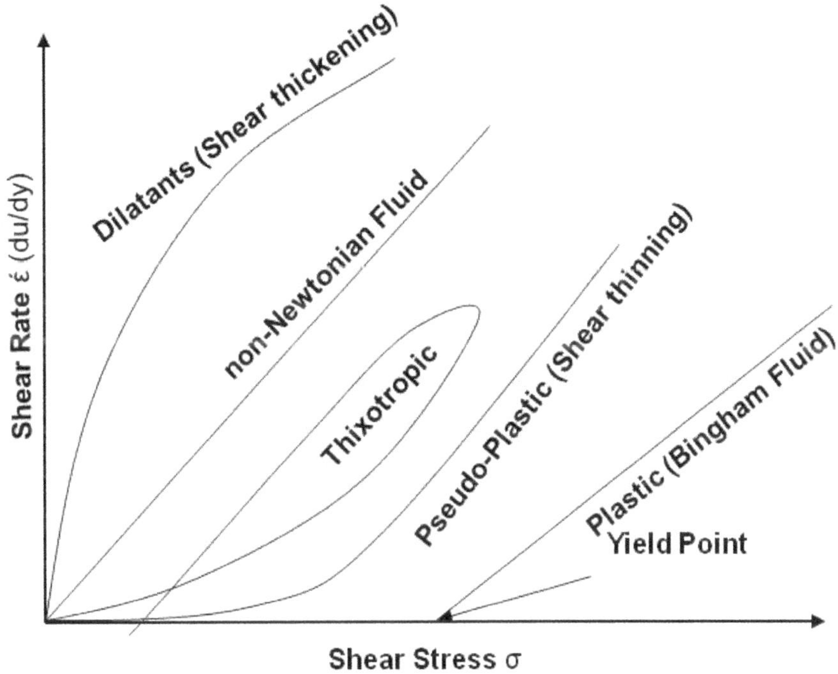

Figure 2-7. Newtonian and non-Newtonian fluids typical shear rate vs. shear stress relationships (adapted from [15])

When transporting non-Newtonian fluids such as bitumen and heavy oils, the viscosity has to be carefully considered. Since the shear rate changes with different fluid velocities, the viscosity curve of a specific fluid must be determined at a known fluid velocity along the fluid temperature profile of a pipeline.

Viscosity characteristics of a typical Bitumen/Bitumen Diluent Blend are shown in Figure 2-8.

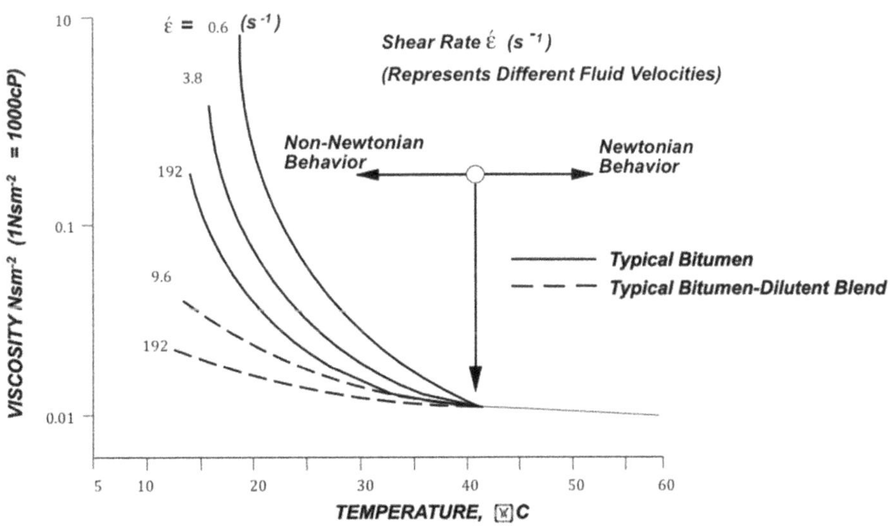

Figure 2-8. Viscosity characteristics of typical bitumen/bitumen diluent blend [12]

2.5.1 Viscosity and Density Relationship

Viscosity and density are not directly related, even though there is a certain correlation between the two for hydrocarbon liquids.

Viscosity and density account for most line pressure changes. Viscosity is the main cause of friction losses in the pipeline, whereas density determines the power requirements and pressure outputs of the pump units.

2.5.2 Viscosity of Blended/Diluted Liquids

Often, dilution occurs in a pipeline system when one fluid stream is injected with another primarily for the purpose of making the final products transported lighter or in the case of product batching, through full or side stream injection or straight injection and delivery. The following technique can be utilized to establish specific gravity and diluted viscosity:

2.5.2.1 (A) New Volume from Current Volume, Current SG, and Target SG

$$V_{\text{new}} = \left(\left(\text{SG}_c - 1.0\right) / \left(\text{SG}_t - 1.0\right)\right) * V_{\text{cur}} \tag{2-26}$$

where
V_{cur} = current volume
SG_c = current SG
SG_t = target SG

New SG from current SG, current volume, and target volume

$$\text{SG}_{\text{new}} = \left(\left(\text{SG}_c - 1.0\right) * \left(V_{\text{cur}} / V_{\text{tar}}\right)\right) + 1.0 \tag{2-27}$$

where
SG_c = current SG
V_{cur} = current volume
V_{tar} = target volume

2.5.2.2 (B) Viscosity Blending Calculation

When two or more liquids are blended, it is also important that the viscosity of the blend is determined to assess pipeline transportation options such as the location of blending and/or injections and as well proper system capability determination. For this purpose, the Refutas viscosity blending index is generally used by the industry. This equation requires input of mass fractions. Often, in error, volume fractions are used which will provide substantially incorrect results if the densities of the two blend crudes are dissimilar.

Calculating the viscosity blending index of a liquid consisting of two or more liquids having different viscosities (using the Refutas equation [16]) is a two step procedure. The first step involves calculation of the Viscosity Blending Index (VBI) of each component of the blend using the following equations:

$$\text{VBI} = 14.534 \times \ln\left[\left(\upsilon + 0.8\right)\right] + 10.975 \tag{2-28}$$

where μ is the viscosity in centistokes and is the natural logarithm (Log_e).

The second step involves using:

$$\text{VBI}_{\text{Blend}} = \left[W_A \times \text{VBI}_A\right] + \left[W_B \times \text{VBI}_B\right] + \ldots + \left[W_X \times \text{VBI}_X\right] \tag{2-29}$$

where W is the weight fraction of each component of the blend. In using the above blending equation, it is necessary that all viscosities are determined at the same temperature.

Once the viscosity blending number of a blend is obtained using Eq. (2-29), the viscosity of the blend can be determined by using the inverse of Eq. (2-28).

$$\upsilon = e^{\left[(\text{VBN}-10.975)/14.534\right]} - 0.8 \qquad (2-30)$$

where VBN is the viscosity blending number of the blend and e is the transcendental number 2.71828, also known as Euler's number.

It may be noted that μ and υ are sometimes used interchangeably but in most pipeline literature, μ is often used for absolute viscosity and υ for kinematic viscosity.

Another method for calculating the viscosity of a blended product is the use of an ASTM equation (Eq. (2-31)) which provides an estimate of the viscosity of a blended product.

$$\text{Log Log}(\upsilon + 0.7) = A - B * \text{Log}(T) \qquad (2-31)$$

where
υ = viscosity of liquid, cSt
T = absolute temperature, K

To calculate the viscosity of a blended product, the following procedure can be followed:

1. Calculate A and B of each individual product from the viscosities of the product at two different temperatures using the ASTM equation.
2. Calculate A and B of the blended product as follows:
   $B_b = \Sigma(Q_i)/\ \Sigma(Q_i/B_i)$
   $A_b = \Sigma(A_iQ_i/B_i)/\ \Sigma(Q_i/B_i)$
   where
   Q_i = flow rate or composition of each individual product
3. Insert the blended coefficients A_b and B_b into the ASTM correlation:

$$\text{Log Log}(\upsilon_b + 0.7) = A_b - B_b * \text{Log}(T) \qquad (2-32)$$

2.5.3 Hydrocarbon Liquids Blending and Volume Shrinkage

Shrinkage occurs when two or more petroleum products are blended. The mixture will experience this volume reduction (shrinkage); however, losses are usually assessed to the light components.

Such a volume reduction is a function of the gravity/density differential between the light and heavy components. Shrinkage decreases as the percentage (%) in volume of the light component in the mixture increases.

It may be noted that the final amount of shrinkage is independent of sequence of injecting light components into the crude.

The following procedure can be used to estimate the volumetric shrinkage percentage [17].

The shrinkage is expressed as:

$$S_v = 2.69 * 10^4 * C(100 - C)^{0.819}(1/\Delta L - 1/\Delta H)^{2.28} \quad \text{Si Unit} \ (2-33)$$

Or

$$S_v = 4.86*10^{-8} C(100-C)^{0.819} \Delta_{API}^{2.28} \quad \text{Imperial Unit} \quad (2-34)$$

where
S_v = volumetric shrinkage, as percent of total mixture ideal volume
C = concentration in liquid volume percent of light component
Δ_{API} = gravity difference, in °API
$(1/\Delta L - 1/\Delta H)$ = inverse density difference of light (ΔL) and heavy (ΔH) components, in m^3/kg

The above equation is only applicable to a pressure range of 100 to 700 kPag (7 to 100 psig), and 15°C (60°F) temperature.

2.5.4 Viscosity Determination

The kinematic viscosity (υ) is defined as the absolute viscosity of a fluid μ divided by its density (ρ) at the same temperature.

$$\upsilon = \mu/\rho \quad (2-35)$$

where
υ = kinematic viscosity, stoke or m^2/s (Centistoke (mm^2/s) mostly used in liquid pipeline industry)
μ = absolute viscosity, Pascal-s
ρ = fluid mass density

For Newtonian fluids, if the viscosities at two different temperatures are known, the viscosity at another temperature can be estimated.

Two viscosity correlations that are often used are the Andrade and the ASTM method. The Andrade correlation shows that the variation of viscosity with temperature is logarithmic:

$$\text{Ln}(\upsilon) = A - B \times T \quad (2-36)$$

where
υ = viscosity of liquid, cSt
T = absolute temperature, K
A, B = constants

2.6 POUR POINT AND VISCOSITY RELATIONSHIP

The pour point of a liquid is the lowest temperature at which it will flow under prescribed conditions. It is a rough indication, but an important one in pipeline design and operation.

In general, hydrocarbon liquids like crude oils have high pour points. As with viscosity, pour points are very much a function of chemical composition for complex mixtures such as crude oils and some distillate products. The pour point temperatures of such mixture are influenced by the precipitation (or solidification) of certain components, such as paraffins.

Crude oils that have significant paraffin or asphalt content (i.e., bitumen or heavy oil) have high pour points. Usually, most light and intermediate crudes have low pour points.

The pour point is one of the critical parameters for heavy or high wax crude pipeline design and operation because extra facilities are generally required if the pipeline flowing temperature falls below the pour point.

The pour point for oil can be determined under protocols set out in the ASTM D-97 pour point test. This protocol requires a hydrocarbon liquids specimen to be cooled inside a cooling bath to allow the formation of paraffin wax crystals. At about 9°C above the expected pour point, and for every subsequent 3°C, the test jar is removed and tilted to check for surface movement. When the specimen does not flow when tilted, the jar is held horizontally for 5 sec. If it does not flow, 3°C is added to the corresponding temperature and the result is the pour point temperature.

It may noted that failure to flow at the pour point may also be due to the effect of viscosity or the previous thermal history of hydrocarbon liquid specimen. Therefore, the pour point may give a misleading view of the handling properties of the oil. It is for these reasons that the pour point is only a rough indicator of the temperature at which the liquid may not flow.

The pour point of crude oil is determined using ASTM D5853-11. This is the only pour point method determination specifically designed for crude oils and provides an index of the lowest temperature of handle-ability for certain applications. The test method can be used to supplement other measurements of cold flow behavior. It is especially useful for the screening of the effect of wax interaction modifiers on the flow behavior of crude oils.

2.6.1 Reasons for Pour Point Determination

Once temperatures of hydrocarbon liquids fall below their respective pour points, these liquids start to show non-Newtonian behavior and therefore conventional pipeline design and operation will have to be modified to be effective. However, there are several options available for design and operating a pipeline transporting high pour point hydrocarbon liquids at temperatures below the pour point—the most frequently used are as follows:

- Heating the hydrocarbon liquid and/or insulating the pipeline to keep the materials above their pour point temperature until they reach their destination.
- Injecting lightweight hydrocarbon liquids (such as natural gas condensate(s)) that are miscible with the heavier hydrocarbon liquid, thereby diluting and lowering both its effective viscosity and pour point temperature.

Other options include the following:

- Partial upgrading, removing those components that will be first to precipitate as the temperature is lowered.
- Water emulsion to lower viscosity and pour point temperature.
- Core annular flow: Introducing water that will preferentially move to the inner walls of the pipe, serving to reduce the effective coefficient of drag exhibited by the viscous petroleum product.
- Use of surfactants/flow improvers (use of additives as a pour point depressant).
- Viscosity reducers.
- Slurry transportation.

Generally, with high pumping power, waxy crude can be pumped below its pour with no sudden change in fluid characteristics at the pour point. However, should pumping be stopped, more energy will be required to put the pipeline back into operation and to keep it flowing. When flow is stopped, wax crystals form, causing the waxy crude to gel in the pipeline. If such a gelling occurs, the waxy crude behaves as if it had a much higher effective viscosity (reminiscent of non-Newtonian behavior) and consequently it would take much higher pumping power (five to ten times more) to reestablish steady state design flows in the pipeline than it did to support the operation when the crude's temperature was above its pour point.

For some products, such as diesel fuels that still contain some waxy components (i.e., saturated, long-chain hydrocarbons), "gelling" may also occur as temperatures are lowered; however, such gelling problems are commonplace in storage tanks and vehicle fuel tanks where the fuel sits motionless for long period of time, but rarely materialize in pipelines where the materials are virtually in constant motion and where their passage through pumps typically imparts some amount of heat. Nevertheless, precipitation or gelling of products contained in pipelines can cause significant operational difficulties. A properly designed pipeline must allow for startup pressures that might be necessary to reestablish pipeline flow during these gelled conditions.

For details refer to chapter 6: "Non-conventional Hydrocarbon Production and Transportation."

2.7 VAPOR PRESSURE

Vapor pressure is an important physical property of hydrocarbon liquids subjected to vaporization. It is the pressure that maintains a liquid in equilibrium at a given temperature and is defined as the absolute vapor pressure exerted by a liquid at 37.8°C (100°F) having an initial boiling point above 0°C (32°F). It is a measure of volatility.

Vapor pressure is an important parameter relating to the design, function, and operation of hydrocarbon products pipeline and storage systems. Vapor pressure of crude oils is of importance to the crude producer and the refiner for general handling and initial refinery treatment. Oil refiners manipulate the Reid Vapor Pressure seasonally specifically to maintain gasoline engine reliability.

Pipeline transportation of hydrocarbon liquids requires that a minimum pressure greater than the vapor pressure be maintained throughout the pipeline to avoid slack flow/two-phase flow conditions, even under transient states (see Section 5.1 for more details). Additionally, in liquid pipeline pumping systems, the pressure at pump suctions must be kept higher than the vapor pressure to avoid cavitation of pumps. Cavitation occurs at the inlet of a pump when the available Net Positive Suction Head (NPSH) drops below the required NPSH of the pump or at area where flow restriction causes a pressure decrease. See Chapter 4 for details.

The vapor pressure of a liquid increases with temperature. Table 2-3 and Figure 2-9 and illustrates the vapor pressure of hydrocarbon liquids commonly transported by pipelines and also stored in storage tanks [18].

2.7.1 True Vapor Pressure

True Vapor Pressure (TVP) is a common measure of the volatility of petroleum distillate fuels. It is defined as the equilibrium partial pressure exerted by a volatile organic liquid as a function of temperature as determined by the test method described within ASTM D 2879.

TABLE 2-3. Properties of selected hydrocarbon liquids [19]

Petroleum Liquid	Weight at 60 °F, MV (lb/lb-mole)	Density At 60 °F, (lb/gal)	True Vapor Pressure, PVA (psi)						
			40 °F	50 °F	60 °F	70 °F	80 °F	90 °F	100 °F
Distillate fuel oil No. 2	130	7.1	0.0031	0.0045	0.0065	0.009	0.012	0.016	0.022
Jet kerosene	130	7	0.0041	0.006	0.0085	0.011	0.017	0.021	0.029
Jet naphtha (JP-4)	80.0	6.4	0.8	1.0	1.3	1.6	1.1	2.4	2.7
Residual oil No. 6	190	7.9	0.00002	0.00003	0.00004	0.00006	0.00011	0.0001	0.00019

Evaporation losses in hydrocarbon tankage systems (refer to Chapter 8) are related to the true vapor pressure (TVP) of hydrocarbon liquids at their storage or pipeline transportation temperature. It is measured by a Reid vapor pressure (RVP) test defined by the American Society for Testing and Materials specification ASTM D323-56. RVP test procedure is described in detail in the API document "Measuring, Sampling, And Testing Crude Oil". There are other API publications that show charts relating RVP and ASTM boiling characteristics of hydrocarbon liquids (gasolines and crude oils) to TVP, and a way to estimate RVP of blends, and the relation of RVP to evaporation losses. Steps to determine TVP and application examples are provided by Vasquez-Esparragoza et al. [20].

The Reid vapor pressure (RVP) differs slightly from the true vapor pressure (TVP) of a liquid due to small sample vaporization and the presence of water vapor and air in the confined space of the test equipment. That is, the RVP is the absolute vapor pressure and the TVP is the partial vapor pressure. Conversion between the two measures is depicted in Figure 2-10.

At normal pipeline operating pressure and temperature, crude oils remain liquid, but LPG and NGL can vaporize because their vapor pressures are high. Consequently, pipelines transporting such products must operate at pressure much higher than their vapor pressure to ensure single-phase flow with no liquid separation.

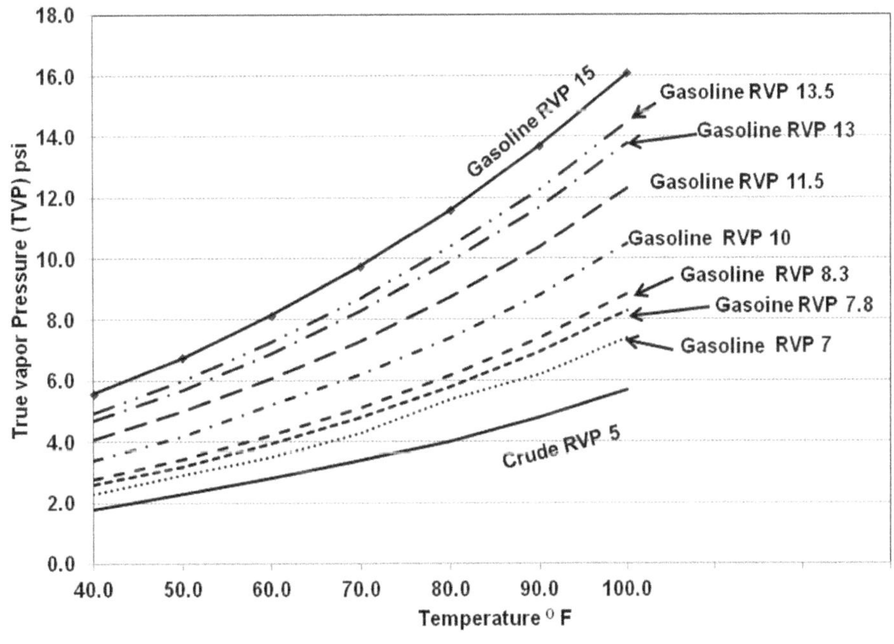

Figure 2-9. Vapor pressure of hydrocarbon liquids commonly transported through pipelines (Source: [18, 19])

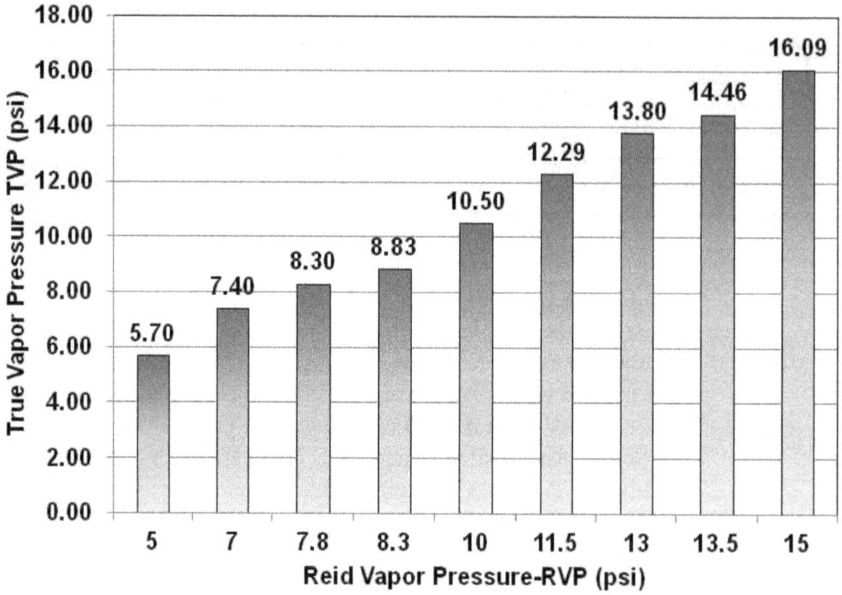

Figure 2-10. Scale comparison of true vapor pressure (TVP) and Reid vapor pressure (RVP)

Generally, pipeline standards have sections related to the design of high vapor pressure pipeline systems. For example, CSA Z662-2011 defines an HVP pipeline system as a pipeline transporting hydrocarbons or hydrocarbon mixtures in the liquid or quasi-liquid state with a vapor pressure greater than 110 kPa absolute at 38°C, as determined using the Reid method. The high vapor pressure (HVP) products include ethylene, ethane, propylene, propane, normal, and iso-butane since pipe flow is almost an isenthalpic process.

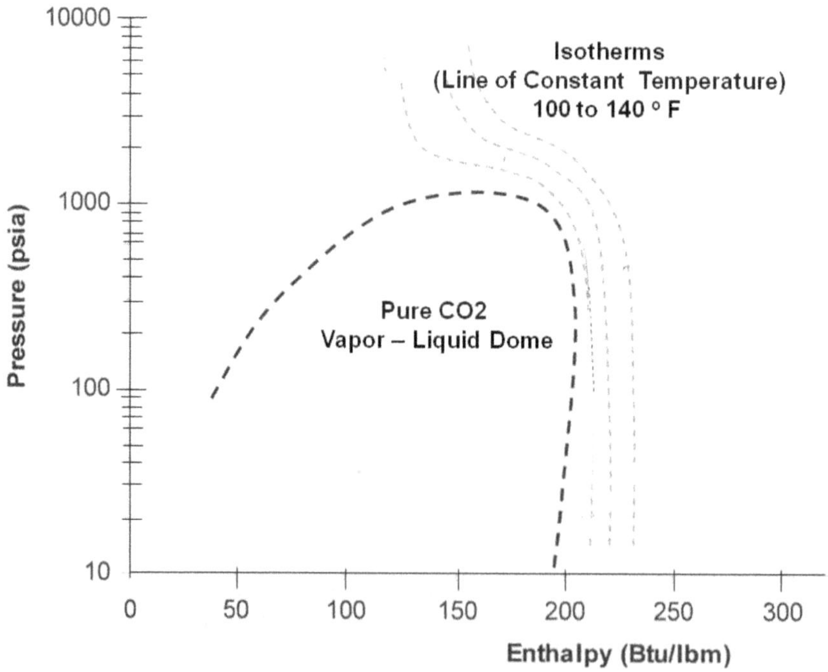

Figure 2-11. Typical pressure-enthalpy diagram (for pure CO_2)

Pressure-enthalpy diagrams are typically used for designing a high vapor pressure (HVP) pipeline. Such diagrams show pressure on the vertical axis and enthalpy on the horizontal axis. Figure 2-11 indicates a typical pressure-enthalpy diagram with isotherms shown for a pure CO_2. Pressure-enthalpy diagrams are often used to determine the minimum pressure for specified operating temperatures for keeping the HVP products in the liquid phase. Therefore, the diagrams are used in locating pipeline operating points in terms of pressure and temperature, and also designing control valves. The diagram may not be needed for the design of low vapor pressure liquids under normal operating conditions because they remain in a liquid phase.

2.8 FLASH POINT

The volatility characteristics of hydrocarbons have an important effect on their safety and performance, especially in the case of fuels. The boiling range gives information on the composition, the properties, and the behavior of the hydrocarbon liquid fuel during transportation, storage, and use.

A fuel's flash point is the lowest temperature at which the hydrocarbon liquid's vapor can ignite momentarily (flash) when exposed to a flame. The lower a fuel's flash point, the more dangerous it is. Some sample flash points for aviation fuels are as follows:

- AVGAS, –50°F;
- JP-4, –10°F; and
- JP-8, 100°F.

These flash points show that fuels give off ignitable vapors at temperatures normally found in vehicles. Aviation-related fuels can ignite even in sub-zero temperatures.

The flash point of a hydrocarbon liquid can be calculated as follows:

$$\text{Flash point}(\text{FPT}) = 1/\left[-0.014568(2.84947 T_{10})0001903\log(T_{10})\right] \quad (2-37)$$

FPT = flash point temperature, °Rankine (°R)
T_{10} – 10% temperature for the material. (Volume %, °R) as per ASTM D86

2.9 HYDROCARBON LIQUID SPECIFIC HEAT CAPACITY

The heat capacity of a liquid is defined as the amount of heat required to increase the temperature of a unit quantity of a liquid by a specific amount. Alternatively:

$$\text{Heat Capacity} = \text{Heat added/Change in temperature}$$

The heat capacity of a hydrocarbon liquid (at constant pressure) can be estimated as a function of specific gravity and temperature as follows:

$$C_p = (1.685 + 0.003391 \times T)/\text{SG} \quad (2-38)$$

where

C_p = heat capacity of the liquid at constant pressure (Isobaric), temperature T (kJ/kg °C)
SG = specific gravity of the liquid at 15°C
T = temperature of the liquid (°C)

It may be noted that the temperature of a liquid with large heat capacity does not rise much for a given amount of heat, whereas the temperature of a liquid with small heat capacity rises significantly when heat is added.

For application to a liquid system, the heat capacity (C_p) at constant pressure (isobaric condition) is used and treated as constant over an applicable temperature interval.

However, the following Lee-Kesler correlation for predicting liquid heat capacities of paraffinic heavy hydrocarbon liquids, such as bitumen and heavy crude oils, provides an accurate estimation of the heat capacity [21, 22].

$$C_p = A_1 + A_2 T + A_3 T^2 \tag{2-39}$$

where

C_p = Isobaric Heat Capacity for Liquid petroleum fraction (BTU/ lb. °R)

A_1 = −1.17126 + (0.023722 + 0.024907 × SG) K_W + [(1.14982 to 0.046535 K_W)/ SG]

A_2 = (10^{-4}) (1.0 + 0.82463 K_W) (1.12172 to 0.27634/SG)

A_3 = (-10^{-8}) (1.0 + 0.82463 K_W) (2.9027 to 0.70958/SG)

T_r = reduced temperature, T/T_{pc}

T = temperature in °Rankine

T_{pc} = pseudocritical temperature in °Rankine

K_W = Watson characterization factor

SG = specific gravity 60°F/60°F

The Watson characterization factor (K_W) denotes the "paraffinic" fraction of petroleum hydrocarbon fractions [23, 24], and, as such, can be expressed as:

$$K_W = \left[(T_b)^{1/3}\right]/\text{SG} \tag{2-40}$$

where

T_b = the mean average boiling point in degrees Rankine (°R)

This is valid from approximately $0.4 < T_r \leq 0.85$, requires significant property knowledge to be applied in practice.

The above approach is indicated to yield significantly improved heat capacity estimates as only the elemental analysis of a material needs to be available to provide accurate heat capacities on a unit mass basis [22].

2.10 THERMAL CONDUCTIVITY

Thermal conductivity is the property of the pipe and the surrounding soil used in heat transfer calculations. Values typically used for cross country pipeline application are:

- K for steel pipe = 50.19 W/m/°C , or 29 Btu/hr/ft/°F
- K for soil = 0.2 to 0.8 Btu/hr/ft/°F, or 0.35 to 1.4 W/m/°C
- K value for insulation may range from 0.02 to 0.09 W/m/°C, or 0.01 to 0.05 Btu/hr/ft/°F.

In SI units thermal conductivity is expressed in W/m/°C. In US Customary units, it is measured in Btu/hr/ft/°F. The thermal conductivity of a material is numerically equal to the amount of heat transferred across a unit area of the solid material with unit thickness, when the temperature difference between the two faces of the solid is maintained at 1°.

The overall heat transfer coefficient is also used in heat flux calculations. Typical value of U may range from 1.7 to 3.4 W/m^2/°C in SI units and 0.3 to 0.6 Btu/hr/ft^2/°F in English units.

2.11 EFFECT OF HYDROCARBON LIQUID PROPERTIES ON MEASUREMENT SYSTEMS

2.11.1 (a) Base Conditions

The base conditions for the measurement of fluids, such as crude petroleum and its fluid products, having a vapor pressure equal to or less than atmospheric at base temperature are:

United States Customary (USC) Units:

Pressure:	101.325 kPaa (14.696 psia)
Temperature:	15.56°C (60.0°F)

International System (SI) Units:

Pressure:	101.325 kPaa (14.696 psia)
Temperature:	15.00°C 59.00°F (59.00°F)

Base conditions may change from one country to the next due to governmental regulations. Therefore, it is necessary that the base conditions be identified and specified for standardized volumetric flow measurement by all parties involved in the measurement. For example, the following is the STP, Standard Temperature and Pressure, for Mexico (SI units)

Pressure:	98 kPaa (14.696 psia)
Temperature:	20.00°C (68.00°F)

For liquid hydrocarbons, having a vapor pressure greater than atmospheric pressure at base temperature, the base pressure must be the equilibrium vapor pressure at base temperature.

2.11.2 (b) Impact of Phase Change

Fluids are classified into four-phase regions, refer to previous Figure 2-2.

- Liquid
- Gas or vapor
- Dense phase or supercritical, and
- Two-phase

A salient point is that fiscal measurement is applicable for single-phase fluids (liquid, gas or dense phase). For phase behavior, refer to Section 2.2 [25].

2.11.3 Properties Important to Measurement Systems

Fluid physical properties are of fundamental importance to measurement and must be ascertained before any serious measurement design or analysis is undertaken. These

TABLE 2-4. Summary of properties important to and required for measurement systems (summarized from ref [25])

Fluid physical properties	Crude Oils	Refined Products	Condensate and LPG
Fluid Density (ρb, ρtp)	√	√	√
Composition (mole percent of C1 – C9, CO2, N2			√
Phase envelope or bubble point prediction			√
Absolute Viscosity (cP), or Kinematic Viscosity (cSt)	√	√	√
API Gravity (API@60)	√	√	
Isentropic Exponent (for orifice meters)			
Pour Point (°F or °C)	√	√	
Cloud Point (°F or °C)	√		
Wax Content (wt %)	√		
Reid Vapor Pressure (psia)	√	√	
Phase envelope or bubble point prediction	√	√	
Sulfur Content (wt % or units of measure specific to fluid application)	√	√	√
H2S Content (ppm wt or units of measure specific to fluid application)	√		√
Sediment and Water Content (vol. %)	√		
TANE or total acid and neutralization levels (mg KOH/gm)	√		
Salt Content (lbm/M Bbl)	√		
Metals in whole Crude and Pitch (ppm wt)	√		
Contaminants (organic chloride, methanol, etc.	√		√
Contaminants (bacteria ,etc.)		√	
Carbon dioxide, nitrogen, oxygen		√	√
Water Content (units of measure specific to fluid application)		√	√

salient properties are summarized in Table 2-4 for crude oils, refined products and condensate/LPG.

2.11.4 Factors Affecting Measurement Accuracy [26–31]

All flow meters are sensitive to various influencing factors due to hydrocarbon properties The designer and operator must therefore be cognizant of the physical principles used in the flow metering technology. The following are typical consideration generally given to the meters that are most used in the liquid pipeline transportation industry:

Positive displacement (PD) meters: factors and associated outcomes which affect the performance of the meter are

- ***fluid viscosity***—mechanical clearances
- ***fluid temperature***—mechanical clearances
- ***fluid pressure***—mechanical clearances
- ***flowrate***—increasing dP with Q
- ***mechanical tolerances***—rotor runout, gear runout, etc.
- ***bearing friction***—due to erosion, corrosion or low lubricity of fluid
- ***deposits***—from solids precipitating out (wax, etc.)
- ***erosion***—due to sand and cavitation
- ***corrosion***—from corrosive contaminants (acid), etc.
- ***cavitation***—from operating at a pressure that is too close to the fluid's true vapor pressure.
- ***accessories affecting torque***—temperature calibrator, register head, packing gland, etc.
- ***surging flows***—large PD meters, due to the mass of the inner mechanism, are

susceptible to damage from fast flow rate changes (power optimization, batch switches, etc.).

Turbine meters: *Two levels of inference* are required to maintain the validity of the volumetric flow rate (which is given by Average Velocity × Area), using turbine meter technology -

1. the flow rate is proportional to the average stream velocity, affected by:
   - ***deposits***—from solids precipitating out (wax, dimerization, polymerization, etc.)
   - ***obstructions***—from filamentary particles ("grass" buildup on the leading edge), etc.
   - ***boundary layer thickness***—from blade surface roughening
   - ***erosion***—due to sand, cavitation, etc.
   - ***corrosion***—from corrosive contaminants (acid), etc.
   - ***cavitation***—from operating at a pressure that is too close to the fluid's true vapor pressure.
2. the average stream velocity is proportional to the rotor's RPM or frequency and is affected by:
   - ***rotor blade angle***—can change if struck by an object.
   - ***rotor stability***—rotor imbalance and poor mechanical/hydraulic bearing conditions negatively impacts the meter's performance.
   - ***fluid velocity profile***—distorted velocity profiles negatively impact the meter's performance.
   - ***fluid swirl***—impacts the boundary layer development at the rotor. When swirl is present, distorted velocity profiles are always present.
   - ***rotor bearing friction***—increased bearing friction impairs the meter's linearity.
   - ***viscous drag on rotor***—the boundary layer development for the blades/rotor is a function of fluid viscosity, rotor velocity and rotor surface finish.
   - ***fluid density***—varying fluid density impacts the rotor driving torque.

REFERENCES

[1] GPSA (Gas Processor Suppliers Association), 1994, *Engineering Data Book*, Tulsa, OK, USA., Vol. II.

[2] Ahmed, T., 2000, *Reservoir Engineering Handbook*, 2nd edition, Gulf Professional Publishing, Houston TX, USA.

[3] Mohitpour, M., Seevam, P., Botros, K. K., Rothwell, B., and Ennis, C., 2011, *Pipeline Transportation of Carbon Dioxide Containing Impurities*, ASME Press, New York, NY, USA.

[4] Young, D., 1998, "Equations of State," http://www.ccl.net/cca/documents/dyoung/topics-orig/eq_state.html.

[5] Young, D., 2001, *Computational Chemistry: A Practical Guide for Applying Techniques to Real World Problems*, John Wiley, ISBN: 978-0-471-33368-5, http://ca.wiley.com/WileyCDA/WileyTitle/productCd-0471333689.html.

[6] ESI (Energy Solutions International), 2012, "Pipeline Studio Version 3.3.1– Liquid Pipeline Simulator TLNET," http://www.energy-solutions.com/products/pipelinestudio/.

[7] Starling, K. E., 1973, *Fluid Properties for Light Petroleum Systems,* Gulf, Publishing Company, USA.

[8] API (American Petroleum Institute), 2001, "API 11.2.2-; Manual Correction of Measured Liquid Volumes to Equivalent Volumes at Reference Conditions."
[9] McCain, W. D., 1973, *The Properties of Petroleum Fluids*, Pennwell Books, ISBN: 296670.
[10] API (American Petroleum Institute), 2000, "Measurement of Petroleum Liquid Hydrocarbons By positive Displacement Meter."
[11] ASTM (American Society of Testing Material), 2005, "D 1298-99 Standard Test Method for Density, Relative Density (Specific Gravity), or API Gravity of Crude Petroleum and Liquid Petroleum Products by Hydrometer Method."
[12] Kung, P., and Mohitpur, M., 1986, "Non-Newtonian Liquid Pipeline Hydraulics Design and Simulation Using Microcomputer". *Proceedings, Pipeline Engineering Symposium, 9th ETCE Conference*, **3**, 73-78.
[13] Withers, V. R., and Mowll, R. T. L., 1982, "How to Predict Flow of Viscous Crude," Pipeline Industry.
[14] Lester, C. B., 1958, *Hydraulics for Pipelines,* Oilden Publishing Co., Houston TX, USA.
[15] Hughes, W. F., and Brighton, J. A., 1967, "Fluid Dynamic" McGraw Hill Co, New York, p. 265.
[16] Maples, R. E., 2000, *Petroleum Refinery Process Economics*, 2nd Edition, Pennwell Books, (ISBN 0-87814-779-9).
[17] API (American Petroleum Institute), 1996, "Manual of Petroleum Measurement Standards MPMS," Chapter 12 *(Calculation of Petroleum Quantities),* Section 3 *(Volumetric Shrinkage Resulting from Blending Light Hydrocarbons with Crude Oils),* 1st edition.
[18] Big Inch Petroleum, 2007, "Hydrocarbon Vapour Pressure," July 17. http://www.eng-tips.com/viewthread.cfm?qid=191613&page=10.
[19] API (American Petroleum Institute), 1969, "Petrochemical Evaporation Loss From Storage Tanks," First Edition, Bulletin No. 2523.
[20] Vasquez-Esparragoza, J. J., Iglesias-Silva, Hlavinka, M. W., and Bullin, J. A. 1994, *How to Estimate Reid Vapor Pressure (RVP) of Blends Encyclopedia of Chemical Processing and Design*, J. J. McKetta ed., Marcel Dekker, Inc., New York, NY, Vol. 47, pp. 415–424. http://www.bre.com/portals/0/technicalarticles/How%20to%20Estimate%20Reid%20Vapor%20Pressure%20(RVP)%20of%20Blends.pdf.
[21] Shaw, J., and Dadgostar, N., 2011, "Developing a Predictive Correlation for the Heat Capacity of Ill -Defined Liquid Hydrocarbons," Chem. Eng. Dept. University of Alberta, Edmonton, Alberta T6G 2G6, Canada. http://www.uofaweb.ualberta.ca/jmshaw/pdfs/2010%20Developing%20a%20Predictive%20Correlation%20for%20the%20Heat%20Capacity%20of%20Ill%20-Defined%20Liquid%20Hydrocarbons.pdf.
[22] Dadgostar, N., and Shaw, J. M., 2011. "A Predictive Correlation for the Constant-Pressure Specific Heat Capacity of Pure and Ill-Defined Liquid Hydrocarbons," *Fluid Phase Equilibria*, **313**, pp. 211–226, Elsevier B.V.
[23] Watson, K. M., and Nelson, E. F., 1933, "Improved Methods for Approximating Critical and Thermal Properties of Petroleum Fractions," *Ind. Eng. Chern.*, **25**(8), pp. 880.
[24] Perry, M. B., and White, C. M.,1985, "New Correlations Between the Watson Characterization Factor (K,) and Properties of Coal-Derived Materials," http://www.anl.gov/PCS/acsfuel/preprint%20archive/Files/30_4_CHICAGO_09-85_0204.pdf.
[25] Gallagher, J. E., 2006, "Effect of Petroleum Properties in Pipeline Measurement," Savant Measurement Corporation, Int School of Hydrocarbon Measurement, Class #2130 http://help.intellisitesuite.com/Hydrocarbon/papers/2130.pdf and www.ishm.info/2012Classes.xls.
[26] API (American Petroleum Institute), 1955, "Measuring, Sampling, And Testing Crude Oil," Bulletin 2500, API, New York, January.
[27] ASME (American Society of Mechanical Engineers), 2002, *ASME ANSI B31.4- Liquid Transportation Systems for Liquid Hydrocarbons and Other Liquids* New York, NY,USA.
[28] ASTM (American Society of Testing Material), 1999, "D323-99a Standard Test Method for Vapor Pressure of Petroleum Products (Reid Method)," ASTM International, PA 19428-2959, United States.

[29] ASTM (American Society of Testing Material), 2011, "D97-11, Standard Test Method for Pour Point of Petroleum Products," ASTM International, PA 19428-2959, United States.
[30] ASTM (American Society of Testing Material), 2011, "D5853-11 Standard Test Method for Pour Point of Crude Oils.
[31] ASTM (American Society of Testing Material), 2011, "D86-11a Standard Test Method for Distillation of Petroleum Products at Atmospheric Pressure."

Chapter 3

SYSTEM HYDRAULICS AND DESIGN

This chapter discusses the fundamentals of liquid pipeline hydraulics and the design and operation (Chapter 5) of hydrocarbon liquid pipeline systems from a hydraulics point of view. Pipeline system design is mainly concerned with line sizing, equipment sizing and location, and flow capacity; while system operation is concerned with pipeline system or facility start-up and shut-down, product receipt and delivery, flow rate changes, emergency shut-down, equipment failure, etc.

A proper pipeline system design requires a system approach taking into account the following design disciplines:

- Hydraulics
- Mechanical design
- Geo-technical design
- Operations and maintenance design

These disciplines are closely interrelated because any decisions or changes in one area of design directly affect or limit the options in another area. Through the hydraulic design, the pipeline route, pipe size, operating pressure and temperature and the number of pump stations are determined. From a hydraulic design, mechanical designs can be developed to meet the criteria of the design basis. The mechanical design is dictated by the relevant codes and standards, resulting in pipe material selection and specifications as well as burial depth requirements. Geo-technical design addresses surface loads, water crossings, buoyancy control and geo-hazard management, which can significantly affect the cost and safety, if the pipeline route traverses challenging environments. The operation and maintenance consideration includes the necessary control systems to operate the system within its design parameters, taking account of the operating tasks while maintaining the functional integrity of the system.

The scope of this chapter includes the governing principles and equations of liquid pipeline hydraulics and their solutions in steady states. The design of any pipeline system is based on various design factors such as flow profile over time and operating pressures.

3.1 FUNDAMENTALS OF LIQUID PIPELINE HYDRAULICS

3.1.1 Pipeline Flow Equations

Pipe flow is dictated by three conservation laws: mass, momentum, and energy conservation. The mass conservation law states that the net change rate of the fluid flow in a segment of pipe is equal to the net packing rate of the fluid in the segment of pipe, while the momentum conservation law states that the momentum applied to a fluid element is conserved, equating the rate of change of momentum to the sum of the applied forces. The energy conservation law holds for fluid flow, so the net rate of energy transport across a pipeline segment is the same as the rate of energy accumulation within the pipeline section. Such energy includes the internal energy, compression or expansion energy (work), and kinetic energy.

The mathematical models used for pipelines are based on equations derived from the fundamental principles of fluid flow and thermodynamics. The hydraulic states of a pipeline can be defined by four independent variables; pressure, temperature, flow rate, and density, and thus four equations are required to relate these four independent variables. These are momentum, mass, and energy conservation equations together with the equations of state appropriate to the fluids in the pipeline. The three conservation laws can be expressed in the form of partial differential equations describing the momentum equation, continuity equation, and energy equation. The one-dimensional form of the conservation equation is adequate to describe the pipeline flow.

3.1.1.1 Continuity or Mass Conservation Equation

The mass conservation equation accounts for mass being conserved in the pipeline. It requires knowledge of the density and compressibility of the fluid in the pipeline together with flows, pressures, and temperatures.

$$\frac{\partial(\rho A)}{\partial t}+\frac{\partial(\rho v A)}{\partial x}=0 \qquad (3-1)$$

where

A = Cross sectional area of the pipe

The cross sectional area can change due to the changes in pressure and temperature:

$$A = A_0\left[1+c_P\left(P-P_0\right)+c_T\left(T-T_0\right)\right] \qquad (3-2)$$

where the subscript zero refers to base or standard conditions. c_T is the coefficient for thermal expansion of the pipe material and its effect on transients is negligibly small. C_P has a large effect on the acoustic speed of a pressure wave and is defined as:

$$c_P = \frac{1}{E}\left(\frac{D}{w}\right)\left(1-\mu^2\right) \qquad (3-3)$$

where

E = Young's modulus of elasticity of the pipe

w = Pipe wall thickness

μ = Poisson's ratio

The first term in the continuity equation represents the change of mass in a pipe segment. It is often called line pack change. The line pack can be increased or decreased due to pressure and temperature changes. The line pack change is useful for gas pipeline operation. It should be distinguished from the line fill volume, which is the quantity of fluid contained in a pipeline. It is also a useful quantity for batch pipeline operation. The second term represents the difference between mass flow into and out of the pipe segment.

3.1.1.2 Momentum Equation

The momentum equation describes the motion of the fluid in the pipeline. It requires fluid density and viscosity in addition to the pressures and flows. Applying Newton's second law of motion to a fluid element together with the Darcy-Weisbach frictional force, the momentum conservation equation, in one dimensional form, is expressed as

$$\rho\frac{\partial V}{\partial t}+\rho V\frac{\partial V}{\partial x}+\frac{\partial P}{\partial x}+\rho g\frac{\partial h}{\partial x}+\frac{f\rho V\,|V|}{2D}=0 \qquad (3-4)$$

where

- ρ = Density of the fluid
- V = Velocity of the fluid
- P = Pressure on the fluid
- h = Elevation of the pipe
- g = Gravitational constant
- f = Darcy-Weisbach friction factor
- D = Inside diameter of the pipe
- x = Distance along the pipe
- t = Time

The first term is a force due to acceleration, and the second term a force due to kinetic energy. These two terms are related to inertial force. The third term is a force due to pressure difference between two points in a pipe segment. The fourth term is a gravitational force, and the last term is a frictional force on the pipe wall, opposing the flow.

The Darcy-Weisbach equation is used to calculate the pressure drop due to the friction of fluid flow against the pipe wall. The friction pressure drop is linearly proportional to the fluid density and the friction factor, squarely proportional to fluid velocity, and inversely proportional to the pipe diameter. The friction pressure drop can be expressed as follows:

$$\frac{f\rho V\,|V|}{2D}=\frac{8f\rho Q^2}{\pi^2 D^5} \qquad (3-5)$$

In terms of flow rate, the frictional pressure drop is proportional to the square of the flow rate and inversely proportional to the fifth power of the pipe diameter. Since the frictional pressure drop and thus pipeline flow capacity depends highly on pipe diameter, it is the most significant design parameter. The friction factor is related to the energy losses resulting from fluid flow. It is a function of the Reynolds number and pipe roughness. Depending on the Reynolds number, the type of pipe flow is classified into three flow regimes: laminar flow, critical flow, and turbulent flow. Turbulent flow can be further divided into partially turbulent, where the smooth pipe law applies, and fully turbulent, where the rough pipe law applies.

The Reynolds number is dimensionless and the ratio of inertial forces to viscous forces. It is defined by

$$\mathrm{Re}=\frac{|V|\rho D}{\mu}=\frac{|V|D}{\nu} \qquad (3-6)$$

where

- μ = dynamic viscosity (kg/m s)
- $\nu = \mu/\rho$ = kinematic viscosity (m^2/s)
- ρ = fluid density (kg/m^3)
- V = flow velocity, m/s
- D = pipe inside diameter, m

The Reynolds number increases as flow rate or flow velocity increases, and is always positive. The kinematic viscosity is frequently used for liquid pipeline design because it is more readily available and is independent of density. A common kinematic viscosity unit is stokes, but centistokes is a practical unit because the viscosities of most hydrocarbon liquids are in centistokes range.

The friction factor is determined empirically and analytically represented by the Colebrook-White correlation for turbulent flow regimes:

$$\frac{1}{\sqrt{f}} = -2\log\left(\frac{k}{3.7D} + \frac{2.51}{\mathrm{Re}\sqrt{f}}\right) \mathit{for}\,\mathrm{Re} \geq 4{,}000 \qquad (3-7)$$

where k is the pipe roughness, D the pipe inside diameter, and Re is the Reynold's number. For laminar flow, the friction factor is:

$$f = \frac{64}{\mathrm{Re}}\,\mathit{for}\,\mathrm{Re} \leq 2400 \qquad (3-8)$$

The critical flow regime is defined between 2,400 < Re < 4,000, in which the flow is unstable. Laminar flow is independent of pipe roughness, while partially turbulent flow is dependent on Reynolds number and pipe roughness, and fully turbulent flow is dependent only on relative roughness being independent of Reynolds number.

The Moody diagram, shown in Figure 3-1, relates the friction factor in terms of Reynolds number and relative roughness.

The Colebrook-White equation is not easily solvable without a computer because the friction factor appears on both right and left sides of the correlation. To facilitate an

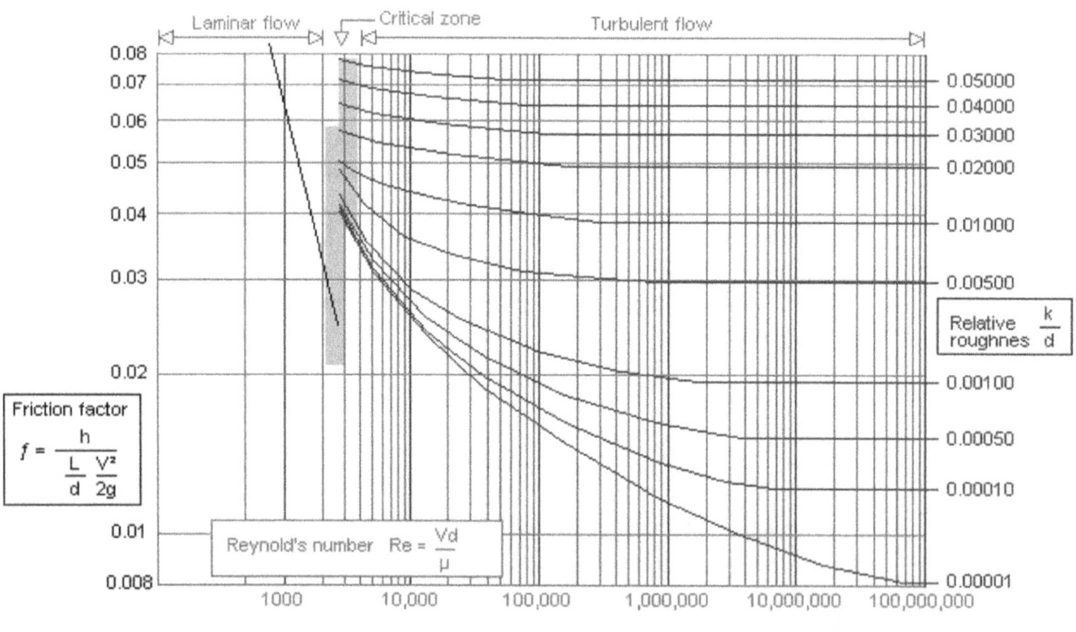

Figure 3-1. Moody diagram for friction factor

explicit calculation, several alternative forms of the correlation have been developed and a few examples are given next:

- Jain's Approximation

$$f = [1.14 - 2 \times \mathrm{Log}(k/D + 21.25Re^{-0.9})]^{-2} \quad (3-9)$$

for $10^{-6} < k/D < 10^{-2}$ and $5000 < Re < 10^{8}$

- Churchill's formula

$$f = 8[(8/Re)^{12} + (A + B)^{-1.5}]^{1/12} \quad (3-10)$$

where $A = \{-2.456 \times \mathrm{Ln}[(7/Re)^{0.9} + 0.27(k/D)]\}^{16}$ and $B = (37{,}530/Re)^{16}$

These equations correlate closely with friction factors on the Moody diagram. The Fanning friction factor f_f is occasionally used and related to the Darcy friction factor as follows:

$$f = 4 \times f_f \quad (3-11)$$

Other pressure drop equations, such as the Shell-MIT equations and Hazen-Williams, are sometimes found in the literature. Since the Darcy-Weisbach equation with the associated Darcy friction factor is most widely used in the petroleum pipeline industry, it will be used throughout this book.

Most liquid hydrocarbon pipelines are operated in partially turbulent flow regimes, with the exception of ethylene and ethane flow which may be in a fully turbulent regime and heavy crude which may be in a laminar flow regime.

3.1.1.3 Energy Equation

The energy equation accounts for the total energy of the fluid in and around the pipeline, requiring information regarding the flows, pressures, and fluid temperatures together with fluid properties and environmental variables, such as conductivity and ground temperature.

$$\left(\rho C_v + \frac{4w\rho_p C_p}{D}\right)\frac{\partial T}{\partial t} + \rho v C_v \frac{\partial T}{\partial x} + T\frac{\partial \rho}{\partial T}\left(\frac{\partial v}{\partial x} + \frac{v}{A}\frac{\partial A}{\partial x}\right) + \frac{f\rho v^2 |v|}{2D} - \frac{4k}{D}\left(\frac{dT}{dz}\right) = 0 \quad (3-12)$$

where

C_v = Specific heat of the fluid at constant volume
T = Temperature of the fluid
ρ_p = Density of the pipe material
C_p = Heat capacity of the pipe material
k = Heat transfer coefficient
z = Distance from the pipe to its surroundings

The first term is the temperature change over time, the second is the rate of temperature change due to the net convection of fluid energy into the fluid element. The third term describes the change rate due to expansion/compression of the fluid including the Joule-Thomson effect. The fourth term represents the heat flow to, or from, the surroundings due to conduction, the fifth term is the effect of work against or by gravity, which will heat the fluid going downhill and cool it going uphill. The last term accounts for heating due to friction, assuming that all the frictional heat is deposited in the fluid. Fourier's law of heat conduction is often used to describe the flow of heat from a pipeline to ground or vice versa. The ground heat transfer takes into account the heat transfer through pipe, insulation, and soil. Ground temperature along the pipeline is not normally measured and thus is not well defined; however, it is an important parameter for designing a heated pipeline system. Therefore, a thorough sensitivity analysis has to be performed to account for ground temperature and soil conductivity.

3.1.1.4 Equation of State

In addition to the three conservation equations, an equation of state is needed to define the relationship between product density or specific volume, pressure, and temperature. The bulk equation of state and API equation were discussed in the previous chapter, so they are briefly summarized here for completeness. For liquid pipeline design, the simple equation of state given below is adequate for heavier hydrocarbon liquid pipeline design:

$$\rho = \rho_0 \left[1 + \left(\frac{P - P_0}{B} \right) - \gamma \left(T - T_0 \right) \right] \tag{3 – 13}$$

where

B = Bulk modulus of the fluid

γ = Thermal expansion coefficient of the fluid

As pointed out in Chapter 2, the API equation of state [1] is a variation of the above equation and is valid for those petroleum products whose density is greater than 635 kg/m^3. For products whose density ranges from 350 kg/m^3 to 635 kg/m^3, the above equation may be inadequate, and API 11.2.2 or NIST equations [2] can produce more accurate results.

3.1.2 Solution Methods

The four equations are solved simultaneously for the four primary variables: flow or velocity, pressure, temperature, and density. These variables are functions of both pipe length and time. In solving these equations, it is assumed that no chemical reaction takes place in the pipeline system and that the fluid remains in a single phase. Multiphase modeling and applications are not addressed in this book.

These partial differential equations involved are coupled non-linear equations. Since these equations cannot be solved directly by an analytical method, approximate solutions may be found using numerical methods instead. With such methods, certain approximations are required such as replacing derivatives in the differential equations with finite differences using averages calculated over distance and time intervals and truncating certain terms in the differential equations.

Since the partial differential equations are expressed in terms of pipe length and time, the solution requires initial conditions for the time variable in order to establish initial pipeline state and boundary conditions to provide boundary values at specific locations. A pipeline state is expressed in terms of four primary variables: flow, pressure, temperature, and density. Initial pipeline states are established in terms of flow, pressure, temperature, and density profiles along the pipeline system. The initial pipeline state can be obtained by either a steady-state solution if there is no known pipeline state or by the previous pipeline state if it is available. At the end of a time interval, the current pipeline state is calculated from the four equations using the initial state conditions and by applying the boundary values. Boundary conditions required to solve for realistic operation analysis are: upstream pressure — downstream pressure boundary, upstream flow — downstream pressure boundary, and upstream pressure — downstream flow boundary.

There are many different ways to solve the difference equations representing the partial differential equations. Three popular solution techniques for pipeline flow simulation are briefly described below. For more details refer to specialized books for solving partial differential equations [3].

3.1.2.1 Method of Characteristics

Streeter and Wylie [4] applied the method of characteristics extensively in solving various pipeline-related problems. The method of characteristics changes pipe length and time coordinates to a new coordinate system in which the partial differential equation becomes an ordinary differential equation along certain curves. Such curves are called characteristic curves or simply the characteristics.

This method is elegant and produces an accurate solution if the solution stability condition is satisfied. This stability condition, called the Courant-Levy condition, requires that the ratio of the discretized pipe length to time increment must be smaller than the acoustic speed of the fluid in the pipeline. In other words, the time increment is limited by the discretized pipe length and the fluid acoustic speed. This is not necessarily a limitation for real-time applications where the time increment is short. However, it can be a severe limitation if applications such as a training simulator require flexible time steps.

The method of characteristics is easy to program and can produce a very accurate solution, it also does not require large computer computational capability.

3.1.2.2 Explicit Methods

In explicit methods, the finite difference equations are formulated in such a way that the values at the current time step can be solved explicitly in terms of the known values at the previous time step [5]. There are several different ways of formulating the equations, depending on the discretization schemes used and which variables are explicitly expressed.

The explicit methods are restricted to a small time step in relation to pipe length in order to keep the solution stable. Just like the method of characteristics, this is not an issue for real-time applications but a severe limitation for applications requiring flexible time steps. For applications extending over a long time, an explicit method could result in excessive amounts of computation.

Explicit methods are very simple for computer programming and can produce an accurate solution. The computer computational capability requirements are relatively light.

3.1.2.3 Implicit Methods

In implicit solution methods [6], the partial differentials with respect to pipe length are linearized and then expressed by finite difference form at the current time step, instead of the previous time as in the explicit method. The values at the current time step are arranged in a matrix, so the solution requires the use of matrix inversion techniques. There are several ways to arrange the numerical expressions, depending on the discretization schemes and whether values are expressed during or at the end of the time interval. Initially, a trial solution is n guessed and then successive changes to the approximated solution are made iteratively until convergence is achieved within a specified tolerance.

The implicit methods produce unconditionally stable solutions no matter what size the time step or pipe length is. Unconditional stability does not mean the solution is accurate. Other errors may make the solution inaccurate or useless. The methods can generate accurate results if the pipe length and time step are short and the specified tolerance is tight. Therefore, they can be used not only for real-time model but also for applications requiring flexible time steps.

The disadvantages are that the methods require matrix inversion software, the computer programming is complex, and the computer computational capability requirement is comparatively high, especially for a simple pipeline system. However, the absence of a restriction on the size of time step generally outweighs the increase in the extra requirements, particularly for large pipeline systems.

There are other solution techniques such as variational methods [7], a hybrid explicit-implicit scheme, and succession of steady states. These are not discussed here.

3.1.3 Steady-State Solutions and Design Equations

A steady state is a condition of a pipeline system that does not change much over time. Under a steady state, pressure and flow remain constant from one instant to another, being considered independent of time. A pipeline system design can be based on a steady-state assumption. In general, the assumption is valid when the system is not subject to sudden changes in flow rates or other operating conditions over a short period of time. However, a steady-state assumption is invalid for short-term operation analysis and even for designing control systems, testing the level of safety under abnormal operating conditions, etc, because these behaviors are time-dependent.

Steady-state equations are good approximations of fluid behaviors for pipeline design. Steady-state solutions can address design issues because a system design is concerned with long time horizons. They are simpler and thus faster to get a solution for each design case. In addition, time-dependent data may not be fully available during a design phase, so transient equations may not be usable. A steady-state solution can generate pressure, flow, temperature, and density profiles along with a list of station suction and discharge pressures. Such a solution is generally adequate for pipeline system design, excluding a control system design, because it can:

- Determine liquid pipeline capacity,
- Determine an efficient operating mode by selecting appropriate units if the line pack changes or transients in the pipeline network are relatively small compared to the system line pack,
- Calculate power or fuel usage and pump or compressor efficiency,
- Identify pipeline operations and an alternate configuration.

In this section, the concept of hydraulics is summarized and a calculation method is presented for design and operation analysis. For detailed hydraulic analysis and calculation, the readers may refer to other books on hydraulics or computer software. In general, the following parameters are required to calculate pipeline hydraulics:

- Pipe grade, size, wall thickness, and pipe roughness,
- Pipe length,
- Elevation profile,
- Fluid properties such as density and viscosity,
- Number of products for batched pipelines,
- Discharge pressure and temperature,
- Delivery or suction pressure,
- Ground temperature and thermal conductivity

3.1.3.1 Solution of Continuity Equation and Volume Correction

Under a steady-state condition, the continuity and momentum equations can be easily solved. The continuity equation is reduced to a total differential equation under a steady state as

$$\frac{d(\rho V)}{dx} = 0$$

From this steady-state form of continuity equation, we get

$$\rho(P,T)V(P,T) = \rho(P_0,T_0)V(P_0,T_0)$$

or

$$\rho(P,T)Q(P,T) = \rho(P_0,T_0)Q(P_0,T_0) \qquad (3-14)$$

This relationship is the basis of converting volume or flow rate from one pressure and temperature condition to another including volume correction to base conditions. Its application is illustrated with the following base design example (this example will be extended further to a realistic design case):

Example: Base Case

A crude oil pipeline from CE to QU is 200 km long and is 20″ in nominal diameter, with a 0.281″ wall thickness. It is constructed of 5LX-56 electric resistance welded steel pipe. At the injection point, crude oil of 32°API gravity and ambient pressure enters the pipeline at an initial flow rate of 18,000 m^3/d at 15°C. The average operating pressure and temperature are 4000 kPag and 4°C. Calculate the flow rate at the operating conditions.

Figure 3-2 illustrates this pipeline configuration, which will be used for the subsequent example problems in this chapter. CE is the initial injection station, QU is the final delivery station, and TO a side stream delivery point.

Solution:

It is assumed that the API correction equation or equation of state (Refer to Chapter 2) is applicable to convert the density at the base condition to the density at the operating pressure and temperature.

Step 1. To determine the flow rate at the operating conditions, the crude density at the same conditions should be determined.

The density equivalent to 32°API gravity is calculated by applying the API gravity and the specific gravity relationship, thus the specific gravity is $\gamma = 141.5/(131.5 + °API) = 0.8654$, and the density is

$$\rho = \gamma \times 1000 = 865.4 \text{ kg/m}^3 \text{ at } 15°C$$

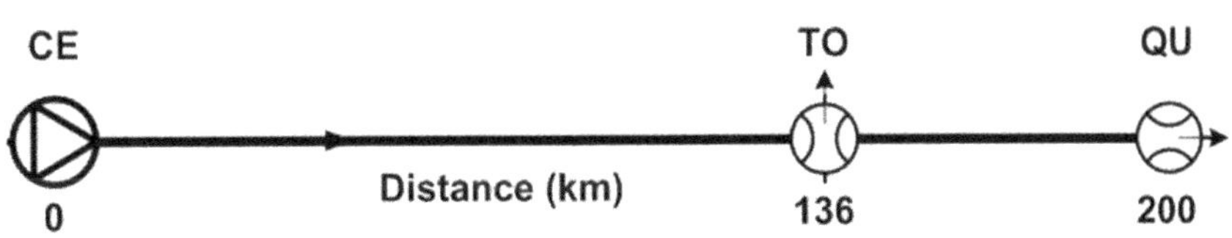

Figure 3-2. Pipeline configuration

Step 2. Since the operating conditions are different from the base conditions of the fluid, it is necessary to convert the density in order to determine the flow rate at the operating condition, by applying the API equation of state:

- Apply the API pressure correction at 4000 kPag: $C_f = 0.6476 \times 10^{-6}$ and $C_P = 1.0026$
- Apply the API temperature correction at 4°C: $C_T = 1.0090$,
- Therefore, the density at 4000 kPag and 4°C = 865.4 × 1.0026 × 1.0090 = 865.4 × 1.0116 = 875.4 kg/m^3

Step 3: Calculate the flow rate at the operating conditions by applying the steady-state mass balance equation.

- Pressure and temperature adjusted flow rate = 18,000 /1.0116 = 17,794 m^3/d
- This volume flow rate is lower because the density at the operating conditions is higher. This is the consequence of mass conservation.

3.1.3.2 Solution of Momentum Equation and Pressure Profile Calculation

The momentum equation can also be simplified under a steady state. Since the kinetic energy or velocity head term for long pipeline systems is negligibly small compared to the total pressure requirement, the momentum equation can be simplified to a total differential equation as shown below.

$$\frac{dP}{dx} + \rho g \frac{dh}{dx} + \frac{f\rho V|V|}{2D} = 0 \qquad (3-15)$$

It can be assumed that the liquid density and velocity are constant between two points along the pipeline. This assumption is valid as long as the distance between two points is not long. Therefore, the pressure-flow equation can be obtained by integrating the above steady-state momentum equation:

$$P_x = P_0 - \rho g(h_x - h_0) - f\rho V^2(X - X_0)/2D \qquad (3-16)$$

The left hand side is the pressure at the downstream point. The first term on the right hand side is the pressure at the upstream point, the second the static pressure or elevation head, and the third the friction head. The total pressure requirement in a pipeline system consists of the following components:

- Pressure changes due to elevation changes, depending only on the product density and difference between the elevations between two points on the pipeline;
- Friction pressure drop due to flow rate or velocity, fluid density and viscosity, and pipe diameter;
- Pressure changes due to changes in pipe diameter and subsequent changes in flow velocity.

For a given flow rate, the above pressure-flow equation allows us to calculate the downstream pressure if the upstream pressure is known, and the upstream pressure if the downstream pressure is known. Also, the flow rate can be calculated if the upstream and downstream pressures are known.

If the static pressure term is set aside, the above equation can be arranged as

$$(P_x - P_0)/(X - X_0) = f\rho V^2/2D \qquad (3-17)$$

This is an expression of the pressure gradient with units of kPa/km or psi/mile; frictional pressure drop per unit pipe length. The pressure gradient is frequently used in liquid pipeline design and operation because the unit pressure drop due to friction is almost constant and the pressure gradient is a straight line for a pipeline with constant pipe size, density, and viscosity. If the pressure is expressed in terms of head, the gradient can be superimposed with the elevation profile, graphically displaying frictional head loss together with the elevation profile. Therefore, it is used to locate hydraulic control points on a pipeline.

Example: Base Case Extension 1

A crude oil pipeline from CE to QU is 200 km long and is 20″ in nominal diameter, with a 0.281″ wall thickness. It is constructed of 5LX-65 electric resistance welded steel pipe. At the inject point, crude oil of 32°API gravity enters the pipeline at an initial flow rate of 18,000 m^3/d. The average operating temperature is 4°C. Calculate the total pressure requirement at the operating conditions. Use the following data:

- Density: 865.4 kg/m^3 at 15°C and 875.4 kg/m^3 at the operating temperature
- Viscosities at 4°C: 43.5 cSt
- Pipe roughness: 0.0457 mm
- Delivery pressure: 350 kPag

Solution:

It is assumed that the elevation profile is flat and the pipe flow is isothermal so that the temperature remains constant throughout the pipeline. The isothermal flow assumption is valid in most liquid pipeline hydraulic studies where the liquid flows near ground temperature.

Step 1. Determine the design flow rate at the operating conditions.

In the previous example, the average flow rate was calculated at 17,794 m^3/d at operating conditions. By taking into account the load factor (refer to Section 3.2.2.3.2 Operating Parameters), the maximum design flow rate is estimated at 20,000 m^3/d.

Step 2. Convert the maximum design flow rate of 20,000 m^3/d into the equivalent flow velocity: $V = 1.21$ m/s

Step 3. Determine the friction factor.

- Pipe inside diameter = (20 – 2 × 0.281) × 0.0254 = 0.494 m
- Reynolds number = $v \times D/\nu$ = 1.21 × 0.494/(43.5 × 10^{-6}) = 13,730
- Relative roughness = 0.0457/(0.494 × 1000) = 9.26 × 10^{-5}

Applying the Jane's friction factor formula, $f = 0.0286$

Step 4. Calculate the pressure gradient at the operating conditions.

- Pressure gradient = 0.0286 × 873.1 × 1.212/(2 × 0.494) = 37.0 Pa/m = 37.0 kPa/km
- Injection pressure = 350 + 37.0 × 200 = 7750 kPag.

Figure 3-3 shows the graphical relationship between the delivery pressure, pressure gradient, and discharge pressure at the injection point.

Example: Base Case Extension 2

Table 3-1 contains an elevation profile. The elevation profile is added to the base design data. Calculate the total pressure requirement at the operating conditions.

Solution:

It is assumed that the elevation changes are gradual between two points and the delivery pressure remains the same as the previous case, the only change is in the elevation.

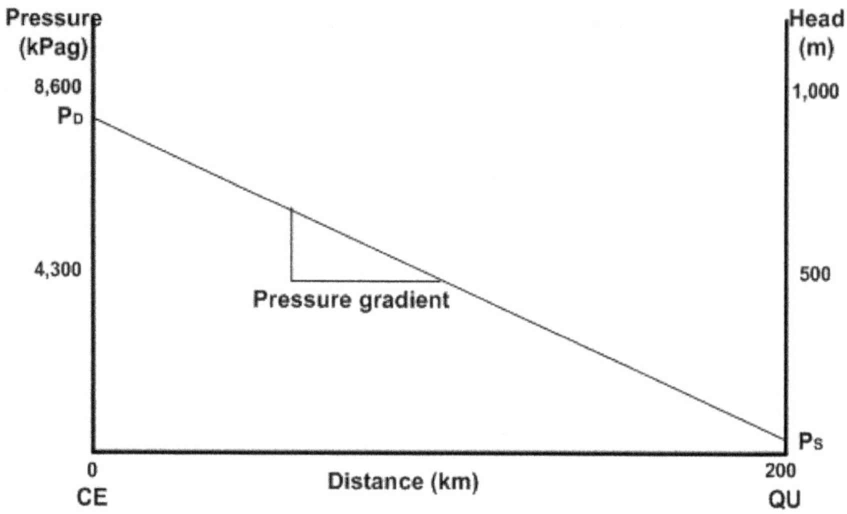

Figure 3-3. Pressure profile and gradient

Note that the frictional pressure drop remains the same even though the elevation changes.

Step 1. Use the same pressure gradient as obtained in the previous example.

Step 2. Calculate the pressures at the above profile points by adding the static pressure difference to the frictional pressure drop.

KMP (km)	Elevation (m)	Pressure (kPag)	KMP (km)	Elevation (m)	Pressure (kPag)
0	30	8605	100	130	4049
20	55	7650	130	100	3196
30	45	7366	150	60	2799
60	30	6384	160	110	2001
80	70	5302	180	150	919
90	100	4676	200	130	350

Step 3. Assess the pressure profile.

The elevation difference between point KMP = 0 km and point KMP = 200 km is 100 m. The static pressure difference is $P_h = 873.1 \times 9.8 \times 100/1000 = 855$ kPa or $8605 - 7750 = 855$ kPa. Since the elevation at the delivery point is 100 m higher than the elevation at the inlet point, the total pressure required at the inlet point is 8605 kPag, which is 855 kPa higher than the previous case for flat elevation.

As shown in Figure 3-4, the pressure profile is shifted by the elevation difference from a reference point, which is in this case the delivery point. Note that the left *y*-axis is represented in pressure and the right *y*-axis in head. Since the elevation profile is

TABLE 3-1. Elevation profile

KMP (km)	Elevation (m)	KMP (km)	Elevation (m)
0	30	100	130
20	55	130	100
30	45	150	60
60	30	160	110
80	70	180	150
90	100	200	130

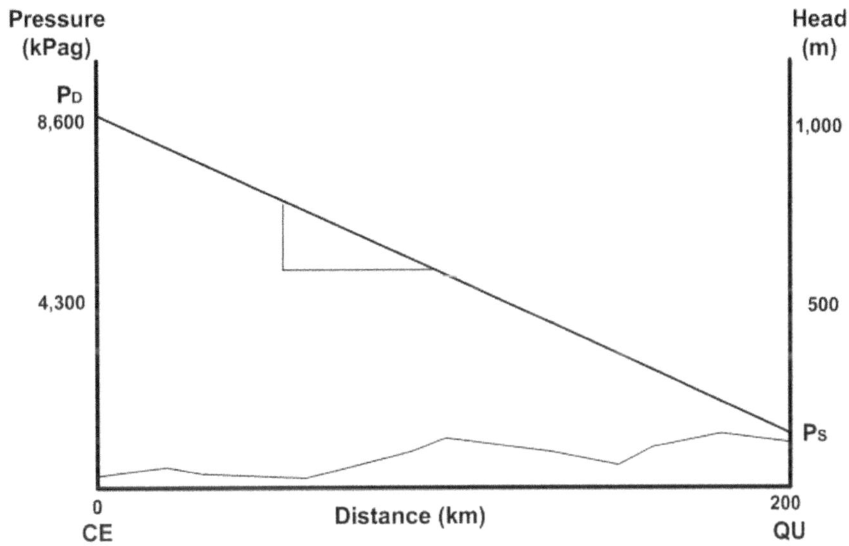

Figure 3-4. Pressure profile with elevation profile

represented in head (m or ft), it is sometimes more convenient to graphically display the pipeline pressure profile in head.

3.1.3.3 Solution of Energy Equation and Temperature Profile Calculation

In the previous examples, an isothermal assumption was made to calculate the pressure profile. The isothermal flow assumption can be justified for fluid which is transported near ground temperature. It is especially valid for a long transmission pipeline with multiple pump stations, because the temperature approaches close to the ground temperature within the first section and the temperature increases at the subsequent pump stations are in the order of a few degrees. However, large changes in liquid temperature can affect liquid density and/or viscosity, which will subsequently affect pressure drop. Therefore, the following hydraulic problems should be treated as temperature dependent flow:

- Heavy hydrocarbon liquids or waxy crudes whose viscosity changes significantly with temperature
- Light hydrocarbon liquids whose density changes significantly with temperature
- Injection temperature is significantly higher than the soil temperature
- Pipelines with a large pipe size running in a hot ambient temperature condition

The liquid temperature rises or falls along a pipeline and rises through a pump station. Temperature profile along the pipeline is influenced by external factors such as ground temperature and soil conductivity as well as heat generated by friction. Fluid temperature rises through a pump station mainly because of the inefficiency of the pump and the small temperature drop through station piping. The temperature change along a liquid pipeline consists mainly of the following components:

- Temperature rise due to volume expansion in an isenthalpic process, raising liquid temperature as the pressure drops;
- Temperature change due to heat conduction with the surrounding ground and ambient temperatures.

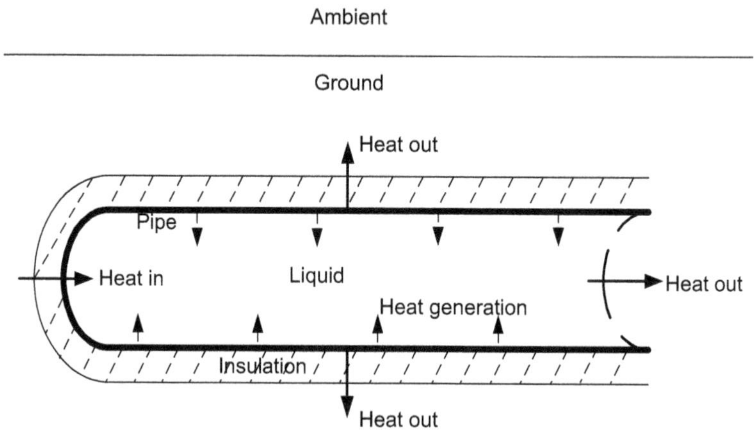

Figure 3-5. Heat balancing mechanism

Some pipelines may be partially or wholly installed aboveground to save construction or maintenance cost. However, transmission pipelines are generally buried in order to:

- Minimize land use disturbance,
- Provide longitudinal restraint along pipeline length,
- Protect pipe from possible pipe material fatigue due to stress changes caused by fluctuations in ambient temperature,
- Minimize effects of changes in ambient temperature on fluid viscosity and density,
- Protect pipe from intentional or accidental damage, and
- Use the pipeline right of way.

The temperature calculation from the energy equation is not simple even under a steady-state condition. The steady-state energy equation can be derived by balancing heat entering and leaving a pipe section, heat transferred from/to the pipe section, to/from surrounding soil or ambient, and heat from friction. The heat balancing mechanism can be shown in Figure 3-5, and the heat balance is expressed as:

$$H_{in} - H_{out} - H_{con} + H_w = 0 \qquad (3-18)$$

where

H_{in} = Heat entering a pipe section (w)
H_{out} = Heat leaving a pipe section (w)
H_{con} = Heat transferred from/to the pipe section to/from surrounding soil or ambient (w)
H_f = Heat from friction (w).

Described below is a temperature calculation procedure. Another method for calculating temperature profile is presented in Addendum 3.1, which includes a temperature calculation method for above-ground pipelines.

1. Assuming that the specific heat of the fluid remains constant at the entering and leaving conditions, the heat entering and leaving a pipe section can be expressed in terms of temperatures and engineering quantities as follows:

$$H_{in} - H_{out} = \rho Q C_p (T_i - T_0)/3600 \quad (3-19)$$

where

ρ = liquid density (kg/m^3)
Q = flow rate (m^3/hr)
C_p = specific heat of liquid, kJ/kg/°C
T_i = temperature of liquid entering the pipe section, °C
T_o = temperature of liquid leaving the pipe section, °C

2. As the liquids flow through the pipe, the pipe pressure drops by friction, liquid flows undergo an isenthalpic process, and as a result the pressure dissipated by friction becomes heat in the flowing fluid. The temperature of liquids rises in frictional heating due to their volumetric properties as they are expanded in an isenthalpic process. The effect of friction heating generally increases with flow rate, viscosity, insulation, and line length. For large diameter pipelines and high flow rates, heat generated by friction loss should be included in the temperature profile.

   Heat of friction should be considered at high flow rates in large pipelines to ensure that overheating does not occur. Pump stations operating on flow control may experience increasing or decreasing discharge pressures as the temperature of the fluid in the pipeline rises or falls after leaving the pump station. As the temperature increases, the fluid expands. As expansion continues in the pipeline, the local pressure and volumetric flow rate increases. The heat generated by frictional pressure drop is expressed as

$$H_w = q\, \Delta P_f = 0.278 Q \times (\Delta P_f/\Delta x) \times L \quad (3-20)$$

where

H_w = frictional heating, w
q = liquid flow rate, m^3/sec
ΔP_f = frictional pressure drop, Pa
Q = liquid flow rate, m^3/hr
$\Delta P_f/\Delta x$ = frictional pressure gradient, kPa/km
L = length of the pipe section, km

3. Even though ground temperature along the pipeline is not normally measured on a daily basis, it is an important parameter for designing a pipeline system. Significant temperature changes can occur due to heat transfer through conduction between the liquid and surrounding soil. In describing the flow of heat from pipeline to ground, Fourier's law of heat conduction is applied by taking into account the heat transfer through pipe, insulation, and soil. The conduction heat transfer can be expressed as:

$$H_{con} = U \times A \times \Delta T_m = 2\pi\, D_T \times L \times U \times \Delta T_m \quad (3-21)$$

where

U = overall heat transfer coefficient ($w/m^2/°C$)
A = surface area of the outside of the pipe (m^2)
D_T = outside pipe diameter or insulated pipe diameter (m)
L = length of the pipe section (m)
ΔT_m = $T_m - T_g$ = log mean temperature difference between the liquid in a pipe section and its surrounding soil (°C)
T_g = ground temperature (°C)

In heat transfer calculations, the log mean temperature can be used, because theoretically it produces a more accurate result in temperature calculation. In practice, there are many factors that prevent the calculation of temperature accurately; these factors include ground temperature, soil conductivity, etc.

In the above heat transfer equation, the overall heat transfer coefficient and log mean temperature difference need to be determined. As shown in the figure below, the overall heat transfer for pipe flow includes the heat transfer effects due to the boundary layer, pipe wall, surrounding soil, and insulation if the pipe is insulated. Therefore, the overall heat transfer coefficient is defined as

$$U = 1/(R_{if} + R_p + R_{ins} + R_s) \qquad (3-22)$$

where

R_{if} = thermal resistance due to the boundary layer that builds up on the inside of the pipe wall ($m^{2\circ}C/w$)
R_p = thermal resistance of the pipe wall ($m^{2\circ}C/w$)
R_{ins} = thermal resistance of insulation ($m^{2\circ}C/w$)
R_s = thermal resistance due to the surrounding medium ($m^{2\circ}C/w$)

However, the heat transfer effects due to the boundary layer and pipe wall are much smaller than those due to surrounding soil or insulation. Therefore, these two terms are usually ignored, and only the last two terms are considered in the overall heat transfer calculation.

Pipelines are not frequently insulated unless the fluid viscosity is so high that it can be significantly reduced by heating the fluid. If the fluid such as heavy crude is heated, certain parts of the pipeline are insulated. For an insulated pipe, the heat resistance can be determined by,

$$R_{ins} = (D_T/k_{ins})\, \mathrm{Ln}(D_T/D) \qquad (3-23)$$

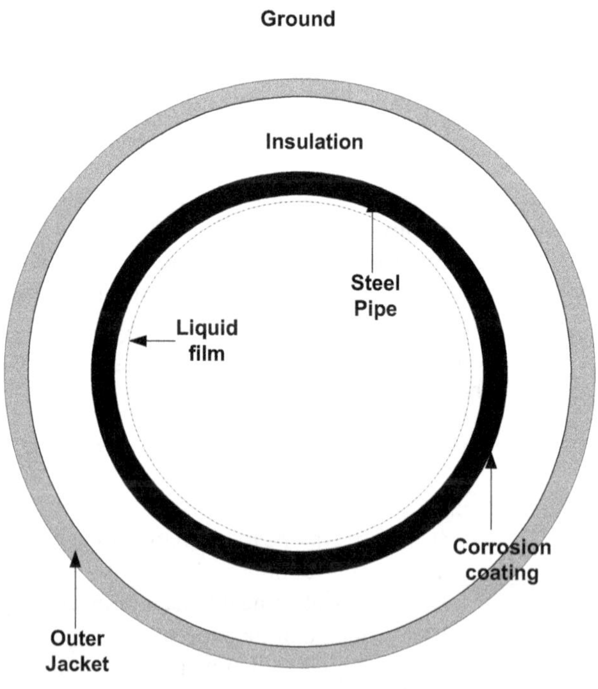

Figure 3-6. Cross section of insulated pipe

where

Ln = natural log

D_T = the outside diameter of the insulated pipe in m ($D_T = D + 2 \times T$),

k_{ins} = thermal conductivity of the insulation,

T = the insulation thickness in m.

In general, the thicker the better; however, insulation efficiency is not proportional to the thickness. Although greater thickness reduces conductive heat transfer, it may not offset the cost of the extra insulation nor reduce the overall heat transfer. The outer jacket is intended to prevent water from making direct contact with insulation material, thereby limiting or even destroying the insulation properties of the insulation. It should be noted that pipeline insulation to reduce heat loss during cold weather may contribute to overheating in summer, particularly for large diameter pipelines. Normally, pipes are coated under the insulation layer.

As discussed earlier, most pipelines are buried along their entire length or at least almost all of their length. The thermal conductivity of insulation can be ten or more times lower than that of soil, but the depth of burial is much deeper than the insulation thickness. In general, heat resistance of a buried pipe is greater than that of insulation, and thus most heat transfer is concerned with heat conduction through the surrounding soil. The heat resistance can be determined by:

$$R_s = (D_T/k_s)\ \mathrm{Ln}\{[2X_c + (4X_c^2 - D_T^2)^{0.5}]/D_T\} \qquad (3-24)$$

where

D_T = D if the pipe is not insulated (m)

X_c = burial depth to the center line of the pipe (m)

= burial depth to the top of the insulation = $D_T/2$

k_s = thermal conductivity of the soil (w/m °C)

The thermal conductivity is a measure of how easily heat conducts through the material. It appears in Fourier's law of heat conduction. Generally, the thermal conductivity can be nearly constant over the temperature range normally encountered in pipelines. Thermal conductivity is measured in units of W/(m°C) (Table 3 2).

Certain portions of a pipeline may run above-ground, even for heated liquids, in order to reduce the construction and other costs. Above-ground pipelines are usually insulated. If the above-ground pipe length is long enough to affect the temperature profile, the heat transfer between the liquid and ambient air needs to be calculated. Since the ambient air conditions can change significantly in a short time, their effects need to be evaluated for design based on the average and worst conditions but are difficult to assess for operation.

In heat transfer calculations, the log mean temperature difference between the liquid in a pipe section and the surrounding soil is often used. This is because the fluid

TABLE 3-2. Thermal conductivity

Substance	Thermal conductivity (W/m°C)
Sandy soil, dry	0.45–0.70
Sandy soil, moist	0.85–1.05
Sandy soil, wet	1.90–2.25
Clay soil, dry	0.35–0.50
Clay soil, moist	0.70–0.85
Clay soil, wet	1.05–1.55
Insulation	0.02–0.05

temperature drop in the pipe section shows an exponential behavior (Figure 3-7). The log mean temperature is defined as:

$$T_m = T_g + (T_i - T_0)/\text{Ln}[(T_i - T_g)/(T_0 - T_g)] \quad (3-25)$$

where

T_i = temperature of liquid entering the pipe section (°C)
T_0 = temperature of liquid leaving the pipe section (°C)
T_g = ground or surrounding medium temperature (°C)

Therefore, the log mean temperature difference is determined by the equation:

$$\Delta T_m = T_m - T_g = (T_i - T_0)/\text{Ln}[(T_i - T_g)/(T_0 - T_g)] \quad (3-26)$$

Note that a log mean temperature is similar to a simple arithmetic average temperature for short pipe lengths over which the temperature is calculated, and that both the log mean temperature and arithmetic average temperature contain the downstream temperature that has to be calculated in the temperature profile calculation. Therefore, an iterative technique is used to calculate either the log mean or arithmetic average temperature and this can be easily implemented in software. A manual calculation can also generate a reasonable temperature profile to the known upstream temperature instead of using the log mean temperature.

Combining the above equations for temperature, we have

$$T_0 = T_i + P_f/(\rho C_p) - H_{con}/(\rho Q C_p) \quad (3-27)$$

where

T_0 = Outlet temperature (°C)
T_i = Inlet temperature (°C)
ΔP_f = frictional pressure drop, Pa
ρ = density (kg/m^3)
Q = flow rate (m^3/sec)
C_p = specific heat (J/kg °C)

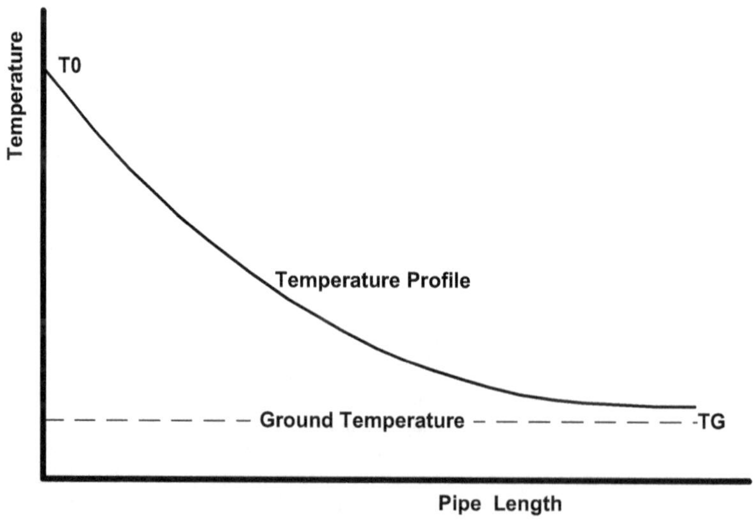

Figure 3-7. Temperature profile

The heat conduction term, H_{con}, includes T_0. In other words, the above temperature equation contains the term T_0 on both sides of the equation. Therefore, it requires an iterative process to calculate T_0 accurately. Except for heavy crudes, the friction heating term is small compared to the heat conduction term, so the above temperature equation can be simplified to:

$$-\rho QC_p dT = UA\Delta T_m dx \qquad (3-28)$$

where

A = pipe surface area
dx = differential in distance

This equation can be integrated to obtain

$$T_x = T_g + [T_0 - T_g] \times \exp[-(2\pi \times UD_T X)/(\rho QC_p)] \qquad (3-29)$$

This equation shows that the temperature profile decays exponentially and that the delivery temperature drops closer to the ground temperature. If the frictional heating term is included, the overall temperature profile is elevated. The temperature equation indicates that, assuming the ambient temperature is lower than the liquid temperature, the liquid cools faster and its viscosity increases as flow rate decreases.

Note that the effect of friction heating increases with flow rates and viscosity because the frictional pressure drop is high. Therefore, a frictional heating term should be included for the case of high flow rates and/or high viscosity liquid. Also, the calculation of a temperature profile is so complex and prone to error that it is beneficial to use a computer software package to obtain quick and accurate results. Temperature-related problems are more severe for larger pipelines because the conduction heat loss is proportional to pipeline surface area.

The surrounding environment is the key factor in the overall heat transfer coefficient, which is most critical in calculating the temperature profile along the pipeline. Table 3-3 shows the range of overall heat transfer coefficients for an on-shore pipeline's surrounding environment [14].

Example: Base Case Extension 3

The previous base case is extended to include the temperature profile by removing the isothermal assumption. Assuming that the pipeline is not insulated, calculate the pressure and temperature profiles using the following data:

- Oil inlet temperature: 35°C
- Average soil temperature: 4°C
- Depth of cover: 1.2 m
- Soil thermal conductivity: 0.5 W/m°C

TABLE 3-3. Environment vs. overall heat transfer coefficients

Environment	U Value (W/m^2 °C)
Buried, dry soil (uninsulated)	0.85–3.69
Buried, dry soil (2" thick insulation)	0.28–0.85
Buried, wet soil (uninsulated)	1.70–4.54
Buried, wet soil (2" thick insulation)	0.57–1.14
Above-ground, exposed to atmosphere (uninsulated)	3.97–8.52
Above-ground, exposed to atmosphere (2" thick insulation)	0.57–1.15

- Specific heat: 1880 J/kg°C
- Viscosities: 9.5 cSt at 35°C and 43.5 cSt and 4°C
- Pour point: 0°C

Solution:

It is assumed that the viscosity of this product is Newtonian and that the density and viscosity depend on temperature. The fluid density and viscosity are calculated at the starting point temperature in the segment between two profile points. Let the inlet pressure be 8605 kPag, the same as for the isothermal case.

Step 1. Since the density and viscosity change with temperature, the temperature relationships of density and viscosity need to be established to calculate these quantities as the temperature profile is calculated.

- Applying the API temperature correction term, we get
  $\rho(T) = \rho(15) \times \text{Exp}[-0.00082 \times (T - 15) \times (1 + 0.000656 \times (T - 15))]$
- Applying the ASTM viscosity correlation, we get
  $\text{Log}\,(v + 0.7) = 11.4667 - 4.6062\,\text{Log}(T + 273)$

Step 2. Calculate the density and viscosity at the inlet conditions; $\rho(35) = 851.0\ kg/m^3$ *and* $(16) = 9.5\ cSt$.

Step 3. Use the inlet temperature of the first segment to calculate the friction factor of 0.0201 and the frictional pressure drop of 508 kPa.

Step 4. Calculate the temperature at the downstream point of the first segment.

- The temperature increase due to the frictional pressure drop is 0.32°C
- To calculate the temperature drop due to conduction, the following values are calculated iteratively:
- the heat transfer coefficient, 0.324 $W/m^{2\circ}C$;
- the log mean temperature, 34.1°C;
- the temperature drop at downstream temperature 2.1°C;
- hence the downstream temperature is 35 + 0.32 – 2.10 = 33.2°C.

Step 5. Calculate the pressure and temperature at the other profile points by repeating the above steps.

KMP (km)	Elevation (m)	Pressure (kPag)	Temp (°C)	KMP (km)	Elevation (m)	Pressure (kPag)	Temp (°C)
0	30	8605	35.0	100	130	5152	27.3
20	55	7889	33.2	130	100	4586	25.4
30	45	7714	32.4	150	60	4368	24.3
60	30	7060	30.0	160	110	3666	23.9
80	70	6195	28.6	180	150	2765	22.8
90	100	5674	27.9	200	130	2364	21.9

It is expected that the total pressure requirement is lower than the pressure requirement under the isothermal assumption, because the operating temperature would be higher and thus the values of density and viscosity are lower. Indeed, the delivery pressure turns out to be much higher than the delivery pressure for the isothermal case, and so the total pressure requirement is less by 2014 kPa. It is concluded that the temperature effects have to be included in hydraulic calculations if the liquid injection temperature is much higher than the ground temperature.

In summary, the following data are required to determine the steady-state pressure and temperature profiles:

- Flow rates
- Fluid properties — density, viscosity, bulk modulus and thermal expansion coefficient, heat capacity, vapor pressure, and pour point
- Pipe grade, and pipe size, wall thickness and roughness
- Pipe length and elevation profile
- Injection and delivery locations
- Delivery pressure or discharge pressure
- Injection and ground temperature
- Depth of burial and soil conductivity
- Operating data such as operating or ground temperatures, batch operation, etc.

3.2 DESIGN PROCESS

A design is a plan developed according to a set of given factors. Included in these factors are parameters and criteria used to provide boundaries to the plan. A parameter is a physical property whose value affects the behavior or characteristic of a related system, while criteria are a set of conditions or guidelines on which a decision can be made. An optimal design provides a plan that optimizes a set of these factors. In pipeline design, these factors may include, supply/demand profiles over time, pipeline route and topology (including supply and delivery locations), fluid properties, pipe parameters, operating pressures and temperatures, and economic parameters. Further restraints are imposed by the codes and standards that apply to the industry. This section discusses key design factors and the design procedure for proper pipeline system design. Refer to ref. [9] for a further discussion of the design factors.

3.2.1 Codes and Standards

The primary responsibility of a pipeline engineer is safety. Therefore, the prime consideration in the design, operation, and construction of a pipeline system is safety. Pipeline standards have been developed to help engineers attain safety consistently in their designs and are normally administered by organizations such as ASME, ISO or Canadian Standards Association (CSA). Standards become codes when they are incorporated into a set of government regulations where they then have the force of law.

Codes and standards are considered to be criteria, and they are essential for pipeline system design and operation. They represent a category that is constant across the industrial activity. Due to the inherent risk of high pressure fluid transmission, codes and standards have been developed to minimize such risk. They set down requirements for design, operation, and construction of pipeline systems with the intent of ensuring good engineering practices and public safety and health. It should be noted that in any given jurisdiction, the government has the ultimate authority, issuing regulations defining minimum requirements. These regulations are legally binding for the design, operation and construction of pipeline systems.

CFR 195 is required in the US. However, ASME B31.4 Code, titled "Pipeline Transportation Systems for Liquid Hydrocarbons and Other Liquids," is applied to liquid pipelines in many parts of the world. B31.4 covers both onshore and offshore liquid pipeline systems, but not gas pipeline systems. Petroleum liquids covered by B31.4 include crude oils, refined products, natural gas liquids, oil-water emulsions, ammonia, alcohols, carbon dioxide, etc. The design section includes design criteria, design and selection of piping components, piping joints, supports and restraints, and auxiliary and other specific piping. The standard also specifies the following subjects:

- Acceptable materials and limitations;
- Dimensional requirements for piping components and threads;
- Construction, welding, and assembly of components, equipment and facilities;
- Inspection and testing, including repair of defects and test pressure;
- Operation and maintenance procedures of pipeline, equipment and facilities, right of way, communications, etc.
- Internal and external corrosion control and monitoring.

The CSA pipeline standard Z662 is more comprehensive than B31.4 in its scope and covers the following:

- Petroleum liquids and gases including sour gas and oil field steam;
- Onshore and offshore liquid and gas pipelines;
- Steel pipe, reinforced composite and polyethylene pipes, and aluminum pipe.

There are many differences between Z662 and B31.4 in design specifications, materials, welding and in other areas. However, the discussion of the differences is beyond the scope of this book. A summary of the differences can be found in [8]. In this book, ASME B31.4 and if necessary, the Canadian standard, CSA Z662, are referenced whenever they are used. Other standards referenced include ASME B16.5 for pipe flanges and flanged fittings, ASME B16.34 for valves, and API 5XL for specifications for line pipe.

3.2.2 Design Factors [9]

3.2.2.1 Supply and Demand

The need for a pipeline system has to be identified before the pipeline system is built. This need results from actual or anticipated requests for transportation of petroleum products.

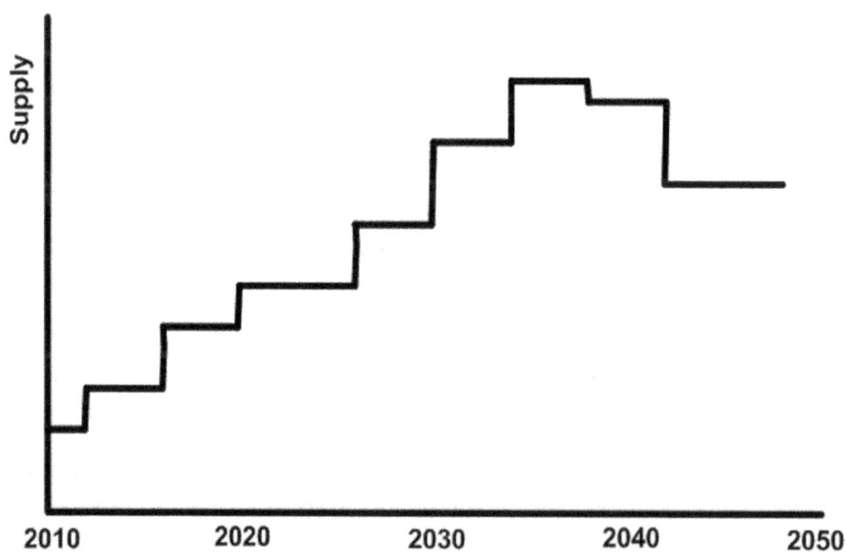

Figure 3-8. Supply profile

The need can be a new pipeline or an increase in the capacity of an existing line, depending on the supply and/or demand locations and volumes specified in the requests.

As shown in Figure 3-8, the flow rates are initially low and increase to a future flow rate. The flow rates can be decreased during the life of the project, and the supply and demand locations may also change. Therefore, an optimum design includes pipeline system growth in terms of pipe and facilities requirements, taking into account future incremental flow rate increase and eventual decrease.

The first step in identifying the need is to determine the supply and demand as well as their respective locations. In general, the demand profile drives the pipeline capacity for petroleum products in consuming areas or oil importing countries, while the supply profile drives the pipeline capacity for producing areas. However, the supply and demand change over time, and their build-up patterns in terms of volume and time greatly influence the determination of the economically optimum size of the pipeline and facilities required for the entire range of flow rates. In other words, the supply/demand projection into the future is required to determine the optimum pipe size, facilities, timing of system expansion, and other requirements. The locations of supply and delivery points strongly influence the selection of the pipeline route and subsequently the locations of facilities and control points.

The supply information includes the oil reserves or production capacities (refinery capacities) estimated at a given time as well as the locations where these volumes will grow or shrink over time. Depending on the particular pipeline system under consideration, supply may or may not be a major factor. If the pipeline system is to be supplied by a large supply source, it may be assumed that the supply will satisfy the demand over the life of the project. On the other hand, if the pipeline system transports fluid from many supply sources, demand may dictate the pipeline system design instead. Therefore, transportation facilities should be designed and built to accommodate these volume forecasts and the accuracy of the supply and demand forecasts reduces the risk of over or under design of the system. Figure 3-8 shows an example of a supply profile over time.

The demand is forecasted on the basis of average annual flows over the period of the project; the yearly volume increases or variations are important for system design. Seasonal variations in the demand also need to be taken into account in design. If the pipeline system transports petroleum products such as gasoline to a large consuming area, seasonal variations in the demand can be more important than the annual increase. In addition, the storage capacities around the consuming areas are also important not only to offset some of the peak requirements but also to avoid over design. If the pipeline system has no storage facilities available, the peak requirements must be transported and the facilities must be sized accordingly to accommodate these requirements.

3.2.2.2 Pipeline Route and Environmental Issues

The routing of the pipeline system is directly related to supply and demand locations. The routing selection is important especially for new pipeline systems. A preliminary route is selected using a combination of immersive video, aerial photography, LIDAR (laser interferometry and distance ranging), and geographical information system (GIS). The latter provides detailed geographical information such as major locations, roads, rivers and lakes, mountains, and even existing pipelines [10]. If major obstacles are located along the preliminary route, the route may be modified before hydraulic studies are performed. In later phases of design, the preliminary route can be modified as more detailed information is made available. For existing systems, the routing considerations may be as simple as paralleling the existing system. However, a new routing may offer significant benefits such as cost savings or additional volume pickups or deliveries over the paralleling option.

The routing selection factors may include terrain, supply sources, population centers, environmental constraints, and other impediments. The weighting of these factors can vary from location to location, but cost and timing are the major considerations along with environmental impacts. The following factors should be taken into consideration in selecting the pipeline route because of the significant impact they may have on the pipeline economics and permitting requirements:

- Pipeline right of way affects construction and land acquisition costs
- Compliance with environmental regulations affect construction timing/methods and hence costs
- Elevation profile directly affects hydraulics and pumping requirements as well as construction cost
- Depth of cover or burial affects hydraulics due to heat conduction and the integrity of pipe as well as construction cost
- Soil types along the route affects construction cost and heat conduction
- Water crossing including rivers affects construction cost, requiring extra valves and overcoming other environmental restrictions
- Geotechnical considerations such as slope stability, earthquake, permafrost, muskeg, etc.

Environmental assessments help pipeline operators develop the guidelines for the pipeline system during the design, construction, and operation phases. They are intended to protect the possible varied environments along the pipeline route. The following environmental issues may arise along the proposed pipeline route:

- Soil resources/farm land
- Protected areas
- Areas of potential archaeological value
- Wildlife, endangered species, etc.

3.2.2.3 Operating Parameters

Since the final purpose of the design is satisfactory system operation, the operating parameters have to be defined in an early phase of the design. They may include operating flow range, operating pressures and temperatures, fluid properties, and ambient conditions.

For optimum design and operation, required factors are not only the future growth of the system throughputs, as discussed in Section 3.2.2.1, but also maximum and minimum daily or annual throughputs. The pressure drop is almost proportional to the square of flow rate or flow velocity. Liquid velocity in a pipeline is the velocity averaged across the cross section of the pipe and is calculated as follows:

$$V = Q/A \qquad (3-30)$$

where:

V = Liquid velocity
Q = Flow rate
A = pipe cross sectional area

It may be noted that there are a number of situations where selecting a pipe size based on the optimum fluid velocity is not appropriate and a detailed analysis will be required

The pressure gradient or pressure drop per unit length of pipe is an important measure for designing a safe and economic pipeline system. Since the liquid velocity is directly related to the frictional pressure drop, the maximum velocity is used as a guideline for an optimum system design. In other words, the required facilities such as pipeline and pump station and operating costs can be minimized by keeping the velocity around an optimum velocity. The maximum velocity can be different for fluids with different density and viscosity. It also depends on surge conditions, potential erosion, facility limits, and economics. Refer to Addendum 3.2 for the discussion of erosional velocity.

Pipeline and piping a major proportion of a pipeline and facilities costs (for example petrochemical plants, piping makes up 20% to 30% of the total capital costs). Therefore, optimizing the pipe size is a key to reducing capital costs.

The optimum pipe diameter is a balance between two opposing factors: material costs and pumping (energy) costs. To obtain an exact optimum size would require a rigorous analysis taking into account: energy costs and capital costs of pumps/piping. These factors will change over time and several of them may be difficult to determine accurately [9]. The following provide fluid velocity ranges that typically provide optimum velocity and hence pipeline diameter operation:

3.2.2.3.1 Low-Viscosity Liquids For low-viscosity liquids, (i.e., with a viscosity of less than 10 cSt — e.g., water, light oils, caustic solutions),

Pipe diameter	Suggested velocity
Below 75 mm NB (Nominal Bore)	0.9 m/s to 2.0 m/s
75 mm NB to 150 mm NB	1.5 m/s to 3.5 m/s
100 mm NB to 200 mm NB	1.8 m/s to 4.0 m/s
Above 200 mm NB	2.4 m/s to 4.5 m/s

These figures approximate only but generally provide an economic pipeline and piping design.

3.2.2.3.2 High Viscosity Fluids As the liquid viscosity increases above 10 cSt, the suggested velocities are lower than those listed above. However, for high viscosity liquids (i.e., these with viscosities approaching 1000 cSt and higher), pipeline and piping design would not be based purely on economic factors. For high viscosity liquids, keeping the pressure drop to within acceptable limits is likely to be the key. It may be noted that there are a number of situations where selecting a pipe size based on the optimum fluid velocity is not appropriate and a detailed analysis will be required.

No pipeline systems can operate continuously for a full calendar year due to operational restrictions such as system maintenance or other reduced capacity operations. The average daily flow is obtained by dividing the annual throughput by 365 (yearly calendar days), and the actual maximum daily flow by the actual number of operating days. The ratio of operating days to calendar days is called load factor, so the load factor can be defined as the average daily flow divided by the actual maximum daily flow. Normally, the maximum daily flow is used for design in order to compensate for the downtime. In the design procedure, a load factor of up to 95% is used for a simple pipeline, while it may be as low as 85% for more complex systems or pipelines operated with expected large flow variations.

The minimum flow rate has to be defined for system design and operation, because all equipment has maximum and minimum operational limits in capacity and efficiency. For example, a pump can only operate within a flow bound between the maximum and minimum capacity. In a highly mountainous terrain, slack flow conditions may occur at low flows so that extra equipment specifications are required to operate the pipeline safely. Refer to Section 3.3.3 for a detailed discussion of slack flow conditions.

Choice of operating pressures directly affects pipeline safety and operating requirements. The requirements include shipping capacity and volume demands, location and method of installation, and the type of pipe material selected. The operating pressure of a pipeline must be maintained within minimum and maximum pressures. These pressure limits are critical for safe and efficient operation. The maximum operating pressure in a liquids pipeline is constrained by the yield strength of the pipe material, pipe diameter and wall thickness, the fluid density and the elevation of the lowest point of the pipe, while the minimum pressures are determined by vapor pressures of the liquids along the pipeline. The elevation affects the operating pressure due to high static head for liquid pipelines.

The delivery pressure is generally defined in the contract between the pipeline company and the shippers or third party pipeline to which the fluid is delivered. The determination of the delivery pressures is influenced by the terminal equipment such as tank and control valves as well as the elevation profile upstream of the terminal. A peak elevation can dictate the pressure required, which can result in higher delivery pressure at the terminal. The delivery pressure is determined by the fluid vapor pressure, pressure rating of the equipment at the delivery site, and pressure requirements imposed by the delivery facilities such as a tank or connecting pipeline. Therefore, the delivery pressure requirement dictates the operating pressure for a given flow rate.

As noted earlier, temperature affects viscosity, density, and specific heat in liquid lines. A temperature rise is beneficial in liquid pipelines as it lowers the viscosity and density, thereby lowering the pressure drop. The cooling effect on non-Newtonian or viscous fluids can be significant because their viscosity can increase significantly and subsequently the pressure drop can be very high. To reduce the effect of temperature cooling, the pipeline can be insulated and/or operated at high temperature. The viscous fluids can be blended with light hydrocarbon liquids such as condensate. The temperature along the pipeline is least controllable due to its dependency on variable soil thermal conductivity and ambient temperature.

The maximum temperature limit for buried pipe is determined by a combination of the following three factors:

- Ground conditions
- Stress level the pipe material can withstand without buckling
- Economics of pipeline flow (the liquid flows most efficiently at high temperature)

The minimum temperature limit is normally determined by the metallurgical (fracture toughness) properties of the pipe material or by the ground conditions.

Fluid properties were fully discussed in the previous chapter. Summarized below are fluid properties that directly and indirectly affect the design and operation of liquid pipeline systems.

- Density or specific gravity — the higher the fluid density, the higher the pressure drop. The pressure drop due to friction is directly proportional to the fluid density.
- Compressibility or bulk modulus is not important for liquid pipeline capacity calculation, but important for controlling pressure surges and determining line pack changes.
- Viscosity is important in calculating line size, hydraulics, and pumping requirements for liquid pipelines.
- Vapor pressure determines the minimum pressure in the pipeline. It must be high enough to maintain the fluid in a liquid state and to avoid cavitation at inlet to a pump.

- Pour point is the lowest temperature at which oil flows and around which it starts behaving more like a non-Newtonian fluid. Oil can be pumped below the pour point, but here the design and operation require special consideration and pumping equipment. It should be noted that the change in fluid characteristics occurs gradually at a higher temperature than the pour point.
- Specific heat affects heat transfer rate through conduction processes between fluid and surrounding soil.

The ambient parameters include ambient air temperature and ground conditions. These parameters play a critical role in design and operation, particularly for long pipelines or for pipelines in extreme environments such as a desert or the Arctic. In permafrost areas, for example, the fluid has to be chilled to a few degrees below 0°C to avoid melting the surrounding frozen soil. The ambient air temperature affects turbine driver thermodynamic performance as well as the fluid properties due to conduction. Since ambient conditions change daily and seasonally, these variations have to be taken into consideration in design and operation.

Most pipelines are buried for various reasons. Even though it is costly to bury pipe, buried pipelines offer significant advantages over aboveground pipelines:

- Limited changes due to ambient temperature and minimum effects on fluid properties such as viscosity
- Pipeline is restrained by the soil along its length
- Protection from intentional or accidental damage as well as against expansion and contraction from ambient temperature changes
- Allows surface use of pipeline right of way.

The greater the depth of burial, the lower the rate of heat transfer. The effect of soil thermal conductivity on the fluid depends on the differential temperature between the fluid and the surrounding soil. If the soil temperature is colder than the fluid temperature, the fluid temperature drops. This results in higher viscosity and a higher pressure drop. If the soil temperature is hotter than the fluid temperature, the opposite results occur. The following parameters are required to determine the temperature profile due to heat transfer along the pipeline:

- Receipt temperature — determine the temperature profile along the pipeline
- Soil temperature
- Thermal conductivity
- Depth of cover
- Thermal insulation properties
- Ambient temperature — has a direct impact on soil temperature and turbine performance

3.2.2.4 Pipe Parameters

Most liquid transmission lines are constructed of steel pipes. Steel pipes are structurally strong and ductile; they do not fracture easily. Steel pipes are made of various grades of steel with yield strength in the range of 30,000 to 120,000 psi. In the hydraulic design, line size is initially based on a preliminary choice of pipe grade, diameter, and wall thickness from experience. Further calculations are needed to finalize the system design based on the code requirements, project cost, and material availability.

The profitability of a pipeline operation is directly related to how much volume is delivered from sources to destinations, and the maximum throughput is mostly determined by pipe size and pressure. Pipe grade, diameter, and wall thickness are the largest factors in determining the throughput capacity. They also affect pipeline operating pressure and thus overall economics:

- Pipe size — the larger the inside diameter of the pipeline, the more fluid can be moved through it and the smaller the pressure drop per unit length.
- Pipe wall thickness — determines the steel and construction cost and operating pressure.
- Pipe grade — determines the steel strength and the operating pressure affecting pipe construction and operating costs.
- Pipe roughness — affects pressure drop and cleaning pig run frequency.
- Pipe coating — protects against corrosion and other damage by inhibiting the flow of electric current from the pipe to the surrounding soil.
- API 5LX specifications are often applied to the acquisition of high pressure steel pipeline in Grades X42 through X80.

Pipe size is the largest factor in determining the throughput and one of the most important parameters in the design and operation of pipeline system to meet a set of projected flow profiles. The minimum size may be selected based on the maximum input pressure and the minimum output pressure for short pipelines, while the pipe size together with other factors including pumping facilities have to be optimized for longer systems.

Pipes are designated in pipe size, pipe wall thickness, and weight. A common designation of pipe size is the nominal pipe size (NPS), which indicates the outside diameter of a pipe. The internal diameter of the pipe defines the cross-sectional area available for the flow of fluids. It is obtained by subtracting twice the pipe wall thickness from the outside diameter. For a given pipe diameter, several different wall thicknesses are available to satisfy different levels of the maximum design pressure. For a specified pipe design pressure, the pipe wall thickness varies with the pipe grade and its elevation changes. Nominated pipe sizes and wall thicknesses are intended to standardize pipelines and associated facilities.

The nominal pipe size outside diameter is expressed in millimeters or inches. The weight of a unit pipe length is determined by the actual pipe size and wall thickness. Table 3-4 lists pipe sizes and wall thicknesses with their corresponding weights.

3.2.2.4.1 Design Pressure The three pipe parameters determine the level of internal pressure that a pipe can withstand. Associated with these parameters together with a design or "safety" factor is the maximum design pressure. The design pressure sets the maximum limit that the pipeline is allowed to be pressurized safely. The design

TABLE 3-4. Pipe Size, standard wall thickness, and weight

Nominal pipe size (NPS)	Pipe OD (mm/in)	Standard wall thickness (mm/in.)	Weight (tons/km)
6	168.3 / 6.625	7.11 / 0.280	28.27
8	219.1 / 8.625	8.19 / 0.322	42.54
10	273.1 / 10.752	9.27 / 0.365	60.32
12	323.9 / 12.752	9.53 / 0.375	73.84
14	355.6 / 14	9.53 / 0.375	81.31
18	457.2 / 18	9.53 / 0.375	105.18
20	508.0 / 20	9.53 / 0.375	117.11
24	609.6 / 24	9.53 / 0.375	140.98
32	812.8 / 32	9.53 / 0.375	188.72
36	914.4 / 36	9.53 / 0.375	212.59
42	1066.8 / 42	9.53 / 0.375	248.40
48	1219.2 / 48	9.53 / 0.375	284.20

pressure is determined by modifying Barlow's formula for a given pipe grade, pipe size, and wall thickness to include a design safety factor:

$$P_{design} = (2S \times t/D_0) \times F \times L \times J \times T \qquad (3-31)$$

where

- S = specified minimum yield strength (SMYS) of pipe, kPag, or psig
- t = pipe wall thickness, mm or in.
- D_0 − outside pipe diameter, mm or in.
- F = design factor or safety factor
- L = location factor ($L = 1$ for liquid pipelines)
- J = joint factor (to reflect the method of pipe joining generally taken to have a value of 1)
- T = temperature derating factor, to account for the effect of higher temperatures on yield stress

The SMYS is a standard measure of the specified minimum yield strength for steel pipe. Standards that are frequently used by the pipeline industry are API 5L: Specifications for Line Pipe, which includes API 5LX and 5LS. API 5LX specifies various strength grades, ranging from Grade B, rated at 42,000 psig (289 MPag) to Grade X120, rated at 120,000 psig (827 MPag), where the Grade X120 refers to the SMYS in 1000 psi. Pipes are manufactured to these specifications.

ASME B31.4 does not define the location factor. The design factor, F, specified in ASME B31.4 is 0.72 for liquid pipelines regardless of the location of the pipeline, while other codes such as CSA Z662 define the design factor differently depending on the locations. The joint factor is 1 for all types of pipe manufactured to 5LX and 5LS specifications. The temperature derating factor is generally taken as 1 for transmission pipelines, because transmission lines are seldom operated beyond the temperature derating range. Several mechanical design aspects are discussed in the next chapter.

Effective pipe roughness is a pipe parameter that affects frictional pressure drop and pipeline efficiency. It includes pipe roughness as well as other pressure loss terms such as bends, welds, fittings, etc. It directly influences the friction factor of the fluid flow; the larger the pipe roughness, the higher the frictional resistance. To reduce roughness, pipes are internally coated or cleaned by pigging. Several examples of pipe conditions and their corresponding roughness are listed in Table 3-5, showing also that pipe roughness varies with pipe conditions.

3.2.2.4.2 Maximum Allowable Operating Pressure (MAOP) In the design of pipelines and their components, the design engineer must ensure that the design pressure at any point along the pipeline is lower than or equal to the maximum design pressure or maximum allowable operating pressure (MAOP). As discussed earlier, the design pressure is proportional to pipe strength and the MAOP defines the maximum pressure permitted for steady-state pipeline operations which relates to the pipe's ability to withstand internal pressure. The MAOP is the sum of the pressure required to overcome friction losses, static head pressure, and any required back pressure or

TABLE 3-5. Pipe roughness

Pipe conditions	Roughness (in.)	Roughness (mm)
New clean bare pipe	0.0005–0.0008	0.0127–0.0200
Scraper burnished pipe	0.0003–0.0005	0.0076–0.0127
Internally coated pipe	0.0002–0.0003	0.0051–0.0076
Pipe after two years of atmospheric exposure	0.0018–0.0020	0.0445–0.0508

delivery pressure. Therefore, the values of a point specific MAOP along the pipeline vary with elevation changes.

In many jurisdictions, MAOP is obtained by choosing the lowest of the following four values in a pipeline section:

- Design pressure determined by Barlow's formula,
- Pressure established during hydrostatic testing of pipe with hydrostatic pressure limit equal to 80% of hydrostatic test pressure (hydrostatic test pressure results in 90% of SMYS for new pipe), which is illustrated in Figure 3-9. Note that in Canada and a number of other jurisdictions test pressures causing the pipe to reach or go slightly above yield are permitted,
- Flange rating: B16.5 based on grade, material and operating temperature,
- Documented historical operating pressure.

Figure 3-9 shows the MAOPs determined after a hydrostatic test is performed, assuming that the pipe grade, diameter, and wall thickness are uniform. Hydrostatic testing must be performed on new pipelines, as specified in ASME B31.4 and other standards, prior to in-service use. Hydrostatic testing is also used on operating pipelines to assess their structural integrity. For testing a new pipeline, the pipeline is divided into multiple pipe segments, which are tested individually. The length of each segment and hence the overall number of test sections is determined on the basis of acceptable elevation changes within the segment.

After a certain period of operation, some segments of pipe may have corroded internally or externally, and thus effective pipe wall thickness is reduced. In such cases, the new pipe wall thicknesses have to be determined and the pipe repaired or else the MAOP of the segments must be lowered.

3.2.2.4.3 Pipe Wall Thickness Pipe wall thickness seldom remains uniform along the pipeline. ASME B31.4 requires that an allowance of 10% over the internal design pressure or 80% of specified minimum yield strength (SMYS) of the pipe is made to take into account surges and other operational changes in pressure. After an optimum pipe wall thickness is determined, a thorough transient analysis is performed using potentially worst case operation scenarios. Based on this analysis, the pipe wall thicknesses need to be increased to satisfy local transient pressure requirements or can be

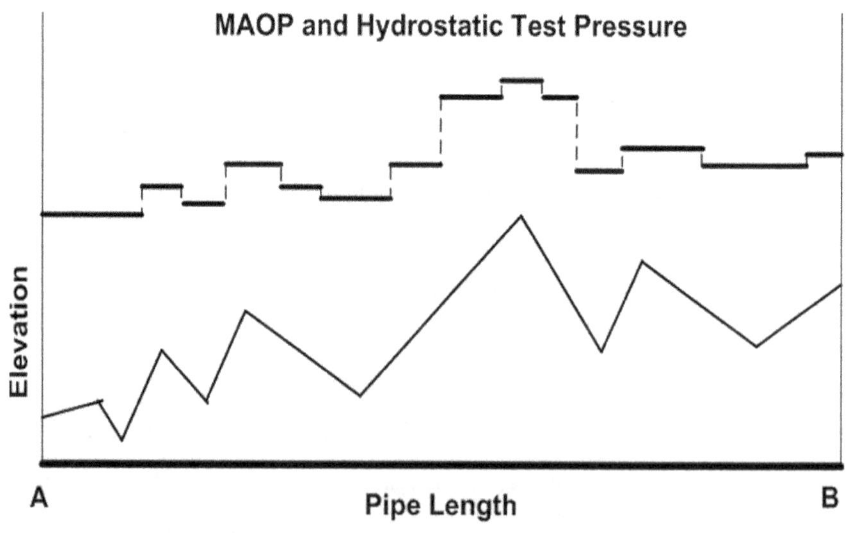

Figure 3-9. Hydrostatic test and MAOP

decreased not only to satisfy safe pressure requirement but also reduce pipe cost. As a rule of thumb, pipe wall thickness tends to be larger than the optimum thickness around river crossings or in deep valleys, while it is smaller at the highest elevations.

Section 401.2.3 of B31.4 specifies that a component of the pipeline system shall be designed to withstand the maximum differential pressure between external and internal design pressures. External surface loading on the buried pipe at road and railroad crossings, or caused by heavy agricultural equipment may require extra pipe wall thickness.

3.2.2.5 Pumping Parameters

All liquid pipeline systems have one or more pump stations in order to boost the pressure level of the liquid. In the early phase of the pipeline system, the number of pump stations may be small due to low flow rate. As the flow rate requirements increase, one way of addressing the system growth is to add more pump stations.

Pump characteristics and station design are detailed in the next chapter. Summarized below are the pumping parameters required for the selection of pumps and the design of pump stations:

- Pump Capacity
- Performance curves
- Operating ranges (flow, pressure and temperature)
- Pump efficiency
- Cooler/heater requirement parameters
- Station auxiliary equipment requirements and specifications
- Energy/Power requirements and specifications
- Driver requirements and specifications
- Piping requirements and specifications

3.2.2.6 Economic Factors

Several stakeholders are involved in building and operating a pipeline including both users and non-users of the pipeline system. Either directly or indirectly, these stakeholders have an interest in the pipeline system. The users may include the shippers on the pipeline system as well as the owner and operating company. Non-users of the system are land owners, the general public, environmentalists and multiple levels of governments. Other non-users may include users of other transportation modes, such as trucking and railroad companies, whose business could be directly affected by the pipeline system. Some of the non-users such as land owners have an economic interest, but others such as the general public may not be directly involved in the development of the system. However, labor unions and/or environmentalists might show opposing interests to the project; citing economic impacts vs. potential adverse consequences due to changes in the socio-economic and natural environment. Governments, through their regulatory agencies, make a decision by balancing the views of all of the stakeholders based on sound engineering and economic merits. Therefore, an unbiased economic study including an environmental assessment is necessary to satisfy all the stakeholders.

For a new pipeline system, an economic study is necessary to provide a measure of economic benefits for not only shippers and pipeline companies but also other key stakeholders. The study must justify the need of a new pipeline system to satisfy the energy requirements in new markets. The study assesses the project feasibility, financing requirements, and optimum system design and operation. If the pipeline is of strategic importance for a country or a certain region, the assessment of the project feasibility may not be critical. However, the need for a new pipeline or a major expansion of an existing system can be justified through an economic analysis. The economic study covers the financing requirements that may include the project profitability, amounts of financing and their payment schedule. It also includes preliminary design and operation, all costs, and comparative analysis of the capital costs along with the operating costs as well as the proposed tariff structure in the case of a cost recovered public utility.

A pipeline economic analysis includes a process of optimizing the pipeline system, determining an optimum pipe size and pumping requirements over the life of the project life. The economic study may include key, not necessarily all, design factors discussed above. The optimizing process involves achieving a desired level of profitability, balancing the capital costs including material and construction against the operating costs. During the process, due considerations should be given to design factors that are suitable for operating the pipeline system safely and economically.

The performance of an economic study is beyond the scope of this book, so no attempt is made to discuss an economic analysis and tariff structures. However, some of the major cost factors are discussed in this section because they influence pipeline system design greatly and will be referred to again in the subsequent chapters:

- Mechanical factors
  1. Pipe grade, pipe size or diameter, and wall thickness
  2. Pipeline route and depth of cover
- Capacity factors
  1. Operating parameters
  2. Station spacing and pumping costs
- Reliability and safety factors
  1. Valve spacing
  2. Other valve-related costs

3.2.2.6.1 Pipe Grade, Size, and Wall Thickness It is critical to optimize the pipe grade, diameter, and wall thickness to minimize the project cost. The pipe cost is based on the grade, diameter, and wall thickness. For most pipeline systems, the pipe cost is the highest material cost. In addition, these three factors have a direct effect on the cost of installation. Pipeline economics begins with the selection of the pipe material. Since pipe material for transmission lines is steel, it boils down to the selection of pipe grade. Higher grade steels are more costly to produce and because of their chemical composition require specific welding procedures. Nevertheless they do result in thinner pipe wall hence less steel tonnage, lower transportation costs, and reduced amounts of welding. A case specific study is needed to determine if such steels are the optimal solution to a given project.

One common economic decision is whether to construct a large line initially, or put in a smaller line first and parallel it or add pumps at a later time. Once the need for a pipeline system is recognized, the maximum pipe size is determined such that it can be economically optimized. The larger the pipe size, the larger the carrying capacity and the lower unit shipping costs. The pipeline capacity increases approximately by 5/2 power for a fixed pressure drop, but the pipe material cost increases significantly and construction costs increase almost linearly as the size is increased.

The design pressure is directly proportional to pipe wall thickness for the same grade and size. The larger the wall thickness for a given pipe size, the higher the design pressure. The larger the wall thickness, the higher the pipe and construction costs. Higher grade pipe requires thinner pipe wall for the same design pressure, resulting in lower steel weight and reduced cost even though higher pipe grade costs more per ton. Cost savings can also result from reduced construction costs.

3.2.2.6.2 Pipeline Route Both direct and indirect costs due to time delays have to be taken into account in selecting a pipeline route. As noted earlier the costs of selecting a pipeline route are related to pipeline length, terrain features, intermediate supply and delivery locations, cost and restrictions on facilities and land, and permitting requirements. If possible, a straight line is selected to minimize the pipe cost, and severe mountainous terrains are avoided because of high construction, pumping and maintenance cost requirements. Obtaining right-of-ways for certain portions of the route can be difficult or even impossible due to environmental restrictions or land claims.

The determination of pipeline location must take account of population density, as well as the proximity of features such as roads, railways, rivers, lakes, unusually sensitive areas, etc. The route should be evaluated in terms of the safety and environmental issues, accessibility, extra material requirements, land claims, etc. Also, the locations of facilities have a direct influence on construction cost.

The minimum depth of cover from a safety standpoint is specified in the applicable codes and standards. However, the operational requirement depends on the temperature condition and thus varies along the pipeline route, particularly for long pipelines. The effect of depth on the installation and labour cost component is largely dependent upon the burial depth, soil conditions and location. Extra labour, material and/or equipment costs are incurred for conditions such as rocky ground, soft ground, e.g., muskeg, river beds, roadbeds, railway crossings, etc.

3.2.2.6.3 Operating Parameters No extra cost is associated with the flow rate because the design is based on it. Since operating pressures are based on maximum allowable operating stress levels of pipe grade, pipe size and wall thickness, and class location factors, a range of design pressures is available in the design phases. If higher operating pressure is selected, the station spacing is increased, resulting in lower material and energy costs.

If the fluid viscosity is sensitive to temperature, the major cost items could be the provision of heaters and heating, pipe insulation, and/or a blending operation.

3.2.2.6.4 Station Spacing and Pumping Costs Station spacing is determined by factors such as pipe size, flow profiles, hydraulics and elevation profile, and capital and operating costs. In an environment of high energy cost or rapid increase in flow, the option with a larger pipe size is preferred, even though its capital cost is higher than that of a smaller size. For a given flow profile, the larger the pipe size, the longer the station spacing. The longer the station spacing, the lower the capital costs associated with station construction and the pumping cost associated with power and energy.

3.2.2.6.5 Valve Spacing Valves are significant cost items. Placement of valves provides for effective control of pressure or flow; sectionalizing the system in case of emergency, isolation of components of the system, etc. The minimum valve spacing and operation requirements are specified in the applicable codes and standards. The number and locations are determined by such factors as system layout, product, adjacent population density, proximity to river crossings, etc.

3.2.2.6.6 Other Valve-Related Costs Other valve related costs have to be considered for safety in certain designs: namely, the need for and location of pressure-reducing valves and pressure relief valves. The latter is discussed in Section 5.1.3 Surge Control.

3.2.2.6.7 Pressure-Reducing Station (PRS) A pressure-reducing station (PRS) is usually installed to reduce the back pressure of a pipeline if the pipeline is sloping down severely. This is due to the static pressure increase beyond the MAOP caused by the elevation gain on the downstream side (refer to Section 3.3.3). A PRS is installed to maintain the downstream pressure below the MAOP, independent of the upstream pressure, unless the upstream pressure becomes less than the downstream pressure set point. Through the downstream pressure controlling process, the upstream pressure can be increased. The installation of a PRS has both cost and operation implications; a PRS requires not only various types of valves including a relief valve and relief tank but also a pig trap and launcher pair. An example of a PRS operation is discussed in Section 5.1.4.

Major capital costs are 30% to 40% of the total capital cost in material, 35% to 50% in labor and construction, 5% to 10% in right of way, and 12% to 15% in miscellaneous items. The materials include pipe, pump stations, valves and fittings, meter stations, SCADA and telecommunication equipment, and tanks and manifold piping, while the miscellaneous items include engineering, surveying, administration, regulatory filing, freight, taxes, etc. Among the major operating costs, general and administration costs such as payroll is the largest, and power and energy cost the next largest. The rest are SCADA and telecommunication costs, utility costs, lease costs such as ROW easements, office buildings, etc.

3.2.3 Hydraulic Design Procedure

A pipeline design process describes a way of combining the design considerations with appropriate codes and standards. Of course, all of the design factors discussed in the previous section are not always required; a certain type of pipeline design requires a certain set of factors and another type requires a different set of factors. For example, consideration of the pour point of light crude may not be required in a warm temperature environment, but may be required for a viscous heavy crude in a cold temperature environment.

The system design is done in several phases; conceptual design, system planning, and detailed engineering design. In the conceptual design phase, the following data is available with which a preliminary hydraulic study is performed:

- Product properties such as gravity and viscosity
- Flow profile over the life of the project
- Pipeline length and preliminary route with the points of injection and delivery
- Macro-economic data such as trends of economic growth, demographic changes, etc.

The conceptual design may include hydraulic and economic studies, which result in overall system and financial requirements.

There are several types of pipeline system design; a new pipeline system, increasing the design capacity of an existing pipeline, and delivery from and/or injection to other points outside the existing system. The increases in the capacity of the existing system may require additional pump stations, a parallel line, or replacement of the existing pipeline with a larger pipe size. Also, the route of the existing pipeline can be moved due to significant supply/demand changes, or some existing pump stations may be relocated to other sites to improve the operational efficiency and subsequently to increase capacity.

After the conceptual design is approved, the pipeline system design is done to achieve the minimum combined capital and operation costs. In the system planning phase, the hydraulic and economic evaluation studies are performed in relative detail, by taking into account the product properties and volumes to be transported, pipeline route and terrain data, operating temperature ranges and possibly preliminary pressure ranges, economic and financial data, and other factors such as environmental conditions and restrictions. Described below is a process for performing hydraulic and economic studies:

1. *Gather data*
   - Receive the commitment from shippers for the proposed pipeline
   - Forecast the supply/demand volumes
   - Select a preliminary route for the pipeline

2. *Prepare a set of design criteria and parameters.*
   - Define the ranges of operating pressure and temperature, pipe size, flow velocity, and other parameters if necessary. Refer to Section 3.2.2.3 for estimating an optimum flow velocity.
   - Define the economic evaluation criteria such as profitability measures.
3. *Develop a base design using the following steps:*
   - Step 1: Determine the design pressure for a pipe size and wall thickness and the flow velocity for the maximum flow rate. Select the pipe size and wall thickness combinations that fall within the operating pressure, velocity, and pipe size.
   - Step 2: Calculate the required total pressure for the combinations over the flow ranges determined from supply or demand forecasts.
   - Step 3: Determine the number of pump stations and power requirement as well as energy cost for each combination over the flow ranges.
   - Step 4: Calculate the capital costs, which include the cost of the pipe and the cost of pump stations, and annual operating costs for each combination.
   - Step 5: Determine the total annual cost by adding the operating costs to the annualized capital cost for each combination
   - Step 6: Select the best case to be the base design after reviewing the results.
   - Step 7: Analyze the operability of the selected design to finalize the base design.
4. *Evaluate alternative design cases to the base design.*
   - Develop alternative design cases to the base design
   - Evaluate them with some or all of the above six steps
   - Compare each alternative against the base design and select the best design.
5. *Perform a sensitivity study*
   - Perform a set of sensitivity study for forecasted flows, product properties, pipe cost, pump station cost, fuel cost, discount rate, and other key parameters.
   - If the sensitivity study reveals a significant flaw in the selected design, then another alternative design has to be considered.
   - If a minor deviation is found, the selected design may be modified.

A hydraulic study is a process to identify a viable combination of pipe parameters, pumping power, and flowing temperature. As the result of the hydraulic studies, the following values are determined:

- Pipe grade, and optimum pipe size and wall thickness as well as their subsequent pump station and power requirements,
- Operating pressure ranges, including frictional pressure drop and static pressure change due to elevation difference,
- Facility requirements such as tank and meter stations,
- For heavy or waxy crudes, temperature to which the fluid would be treated or heated, or blending ratio if blended with diluent,
- Future plans for capacity increase of the pipeline system.

When combined with an economic evaluation, the system planning effort yields the following economic information:

- Annual average energy cost
- Annualized capital cost including costs for pipe, pump station and facilities over the life of the pipeline
- Total annualized cost over the life of the pipeline

The optimum pipeline system is selected through an economic evaluation process of comparing several alternative designs against a base design on the basis of the profitability index such as net present value or internal rate of return. The hydraulic design at a system planning level is within the scope of this book, but not the economic evaluation process.

A detailed engineering design phase covers the hydraulic, mechanical, and operational aspects of design as well as construction and maintenance. The detailed design results in engineering specifications, which will be issued for procurement. Detailed engineering design is beyond the scope of this book.

3.3 LIQUID PIPELINE DESIGN

This section illustrates the hydraulic design process with an example, focusing on hydraulic studies. Common hydraulic studies for pipeline system design and/or operation include the following:

- Determine the unknown upstream pressure from the known downstream pressure
- Determine the unknown downstream pressure from the known upstream pressure
- Determine the unknown pipe flow from the known upstream and downstream pressures
- Size the pipe diameter from the known flow and upstream and downstream pressures
- Determine pump station spacing from the known flow and pressures
- Perform sensitivity analysis

Hydraulic design of liquid pipelines can be viewed from several angles:

- Temperature dependence of liquid properties
- Batch or segregated pipeline operation
- Pipeline configuration
- Terrains where pipeline system is constructed and operated
- Weather conditions

Changes in liquid temperature affect liquid density and/or viscosity, and subsequently pressure profile. In most liquid pipelines, in which the liquid is transported near ambient temperature, the temperature effect on hydraulics is minimal. As discussed earlier, an isothermal assumption is valid in such a condition and hydraulic analysis can be greatly simplified.

However, the temperature effects for certain types of fluid can be so significant that the temperature changes should be taken into account to determine an accurate pressure profile, while certain products cannot be transported without being heated or blended with lower viscosity product. Depending on the temperature behaviors of petroleum liquids, liquid pipeline design problems can be classified into three categories: light hydrocarbon liquids, intermediate hydrocarbon liquids, and heavy hydrocarbon liquids.

- Light hydrocarbon liquids include ethylene, ethane, propylene, propane, and natural gas liquid (NGL). These liquids are characterized by their high vapor pressure and high density variations due to temperature changes. The density of these high vapor pressure (HVP) liquids changes significantly with temperature.

Section 3.3.6 describes a HVP pipeline design process with an ethane pipeline as an example.

- Although the heating effect on viscosity is inherent to all real fluid flow situations, the temperature effect on viscosity of heavy and waxy crudes is significant. Temperature of the highly viscous fluids at the entrance to the pipe can be quite different from the temperature of the soil surrounding the pipeline system. Viscous liquids such as heavy oil or waxy crude may be heated or blended with lighter hydrocarbon liquids to reduce the viscosity for pumping. Section 3.3.7 describes a heavy oil pipeline design process as an example.
- Intermediate hydrocarbon liquids such as light or medium crude and refined products such as diesel or gasoline are not as sensitive to temperature in terms of density and viscosity. Also, frictional heating is negligibly small for these products. Therefore, the assumption of isothermal flow is reasonable if an adequate average temperature is used for the operating temperature. However, the design consideration should include the vapor pressure because it depends on temperature. Section 3.3.1 begins with an isothermal pipeline system design example, demonstrating the hydraulic design process. An average flow profile is added to the base design problem in order to demonstrate the above design process for a realistic design problem. The last three steps are not included in these examples because as mentioned quantitative economic analysis is beyond the scope of this book.

3.3.1 Crude Oil Pipeline System — Isothermal Flow

Example: A crude oil pipeline from CE to QU is 200 km long. Refer to Figure 3-2 for the pipeline configuration. At the injection point, crude oil of 32°API (specific gravity of 0.8654) and ambient pressure enters the pipeline at an initial flow rate of 18,000 m^3/d at 15°C. The operating temperature in winter and summer is 4°C and 14°C, respectively. Design a crude oil pipeline to transport the amounts defined in the flow profile, using the data listed below:

- Density: 865.4 kg/m^3 at 15°C and 875.4 kg/m^3 at the operating temperature
- Viscosities at 4°C: 43.5 cSt
- Pipe roughness: 0.0457 mm
- Delivery pressure: 350 kPag
- Load factor: 90%

The average flow profile is as follows:

Year 1 — 18,000 m^3/d
Year 4 — 20,000 m^3/d
Year 10 — 27,000 m^3/d

Design an optimum pipeline system. Assume that the design factor of 0.72 as specified in ASME B31.4 Codes is applicable and that the elevation profile is flat and flow is isothermal.

Solution: The design considerations for this type of design problem are:

- Satisfy the delivery pressure requirement that must be greater than the vapor pressures of the delivered products,
- Find an optimum solution in terms of not only capital and operating costs but also hydraulics for flexible operation.

Discussed below is the solution procedure described in Section 3.2.3, except an economic analysis.

1. *Gather data*
   It is assumed that the shipper commitments have been received, the approximate volume forecasts are made, and a preliminary route of the pipeline is selected.
2. *Prepare a set of design criteria.*

- Range of maximum operating pressure: from 8100 kPag to 9500 kPag based on common practice for liquid pipelines.
- Operating temperature: winter operating temperature of 4°C is used.
- Minimum operating pressure: 250 kPag
- Pipe grade: X70 (483 MPag) and X65 (448 MPag)
- Pipe sizes: 18″ (457.2 mm), 20″ (508.0 mm), and 22″ (558.8 mm)
- Pipe wall thickness: 0.25″ (6.35 mm) and 0.281″ (7.14 mm)
- Maximum liquid velocity: 2 m/s

3. *Develop a base design.*

Step 1. Calculate the flow velocity and design pressure for each pipe grade, size and wall thickness.

In the flow profile, the largest flow is scheduled from the 10th year on, and the design flow rate is obtained by dividing the flow rate by the load factor: 27,000/0.9 = 30,000 m^3/d. For the design flow rate, the table below gives the flow velocity for each combination of the pipe size and wall thickness:

Pipe size (in/mm)	Wall thickness (in/mm)	Velocity (m/s)	Wall thickness (in/mm)	Velocity (m/s)
18/457.2	0.250/6.35	2.24	0.281/7.14	2.26
20/508.0	0.250/6.35	1.80	0.281/7.14	1.81
22/558.8	0.250/6.35	1.48	0.281/7.14	1.49

The 18″ pipe is excluded from further consideration because the velocity exceeds the velocity limit by more than 10%, so the pipe sizes to be considered further are 20″ and 22″. Next, calculate the design pressure for X65 and X60 grade pipes, respectively.

	X65		X70	
Pipe size (in/mm)	Wall thickness (in/mm)	Design pressure (psig/kPag)	Wall thickness (in/mm)	Design pressure (psig/kPag)
20/508.0	0.250/6.35	1180/8132	0.250/6.35	1260/8688
20/508.0	0.281/7.14	1315/9067	0.281/7.14	1417/9765
22/558.8	0.250/6.35	1064/7334	0.250/6.35	1145/7897
22/558.8	0.281/7.14	1196/8246	0.281/7.14	1288/8880

Since the operating pressure range is between 8100 kPag and 9500 kPag, the design pressures far outside of the range are removed from further consideration. Therefore, the selected combinations for X65 pipe are 20″ with wall thicknesses of 0.25″ and 0.281″, 22″ with wall thickness of 0.281″, and those for X70 are 20″ with wall thicknesses of 0.250″ and 0.281″, and 22″ with wall thickness of 0.281″. The allowable design pressure for the 20″ with 0.281″ wall thickness exceeds the maximum operating pressure, but the combination is selected for further consideration because it is within a tolerance level.

Step 2. Calculate the required total pressure drop and total pressure requirement or inlet pressure for the design flow rate of 20,000m^3/d during the first three years.

The design flow rate is obtained by dividing the given flow rate, 18,000m^3/d, by the load factor, 0.9. The total pressure drop is added to the delivery pressure to get the total pressure requirement. The pressure calculation is based on the worst condition, which is the winter temperature.

Pipe grade	Pipe size (in/mm)	Wall thickness (in/mm)	Design pressure (kPag)	Total pressure drop (kPag)	Total pressure req. (kPag)
X65	20/508.0	0.250/6.35	8132	7288	7638
X65	20/508.0	0.281/6.35	9067	7288	7638
X65	22/558.8	0.281/7.14	8246	4649	4999
X70	20/508.0	0.250/7.14	8688	7400	7750
X70	20/508.0	0.281/7.14	9765	7400	7750
X70	22/558.8	0.281/7.14	8880	4649	4999

All six combinations satisfy the total pressure requirements for 20,000 m^3/d flow. For the same combinations as above, calculate the required total pressure drop and inlet pressure for 22,300 m^3/d from the fourth year to the tenth year.

Pipe grade	Pipe size (in/mm)	Wall thickness (in/mm)	Design pressure (kPag)	Total pressure drop (kPag)	Total pressure req. (kPag)
X65	20/508.0	0.250/6.35	8132	8811	9161
X65	20/508.0	0.281/6.35	9067	8946	9296
X65	22/558.8	0.281/7.14	8246	5618	5968
X70	20/508.0	0.250/7.14	8688	8811	9161
X70	20/508.0	0.281/7.14	9765	8946	9296
X70	22/558.8	0.281/7.14	8880	5618	5968

Only the combinations of pipe size 22″ with the wall thickness of 0.281″ for X65, and of the pipe size 20″ with the wall thickness of 0.281″ and the pipe size 22″ with the wall thickness of 0.281″ for X70 pipe, satisfy the pressure requirement with no intermediate pump station. It may not be cost-effective to install and operate an intermediate pump station to accommodate a small amount of the flow increase from the fourth year.

For the above three combinations, calculate the required total pressure and inlet pressure for the flow rate of 30,000 m^3/d from the tenth year on. It should be noted that the pumping power requirement for the 20″ pipe at the inlet point is higher by 59% (8946/5618 = 1.59) than the power requirement for the 22″ pipe size. Therefore, the pump units for the 20″ pipe have to produce higher head than those for the 22″ pipe and thus their capital and operating costs are higher. On the other hand, the required pressure for the 22″ pipe is low for the first 10 years, and so the facility usage would be limited unless further flow increase is expected in earlier years.

Pipe grade	Pipe size (in/mm)	Wall thickness (in/mm)	Design pressure (kPag)	Total pressure drop (kPag)	Total pressure req. (kPag)
X65	22/558.8	0.281/7.14	8246	9432	9782
X70	20/508.0	0.281/7.14	9765	15,039	15,389
X70	22/558.8	0.281/7.14	8880	9432	9782

The total pressure requirements for all three combinations are higher than their respective design pressure. Therefore, they require an intermediate pump station to satisfy the total pressure requirement.

Step 3. Determine the number of intermediate pump stations and their power requirements.

Only one intermediate pump station is required for all three cases because the design pressures for all cases are less than half of the total pressure drops. Assuming the suction pressure of the intermediate station is the same as the delivery pressure, the discharge pressure at the inlet and intermediate stations are as follows:

Pipe grade	Pipe size (in/mm)	Design pressure (kPag)	Discharge pressure at inlet point (kPag)	Discharge pressure at intermediate station (kPag)
X65	22/558.8	8246	5066	5066
X70	20/508.0	9765	7870	7870
X70	22/558.8	8880	5066	5066

The capital cost due to the extra pumping power requirement for the 20″ pipe is higher than the cost for the 22″ pipe size, while the pipe cost for X70 with 20″ diameter may cost less than the other two options. The extra capital cost for the 22″ line is more than 20% and is incurred by the extra pipe material and construction expenses. However, the extra capital cost of the 22″ diameter pipe might be partly compensated by lower unit pumping cost. Assuming that the annualized cost for the 20″ pipe case is lowest, it is selected as the base design.

The facilities such as the initiating pump station for the selected base design would be designed to accommodate the capacity until the capacity increases in the 10th year. In the 10th year, the additional facility increases include the pumping capacity at the inlet point for the additional flow and an intermediate pump station with the pumping capacity of 30,000 m^3/d.

4. *Develop alternative design cases and perform comparative studies against the base design*

- **Alternative 1:** This alternative design is to use a pipe wall thickness larger than 0.281″ in order to increase the design pressure slightly higher than the total pressure requirement. No intermediate pump station is required if the design pressure is slightly higher than the total pressure requirement. Note that the required total pressure will be increased due to slightly smaller inside pipe diameter.

  The design pressure for the X70 22″ pipe is lower than the total pressure requirement, which in turn is lower than the maximum operating pressure range. The next largest nominal wall thickness is 0.312″ or 7.92 mm, and its design pressure is 9857 kPag or 1430 psig, but the required total pressure is 9914 kPag for a flow rate of 30,000 m^3/d. Therefore, the wall thickness is not sufficient to meet the total pressure requirement without an intermediate pump station.

  The next largest nominal wall thickness is 0.344″ or 8.74 mm, which can allow the design pressure to increase up to 10,870 kPag. For this wall thickness, the required total pressure turns out to be 10,051 kPag. Since this design pressure is higher than the required total pressure, no intermediate pump station is required for the flow rate expected beyond the 10th year, and thus the capital and operating costs due to an extra pump station can be saved. However, two points should be evaluated; the required pressure is very high for a liquid pipeline and the extra capital cost. The required pressure in this case is much higher than the maximum operating pressure, and normally crude oil pipelines are not operated at such a high pressure. Extra pipe and construction costs will be incurred due to the extra pipe material needed.

Therefore, these extra capital cost should be compared against the costs of the base design in terms of annualized cost. The base design may a better choice in terms of the overall cost and its pipeline system operation due to its lower operating pressure.

- **Alternative 2:** This alternative design is to use X80 grade pipe to increase the design pressure, also allowing the operating pressure limit to be raised. If the design pressure for this pipe grade is higher than the required total pressure of 9782 kPag, then no intermediate pump station is required even for the maximum flow rate.

  For this grade, the design pressure is 10,149 kPag, which is higher than the total pressure required from the 10th year on. Therefore, an intermediate pump station is not needed. Also, the pipe material cost for X80 pipe is only slightly higher than X70 pipe cost.

  When compared against the base design, both designs are comparable, because this alternative design offers the lower cost solution even though its operating pressure range for a crude line seems to be high. To finalize the design, it is necessary to perform sensitivity studies for these two designs.
- **Alternative 3:** The base design is modified by adding storage tank capacity to allow the system to transport more in summer during which time the transportation capacity is higher than during the winter when capacity decreases due to lower viscosity and density. It costs much less to add the extra tank capacity than to increase pipe diameter or wall thickness, but the tank operation does add costs.

  The summer capacity listed in the table below can be found by setting the operating temperature at 14°C and the discharge pressure at the maximum operating pressure of 9500 kPag with and without an intermediate pump station; 24,580 m^3/d without and 36,300 m^3/d with an intermediate pump station. For the same inlet pressure as for the base design, the capacity increases to 23,280 m^3/d without an intermediate pump station. Therefore, assuming that the pipeline operates in summer condition for the first half of a year and in winter condition for the rest of the year, this alternative design allows an increase in the transportation capacity on a yearly basis.

  This alternative design offers a more flexible solution than the base design, even though it costs more. Also, its transportation capacity is 10% higher that the base design capacity. To finalize the design, it is necessary to perform sensitivity studies for these three designs.

5. *Perform a sensitivity study with respect to the flow profile.*

- Sensitivity on the modified flow profile: The flow rate is expected to gradually increase by approximately 1% yearly over the initially projected flow profile; 21,000 m^3/d or design flow rate of 23,300 m^3/d for year 9 and more than 27,000 m^3/d or design flow rate of 30,000 m^3/d beyond year 10. For these flow rate changes, the required total pressures are calculated for the three cases:

Design	Design pressure	Inlet pressure (9th year flow)	Inlet pressure (beyond 10th year)	Discharge pressure at intermediate station
Base	9765	9720	18,133	9242
Design 2	10,149	6234	11,499	6275
Design 3	9765	9300	9300	9300

The base design barely satisfies up to the seventh year transportation requirement without an intermediate pump station, while Alternative 3 provides more than the ninth year requirement within the design pressure. Also, Alternative 3 fully utilizes the facility, but Alternative 2 does not. Still, both alternatives need an intermediate pump station from the 10th year on, and the pumping capacity at the inlet station has to be increased at the same time. The intermediate pump station will be located at 100 km from the inlet station, because the pumping head at both stations is the same (the criteria for locating pump stations are discussed in the next chapter). With the discharge pressure of 9300 kPag, Alternative 3 has a higher flow capacity than needed. In summary, Alternative 3 is selected as the best design under this flow condition, because:

- With slightly more capital and operating costs than the base design, Alternative 3 offers more flexible operation,
- If needed, the flow capacity can be increased significantly.

- Sensitivity on a fast flow growth: The flow rate is expected to grow at the yearly rate of 1000 m^3/d from the first year on and to level off at 32,000 m^3/d; 18,000 m^3/d or design flow rate of 20,000 m^3/d in the first year, 19,000 m^3/d or design rate of 21,100 m^3/d in the second year, etc. For these flow rate changes, the required total pressures are calculated for these three cases:

Design	Design pressure	Inlet pressure (kPag)	Year of pump installation
Base	9765	9436	3rd year
Design 2	10,149	9782	10th year
Design 3	9765	9440	4th year

Alternative 2 does not require an intermediate pump station until the 10th year, while the other two require it in 3rd and 4th year, respectively. Alternative 2 needs higher initial capital cost due to the higher pipe grade and larger pipe size. However, Alternative 2, using a large pipe size, offers a better option in terms of the operating cost for such a high flow growth rate.

3.3.2 Pipeline Configurations

This section describes the key design and operation issues on different pipeline configurations. In addition to pipe, a pipeline network is composed of the following facilities:

- Injection points, also known as receipt or inlet stations, these are where the products are lifted or injected into the line. Storage facilities such as tanks and booster pumps are usually located at these locations.
- Delivery point, also known as terminal, is where the product will be delivered to the final consumer or to another pipeline.
- An intermediate station can provide a side stream injection or delivery point.

These stations allow the pipeline operator to inject or deliver part or all of the product being transported.

- Pump stations are located along the line to move the liquid through the pipeline.
- Block Valve Stations are the first line of consequence mitigation for pipelines. With these valves the operator can isolate any segment of the line to perform some specific maintenance work or isolate a rupture or leak. Block valve stations are usually located every 20 to 30 km, depending on the type of pipeline and applicable standards.
- Regulating station is a special type of valve station, where either pressure or flow is controlled. Pressure regulators are usually located on the downhill side of a peak, while flow regulators are installed at delivery stations.

Depending on the requirements and arrangements of these facilities, liquid pipeline networks can be diverse; some are short and straight, some are long with multiple pump stations, or some are complex with multiple injection and delivery points. The pipeline system design and operation has to comply with the required system network. A simple pipeline consists of one inlet with a pump station and one delivery. In addition to simple networks, the following types of pipeline networks can be built and are frequently encountered:

- Pipelines including one injection and one delivery with multiple pump stations
- Pipelines including multiple injection and multiple delivery points with multiple pump stations
- Pipelines with branch or lateral lines that connect to/from other pipelines or facilities from/to the main line
- Series pipelines of partial or entire length, referring to the connection of pipes of the same or different diameters in series.
- Parallel Pipelines of partial or entire length to increase throughput by reducing pressure drop.

Note that the pressure gradients for these networks, except the first type, vary because the flow rate of each segment is different, and so is the pumping requirements. If such a network is anticipated in the initial design phase, the pump station locations are determined accordingly. If the existing network has to be modified to meet the new requirements, additional pump stations are added and/or certain stations need to be modified.

3.3.2.1 Side Stream Delivery

Liquid may be delivered off the pipeline (stripping) at intermediate locations, thus reducing the main line flow rate while the remainder of the product continues to the main line terminal. The final delivery location is right on the main line or connected through a branch line. Since the downstream flow is lower, the frictional pressure drop is lower. Normally, a holding pressure control valve is installed at the delivery point to maintain the delivery pressure level or a pressure regulator is placed on the branch line. A block valve is installed downstream of the take-off point on the main line and branch line to block the flow when a full stream delivery takes place on either line.

The modes of side stream delivery operation can vary depending on the delivery flow requirements or availability, nomination status, and pipeline operational status. For example, the main line downstream of the take-off point cannot be operational if a line break occurs there, or the branch line should be shut down if no volume is nominated to the branch line delivery site. Therefore, the following modes of side stream delivery operation are possible, and thus have to be included in the design:

- Strip delivery through the branch line or at the delivery point as originally designed,
- Full stream delivery through the branch line due to the main line problem in the segment downstream of the take-off point,
- Full stream delivery through the main line due to a problem in the branch line.

The design considerations for this type of design problem are:

- Satisfying the delivery pressure requirements at both delivery locations, while maintaining sufficiently high pressure at the take-off point. Note that the delivery pressures at both locations can be different because the delivery conditions can be different.
- Using a pipe with a smaller diameter downstream of the delivery point if the side stream delivery volume is large.
- Installing an extra facility such as a pressure regulator or pump at the take-off point on the branch line in order to satisfy the branch line delivery pressure requirement.
- Selecting pumps to meet the maximum and minimum flow requirements. When the main line is shut down downstream of the take-off point, the minimum flow rate along the main line can be as low as or even lower than the design flow rate of the branch line.

Example: A crude oil pipeline from CE to QU is 200 km long. It is constructed of 5LX-70 steel pipe with NPS = 20″ and a 0.281″ wall thickness. At the CE terminal, the crude oil of 32°API (specific gravity of 0.8654) enters the pipeline at the design flow rate of 30,000 m^3/d. Crude oil is taken off at TO, 136 km downstream of CE, where up to 7200 m^3/d is stripped off the pipeline, and the rest is delivered to the final destination, QU. Occasionally, the full flow has to be delivered to QU. At TO, a 50-km branch line is connected to a third party pipeline, which requires a delivery pressure of 3000 kPag. This branch pipeline is constructed with X52 grade pipe, and the pipe size is NPS = 12″ (actual pipe diameter = 12.75″) with a 0.219″ wall thickness.

Determine the pressure requirements of the pipeline system, using the following data:

- Average operating temperature: 4°C
- Density: 865.4 kg/m^3 at 15°C and 875.4 kg/m^3 at the operating temperature
- Viscosity at 4°C: 43.5 cSt
- Pipe roughness: 0.0457 mm
- Delivery pressure at QU: 350 kPag

Assume that the design factor of 0.72 is applicable and that the elevation profile is flat and flow is isothermal. Figure 3-10 shows the configuration of this pipeline system.

Solution:

It is assumed that the Alternative 3 design has been used for the main line in anticipation of flow increase and the intermediate pump station has been operating.

Step 1. Determine the design pressure of the main and branch lines using the Barlow formula with the hoop stress limited to 72% of the SMYS

- $P_{main} = 2 \times 70{,}000 \times 0.281 \times 0.72/20 = 1416$ psig = 9765 kPag
- $P_{branch} = 2 \times 52{,}000 \times 0.25 \times 0.72/12.75 = 1286$ psig = 8868 kPag

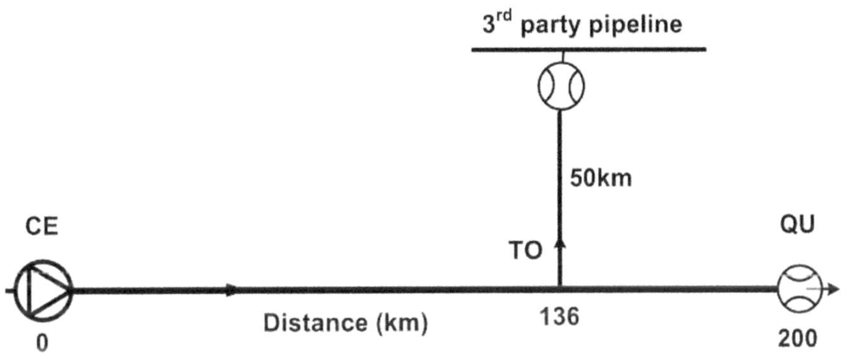

Figure 3-10. Side stream take-off

Step 2. Calculate the required pressures at TO and the discharge pressure at the intermediate pump station.

1. First calculate the pressure required at TO for the design flow rate that can meet the branch line delivery pressure requirement within the design pressure limit of 8868 kPag.
   - Flow velocity = 1.086 m/s
   - Reynolds number = 7820
   - Relative roughness = 0.000146
   - Friction factor = 0.0334
   - Friction pressure drop = 2745 kPa

   Therefore, the pressure required at TO is 3000 kPag + 2745 kPa = 5745 kPag. Assuming that no pump station is installed at TO, the actual pressure required at TO may be around 5900 kPag when minor pressure losses at TO and the delivery site are taken into account (refer to Addendum 3.3 for the discussion of minor pressure losses).
2. Next, determine the discharge pressure required at the intermediate pump station.
   - Distance from the pump station to TO = 136 km – 100 km = 36 km
   - Pressure gradient of the main line = 75.2 kPa/km
   - Total pressure drop between the pump station and TO = 75.2 kPa/km × 36 km = 2707 kPa
   - Discharge pressure required at the intermediate station = 5900 kPag + 2707 kPa = 8607 kPag
   - Discharge pressure difference with and without the branch line = 8607 – 7870 = 737 kPa

Step 3. Determine the total pressure requirement when the branch line is shut down.

When the branch line is shut down, the maximum flow rate along the entire main line reaches the design flow rate of 30,000 m^3/d, and thus the main line can transport the design flow rate, as demonstrated in the previous example.

Step 4. Evaluate this design

This pressure is lower than the design pressure, but 737 kPa higher than the discharge pressure required for the main line pressure. Since the discharge pressure difference is large, there are several options to correct this large pressure difference problem;

- No modification to the existing pump units at the intermediate station,
- Add a pump at the intermediate station, locate TO closer to the intermediate station,
- The installation of a pump station at TO on the branch line.

Discussing these options further,

- Alternative 3 has been selected in anticipation of future flow increases. Therefore, the pump units would have been chosen so as to accommodate such flow increases and thus pump head. If the pump driver has extra power, the pump units may not need to be modified by increasing the pump impeller size. (Refer to the next chapter on pumps.)
- A pump is added to the existing pumps at the intermediate station to provide the extra pumping head. If the extra head required is large, this option may be viable but the 737 kPa head is too small to warrant another pump.
- If there is no restriction in locating the take-off point, it can be the best option to locate TO at 126 km:
  - Distance from the pump station to TO = 126 km – 100 km = 26 km
  - Pressure gradient of the main line = 75.2 kPa/km
  - Total pressure drop between the pump station and TO = 75.2 kPa/km × 26 km = 1955 kPa
  - Discharge pressure required at the intermediate station = 5900 kPag + 1955 kPa = 7855 kPag
  - Discharge pressure difference with and without the branch line = 7855 – 7870 = – 15 kPa

  Even if the branch line gets slightly longer than the original distance, the discharge pressure difference is small enough so as not to require any changes to the existing pump station.
- It is a costly option to install a small pump station on the branch line, because extra capital and operating costs are required.
- It may not be a viable option to use an 18″ pipe downstream of the side stream delivery point, because the pressure drop for the pipe size is so high that the maximum design pressure limit will be violated.

If it is known that the branch line will be added at the time of the main line design, other considerations need to be included in order to optimize the system design:

- Location of the intermediate pump station
- Location of the take-off point
- Pressure requirements
- Selection of pumping units

3.3.2.2 Side Stream Injection

Instead of flow take-off, liquid may be injected from branch lines into the main pipeline, entering the main pipeline at these intermediate locations, adding flow rate to the main line flow downstream of the injection point. Since the flow is lower in the upstream segment of the injection point, the frictional pressure drop there is lower. A block valve is installed upstream of the injection point on the main line side and closed when a full stream injection takes place or a new batch is created at the injection point (refer to Chapter 5).

The modes of side stream injection can vary depending on the injection flow requirements or availability, nomination status, and pipeline operational status. For example, the branch line cannot be operational if a line break occurs downstream of the injection point. Therefore, the following modes of side stream injection are possible, and thus have to be included in the design:

- Side stream injection through the branch line or at the injection point as originally designed,
- The main line upstream of the injection point is shut down due to a problem in the segment, so the operational segment of the pipeline is the branch line connected to the main line at the injection point,
- The branch line is shut down if no volume is available to be injected into the branch line or other operational problems occur.

Unlike the side stream delivery problem, other operational scenarios are available; the same or different product injection and batching or blending operation for different product. If the same product is injected, the product is mixed with that in the main line and there is no operational issue. If a different product is injected into the main line, the following operational issues need to be addressed:

- Two products are blended if a partial injection takes place and the properties of the blended product will be different from the liquid in the main line before they are blended. Then, a new batch is created at the injection point and its volume grows until the side stream injection is finished. The movement of the new batch has to be tracked until it is fully delivered to the shipper.
- If the two products are not allowed to be blended, then the injection should be a full stream injection and a new batch is created at the injection point. The main line flow is stopped upstream of the injection point.

This type of design problem requires the following design considerations:

- The injection pressure on the branch line should be higher than the main line pressure at the injection point. The branch line design is similar to that of the delivery to the third party pipeline discussed in the previous design problem.
- A pipe with a larger diameter can be used downstream of the injection point if the side stream injection volume is large.
- For partial side stream injection, the pumps upstream of the injection point should be designed to accommodate the reduced flow. If the side stream injection rate is high, the upstream flow rate can be lower than the minimum main line flow. If the viscosity of the injection fluid is much higher than the viscosity of the main line liquid, then the pumping units at the downstream of the injection point have to be selected to accommodate high viscosity.
- The side stream injection flow rate can be much lower than the minimum main line flow. The pump stations downstream of the injection point have to be designed to meet the low flow rate during a full stream injection, particularly if the injection flow is lower than the minimum main line flow.
- A block valve is installed on the upstream side of the main line to take into account full stream injections into the main line.
- Injection of the same product: no batch is created. If the injection is a partial injection, the upstream flow is reduced and thus the pump stations have to be designed to accommodate the maximum and minimum flow requirements.
- Partial injection of a different product: blending of two different products occurs and a new blended batch has different density and viscosity. The property

differences have to be taken into account in the design of the pipeline system including pumps and tanks.

- Full stream injection of a different product: a new batch retains the product properties of the injection fluid. For this operation, the effects of the injection fluid have to be taken into account in the selection of the pump units in the downstream segment of the injection point, particularly if its viscosity is much higher than the viscosity of the main line liquid.

Example: Product Blending

A crude oil pipeline from CE to QU is 200 km long. It is constructed of 5LX-70 steel pipe with NPS = 20″ and a 0.281″ wall thickness. At the CE terminal, the crude oil of 32°API enters the pipeline at the design flow rate of 30,000 m^3/d. A 60-km branch line is planned to transport a crude oil of 35 °API (specific gravity of 0.850) from a tank to SI, a side stream injection point where the crude oil enters the main line at the design flow rate of 7200 m^3/d. SI is initially located at 78 km downstream of CE, because it is closest to the flow lifting point, LP. Considered initially are X52 grade pipe and the pipe size is NPS = 10″ (actual pipe diameter = 10.75″) with a 0.219″ wall thickness. Figure 3-11 shows the configuration of this pipeline system.

The product density and viscosity of 32°API gravity are 875.4 kg/m^3 and 43.5 cSt, and the density and viscosity of 35°API gravity are 857.6 kg/m^3 and 21.0 cSt at the operating temperature, respectively. Determine the pressure requirements of the pipeline system, using the following data:

- Average operating temperature: 4°C
- Pipe roughness: 0.0457 mm
- Delivery pressure at QU: 350 kPag

Assume that the design factor of 0.72 is applicable and that the elevation profile is flat and flow is isothermal.

Solution:

It is assumed that the Alternative 3 design has been used as before and the intermediate pump station has been operating.

Step 1. Determine the design pressure of the main and branch lines using the Barlow formula with the design factor of 0.72.

- Pmain = 2 × 70,000 × 0.281 × 0.72/20 = 1416 psig = 9765 kPag
- Pbranch = 2 × 52,000 × 0.25 × 0.72/10.75 = 1525 psig = 10,518 kPag

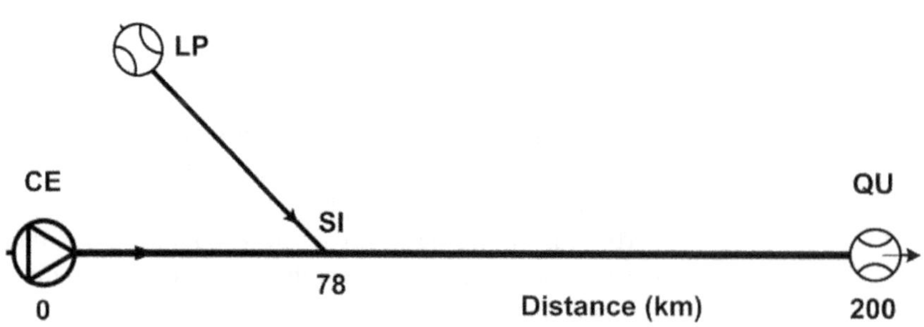

Figure 3-11. Side stream injection

Step 2. Calculate the required pressures at SI on the main line.

1. Calculate the pressure profile of the main line and the pressures at SI and LP for the design flow rate.
   - From the base design example, the pressure gradient is 75.2 kPa/km, the pressure at CE is 7870 kPag, and the suction pressure at the intermediate pump station and the delivery pressure at QU are 350 kPag. Therefore, the minimum pressure required at SI becomes:
     Pressure at SI = 7870 – 75.2 × 78 = 2005 kPag
   - Minor losses in pressure are expected due to facilities like a pressure regulator and block valves installed on the branch line. Taking into account various minor losses, the actual pressure required at SI is assumed to be approximately 2100 kPag. The discharge pressure at CE has to be increased by 95 kPa and the pressure at LP should be determined to satisfy this pressure requirement.
2. Calculate the discharge pressure at the branch line lifting point, LP.
   - Flow velocity = 1.547 m/s
   - Reynolds number = 19,280
   - Relative roughness = 0.000174
   - Friction factor = 0.0265
   - Friction pressure drop = 6222 kPa

   Therefore, the discharge pressure required at LP is 2100 kPag + 6222 kPa = 8322 kPag. This pressure requirement is lower than the design ressure of 10,518 kPag, and thus the side stream injection is appropriate.
3. Check if the suction pressure at the intermediate pump station is adequate. Since the main line pressure is increased by 95 kPa, the suction pressure will be higher than the minimum suction pressure by that amount. This pressure increase is well within the tolerance. Therefore, this design including the injection location is an adequate solution.

Step 3. Calculate the pressure requirement of the segments upstream and downstream of the injection point when the two products are blended.

1. Calculate the pressure profile upstream of the injection point.
   If the design flow rate is injected, the upstream flow rate is 32,000 – 7200 = 24,800 m^3/d, for which the pressure gradient 53.9 kPa/km. Then the discharge pressure at CE is 2100 + 53.9 × 78 = 6304 kPag.
2. Calculate the density and viscosity of the blended crude. These quantities are calculated for the design flow rates of the two lines; approximately 80% of the main line and 20% of the injection flow rate. Actually, they change depending on the percentages of blending of these two products. However, it is assumed here that they remain constant to simplify calculations for other blending percentages.
   - Density of the blended liquid at 4°C = 871.8 kg/m^3
   - Viscosity at 4°C using the ASTM method = 38.2 cSt
3. Calculate the suction and discharge pressures at the intermediate station.
   - Flow velocity = 1.815 m/s
   - Reynolds number = 23,450
   - Friction factor = 0.0250
   - Pressure gradient = 72.9 kPa/km
   - Suction pressure at the intermediate station = 2100 – 72.9 × (100 – 78) = 496 kPag
   - Discharge pressure required at the intermediate station = 72.9 × 100 + 350 = 7640 kPag

Since the pressure difference at the station is slightly reduced due to lower density and viscosity, the pumping requirement is reduced and no modification to the pump units is needed. If the density and viscosity are higher than those of the main line liquid, the pumping requirement will have to be increased.

Step 4. Determine the total pressure requirement when the branch line is shut in.

When the branch line is shut down, the maximum flow rate along the entire main line reaches the design flow rate of 30,000 m^3/d, and thus the main line can transport the design flow rate without any changes to the main line.

Step 5. Evaluate this design

If the injection point is located closer upstream of the main line injection point, the injection pressure has to be high, requiring high discharge pressure at the branch line injection point. If it is higher than the design pressure, potential options include an increase in branch line pipe size, increase in the branch line discharge pressure, and/or locating the injection point further to a downstream point along the main line.

If it is known that the branch line will be added at the time of the main line design, other considerations need to be included in order to optimize the system design:

- Location of the intermediate pump station
- Location of the injection point
- Pressure requirements
- Selection of pumping units

3.3.2.3 Pipeline in Series

Pipelines may include different pipes connected in a series. Such situations occur when different flow rates are transported due to intermediate take-off or injection or different pressures are required along certain pipe segments. Depending on the purpose of arranging pipes in series, there are three types of series arrangement; different pipe sizes, different pipe wall thickness, and different pipe grade. Except for the flow change due to side stream injection or delivery, the same flow rate goes through the pipes connected in series but the flow velocity of each segment is different.

The pressure requirement in a series pipeline for the entire pipeline network is determined by applying the appropriate flow equation for each segment and combining all the segment pressure drops. The total pressure requirement can also be determined by calculating the pressure required for each segment and then adding all the pressures over the entire length.

3.3.2.3.1 Different Pipe Sizes Connected in Series Different pipe sizes are connected in series in two cases: significant change in flow or in elevation. The larger the pipe diameter, the slower the velocity, the smaller the friction factor, and the lower the friction pressure loss. A larger pipe is required as the throughput along a pipeline increases significantly, or vice versa. Therefore, a pipe is connected in series at a junction

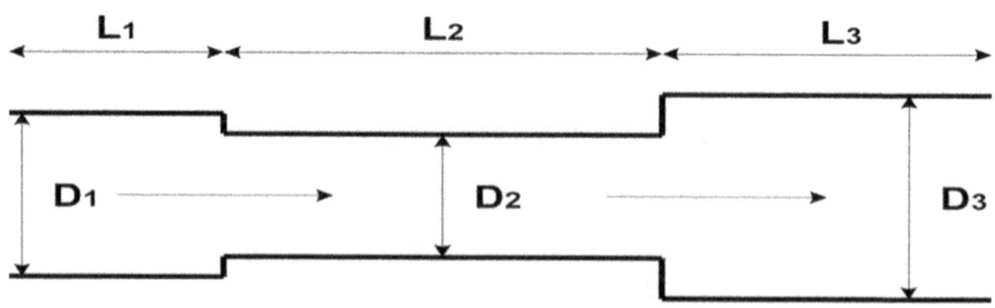

Figure 3-12. Series pipes

where there is a large flow increase or decrease due to side-stream injection or delivery. Figure 3-12 shows a pipeline with different lengths and diameters connected in series; flow is taken off at the end of L_1, requiring a smaller pipe size downstream of the side stream delivery, and flow is added at the end of L_2, requiring a larger pipe size downstream of the side stream injection point.

If the future throughput may not be known beforehand, it is not easy to determine the different pipe sizes for each segment. Therefore, it may be more economical to use the same size pipe over the entire length of the pipeline in case future flow requirements are not well known or show an increasing trend, even if there is intermediate take-off or injection.

Where the pipeline is sloping down significantly, the pipe pressure can be increased due to elevation gain on the downstream side beyond the pressure loss due to friction. As a result, it may be more advantageous to use a smaller pipe size (refer to Section 5.1.4) to increase the frictional pressure drop so that overall pressure gain can be reduced. For the opposite case, it may be safer to use a larger pipe size where the elevation gains significantly, particularly if it is difficult to maintain the peak point pressure above the minimum required level.

The pressure gradients change with pipe sizes, and the total pressure requirement is the sum of the total pressure required in each pipeline segment, including the static pressure due to elevation changes. One method of calculating the pressure drop in a series pipe is to use the equivalent length technique, in which the first pipe is hydraulically equivalent to another pipe if the frictional pressure drop in the first pipe is the same as that in the other pipe with a different length. Refer to hydraulic books that detail this method.

At the connection point, either a reducer or expander may be used to provide smoother transition from one size of pipe to another size. Minor pressure losses occur at each junction, and a pigging station with a pig trap and launch facility has to be installed. Dual diameter cleaning pigs may be required on this type of pipeline.

3.3.2.3.2 Different Pipe Wall Thickness The main reason for connecting pipes with different wall thickness is to reduce the pipe material and construction costs while at the same time maintaining the same level of safety. Unless the pipe pressure increases due to significant elevation gain, the pressure tends to decrease continuously from upstream to downstream and so does the pressure requirement. In other words, the discharge pressure of an upstream pump station is much higher than the delivery or suction pressure of the downstream pump station. The design pressure of a pipe is proportional to the pipe wall thickness. Therefore, a pipe with thinner wall can be used on the delivery or suction side, while a thicker pipe wall on the discharge side. Since the pipe costs less for thinner wall pipe, the overall material and construction costs can be reduced. In practice, different pipe wall thicknesses are used to compensate for different pressure requirements locally in the pipeline system.

When a pump station is shut down, the suction pressure increases greatly due to potential surge and subsequently the surge pressure moves towards the upstream. In addition, when the pipeline reaches a new steady state, the pressure level of the suction side increases substantially in order to maintain the delivery pressure or the suction pressure set point at the next pump station. Therefore, the pipe wall thickness on the suction side of an intermediate pump station has to be high enough to withstand the higher pressures that result from shut-down of the station.

Note it is not uncommon to reverse the flow direction on some pipelines for operational reasons and this capability can be denied if different wall thicknesses are used.

3.3.2.3.3 Different Pipe Grade A different pipe grade may be used instead of using a pipe with different wall thickness to satisfy the different design pressure requirement, not only for reducing the cost but also maintaining the same level of safely. In other words, the high pressure sections are constructed of a high grade pipe, while the lower pressure sections are constructed of somewhat lower grade steel. The same precaution as mentioned for different pipe wall thickness has to be exercised.

The design strategy of using different pipe wall thicknesses and/or pipe grade may not be a good option if more pump stations are added, or flow is reversed, at a later time. When the pipe flow increases, it is an option to add a pump station between two existing pump stations. If the pressure rating on the discharge side of the new pump station is low due to low pipe grade and/or thinner pipe wall, the new pump has to discharge at a low pressure unless the pipe sections with low pressure rating are replaced with thicker wall pipe or higher grade pipe.

3.3.2.4 Pipelines in Parallel

Excessive pressure drop can occur in certain sections of a pipeline system where a bottleneck is formed. As a result, the throughput can be severely limited throughout the pipeline. Pipes are arranged in parallel to reduce the excessive pressure drop in a certain section of the pipeline, and as a result to increase the throughput in the bottleneck and relieve the throughput limitation in the pipeline system.

Two or more pipes are connected at the upstream and downstream points, so that the flow splits among individual pipes at the upstream point and combines into a single pipe at the downstream point as illustrated in Figure 3-13. Such a piping system is referred to as parallel piping or looped piping system. The liquid flowing through AB splits into Pipe 1 and Pipe 2, through which the liquid flows separately into point C. The liquid flows recombine at point C and move to point D. An example is given in the Addendum 3.3.

The flow rate splits in such a way that there is a common pressure across each parallel pipe and the total flow is the sum of the flows across all parallel pipes at the splitting point and at the combining point. The pipe sizes of the parallel piping sections can be determined to meet the overall pressure requirements for the required throughput. The sizes of parallel pipes can be different. If the pipe sizes of the parallel pipes are different, so is the flow velocity through each pipe.

If the pipe sizes are different between the parallel pipes, the flow rate through each parallel pipe is initially unknown. Two principles are used to calculate the flow split and pressure across the parallel pipes:

- Conservation of mass or total flow at the junction
- Common pressure at the end of or pressure loss across each parallel pipe.

Applying the flow conservation principle at B or C,

$$Q = Q_1 + Q_2 \qquad (3-32)$$

where Q represents the flow rate in the base conditions. Applying the common pressure principle, we have

$$P_B - P_C = P_1 = P_2$$

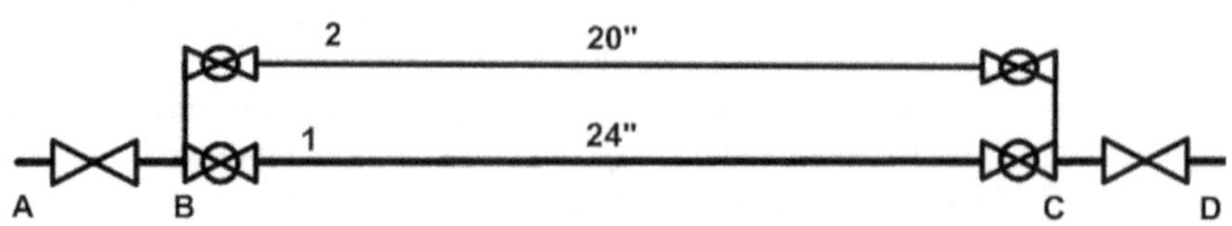

Figure 3-13. Parallel pipes

where P_1 and P_2 are pressure drops between B and C along the parallel pipes 1 and 2, respectively.

A pipeline is looped to increase throughput. Since the frictional pressure drop is lower with a parallel pipe, so is the pumping requirement. However, if the pipe sizes in parallel are different, caution must be exercised for batch pipeline design and operation. Since the flow velocity through each pipe is different, the batch front through a smaller pipe arrives at the other end earlier than the other batch front, allowing the early arriving batch to be blended with the leading batch. This blending increases the mixing volume, thereby increasing slop. A batch controller is installed at the other end of a parallel pipe in order to avoid this blending problem.

3.3.3 Severe Elevation Change — Slack Flow

It is a challenge not only to construct a pipeline in mountainous areas with severe elevation changes but also to operate the pipeline. The difficulties result because the total pressure required to transport in such an area may depend more on the elevation change than on the frictional pressure drop. When the pressure of the liquid drops below vapor pressure, the liquid evaporates or boils forming vapor pockets inside the pipe as shown in the diagram below. This condition is called slack flow and shown in Figure 3-14. Note that there is a free surface between the liquid and vapor, at which significant turbulent mixing can take place. Therefore, batch interface mixing can be significant under a slack flow condition.

A vapor pocket is formed where the elevation drops severely, because the pressure downstream of a peak point must be increased due to the elevation gain but the required pressure there is brought down by the low back pressure setting. Refer to the elevation profile-pressure gradient diagram shown in Figure 3-16. With such severe elevation drops, the slack flow condition can occur downstream of the high points in the profile, if the back pressure is set low. The vapor pockets tend to stay on the downstream side of the high point, and the liquid flow is restricted due to the vapor pockets, resulting in high pressure drop.

The slack flow problem may not occur at a high flow because the frictional pressure drop can overcome the pressure increase due to elevation gain. However, the problem becomes more pronounced at a lower flow rate because the frictional pressure drop at a low flow is so small that the downstream pressure becomes much higher than the pressure at higher flow rate. These points are demonstrated in Figure 3-16 (refer to Slack Flow Design Problem), showing the two pressure profiles.

A slack flow condition disrupts the pipe flow, reducing pipeline transmission efficiency and increasing batch interface mixing sizes. Damage to the interior of the pipe can result if the vapor pocket suddenly collapses. Slack flow operation is difficult to avoid for liquid pipelines, if the elevation drops severely and the back pressure has

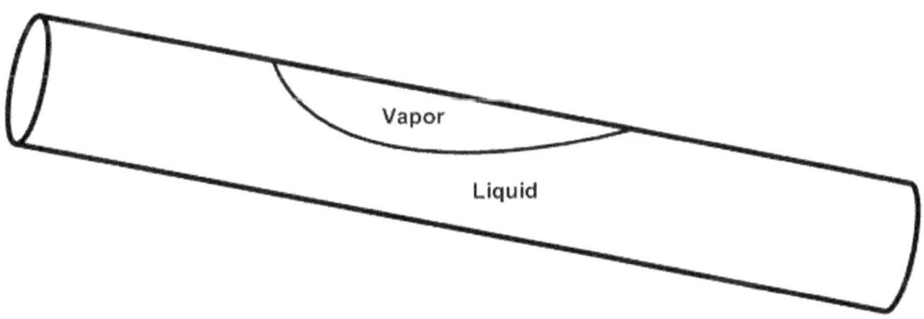

Figure 3-14. Vapor pocket in a slack flow condition

been set low due to pressure limitations on equipment. Even though slack flow is not desirable, pipeline systems transporting low vapor products such as crude oils can be successfully operated in a slack flow condition. However, slack flow operations need to be avoided for batch lines in order to limit the growth of batch interface mixing.

The design considerations for this type of design problem are:

- A minimum pressure, which is sufficiently higher than the vapor pressure, has to be maintained at the peak point to prevent vaporization.
- Since a slack flow condition occurs more frequently at a low flow rate, a thorough hydraulic analysis has to be performed, particularly at low flow rates, in order to fully understand the consequences of the slack flow on the design and to determine the extra facility requirement and pressure rating on the equipment. Valves and flanges in the downstream segment of the peak point should have a high pressure rating if the back pressure is not reduced using the following methods; installation of smaller pipe size and/or pressure-reducing station (PRS).
- Smaller pipe sizes can be used downstream of the peak point to increase the frictional pressure drop and at the same time reduce the pipe pressure. Pipe and construction costs can be reduced significantly. However, separate pig launchers and receivers have to be installed at either end of the pipe segment with smaller pipe size because the pipe size is changed.
- A PRS may be installed to operate the pipeline in a full flow condition by keeping the back pressure low and at the same time maintaining the downstream pressure lower than the MAOP. As an additional benefit, the PRS can help to keep the peak point pressure above the vapor pressure of the liquid. A PRS is needed on batch pipelines to be operated in a full flow. Occasionally, two PRSs may be installed if the elevations change several times and the drops are extremely severe, or a combination of a smaller pipe size and PRS is adopted in the design (refer to OCP pipeline in Section 5.1.4).
- A typical PRS station is shown in Figure 3-15. It is noted from the figure that the number of the pressure control valves is selected depending on the flow rate and their positions are adjusted depending on the downstream pressure. A pig receiver/launcher is not necessarily required if a pig can bypass the pressure-reducing station.

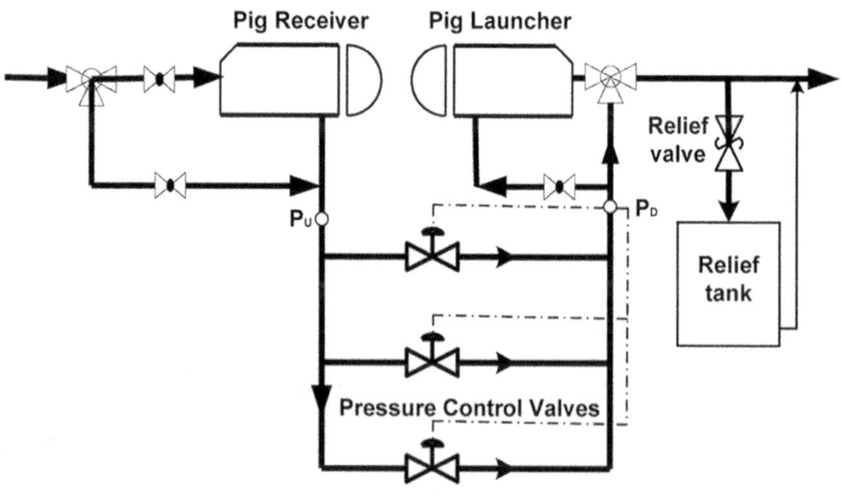

Figure 3-15. Pressure-reducing station

Example: Slack Flow Line

A crude oil pipeline from CE to QU is 200 km long, crossing a mountainous area. The table below shows an elevation profile. At the CE terminal, the crude oil of 32°API gravity enters the pipeline at the design flow rate of 30,000 m^3/d. The minimum flow rate is 9000 m^3/d. Determine the pressure requirements of the pipeline system, using the following data:

- Average operating temperature: 4°C
- Minimum delivery pressure at QU: 350 kPag
- Pipe grade: 5LX-70
- Pipe size: NPS = 20″ and a 0.281″ wall thickness.
- Density at the operating temperature: 875.4 kg/m^3
- Viscosity at the operating temperature: 43.5 cSt

Kilometer post (km)	Elevation (m)	Kilometer post (km)	Elevation (m)
0	30	110	100
20	55	130	300
30	45	150	770
60	30	160	425
80	70	180	150
90	100	200	130

Assume that the design factor of 0.72 is applicable and that the flow is isothermal.

Solution:

It is assumed that the elevation changes are gradual between two profile points, the peak point pressure is kept at 350 kPag, and the minimum suction pressures are the same as the delivery pressure. An intermediate pump station is located at KMP = 110 km. Note that the elevation changes in the first section between CE and the intermediate station are mild, but the changes in the second section are significant.

Step 1. Calculate the pressures for the design flow rate of 30,000 m^3/d at the above profile points. The discharge pressure at CE is 9220 kPag so as to satisfy the minimum suction pressure requirement, and the discharge pressure at the intermediate station is 9091 kPag so as to keep the peak point pressure at 350 kPag. The delivery pressure of 2067 kPag is obtained in order to keep the pipeline flow in a full flow condition. As a result, the pressure profile is determined as shown in the table below.

KMP (km)	Elevation (m)	Pressure (kPag)	KMP (km)	Elevation (m)	Pressure (kPag)
0	30	9220	110	100	350/9091
20	55	7502	130	300	5876
30	45	6836	150	770	350
60	30	4708	160	425	2550
80	70	2862	180	150	3400
90	100	1854	200	130	2067

The discharge pressure at the intermediate station has to be sufficient to overcome the static pressure loss due to the elevation increase and the friction pressure drop and at the same time to keep the peak pressure higher than the vapor pressure of the liquid. Note that the delivery pressure is higher than the minimum required delivery pressure of 350 kPag, because the pressure downstream of the peak point is gained due to the elevation drop while maintaining the required peak point pressure in a full flow condition. If the delivery pressure is set at 350 kPag, then the peak pressure drops below the vapor pressure and vapor pockets are formed to meet the set point pressure.

Step 2. Calculate the pressures for the minimum flow rate of 9000 m³/d at the above profile points. The table below shows the pressure profile for the minimum flow rate.

KMP (km)	Elevation (m)	Pressure (kPag)	KMP (km)	Elevation (m)	Pressure (kPag)
0	30	1975	110	100	350/6456
20	55	1575	130	300	4558
30	45	1567	150	770	350
60	30	1415	160	425	3209
80	70	886	180	150	5375
90	100	536	200	130	5360

The discharge pressure at CE is 1975 kPag so as to satisfy the minimum suction pressure requirement, and the discharge pressure at the intermediate station is 6456 kPag so as to keep the peak point pressure at 350 kPag. The difference in the two discharge pressures are significantly large because the friction pressure drop is small at a low flow rate but the intermediate station has to pump the liquids at a higher pressure to compensate for the large elevation gain up to the peak point.

Therefore, the discharge pressure or head at the intermediate station must be high to satisfy the pressure requirement at the peak point, which is, as shown in Figure 3-16, the control point that dictates the discharge pressure and the downstream pressure too. Note that the delivery pressure is much higher than the minimum required delivery pressure of 350 kPag as well as the delivery pressure for the design flow rate. This is caused by the low frictional pressure drop at the low flow rate while requiring the same pressure gain due to the elevation drop.

Figure 3-16 graphically shows three pressure profiles; pressure profile for the design flow, pressure profile for the minimum flow, and pressure profile for the design flow with the delivery pressure set at 350 kPag. The pressure gradients AB and CD represent a full flow condition for the maximum design flow, while A'B and C'D' are the profiles for another full flow condition, and for the minimum design flow, respectively. The lines EF and E'F show the pressure gradients where the delivery pressure at QU is set at the minimum delivery pressure of 350 kPag. The liquid in the segment between

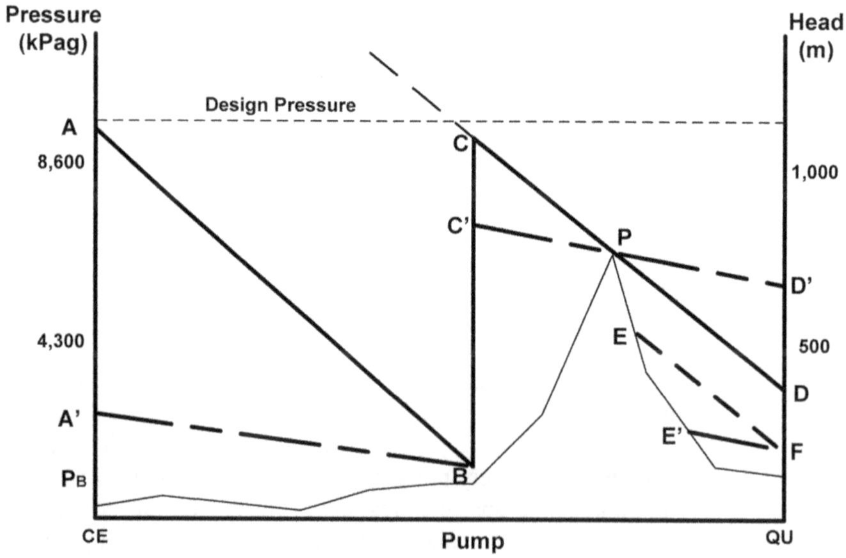

Figure 3-16. Slack flow conditions and pressure gradients

the peak point and E or E' flows in a slack flow condition for the maximum or minimum flow rates. In other words, the pipeline segment is at zero gauge or atmospheric pressure. The segment between EF or E'F remains in full flow. If the back pressure is kept constant at the minimum delivery pressure, the slack flow segment grows larger as the flow rate is reduced. A PRS may be needed at, or preferably upstream of, QU to bring these slack flow lines to the full flow condition.

Step 3. Two alternative designs are available; reduce the pipe size from 20″ to a smaller pipe size and/or install a pressure-reducing station (PRS). If the pipe size is reduced to 14″, the delivery pressure at the minimum flow rate drops to 3218 kPag. A PRS can be installed downstream of the peak point for keeping the peak point pressure high enough while reducing the downstream pressure.

3.3.4 Severe Weather Conditions

Severe weather conditions significantly influence the pipeline design and operation. A severe weather condition can result in extremely hot or cold ambient temperature and have a similar effect on soil temperatures. If a pipeline operates in hot weather conditions, the pipeline system can pick up ambient heat. On the other hand, if a pipeline operates in an extremely cold area, the ground remains frozen and the fluid has to be transported at lower than the freezing temperature in order to avoid melting the ice in the surrounding soil.

3.3.4.1 Pipeline in a Hot Environment

As discussed in Section 3.1.3, the liquid temperature can increase mainly due to pump inefficiency, heat gain through the frictional heating, as well as from the surrounding soil. The temperature increase due to pump inefficiency will be high if the station spacing is short, because the next pump will add more heat before the liquid temperature drops sufficiently to the ground level temperature. The temperature increase due to frictional heating is higher as the flow rate increases. Normally, the temperature rise due to conduction is largest. If the surrounding soil temperature is high due to prolonged high ambient temperature, the liquid in the pipeline absorbs the heat from the soil, raising its temperature. The temperature increase will be greater for larger diameter pipelines, because the larger the pipe surface area the larger the heat conduction. The temperature increase results in a decrease in liquid viscosity and density as well as a decrease in vapor pressure. The decrease in viscosity and density will help to reduce the friction loss. However, the decrease in vapor pressure has the following negative consequences:

- The pipeline pressure drops below the vapor pressure unless the pumps discharge at higher pressure.
- Evaporation of the liquid in the pipeline and storage tanks would increase.

If the temperature increase is high, then cooling facilities need to be installed along the pipeline in order to cool the temperature of the liquid. The best locations for any cooling facilities would be near rivers or other water crossing areas where line temperatures are low.

3.3.4.2 Pipeline in a Cold Environment

In the Arctic, the temperature in winter is very low, but can be hot in summer. However, the ground is permanently frozen in most areas. This condition is called permafrost. It is expensive to construct and operate a pipeline in a permafrost zone. The operating temperature is one of the most critical design parameters in an Arctic pipeline. Therefore, the following considerations must be given when designing a pipeline for a cold climate:

- Selection of pipe – low temperature steel pipes are required to control fracture.
- The pipeline is buried or installed aboveground – the line is installed aboveground in areas where the ground is permanently frozen, to avoid the need to chill the oil.
- If buried, the fluid is chilled. If the liquid temperature enters a pipeline close to or greater than the freezing point of water, the flowing temperature increases, as discussed above, and becomes higher than the freezing point. The liquid will warm the pipeline and eventually the surrounding soil, which will be softened around the pipeline. This may lead to "thaw settlement" resulting in the pipeline being bent and eventual damage to the pipeline. Therefore, for pipelines in permafrost zones, the operating temperature must be lower than the freezing temperature for the soil.
- If a crude oil pipeline is shut down for a prolonged period, the crude oil may congeal in the pipeline. Therefore, the relationship between the crude viscosity and temperature has to be determined and temperature cooling behavior evaluated while the pipeline is shut down. If there is a possibility of congealing during shut-down, special facilities may be needed to restart the pipeline.

A chiller is installed at the liquid injection point in order to reduce the liquid temperature below the freezing temperature. The liquid temperature is reduced to at least –5°C in consideration of temperature increases due to pump inefficiency and heat conduction in summer. In addition to a chiller, a wax removing facility may be required at the injection location, because wax can build up on the pipe wall at low temperatures. The requirement of the wax removing facility is determined after analyzing the viscous behavior of the liquid with respect to the temperature.

3.3.5 Batch Pipeline Hydraulics Design

Since the densities and viscosities of the batching products can be different, the pressure gradients are different and the pipeline capacities vary too. This is the result of the dependence of capacity on pressure drops of the products in the pipeline, the order of the products along the pipeline, and the position of the products with respect to pipeline and pump stations. In addition, the vapor pressure of each batch differs, requiring a different minimum pressure along the pipeline. If the differences of the vapor pressures are significant, the pipeline may be operated with different minimum pressures in order to reduce pumping costs. An example is an ethane-propane batch pipeline, where the ethane batch requires a minimum pressure of around 4500 kPag but the propane batch a minimum pressure of about 1700 kPag. Elevation changes, particularly severe changes, have to be included in the analysis of pressure drops and batch movements to analyze the minimum pressure requirements. The design considerations for a batch pipeline design problem are as follows:

1. Select the fluid that produces the biggest friction loss, i.e., with the largest viscosity and/or the highest density within the operating temperature range. The minimum operating pressure is determined for the fluid with the highest vapor pressure in order to maintain all the batching products in a full flow condition. Selecting these two products ensures that all pumping stations provide adequate pressure and power to sustain the design flow rates and pressure for all the batching products, while keeping the operating pressure within the maximum and minimum pressure limits. If future growth in the pipeline capacity is expected, this design approach is preferred because it provides enough room for future growth in the throughput capacity.

2. Pressure drop averaged over the batching products can be used for a hydraulic design if one of the following conditions is met:
   - Batch sizes are smaller than the volumes of the pipe section between two pump stations,
   - The system load factor is low, or
   - All batching products have similar viscosities and densities.

This approach can result in a tight system design in terms of future system growth, but can be acceptable if the pump station spacing is very long, the future growth in the capacity is limited, or all future batching products have properties similar to the existing products.

Example: Batch Pipeline

A petroleum product pipeline from CE to QU is 200 km long and is 20″ in nominal diameter, with a 0.281″ wall thickness. It is constructed of 5LX-65 electric resistance welded steel pipe. At the injection point, the following three products enter the pipeline at the design flow rate of 30,000 m^3/d in a batch mode:

Product	Density at 4°C (kg/m^3)	Viscosityat 4°C (cSt)	Batch size at 4°C (m^3)	Vapor pressure (kPa)
32°API	875.4	43.5	20,000	10
35°API	857.6	21.0	15,000	15
Condensate	705.0	0.7	25,000	95

It is assumed that these values are measured at the average operating temperature of 4°C. Design the batch pipeline including the delivery pressure.

Solution:

It is assumed that the base design is used; an intermediate pump station is located 100 km downstream of CE and the design pressure is 9765 kPag.

Step 1. Determine the line fill volumes of the two sections of the pipeline. The line fill volume is the volume of liquid contained in a segment of pipe, and is the pipe volume in the ambient conditions, even though actual volume of liquid shrinks under pressure. Addendum 3.4 discusses the effect of pressure and temperature on line fill volume. A section is defined as the pipeline between two pump stations or between a pump station and the delivery point. Therefore, the first section is defined from CE to the intermediate station, where the second section starts, ending at QU.

- Since the length of each section is the same, so is the line fill volume of each section. Assuming that the pipe volume does not change in a pressurized condition, the line fill volume of each section becomes 19,150 m^3.
- Since this volume is smaller than the size of a 32°API batch, the batch covers the whole section when fully lifted at CE or has passed the intermediate station.

Step 2. Select the product with the largest viscosity and the product with the highest vapor pressure among the three batch products.

- The 32°API batch has the highest viscosity among the three products, with a viscosity of 43.5 cSt.
- The condensate batch has the highest vapor pressure of 95 kPa or –6 kPag. Taking into account the minor pressure losses and transient effect, extra pressure of 400 kPa is added to the highest vapor pressure to get the minimum pressure of 400 kPag at the delivery and pump station.

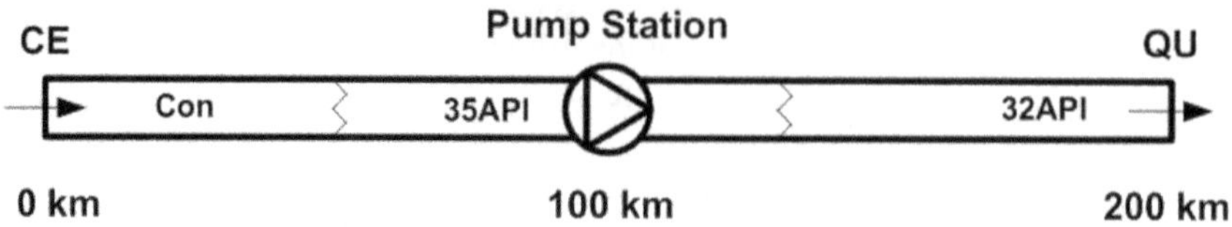

Figure 3-17. Line fill of batches

Step 3. Determine the batch sequence. Here, the batch sequence is given below without describing the sequencing method which is detailed in Chapter 5.

- The batch sequence for minimizing the interfacial mixing is 32°API – 35°API – Condensate, and the same sequence is repeated in the next batch cycle.
- When these batches are placed in the pipeline, the batch line fill profile can be shown in Figure 3-17.

Step 4. Calculate the pressure profile using the 32 °API properties.

- Discharge pressure at CE = 7920 and suction pressure at the intermediate station = 400 kPag.
- Discharge pressure at the intermediate station = 7920 and delivery pressure at QU = 400 kPag.

Step 5. Determine the average pressure profile.

- Calculate the pressure drops P_{32}, P_{35} and P_{con} for 32°API, 35°API and condensate, respectively, using the common minimum pressure of 400 kPag.
- Calculate the total required pressure averaged over the weight of each batch size:

$$P_{avg} = (20{,}000 \times P_{32} + 15{,}000 \times P_{35} + 25{,}000 \times P_{con})/(20{,}000 + 15{,}000 + 25{,}000)$$

- This approach is acceptable if the load factor is low.

3.3.6 High Vapor Pressure (HVP) Pipeline Design

HVP products are defined as the liquids whose vapor pressure at 38 °C exceeds 110 kPa. High vapor pressure (HVP) pipelines are characterized by low density, low viscosity, and the requirement to operate the system at high pressure to maintain the fluid in a single phase in the pipeline. HVP products are highly flammable and heavier than air even when they evaporate into a gaseous form. They expand greatly as the temperature increases, and their vapors are not easily visible. If the HVP liquids leak out of a pipeline, the vapors may creep along the ground or gather in low places, and can explode if they encounter an ignition source. Therefore, extra precautions are necessary to transport and store the products.

The temperature effects on HVP and dense phase fluids (refer to Chapter 2 for the definition of dense phase) are so sensitive that the temperature behaviors in the pipeline should be taken into consideration to determine the pressure profile accurately.

- The density of light hydrocarbon such as ethane or propane changes significantly with temperature. The viscosity of lighter hydrocarbon liquids is small and does not vary with temperature significantly. Therefore, hydraulic design for such fluids is relatively independent of viscosity, because the Reynolds number is so high that the fluids flow in or close to a fully turbulent flow regime. However, the design consideration should include the dependence of their high vapor pressures and phase changes on the operating temperature. This is the subject addressed in this section.

- Viscous liquids such as heavy oil or waxy crude need to be heated or blended with diluent to reduce the viscosity for pumping. Although the viscous heating effect is inherent to all real fluid flow situations, its relative influence on heavy and waxy crudes is very high. The temperature of the highly viscous fluids at the entrance to the pipe can be significantly different from the temperature of the medium surrounding the pipeline system. This subject will be discussed in the next section.

The hydraulic design for fluids such as NGLs or LPGs is relatively independent of viscosity, but more dependent on consideration of vapor pressure. High vapor pressure (HVP) pipelines are characterized by low density, low viscosity, and the requirement to operate the system at high pressure to maintain the fluid as a single phase liquid in the pipeline. Single phase should be maintained throughout the pipeline by keeping the local pressure above the vapor pressure.

The governing design parameters for HVP pipelines are thus the vapor pressure and maximum temperature.

- The vapor pressure is directly related to fluid temperature in the pipeline.
- The maximum vapor pressure occurs at maximum temperature in the pipeline.

The delivery points for HVP liquids require much higher minimum pressures over the vapor pressure of the liquid. Because of the complex dependence of fluid properties on pressure and temperature in the dense phase, pressure and temperature calculations should be performed simultaneously to maintain high accuracy. The delivery point for HVP liquids may be equipped with a pressurized sphere and thus require much higher minimum pressures over the vapor pressure of the liquid.

HVP products can be economically transported in liquid phase, except ethane and ethylene which may be transported in dense phase. In order to avoid vaporization of the HVP liquids, HVP pipelines have to be operated at high pressure, above a minimum pressure greater than the vapor pressures throughout the pipeline. Normally, the minimum pressure is determined by adding extra pressure to the vapor pressure. The extra pressure takes into account the transient effect and elevation difference along the pipeline as well as piping losses through manifold and other equipment at pump stations. If the liquids are delivered to a tank, the delivery pressure should be much higher because the tank is often pressurized at a very high pressure level.

Dependent upon the pressure and temperature conditions, the fluid in a pipeline can exist as a liquid, gas or a mixture of both (two-phase flow). The phase behavior does not play a critical role in designing and operating heavier hydrocarbon liquid pipelines, because their operating ranges are far away from the phase change zone. However, the phase behavior of the HVP liquids has to be taken into account in pipeline design and operation, because their pipelines operate closer to the zone where a phase change occurs.

Some examples of HVP products include pentane, butane, propane, ethane and ethylene. Any pipeline transporting these products in liquid phase is called an HVP pipeline. The vapor pressures of these products are listed in Table 3-6 (these vapor pressures are obtained from GPSA Handbook [11], measured at 40°C instead of 38°C).

TABLE 3-6. High vapor pressure product parameters

Products	Vapor pressure (kPa)	Thermal expansion (/°C)	Critical pressure (kPa)	Critical temperature (K)
I-Pentane	151	–0.00160	3381	460.4
N-Pentane	116	–0.00154	3370	469.7
I-Butane	530	–0.00216	3640	407.8
N-Butane	379	–0.00194	3798	425.1
Propane	1370	–0.00280	4244	369.8
Ethane	6000 (*)	–0.015 (+)	4872	305.3
Ethylene	9700 (*)	–0.025 (+)	5040	282.3

(*) The vapor pressures and thermal expansions of these liquids are highly dependent on the pressure and temperature conditions. Therefore, a representative value does not have a definite meaning for these products. These values are estimated by extrapolating measured values and are presented for an illustrative purpose only.

(+) These values are estimated about 40°C at 9000 kPa and presented for an illustrative purpose only.

As shown in the table, the vapor pressures and thermal expansions of ethane and ethylene are significantly higher than the other HVP liquids. Normally, these two products are transported in dense phase. For a hydrocarbon mixture, there is no clear line dividing dense phase from the liquid phase or other single line dividing the dense phase from the gas phase, but the dense phase lies between critical temperature and cricondentherm if the pressure is above the cricondenbar. Phase change from dense-to-liquid or vice versa is gradual. Ethane (C_2H_6), ethylene (C_2H_4), and carbon dioxide (CO_2), can be "liquefied" in pipelines at temperature and pressures even below the critical point, and treated as liquids in transportation. Dense phase liquid is a highly compressible liquid that shows properties of both liquid and gas; a density similar to that of a liquid, but a viscosity similar to that of a gas. For liquid pipeline design and operation, it is considered that the fluids are in dense phase if the pressure and temperature are around the critical pressure and critical temperature but above the vapor pressure.

Because of the complex dependence of fluid properties on pressure and temperature in or near to the dense phase, pressure, and temperature should be determined as accurate as possible and thus their calculations must be performed simultaneously to achieve the desired accuracy. Reference [12] details the method of calculating pressure and temperature in dense phase and identifies the following key design parameters:

- The critical point is not well defined nor are the properties near the critical point. Therefore, one should try to avoid approaching the critical points too closely.
- Since most ethane or ethylene pipelines are operated in a fully turbulent flow regime, the friction factor is independent of the Reynolds number and depends only on the relative roughness of the pipe. Therefore, the accuracy of the pressure profile is sensitive to the values of the relative roughness. Sometimes, other HVP products flow in a similar fully turbulent regime.
- Both pressure and temperature profiles are relatively sensitive to the specified value of the overall heat transfer coefficient, which in turn depends on soil conductivity. The soil conductivity not only varies along the pipeline but also changes frequently with moisture content.
- The effect of the seasonal variation in the average soil temperature depends on the difference between the fluid temperature and soil temperature. If the

difference is large, the effects on the pressure and temperature profiles will also be large.

- Higher flow rates result in greater friction losses and thus lower pressures, causing lower density and higher velocities. At the higher flow rates, the temperature is high and this, together with the low pressures, results in a further pressure drop with flow rate.
- The elevation effects on the pressure drop in the uphill segments are different from that in the downhill segment. In the uphill segment, the total pressure gradient remains the same, because the decrease in the frictional pressure gradient is compensated by the increase in the static pressure increase rate due to the elevation gain. In the downhill segment, however, the magnitude of the hydrostatic term exceeds the magnitude of the friction term, resulting in less pressure drop.

The magnitude of these effects depends on the rate of change of the fluid properties with pressure and temperature under the particular flowing conditions. In many cases, accurate values for these design parameters are unknown. For example, soil temperature will vary considerably from place to place, adding another uncertainty. In such situations, one should perform calculations for a range of values to examine the overall uncertainty in the calculated pressures and temperatures.

Normally, when designing a liquid pipeline, consideration is given to use the maximum flow that is required at a specific time and a larger pipe size taking into account future volume increase. However, a HVP and particularly dense phase pipeline is designed with the following criteria in mind, , high operating pressure, because of uncertainties in defining the product properties in their operating ranges, and limited accuracy in determining temperature profile:

- Low flow velocity resulting in low pressure drops to operate at high pressures, requiring a larger size pipe,
- A high pressure required at the storage facility of the HVP products, also requiring high delivery pressure,
- Overpressure problem if the pipeline is shut down for a prolonged period, requiring blowdown valves,
- Frequent block valve spacing to reduce spillage and increase safety,
- Installation of blowdown valves on either side of each block valve to relieve an overpressure condition or deal with other emergency conditions. For added safety purposes, the blowdown valves need to be automated and the isolated segment has to be blown down as quickly as possible. Normally, a flaring system may be provided to flare the spillage.

Hydrate problems in HVP pipelines have been reported in the presence of free water. Since it is impractical to control the pressure and temperature conditions for forming hydrates, it may be simpler to reduce the contents of free water.

Example: Ethane Pipeline

A pipeline company plans to build an NPS 12″ pipeline, transporting ethane, with wall thickness of 0.219″. The total length of the pipeline is 200 km and its elevation profile is assumed to be flat. Initially, no intermediate pump station is planned. The yearly throughput is expected to grow to 1,500,000 tons. The inlet pressure is planned to be 600 kPa less than the maximum design pressure and the maximum inlet temperature is 30°C. Refer to the pressure-enthalpy diagram shown in Figure 3-18. Determine the minimum operating pressure, and pressure and temperature profiles using the following data:

- Pipe grade: X56
- Maximum inlet temperature: 30°C
- Ground temperature: 4°C
- Heat capacity: 4.76 kJ/kg°C
- Ethane viscosity: 0.14 cSt
- Soil conductivity: 0.5 W/m°C
- Depth of cover: 1.2 m

Solution:
Refer to the pressure-enthalpy diagram, Figure 3-18, which shows the phase behavior of ethane. The Pressure-Enthalpy diagrams show pressure on the vertical axis and enthalpy on the horizontal axis. The diagrams are used in locating pipeline operating points in terms of pressure and temperature and for designing control valves. Pipe flow is almost an isenthalpic process, so the diagram shows a graph of the enthalpy during various pressures and physical states. The critical point is defined at the critical pressure and critical temperature (point C in the figure), where the liquid phase and vapor phase meet, and either phase cannot be distinguished. The rectangular box in the diagram shows the operating pressure range of an ethane pipeline for an operating temperature range (assuming that the operating temperature ranges from 0°C to 30°C (solid lines in the figure) and the pressure from 4500 kPa to 10,000 kPa). Since the operating temperature range is lower than the critical temperature, the ethane in this operating condition remains in liquid phase. For different operating temperatures, the operating pressure range should be different to avoid vaporization.

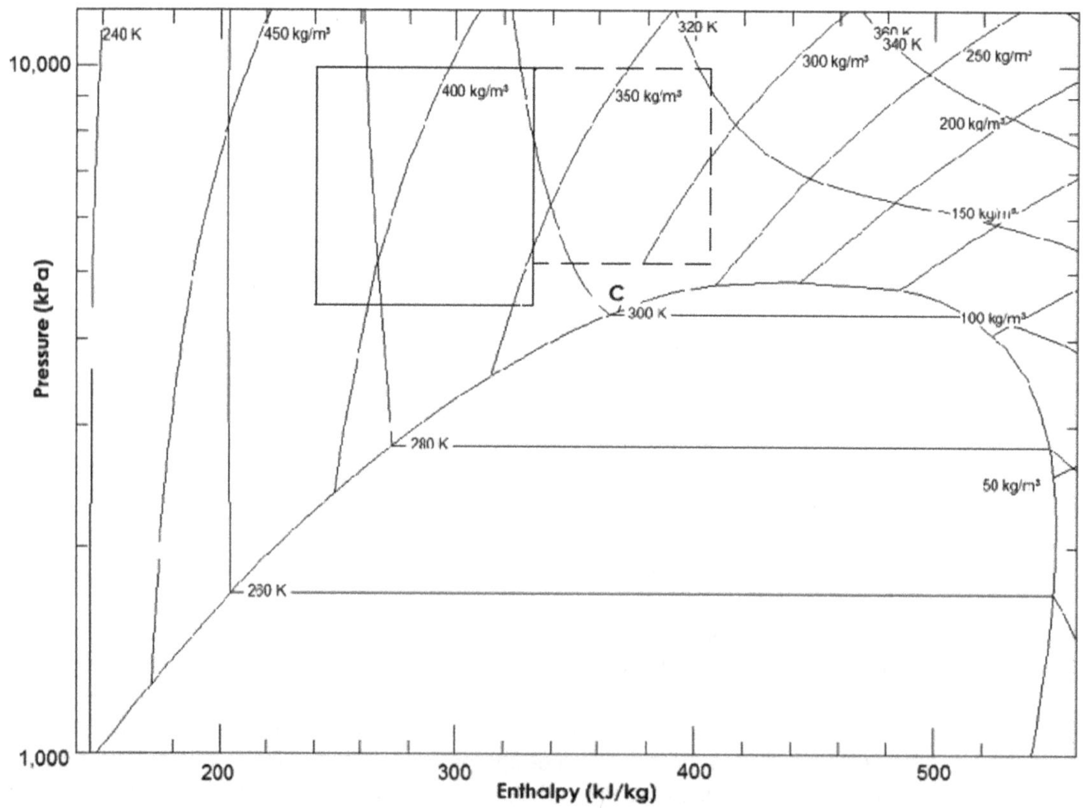

Figure 3-18. Ethane pressure-enthalpy diagram [2]

As shown in broken lines, the ethane will be in dense phase and the minimum pressure has to be increased if the operating temperature is increased to 37°C at the maximum pressure.

It is assumed that the pipe design factor is 0.72 and pipe roughness is 0.0018″ or 0.0457 mm.

Step 1. Determine the maximum design pressure for the X60 grade pipe.

- Applying Barlow formula with the design factor of 0.72, design pressure = $2 \times S \times t/D \times F = 2 \times 56{,}000$ psig $\times$ 0.219″/12.75″ $\times$ 0.72 = 1385 psig = 9550 kPag

Step 2. Determine the density of ethane at the maximum inlet conditions using the Pressure-Enthalpy diagram.

- The maximum operating inlet pressure is 9550 – 600 = 8950 kPag, and the inlet temperature is 30°C.
- From the ethane pressure-enthalpy diagram in Figure 3-18, the ethane specific volume at the inlet conditions is about 0.00267 m^3/kg, or the density is about 375 kg/m^3. However, the pressure and temperature change as the ethane flows along the pipeline, and so does the density.

Step 3. Determine the vapor pressure and minimum delivery pressure for this pipeline design, using the ethane Pressure-Enthalpy diagram.

- Since the pipeline flow is an almost isenthalpic process, the ethane vapor pressure is determined close to 4000 kPa by following down the isenthalpic line to the phase envelope.
- The minimum delivery pressure is obtained by adding to the vapor pressure a safety pressure of 600 kPa: 4000 + 600 = 4600 kPa or about 4500 kPag. The safety pressure includes minor pressure losses due to valves, pump station piping loss, meter station piping loss, and transient effects.

Step 4. Calculate the volume flow rate and velocity at the design flow rate.

- At the inlet conditions, the density is 375 kg/m^3 and thus the volume flow rate 1,500,000/(0.375 × 365 × 24) = 457 m^3/hr. The flow velocity at the inlet conditions is about 1.65 m/s. Note that the local velocity varies somewhat because of mass conservation.
- If the number of yearly operating days is less than 365 days, then the actual number of operating days should be used. Then, the flow velocity is larger because the total amount of yearly shipment is divided by a smaller number of days than 365.

Step 5. Determine the pressure, temperature and density profiles (Figure 3-19).

- The viscosity effect on the friction factor may be negligibly small. However, the density and heat capacity density change with temperature, and the density is a complex function of pressure and temperature. The pressure and temperature profiles may not be determined reliably without their accurate behaviors with respect to pressure and temperature.
- It is time-consuming to manually calculate the pressure and temperature profiles in detail. Therefore, it is suggested to use a pipeline simulator for hydraulic

design work within the boundaries that have been established in the above steps.

- To calculate the pressure and temperature profiles accurately, the total pipeline length is broken down into multiple short pipe lengths, say 5 km spacing for elevation changes or 10 km spacing for flat elevation. The profiles are plotted in the figure below.

Step 6. Analyze the ethane pipeline design.

1. The profiles show the following behaviors:
   - The delivery pressure is set at 4500 kPag and discharge pressure is calculated at 8550 kPag for the flow rate of 457 m^3/hr. The pressure gradient is almost linear at high operating pressures, where the ethane remains in dense phase within the operating range. Since the mass rate has to be conserved, the flow velocities at high temperatures, where the densities are lower, are faster than those at low temperatures. Therefore, the frictional pressure drop is somewhat higher in the upstream segment where the operating temperature is high than that in the downstream segment. This hydraulic behavior is similar to that of a gas pipeline.
   - The temperature drops to the ground temperature about 80 km from the injection point. For a lower flow rate, the temperature drops faster and reaches the ground temperature nearer to the injection point, because the heat conduction is faster at a low flow velocity.
   - Within this operating range, the density varies from 365 kg/m^3 to 415 kg/m^3, about 11% change. Note that the density profile does not necessarily keep

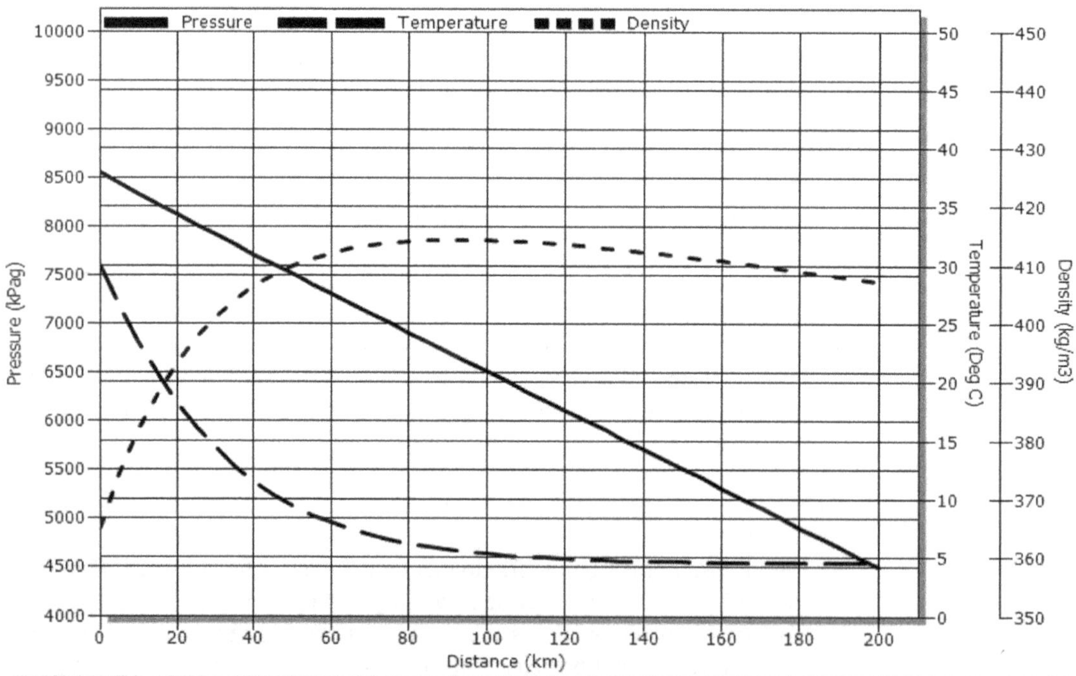

Figure 3-19. Pressure and temperature profiles of an ethane pipeline.

increasing as the pressure and temperature drop, because the density has a non-linear relationship with pressure and temperature.

2. Determine the pipeline capacity.
   - The maximum throughput can be determined by setting the injection pressure at the maximum pressure.
   - The initially planned maximum pressure is 8950 kPag. At the injection pressure, the capacity is 480 m^3/hr, which is about 5% higher than the design flow rate. If the expected flow increases beyond this capacity, the maximum operating pressure can be increased to the MAOP.
   - If the operating pressure is allowed to go up to the MAOP, the capacity increases to 517 m^3/hr, which is about 13% higher than the design flow rate. This capacity is equivalent to the annual rate of 17.0 million tons.
   - If the throughput needs to be increased beyond this limit, an intermediate pump station has to be installed.
3. Calculate the pressure and temperature of the liquid. Since the pipeline pressure of HVP products is very sensitive to temperature changes, it is necessary to understand the temperature and pressure behaviors in order to avoid potential overpressure problems. If the ambient pressure is high while the pipeline is shut in for a prolonged period of time, an overpressure problem can occur because the ethane temperature can increase in the pipeline. It is necessary to install automatic blowdown valves to relieve the pipeline pressure.

3.3.7 Heavy Crude Pipeline Hydraulic Design

Heavy or waxy crudes do not flow easily in normal operating temperature ranges mainly because of high viscosity and their high pour points. The pour points of these viscous crudes are higher than normal operating temperature ranges. Even though light and medium crudes are easy to pump above their pour points, they can exhibit similar behaviors if the crude temperature drops below their pour points, as is possible in cold climates. Hydraulic design for heavy crudes or for hydrocarbon liquids transported below their pour points is influenced largely by the effect of temperature on viscosity and related friction losses. Therefore, the design aspects discussed in this section are equally applicable to not only heavy crudes transportation in normal operating temperature ranges but also light and medium crudes transportation in very cold areas.

Transportation of such crudes through pipelines requires much higher flowing temperature than their pour points or else a reduction of the pour points by blending with diluent. Reference [15], in the five part series, describes various issues of pumping heavy crudes and lists the following methods of transporting high pour point crudes:

- Blending with a hydrocarbon diluent to keep the fluid behavior as Newtonian. The diluents frequently used for bitumen transportation are natural gas condensate and synthetic crudes.
- Heating the crude to a higher inlet temperature to allow it to reach the delivery or intermediate station before cooling to below its pour point.
- Combination of the above two methods
- Mixing hot water with the highly viscous crude to form an emulsion, primarily being used to transport bitumen to its processing plant
- Processing the crude before pipelining to remove the wax and bring down the pour point and viscosity

- Injecting paraffin inhibitors, primarily being used in crude oil production systems to reduce pour point by preventing paraffin deposition and wax crystallization on the pipe wall
- Heating both the crude and the pipeline by steam tracing or electrical heating, which is only applicable to short pipelines due to the poor economics of applying it to long transmission lines

However, before deciding which method is selected, it is necessary to evaluate the physical properties of crude, the temperature behavior in the pipeline, restarting after shutdown, and facilities design.

3.3.7.1 Determine the Physical Properties under Pipeline Conditions

The critical design parameters for heavy oil pipelines are the viscosity and pour point, because the viscosity is directly related to fluid temperature in the pipeline and the non-Newtonian viscosity behaviors appear near the pour point. The following physical properties are important for designing a heavy or waxy oil pipeline system including pipeline hydraulics, pump station, and terminals [14]:

- Density or specific gravity
- Wax content
- Shear stress vs. shear rate for non-Newtonian region
- Yield stress for non-Newtonian region
- Bulk modulus
- Heat capacity

Heavy crude is characterized by high density, high viscosity and high pour point, and may contain a significant amount of wax and/or sulphur. Heavy crude may exhibit non-Newtonian viscosity behavior at normal operating temperature ranges because its pour point can be higher. It is known that the apparent viscosities of non-Newtonian liquids are sensitive not only to temperature changes but also to the shear rate and cooling rates. Laboratory tests should be performed at the pipeline operating conditions to determine the crude's viscosity types and behaviors in terms of the shear stress vs. shear rate and yield stress over the operating temperature ranges including the pour points. The types include Newtonian, dilatant, Bingham plastic, pseudoplastic, and thixotropic (time-dependent) fluid, because heavy crudes show different fluid characteristics.

Another potential engineering problem in dealing with heavy crudes, and sometimes with light and intermediate crudes, is the significant presence of wax. A waxy crude may exhibit Bingham plastic characteristics after gelling, requiring a finite shear stress to initiate flow. Heavy and/or waxy crudes start developing a yield stress near their pour point, which may require additional pressure to restart flow. It is known that wax does not deposit in turbulent flow at high temperatures, certain parts of a pipeline may have wax deposits, and wax deposits could have an insulating effect.

TABLE 3-7. Viscosity, temperature, and pour point [14]

Product	Specific Gravity	Temperature (°C)	Viscosity (cSt)	Temperature (°C)	Viscosity (cSt)	Pour Point (°C)
Bitumen	1.02	65	50,000	120	330	55
Residuals	0.96	65	1000	120	46	32
Crude	0.84	20	11	50	4	13
High wax	0.81	50	7.4	60	3.3	35
Diesel	0.84	-1	2.8	27	1.4	–
Jet fuel	0.78	-1	2.2	27	1.3	–
Gasoline	0.73	-1	0.8	27	–	–
NGL	0.50	-1	0.23	38	0.2	–

The common characteristic of heavy and waxy crudes is their high pour point. Due to non-Newtonian behavior near the pour point, more pressure is required to pump in the non-Newtonian range. No problem may arise in pumping heavy crude below its pour point, if the fluid is kept in motion. However, when the crude temperature is below its pour point, a few unique behaviors are observed:

- If a crude pipeline being pumped below its pour point is shut down, the resulting gelled state will require substantially more pressure to put it into motion.
- This additional restart pressure is substantially less than if a crude pipeline being pumped above its pour point is shut down and allowed to cool down.

Density and bulk modulus of heavy oil are very high compared to other types of crude. The high bulk modulus can result in a large potential surge during pump shut-down or valve closure. The frictional pressure drop of a heavy crude pipeline is significantly high due to the high density and viscosity, and so is the surge pressure due to high bulk modulus. As a result, a heavy oil pipeline tends to be operated at low flow velocity for economic and safety reasons. As usual, heat capacity is used for calculating temperature profile. The yield stress is a parameter used for determining the pumping requirement upon restart. Therefore, yield stresses should be measured over the range of temperatures and shutdown times which are expected in the pipeline.

For extra heavy oil transportation through a pipeline, blending with diluent is most effective. It lowers the pour point of the blended heavy crude and viscosity significantly. The level of blending diluent changes with the temperature and viscosity behaviors of the crudes, and the diluent requirement varies within individual pipeline systems to meet their specifications. When delivered to a third-party pipeline, the pipeline specifications for density and viscosity are 940 kg/m^3 (19°API) and up to 350 cSt in Alberta, Canada [16]. This provides for lower diluent requirements in summer months than in the winter. Typically, summer requirements are about 20% less than maximum requirements in mid-winter. Normally, crudes with an API gravity of 18 or higher may not require any diluent, unless the operating temperature is very low. Even in winter months, diluent requirements may be less than 5%. Crudes lower than 18°API need to be blended to a level that will provide for optimum pumpability and protection from congealing in case of line shutdown [13]. Regardless of density or viscosity before blending, all blended crudes should have a common consistency so that all heavy crude moving in a common carrier pipeline has the same hydraulic characteristics.

It is too expensive and risky to transport bitumen a long distance by means of heating only. Therefore, in long lines it is necessary to blend it with a diluent. The key issue is the availability of diluent at the injection location. If it is not available, then it has to be shipped from other sources through another pipeline or by train. If the availability of diluent is limited, the diluent can be separated from the blended bitumen after it is delivered. It requires a separation plant at the delivery location and then the separated diluent is shipped back to the injection location.

3.3.7.2 Determine the Pressure and Temperature throughout the Pipeline for the Anticipated Flow Rates

Section 3.1.3 discusses the equations, surrounding environment, and procedures for calculating pressure and temperature profiles. The surrounding environments can vary significantly, resulting in different overall heat transfer coefficients. As noted previously, the viscosity of extra heavy crude such as bitumen is very sensitive to temperature change. Therefore, an accurate temperature calculation is necessary. In order to improve calculation accuracy, temperature and pressure profile calculations are performed by dividing the entire pipeline into many short pipe segments and analyzing each segment separately.

As discussed earlier, hydraulic design for heavy crudes is influenced largely by the effect of temperature on viscosity and related friction losses. If the crudes are heated, a thermal analysis is required to predict the performance of the system over its design temperature range and subsequently to determine the pumping requirements in a pipeline system. If the injection temperature is much higher than the ground temperature, the frictional pressure drop accelerates as a result of cooling. As the fluid is cooled, both the density and the viscosity increase, and the frictional pressure drop increases.

Due to the high viscosity of heavy crudes, the frictional pressure drop per unit distance is very high even for low flow velocity, and thus the friction heating and fluid temperature increase. Normally, liquid pipelines operate in the turbulent flow regime and the boundary layer is thin. Therefore, the thermal resistance due to the boundary layer that builds up on the inside of the pipe wall is negligibly small. The contribution of the thermal resistance for extra heavy crudes to the overall heat transfer coefficient turns out to be small, even if the crudes flow in the laminar flow regime. However, the actual temperature drop reduces slightly as a result of the added thermal resistance.

Pipelines for heavy crudes may be insulated to reduce heat loss, if the economics is justified. Insulation thickness is important as the design and operation of a hot oil pipeline depend on the amount of heat lost by the heavy crudes. If the temperature difference between the pipeline and the ground is greater, the insulation thickness can be increased to a certain extent. However, it should not be thicker than the economic and physical optimum insulation thickness. Insulation applied to large diameter pipelines to maintain temperature at low flow rates and low ambient temperature may cause overheating of the line for high flow rates at high ambient temperatures. For any given insulation thickness, the heat loss is greater if it takes longer for the crude to travel between the initial pumping station and terminal or between reheat stations. Therefore, the velocity and viscosity of the oil determine the distance between stations and the number of pump stations and/or reheat stations.

Design considerations for heavy crude pipelines with thermal effects should include the following:

- The temperature behaviors of the environment and its effect on the physical properties of the fluid over the range of operations.
- The temperature effects on shutdown and restart.

In addition to hydraulic design, the operating temperature range affects the mechanical design and design for operations including shutdown and restart. The following types of hydraulic design and operation problems arise for heavy crude pipelines:

- Pressure and temperature profile calculations,
- Pipe line sizing,
- Maximum throughput determination,
- Pump and heater station spacing,
- Heat retention for a certain period, thereby determining insulation thickness and other facility requirements such as extra pumps.

3.3.7.3 Review the Restart after Shutdown

As pointed out above, heavy crudes show unique behaviors near or below their pour points; the gelled state will require more pressure to put it into motion due to the high yield stress below the pour point. Therefore, it is necessary to determine crude temperature throughout a pipeline whose temperature may cool down during the expected shutdown period. During shutdown periods, fluid in the pipeline cools without the heat of friction until flow resumes or the pipeline temperature reaches the ground temperature.

During cooling, the temperature at a certain location may be calculated by analyzing the rate of heat loss of the crude in the pipeline:

$$T(t) = T_g + [T(0) - T_g] \times \exp\{-(4UD_T t)/[\rho C_p d^2)]\} \qquad (3-33)$$

where

$T(t)$ = temperature at time t, (°C)
$T(0)$= temperature at the time when the pipeline is shut down (°C)
T_g = ground temperature (°C)
D_T = pipe outside diameter including insulation thickness for insulated pipe
t = time from start of static cooling, in second
d = inside pipe diameter, m
ρ = liquid density, kg/m^3
C_p = heat capacity of the liquid averaged at the inlet and discharge temperatures, kJ/kg°C

If the heat conduction through the boundary layer and pipe wall is excluded from the overall heat transfer coefficient, the calculated temperature would be lower than the actual temperature, requiring somewhat lower restart pressure. If possible, pipeline start-up or restart can be scheduled during periods of warmest ambient temperatures in order to avoid the difficult problems that may be encountered during start-up.

3.3.7.4 Design Facilities

The effect of the yield stress of heavy crudes is non-trivial below pour point. When a pipeline is shut in and thus the heavy crude cools down below the pour point, it requires an extra pressure to put the crude in motion. This extra pressure requirement has to be provided by a pump to initiate flow. The pressure required to initiate flow is sum of the pressure differentials required to break the gel in each section of the pipeline. Since yield strength is sensitive to temperature, the required pressure has to be determined on each segment to reduce potential calculation error.

When starting up the pipeline after shut-in, the flow rate should be very low to push the gelled crude gently. It is essential to establish the minimum flow rate needed to be maintained during initial start-up, and it may be necessary to include redundant provisions for emergency and planned shutdowns. In selecting a mainline pump, the maximum operating point should be satisfied as usual. If the minimum flow cannot be met by the mainline pump during an initiating period after shut-in, special startup/restart pumps with the capability of high pressure and low flow should be considered. Note that the performance of a centrifugal pump deteriorates for pumping high viscosity fluids, thus requiring a rerate of the pump performance. Refer to the next chapter for pump performance rerating for high viscosity conditions.

Systems to consider would include standby pumps for displacing the crude oil in the pipeline with water, and adding pour point depressant injection facilities. If bitumen is blended with a diluent, a diluent blending and storage facility is required at the lifting point. If the crude is heated for pumping, heaters have to be installed not only at the initiating station but also intermediate pump stations. Depending on the temperature requirements along the pipeline, the number of heaters required at each station can vary. The heater duty can be calculated from:

$$q_h = \rho Q C_p (T_d - T_i)/\eta_h \tag{3 – 34}$$

where

q_h = heater duty required to heat the liquid to the discharge temperature, kJ/hr
ρ = liquid density, kg/m^3
Q = liquid flow rate, m^3/hr
C_p = heat capacity of the liquid averaged at the inlet and discharge temperatures, kJ/kg °C
T_d = discharge temperature of the heater, °C
T_i = inlet temperature of the heater, °C
η_h = efficiency of the heater

Example: Extra Heavy Crude Oil Pipeline

A pipeline company has decided to build a pipeline transporting bitumen to a common carrier pipeline system. The company receives bitumen at 65°C at the initiating station with an expected maximum throughput of 151,000 bbls/day or 24,000 m^3/day. The delivery point of the common carrier pipeline is located 200 km from the lifting point. The pipeline system crosses an unpopulated area, and the elevation profile is almost flat. An X56 grade pipe with 28″ pipe diameter and 0.350″ wall thickness is considered, and the pipeline has to be buried 1.2 m below the ground surface. The maximum shutdown period expected for scheduled maintenance or emergency repair is estimated at 120 hours.

Soil temperatures are 4°C in summer and –4°C in winter and the soil conductivity is 1.0 W/m°C. Assume that the soil temperatures are uniform throughout the pipeline. The bitumen gravity is 8.6°API and its viscosities are 10,000 cSt and 100 cSt at 45°C and 115°C. A series of laboratory tests has found that its pour point is 50°C, below which the yield stress grows significantly exhibiting a Bingham plastic behavior. The common carrier pipeline requires that the viscosity of the delivered product must be maintained at 350 cSt or lower. The heat capacity of the blended bitumen ranges from 2.01 kJ/kg°C to 2.40 kJ/kg°C depending on the temperature and density.

Considering the pipeline length, high viscosity and pour point, and low operating temperature, the company has decided to blend bitumen with condensate as a diluent in order to facilitate easy transportation of bitumen. The bitumen and condensate are blended at the production area before the blended bitumen (dilbit) is lifted. The API gravity of the condensate is 76° and the viscosities of the condensate at 5°C and 45°C are 0.7 cSt and 0.4 cSt. Determine the following:

- diluent requirements in both summer and winter conditions,
- pressure and temperature requirements,
- a heater requirement,
- temperature profile after the maximum shutdown period
- pipe insulation requirement

Solution:

It is assumed that the pour point is low enough to transport the blended bitumen as a Newtonian fluid and that the possible contents of diluent for winter condition are 45%, 40% and 35% and the contents for summer condition are 30%, 25% and 20%.

Step 1. Calculate the base densities of the blended bitumen, and viscosities of and volume requirements for the blended bitumen at the minimum temperatures.

The minimum temperature is the temperature required to satisfy the maximum viscosity requirement of 350 cSt, and the flow requirement is the total volume or daily flow rate of the blended product. Assuming the daily bitumen production remains the same, the flow requirement for the winter blended bitumen is larger because the diluent requirement is higher in winter.

- Density of the bitumen at 15°C and atmospheric pressure: $\rho = 1000 \times 141.5/(131.5 + 8.6) = 1010$ kg/m^3
- Density of the condensate at 15°C and atmospheric pressure: $\rho = 1000 \times 141.5/(131.5 + 76) = 682$ kg/m^3

Contents of Diluent (%)	Density (kg/m^3)	Thermal Expansion (°C)	Minimum Temperature (°C)	Viscosity (cSt)	Flow Requirement (m^3/day)
45	862.4	8.27×10^{-4}	7	321	34,800
40	878.8	7.95×10^{-4}	15	330	33,600
35	895.2	7.66×10^{-4}	23	342	32,400
30	911.6	7.37×10^{-4}	32	330	31,200
25	928.0	7.12×10^{-4}	41	324	30,000
20	944.4	6.89×10^{-4}	50	339	28,800

Step 2. Calculate the pressure and temperature profiles. The maximum design pressure for the selected pipe is 6950 kPag, so the blended bitumen is discharged at 6900 kPag.

The table below summarizes the injection and delivery temperatures and the delivery pressures for different amounts of diluent. The injection delivery temperatures are determined in such a way that the viscosity at the delivery point is kept below 350 cSt. The table shows the temperatures and their corresponding viscosities after the pipeline is shut down for 120 hours.

Contents of Diluent (%)	Injection Temperature (°C)	Delivery Temperature (°C)	Delivery Pressure (kPag)	Temperature after 120 hours (°C)	Viscosity after 120 hours (cSt)
45	15	7	532	0.2	750
40	34	15	1458	3.3	1264
35	52	23	2019	6.4	2138
30	62	32	2236	14.7	1917
25	83	41	2774	18.2	2958
20	104	50	3203	21.8	4359

The required injection temperature decreases as the amount of the diluent increases. As expected, the pressure requirements for the winter condition are higher than those for the summer condition as a result of the operating temperatures in summer condition being much higher than in winter and the flow rates are lower for the summer condition. Also, as the amount of diluent gets smaller, a heating facility has to be installed to raise the injection temperature.

If this bitumen starts showing its non-Newtonian behavior about 2000 cSt, the yield stress has to be measured in order to assess the requirement for extra pumping facilities to dislodge the blended bitumen that was congealed during the 120 hours of the shut-in period.

Step 3. Repeat the same calculations for the case where the pipeline is insulated with 2″ polyurethane insulation material. The insulation conductivity is 0.035 W/m°C.

Contents of Diluent (%)	Injection Temperature (°C)	Delivery Temperature (°C)	Delivery Pressure (kPag)	Temperature after 120 hours (°C)	Viscosity after 120 hours (cSt)
45	10	9.7	390	5.6	410
40	18	16	381	9.8	610
35	29	23	841	15.4	774
30	38	31	1139	23.8	742
25	51	40	1706	30.4	853
20	63	48	2109	36.5	997

Compared to the results of the un-insulated case, the insulated pipeline has the following advantages over the un-insulated pipeline:

- No extra pumping facilities are required even after shutting down for 120 hours,
- There is less need for a heater because the required injection temperature is low, or the heater duty is lower than the duty for the un-insulated pipeline even if a heater is installed, assuming that the same amount of the diluent is mixed,
- The diluent requirement is much smaller than the requirement for the un-insulated pipeline,
- Restarting after the shut-down is much easier due to low viscosity.

Step 4. Finalize the pipeline system design

- The 28″ pipe with wall thickness of 0.35″ and pipe grade X56 satisfies the pressure requirements for both winter and summer conditions.
- The insulated pipeline can save both capital and operating costs by reducing the requirements for extra facilities such as a heater and an extra pump to deal with the congealed non-Newtonian crude.
- The selection of the diluent requirement vs. heater installation is based on the cost comparison of the diluent costs against the heater costs. If the pipeline is insulated, 35% of diluent and 65% of bitumen blending can be sufficient in winter. If the temperature of the blended bitumen is higher than 63°C, a heater is not required and thus 20% of diluent may satisfy the summer transportation requirement.

3.4 LOCATING PUMP STATIONS

The initiating point of the pipeline system, into which petroleum products are lifted, must have a pump station. Also, a long pipeline may require multiple pump stations along the mainline. The pumping requirements should be considered in terms of the number and locations of the stations. The number of pump stations is dictated by the installation and operating costs as well as the flow velocity and controllability of the pipeline system and pump station. If the number of stations increases, the costs and flow velocity increase while making the system control difficult due to large surge pressure and its fast response. Refer to Section 5.1.3 for controlling surge.

The key criteria of initially locating mainline pump stations are that the MAOP should not be violated downstream of each pump station and each station has the same differential pressure or head. Here, the differential pressure includes all minor pressure losses due to station piping, bends, fittings, and various valves including control valve. The second criterion offers the following advantages:

- The total energy or power consumption is reduced by adding the same amount of energy to the liquid at each pump station.
- The pump maintenance and spare part inventory costs can be minimized, because the equipment can be identical.

However, these advantages should be compared against potential extra costs to design a pipeline system as such. For example, the power line may be too far from an optimum location to satisfy the above criterion.

This criterion is applicable to the design of all new pipelines in locating pump stations. However, the procedure of locating stations can be different for different terrains or pipeline configuration:

- Relatively flat terrain,
- Complicated terrain in terms of elevation profile,
- Simple pipeline system with one injection and one delivery,
- Complex pipeline system with multiple injection and delivery points.

For a simple pipeline system with relatively flat terrain, the criteria for locating stations results in almost equal station spacing along the pipeline, and the number of pump stations can be determined by dividing the total required pressure by the difference between the MAOP and the minimum pressure;

No. of stations = Total required pressure / (MAOP – minimum suction pressure)

The procedure of locating pump stations is to start from the delivery pressure, drawing the pressure gradient upstream to the intersection of the maximum design pressure, which is superimposed on the elevation profile. If the discharge pressure of the initiating station is smaller than the design pressure, then reduce the design pressure and move the initial locations to further downstream locations. The same differential pressure can be calculated by dividing the total pressure requirements by the number of pump stations.

Example 1: Simple Pump Station Location

Refer to the design example described in Section 3.3.1. The total required pressure is 15,389 kPag, maximum design pressure 9765 kPag, and minimum delivery pressure 350 kPag. It is assumed that the minimum suction pressure is 350 kPag. Since the elevation profile is flat, the number of pump stations is obtained from the above formula:

$$15{,}389 / (9765 - 350) = 1.6$$

Therefore, the total number of pump stations required is 2; one at the initiating station and the other at an intermediate location. Applying the station location criteria, the intermediate station is located at the mid-point of the pipeline as shown in Figure 3-20.

For a simple pipeline system with severe elevation changes, the station locations can be determined by applying these criteria through trial and error on a graph. The procedure of locating intermediate pump stations is as follows:

Step 1. Using the maximum design pressure as the discharge pressure at the initiating station, the first intermediate station is found at a location where the pressure reaches the minimum suction pressure by drawing the pressure gradient on the elevation profile. In practice, a pressure allowance of 200 kPa to 300 kPa at the intake of the pump station is required to account for the losses due to station piping, valves, fittings, and other equipments.

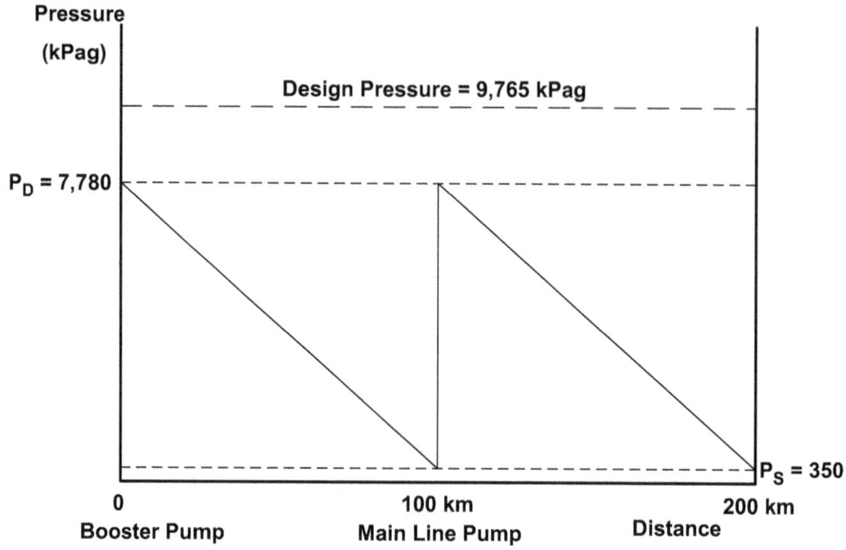

Figure 3-20. Locating intermediate pump stations for flat elevation

Step 2. Progressing downstream from the maximum design pressure at the intermediate station, the next intermediate station is located in the same way as above. Repeat these steps until the minimum suction pressure of the last section is greater than or equal to the delivery pressure.

Step 3. If the suction pressure is much greater than the delivery pressure, reduce the discharge pressure equally at each pump station and then repeat the second step to move the initial locations to upstream locations.

Step 4. If the discharge pressure has to be reduced significantly, the maximum design pressure can be lowered by selecting lower grade pipe or thinner pipe wall thickness.

Figure 3-21 shows that the total pressure requirement is greater than the design pressure. This pressure requirement can be met by installing an intermediate pump station or choosing a thicker pipe in the upstream segment where the required pressure is

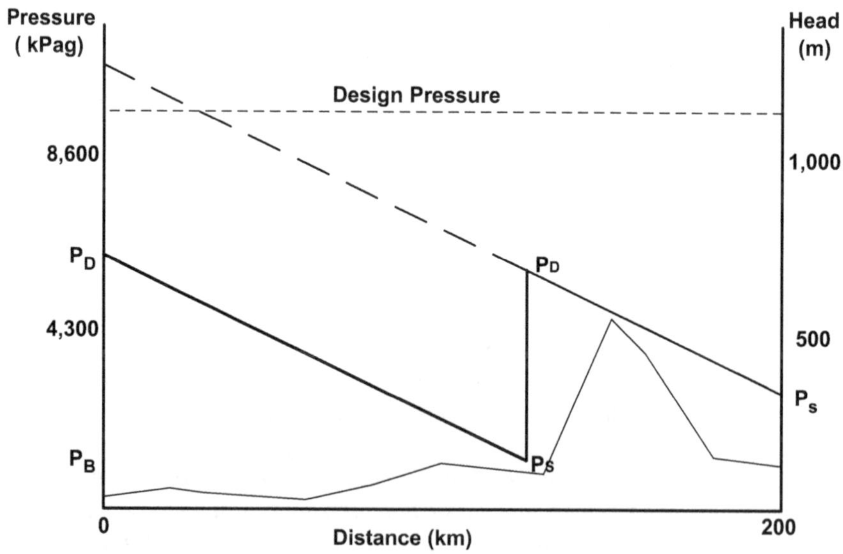

Figure 3-21. Locating intermediate pump stations

violated. Assuming an intermediate pump station is installed, it can be located in such a way that the differential pressure, $P_D - P_B$, at the initiating station is the same as the differential pressure, $P_D - P_S$, of the intermediate station. In this case, the station location is shifted toward the high elevation side. The shift depends on the elevation profile and site conditions.

Example 2: Pump Station Location in Changing Elevation Profile

A pipeline company plans to build and operate a crude oil pipeline, delivering to a tank farm. The pressure rating of the tank equipment is designed at 700 kPag. The average flow rate is 1175 m^3/hr and the pipeline system is expected to operate at least 345 days a year. The average operating temperature is 4°C. The density and the viscosity of the crude at the operating temperature is 870 kg/m^3 and 40 cSt, respectively. The vapor pressure is 80 kPa or –21 kPag, and a slack flow condition has to be avoided. Analyze the pressure profile for the minimum design flow rate of 500 m^3/hr. Assume the suction pressure at each pump station is 350 kPag and the pump pressure differential should be less than 8000 kPa. The pipe specifications are as follows:

- Pipe sizes – NPS = 20″
- Wall thickness – 0.281″
- Pipe roughness – 0.0018″
- Pipe grade – API X60
- Design factor – 0.72

The pipeline length is 350 km and the elevation profile is given below.

KMP	0	50	150	190	230	290	320	350
Elevation	10 m	250 m	250 m	250 m	250 m	310 m	460 m	10 m

Solution:

Step 1. Determine the design flow rate and the maximum design pressure.

- Since the number of yearly operating days is 345 days, the load factor is 345/365 = 94.5%, and thus the design flow rate is 1175/0.945 = 1243 m^3/hr, or rounding up to 1250 m^3/hr.
- The design pressure is obtained by applying the Barlow formula and the design factor for the X60 pipe grade; 8370 kPag.

Step 2. Calculate the pressure gradient.

- Flow velocity = 1.82 m/s
- Reynolds number = 1.82 × 0.494/0.00004 = 22,500
- Relative roughness = 0.00009
- Friction factor = 0.0377
- Pressure gradient = 0.0377 × 870 × 1.822/(2 × 0.494) = 110.0 Pa/m = 110.0 kPa/km

Step 3. Determine initial station locations and calculate the pressures at the locations.

- Assuming that the discharge pressure at the initiating station is 8230 kPag, the first pump station will be located approximately 53 km downstream with a suction pressure of 354 kPag.
- Since the elevation difference is zero for about 180 km downstream of the first intermediate station, the next pump station can be located with a similar

differential pressure to the initiating station; 7920/110 = 72 km. Therefore, the first and second pump stations are located at 125 km and 197 km, where the discharge pressures are 8274 kPag and 8270 kPag, respectively.

- The fourth station is located at 269 km if the elevation were 250 m. Since it is higher than 250 m at 269 km point, the spacing will be shorter than 72 km and the elevation is determined by prorating elevations between two adjacent known locations. At the 266 km post, the elevation is prorated at 286 m and the suction pressure becomes 374 kPag.
- If the discharge pressure of the fourth station is set at 8270 kPag, then the pressure at KMP = 290 is 5596 kPag and the pressure at KMP = 320 is 1021 kPag. Note that the KMP = 320 is the peak point in this pipeline system. Since the peak point pressure is higher than the vapor pressure by 1042 kPa (1021– (–21) = 1042), theoretically the discharge pressure can be reduced by about 1000 kPa. However, considering the transient effect on the pressure, an extra allowance of about 300 kPa has to be added to the vapor pressure.

Step 4. Since the discharge pressure of the last station can be reduced by (1000–300) kPa, that is to 700 kPa, the initial station locations can be adjusted.

- First, locate the first intermediate station at 52 km, where the suction pressure is set at 350 kPag, the discharge pressure and differential pressure at the initiating station are 8116 kPag and 7766 kPa.
- Using the similar differential pressure, the station spacing of the next two stations is 70.5 km and the second and third station locations are 122.5 km and 193 km, respectively. Then, setting the suction pressures of the second and third stations at 350 kPag, the discharge pressures of the first and second stations are 8105 kPag and the differential pressures are 7755 Pa. This differential pressure is very close to the differential pressure at the initiating station.
- The fourth station was located initially at 266 km. By taking into account the elevation difference and locations due to the location shift, the new location is determined at KMP = 263 km. Then, the discharge pressure at the third station is 8160 kPag. If the discharge pressure at the fourth station is set at 8079 kPag, the peak point pressure is calculated at 300 kPag.
- If a surge analysis shows that the peak point pressure is too low, the pump stations need to be moved slightly toward downstream locations.

Step 5. Determine the delivery pressure when the station locations are finalized.

- The hydraulic pressure gain from the peak point to the delivery point is 0.87 × 9.8 × 450 = 3837 kPa, but the friction pressure loss is 3300 kPa. Thus the delivery pressure is 887 kPag, which is greater than the tank equipment pressure rating. Therefore, a pressure control valve (PCV) is needed upstream of the tank farm.
- Since the MAOP is much greater than the delivery pressure, a pressure-reducing station (PRS) is not required as long as the peak point pressure is maintained above the vapor pressure.

Step 6. Analyze the pressure profile for the minimum design flow rate.

- Flow velocity = 0.727 m/s
- Reynolds number = 0.727 × 0.494/0.00004 = 8980
- Relative roughness = 0.00009

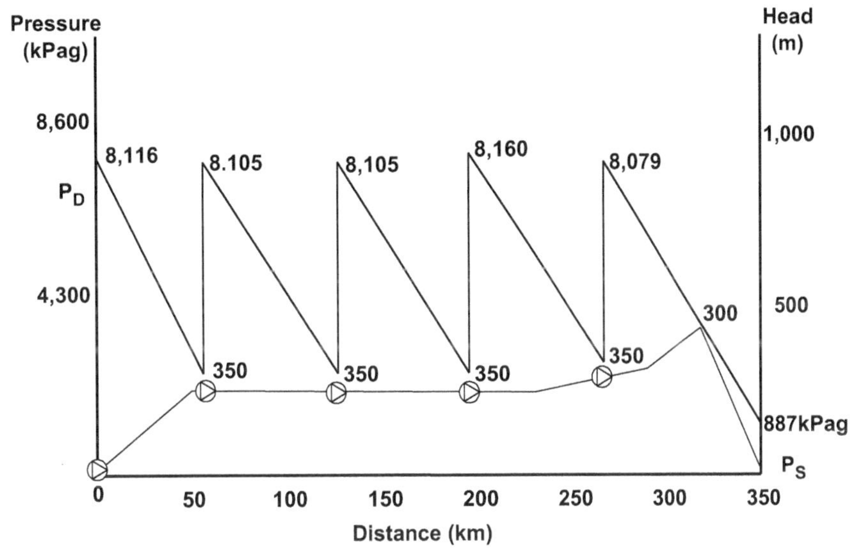

Figure 3-22. Station locations with elevation and pressure profiles

- Friction factor = 0.0505
- Pressure gradient = 0.0505 × 870 × 0.7272/(2 × 0.494) = 23.5 Pa/m = 23.5 kPa/km
- Since the pressure gradient is low, the first and second pump stations can be bypassed. Assuming the suction pressure is set at 350 kPag at the third pump station, the discharge pressure required at the initiating station is 6952 kPag. If the discharge pressure at the third station is 4844 kPag, then the pressure at the peak is 350 kPag and the delivery pressure becomes 3482 kPag.
- Since this pipe pressure is much higher than the tank equipment pressure rating, the PCV must have the capacity to reduce pressure by 3482 – 700 kPa = 2782 kPa.

Figure 3-22 shows the pump station locations with elevation and pressure profiles for the design and minimum flows. Note that the pressures at the delivery gate for low flows are higher than those for high flows in order to keep the minimum pressure required at the peak point.

In general, the same criteria are applied to more complex pipeline systems for locating intermediate pump stations by a trial and error method. Through this hydraulic analysis, the approximate pump station locations are determined that would meet the design and operating parameters. However, the same differential pressure at all pump stations cannot always be achieved.

Example 3: Pump Station Location with a Branch Line

The pipeline from CE to QU is 214 km long and is 20″ in nominal diameter, with a 0.281″ wall thickness. It is constructed of API X-60 grade steel. At CE, diesel enters the pipeline at the design flow rate of 1800 m^3/hr. The booster pumps at CE discharge into the main line pump at 350 kPag, and the minimum delivery pressure required at QU is 350 kPag.

The diesel is taken off at TO, 176 km downstream of CE, where up to 600 m^3/hr is stripped off the pipeline, and the rest is delivered to the final destination, QU. Occasionally, the full flow has to be delivered to QU. At TO, a 50-km branch line is connected to a third party pipeline, which requires the delivery pressure of 3000 kPag.

The pipeline is constructed with X52 grade pipe, and the pipe diameter is 16″ with a 0.25″ wall thickness.

Locate the pump stations along the main pipeline, using the following data:

- Average operating temperature: 15°C
- Density: 850.0 kg/m^3 at 15°C at the operating temperature
- Viscosities at 15°C: 10.0 cSt
- Pipe roughness: 0.0018″
- Delivery pressure at QU: 350 kPag

Assume that the design factor of 0.72 is applicable and that the elevation profile is flat and flow is isothermal.

Solution:

Step 1. Calculate the design pressure (MAOP) of the main and branch lines.

- MAOP of the main line = (2 × 60,000 × 0.281 × 0.72/20) × 6.895 = 8470 kPag
- MAOP of the branch line = (2 × 52,000 × 0.250 × 0.72/16) × 6.895 = 8067 kPag

Step 2. Calculate the pressure required at TO on the branch line side.

- Flow velocity = 1.37 m/s
- Reynolds number = 1.37 × 0.394/0.00001 = 54,000
- Relative roughness = 0.0001125
- Friction factor = 0.0208
- Pressure gradient = 0.0208 × 850 × 1.372/(2 × 0.394) = 42.3 Pa/m = 42.3 kPa/km
- The pressure required at TO = 3000 + 42.3 × 50 = 5115 kPag, which is the minimum pressure required at the take-off point.

Step 3. Calculate the pressure at CE.

- Flow velocity upstream of TO = 2.62 m/s
- Reynolds number = 2.62 × 0.494/0.00001 = 129,400
- Relative roughness = 0.00009
- Friction factor = 0.0175
- Pressure gradient = 0.0175 × 850 × 2.622/(2 × 0.494) = 103.3 Pa/m = 103.3 kPa/km
- The pressure required at CE = 5115 + 103.3 × 176 = 23,296 kPag
- Since this pressure is much higher than the main line design pressure, pump stations should be installed along the main line.

Step 4. Find the minimum number of pump stations and locate the required pump stations along the main line.

- Making a small allowance of 370 kPa in the discharge pressure, the discharge pressure is set at 8100 kPag.
- Since equal pumping head reduces overall cost, the equal spacing in flat terrain can achieve the equal pumping head. Also, too short a spacing should be avoided to minimize capital and operating costs. It can then be safely assumed that the suction and discharge pressures at each pump station are 350 kPag and 8100 kPag, respectively.
- The station spacing is determined by (8100 – 350)/103.3 = 75 km, which is the maximum station spacing. Therefore, 214/75 = 2.85 or three pump stations are

required along the main line. In order to maintain equal pump head for each station, the spacing is 214 km/3 = 71.3 km or 72 km, if the pressure requirement of 5115 kPag at TO is satisfied. Therefore, the mainline pump stations are temporarily located at 72 km and 144 km.

Step 5. In order to justify the selection, we need to prove that the pressure requirements at TO and QU are satisfied with the pump stations. Since the full flow can be delivered to QU, we need to study the hydraulic behaviors of both operations.

- TO is located at 32 km downstream of the third mainline pump station, and the pressure required at TO is 5115 kPag. When the pump station discharges at 8100 kPag, the pressure at TO is 8100 – 103.3 × 32 = 4794 kPag. This pressure does not satisfy the pressure required on the mainline side of TO. The upstream pump has to be located at (8100 – 5115)/103.3 = 28.9 km from the TO or 176 – 28 = 148 kmp.
- Dividing this distance in two stations, the station spacing is 148 km/2 = 74 km, which is still less than 75 km. Therefore, the new station locations become KMP = 74 and KMP = 148. When the second pump is located at 148 km and discharges at 8100 kPag, the pressure at TO is 5208 kPag, which is higher than the required pressure there.
- If the pressure is maintained and other pressure losses are less than 93 kPa (5208 – 5115), no pumping is required along the branch line. Instead, a pressure control valve is required at TO on the branch line side to regulate the delivery pressure for low flow rate.
- The delivery pressure at QU for full flow delivery – The distance between the third station and QU is 66 km. When the pump station discharges at 8100 kPag, the full flow delivery pressure is 8100 – 103.3 × 66 = 1282 kPag. This pressure falls outside the acceptable delivery pressure range, and thus a pressure regulator is required at the delivery point.
- The delivery pressure at QU for partial flow delivery – The partial flow rate is 1800 m^3/hr – 600 m^3/hr = 1200 m^3/hr.
  - Flow velocity between TO and QU = 1.75 m/s
  - Reynolds number = 1.75 × 0.494/0.00001 = 86,450
  - Relative roughness = 0.00009
  - Friction factor = 0.0189
  - Pressure gradient = 0.0189 × 850 × 1.752/(2 × 0.494) = 49.8 Pa/m = 49.8 kPa/km
  - The delivery pressure at QU = 8100 – 103.3 × 28 – 49.8 × 38 = 3315 kPag. This pressure is much higher than the required delivery pressure range, and thus a pressure control valve has to be installed at or upstream of the delivery location.

It should be noted that the differential pressure at the third pump station is different from the pressure at the other stations.

As a final step of locating the pump stations, the best pump station locations are adjusted on the basis of the following criteria at the time of detail design and construction:

- Site terrain conditions
- Availability of power infrastructure
- Availability of access roads
- Potential impact to environment and habitat
- Potential impact to the local land owners due to noise, etc.

ADDENDA TO CHAPTER 3

A3.1 Temperature Calculation

Temperature has considerable influence on the design of pipelines and related facilities, including the establishment of facilities sizing and optimization, economic and technical evaluation, etc.

Temperature and pressure influence all fluid properties. In fluid transmission pipelines, both pressure and temperature vary along the pipeline length. In long-distance-transmission pipelines traversing varied terrain, from permafrost regions to moderate climate conditions, pipelines experience significant temperature changes. Temperature change affects viscosity, density, and specific heat in liquid lines, particularly in crude oil pipelines.

In any pipeline segment, the significant overall temperature change (ΔT) is due to conduction and convection (ΔT_c). However, there are other factors that affect the overall temperature change. These are (ΔT_e) due to isentropic expansion caused by elevation change and due to isenthalpic expansion caused by friction (ΔT_f) [17]. Therefore, the overall temperature change in a pipeline segment is:

$$\Delta T = \Delta T_c + \Delta T_e + \Delta T_f \tag{A3 – 1}$$

The following illustrates a method of overall temperature change due to conduction and convection, ΔT_c. For a pipeline (Figure A3-1) buried at a finite depth (h^o) with insulation, the following expression for computing fluid flow temperature T_o is applicable, Holman [18].

Nomenclature:

C_p = Fluid isobaric specific heat
D_p or D = Pipe outside diameter

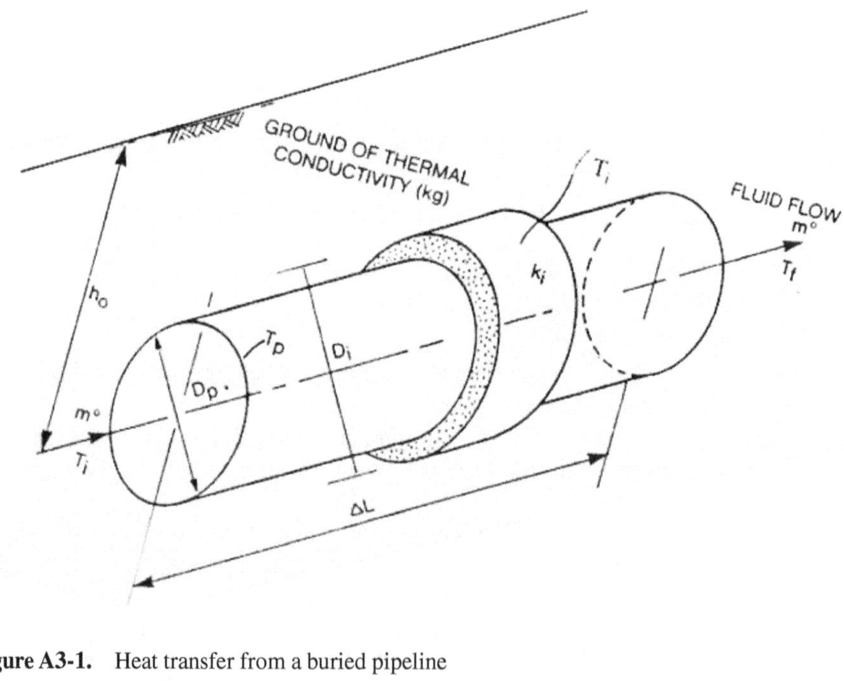

Figure A3-1. Heat transfer from a buried pipeline

D_i	= Pipe diameter with insulation
d	= Pipe inside diameter
h_a	= Air film coefficient
h_f	= Fluid film coefficient
Δh	= Elevation change
K_g	= Soil/surrounding ground thermal conductivity
K_i	= Insulation conductivity
K_p	= Pipe thermal conductivity
ΔL	= Pipe segment length
Q	= Fluid flow rate
T_i	= Inlet fluid temperature
T_o or T_f	= Outlet/Fluid temperature
V_a	= Ambient air/surrounding fluid velocity

$$\rho Q C_p (T_i - T_o) = \frac{2\pi \times \Delta L k_i (T_o - T_g)}{\ln(D_i / D_p)} \qquad \text{(A3 – 2)}$$

By introducing the shape factor, S, and rearranging the above equation, we have

$$T_o = \frac{\rho Q C_p T_i + \dfrac{k_g S}{1+\alpha} T_g}{\rho Q C_p + \dfrac{k_g}{1+\alpha}} \qquad \text{(A3 – 3)}$$

where

$$\alpha = \frac{k_g}{k_i} \ln\left(\frac{D_i}{D_p}\right) \qquad \text{(A3 – 4)}$$

and the shape factor is defined as

$$S = \frac{2\pi \times \Delta L}{\ln\left[\dfrac{2h_o}{D_p} + \sqrt{\left(\dfrac{2h_o}{D_p}\right)^2 - 1}\right]} \qquad \text{(A3 – 5)}$$

For un-insulated pipeline, $\alpha = 1$

For an above-ground or offshore pipeline (Figure A3-2) the corresponding fluid flow temperature is:

$$T_o = T_a + (T_i - T_a)\,\mathrm{Exp}\left(\frac{-\Delta L}{\rho Q C_p U}\right) \qquad \text{(A3 – 6)}$$

Where U = overall heat transfer coefficient and is given by:

$$U = \frac{1}{\pi}\left[\frac{1}{h_f d} + \frac{1}{2k_p}\ln\left(\frac{D}{d}\right) + \frac{1}{2k_i}\ln\left(\frac{D_i}{d}\right) + \frac{1}{h_a D_i}\right] \qquad \text{(A3 – 7)}$$

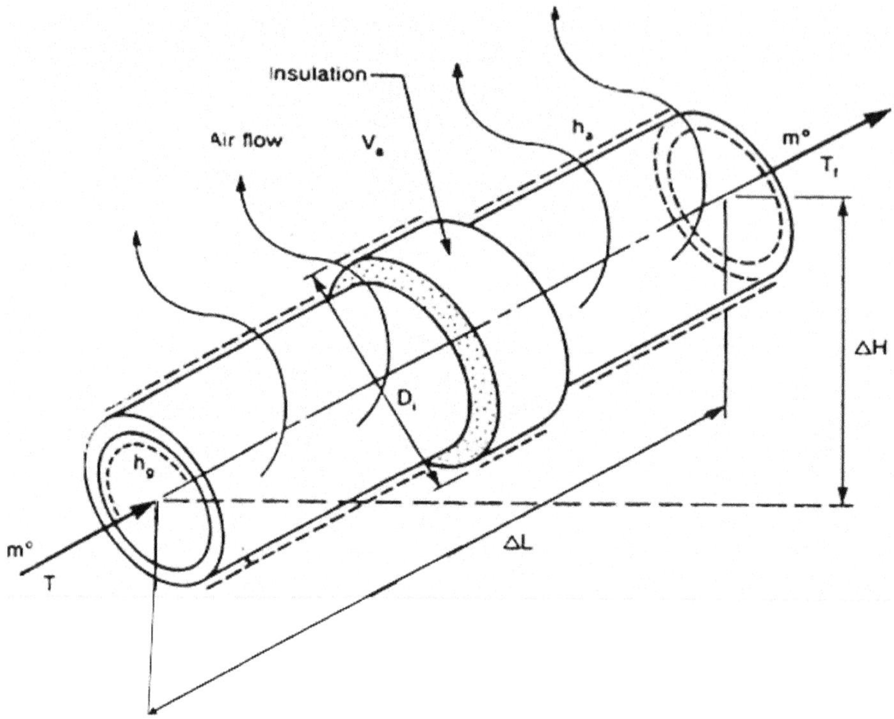

Figure A3-2. Heat transfer from an above-ground or offshore pipeline

In Eq. (A3 – 7), radiation heat losses are ignored as they are small at most normal pipeline operating temperatures. When the pipe is not insulated, the third term in Eq. (A3 – 7) is reduced to zero and Di in the fourth term is set equal to D (i.e., outside diameter of the pipe).

For above-ground pipeline, the film coefficient (h_a) for air can be calculated from the following equation recommended by Dittus and Boelter, Holman [18].

$$N_u = 0.023(R_e)^{0.8}(P_r)^n \qquad (A3-8)$$

Where the Nusselt number, N_u, Reynolds number, R_e, and Prantl number, P_r, are defined as follows:

$$N_u = \left(\frac{h_a D}{k_a}\right)$$

$$R_e = \left(\frac{\rho_a V_a D}{\mu_a}\right)$$

$$P_r = \left(\frac{-\mu_a C_{pa}}{k_a}\right)$$

Fluid film coefficient (h_f) for fluid flowing at velocity V_f through the pipe segment is given by:

$$h_f = 0.023 k_f^{0.6} \left(\frac{C_{pf}}{\mu_f} \right)^{0.4} \frac{(\rho V_f)^{0.8}}{d^{0.2}} \quad \text{(A3 – 9)}$$

In the above equations, C_p, μ, ρ, and k, respectively, refer to isobaric (constant pressure) specific heat, viscosity, density, and conductivity of the flowing medium in the either denoted as suffix (a) for air and (f) for fluid.

For wind blowing over a pipe segment at a velocity of (V_a), the film coefficient (h_a) can be calculated from the following equation:

$$h_a = C \left(\frac{k_a}{D_i} \right) \left(\frac{\rho_a V_a D_i}{\mu_a} \right)^n \quad \text{(A3 – 10)}$$

Properties of air are provided elsewhere [18]. Values of constant C and exponent n are dependent on the Reynolds number and are also given elsewhere [18].

For an offshore pipeline, h_a can be calculated from Eq. (A3-10) with appropriate values of C_p, ρ, k, and μ for sea water, and knowing the current velocity.

The following expressions summarize the computation of ΔT_e and ΔT_f.

$$\Delta T_e = -E_s P_h = \left(\frac{\partial T}{\partial P} \right)_s = \frac{T}{C_p} \left(\frac{\partial V}{\partial T} \right)_p P_h \quad \text{(A3 – 11)}$$

And

$$\Delta T_f = -J P_f = -\left(\frac{\partial H}{\partial P} \right)_T \Big/ \left(\frac{\partial H}{\partial T} \right)_P P_f = -\frac{1}{C_p} \left(\frac{\partial H}{\partial P} \right)_T P_f \quad \text{(A3 – 12)}$$

Where E_s and J are elevation sensitivity and Joule Thompson coefficients, respectively, P_h is pressure loss due to elevation change, and P_f is pressure loss in overcoming friction. E_s can be computed from graphs of pressure (P) and temperature (T) at constant entropy (s), and P_k can be calculated from graphs of enthalpy (H) versus pressure (P) at constant temperature (T).

The sign of Joule Thompson coefficient J indicates whether fluid expansion or compression will cause an increase or decrease in the temperature. As an example, in an expanding gas if J is positive, the gas will cool. A negative J in an expanding gas indicates temperature rise, and is observed in expansion of some special gases, e.g., hydrogen. Methods for calculating E_s and J are given elsewhere [19].

The above procedure outlined above provides an accurate prediction of fluid flow temperature under steady-state condition for buried and exposed pipelines. Sample plots of temperature profiles for a liquid pipeline (carrying bitumen/condensate mixture) is shown in Figure A3-3.

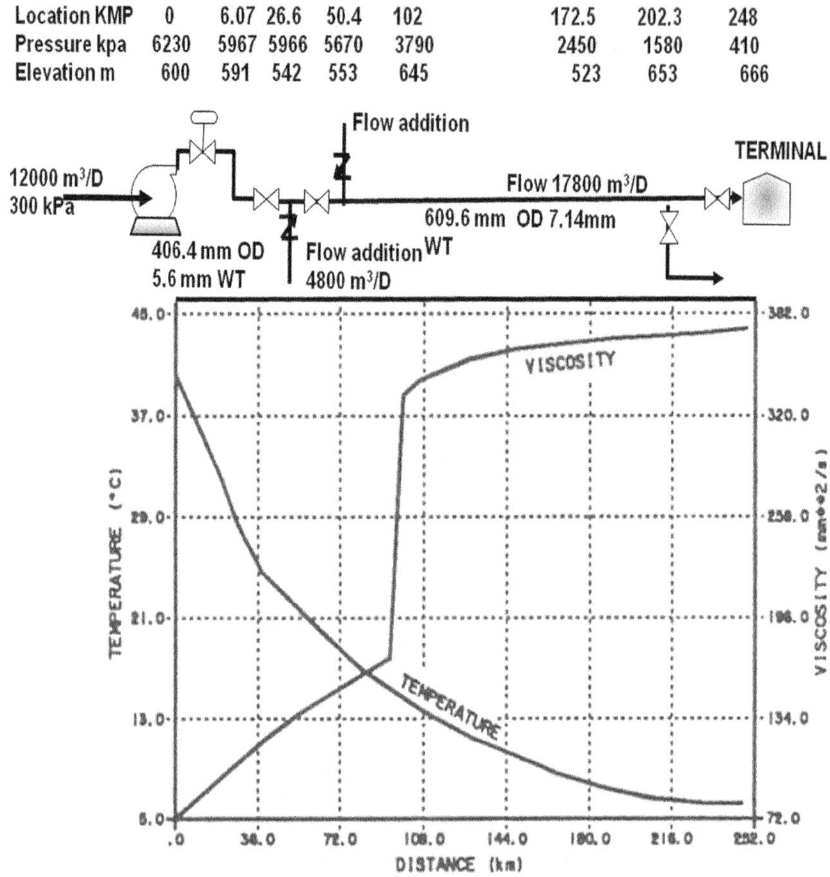

Figure A3-3. Temperature profile in a buried 12″ pipeline transporting non-Newtonian fluid [17]

A3.2 Erosional Velocity of Fluid

Liquid pipeline operations are limited by the following factors that impact the fluid flow velocity:

- Static electricity (affecting both low and high vapor pressure (HVP) products pipeline)
- Erosional velocity (affecting oil and low vapor pressure (LVP) pipeline)

Generation of static electricity is of a concern for pipeline transporting high vapor pressure products such as LPG (propane, butane etc.). Industry's practice is to limit pipeline fluid velocity to 3 to 5 m/s (< 16 ft/s), depending on liquids.

Erosion occurs due to high velocity, especially in the presence of sand or bubbles of particles (Figure A3-4). Erosion is particularly severe when corrosive agents also exist in the fluid. Erosion can be best controlled by proper design and operational limits.

Erosional velocity limits in liquid pipelines are based on gas/liquid density at the operating pressure and temperature and the likely entrainment of particulates such as sand in the pipeline in gathering and injection lines.

Erosional velocity can be calculated from equation, $V_e = C/\rho^{0.5}$. In this case density ρ is replaced by ρ_m (density mean value), representing the density at initial and final flowing conditions:

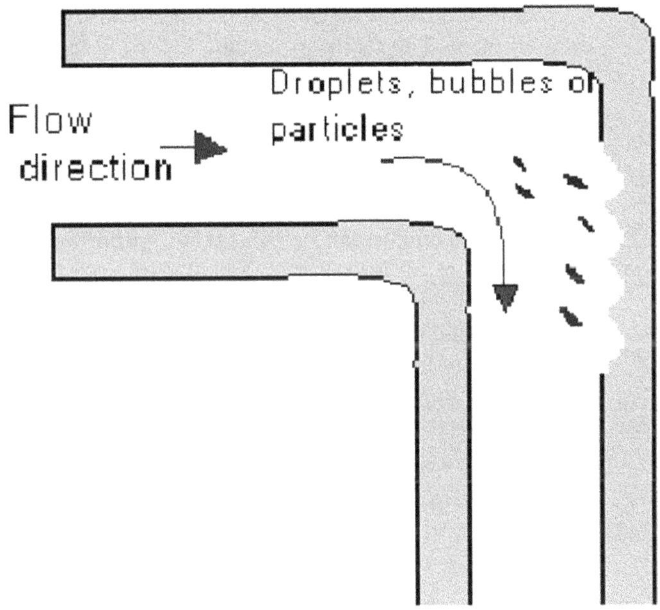

Figure A3-4. Erosion in a pipeline

$$\rho_m = \frac{12{,}409 S_i P + 2.7 R S_g P}{192.7 P + R_a T Z} \quad \text{(A3 – 13)}$$

where

V_e = Erosional velocity, ft/sec
C = Constant defined as 75< C <150
P = Operating pressure, psia
S_i = Liquid specific gravity (water = 1; use average specific gravity for hydrocarbon mixes such as interface) at standard conditions.
R_a = gas/liquid ratio, ft^3/bbl at standard conditions
T = Operating Temperature °R
S_g = Gas specific gravity (air = 1) at standard conditions
Z = compressibility factor at the specified pressure and temperature,

Depending on gas/liquid ratio and the amount of entrainment of particulates, it is a standard practice to limit the fluid velocity about 5 m/s. However, in two-phase flow lines minimization of slugging in pipeline and separation equipment set the maximum velocity. In this case the velocity is set to about 10 ft/sec (3 m/sec), ANSI/API RP-14E [20]. This is particularly important in long lines with elevation changes.

A3.3 Minor Pressure Losses

A liquid pipeline system is composed of pipe, valves, flanges and fittings, and includes facilities such as pump stations, meter stations, tanks, and pig launchers and traps. A heavy crude pipeline may require heaters, while a pipeline in a permafrost zone needs chillers.

All pipeline components and facilities cause pressure losses as the liquid passes through them. The facilities include valves, joints and fittings, pipe bends and elbows, pipe enlargers or contractors in addition to pipes. In most long-distance pipelines, the pressure losses due to valves and joints are comparatively small, so such pressure losses are called minor pressure losses. Since the number of components and facilities is small, the overall minor pressure losses are so small that they can be approximated without significant errors. On the other hand, the pressure loss due to these components can be substantial within facilities such as stations and tank farms.

This addendum discusses the calculation of minor pressure losses that would occur when liquid flows through some valves, pipe and fittings. Depending on the components, minor losses can be calculated in two ways:

- Minor losses are expressed as a kinetic energy term in terms of the pressure loss coefficient, *K*:

$$\Delta P_{\mathrm{m}} = K\Delta V^2/2 \qquad \text{(A3 – 14)}$$

- Minor losses in terms of an equivalent length of straight pipe, L_e:

$$P_{\mathrm{m}} = \mathrm{f}\rho V^2 L_e/(2\mathrm{D}) \qquad \text{(A3 – 15)}$$

where

ΔP_{m} = Minor pressure loss
K = pressure loss coefficient (dimensionless).
V = velocity of liquid through valve, joint or fitting
ρ = liquid density
f = friction factor
D = pipe inside diameter
L_e = equivalent length of straight pipe

The value of *K* is determined mainly by the flow geometry or the shape of the device. *K* is analogous to the term fL/D for a straight length in the Darcy equation.

It is more convenient to use *K*, because *K* is less dependent on the Reynolds number and relative roughness. *K* and L_e values are obtained from Reference [21]. Some representative values of *K* and L_e are shown in Table A3-1. Since actual values depend on the design and fabrication of the components, actual manufacturer's data shall be consulted in the final design.

Example of Equivalent Length Method: A piping system of a pump station is 140 m of NPS 24 pipe that has two 24″ gate valves, three 24″ ball valves, on swing check valve, and two 90° standard elbows. Using the equivalent length concept, calculate the total pipe length of the station.

Solution:

Convert all valves and fittings in terms of 24″ pipe as follows:

- Two 24″ gate valves = 2 × 24″ × 8 × 0.0254 = 9.75 m
- Three 24″ ball valves = 3 × 24″ × 3 × 0.0254 = 5.49 m
- One 24″ swing check valve = 1 × 24″ × 100 × 0.0254 = 60.96 m
- Two 90° elbows = 2 × 24″ × 30 × 0.0254 = 36.58 m

TABLE A3-1. Loss coefficients and equivalent lengths

Types	Loss coefficient (K)	Types	Equivalent length (L_e/D)
Pipe entrance:		Valves (fully open):	(*)
• Sharp edged	0.5	• Gate valve	8
• Well-rounded	0.03	• Globe valve	340
Pipe exit:		• Plug valve	18
• Sharp edged	1.0	• Angle valve	150
• Well-rounded	1.0	• Ball valve	3
90° Bends:		• Globe lift check valve	600
• Smooth threaded	0.9	• Angle lift check valve	55
• Smooth flanged	0.3	Standard elbow:	
• Miter bend	1.1	• 45°	16
• Miter bend with vanes	0.2	• 90°	30
Tee Branch flow:		Standard tee:	
• Flanged	1.0	• Flow through run	20
• Threaded	2.0	• Flow through branch	60
Tee line flow:			
• Flanged	0.2		
• Threaded	0.9		

(*) *K* values for valves vary with their type and size.

- Total equivalent length of straight pipe and all fittings = 140 m + 112.78 m = 252.78 m
- The pressure drop due to friction is calculated based on 252.78 m of pipe.

Gradual Enlargement and Reduction

Consider liquid flowing through a pipe of diameter D_1. If the diameter enlarges or reduces to D_2, the pressure loss can be calculated as follows:

$$\Delta P_m = K\rho(v_1 - v_2)^2/2 = K\rho(A_2/A_1 - 1)^2\, v_2^2/2 \qquad (A3-16)$$

where v_1 and v_2 are the velocity of the liquid in D_1 and D_2 pipes and A_1 and A_2 the areas. The value of *K* depends on the diameter ratio and the different angle due to the enlargement or reduction. The figure below shows a gradual pipe enlarger and a reducer.

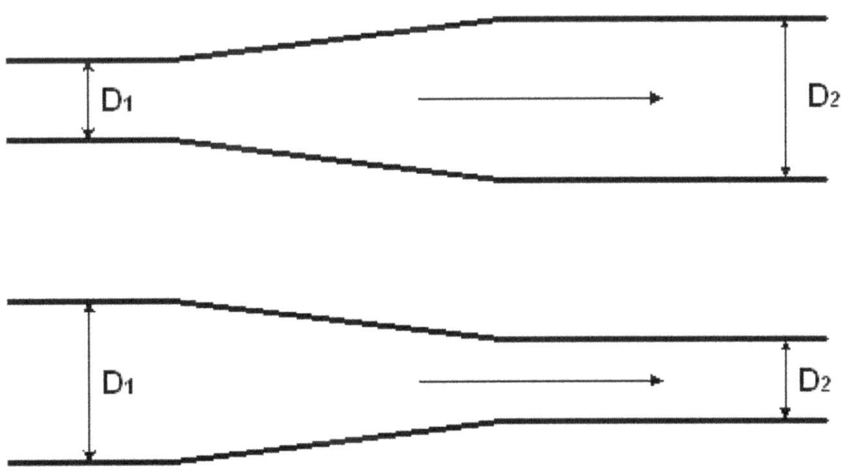

Figure A3-5. Gradual pipe enlargement and reduction

The loss coefficients for an enlarger are given in Tables A3-2 and A3-3 below:

TABLE A3-2. Loss coefficients for enlarger

Angle	20°	45°	60°	90°	180°
$A_2/A_1 = 2.25$	0.45	1.06	1.19	1.11	1.00
$A_2/A_1 = 9$	0.43	0.88	1.02	1.04	1.02

The loss coefficients for a reducer are given in the table below:

TABLE A3-3. Loss coefficients for reducer

Angle	20°	45°	60°	90°	180°
$A_2/A_1 = 0.5$	0.05	0.06	0.06	0.12	0.26
$A_2/A_1 = 0.25$	0.03	0.07	0.07	0.17	0.41

Enlarger Example: Calculate the pressure loss due to a gradual enlargement in a pipe that flows 1800 m^3/hr of diesel from 8″ diameter to 12″ with an angle of 60°. Both sizes are internal diameters.

Solution:

Solution: The liquid velocity in the 12″ pipe size is v_2 = 1.714 m/sec, and diameter ratio = 12/8 = 1.5. From the diagram, the value of K is 1.19 for area ratio = 2.25 and angle = 60°. Therefore, pressure loss due to gradual enlargement is

$$\Delta P_m = 1.19 \times 850 \times (2.25 - 1)^2 \times 1.714^2/2 = 6037 \text{ Pa} = 6.037 \text{ kPa}$$

Sudden Expansion: Minor loss for the sudden expansion of a pipe can be expressed as

$$\Delta P_m = \rho(1 - D_1^2/D_2^2)^2\, V_1^2/2 \qquad \text{(A3 – 17)}$$

Sudden Contraction: Minor loss for the sudden contraction of a pipe can be expressed as

$$\Delta P_m = \rho(1/C_c - 1)^2\, V_2^2/2 \qquad \text{(A3 – 18)}$$

where C_c is given in Table A3-4 below.

TABLE A3-4. Sudden contraction parameters

A_2/A_1	0.0	0.2	0.4	0.6	0.8	1.0
C_c	0.585	0.632	0.659	0.712	0.813	1.0

Example: A tank open to the atmosphere is filled with 40°API (specific gravity of 0.825) oil to a height of 10 m from the bottom. A tap at the bottom of the tank is opened, and oil flows out from the outlet. Assuming that the flow is steady and incompressible, determine the tank discharge pressure at point 2.

Solution:

Assume that as the oil is discharged out of the tank, the tank level drops so slowly that the flow velocity at point 1 is negligibly small. The data for this example are:

- Pipe = 16″ OD × 0.25 wt: inside diameter = 0.394 m
- Pipe roughness = 0.0018″ = 0.0457 mm
- Crude viscosity = 3 cSt of 40°API crude

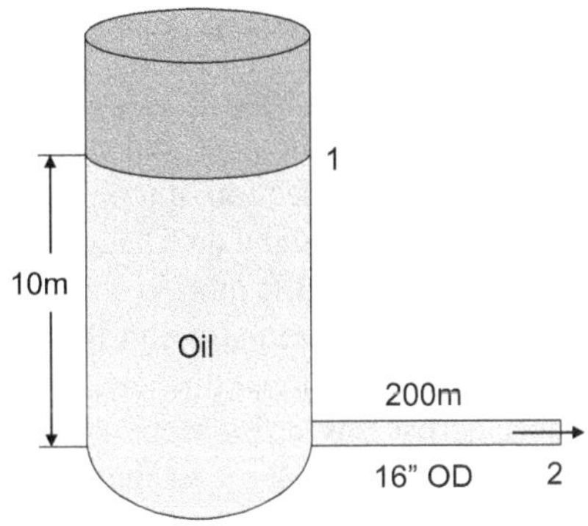

Figure A3-6. Example of Oil Discharge from a Tank

- Density = 825 kg/m^3
- Flow rate = 1000 m^3/hour or velocity = 2.28 m/sec

Step 1. From the momentum conservation in steady state, we have

$$(P_1 + \rho v_1^2/2 + \rho g z_1) - (P_2 + \rho v_2^2/2 + \rho g z_2) = f\rho v_2^2/2 \times L + K\rho v_2^2/2$$

where f is the Darcy friction factor and L the pipe length, and $K = K_{entrance} + K_{exit}$ are the pressure loss coefficients for pipe entrance and exit, respectively.

Step 2. P_1 is atmospheric pressures, $v_1 = 0$, $v_2 = v$, $z_1 = 10$ m, and $z_2 = 0$. Therefore, the equation becomes

$$P_2 = \rho g z_1 - \rho v^2/2 - f\rho v^2/2 \times L - (K_{entrance} + K_{exit})\rho v^2/2$$

where $K_{entrance} = 0.5$ and $K_{exit} = 1.0$.

Step 3. Calculate friction factor and pressure drop for the 200 m long 16″ pipe

- $Re = 2.28 \times 0.394/0.000003 = 299{,}000$: partially turbulent
- Relative roughness = 0.0000457 m/0.394 m = 0.000116
- $f = 0.0156$
- P_f/L = 0.0156 × 825 kg/m^3 (2.28 m/sec)^2/(2 × 0.394 m) = 84.9 Pa/m = 0.0849 kPa/m
- Total pressure drop for 200 m long pipe = 0.0849 × 200 = 17.0 kPa

Step 4. Determine the tank discharge pressure

- Static pressure at the tank outlet, $\rho g z_1 = 825 \times 9.8 \times 10/1000 = 80.9$ kPa
- Velocity pressure, $\rho v^2/2 = 825 \times (2.28)^2/2 = 2144$ pa = 2.14 kPa
- Pressure losses due to entrance and exit, $(K_{entrance} + K_{exit})\rho v^2/2 = 1.5 \times 2.14 =$ 3.21
- Pressure at the pipe discharge point = 80.9 – 2.14 - 17.0 – 3.21 = 58.6 kPa

Step 5. Discuss the result of this example

The pressure losses due to pipe entrance and exit can be relatively significant for low pressure system, whereas they are negligibly small for large pressure system such as a transmission pipeline.

A.3.4 Effect of Pressure and Temperature on Pipe Volume

When a pipeline transporting hydrocarbon liquids experiences a change in pressure and temperature, both the pipe and the liquid contained therein are affected. For a given segment or cross section of a pipe (Figure A3-7), the change in pressure can be due operational variables and the change in temperature due to environmental/ambient conditions (surrounding the pipe) at a given time.

Considering a uniform stress in the pipe wall thickness and assuming that variations in pipe cross sectional diameter are negligible and that there is no variation in temperature across any cross section and as well pipe ends are considered as fully closed, so the internal volume of the pipe is equal to the volume of contained fluid, then the following illustrate the relationship between pressure, temperature and volume in a section of a pipeline filled with a hydrocarbon liquid [9].

As indicated previously both the pipe and fluid are affected by changes in pressure (δP) and temperature (δT). Therefore, any incremental increase in volume of the pipe (δV_{pipe}) must be related to change in liquid volume (δV_{fluid}) caused by change in temperature or pressure. Thus,

$$\delta V_{\text{pipe}} = \delta V_{\text{fluid}} \qquad \text{(A3 – 19)}$$

By rearranging the expressions for both the bulk modulus and the coefficient of volume thermal expansion of the fluid, it can be written that

$$\delta V_{\text{fluid}} = BV\delta T - V\delta P / K \qquad \text{(A3 – 20)}$$

where

B = coefficient of volumetric thermal expansion of the fluid.
K = bulk modulus of fluid

The determination of the incremental increase in volume of the pipe (δV_{pipe}) is dependent upon the degree to which the pipeline is free to move. Consider first the case where a welded pipeline has unrestrained movement such that it is free to expand in both the radial and longitudinal directions.

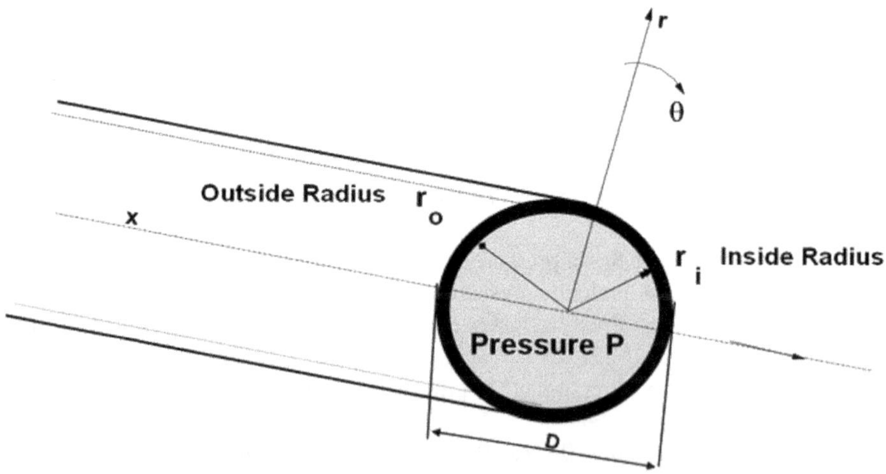

Figure A3-7. Generalized view of a liquid pipeline under internal pressure

The incremental volume increase in the pipe will be due to the effects of both pressure and temperature. In the case of pressure change, the pipe wall experiences a state of biaxial stress, a circumferential or hoop stress S_θ, and an axial stress S_x. The respective hoop and axial strains are:

$$\varepsilon_\theta = \left(\frac{S_\theta - \nu S_x}{E}\right) = \frac{D\delta P}{4tE}(2-\nu) \tag{A3 – 21}$$

and

$$\varepsilon_x = \left(\frac{S_x - \nu S_\theta}{E}\right) = \frac{D\delta P}{4tE}(1-2\nu) \tag{A3 – 22}$$

where

ν = Poisson's ratio
ε_θ = Hoop strain (i.e., strain in radial direction)
ε_x = Axial strain (i.e., strain along the length of the pipeline)
S_x = Axial stress = $D\delta P/4t$
S_θ = Hoop stress = $D\delta P/2t$
E = Modulus of elasticity
t = Pipe wall thickness

and the new pipe length L^* and diameter D^* become

$$L^* = L(1 + \varepsilon_x) \tag{A3 – 23}$$

$$D^* = D(1 + \varepsilon_\theta) \tag{A3 – 24}$$

The new volume of the pipe section is given by $\pi L^*(D^*)^2/4$. Upon substitution of the expressions for D^* and L^*, and ignoring the cross products of strain as being negligible, the new pipe volume V^* is given by:

$$V^* = \frac{\pi D^2 L}{4}(1 + 2\varepsilon_\theta + \varepsilon_x) \tag{A3 – 25}$$

The incremental increase in volume, δV, due to incremental change in pressure, δP, can be found by substituting the expressions for ε_θ and ε_x into Eq. (A3 – 25), and noting that $\delta V = V^* - V$, where V is the original volume of the pipe. Hence, for fluid and pipe:

$$\delta V_{pipe} = \frac{VD\delta P}{4tE}(5-4\nu) \tag{A3 – 26}$$

The incremental volume change due to a temperature change δT is calculated using

$$\delta D = D\alpha\delta T \tag{A3 – 27}$$

and

$$\delta L = L\alpha\delta T \tag{A3 – 28}$$

where α is the coefficient of linear expansion of pipe material, and δD and δL are changes in diameter D and length L, respectively. By neglecting the cross products of the coefficient of thermal expansion when evaluating the new volume, the incremental volume is:

$$\delta V = 3\alpha V \delta T \qquad (A3-29)$$

The combined effect of incremental pressure δP and temperature δT change can be found by substituting Eqs. (A3 – 20), (A3 – 26) and (A3 – 29) into Eq. (A3 – 19):

$$V\left(\frac{D\delta P}{4tE}(5-4\nu)+3\alpha\delta T\right)=V\left(B\delta T-\frac{\delta P}{K}\right) \qquad (A3-30)$$

Rearranging the terms and eliminating V:

$$\delta P = \frac{(B-3\alpha)\delta T}{\frac{1}{K}+\frac{D(5-4\nu)}{4tE}} \qquad (A-31)$$

Equation (A3 – 31) provides the change in pressure due to a temperature change for an unconstrained pipe with closed ends. It may be noted that, in this circumstance, it is independent of pipe length. Similar expression can be derived for buried pipeline which are considered as restrained. Refer to ref. [9].

For a buried pipeline change in pressure δP caused by change in temperature δT and thence the change in volume δV due to the change in pressure can be computed from the following expressions:

$$\delta P = \frac{\left[B-2\alpha(1+\nu)\right]\delta T}{D\left(1-\nu^2\right)/tE+1/K} \qquad (A3-32)$$

The effect of axial thermal stress on restrained pipelines capped at the ends can be neglected [22]. Ignoring the temperature increase effects simplifies Equation (A3.4.14) such that, for a restrained pipe, the theoretical volume required to pressure (squeeze) the pipeline section is given by:

$$\delta V = V\delta P\left(\frac{D}{tE}\left(1-\nu^2\right)+\frac{1}{K}\right) \qquad (A3-33)$$

Example:

To illustrate the effect of a change of temperature on pressure consider a pipe of outside diameter 114.3 mm and wall thickness 3.18 mm transporting water under pressure. Neglect the effect of axial thermal stress on a restrained pipeline Eq. (A3 – 32) can be simplified to become [22]:

$$\delta P = \frac{B-2a}{\frac{D}{Et}\left(1-\nu^2\right)+C}$$

where $a = 1.17 \times 10^{-5}$

$B = (-64.268 + 17.105\,T - 0.20396\,T2 + 0.016048\,T3) \times 10^{-6}$

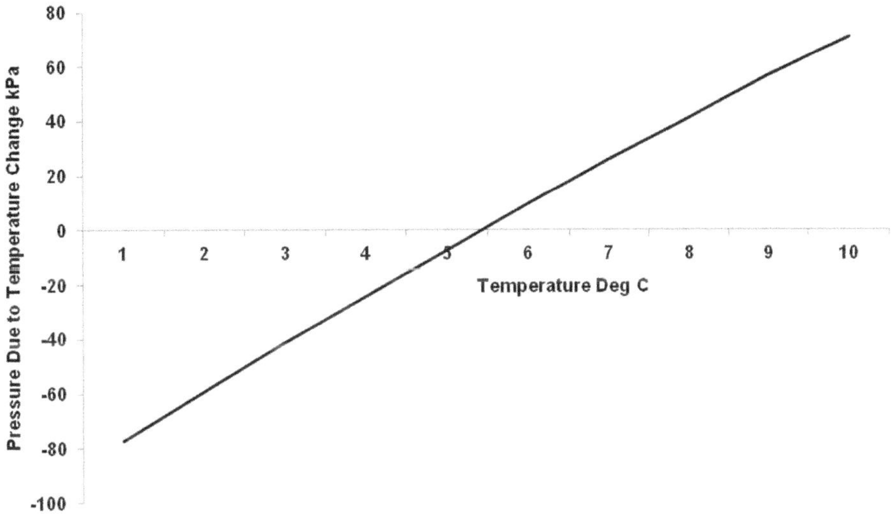

Figure A3-8. Pressure change due to temperature change (medium water in 48″ pipeline, no flow condition)

$E = 2.0 \times 10^8$
$\nu = 0.3$
$C = 1/K = 4.86 \times 10^{-7}$ (K= 2.057×10^6 kPa assumed for water at 10°C)

Note that C (the reciprocal of the bulk modulus for liquid, in this case water) is a function of the pressure and temperature. Assuming the temperature T is 10°C then upon substituting, B has the value 81.0728×10^{-6} and the pressure increase due to a 1°C temperature change is:

$$\delta p = \frac{81.0728x10^{-6} - 2\times 1.17\times 10^{-5}}{\dfrac{114.3}{200\times 10^{6}\times 12.7}(1-0.3^{2}) + 4.86\times 10^{-7}}$$

That is a pressure increase of 98.9 kPa.

In a similar manner, the change in pressure due to a temperature increase over a range of temperatures for a large restrained transmission pipeline whose outside diameter is 1219 mm (48″), having a wall thickness of 12.7 mm (1/2″) and containing water, can be calculated. The resulting pressure change due to temperature change is shown in Figure A3-8.

If it is assumed that water temperature increases from 1 to 3°C during a period of time then the end result is a pressure decrease of 100 kPa would be obtained by summing the pressure decreases in each time interval, that is 59 kPa between 1 and 2°C, and 41 kPa between 2 and 3°C. It may be noted that had the temperature decreased from 3 to 1°C the pressure would have increased by 100 kPa.

REFERENCES

[1] American Petroleum Institute, 1984, "Petroleum Liquid Volume Correction," API Publication 1101.

[2] NIST Standard Reference Database 4, Supertrap version 3, National Institute of Standards and Technology, Gaithersburg, MD 20899, U.S.A.

[3] Finlayson, B. A., 1980, *Nonlinear Analysis in Chemical Engineering*, McGraw-Hill, New York, N.Y.
[4] Wylie, E. B., and Streeter, V. L., 1983, *Fluid Transients*, FEB Press, Ann Arbor, MI.
[5] Carnahan, B., Luther, H. A., and Wilkes, J. O., 1969, *Applied Numerical Methods*, John Wiley & Sons, Inc. New York, N.Y.
[6] Wylie, E. B., Stoner, M. A., Streeter, V. L., 1971. "Network System Transient Calculations by Implicit Model," *Soc. Pet. Eng. J.*
[7] Rachford, H. H., and Dupont, T., 1974, "A Fast, Highly Accurate Means of Modeling Transient Flow in Gas Pipeline Systems by Variational Methods," *Soc. Pet. Eng. J.*
[8] Meyer, K. J., Smith, R. W., Murray, A., 2008, "Comparison of US and Canadian Transmission Pipeline Consensus Standards," Proc. of IPC 2008, Sep. 29–Oct. 3, pp. 1–7.
[9] Mohitpour, M., Golshan, H., Murray, A., 2007, *Pipeline Design and Construction – A Practical Approach*, 3rd Ed. ASME Press, New York, NY, USA.
[10] Adam, S., and Davis, K., 2009, *Pipeline Geomatics*, ASME, New York, NY.
[11] 1998, *Engineering Data Book (SI Version)*, 11th Edition, Gas Processors Suppliers Association (GPSA), Tulsa, OK, USA.
[12] Gregory, G. A., Aziz, K., and Moore, R. G., 1979, "Computer Design of Dense-Phase Pipelines," *J. Petroleum Technology*.
[13] Urquhart, R. D., 1986, "Heavy Oil Transportation – Present and Future," *J. Can. Pet. Eng.*, pp. 68–71.
[14] Arnold, C. L., 1981, "How Temperature Affects Pipeline Hydraulics Design for Heavy Crudes, Resid," *Oil & Gas J.*, pp. 104–120.
[15] Smith, B., 1979, "Pumping Heavy Crudes – 1: Guidelines set out for pumping heavy crudes," Oil & Gas J., pp 111 – 114, May 28, 1979, "Pumping Heavy Crudes – 2: Steps for finding crude properties," Oil & Gas J., pp 150 – 152, June 4, 1979, "Pumping Heavy Crudes – 3: Heat transfer explored in pipelining high-pour-point crude oil," Oil & Gas J., pp 110 – 111, June 18, 1979, "Pumping Heavy Crudes – 4: Restart of heavy crude lines probed," Oil & Gas J., pp 105 – 106, July 2, 1979, "Pumping Heavy Crudes – 5: Design of heavy crude facilities explored," Oil & Gas J., pp 69 – 70, July 16, 1979.
[16] Rahimi, P., et al., 2009, *5th National Centre for Upgrading Technology (NCUT) Upgrading and Refining Conference*, Sep. 14–16.
[17] Mohitpour, M., 1991, "Temperature Computation in Fluid Transmission Pipelines," *Proc. of ETCE, Pipeline Engineering Symposium*, ASME, Houston, TX, PD-Vol. 34.
[18] Holman, J. P., 1981, *Heat Transfer*, 5th Ed, McGraw Hill Book Co., New York, NY, USA.
[19] Edminster, W. C., 1972, *Applied Hydrocarbon Thermodynamics*, Gulf Publishing Co., Houston, TX, USA. Vol. 2.
[20] ANSI/API RP 14E 2007, "Recommended Practice for Design and Installation of Offshore Products Platform Piping Systems".
[21] Fox, R., McDonald, A., and Pritchard, P., 2004, *Introduction to Fluid Mechanics*, 6th Edition, John Wiley & Sons, Inc. NJ., USA.
[22] Gray, J. C., 1976, "How Temperature Affects Pipeline Hydrostatic Testing," *Pipeline and Gas J.*, pp. 26–30, Aug.

CHAPTER 4

PUMPS AND PUMP STATIONS

4.1 INTRODUCTION

The main objective of liquid pipeline operations is to transport liquid petroleum products from the producers to the customers. In order to achieve this objective, energy is added to the products to increase the pressure at the pump stations for offsetting the pressure loss in the pipeline. In addition, measurements of pressure and flow are required for facility control and custody transfer. There are other tasks required to operate pipeline systems. This chapter discusses such key subjects as pump selection and sizing, pump operating points, pump station design, and station control.

A pump transforms energy to increase pressure of a liquid and is used extensively to transport liquid through a pipeline system. The pressure of a liquid has to be increased either to overcome frictional losses or to raise the liquid from one elevation to a higher elevation. As the flow rate increases, more pumps are required to produce the required pressure along the pipeline (Figure 4-1).

Depending on the method of adding energy to the liquid, pumps are classified into two types; centrifugal pumps and positive displacement (PD) pumps. Centrifugal pumps add kinetic energy to the liquid by increasing the liquid flow velocity, while PD pumps add energy periodically to the liquid by the direct application of a force to movable volumes of liquid. The two types of pumps can be compared in general terms, as listed in Table 4-1.

As shown in this table, centrifugal pumps are most suitable for transmission pipelines transporting most petroleum products. Therefore, they are extensively used in liquid transmission pipelines and thus this book only discusses the design and operation of centrifugal pumps and pump stations.

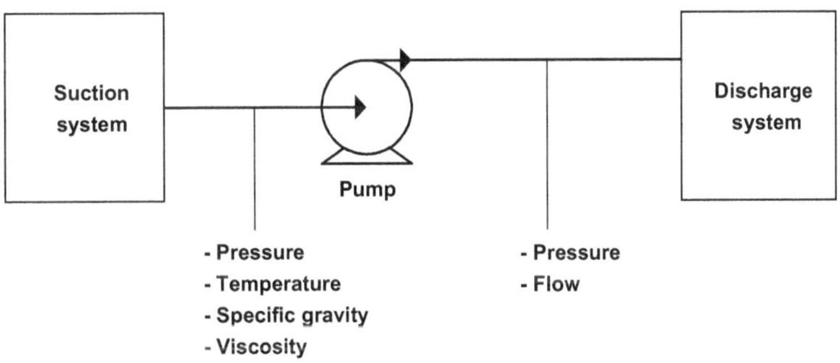

Figure 4-1. Pump suction and discharge systems

TABLE 4-1. Centrifugal pumps vs. positive displacement pumps

	Centrifugal Pump	PD Pump
Operating flow range	Flexible	Limited
Operating pressure range	Low–medium	High
Control of pressure and flow	Flexible	Limited
Pumping efficiency	Low–medium	High
Viscosity range	Low–medium	High
Capital cost	Low–medium	High
Maintenance requirement	Low	High
Physical size	Small	Large

4.2 CENTRIFUGAL PUMPS

Centrifugal pumps have prevalent application in liquid hydrocarbon pipeline transmission systems as they are capable of handling variable heads and flow rates [1]. These pumps can handle multiple products over a wide range of viscosities and other properties [2].

Figure 4-2 shows the cross-sectional and axial views of a centrifugal pump with an end suction impeller. The main components of a centrifugal pump include an impeller, casing, housing or frame, and shaft and stuffing box/mechanical seal. A centrifugal pump uses the centrifugal action through the impeller within the pump to transfer energy. As the pump shaft is rotated by a pump driver, the impeller rotates inside the pump casing. Liquid flows from the suction piping into the impeller through its eye, and the rotating impeller imparts kinetic energy to the liquid. As the liquid slows down while passing through the volute, the liquid kinetic energy is converted into potential energy or pressure to conserve the total energy.

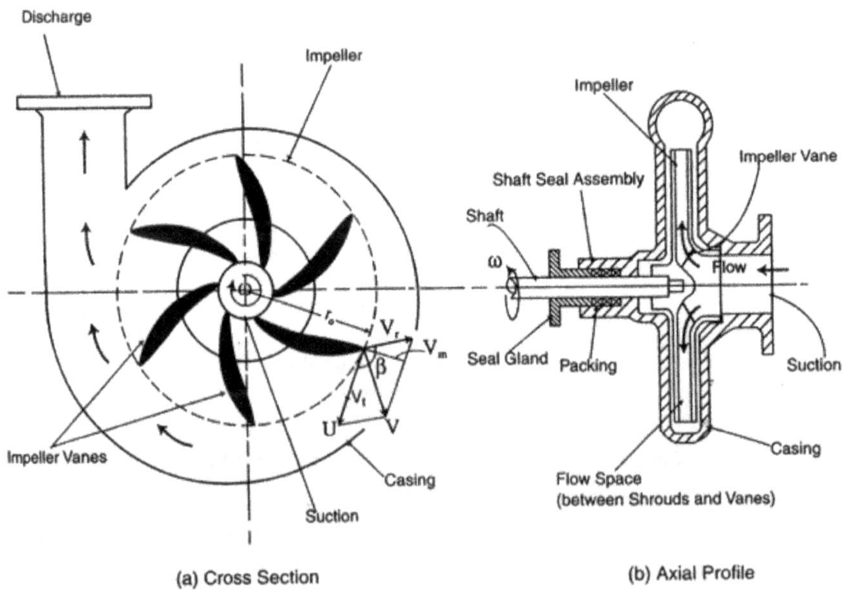

Figure 4-2. Centrifugal pump with impeller

4.3 CENTRIFUGAL PUMP TYPES

Mainline centrifugal pumps are usually designed to ANSI/API 610 or ISO 13709 (Identical Standards) – Centrifugal pumps for petroleum, petrochemical and natural gas industries.

Pump types typically used in liquid hydrocarbon process, refining and pipeline transportation include:

4.3.1 End Suction Single Stage Pumps

These pumps have limited application in pipeline systems but are mostly used in process industries (Figure 4-3).

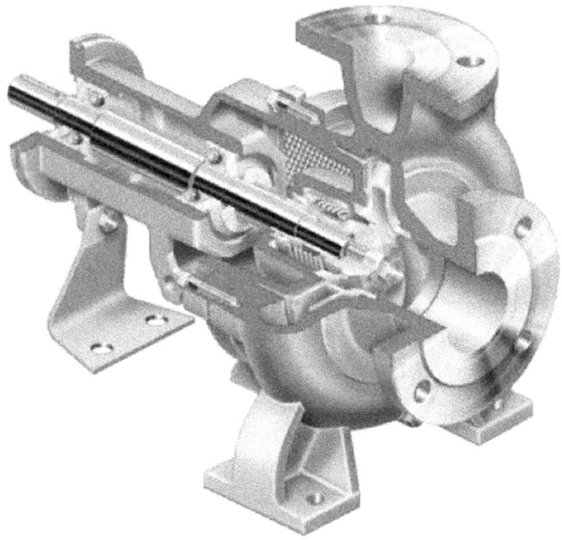

Figure 4-3. API 610 end suction pump — Courtesy of Flowserve Corporation, all rights reserved.

4.3.2 Vertical In-Line Single Stage Pumps

These pumps are typically used for product transfer within terminals or as booster pumps for smaller mainline pumps at initiating stations. API 610/ISO 13709 versions are used extensively in petrochemical and refinery service (Figure 4-4).

4.3.3 Horizontal Axially Split Between-Bearing Single-Stage Pumps

These pumps are typically used as mainline pumps. These high volume pumps are often piped in series configuration to provide the high pressures required for pipeline transmission lines. These double suction, double volute pumps provide high efficiency over a large range of flow. They are inherently balanced with minimal axial thrust issues at flows well off the best efficiency point due to the double volute design (Figure 4-5).

4.3.4 Horizontal Axially Split Between-Bearing Multi-Stage Pumps

These pumps have application where higher pressures and lower volume capacity than the single state pumps are required. This pump design, along with the single stage

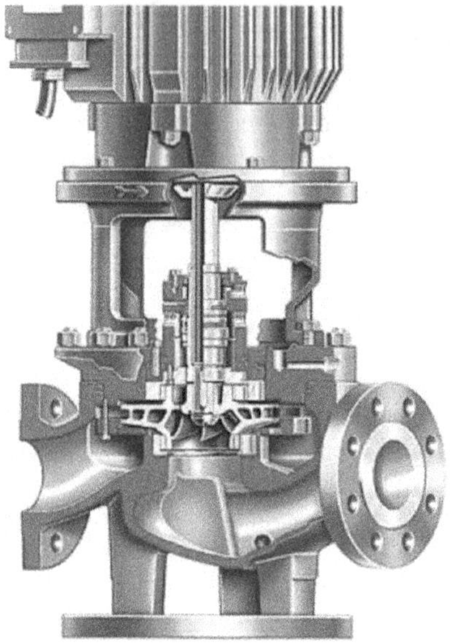

Figure 4-4. Vertical in-line. Courtesy of Flowserve Corporation, all rights reserved.

version, allows for ease of maintenance as the pump rotating element can be serviced or removed for impeller modifications without disturbing the station pressure piping. These pumps are robust in design and provide long service life. API 610/ISO 13709 is the guiding standard for design, manufacturing and performance testing requirements of these pumps. API 610/ISO 13709 designates these pumps as BB1 — axially split single stage between-bearing pumps and BB3 — axially split multistage between-bearing pumps (Figure 4-6).

4.3.5 Double–Case (Can) Vertically Suspended Volute Pumps

These pumps are used where there is limited suction pressure available such as in tankage terminals and where higher viscosity product is transported. Tank farm de-

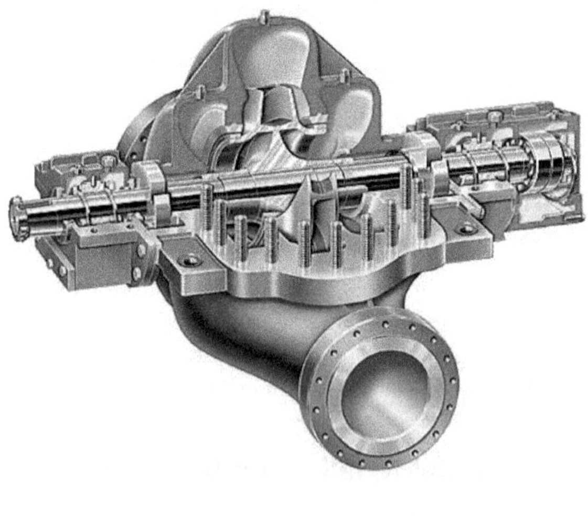

Figure 4-5. Horizontal axially split BB single stage pump. Courtesy of Flowserve Corporation, all rights reserved.

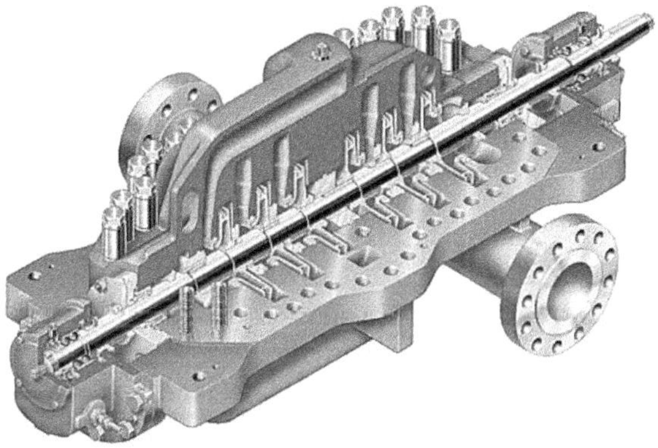

Figure 4-6. Multi-stage horizontal axially split pump. Courtesy of Flowserve Corporation, all rights reserved.

signs usually provide manifolds located near the tanks and incorporate booster pumps with low NPSH requirements. These pumps are typically can-type vertical multi-stage centrifugal pumps that accommodate the low suction head available from the tankage. These booster pumps are designed to provide sufficient pressure to overcome frictional losses of valves, piping, and fittings throughout the station and to meet the NPSH requirements of the mainline pumps (Figure 4-7).

Impeller hydraulics can be optimized for individual system requirements with these pumps. Head rise from design point to shutoff can be as low as 15% over design

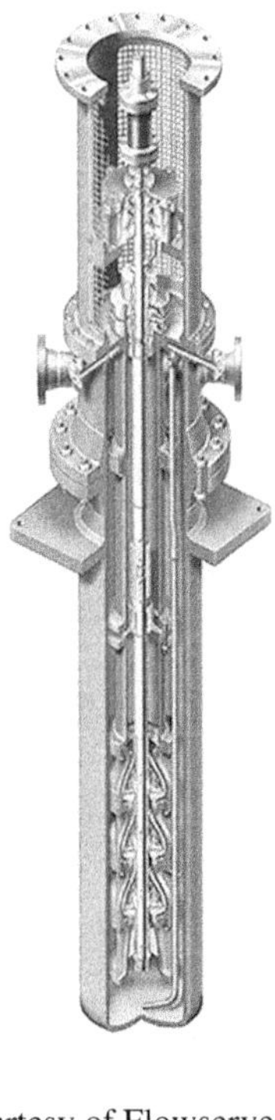

Figure 4-7. Vertical Can pump. Courtesy of Flowserve

head. This can result in considerable energy savings particularly under partial loads or off-peak operating conditions.

The can length will determine the NPSH requirements for the pump and is specific to the particular application. API 610/ISO 13709 designates this type of pump as VS-7 — Double casing volute vertically suspended pump.

4.4 PUMP SELECTION AND SIZING

The pump performance is the basis of pump selection. To select the pump, it is most critical to determine the operating range of the pump and the system curve in the pipeline system. The actual procedure to be followed is:

- Determine the required flow range of pump: It is important to define the maximum, normal and minimum flow requirements. The normal flow is the flow at which the pump or pumps will usually operate. The rated flow is likely to be the maximum flow under current or near term conditions, whereas the minimum flow limit should be established to accommodate temperature rise and low efficiency or to install recirculation facility or multiple pumps. It is also important to clarify the number of days per year of service at which the pump will operate at maximum, normal and minimum flow rates.
- Determine the system curves for the maximum, normal, and minimum flow rates so that the operating envelope of one or more pumps is established. It is preferred to express the system curves in terms of head because the pump curve is expressed in head.
- Select pump type, size, and arrangements, which are interrelated. For example, two small pumps are arranged in parallel for wide range of flows between the minimum and maximum flows.
- Prepare pump data sheets showing pump performance requirements and service conditions.
- Solicit for bids and make final selection

4.4.1 Pump Performance

Normally, centrifugal pumps are characterized by efficient performance over a wide range of pressures and flow rates. Their size is relatively small. They cost less than other types of pumps and operate reliably. In addition, centrifugal pumps are capable of pumping high throughput and various products with different liquid densities and viscosities. They can handle products with viscosities up to 350 cSt depending on pump size and speed before efficiency begins to fall off significantly.

Pump characteristics are represented graphically to describe pump performance. The pump performance curves are normally provided by the manufacturer. The pump manufacturer tests the performance with water and only guarantees the performance at the rated point. Therefore, the pump performance curve, including the shut-off pressure, can vary and this variation must be specified or tested.

Pump performance curves show graphically the relationship of flow rate with head, efficiency, net positive suction head (NPSH) required, and power requirement for several impeller diameters. If the driver connected to the pump can vary its speed, these pump curves are produced for different speeds. The information is used for pump selection.

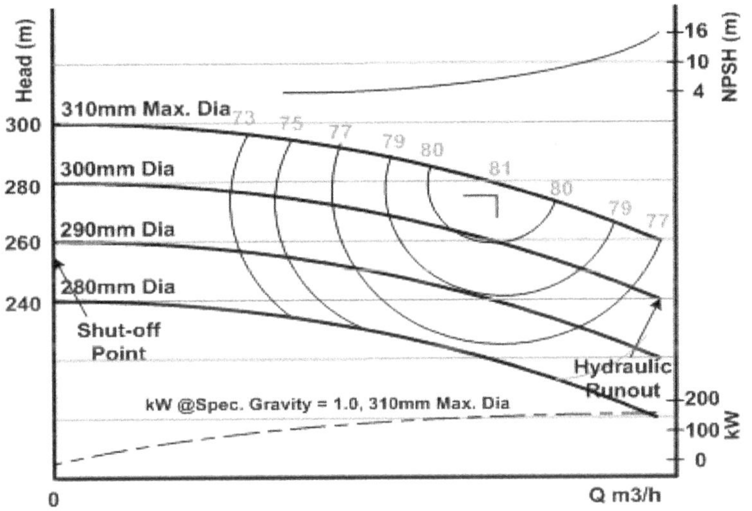

Figure 4-8. Typical pump performance curve

4.4.1.1 Pump Performance Curves

Pump manufacturers supply performance curves for their pumps with information on pump performance over a range of flows. The following diagram (Figure 4-8) demonstrates typical pump curves depicting pump head delivered by the impeller diameter chosen over the recommended flow rate for that pump. The Best Efficiency Point (BEP) is shown and usually forms the basis of the design flow rate. These pump curves typically provide the following information:

- Range of impeller diameters available
- Head vs. flow rate
- Pump efficiency vs. flow rate
- Brake power to drive the pump vs. flow rate
- Net Positive Suction Head Required (NPSHR) for the pump vs. flow rate

These curves are plotted for the rated speed and different impeller diameters. For variable speed drivers, the curves are shown at various pump speeds.

Some pumps are equipped with double suction impellers with the impeller eyes located on both sides. Pump shut-off is the head developed at zero flow, while hydraulic runout is the pump capacity above which the pump should not be operated due to instability and other operational problems. This point is usually defined at 120% of the best efficiency point (BEP).

4.4.2 Service Conditions

There are many factors that must be considered in the selection and sizing of pumps for a liquid hydrocarbon pipeline system. In order to select the proper pump, the following parameters should be known:

- Liquids to be pumped — clear liquids or liquids containing solids or vapor
- Liquid specific gravity/density
- Liquid vapor pressure at the pumping temperature

- Viscosity— pump performance drops more rapidly above 110 cSt
- System throughput — desired pump capacity and expected future changes to the desired capacity are functions of pump speed and size
- Pump head requirements — pump head depends on speed and impeller diameter
- Pressure conditions — suction and discharge pressures, future pressure condition, and series/parallel operation conditions
- Suction requirement — NPSHA must be greater than NPSHR of the pump
- Pump unit efficiency — energy usage is dictated by pumping efficiency
- Type of service — number of operating hours and criticality of the service
- Preferred pump and driver type
- Operability and maintainability
- Equipment life cycle cost including the initial purchasing and installation costs, operating cost, and maintenance cost
- Specific site conditions and space limitations: topography and elevation above sea level and NPSHA
- Other considerations such as codes and regulations

A rule of thumb of sizing and selecting centrifugal pumps is to choose the physically smallest pump that will satisfy the service requirements. Centrifugal pumps are sized on the following basis:

- Impeller diameter: The pumping head is proportional to the square of the impeller diameter, while the flow rate varies linearly with the diameter. Therefore, the larger the impeller diameter, the higher the head and the throughput. Normally, pump vendors provide a range of impeller diameter suitable for a pump. Impeller diameters are determined based on required head at design point. The pump manufacturer will then trim the impeller to the required diameter.
- Impeller speed: The head and flow varies in a similar manner to the impeller as described above. However, because of dynamic forces on the impeller, speed limits impeller size. The speed ranges from 1200 RPM up to 5500 RPM. Refer to Section 4.6.3 for the Affinity Laws.
- Suction pressure: The NPSHR of a pump is the limiting factor that affects size, speed, and capacity. This topic is discussed in Section 4.4.3.
- Suction and discharge nozzle sizes: Suction nozzles are usually larger than discharge nozzles. The larger the nozzle size, the higher the flow capacity of the pump. Nozzles sizes are determined by the pump manufacturer.

Pipeline system hydraulic requirements will determine the selection of pumps. If there are batching requirements with products of varying density and viscosity, it may be necessary to provide multiple pumps operating in series or parallel, or to use some combination of series/parallel configuration. Variable speed pumps can simplify the selection process. The pump configuration should be selected to maintain the maximum pump efficiency over the range of flows expected for the pipeline system. Initial and future flow rates should be evaluated so that the selected pumps have sufficient flexibility to handle any anticipated flow and product properties. For comparison purposes, pump efficiencies should be evaluated for both series and parallel pump configurations.

Once the pump configurations with head requirements at various flow rates have been determined, it will be necessary to select actual pumps from manufacturers' published pump performance data. Pump manufacturers publish pump performance maps that depict pump performance for a specific family of pumps over a large range of

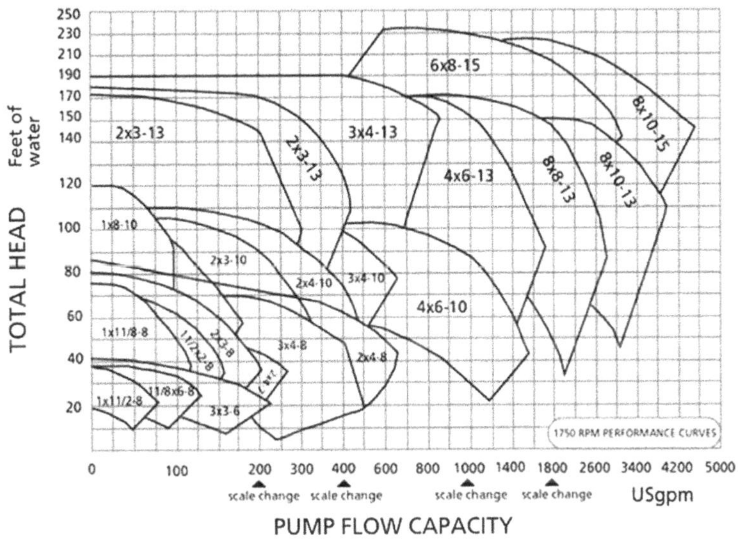

Figure 4-9. Pump selection map

flows and differential heads. Pump manufacturers produce charts to show the capacities of different sizes of pumps. Figure 4-9 shows a typical performance map for centrifugal pumps suitable for pipeline service.

4.4.3 Net Positive Suction Head (NPSH)

Net positive suction head (NPSH) is the total absolute suction pressure at the impeller eye less the absolute vapor pressure of the liquid pumped. NPSH must be of a magnitude to avoid vapor formation in the liquid and hence cavitation. Another way to describe NPSH is that it represents the amount of head required to push the fluid into the pump to suppress cavitation [3]. Available and required NPSH are calculated as follows:

4.4.3.1 Net Positive Suction Head Required (NPSHR)

The NPSHR specified in the pump performance curves is another important performance parameter. All centrifugal pumps accelerate fluid and possess a corresponding internal friction loss. Therefore, every pump requires a certain amount of positive suction pressure in order to avoid cavitation. The pump NPSHR is determined by actual tests conducted by the pump manufacturer using procedures established by the Hydraulic Institute.

Cavitation is a critical problem in controlling pumps. When a liquid enters into the pump impeller in such a manner that the local pressure drops below the liquid's vapor pressure, the liquid transforms into the vapor phase and bubbles are formed in and around the impeller. If the local pressure recovers due to centrifugal action, the vapor bubbles collapse into the liquid phase. Since the bubbles occupy much larger volume than the liquid, collapsing bubbles release a huge amount of energy. This energy will hit the surrounding metal, causing physical damage to the impeller and casing. The other effects of cavitation include vibration and noise and dramatic reduction in pump performance.

4.4.3.2 Net Positive Suction Head Available (NPSHA)

Net Positive Suction Head Available (NPSHA) is calculated by the following expression:

$$NPSHA = h_a - h_{VP} + h_{st} - h_{fs}$$

Or

$$NPSHA = h_{st} + (P_a - P_{VP})\frac{2.31}{G} - h_{fs}$$

where

P_a = absolute pressure at the surface of the liquid supply level
P_{VP} = vapor pressure of the liquid at the temperature being pumped
h_a = P_a expressed in equivalent head
h_{VP} = P_{VP} expressed in equivalent head
h_{st} = static elevation of the liquid supply above or below the pump inlet centerline.
h_{fs} = suction line losses including entrance losses and friction of the piping
G = specific gravity

Figure 4-10 shows that NPSHA must be greater than NPSHR for stable operation and that NPSHR increases and NPSHA decreases as the flow rate increases.

The NPSHR stated by the pump manufacturer is at a point where the pump is in full cavitation. Therefore, it is important to allow a margin between NPSHR and NPSHA. As a rule of thumb, the NPSHA should be at least 10% greater than the NPSH required by the pump [1].

If the requirement for stable operation cannot be satisfied, either NPSHR should be reduced, NPSHA increased, or both. The NPSHR reduction can be accomplished by using double suction impellers or by impeller design with a larger impeller eye area. Also, smaller

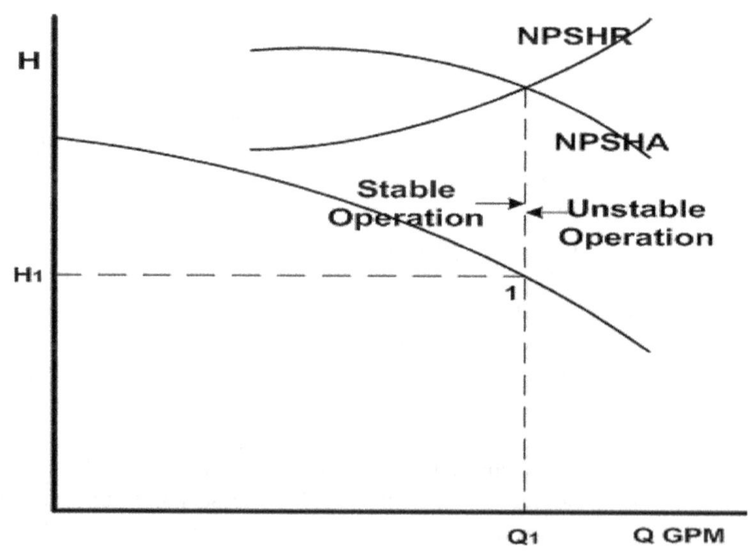

Figure 4-10. Stable operating point

pumps can be installed in parallel or a larger suction pipe size can be used to reduce frictional pressure losses in suction piping. NPSHA can be increased by installing a booster pump in front of the pump or by reducing the friction pressure losses in suction piping.

4.4.4 Specific Speed

Specific speed (N_S) is used to predict pump characteristics for the purpose of classifying pump impellers according to type, proportions and performance. Specific speed is often used for comparison purposes on selection of pumps from different manufacturers [3].

Specific speed is expressed as:

$$N_s = \frac{N\sqrt{Q}}{H^{\frac{3}{4}}}$$

where

N_s = pump specific speed, dimensionless
N = pump speed, RPM
Q = capacity at best efficiency point, USGPM
H = total head per stage at the best efficiency point, feet

For double suction impellers, one half of the flow is used to calculate the specific speed. The specific speed (N_s) determines the general shape or class of the impeller. As the specific speed increases, the ratio of the impeller outlet diameter, D_2, to the inlet or eye diameter, D_1, decreases. This ratio becomes 1.0 for a true axial flow impeller.

Radial flow impellers develop head mainly through centrifugal force. Pumps of higher specific speeds develop head partially by centrifugal force and partially by axial force. A pump with a higher specific speed generates head more by axial forces and less by centrifugal forces. An axial flow or propeller pump with a specific speed of 10,000 or greater generates its head exclusively through axial forces.

Typical values for specific speed (N_s) for different designs in US units (*gpm, ft*)

- radial flow — $500 < N_s < 4000$ typical for centrifugal impeller pumps with radial vanes — double and single suction. Francis vane impellers operate in the upper range
- mixed flow — $2000 < N_s < 8000$ — more typical for mixed impeller single suction pumps
- axial flow — $7000 < N_s < 20{,}000$ — typical for propellers and axial fans

To convert between US units (*USgpm*) and Metric units (m^3/h)

$$N_{s(US\ gpm,\ ft)} = 0.861\ N_{s(m^3/h,\ m)}$$

The specific speed of an impeller can provide a wide variety of information about its performance:

Impellers with low specific speed are long and thin and are used for low-flow, high-head applications. Impellers with high specific speed are short and stubby and are used for high-flow, low-head applications (Figure 4-11).

Efficiency is determined by considering the losses through pump impeller friction, ring leakage, and mechanical losses, as well as losses incurred by movement of the liquid within the pump, referred to as hydrodynamic losses. Specific speed affects pump efficiency. The lower the specific speed, the lower the efficiency. The reason is

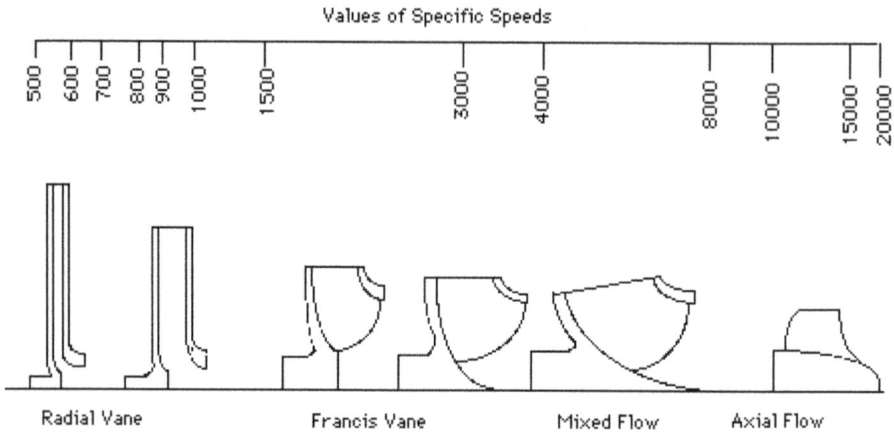

Figure 4-11. Impeller design vs. specific speed — Hydraulic Institute

that a higher percentage of energy is lost to overcome the impeller disk friction that is necessary to generate high heads (Figure 4-12).

Once an impeller is designed for a certain specific speed, it will produce a typical head capacity curve and efficiency curve shape. A low specific speed impeller has a flat curve with a wide efficiency range. A high specific speed impeller produces a steep curve with a narrow efficiency range.

The major use of the specific speed number is to help specify pumps to be as efficient as possible for the service intended. Maximum pump efficiency is obtained in the specific speed range of 2000 to 3000. Pumps for high head low capacity occupy the range 500 to 1000 while low head high capacity pumps may have a specific speed of 15,000 or larger.

4.4.5 Suction Specific Speed

Suction specific speed is an index that describes the characteristics of the suction side of the impeller. It is calculated at the pump's best efficiency point and maximum impeller diameter. The equation for suction specific speed (designated Nss or S) is:

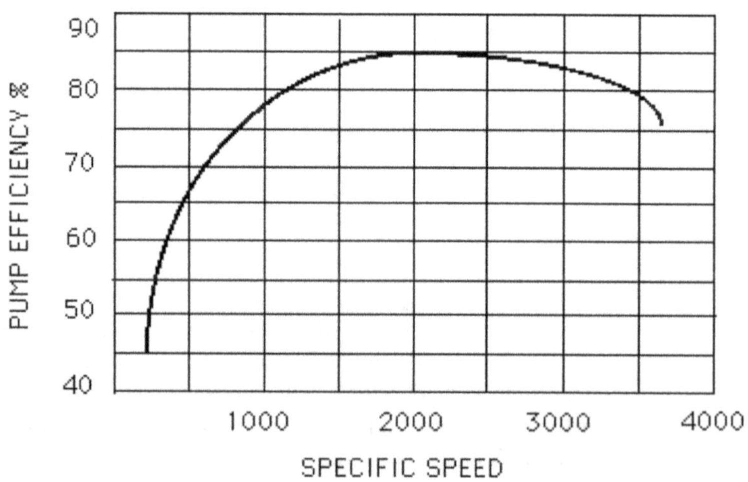

Figure 4-12. Pump efficiency vs. specific speed

$$Nss = \frac{N(Q)^{0.5}}{(NPSHR)^{0.75}}$$

where

N = rotating speed (rpm)
Q = flow per impeller eye (m^3/second)
$NPSHR$ = Net Positive Suction Head Required (see Section 4.4.3)

For double suction impellers, Q is one half of total flow. Nss derived using SI units can be converted to US Customary Units by multiplying by a factor of 51.64 (Ref. API 610)

From the equation, we can see that the lower the NPSHR for a pump the higher the Nss. Nss values for many standard impellers typically range from 7000 to 9000, but some designs may have an Nss as high as 18,000 to 20,000.

It is important to consider that increasing the Nss of a pump has been shown to shift the onset of suction or discharge recirculation closer to the best efficiency point (BEP) flow of the pump. This effectively decreases the window of stable operation for the pump. Suction recirculation is a reversal of flow in the impeller eye that can lead to increased noise, surges, and cavitation-like damage to the impeller vane. Discharge recirculation is a similar reversal of flow occurring at the discharge of the impeller vane. It is recommended that pumps should have an Nss of no more than 9000 for water and 11,000 for hydrocarbons [3].

4.4.6 Pump Performance Curve Characteristics

There are a number of pump head-capacity (H-Q) curve shapes that are shown in Figure 4-13 below. These characteristic curves are listed below and described thereafter:

- Rising
- Drooping
- Steep
- Flat

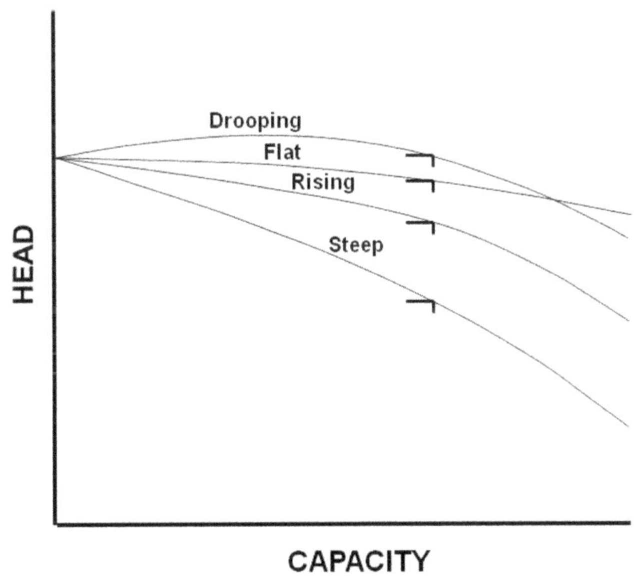

Figure 4-13. Pump performance curve characteristics

- Stable
- Unstable

Rising characteristic — the head rises continuously as the capacity is decreased to shut-off.

Drooping characteristic — the head developed at shut-off is less than at some of the other capacities.

Steep characteristic — the head developed at shut-off is significantly larger than that developed at the design capacity.

Flat characteristic — the head developed at shut-off is approximately that developed at the design capacity. The curve can be a slightly rising or drooping.

Stable curve — is a rising curve where only one capacity can be obtained at any one head. A curve with a rising characteristic would be an example of a stable curve.

Unstable curve — is a drooping curve where more than one capacity can be obtained at one head. A curve with a drooping characteristic would be an example of an unstable curve.

Unstable curves where the maximum developed head is at some flow greater than zero are undesirable in applications where multiple pumps operate in parallel. In such applications, zero flow head may be less than system head, making it impossible to bring a second pump on line. It is also possible for pumps in this configuration to deliver unequal flow with the discharge pressure from one pump determining the flow rate from another [1].

4.4.7 Centrifugal Pump Power and Efficiency

The ideal hydraulic power to drive a pump depends on the flow rate, the liquid density, and the differential head generated by the pump. This can be calculated as:

$$P_h = \frac{q\rho gh}{(3.6\times 10^6)}$$

where

P_h = hydraulic power (kW)
q = flow capacity (m^3/h)
ρ = density of fluid (kg/m^3)
g = acceleration due to gravity (9.81 m/s^2)
h = differential head (m)

Power is more commonly expressed as kilowatts (kW) or horsepower (hp = kW × 0.746).

The shaft power is the power required transferred from the motor to the shaft of the pump, and it depends on the efficiency of the pump. Shaft power can be calculated as:

$$P_s = {P_h}/{\eta}$$

where

P_s = shaft power (kW)
η = pump efficiency

4.4.8 Performance Modifications For Varying Pipeline Applications

The performance of pipeline pumps often needs to be altered to accommodate varying liquid transmission conditions. These generally require that the pump be physically modified to meet the new conditions. The following are considered by the pipeline industry:

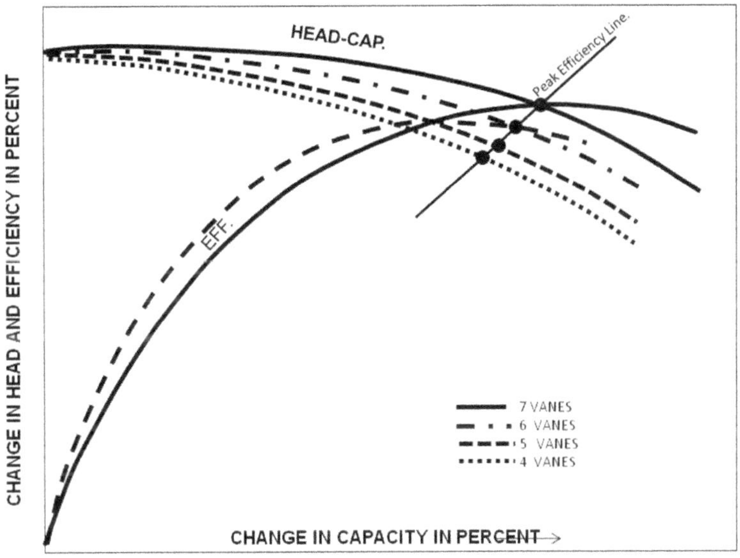

Figure 4-14. Changing performance by impeller vane number

Impeller change — In order to change the specific speed, impeller size may be changed to meet the new demand on performance. Pump manufacturers can usually offer several different impeller diameters and vanes that will fit the pump casing without any further internal modifications. The effect of changing the impeller characteristics is illustrated in Figures 4-14 and 4-15.

Restaging — Pumps with multi-staging capabilities can be restaged (up or down) to meet the change in pressure, head or flow requirements. For example, if an entire pressure range is not needed for a particular period of time, a number of impellers can be removed to meet the required conditions. Manufacturers provide de-staging kits to block off the unused pump impeller areas to maintain efficiency (Figure 4-16).

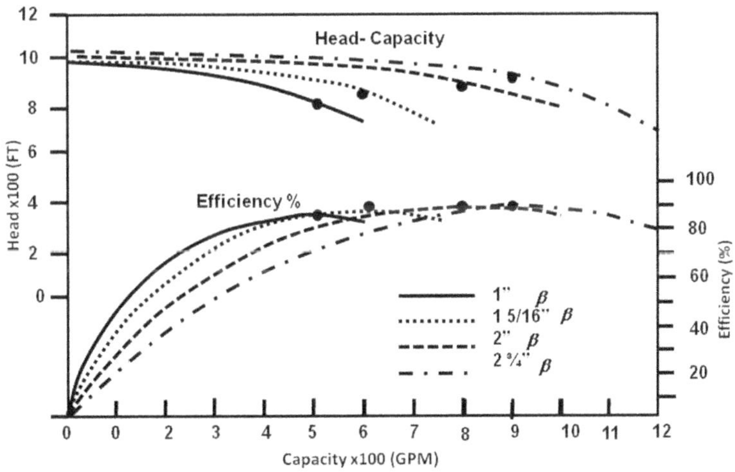

Figure 4-15. Changing pump performance by impeller vane angle

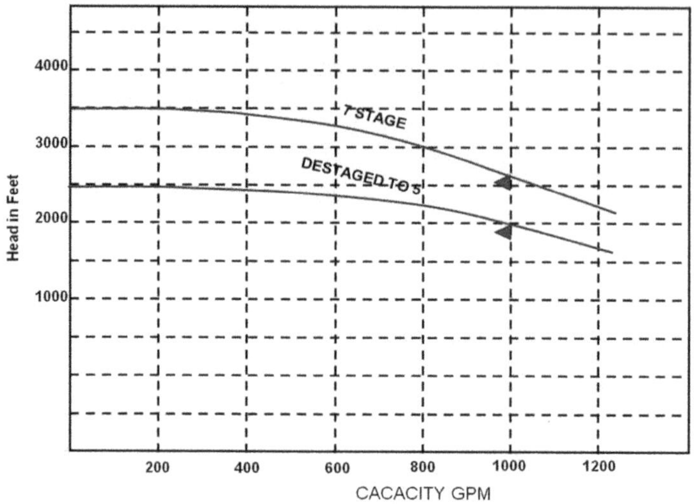

Figure 4-16. Changing performance by re-staging

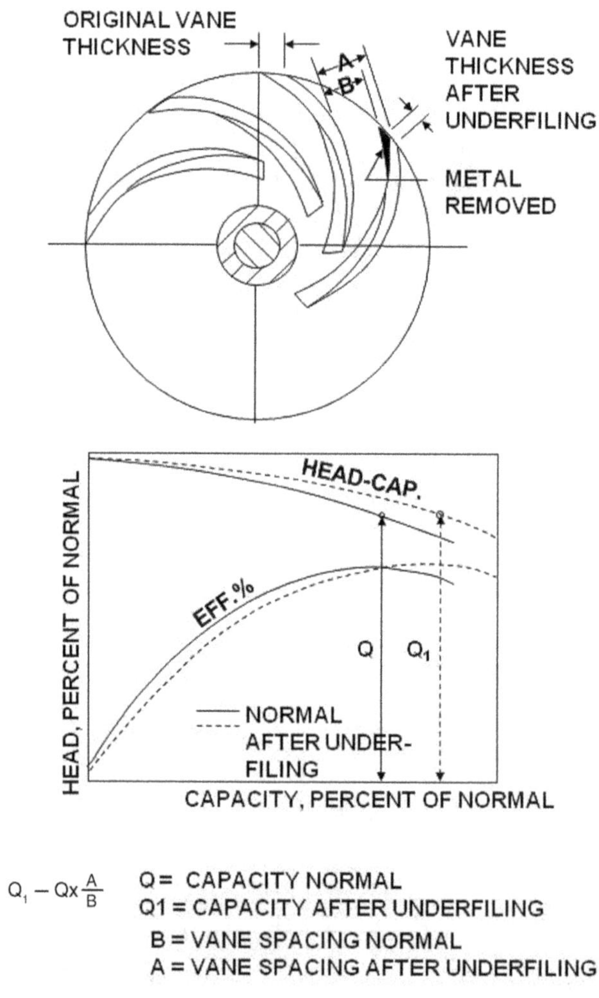

Figure 4-17. Changing performance with underfiling

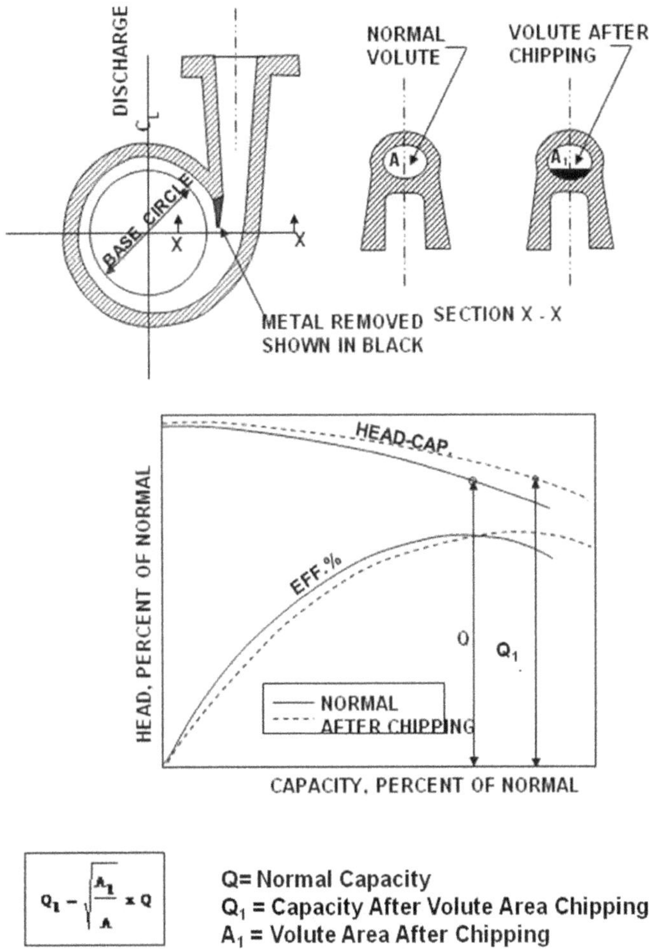

Figure 4-18. Changing performance with volute chipping

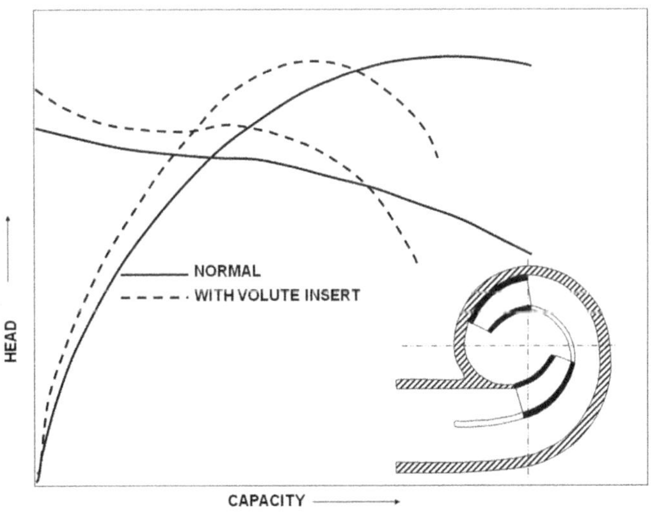

Figure 4-19. Changing performance with volute inserts

Impeller underfiling and overfiling — This is undertaken to alter the performance of a pump. It involves modifying the flow area of the impeller by grinding metal off the impeller outlet vanes (Figure 4-17).

Impeller volute chipping — This is a technique that is used to alter the outlet flow area of the pump casing in order to modify performance (Figure 4-18).

Impeller volute inserts — This technique (Figure 4-19) involves inserting special removable volutes into the pump to allow for a wider performance range. It also allows for more accurate and close control over the performance.

4.4.9 Cavitation [6–10]

Cavitation is the rapid formation and collapse of vapor bubbles that form in a pump inlet whenever the local absolute pressure of the liquid falls below its vapor pressure. These bubbles collapse rapidly and violently when the local absolute pressure increases due to kinetic forces being imparted by the impeller (see Figure 4-20 below). Cavitation is the rapid formation and collapse of these vapor bubbles.

Figure 4-20. Cavitation bubble implosion, arrows indicate fluid pressure [16 modified]

Collapsing vapor bubbles cause noise, vibration, and erosion of material from the damaged impeller as shown below in Figure 4-21 [4].

Cavitation control is a very important consideration in any liquid system and thus any cavitation induced conditions must be avoided when operating centrifugal pumps. If a liquid is accelerated in such a manner that the local pressure falls below the liquid vapor pressure, the liquid will transform into the vapor phase, which results in the formation of bubbles. If the local pressure recovers, the vapor bubbles will transform themselves back into a liquid. There is a tremendous volume change during transformation, because collapsing bubbles release a large amount of energy. Because the bubbles are very small, the resulting impact loads on the surrounding metal can be significant. This can result in the creation of high noise levels and physical damage to the metal [6].

Some liquids (such as water) are more difficult to handle from a cavitation point of view. When the vapor pressure of a homogenous fluid such as water is reached, the entire fluid begins to change phase, resulting in the formation of a large number of damage-causing bubbles. For a non-homogeneous fluid such as a hydrocarbon, only the light ends (such as condensates) are affected and the impacts of cavitation are reduced (Figure 4-22).

In a centrifugal pump, the fluid is accelerated by the impeller. The area of lowest pressure in the pump suction system, as shown in Figure 4-23, is the eye of the impeller at cross section A-A. If the pressure falls below the vapor pressure of the liquid, vapor bubbles form. As the mixture of liquid and bubbles continue through the pump, the pressure increases and the bubbles return to the liquid state. Damage to the impeller

Figure 4-21. Pump impellers destroyed by cavitation [10,8]

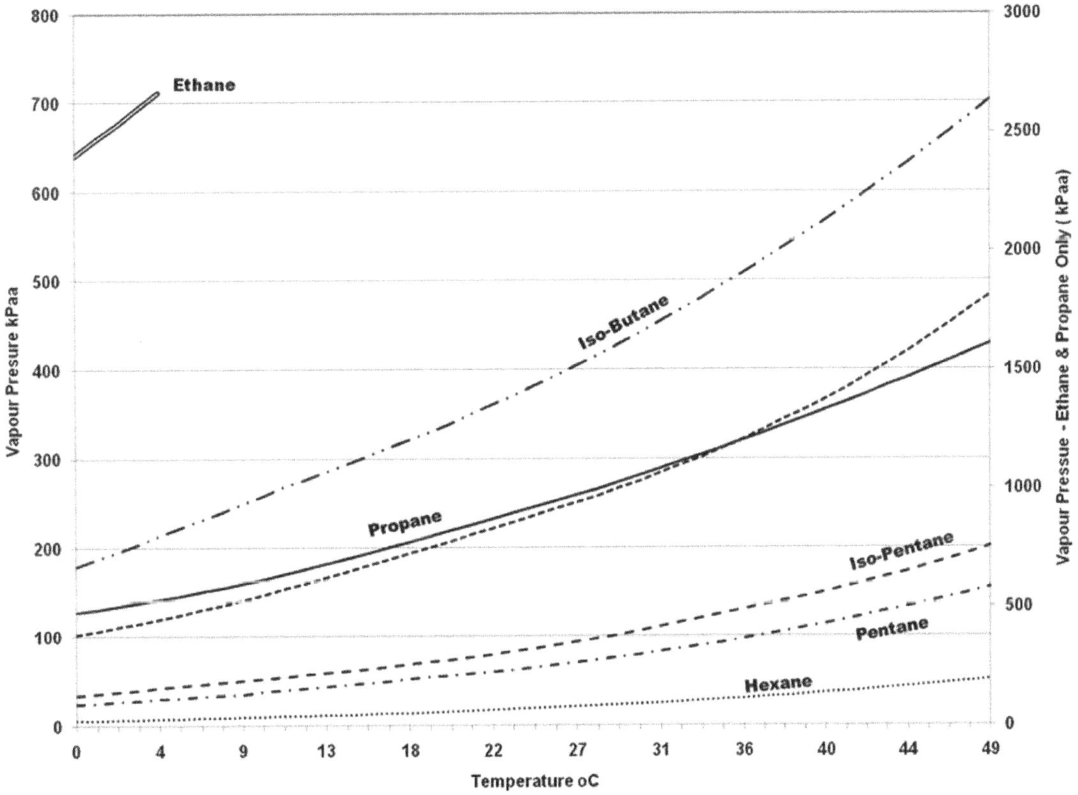

Figure 4-22. Vapor pressure of some fluids

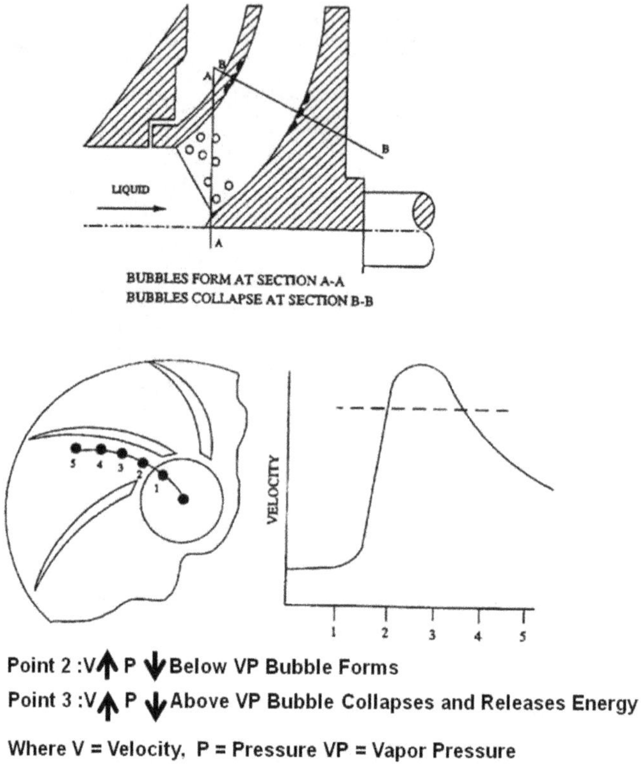

Figure 4-23. Cavitation formation

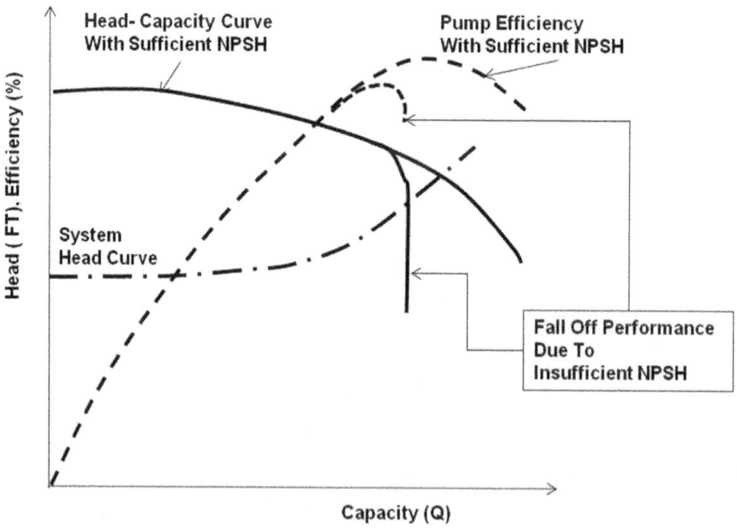

Figure 4-24. Result of cavitation on pump performance

occurs where the bubbles collapse as shown at cross section B-B. This location varies for different impellers and different suction conditions:

The effects of cavitation include:

- Noise and vibration
- Pump damage (e.g., pitting of the impeller)
- Fall off of pump performance and efficiency

Cavitation in centrifugal pumps can be recognized by a characteristic noise, which sounds just like it is trying to pump gravel. A typical break-off in the performance curve of a pump due to cavitation is shown in Figure 4-24.

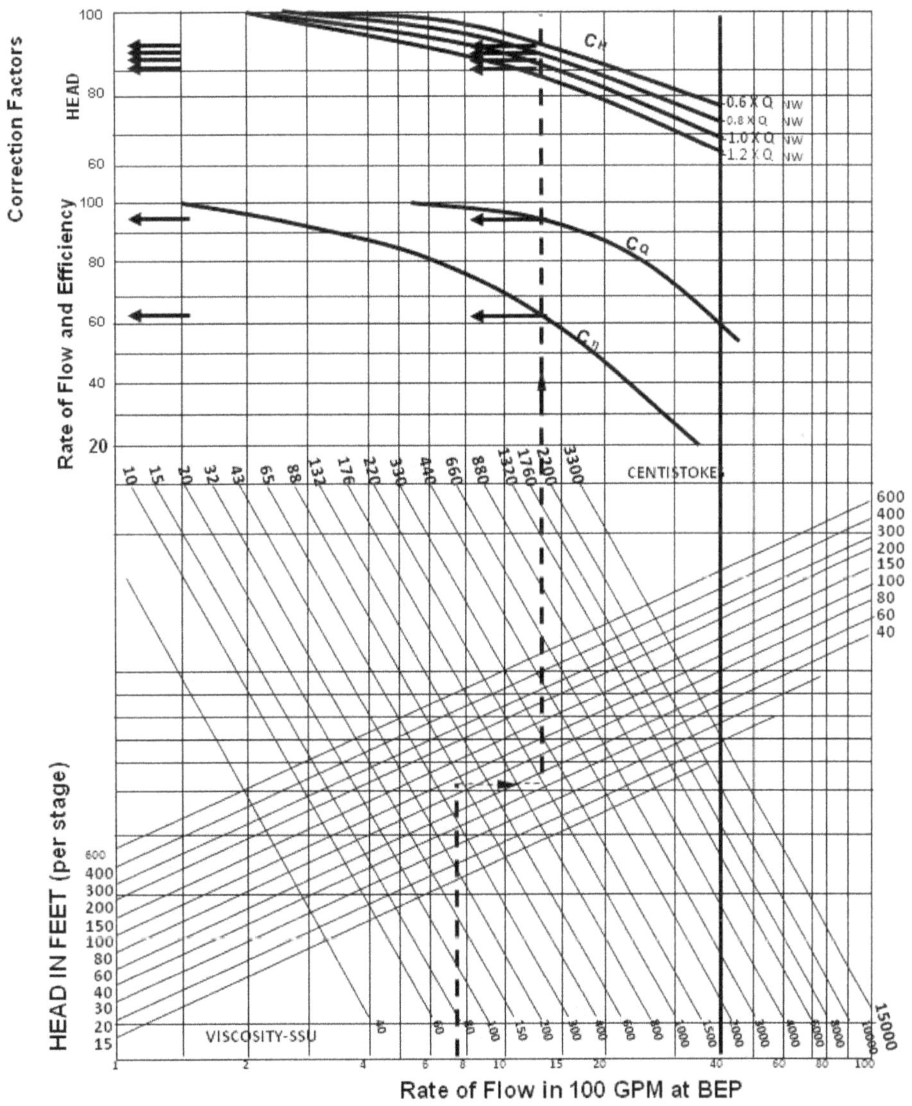

Figure 4-25. Performance correction chart for viscous liquids — with permission of Hydraulics Institute

4.4.10 Viscous Hydrocarbon Behavior in Pumps [11]

The performance of centrifugal pumps is affected by fluids with higher viscosities including fluids that behave in a non-Newtonian manner. An increase in power required a reduction in head generated, a loss in efficiency and in some cases a loss of capacity can be expected with the transmission of high viscosity fluids.

Performance correction factors, as depicted in Figure 4-25, can be applied to estimate the actual performance of a pump handling viscous fluids.

Figure 4-25 is to be used only within the scales shown and only for pumps of conventional design. The chart is applicable only for conventional centrifugal pumps operating close to their best efficiency point (BEP), having sufficient net positive suction head available (NPSHA) and transferring Newtonian fluids (such as crude oil). For description on non-Newtonian fluids refer to Chapter 1.

The following equations are used for determining the viscous performance when the water performance of the pump is known:

$$Q_{vis} = C_Q \times Q_W$$

$$H_{vis} = C_H \times H_W$$

$$E_{vis} = C_E \times E_W$$

$$BHP_{vis} = \frac{Q_{vis} \times H_{vis} \times sp\ gr}{3960 \times E_{vis}}$$

where

Q_{vis} = viscous capacity (USgpm)
H_{vis} = viscous head (ft)
E_{vis} = viscous efficiency (%)
BHP_{vis} = viscous brake horsepower

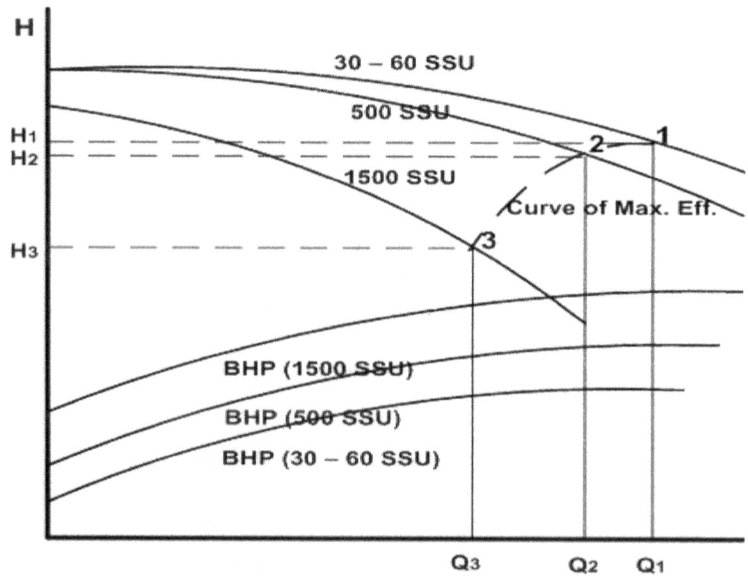

Figure 4-26. Viscosity increase vs. pump performance

Q_W = water capacity (USgpm)
H_W = water head (ft)
C_Q = Capacity correction factor
C_H = Head correction factor
C_E = Efficiency correction factor
Q_W = Water Capacity at which maximum efficiency is obtained

Figure 4-26 illustrates graphically the consequences of viscosity increases on the head, capacity, and brake power requirements. As the liquid viscosity increases, the head generated by a centrifugal pump is reduced, the pumping efficiency drops, and as a result the brake power required for pumping increases. Note that the viscosity is greater than 500 SSU or about 100 cSt, the pump performance drops rapidly.

4.4.11 Temperature Rise

Fluid temperature rises across the pump for two reasons. First, there is an adiabatic temperature rise due to compression of the hydrocarbon fluid across the pump. As well, heat is generated due to frictional forces generated within the pump and by recirculation caused by leakage through clearances within the pump. This major component of the temperature rise is directly related to the efficiency of the pump at the operating flow rate. The amount of heat generated is the difference between the input energy to the pump and the delivery energy.

Temperature rise in a pump with a closed discharge valve can occur quickly if the pump generates high head. Power losses are equal to the shutoff input power, and all this power goes into heating the small quantity of fluid contained within the pump casing. Care must be taken to prevent extended shut-in head conditions during pump startup.

Since the heat is generated as the liquid passes through the pump, the pump discharge temperature is increased and is calculated using the fundamental concept that the mechanical energy lost in the pump due to mechanical efficiency is converted to heat energy.

$$T_d = T_s + \frac{\Delta P}{\rho C_p}\left(\frac{1-\eta}{\eta}\right)$$

where

T_d = discharge temperature (°C)
T_s = suction temperature (°C)
C_p = liquid heat capacity (kJ/kg °C)
η = pump mechanical efficiency
ρ = liquid density (kg/m^3)
P = discharge and suction pressure difference (kPa)

When the pump is running normally, the temperature increase is small, in the order of a few degrees Celsius. If the pump discharge is shut off or the flow is too slow, energy is converted to heat and the heat cannot be carried away quickly. The liquid in the pump will heat and eventually vaporize. This can result in dramatic failures, particularly for large multi-stage pumps. Such a situation can be avoided by automatically shutting down the pump as the flow rate drops below the pump's Minimum Continuous Stable Flow (see Section 4.4.12) or by providing a recirculation system (see Section 4.9.4).

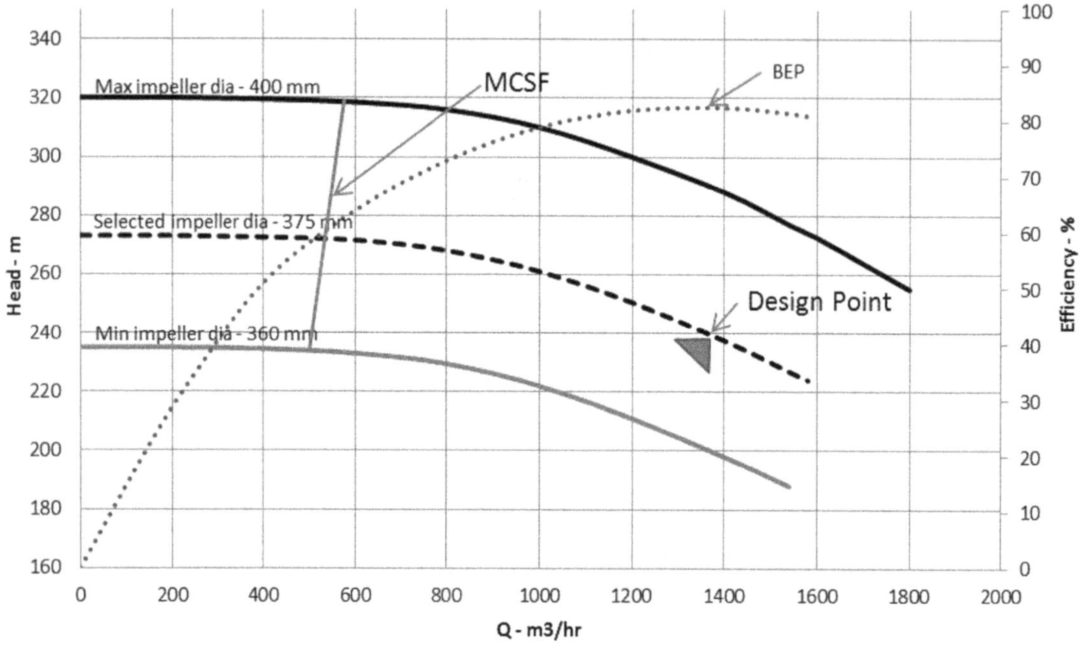

Figure 4-27. Performance curve showing MCSF

4.4.12 Minimum Flow

Minimum Continuous Stable Flow (MCSF) is usually provided by the pump manufacturer and is the lowest flow rate at which a pump can operate continuously without exceeding the vibration and/or noise limits specified. These limits are usually referenced to industry standards such as ANSI, API, ISO, and ASME or, in some cases, to the customer's own pump specifications (Figure 4-27).

There are many ways by which a manufacturer determines its recommended MCSF for a specific pump. This can be based on actual test results, historical experience or the specifics of the pump design. MCSF can be influenced by the properties of the pumped liquid as well [5].

4.5 PUMP SPECIFICATION AND PURCHASE

A pump purchase requisition must be prepared and should consist of at least two parts:

- Completed API 610/ISO 13709 Pump Data Sheets
- Clarifications/Supplements to API 610/ISO 13709 Standards, if required

4.5.1 Pump Data Sheets

Appendix B of API 610/ISO 13709 contains sample data sheets that should be used when purchasing an API 610/ISO 13709 pump.

It is important to complete all sections of the pump data sheets. There are many options in the API 610/ISO 13709 standards and the data sheet will clarify which options are required and ensure that all vendors are quoting to the same requirements. Important aspects that the data sheets clarify are:

- Service conditions (see Section 4.4.2)
- Liquid properties (see Chapter 2)
- Site conditions
- Preferred coupling type
- Materials of construction
- Mechanical seal type
- Mechanical seal flush piping
- Bearing lubrication system requirements
- Instrumentation requirements
- Painting requirements
- Electrical power available and hazardous area classification
- Pump testing requirements
- Material inspection requirements
- Spare parts requirements

4.6 RETROFITTING CENTRIFUGAL PUMPS FOR CHANGING SERVICE CONDITIONS

4.6.1 Reduced Pipeline Throughput

When pipeline throughput is reduced, the pumps in effect become technically oversized and are therefore operating inefficiently. These pumps then become subject to problems that are associated with low flows, such as vibration or seal and bearing failures. There are a number of retrofit solutions:

Change the pump speed to adapt to the desired flow conditions in the pump,

- reduce the pump impeller diameter to better suit the lower flow rate,
- install volute inserts; or,
- combination of the above solutions.

4.6.2 Increased Pipeline Throughput

When pipeline throughput above current system design is required, it will be necessary to add intermediate pump stations to the pipeline system. By adding stations, the system hydraulic performance curves are modified to allow increased flow rates while staying within system Maximum Allowable Operating Pressures.

These increased flow rates through existing pump stations may change the pump operating conditions to the point that pumps operate inefficiently or overload the pump drivers. Depending on the actual flow rate increases required, there are a number of modifications that can be adopted to have the existing pumps perform well under the new operating conditions. Retrofit solutions include:

- Trimming the pump impeller to allow a constant speed pump driver to remain within its power limits.
- Reducing the speed of a variable speed driver to limit the power requirements of increased flow through the pump.

- Adding a parallel pump unit capable of delivering the same pump head as the existing pump/pumps that operate in series. This approach then reduces the flow to the existing pump/pumps to allow pump performance to be maintained as originally designed.

The following section on affinity laws outlines the method of calculating impact on capacities, pressure rise and pump speeds by changing certain parameters.

4.6.3 Affinity Laws [4]

The flow rate and head generated by a centrifugal pump may be changed by varying either the pump speed or changing the impeller diameter. This results in a change to the impeller tip speed or velocity of its vanes, which causes a change in the velocity at which the liquid leaves the impeller. Usually, impellers can be cut down to 80% of their original diameter without lowering their efficiency significantly.

For centrifugal pumps with radial impellers, the relationships are approximated as follows:

For diameter change only:

$$Q_2 = Q_1\left(\frac{D_2}{D_1}\right), \quad H_2 = H_1\left(\frac{D_2}{D_1}\right)^2, \quad BHP_2 = BHP_1\left(\frac{D_2}{D_1}\right)^3$$

For speed change only:

$$Q_2 = Q_1\left(\frac{N_2}{N_1}\right), \quad H_2 = H_1\left(\frac{N_2}{N_1}\right)^2, \quad BHP_2 = BHP_1\left(\frac{N_2}{N_1}\right)^3,$$

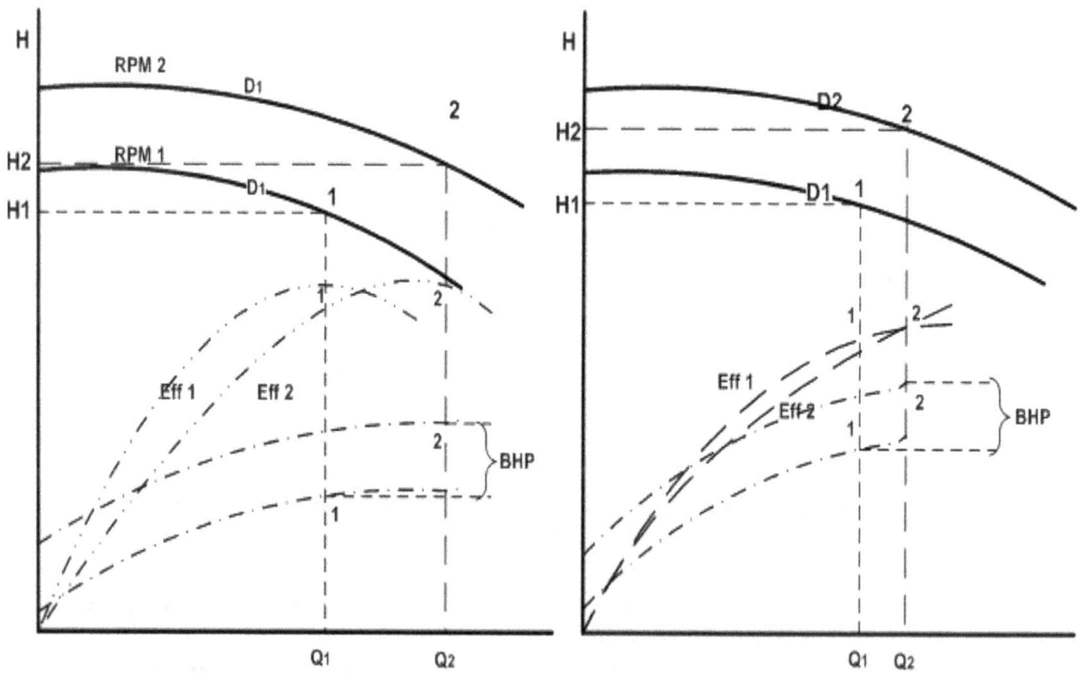

Figure 4-28. Affinity laws

For diameter and speed change:

$$Q_2 = Q_1\left(\frac{D_2}{D_1}\times\frac{N_2}{N_1}\right),\quad H_2 = H_1\left(\frac{D_2}{D_1}\times\frac{N_2}{N_1}\right)^2,\quad BHP_2 = BHP_1\left(\frac{D_2}{D_1}\times\frac{N_2}{N_1}\right)^3,$$

where

D = impeller diameter (in.)
H = head (ft)
Q = capacity (USgpm)
N = speed in RPM
BHP = brake horsepower
1 = original conditions subscript
2 = new design conditions subscript

The affinity laws can be presented graphically as shown in Figure 4-28. The pumping capacity and efficiency increases as the impeller diameter and/or speed increases. Since the head and flow capacity are higher, the power needed for higher speed and/or larger diameter is greater. Figure 4-30 illustrates the capacity changes with speed change on the left and impeller diameter change on the right hand side.

4.7 PIPELINE HYDRAULIC REQUIREMENTS

4.7.1 System Head Curves and Pump Operating Points

A system curve or system head curve for a pipeline demonstrates the head required at that location for the flow rate range. As demonstrated in Chapter 3, the pressure or head required to overcome frictional losses increases with the flow rate. The system curve depends on the following variables:

- Flow rate
- Liquid density and
- Liquid viscosity
- Pipe diameter, wall thickness, and
- Pipe roughness
- Elevation difference

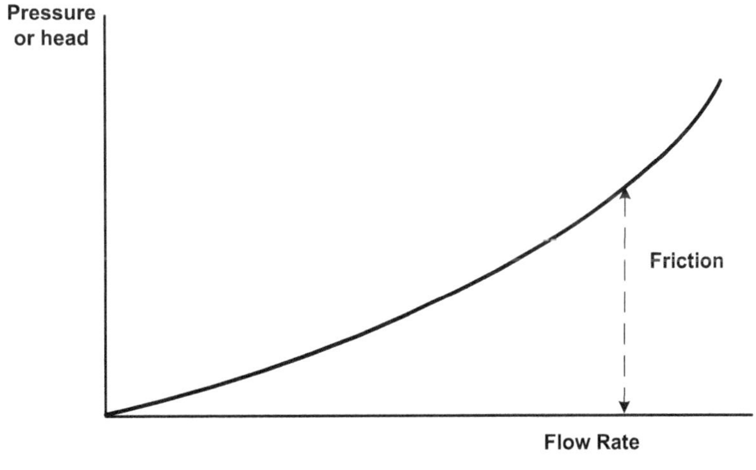

Figure 4-29. System curve

TABLE 4-2. System head vs. flow rate

Flow Rate (m^3/hr)	Friction Head (m)	Static Head (m)	Total Head (m)
250	110	100	210
625	524	100	624
835	865	100	865

As detailed in Chapter 3, the friction pressure loss can be calculated for a given liquid with its known product properties and pipe parameters. Since pump curves and elevation differences are given in head, the friction pressure loss is usually expressed in head. Using the example given in Section 3.1.3.2, a complete system curve can be built for the entire operating flow range. Assuming that the operating flow ranges from 250 m3/hr to 835 m3/hr, the friction pressure loss is calculated in the same way as described in the section and the system head required for each flow rate is determined from the relationship of $\Delta h = \Delta p/(\rho \times g)$, where Δh is the head, Δp the friction pressure loss, ρ the fluid density, and g the gravitational constant (Figure 4-29).

In general, ρ the friction pressure loss approximately varies as the square of the flow in a pipeline. This approximation works well if the operating flow range is narrow or the friction factor does not vary with the flow rate significantly. Since the friction factor change is sensitive to flow rate change for high viscosity liquid, the friction factor and friction head has to be calculated for each flow rate. Once the friction heads are calculated for various flow rates, the total head can be determined by adding the static head to the friction head. Assuming that the head increase due to elevation gain is 100 m, the total head requirements are summarized in Table 4-2. This table is used to generate the system head curve over the operating flow range.

In addition, the frictional components of the system curve can vary, sometimes over a long time period, with the following changes:

- Density due to batch movement,
- Temperature,
- Wax buildup on the pipe wall,
- Side stream delivery or injection,
- Throttling of a control valve.

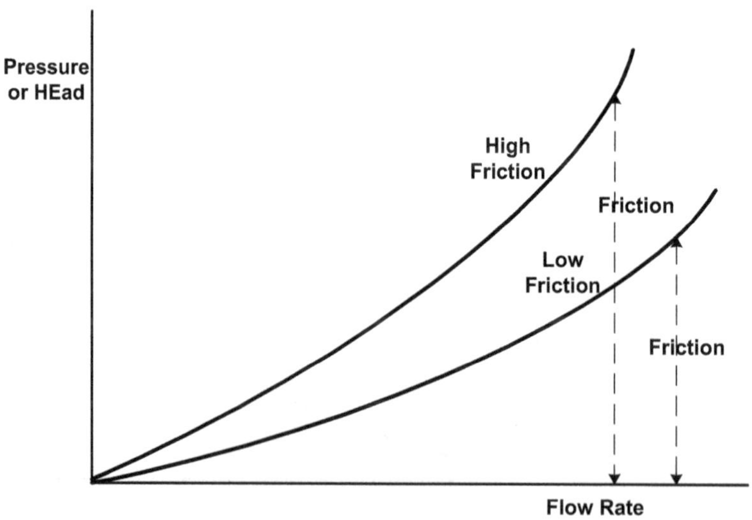

Figure 4-30. System curves with two products

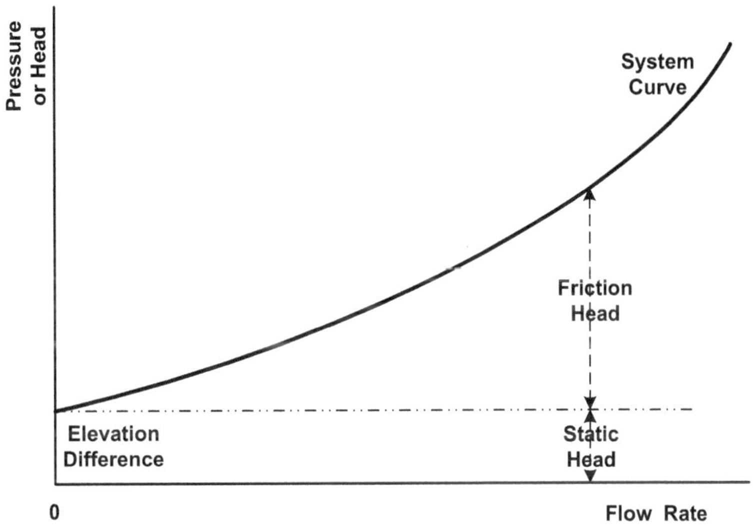

Figure 4-31. System curve with elevation difference

These changes could move the system curve and consequently pump operating point. Therefore, the range of these changes in operating point and their consequences in the power requirement, capacity and NPSH has to be considered in designing and selecting pumps.

A system curve is a graphic representation plotted on an *x–y* graph, where *x*-axis represents the flow rate and *y*-axis the pressure or head caused by the frictional pressure drop along the pipeline. Figure 4-31 shows a typical system curve for a flat pipeline system in which a single product is transported.

Figure 4-30 below shows system curves for a level pipeline system in which two products are transported, assuming that the two products have different densities and/or viscosities. The higher specific gravity and viscosity product requires greater pressures compared to the liquid with lower gravity and viscosity, generating different system curves. Figure 4-30 shows two examples of system curves; one for heavier product

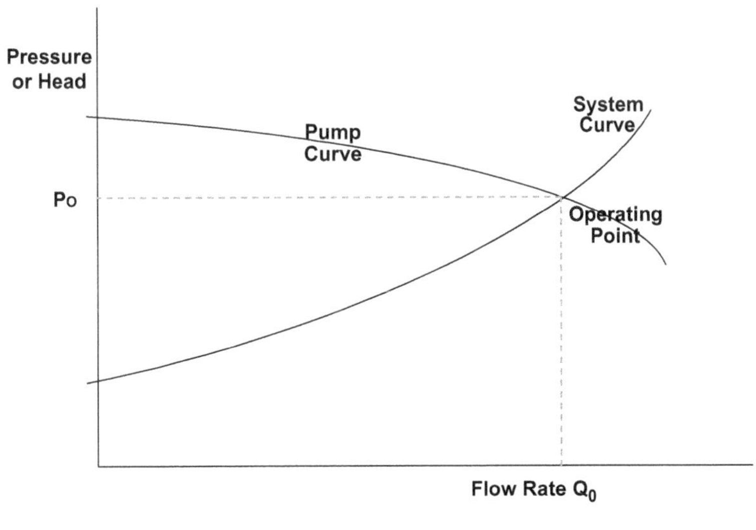

Figure 4.32. Pump operating point

causing higher frictional pressure drop and the other for lighter product causing lower frictional pressure drop. It is assumed that the pipe size remains constant and the elevation profile is flat. If the pipe size changes, so does the slope of the system curve where the pipe size changes. If the elevation changes at two different locations, the system curve simply moves up or down as shown in Figure 4-31.

If the next pump station or delivery point is higher in elevation, the system curve is shifted upward as shown in Figure 4-33.

The pump operating point is the point where the pump curve meets the system curve, as shown in Figure 4-32 below.

4.7.2 Hydraulic Performance in Batched Pipeline Systems with Constant Speed Pumps

If more than one product is batched in a pipeline, the operating points will change depending on their positions with respect to pipeline system and pump curves. As an example, refer to Figure 4-33, showing the system curves for the heavy and light oil. The operating points of these two liquids are located at H for the heavy oil and L for the light oil for the pump. Even though these two products have different densities, the pump curve in head is the same. With all heavy oil in the pipeline, the operating point is located at H, the flow rate being Q_H and the head being H_H. When batching products, the positions of the batched products shift with respect to a pump and pipeline, filling the pipeline with the heavy oil and the rest with the light oil. Therefore, the flow rate will be at some point between Q_H and Q_L, and the operating point is determined with a new system curve. In this example, as the light oil moves into the pump and the pipeline, the operating point moves to L, eventually reaching equilibrium at an operating point where the flow rate is located at Q_L and the head at H_L.

Figure 4-34 below shows the pump and system curves plotted in pressure instead of head. Since the pressure changes with density, two pump pressure curves are displayed with four operating points. If all heavy oil is in the pipeline and passes through

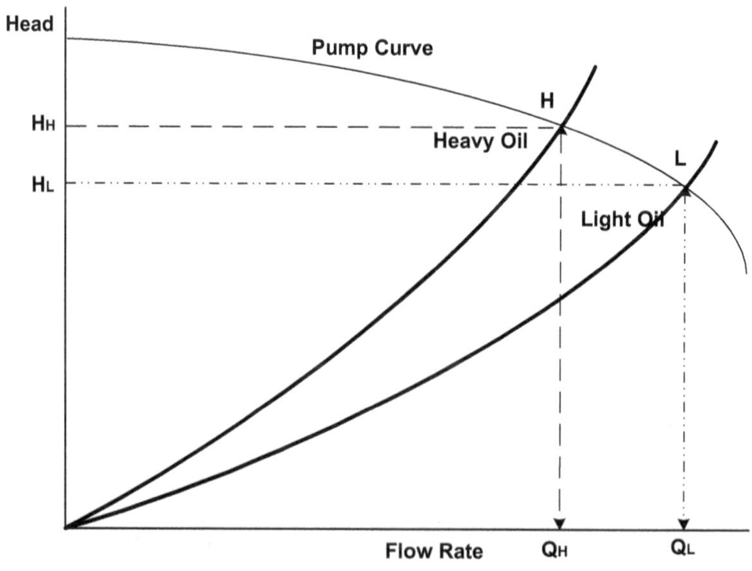

Figure 4-33. Operating points in head for multiple products

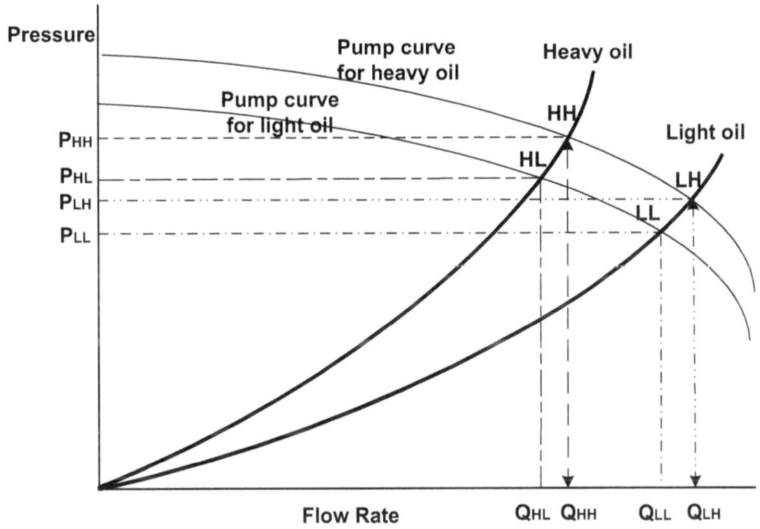

Figure 4-34. Operating points in pressure for multiple products — constant speed pumps

the pump, the operating point is located at HH in the figure. As the light oil enters the pump, while the heavy oil is still flowing in the pipeline, the operating point slowly moves from HH to HL, but with reduced flow rate. In reality, this movement does not occur abruptly because the batch interface enters first and the density gradually changes before the light oil completely replaces the interface. The flow reduction is caused by the low density of the light oil passing through the pump and thus reducing the differential pressure. As the light oil moves into the pipeline, the operating point gradually moves to the new operating point, LL, because the light oil requires lower system curve. As the light oil fills the pipeline, the flow rate increases but the pressure decreases. The new point represents the operating point with the light oil in both the pipeline and the pump.

In the same process as described in the previous paragraph, as the heavy oil moves into the pump and the pipeline is filled with light oil, the operating point moves to LH, and eventually the operating point moves back to HH as the heavy oil fills both the pump and the pipeline. In batch operations, the flow rates vary between Q_{HL} and Q_{LH}, so does the pressure assuming the pump is not throttled. Note that the positions of these four operating points may be altered depending on the viscosity of the heavy oil because pump performance curves are changed with higher viscosity liquids (refer to Section 4.4.10).

4.7.3 Hydraulic Performance in Batched Pipeline Systems with Variable Speed Pumps

With the introduction of economically available variable speed drivers for centrifugal pumps (see later descriptions in this chapter); it is possible to provide more capacity to batched product pipelines very efficiently. By increasing pump speed in the example shown above (Figure 4-34) as light oil enters the pump with the pipeline full of heavy oil, it is possible to maintain pressures and thus flow rates at Q_{HH} with more rapidly increasing flow rates to Q_{LH} than in the previous example. In this case, the design point for the pumps would be at LH with heavy oil density and viscosity determining the power requirements of the variable speed pump.

4.7.4 Pump Configurations

4.7.4.1 Parallel Operation

If a station has more than one pump unit, it is operated in one of two modes: parallel operation and series operation. Multiple pumps in a pump station can be arranged in parallel, series, or combination of parallel and series arrangement. Some combination of these arrangements is needed to increase both flow and pressure.

The primary purpose of operating units in parallel is to allow a wider range of flow than would be possible with a single unit for systems with widely varying flows. Generally, pumps are arranged to operate in parallel to increase system flow rates by maintaining discharge head through a much greater pipeline throughput. The following alternatives should be considered in selecting pump configurations:

- One operational pump with one spare unit in parallel;
- Two operational pumps in parallel with one parallel pump as spare;
- Two pumps in parallel without a 100% spare, if less capacity can be tolerated at times.

When pump units are arranged in parallel, loads are shared by more than one unit. The following load sharing strategies are available:

- Base loading: one or more units may be operated at a constant load while other units are operated to handle additional flow rate requirements;
- Optimum load sharing: set points for each unit are determined based on knowing the individual operating curves and allocating load to optimize pump efficiency and minimize overall energy consumption;
- Equal load sharing: each unit shares load equally. This strategy is used for identical units which are arranged in parallel.

There are several areas of concern when operating centrifugal pumps in parallel. If adequate flow control is not provided, hydraulic imbalance could occur resulting in high flow rates in one of the parallel pump units and low flow rates in the other unit. Possible issues include:

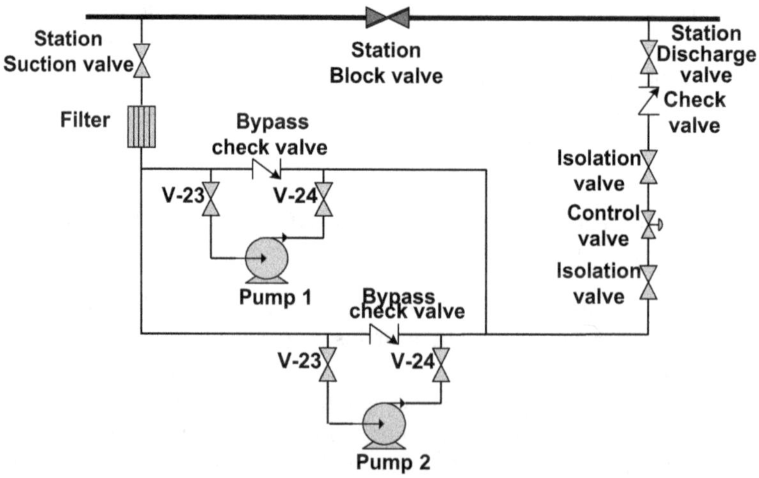

Figure 4-35. Parallel operation

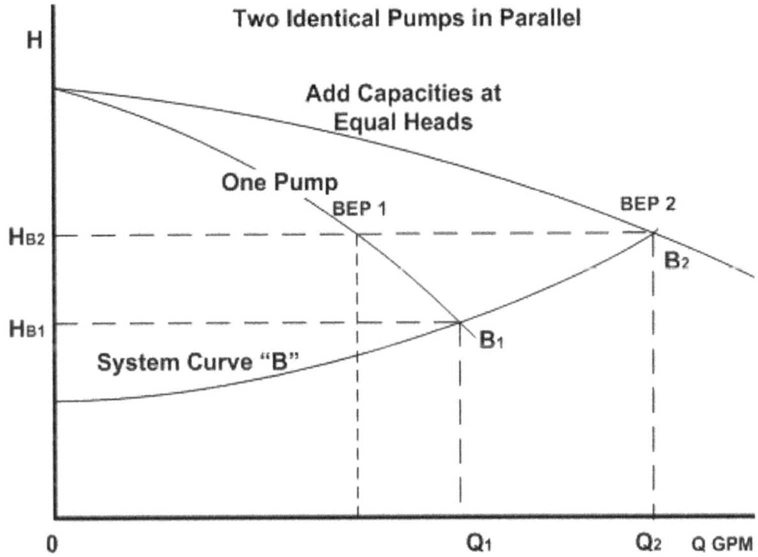

Figure 4-36. Operating points for parallel operation

- High flow rate induced cavitation;
- Pump driver overloading;
- Prolonged operation of one of the pumps at a flow below its minimum acceptable continuous flow rate.

Selection of appropriate control systems is essential with pumps operating in parallel configuration.

Figure 4-35 shows a parallel arrangement of two pumps, where more than one unit can be operated at the same time. Figure 4-36 demonstrates the operating points of one pump and two identical pumps when they are arranged in parallel. When two or more units operate in parallel, all units have common suction and discharge pressures.

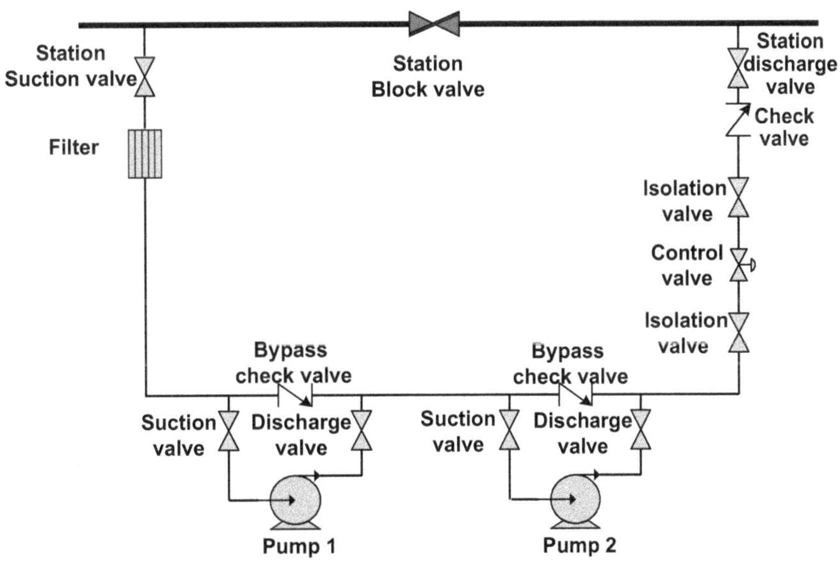

Figure 4-37. Two pumps in series

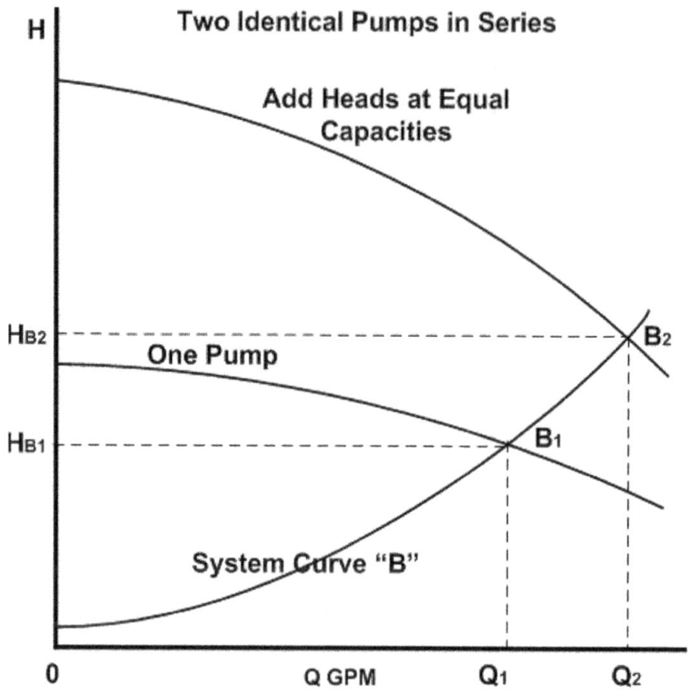

Figure 4-38. Operating points for series operation

4.7.4.2 Series Operation

The main reason for operating units in series is to increase the pumping head from what would be possible with a single unit. Pumps operating in series add head and capacity to the system output. In areas of large pipeline elevation rise, if two pumps are arranged in series and one shuts down, the remaining pump alone may not be able to provide the necessary head for the static lift necessary to maintain pipeline flow. In this case, appropriate station design would require that a spare pump unit be installed to be activated in the event of a loss of a pump unit operation.

Figure 4-37 shows a series arrangement of two pumps, and Figure 4-38 demonstrates the operating points of one pump and two identical pumps when they are arranged in series. In series operation, the flow through all of the units is equal and the discharge of one pump feeds the suction of the next unit.

4.8 PUMP DRIVERS

A mainline horizontal centrifugal pump can be driven by an electric motor, gas turbine or a diesel engine. Liquid hydrocarbon transmission pipelines are typically driven by electric motors where electrical power is available. This is primarily due to their lower initial capital cost and inherent reliability. As well, electric motor drivers have benefits over gas turbine and diesel drivers due to ever more stringent emission level limits.

This section will consider pump station operation using both constant speed electrically driven pumps and variable speed electrically driven pumps. Fixed-speed electric motors provide a cost-effective solution for base load applications where electrical power is available and reliable. They have the advantage of low-maintenance costs and are simple to operate. Variable speed drive (VSD) motors are becoming the standard

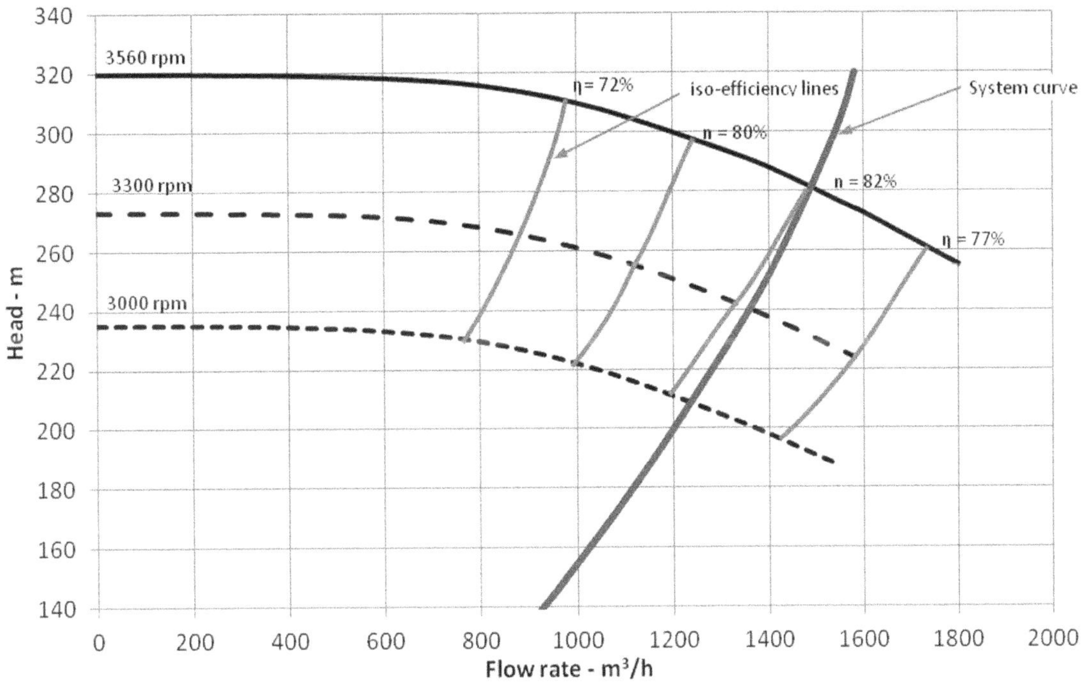

Figure 4-39. Variable speed pump performance curve

for pump stations that have varying flow or product density requirements such as on batched product pipelines. Despite their control systems being more complex than for a constant speed motor, variable speed motors are much more energy efficient. This is because pump capacity can be controlled without the disadvantage of pressure loss incurred by the throttling through a discharge control valve.

Variable speed pumps control the flow and pressures by varying the speed of the drivers with maximum power override. For a pump station that contains both fixed-speed and variable speed motors, the control strategy is to run the fixed speed units at a base load with minimal throttling and use the variable speed units to adjust for the required station set point. Figure 4-39 exhibits the performance curves of a variable speed pump.

In applications that require flow or pressure control, the most energy efficient option is an electronic VSD, referred to as a Variable Frequency Drive (VFD). The most common form of VFD is the voltage-source, pulse-width modulated frequency converter [12]. The converter develops a voltage directly proportional to the frequency which produces a constant magnetic flux in the motor. This type of speed control can be driven by set points of discharge pressure or flow rate.

As energy cost increases, VFDs and thus variable speed pumps are becoming more cost-effective. This is because they allow pressure and flow control without wasting energy incurred by throttling a control valve to control the discharge pressure of the pipeline. Even though variable frequency drives are more expensive, they have the advantage of reduced energy consumption and are efficient over a wide range of flow.

Energy savings of between 30% and 50% have been achieved in many installations by installing VFDs [13].

The following simplified graph, Figure 4-40 shows the significant reduction of power required by variable speed pumps as compared to fixed speed pumps through a range of flows.

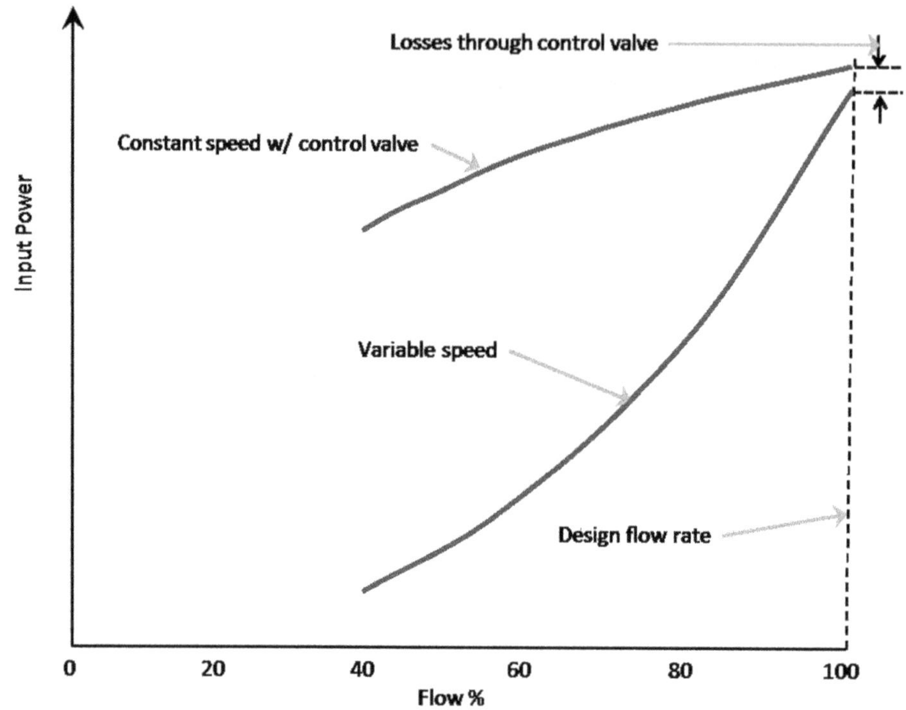

Figure 4-40. Power requirements — constant speed drive vs. variable speed drive

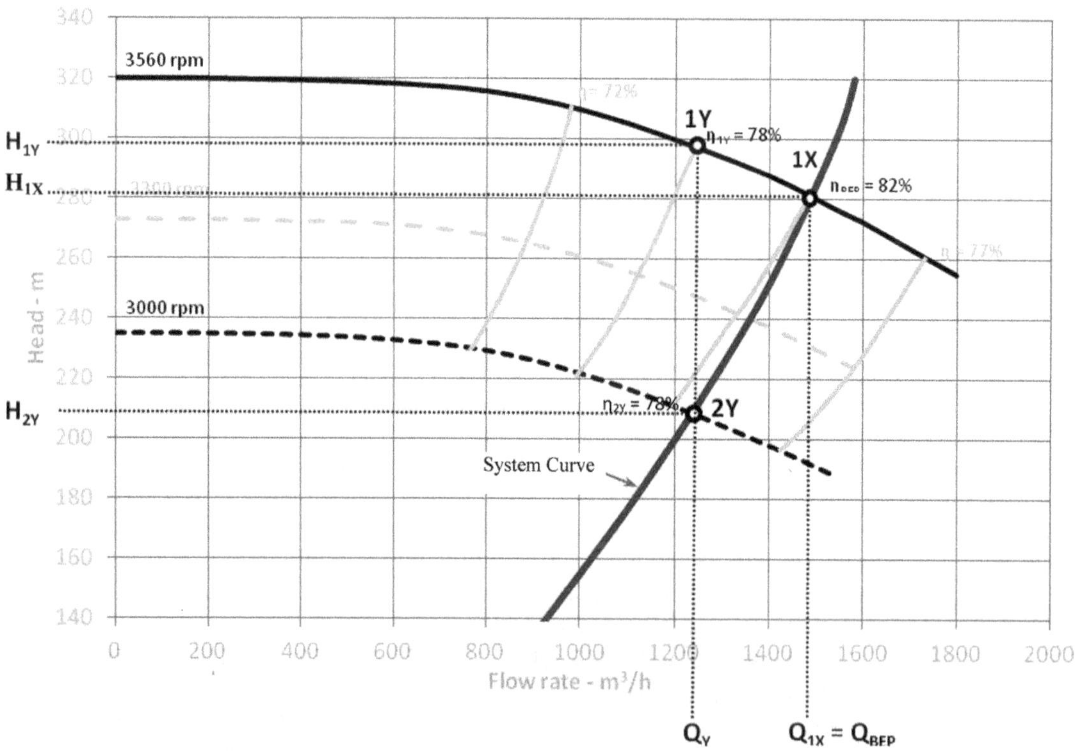

Figure 4-41. Constant vs. variable speed power requirements

Figure 4-41 below shows a pump operating at the operating point 1X, where the flow rate and head are Q_{1X} and H_{1X}, respectively. Here, for illustrative purposes, it can be assumed that the operating point 1X is the best efficiency point (BEP), where the efficiency of the pump is highest. Then, the power required by the pump is $PW_{1X} = Q_{1X} \times P_{1X}/(\text{dens} \times \eta_{BEP})$, where η_{BEP} is the pump efficiency at the BEP.

To achieve a lower flow rate Q_Y, the control valve is partially closed for the fixed speed pump or the pump speed is reduced for the variable speed pump. These are illustrated in Figure 4-41. 1Y is the operating point of the fixed speed pump and 2Y is that of the variable speed pump for the lower flow rate. At 1Y, the flow rate is reduced to Q_{1Y} but the pump head increases to H_{1Y} for the fixed speed pump where the pump efficiency, η_{1Y}, is considerably lower than the pump efficiency at BEP. In the case of the variable speed pump, the operating point 2Y for the same flow rate generates only the head required for the system curve, H2Y and the pump efficiency, η_{2Y}, is only marginally lower than the best efficiency point.

In summary, energy requirements are directly proportional to head generated and are factored by changes in pump efficiency. As demonstrated by Figure 4-41 above, there is considerable energy savings by use of variable speed drives. The above example also does not take into account the continuous pressure loss that occurs in a fully open control valve and therefore is conservative in its energy savings by use of variable speed motors and elimination of the control valve.

See Section 4.10.6.2 for further discussion on the energy savings of variable speed pumps.

4.9 PUMP STATION DESIGN

As described in Section 3.4, intermediate pump station locations on a pipeline system are determined by the hydraulic design of the transmission pipeline and the frictional pressure losses that occur with the product proposed for transportation. Adjustments to station location from its ideal hydraulic location are often required because of such

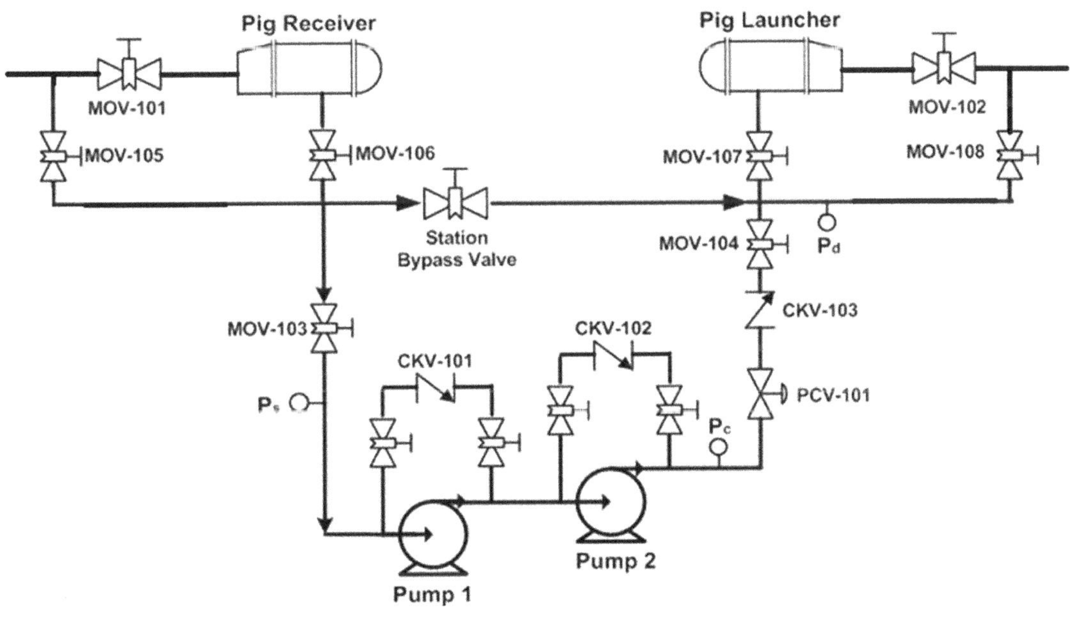

Figure 4-42. Typical pump station diagram

factors as land access issues or topography constraints. Final locations for intermediate pump stations may result in some slight reduction in pipeline capacity because of some hydraulic imbalance in the final station sites. This section addresses several key equipment and features required for designing a pump station.

4.9.1 Pump Station Diagram

A simple pump station diagram is shown in Figure 4-42. In this diagram, the main components include two pump units, station piping connecting various components, station isolation valves, bypass valves, pressure control valve, and check valves. In addition, many intermediate pump stations are installed with a pig receiver and launcher set. Pump stations pumping heavy crudes may be equipped with heaters.

Figure 4-42 shows a typical pump station, which is composed of the following equipment and instruments:

- Station isolation valves are MOV-103 and MOV-104. These valves are used to isolate the pump station for safety, maintenance, or other operating purposes.
- Check valves are shown as CKV-101, CKV-102, and CKV-103.
- Bypass valves include the station bypass valve along the main line, MOV-103 on the suction side, and MOV-104 on the discharge end. The station bypass valve is open and the other two valves are closed when the pump station is shut down, so that the fluid flows through the station bypass valve.
- Instrumentation for pressure measurement (P_s, P_c, and P_d) is essential for station control. Certain pump stations are equipped with a flow meter for control purposes, not for custody transfer.
- A pressure control valve (PCV-101) is installed on the station discharge piping to control the discharge pressure and flow rate.

Pumps are driven by pump drivers which can be either fixed speed type or variable speed type. In some cases, combined fixed and variable speed drivers have been used to take advantage of low cost fixed speed drivers for base load and of low energy cost variable speed drivers for extra load.

4.9.2 Pump Station Piping

Station suction piping design is important to ensure that NPSHA be maintained above the NPSHR for the pump units. Piping must be designed with sufficient pressure to withstand potential surge pressure changes during pump start-up and shut-down operations. Piping sizing should be determined based on minimizing pressure losses to pump suction. As a rule of thumb, flow velocity on the suction side should be in the 1.5 m/sec to 2.5 m/sec range. Higher velocities increase the frictional pressure loss and potential surge pressure. Due to potential surge when a pump shuts down, the suction pressure can suddenly increase significantly. Therefore, the suction line pipe must be able to withstand the increased pressure.

Since centrifugal pumps generate a performance curve that rises as the flow decreases, the discharge pressure should be determined at shut-off.

If the liquid is fed from a tank, the amount of entrained vapor or air must be kept to an absolute minimum. Entrained vapor causes not only vibration and possibly cavitation but also reduced capacity and efficiency. Normally, a booster pump is installed to avoid this condition.

Station piping in pipeline systems that are intended for batched products should be designed to minimize any areas of potential product contamination. All fittings should be close-coupled and piping should be designed to eliminate any connections that could include isolated pockets of product that would contaminate a batch of different products.

For pipelines transporting batches of different product, pump units that are not in operation should have their suction and discharge valves left in the open position so that all products flowing through the station piping are continuously being flushed. When there is a need to put the pump unit into operation, the control logic for the unit start should first close the discharge valve before the pump unit is to be started. Once the pump is on-line, the discharge valve is opened and the flow through the pump is re-established with its added head.

In addition, a control valve is required for fixed speed pumps and some variable speed pumps. Some initiating, intermediate pump and pressure reducing stations are equipped with a pig launcher and receiver. In certain situations, heaters are installed to heat heavy crude oil. In cases where the flow rate through a pump can drop below the minimum continuous flow limit, recirculation piping and valves may need to be installed (See Section 4.9.4 Station Flow Recirculation).

4.9.3 Control Valve and Sizing

The proper selection of a valve for petroleum liquid pipelines depends on factors such as liquid properties, the system curve, the pressure drops to be controlled, and cost. The liquid properties to be specified are its phase (whether or not it contains solids or vapor) and its corrosion/erosion property. Temperature may not be a critical factor for most liquid pipeline applications. The most suitable type of valve required for throttling the flow or controlling the pressure includes globe, control ball, butterfly, and rotary plug valves.

Valve size is obtained from a basic liquid sizing equation, which can be written as follows:

$$Q = C_V \sqrt{\frac{\Delta P}{\gamma}}$$

where

Q = flow rate

C_V = valve sizing coefficient determined experimentally for each type and size of valve

ΔP = pressure differential

γ = specific gravity of liquid

To calculate the expected C_V for a valve controlling liquid flows, the above equation is re-arranged for C_V in terms of the flow rate and pressure differential.

$$C_V = Q\sqrt{\frac{\gamma}{\Delta P}}$$

We can obtain the valve's C_V requirements by inputting into this equation the minimum and maximum flow rates together with valve upstream and downstream pressures. This value is used to select from the manufacturer's data the size of the valve required. The valve size should be compared to the pipe size in which the valve is to be installed. As a rule of thumb, the control valve should not be smaller than two nominal pipe sizes below the nominal pipe size.

4.9.4 Station Flow Recirculation

Pumps can be operated at or near zero flow for a short time at start-up without adverse consequences to the pumping system. To avoid recirculation problems, at least 20% of BEP flow is required for small pumps, and 50% or even 60% of BEP flow for large pumps.

However, if centrifugal pumps operate at low or no flow conditions frequently and the duration of such operation is long, most pumps will need to be provided with a flow recirculation system to protect them from potential damage or unstable flow conditions in a low flow operation. The following conditions can develop at low flow rates:

- Temperature in the pump rises significantly due to low pump efficiency;
- Unstable flow conditions occur, resulting in surging pulsations and pipe vibration.

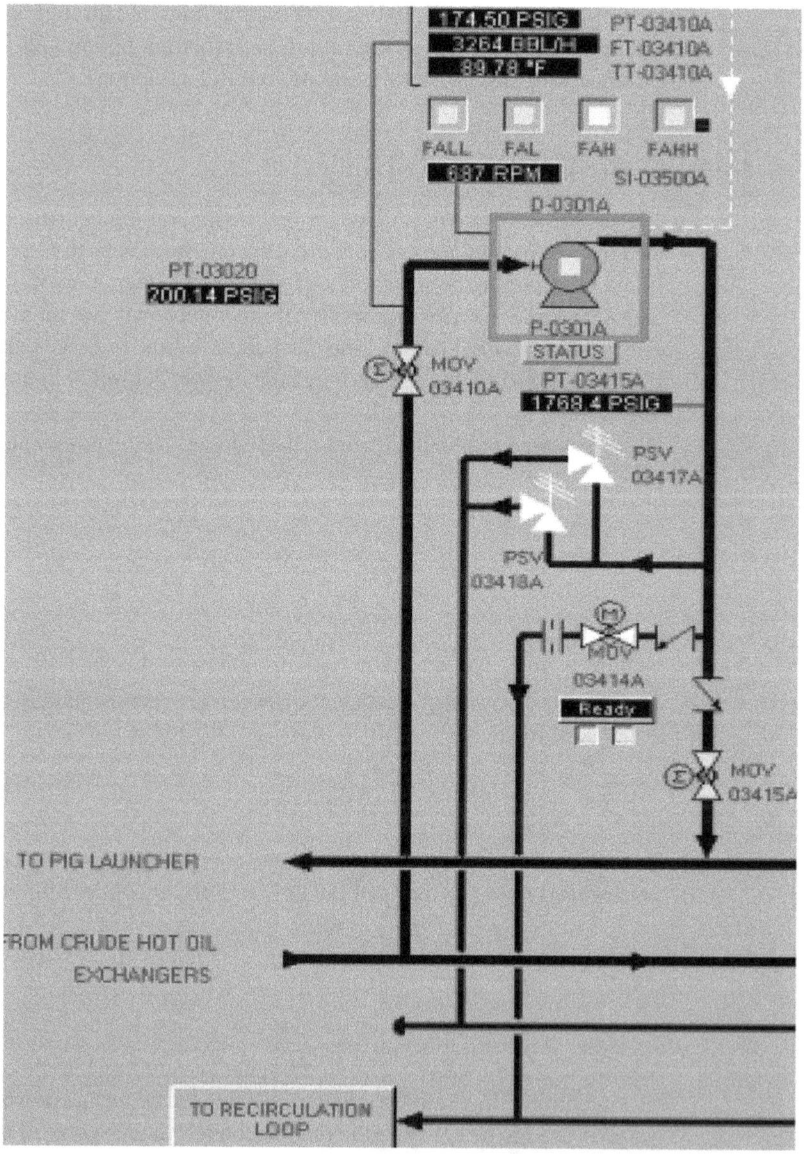

Figure 4-43. Pump station with a recirculation system

If the low flow conditions persist for a prolonged period or occur frequently, pumps and other equipments can be damaged. If a pump station has to pump low flows frequently, there are several options to address this problem:

- Install two or more small pumps in parallel;
- Install variable speed pumps;
- Install a recirculation system with flow measurement, recirculation valve and piping.

The overall economics will dictate the choice of these options.

Several alternative recirculation systems are available. Figure 4-43 shows one such alternative recirculation system, which includes a flow meter, recirculation control valve and piping. When the flow rate gets closer to a set minimum flow, a bypass control valve is activated and the flow downstream of the pump is recycled back to the pump suction through the recirculation piping. As shown in this figure, the recirculation system consists of a flow sensing device, a minimum flow control valve, a check valve, and a bypass valve. Note that a pressure safety valve (PSV) is installed downstream of each pump to prevent the discharge piping from over-pressuring.

Recycling flow rate is controlled by a control valve with the flow meter located in the suction line. As the flow rate decreases, the flow meter sends a signal to open the recirculation control valve to keep the combined flow rate at or above the minimum required flow rate. In general, this system works well in maintaining the minimum flow rate required by the pump. Note that the take-off for the recirculation line is upstream of the check valve.

4.9.5 Pig Launcher and Receiver

Pipeline pigs are extensively used in petroleum pipelines. They are intended to clean and/or inspect the inside of pipelines; or, in some cases, to separate batch interfaces in a multiproduct pipeline to reduce interface mixing. So-called smart pigs can detect pipe corrosion by measuring pipe wall thickness and cracks in the pipe wall by means of ultrasonic responses.

The pig is inserted into the pipeline through a pig launcher. After a pig is loaded into the pig launcher, the MOV-107 valve connected to the pig launcher (refer to Figure 4-42) is opened. When MOV-108 is closed, the flow pushes the pig out of the launcher and the pig travels along the pipeline until it reaches the next pig receiving station or pig trap where the pig is retrieved. The valves along the

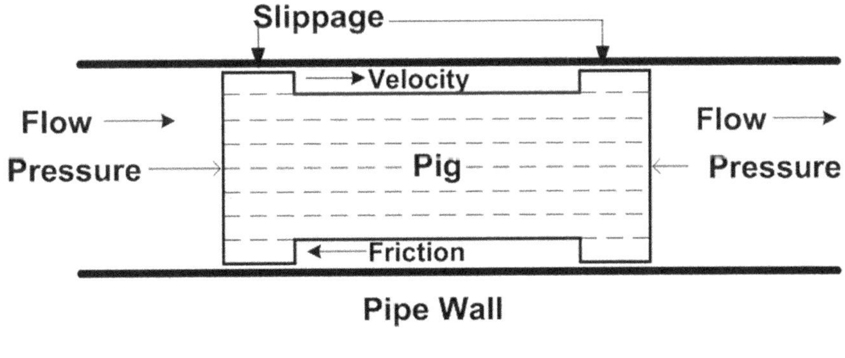

Figure 4-44. Pig movement

pipeline from the launcher to the trap must be full bore valves to allow the pig to pass through them.

Pigging operations are not usually automated because they do not take place frequently. They seldom interrupt normal transportation, but some liquids can be spilled into the sump tank when the pig is retrieved from the pig trap. When a pig is launched, a launch signal can be generated to track its movement until it is received at the designated pig trap.

Pigs usually do not travel at the same velocity as the liquid flow in the pipeline, slipping gradually as the liquid passes by the pig. Some SCADA systems estimate the location and the estimated time arrival to the receiving trap based on the pig slippage factor and current pipeline conditions such as flow velocity. The pig tracking function is useful for the pig retrieving schedule. Figure 4-44 shows a pig movement with pig slippage.

When a pig arrives at the upstream location of the pig receiver, MOV-106 is opened to equalize pressure across MOV-101. MOV-101 is then opened and MOV-105 is closed and the pig moves into the trap. Once the pig arrival has been confirmed, MOV-105 is re-opened and MOV-101 and MOV-106 are closed. The trap is then drained into sump and the pig is removed.

Where batching pigs are utilized, there is a trend toward fully automating the pig operations. Full automation can be justified because the batching operations are required frequently and manpower involvement and operating costs are reduced.

4.9.6 Pump Station at a Tank Farm

A tank farm includes not only multiple tanks and manifolds but also booster pumps and a meter station with a prover. If it is the initiating lifting point, mainline pumps are installed there. Figure 4-45 shows the tanks, booster pumps, meter station and mainline pump station. Normally, a pig launcher is also installed.

The operation of a tank farm and booster pump start-up are relatively simple, unless the tank farm is very large and complicated. Assuming that the pipeline is initially shut-in, the tank farm control system opens the valve connected to the tank containing the desired batch. The density of the batch is measured to confirm that it is the correct batch. After the confirmation, the suction valve and then the discharge valve are opened to prime the pump. When the pump is primed, it starts rotating and the batch

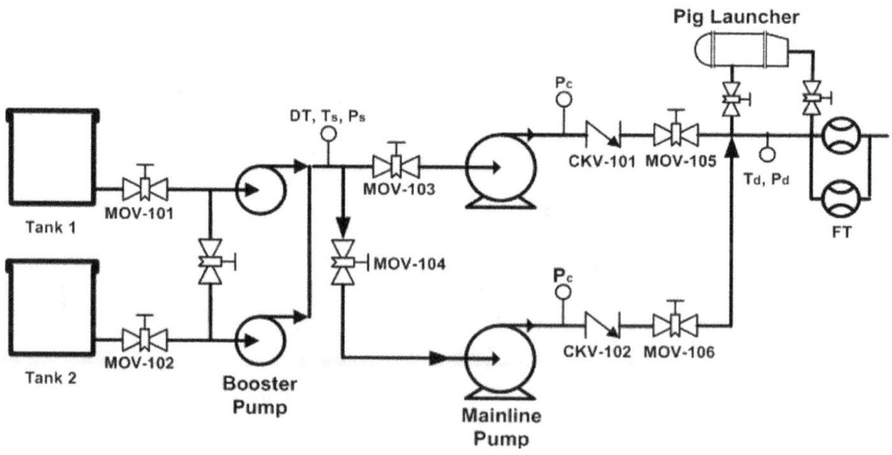

Figure 4-45. Tank farm booster and mainline pumps

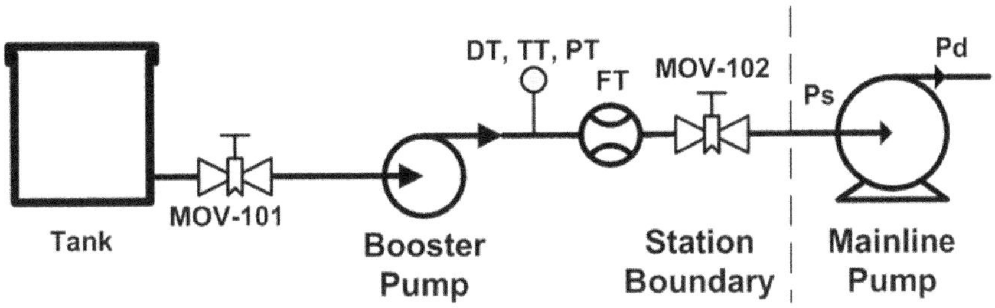

Figure 4-46. Side stream injection

starts flowing. The discharge pressure of the booster pump should be higher than the suction pressure of the mainline pump(s), Ps, which should be greater than or equal to its NPSHR.

A side stream injection takes place on the suction side of an intermediate pump station or in the middle of the pipeline section. Usually, a booster pump and flow measurement system including density, temperature, and pressure measurement are installed in the side stream injection location, but it may not be equipped with a pig launcher. The figure below illustrates a simple side stream injection facility, injecting into the suction piping of an intermediate pump station (Figure 4-46).

4.9.7 Pump Station Heater

As discussed in Chapter 3, heavy oil is often heated to reduce its viscosity and lower frictional pressure drop. If the liquid needs to be heated, heaters are installed at the lifting tank farm and at certain intermediate pump stations along the pipeline. They are installed downstream of the tanks and booster pumps, but on the suction side of the mainline pumps.

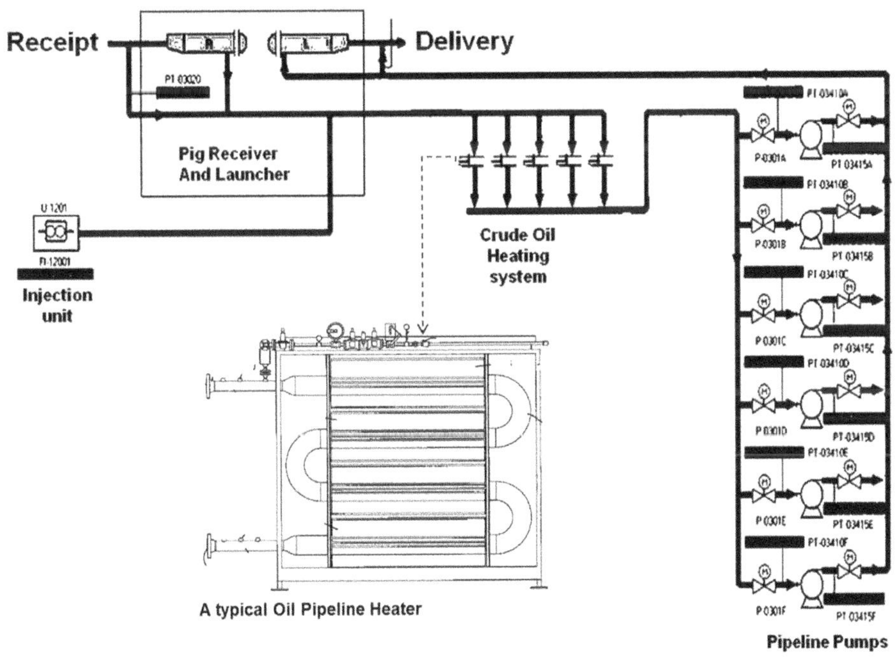

Figure 4-47. Heater, pump and pig launcher/receiver

Figure 4-47 shows a SCADA display including the heating units, pumping units and pig launcher/receiver arrangement at an intermediate pump station. These units can be controlled from this screen. Under certain operating conditions, the heater operations may not be required, particularly at intermediate stations. The operating conditions include high ambient temperature and/or low flow rate:

- If the ambient temperature is high, the fluid temperature may remain high enough to keep the viscosity low.
- The liquid may be able to flow within the pipeline pressure limits without further reducing the viscosity at low flow rates, because the frictional pressure drop is low at the low flow rate.

Normally, heater operations take place before pump operations. In other words, crude oil is heated to increase the pumping efficiency before pumping into the pipeline starts, while heaters are shut down before the pumps are shut down to save the heating cost.

4.10 PIPELINE SYSTEM CONTROL

Pipeline system control requires the selection of a control strategy. The strategy depends on the type of liquid, type of prime mover (fixed vs. variable speed), type of controlling station (meter station, pump station, backpressure controller, etc.), and its location and the pipeline system (delivery junction, steep terrain, permafrost zone, etc.). The variables for controlling pump stations are pressures, flow, and possibly temperature. Temperature control is not required for most liquid pipelines except when the fluid is heavy crude with high viscosity or when the pipeline runs along a permafrost zone. For batch operation, product density and the interface is a control variable.

For a pump with a constant speed prime mover, the pump head is fixed for a given flow and thus the discharge pressure can be reduced by throttling the flowing liquid. The throttling action is performed by a pressure control valve, installed downstream of the pumps. The pressures discharged from pumps with variable speed drivers are controlled by the speed of the drivers with maximum power override. A pump does not require flow control as long as the flow is within the pump's capacity.

Side stream delivery may disrupt the main line pressure. To avoid potential pressure disruptions, the main line pressure is controlled by holding the delivery pressure. If a liquid pipeline runs in terrain with a steep elevation drop, the pressure around the peak elevation point will drop below the vaporization point, creating a slack flow condition, unless a high backpressure or high pressure downstream of the peak point is allowed. A backpressure controller, restricting flow, is needed at a location downstream from the peak point to avoid this condition.

Pipeline system control is accomplished by means of a set point mechanism. In other words, the dispatcher sets pressure, flow or temperature at the desired level and the control system responds to reach the set point. Since pressure is the primary control variable, several pressure set points are discussed below. The controlling pressures, that can be monitored and changed by the dispatchers through the SCADA system, are:

- Suction set point: the desired suction pressure at the station. During normal operation, the suction pressure is equal to or higher than the suction set point. The control system will not function properly if the suction pressure is less than the set point, unless the pressure measurement is erroneous. For liquid pipelines, suction pressure control with discharge pressure override is commonly used to maintain the pressure above the vapor pressure and at the same time keep the pressure below the maximum allowable operating pressure (MAOP). Normally, the minimum suction set point is higher than the station trip pressure below which the station automatically shuts down.

- Discharge set point: the desired discharge pressure at the station. The discharge set point is the pressure that the station control system tries to maintain as a maximum value. No control action takes place if the discharge pressure is below the discharge set point. For a pump with a constant speed driver, a control valve is used to control the discharge pressure. The discharge pressure is equal to or lower than the pump casing pressure, and the difference between the casing and discharge pressures is called the throttle pressure. Normally, the maximum discharge set point at a station is lower, say by 200 kPa, than the maximum operating pressure, in order to avoid an accidental station shutdown.
- Holding pressure: the holding pressure is set to maintain a desired main line pressure at the junction where a side-stream delivery may take place or where there is a delivery point but no pump station. The holding pressure for the latter case is called the delivery pressure, while for the former case the pressure measurement device is installed on the suction side of a station where delivery takes place.

4.10.1 Pump Station Operation

This section presents an overview of the key aspects of the pump station control. The following basic control operations are involved in controlling a pump station properly:

- Control of pumping units including their driver,
- Control of station valves and possibly line valves,
- Alarm annunciation to correct an operation for limit violation,
- Emergency shutdown to prevent possible damage to facilities.

The first two control operations are usually interlocked for starting and stopping operations of stations and/or pipeline system.

Nowadays, pipeline system control is automated to provide the capabilities of operating pipeline systems reliably, efficiently and thus economically. Automated system control enables monitoring and control of pipelines, pump stations, metering stations, and other facilities through a SCADA system. In addition, it can be extended to other applications such as storage management, energy optimization, volume accounting, leak detection, etc.

It is now generally an accepted practice that stations are automated and operated under remote control from a central SCADA control center. This is possible because a centralized system provides the capability to monitor the complete pipeline system and control the entire pipeline system in a safe and efficient manner. Only under abnormal conditions or during some maintenance tasks will the station be under local control. Some stations may be completely unmanned whereas others will have maintenance staff on site but who will not normally be in control of the station equipment. Figure 4-48 shows a typical implementation of a pipeline system and station control.

The key criteria of deciding if the station should be unmanned are that the station can be operated reliably and robustly and the control system has a high level of availability. In addition, transfer of control from "remote" to "local" must be easily supported in the event of an abnormal situation.

The following factors are usually taken into account in deciding if the station is to be unmanned:

- Safety can be enhanced by minimizing potential human errors;
- Reliable and quick control response can be achieved;
- Stations can be more efficiently operated remotely to reduce the operation cost.

Overall, the costs of installing the automation system and of maintaining the unmanned station must be compared against the costs of manning the station.

A typical station control system consists of several components; station control, unit control, driver control, and other auxiliary unit control. A station can be operated locally as well as interfaced to a SCADA system to enable remote control from a central control center. References [14, 15] discuss the control in more detail.

A properly designed remote control system will provide the ability to:

- Monitor all equipment associated with the station including station auxiliary systems;
- Provide two-way communication between the station and the host;
- Monitor the starting and stopping sequence of the drivers and pump units;
- Control and monitor sequencing of station valves;
- Initiate an emergency shutdown of the station or unit.

These extra system control capabilities can meet the following objectives for station operation:

- Operate the station safely and reliably, while maintaining cost efficiency;
- Allow constant monitoring of critical components of the station;
- Shorten response time to potential problems;
- Eliminate mundane tasks for the station operators.

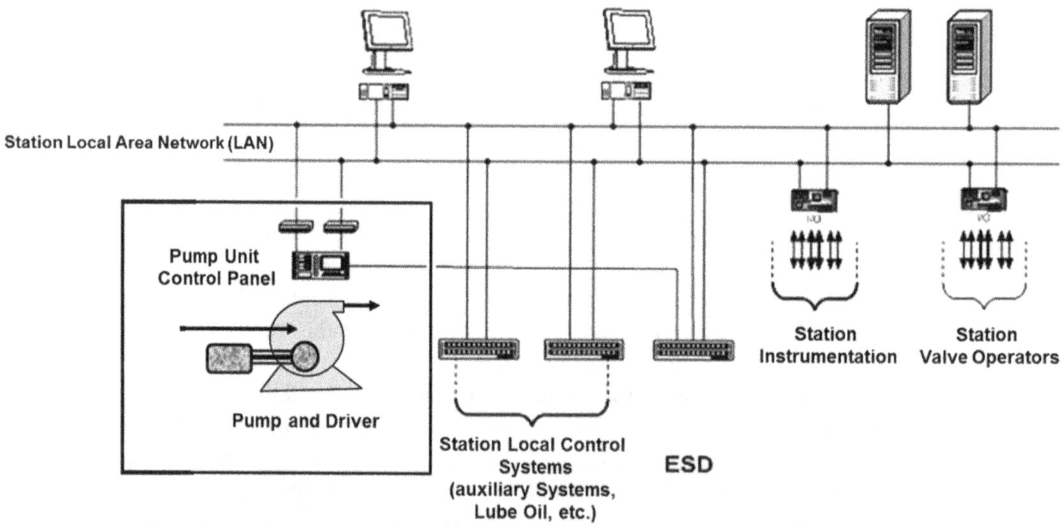

Figure 4-48. Implementation of station control system

Unplanned outages can cost a pipeline company tens of thousands of dollars. Therefore, the station control system must be reliable, robust and have a high level of availability in order to minimize business interruptions and maintain a safe environment for personnel and equipment. It must also be able to transfer control from "remote" to local in the event of an emergency or an abnormal situation. From the perspective of the overall operation of a pipeline, a pump station can be viewed as a "black box" that maintains product flow by offsetting pressure losses in the pipeline. The pipeline operator may only be interested in setting pressures at the various stations and not be concerned with the control of the individual units. In this situation, the station control system would receive station set points rather than individual unit set points from the SCADA system. It would then determine how many units should be operating and the set points for each unit.

An alternate control scheme is to include the station control system within the SCADA system. The system operator would then be initiating start/stop commands and relaying them to individual units as well as sending them the required set points. Through the SCADA system, the following data required for monitoring and controlling pump units are displayed:

- Pump unit status
- Flow rate
- Product name and density or API gravity
- Suction pressure and its set point
- Discharge pressure and its set point
- Throttle pressure or difference between the pump casing pressure and the station discharge pressure
- Holding pressure
- Station electrical load
- Communication status
- Alarms

In practice, the local station control system is implemented with a programmable logic controller (PLC). PLCs are now the heart of station control for station equipment (pumps, drivers, lube oil systems). PLCs consist of a programmable microprocessor unit, communication modules, and input/output modules for connection to field devices. The PLC has overall control capability for the station. This includes all equipment not under the direct control of a unit control system. It ensures that the station operates within the parameters for the station and mainline piping (above minimum inlet pressure, below maximum allowable operating pressure, etc.). In addition, the PLC determines the required set points for the operating units based on the required station set points received from the pipeline operator via the SCADA system. The individual set points sent to each unit will be determined based on a load sharing strategy. This will vary depending on the type of units installed and the overall pipeline operating strategy. They may include such strategies as:

- Base loading: one or more units may be operated at a constant load while other more efficient units are used to compensate for small changes.
- Optimum load sharing: set points for each unit are determined based on knowing the individual unit operating curves and allocating load to minimize overall energy consumption.

With the increase in computing capability, it is now more common for pipeline companies wanting to optimize their pipeline operations to consider having a system optimizer that would optimize pumping usage on the entire pipeline. This is discussed in more detail in Reference [14].

4.10.2 Pump Control Strategy

An optimum design of a station control system includes factors such as pump costs, spare parts requirement, load, operation and maintenance, and flexibility of operation. The pump control strategy must incorporate the following criteria:

- Pump suction pressure must be above the minimum Net Positive Suction Head (NPSH) for the pump in order to prevent cavitation of the pump.
- Pump discharge pressure must be below the maximum allowable operating pressure (MAOP) of the station discharge piping to avoid pipe and associated equipment damage.
- Station discharge pressure must be below the MAOP for the pipeline to avoid damage and to ensure the pipeline is operating within the acceptable limits approved by the regulatory agency.
- Station suction pressure must be above the minimum allowable operating pressure to meet contractual requirements and in the case of liquid lines to avoid "slack line flow" or column separation upstream of the pump station.
- Driver power must be kept within acceptable limits to avoid tripping of the driver.

A pump driven by a constant speed electric motor driver requires a discharge control valve to control pump throughput; the system controlling this valve must have a station suction pressure (or station discharge pressure) control loop. Set points for maximum station discharge pressure, minimum station suction pressure, and maximum motor power are set on the controller. The controller will satisfy the set point for station discharge until the suction pressure or driver power limits are reached then these will override the discharge pressure set point.

Pressure switches are set to provide a trip signal in the event of controller failure. The final backup is a pressure relief valve in the event of a complete control system failure.

For a pump station, that contains both fixed-speed and variable speed motors, the control strategy is to run the fixed-speed units at a base load with minimal throttling and utilize the variable speed unit(s) to adjust for the required station set points.

Generally, there are three major levels of monitoring and station control in the hierarchy of automated pipeline stations, namely:

Local:	In this mode, command control of all local devices and skids is passed to the station control system. This allows a local operator to control the complete station and all auxiliary equipment from a single location at the station. In this mode, command control is limited to the local device or the skid control panel.
Remote:	If the station control system is in "remote" mode, then in effect all control is from the SCADA control center.
Control center/SCADA:	In this mode, command control of the station is passed to the central control center via the SCADA system. No local control is possible. This is essentially the "remote" mode for the station control system. Process values and status may still be sent to SCADA for monitoring and logging purposes.

It is important to realize that the control levels described here affect the state of operator control. In all modes, the local device is always being controlled by its control equipment. The change of mode describes from where, for example, set points or commands to the local controller will originate.

4.10.3 Station Control

An automatic station controller has the ability to start, stop, and adjust units based on the corresponding command sent by the pipeline system operator usually at the control center or an operator occasionally at a station if communication is lost. In addition, it can control pressure or load sharing as flow rate changes, based on the set points entered by the operator. Proper control logic including operation sequence and time is essential to control stations and to reduce pipeline transients. In general, a care must be exercised when a valve is closed. If the valve is closed too rapidly, large pressure surge can be generated and the surge can damage the pump and other pipeline components. Therefore, the surge effect should be minimized by closing the valve slowly.

4.10.3.1 Pump Station Valve Control

One of the key aspects of the pump station control system is the station valve control.

Each valve in the station is controlled remotely, but if necessary can be controlled locally by a field operator. The valve control logic incorporates interlocks with motor-operated valves to ensure proper sequencing and to avoid damage to equipment. Some sequencing scenarios that the control system contains include:

- Station start up and shutdown;
- Scraper launching and receiving;
- Station by-passing;
- Batch receiving and batch launching.

In addition, there may be some control logic required to help minimize or reduce pipeline surges (transients) depending on the results of the pipeline hydraulic studies.

The station controller directs the sequences of station valves by changing their position for the following station operations.

- Open and/or close valves in pre-arranged sequences for starting a pump station or bringing additional units on line;
- Close and/or open valves in pre-arranged sequences for shutting down a pump station, by-passing a station, or bringing additional units off line;
- Close and/or open valves in pre-arranged sequences for launching and receiving a pig or scraper;
- Partially close or open a control valve to adjusting the station discharge pressure for fixed speed pumps. Control valve operation may not be needed for a variable speed pump, unless the station is installed with both types of pumps (Figure 4-49).

Described below are the station start-up and shut-down sequences (refer to Figure 4-51):

- Start the station from an initially shut-in state.
- Assuming the station isolation valves, MOV-101 and MOV-102, are closed, they are opened and Pump 1 is primed.
- The pump starts rotating; the suction and discharge valves are opened as the pump casing pressure increases, and the liquid starts flowing.

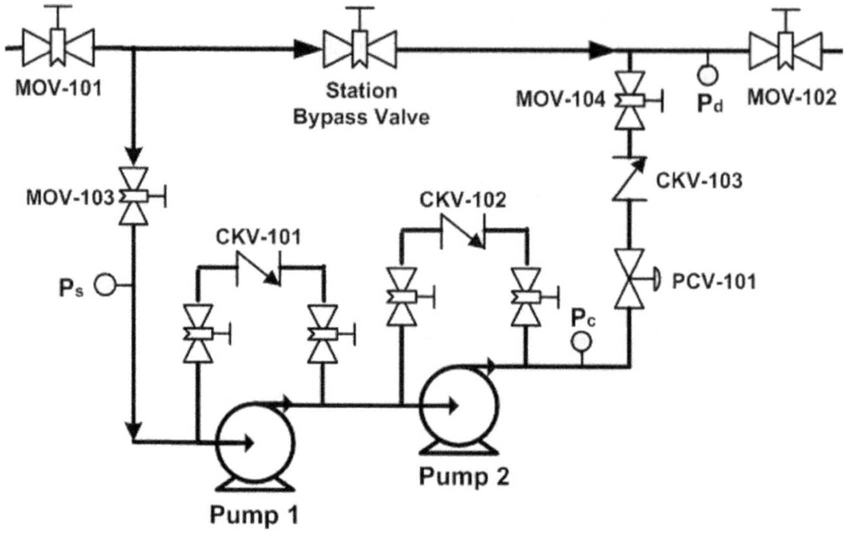

Figure 4-49. Pump station operation

If the flow rate needs to be increased, the same sequence is followed for Pump 2.

- MOV-101 starts closing.
- MOV-103 and MOV-104 start closing, as the pump slows down or shuts down.
- MOV-102 is closed and the station is shut-down.

4.10.4 Injection/Delivery Station Control

Injection of products from and delivery to tank farms or other pipelines requires valve operation to properly control pressure level and flow direction. The pressure control is required to avoid damage to equipment, while flow control is needed to deliver products to their correct destinations. Valve sequencing operation is also necessary for launching and receiving batches.

4.10.5 Pump Unit Control

Pump units are controlled in one of two ways: discharge pressure control by throttling a control valve and pump speed control. Throttling is a common way of controlling the discharge going through a centrifugal pump and the control valve is installed on the discharge side of the pump. Another way to control the discharge flow is to control the speed of the pump, which in turn is controlled by the speed of the pump driver. Unlike valve closing, reducing the flow by reducing the pump speed does not waste energy. This controlling action results in changes in flow through the pump and pressure in the pipeline.

In operating a pump unit, several problems occur: the violation of the required minimum flow and the presence of vapour in the pump. The minimum flow requirements of the pump must be carefully taken into account during the design and operation of pumping units. If the flow is slow, energy is converted to heat due to low pumping efficiency and the heat cannot be carried away quickly. The liquid in the pump will heat and eventually vaporize. Typically, the pump manufacturer will place a minimum flow requirement of about 40% of design flow for pumps associated with the pipeline industry. For most of the time, this does not limit operations but care must be taken during the start-up of the line.

Typically, at the inlet station, some method of recirculation is provided so that the inlet pumps can be brought on-line safely. When the pipeline is running at a large base flow, this recirculation valve may be manual, but for lines where the product is stopped and started, it may be controlled automatically by the unit control logic so that it is open until the minimum flow requirement down the line has been established. The goal is to have the pump operate at or near the most efficient point-labeled Best Efficiency Point (BEP) in Figure 4-50.

The pump may be allowed to continue to run for a short time (at most a few minutes) after the valve is closed and thus the discharge flow is zero. If the valve is closed or even slightly opened, but the pump keeps running, energy is wasted, resulting in an overheated and highly pressurized pump and subsequently shortening the life of the pump.

Another problem is the presence of vapor in a pump. If the pressure drops below the vapor pressure of a liquid at the pump suction, the liquid vaporizes and the vaporized liquid forms bubbles. These bubbles move with the flow into the pump impeller and volute where the pressure of the liquid increases sharply. Then, the bubbles collapse in the high pressure area and the collapsing bubbles can generate localized high-pressure, causing damage in the form of surface pitting. The problem is normally avoided by increasing the pressure of the flow on the suction side of the pump — making the available NPSH higher than the required NPSH.

However, when a pump is shut down, vapor can fill in the pump unit and station piping. If the pump is started under such a condition, the pump impeller will be spinning without liquid flowing through the pump, and thus the liquid cannot be drawn into the pump fully and the flow is slow. As a result, the pump can be overheated if such an operation lasts a long time. To prevent this from happening, the pump must be primed with liquid before starting. After the priming is done, the flow is allowed to increase until it reaches the desired flow level. If a control valve is installed for a fixed speed pump, the control valve should be opened gradually after the flow starts flowing through the pump.

4.10.6 Throttling vs. Speed Controls

This section will consider pump station operation using both constant speed and variable speed electrically-driven pumps. Fixed-speed electric motors provide a cost-effective

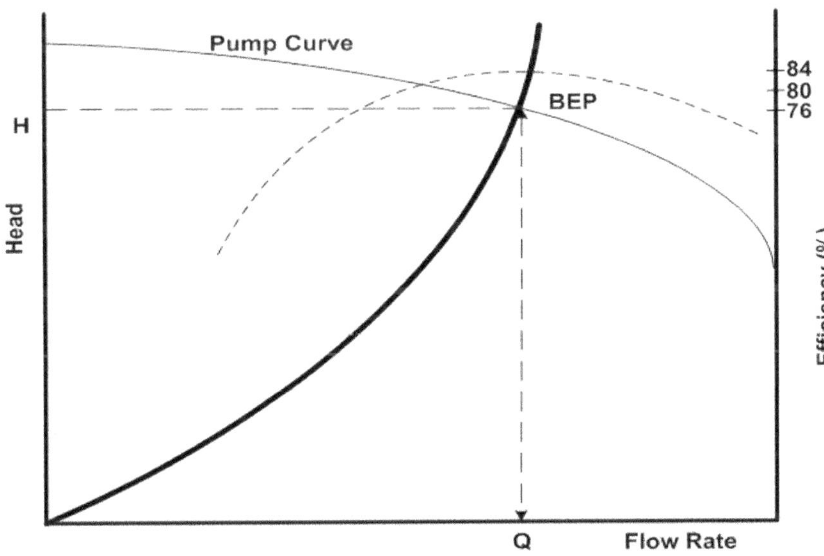

Figure 4-50. Best efficiency point

solution for base load applications where electrical power is available and reliable. They have the advantage of low maintenance costs and are simple to operate. Variable speed drive (VSD) motors are becoming the standard for pump stations that have varying flow or product density requirements such as on batched product pipelines. Despite their control systems being more complex than for a constant speed motor, variable speed motors are much more energy efficient. This is because pump capacity can be controlled without the disadvantage of pressure loss incurred by the throttling through a discharge control valve.

Variable speed pumps control the flow and pressures by varying the speed of the drivers with maximum power override. For a pump station that contains both fixed-speed and variable speed motors, the control strategy is to run the fixed speed units at a base load with minimal throttling and use the variable speed units to adjust for the required station set point. Figure 4-58 exhibits the performance curves of a variable speed pump.

In applications that require flow or pressure control, the most energy efficient option is an electronic VSD, referred to as a Variable Frequency Drive (VFD). The most common form of VFD is the voltage-source, pulse-width modulated frequency converter. The converter develops a voltage directly proportional to the frequency which produces a constant magnetic flux in the motor. This type of speed control can be related to set points of discharge pressure or flow.

4.10.6.1 Throttling for Fixed Speed Pumps

As discussed in Section 4.7.1, there is only one operating point for a fixed speed pump. As shown in the figure below, a throttling action is required to match the system head curve to the pumping head curve of a fixed speed pump at a particular flow rate other than the design flow. Fundamentally, throttling changes the system curve. A control valve is used to throttle the fluid flow, and is installed downstream of the pump.

Figure 4-51 shows that the pump operates at H_1 for the design flow rate Q_0. If a throttle valve is partially closed in the pump discharge line, the throughput drops from Q_0 to Q_T, and additional friction pressure drop occurs through the partially closed valve. As a result, the pump will operate at a new operating point, H_2.

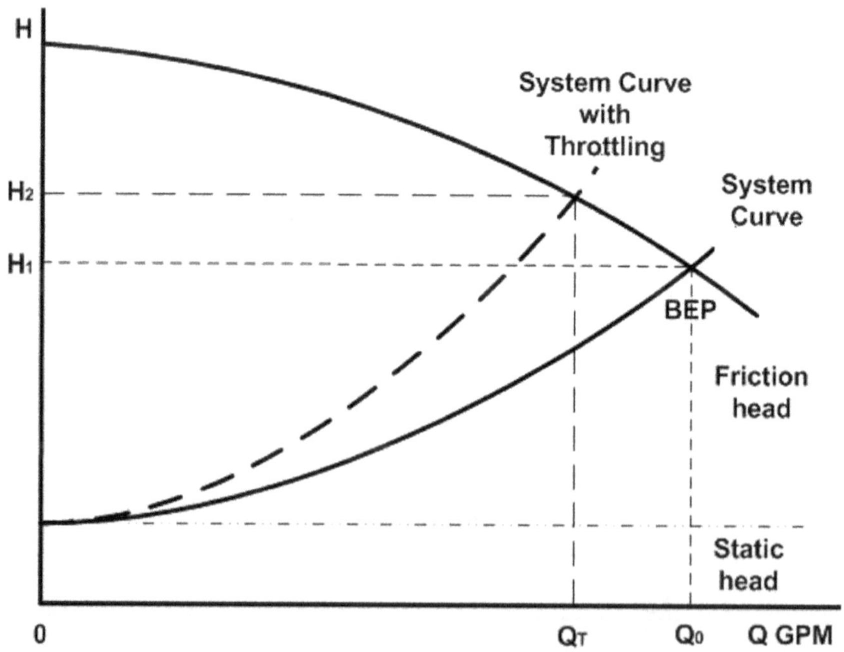

Figure 4-51. Capacity change with throttling

The throttle pressure is the difference between the case pressure and the discharge pressure. The casing pressure of a pump is the available pressure generated by a pump, and the discharge pressure is the pipeline pressure on the discharge side of the pump station. The discharge is the pressure required to transport the liquid to the next pump station or terminal. The throttle pressure is unused pressure developed by the pump and thus results in wasted power. Figure 4-52 illustrates the energy losses caused by throttling; the greater the throttle pressure, the greater the energy loss.

4.10.6.2 Speed Control for Variable Speed Pumps

Pipeline systems operate at flow rates different from the design conditions, because supply or demand changes, liquid properties also can vary as in the case of batch operation, or other operating conditions. Such varying conditions demand flow control. Depending on the varying conditions, there are several ways of controlling flow rates:

- Install a control valve at each pump station to throttle the flow rate;
- Install multiple pumping units to provide sufficient discharge head that can be matched to the flow requirement;
- Install variable speed pumps.

Variable speed pumps control the flow and pressures by varying the speed of the drivers with maximum power override. For a pump station that contains both fixed-speed and variable speed motors, the control strategy is to run the fixed speed units at a base load with minimal throttling and use the variable speed units to adjust for the required station set point.

Even though variable speed pumps are more expensive, it is advantageous to install and operate variable speed pumps because energy cost can be saved and it they easily applicable to a wide range of flow changes.

Compared to fixed speed pumps, variable speed pumps can produce significant energy or power savings as illustrated in Figure 4-52 [16]:

Figure 4.53 shows a pump operating at the operating point B, where the flow rate and pressure are Q_H and P_B, respectively. Here, for discussion purposes, it can be assumed that the operating point is the best efficiency point (BEP), where the efficiency

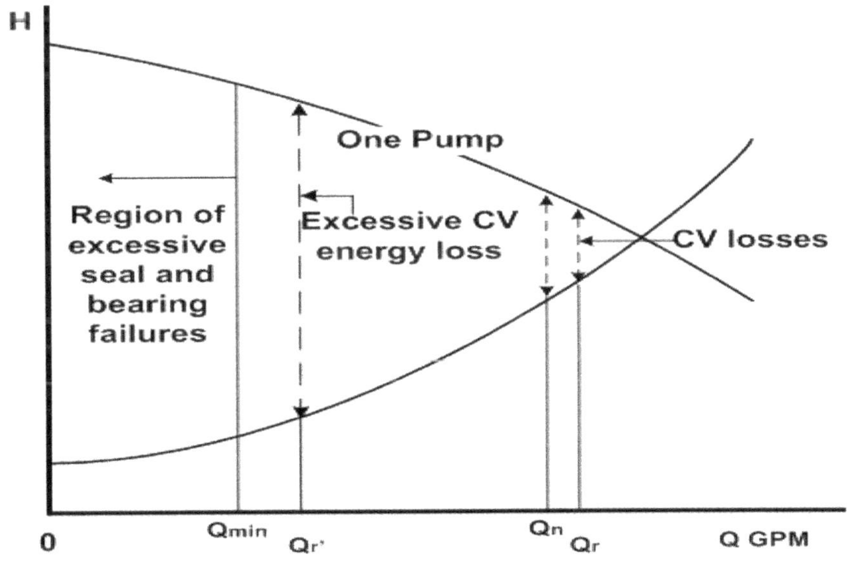

Figure 4-52. Energy losses due to throttling

of the pump is highest. The power required by the pump for the high flow condition is $PW_H = Q_H \times P_B/\eta_{BEP}$, where $\eta_{BEP} = 84\%$, the pump efficiency at the BEP, and is represented by the area of rectangle, P_B0Q_HB, in the figure.

To achieve the lower flow rate Q_L, the control valve is partially closed for the fixed speed pump or the pump speed is lowered for the variable speed pump, as illustrated in the figure; *C* is the operating point of the fixed speed pump and A that of the variable speed pump for the lower flow rate. At *C*, the flow rate is reduced to Q_L but the pump pressure is increased to P_C for the fixed speed pump, and thus the power required by the fixed speed pump is $PW_C = Q_L \times P_C/\eta_C$, where the pump efficiency, η_C is lower than the pump efficiency at BEP. On the other hand, the operating point of the variable speed pump is A for the same flow rate. There, the power is $PW_A = Q_L * P_A/\eta_A$, where the pump efficiency, η_A may be lower than the pump efficiency at BEP but will be higher than η_C. As shown by rectangles in the figure, P_A0Q_LA, the power required by the variable speed pump is lower than the power required by the fixed speed pump, because the pressure requirement is lower and the pump efficiency is higher.

In summary, the fixed speed pumps waste energy by throttling the flow to achieve a lower flow rate, because:

- The pump operates at a reduced efficiency,
- The pump is required to produce an increased pressure.

Energy savings can result from using variable speed pumps, and thus it is advantageous to use them from the viewpoint of reducing the energy cost. In addition, variable speed pumps offer the following advantages:

- Pressure surge can be small, particularly during pump start-up and shut-down operations, because changes in flow and pressure occur gradually.
- They provide flexibility of controlling flow over a wide range.

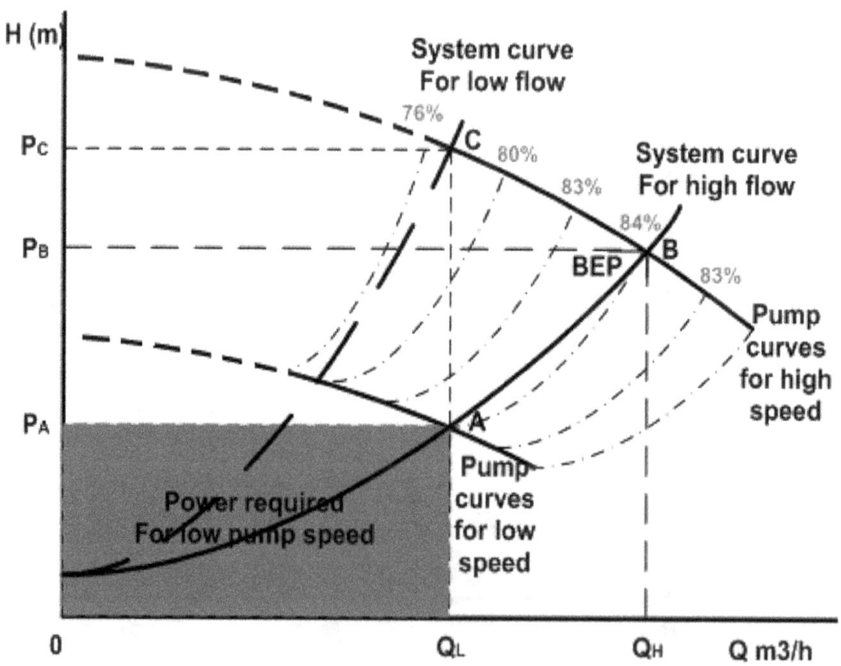

Figure 4-53. Power required at operating points A, B, C

4.11 STATION ELECTRICAL CONTROL

A pump station using electric motor drivers requires a reliable source of electricity. This may be supplied from a commercial source or generated at the station. Economic and reliability considerations usually determine the choice of power source.

The electrical supply usually will have high voltage feeders, voltage reduction equipment, and be a multi-bus operation with its associated transfer equipment. All the bus and equipment protection required to support such a system is normally provided with the electrical equipment. Controls for this equipment may be incorporated into stand-alone control equipment or they may be part of the station control system.

The electrical protection is always contained in stand-alone, specialized equipment package that will protect against:

- Over and under voltage
- Over and under frequency
- Over current and short circuits
- Ground fault
- Voltage imbalance
- Phase reversal
- Transformer gas and high temperature

The electrical supply control system monitors the electrical system and sends the following information back to the station control system:

- Voltage and current values
- Real power, power factor
- Electrical energy consumption
- Circuit breaker and disconnect position
- Frequency

4.11.1 Station Auxiliary Systems

The station control system controls and monitors the functioning of all station auxiliary systems common to the operation of all units. These systems include some or all of the following equipment, depending on the specific station requirements:

- Auxiliary (emergency) electrical generator
- DC Battery charger(s)
- Inverter
- Security system
- Boiler (if required)
- Air-conditioning
- Commercial AC power monitor
- Ground fault detection
- Starting Air System(s) for a gas turbine driver
- Fuel system (for non-electric drivers)
- Vent fans and louvers
- Inlet air filter system
- Central lube oil conditioning (filter/cooling)
- Fire and gas detection system
- Station Emergency Shutdown Device (ESD)

Generally, the design of the station control system allows for the complete control of the station to be from a local control room, with the option of passing control to the pipeline controller via the SCADA system. This would allow the station to be operated remotely and be unattended.

4.11.2 Shutdown Modes

A typical arrangement for station controls is to have different levels or severity of shutdowns such as:

Normal Shutdown:	This will shut down the equipment through a normal shutdown sequence. The unit can be restarted normally. This would be initiated by an operator command or may be required if process conditions exceed limits. Once process conditions have been restored, the unit can be restarted.
Shutdown Lockout:	This is activated to stop a unit due to a serious problem such as loss of lube oil, etc. "Lockout" means the unit cannot be restarted until manually reset locally. This ensures that the site is visited by a technician/operator, who must evaluate the situation before the unit can be restarted. This can apply to individual units or the complete station. Wherever possible, the shutdown will follow normal shutdown procedures to minimize hydraulic disturbances.
Emergency Shutdown (ESD):	This condition requires immediate shutdown of all units and will initiate a hydraulic isolation of the station. In a natural gas pipeline, this will also result in the activation of associated blow-down valves. Following an emergency shutdown, all controls will be in a lockout state and require local resetting.

4.11.2.1 Emergency Shutdown System

The purpose of an Emergency Shutdown System (ESD) is to provide a fail-safe independent control system that can shut down a station and isolate it in the event of a pipeline rupture, station piping rupture or a fire at the station.

From a design perspective, ESD systems should be hardened against the explosive forces and fire associated with this type of system failure. Indeed, to be fail-safe, the ESD feature should be capable of automatically isolating the flow of product to an accident site until it has been verified that it is safe to reactivate normal operations. The ESD system overrides any operating signals from the station or local controls and its design therefore, needs to meet the requirements of both the regulatory regime and the owner's own design philosophies and criteria.

The ESD is the last line of defence to shut down a station and must be able to perform its function even if the station has lost normal power supply, has lost the ability to communicate with SCADA or in the case of local control, system failure. Normal designs of an Emergency Shutdown Controller (ESD) provide for them to be independent of the station controller itself. It should also be possible to test the ESD system on a regular basis without interrupting normal operations. They will typically include redundancy control capability to ensure that no single point of failure in the ESD system will disable the capability to properly detect and execute an ESD action.

A station ESD system has associated shutdown valves to isolate the station. If ESD valves close too quickly a pressure transient can be generated that could damage facilities. Hydraulic studies are usually undertaken to determine ESD valve closure times in order to limit pressure transients along the pipeline from the station.

4.12 APPLICABLE CODES AND STANDARDS

- Hydraulic Institute Standards
- American Petroleum Institute (API) 610 11th Edition, September 2010 — Centrifugal pumps for petroleum, petrochemical and natural gas industries
- ISO 13709:2009 — Centrifugal pumps for petroleum, petrochemical and natural gas industries (Identical to API 610)
- API 614, Fourth Edition: General Purpose Lube Oil System components For Rotating Process Equipment.
- API RP 1110, 2007: Pressure Testing of Steel Pipelines for the Transportation of Gas, Petroleum Gas, Hazardous Liquids, Highly Volatile Liquids or Carbon Dioxide. This RP applies to all parts of a pipeline or pipeline facility including line pipe, pump station piping, terminal piping, etc.
- BS 5136: Acceptance Tests for Pumps for Centrifugal Mixed Flow and Axial Pumps
- ISO 13709:2003: Centrifugal pumps for petroleum, petrochemical and natural gas industries, (Metric)
- ISO/TR 17766:2005: Centrifugal pumps handling viscous liquids — Performance corrections
- ISO 21049:2004: Pumps — Shaft sealing systems for centrifugal and rotary pumps

REFERENCES

[1] Lobanoff, V. R. R., 1992, *Centrifugal Pumps: Design and Application*, 2nd Edition, Butterworth, Heinemann, 1992.

[2] Karassik, I., 1976, *Pump Handbook*, McGraw Hill, New York, NY.

[3] Mohitpour, M., Golshan, H., and Murray, M., 2007, *Pipeline Design & Construction, A Practical Approach*, 3rd Ed., ASME Press, New York, NY.

[4] Chaurette, J., 2004, "Centrifugal Pump Specific Speed Primer and the Affinity Laws," *Pump-Flo*, pp. http://pump-flo.com/pump-library/pump-library-archive/jacques-chaurette/centrifugal-pump-specific-speed-primer.aspx, November.

[5] Karassik, I., 1994, "Setting the Minimum Flows for Centrifugal Pumps," *Pumps and Systems Magazine,* March.

[6] Andrews, D., 2004, "Cavitation - Intelligent Maintenance of Pumps," *Run Times - Lawrence Pumps Inc.,* October.

[7] A. D. B., 2007, "Discharge Recirculation and Cavitation," *Run Times - Lawrence Pumps*, April.

[8] Andrews, D., 2007, "Intelligent Maintenance Management of Pumps," *Run Times - Lawrence Pumps Inc.,* April.

[9] Andrews, D. B., 2004, *Cavitation*, Run Times - Lawrence Pumps Inc., http://www.lawrencepumps.com/Newsletter/news_v01_i5_oct.html, October.

[10] Lohrberg, H., Stoffel B., 2000, "Avoiding Cavitation Erosion," *Pump Users International Forum*, Karlsruhe, Germany.

[11] Andrews, D., 2007, *Viscocity and Pump Performance*, Run Times Vol 4 Lawrence Pumps Inc., February.

[12] Hydraulics Institute, 2004, *Variable Speed Pumping - A Guide to Successful Applications - Executive Summary*, ISBN 1-85617-449-2.

[13] Hydraulics Institute, 2000, *Pump Life Cycle Costs: A Guide to LCC Analysis for Pumping Systems* ISBN 1-880952-58-0.

[14] Yoon, M., Warren, C., Adam, S., 2007, *Pipeline System Automation and Control*, ASME Press, New York, NY.

[15] Mohitpour, M., Van Hardeveld, T., Peterson, W., and Szabo, J., 2007, *Pipeline Operation and Maintenance — A Practical Approach*, ASME Press, New York, NY.

[16] Lohrberg, H., Stoffel B., 2000 Intelligent Maintenance Management of Pumps; Pump Users International Forum.

Chapter 5

PIPELINE OPERATION AND BATCHING

5.1 PIPELINE OPERATION

The scope of pipeline operations is very large. It includes both business and physical operations of a pipeline system. The scope of the physical operations covers not only the product movements by means of daily operations of equipment such as pumps and meter stations but also other operations such as pipe cleaning or pigging and integrity management. The process of transportation service is further discussed in Control Valve Handbook [1]. This section discusses only the daily operations directly related to hydraulics.

5.1.1 Pipeline System Operation

Petroleum products, including crude oil and refined products, are gathered to central points. The gathered products are then scheduled and dispatched to various destinations. The scheduling activity begins with the preparation of a product pumping schedule based on pipeline capacity and shippers' requirements, while the dispatching activity involves organizing various operations of the pipeline facilities to conform to the schedule. In scheduling and dispatching, the pipeline system needs to be operated in a safe and efficient manner. This section briefly discusses three operational issues; pipeline system operation, batch operation, and station operation.

Broadly, pipeline operations involve transporting products from the lifting points to the delivery points; lifting products into the pipeline and delivering the received products to the designated delivery points. The receiving and delivery points may be tank farms, refineries or another pipeline. For transporting products, the pipeline operator monitors product movements and pipeline states to ensure that products are adequately supplied from the lifting points and delivered to the designated points, while controlling flow and pressure at pump and regulator stations for safe and reliable transportation.

The control of pipeline pressure is crucial to ensure safe, reliable, and economical operations because all pipelines are subject to minimum and maximum operating limits. Maintaining these limits is essential to preserve the integrity of the pipeline and pumping equipment, and at the same time to operate the system economically. The operating limits include:

- Maximum allowable operating pressure (MAOP) for safety
- Maximum discharge pressure at pump stations to protect station piping and equipment
- Minimum suction pressure at pump stations to avoid cavitation
- Minimum or maximum pressure at a control point
- Minimum and maximum delivery pressures to satisfy the contractual obligations and protect equipment

- Minimum and maximum flows through pumps for efficient pump operation within the pump design limits
- Maximum power of pumps to operate below the pumping power limit
- Minimum discharge temperature for heavy crude
- Maximum discharge temperature for pipelines in permafrost

The operating pressure must be maintained below the maximum allowable operating pressure (MAOP) and above the vapor pressure of the liquid. Pressures which are too low will result in low flow and inefficient operation. The maximum station discharge pressure is selected as the lower of either the MAOP of the downstream section of the pipeline or the maximum allowable pressure rating for valves and fittings.

A control point is a point in a pipeline section where the pipeline pressure drops below the vapor pressure or exceeds the MAOP. The pressure control at these points is critical to avoid damage to the pipe or pipe rupture or to improve pipeline efficiency. High-pressure control points occur where the MAOP limits are violated, while low-pressure control points occur at the high elevation sections of the pipeline and at pump suction inlets. The pipe section can be damaged around low-pressure control points through a column separation and the subsequent collapse of the accompanying vapor pocket, or through cavitation in the pumps. Refer to Section 5.1.4 for an example of a pipeline operation including the operation of the control points.

In addition to the control point, holding pressure is required at certain stations to regulate the delivery pressure. Pressure control facilities are installed to hold the pressure on the suction side of a station where deliveries take place at the same time as the station is pumping out of a tank farm. In this case, the holding pressure represents the pipeline pressure and the suction pressure represents the pressure at the station suction. Holding pressures are available at certain locations where deliveries take place without any pump station involvement.

The initiating pump station in a tank farm receives liquids from tanks. Since the tanks do not provide sufficient suction head, booster pumps are installed between the tanks and the pump station. Refer to Chapter 9 for tank farm operation. Pump station operation will differ depending on the size and type of equipment as well as the pipeline configuration. If a pipeline is relatively short, the initiating pump station draws from tankage and pumps directly into the tankage at the next station. However, most long pipelines have multiple intermediate pump stations, where one station pumps directly into the suction side of the pumps at the next station. This method is called tight line operation and offers several advantages over the previous operation:

- Tanks are not necessary at the mainline stations, reducing tank costs and simplifying operations,
- Extra operation of pumping into and out of tankage is not required,
- Interfacial mixing is minimized for batch operation.

Before starting mainline stations, the pipeline operator should make sure that the dispatching schedule is planned, the required pump stations and units are in place, and all equipment and protective devices such as valves are ready for service. After such a check is satisfactorily completed, the operator proceeds according to the operation plan. An example of starting up a pipeline system is given in Section 5.1.4.

Since a pump station is the facility used for boosting pressure, there are several pressure monitoring and controlling points at each pump station. Two pressures to be controlled are:

- Suction set point: The pipeline operator can change the suction set point which is the pressure level to be achieved at the station. The actual suction pressure may be close to the set point measured upstream of the pump unit within the station and represents the pressure available in the pipeline to push the liquid into the pumps. There are suction pressure limits; the lower limit below which the pump will not operate and will shut down, and an upper limit above which the suction pressure has to be reduced to avoid over-pressuring the discharge pressure at the upstream station.
- Discharge set point: The discharge set point is the pressure that the station control system tries to maintain as a maximum pressure. No control action takes place if the actual discharge pressure remains below the discharge set point. If the two pressures are equal or the actual discharge pressure exceeds the set point, the pressure regulator activates to reduce the actual pressure level.

In tight line operation, liquid density and/or viscosity changes, which can occur in batch or blending operation, result in pressure and/or flow variations. If the stations are equipped with centrifugal pumps and are operating at maximum pressure, the flow will decrease as the density and/or viscosity increase. If the flow is controlled and pumps discharge at a pressure lower than the maximum, the discharge pressure will increase as the density and/or viscosity increase. The converse is also true. Pump station and pump unit operations are discussed in Chapter 4.

A batch may be injected from or delivered to multiple locations anywhere in the pipeline system at which there are suitable facilities. If a batch is injected at an intermediate location, the injection can be full stream or side stream injection. For full stream injection, the upstream section of the injection location is shut down, producing zero upstream flow and the downstream flow rate is the same as the full stream injection rate. For side stream injection, a batch is injected into the flowing product, resulting in blended product if the two products are different. The downstream flow rate is the sum of the upstream flow and side stream injection rate. Full stream or strip delivery can be made at some points along the pipeline. For full stream delivery, the upstream flow of the delivery location is the same as the delivery rate and the downstream flow is zero. For strip delivery, the upstream flow is the sum of downstream flow and the delivery flow.

All batches should be pumped in a sequence during a fixed period, called a batching cycle. There may be more than one batch cycle per month, the number depending upon tank sizes or capacities at the terminals. The batching sequence is not always fixed, but practically speaking it may be fixed for every cycle as long as the same products are lifted. The batching sequence is arranged in such a way that is likely to result in the least amount of batch interface or mixing size. The batching operations including the method of determining a batch sequence is fully discussed in Section 5.2.

Batching is done either with or without a pig, called a sphere, separating the two adjacent products. A sphere can be inserted at the injection point or at the running pump stations and received at the delivery point or at the running pump stations further along the pipeline. Without a separation pig, interfacial mixing or transmix takes place at the interface boundaries between two adjacent batches. The interfacial mixing sizes depend on the pipe length and Reynolds number; the larger the Reynolds number, the smaller the mixing size. Therefore, it is not advised to design a batch pipeline that is to be operated in a laminar flow regime. Batch movements together with the separation pig or interfaces call for careful monitoring and control.

Common carrier pipeline companies have rules, on which they govern their batch operations for the shippers. The rules regarding the handling of transmix may include the following options:

- Depending on the product specifications, the interface mixture may be cut into one or the other product, or divided between the two adjacent products at the mid-gravity point. However, shipping certain products such as jet fuel does not allow any transmix to be included in the batch, because its specifications are very stringent due to its high flash point and the obvious dangers inherent in fuel control during engine combustion.
- The transmix that does not meet the shipper's product specifications is called slop. This off-spec product is accumulated in a slop tank and then sent to a refinery for reprocessing or blended into other tolerable product.
- If the size of a transmix is larger than the specified amount and cannot be blended into other tolerable product, the pipeline company may have to pay the costs of disposing it.

Batch tracking monitors the volume in each batch, its origin, its destination, its current location, and its estimated time of arrival to delivery locations. When batches are lifted, they can be launched either manually or automatically. Automatic batch launchers are economical and generate accurate and timely batch launch information. The launch information includes batch launch time and batch ID. The batch ID identifies the product of the lifted batch, batch number, size or quantity of the batch, and shipper or owner of the batch. Batch tracking and interface detection is needed to deliver to correct locations at accurate times.

Batch tracking begins when a batch is launched and a launch signal is generated by a batch launcher or the host SCADA. The launch is detected based on a change in density and/or valve status or on a batch schedule. Batch volumes are updated based on injection and delivery volumes obtained from metering locations. The interface positions are determined, given the order and volume of each batch in the pipeline. Batch tracking functions include estimated time of arrival (ETA) to the designated downstream locations.

Batch interface detection is necessary to notify the dispatcher of the batch arrival and to take subsequent operational actions. Batch interfaces can be detected by a densitometer if batches have different densities or by a dye detector if the batch densities are similar. If batches are separated by a sphere, the sphere has to be inserted at the time of batch launch and removed when it arrives at the delivery point. Upon completion of a batch delivery, the batch is removed in the batch tracking list and an over/short volume is calculated, reflecting the difference between metered injections and deliveries.

5.1.2 Concepts of Pipeline Transient Flow

Pipeline system design is mainly concerned with line sizing, route selection, equipment sizing and facility location. System operation is concerned with steady-state operations at different flow rates, pipeline system or facility start-up and shut-down, product receipt and delivery, flow rate change, emergency shut-down, equipment failure, etc. A pipeline system design can be based on a steady-state assumption. In general, the assumption is valid when the system is not subject to sudden changes in flow rates or other operating conditions over a short period of time. Refer to Section 3.1.2 for the discussion of solution methods.

While pipeline systems are being operated at or near steady-state conditions, it is inevitable that the operation conditions change even in normal operations, resulting in a transient state. When an operation change takes place, the flow rate and pressure change immediately, and consequently the change will have an impact on the pipeline system. Therefore, the steady-state assumptions are not valid for analyzing short-term operation problems and for designing control systems, locating facilities, etc because these scenarios are strongly dependent upon time. As a result, the surge effects have to be included in the design and operation assessment.

The solutions of the four fundamental equations governing pipe flow result in time-dependent solutions that can describe transient flows. Transient flows are in transit from one steady state to another. Transient solutions provide hydraulically realistic results. In general, a transient solution is more complex and difficult to use, as well as taking longer to find the solution than a steady-state solution. The transient solutions require extensive data, particularly equipment and control data, which are often unavailable. However, a transient model is essential for the efficient operation of the pipeline. Refer to ref. [2] for a detailed analysis of fluid transients in pipeline systems.

A transient state is an unsteady condition that changes with time between two steady states, while a steady state is a condition of a pipeline system that does not change significantly over time. Transient flow occurs when the flow is disturbed or changed in the pipeline system; these are often caused by changes in hydraulic facilities such as flow or pressure control equipment. Some transients are gradual, in quasi-steady state, and their magnitudes are small, while some are sudden, and their magnitudes are significantly large. Gradual and small transients or quasi-steady states are almost always present in the pipeline system and are easily controllable and thus not destructive in normal operations. An example of such a mild transient occurs while multiple batches move along the pipeline. Therefore, such mild transients are seldom a concern for pipeline design and operation. The distinction between small and large transients depends on how suddenly flow or the response of equipment changes, e.g., sudden valve or pump operation.

Large transients can occur suddenly in a short time when the fluid flow is interrupted because the fluid before the stoppage is still moving forward with its fluid velocity, building up a high pressure. A large transient, often called a pressure surge, is a change in pressure in the pipeline that occurs abruptly during a change from either a normal steady state or another transient state flow in the pipe. All surges travel at acoustic speed through the flowing liquid in the pipeline. If the pipeline pressure increases above the normal operating pressure of the pipeline, the surge is called an upsurge, while a pressure decreasing condition is called a down-surge. Excessively large pressure increases resulting from transients can cause damage to pipe and/or other equipment, causing pipeline systems to fail if the pressure is high enough. On the other hand, when an upstream flow in a pipe is suddenly stopped, the fluid downstream will attempt to continue flowing, creating a vacuum that may cause the pipe to collapse. This problem can be particularly acute if the pipe is on a downhill slope. Methods for mitigating both of these effects will be discussed later in the chapter.

We next describe the behaviors of a surge wave, assuming that a pipe with length L is attached to a liquid tank or reservoir at one end and a valve to the other end. When the valve is suddenly closed, the following sequence of changes in pressure and flow takes place in the pipe line [2]:

- At the valve, the fluid velocity stops instantaneously, and the pressure or head at the valve quickly increases by the amount of potential surge, ΔP.
- The pressure increase immediately results in the slightly enlarged pipe and also a density increase in the fluid. The amount of the pipe enlargement depends on the pipe size and wall thickness and the pipe elasticity, and on the compressibility of the liquid. The pressure increase also causes a sharp-fronted pressure wave to propagate upstream at acoustic speed.

- At time L/a seconds (where a is the speed of sound in the fluid) after the valve closure, the wave front reaches the end of the pipe. At that instant, the velocity is zero throughout the pipe, the pressure is $P + \Delta P$ throughout the pipe, the pipe is enlarged and the fluid is compressed. Right at the exit point of the reservoir, fluid begins to flow toward the reservoir, because the pipe pressure is higher than the reservoir pressure.
- At time $2L/a$ seconds, the pressure throughout the pipe has returned to its original value, but the velocity has reversed from its original direction, undergoing a reflection of the pressure wave. This time is called the critical period. The pressure decreases below the original steady state, causing the pipe cross section to shrink and the liquid to expand. For real liquids, the original transient has significantly died down because of their viscosity and its magnitude is practically negligible.
- At time $3L/a$ seconds, this negative wave has reached the reservoir, and the velocity is zero all along the pipeline. At time $4L/a$, the wave has reached the valve, returning to the original steady state that existed before the valve was closed.

Transients are basically manifested in two types: pressure transients and flow transients, which are different aspects of the same phenomena. Pressure transients occur when a change in energy occurs in the pipeline which adds or remove energy from the pipeline, while flow transients occur when there is a change in flow rate by a change in energy. The main causes of transients in a pipeline are:

- Change in valve settings including open or close status change
- Starting or stopping of pumps
- Changes in pump speed or head
- Pipeline rupture or large leak
- Collapse of column separation or trapped air
- Arrival of a batch interface at the pump

Consequences of a transient vary; flow movements are unsteady and pressures are unstable. Due to increase or decrease in line pressure during a transient, the fluid volume of the pipeline increases or decreases — such a volume increase or decrease is often called line packing or unpacking. If the magnitude of the transient is very large, control capability can be limited resulting in a pump trip or even pipeline system shutdown. In the worst case, the pipeline can be damaged or even ruptured if no form of pressure relief is provided.

A transient or surge is a pressure wave. In this book, transient and surge are used interchangeably. As noted earlier, pressure waves propagate from the source at the acoustic speed of the fluid along the upstream and downstream directions of the pipe. The wave also reflects back at a boundary point, and the reflected wave has negative pressure. The magnitude of an initial pressure wave, called potential surge, is proportional to acoustic speed and fluid velocity. The magnitude attenuates as the pressure wave moves away from the source of the transient. The acoustic speed remains constant in the absence of vapor in a liquid pipeline, but a small amount of vapor in the liquid can significantly reduce the acoustic speed.

The acoustic speed in a buried pipe can be calculated from

$$a = \frac{\sqrt{B/\rho}}{\sqrt{1+(B/E)(D/t)(1-\mu^2)}} \tag{5 – 1}$$

TABLE 5-1. Examples of acoustic speed at base conditions (494 mm diameter and 7 mm wall thickness)

Liquids	API gravity	Base density (kg/m^3)	Bulk modulus (MPa)	Acoustic speed (m/sec)
Heavy crude	20°	934	1720	1099
Medium crude	26°	898	1580	1089
Light crude	32°	865	1440	1075
Diesel	35°	850	1380	1068
Jet fuel	45°	800	1170	1038
Gasoline	65°	720	830	959

where

a = acoustic speed
B = bulk modulus of fluid
ρ = fluid density
E = Young's modulus of the pipe elasticity
D = pipe inside diameter
t = pipe wall thickness
μ = Poisson's ratio of strain (0.3 for buried pipe)

From this equation, the following aspects can be observed:

- The liquid density and bulk modulus or inverse of compressibility have the main effects on the acoustic speed; that is, it is proportional to the square root of the bulk modulus and inversely proportional to the square root of the density.
- The pipe elasticity and pipe size and wall thickness have minor effects on the acoustic speed, (in the order of 20% of the main effects).
- If multiple products with different compressibility and/or density are transported in a batched pipeline, the acoustic speed changes at each batch interface point.
- If the liquid compressibility is assumed zero, the acoustic speed is infinite and the flow is considered to be in a steady state. Therefore, transient phenomena are the consequences of finite compressibility.

See Table 5-1 for the acoustic speed of various hydrocarbon liquids at base conditions.

The initial pressure increase following sudden flow stoppage is referred to as the potential surge. The potential surge occurs upstream of the flow stoppage point or closed valve. If flow stoppage occurs instantaneously, the magnitude of the potential surge is determined by the formula:

$$\Delta P = \rho a v \text{ or } \Delta H = av/g \tag{5-2}$$

where

ΔP = pressure increase
ΔH = head increase
a = acoustic velocity
ρ = density of fluid
v = fluid velocity before valve closure

The magnitude of the pressure increase is reduced as the potential surge travels upstream along the pipeline, because the surge pressure wave attenuates. If the liquid keeps flowing into the upstream segment of the flow stoppage point, the pressure in the segment keeps increasing, and as the pressure increases, the pipe wall expands, the fluid is compressed, and so the line pack increases in the segment. Therefore, the pressure increase at the flow stoppage point is the sum of the potential surge pressure and the pressure rise due to line packing.

Example: base case extension 4 (refer to Chapter 3 for details)

The base case crude oil pipeline runs from the lifting station to the delivery station. The length of the pipeline is 200 km long and is 20" in nominal diameter, with a 0.281" wall thickness. An intermediate block valve is located 100 km downstream of the lifting station.

- Operating temperature: 4 °C
- Flow rate: 830 m^3/hr
- Density: 893.0 kg/m^3 at the operating temperature and delivery pressure
- Bulk modulus: 1,500,000 kPa
- Young's modulus: 200,000,000 kPa
- Viscosity at 4 °C: 43.5 cSt
- Pipe roughness: 0.0457 mm
- Injection pressure: 8040 kPag
- Delivery pressure: 350 kPag
- Maximum design pressure: 8370 kPag

Plot the pressure profiles and describe hydraulic behaviors of this pipeline, when the intermediate block valve is almost instantaneously closed or tripped.

Solution:

Based on the assumption that the flow is isothermal, the pressure profiles in steady and transient states are obtained using the above data. Using the steady-state profile as the initial condition, the potential surge is calculated and its subsequent behaviors are determined when the intermediate valve is almost instantaneously shut off. Since the time-dependent behaviors are complex, a pipeline simulator is used for this analysis. This transient simulation is performed in the following sequence of events:

- The inlet flow rate remains constant at 830 m^3/hr, and the delivery pressure is maintained at 350 kPag.
- The intermediate valve is closed over a one-second interval.
- The transient responses are described at 20, 40, 60, 80, 100, and 120 seconds after the intermediate valve is closed. Since the acoustic speed for crude oil is expected to be about 1 km/s, the transient pressure wave would arrive at the injection and delivery points about 100 seconds after the valve is closed.

Figure 5-1 shows the pressure profiles in both steady and transient states and MAOP. As expected, the upstream section is packed due to the upsurge and the downstream section unpacked due to the down-surge after the valve is closed.

1. Calculate the flow velocity, acoustic speed, and potential surge using the above formula when the valve is almost instantaneously shut off.
   - Flow velocity = 1.17 m/s
   - Acoustic speed = 1063 m/s
   - Potential surge = $893.0 \times 1.17 \times 1063 = 1110$ kPa

   This value of potential surge is a theoretical maximum. In practice, no equipment can be closed instantaneously, and the actual surge pressure will be lower than this surge pressure. In the plot below, the potential surge is shown as a sudden pressure increase at the valve; 700 kPa increase upstream and about the same pressure decrease downstream of the valve.

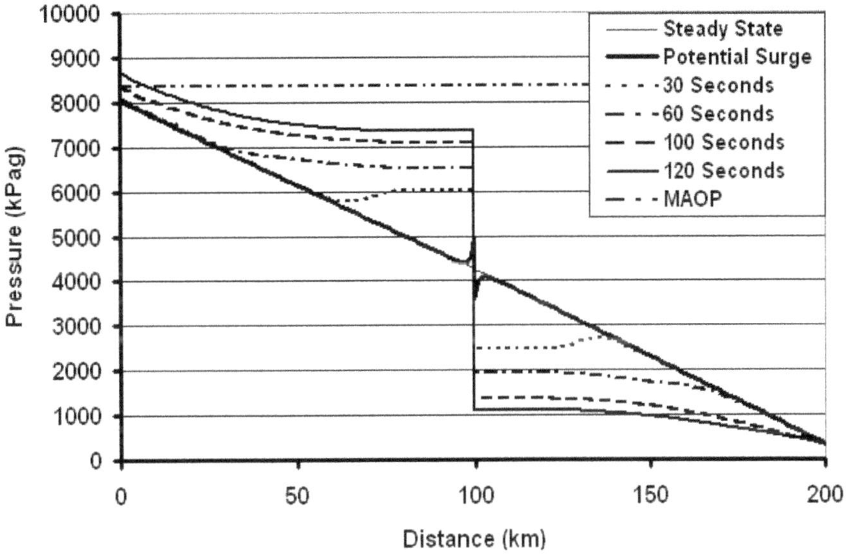

Figure 5-1. Line packing and unpacking in a crude oil pipeline due to valve closure at KMP = 100

2. 20 seconds after the shutdown, the pressure just upstream of the valve increases by 960 kPa and decreases by the similar amount downstream of the valve. The upsurge pressure wave front travels at the acoustic speed of 1063 m/s to 79 km upstream of the valve and the down-surge to 121 km. Beyond the wave fronts, the pipeline pressure and flow are not affected.
3. 40, 60, and 80 seconds after the shutdown, the upstream pressure keeps increasing and the downstream pressure decreasing, while the pressure wave fronts are moving closer to the inlet and delivery points at the same acoustic speed as before.
4. However, 100 seconds after the shutdown, the pressure at the inlet point exceeds the initial injection pressure, approaching closer to the MAOP of 8370 kPag, because the upsurge pressure wave has arrived at the inlet point earlier than 100 seconds and the line pack in the pipeline has kept increasing. In practice, as the pressure reaches closer to the MAOP, the flow rate is ramped down, eventually resulting in stoppage of the flow and pipeline shutdown. The delivery pressure is maintained at the initial pressure level, and the line pressure can be kept above the vapor pressure as long as the elevation is flat.
5. Two minutes after the shutdown, the MAOP is violated if the same injection flow rate is maintained. Modern control systems are able to detect this violation even before the pressure is violated, and shut down the pipeline if the closed intermediate valve cannot be opened.

The potential surge and subsequent line pack increase are shown in Figure 5-1 over time along the entire crude oil pipeline. Since the magnitude of line packing is proportional to the volume of the pipeline, the pressure at the flow stoppage point can increase significantly in a long pipeline, particularly if the stoppage point is located in a downhill segment.

As this example indicates, the actual pressure buildup depends on the total line pack in the pipeline, the velocity at which the liquid flows, and how fast the liquid is stopped. Surge pressure can be much greater than operating pressure, and even static pressure in the pipeline. Surge is a critical factor that must be addressed in high-pressure transmission lines transporting heavy hydrocarbon liquid.

Figure 5-2 shows the potential surge and line packing and unpacking in an ethane pipeline. This profile plot can be compared against the previous plot, which is obtained by simulating the intermediate valve closure in the same manner as the above. The pipeline configuration is also similar to the heavy oil pipeline, except the pipe size, flow rate and operating pressure are different. Two minutes after the valve is closed, line packing and unpacking are very small. From these two plots, it is obvious that the transient responses in an ethane pipeline are very small and the pressure builds up very slowly, while the responses in a heavy oil pipeline are large and the pressure build-up is fast. As a result, it is relatively simple to control transients in light hydrocarbon pipelines and thus a surge analysis is seldom necessary for this type of liquid pipeline.

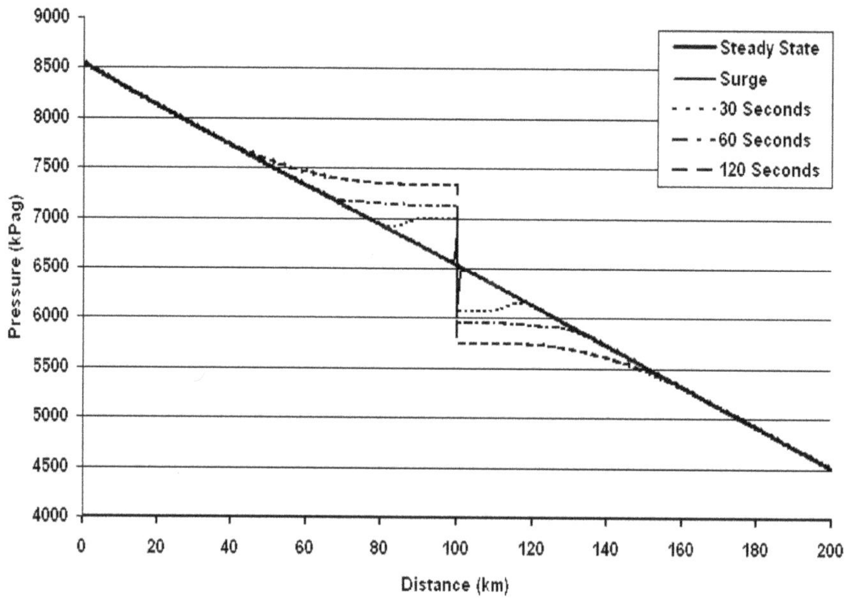

Figure 5-2. Line packing and unpacking in an ethane pipeline due to valve closure at KMP = 100

When transients in a pipe system cause the local pressure to drop below the vapor pressure of the liquid, the liquid vaporizes, forming vapor pockets and splitting the liquid. This phenomenon is called column separation. The size of vapor pockets varies with the local pressure and temperature and even the terrain shape. During the period of separating liquid columns, the flow behaviors become unpredictable due to significant acoustic speed drops in the presence of vapor pockets and the unstable nature of separated columns. Column separation can occur when the back pressure is decreased on the downstream side by starting a pump or opening a valve quickly. It can also occur within a pipeline system when the back pressure is such that the pressure at an upstream point is reduced below the liquid's vapor pressure. Practically speaking, a column separation is likely to occur near the peak point upstream of the sloping down section of a pipeline (see Figure 5-3).

When a transient causes the pressure to drop quickly below the vapor pressure of the fluid, vapor pockets can be formed inside the pipe and column separation occurs. In other words, column separation occurs when the pressure downstream of a valve drops below the vapor pressure upon sudden closure of the valve. Column separation is a phenomenon that often accompanies surge. It happens when a portion of the pipe is subject to low pressure. Column separation is the most serious consequence of down-surge. It is more likely to occur at high points or knees (sharp changes in slope)

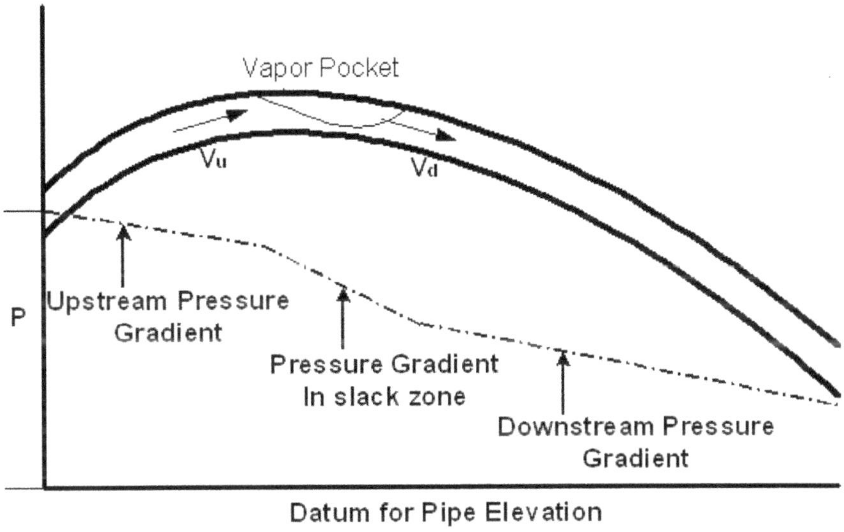

Figure 5-3. Pressure gradient change due to vapor pocket

in the pipeline. Column separation can disrupt the operation of pipelines and should be prevented from happening through proper design and operation.

Figure 5-3 illustrates a vapor pocket around a high elevation point due to column separation often called a slack flow condition. At the onset of the vapor pocket, the upstream liquid velocity is slower than the downstream velocity, and thus the upstream pressure gradient is lower than the downstream pressure gradient. When the back pressure is raised slowly, the downstream flow velocity slows down and the vapor pocket is reduced, eventually disappearing. If the back pressure is raised quickly, the vapor pocket can collapse, generating a high increase in pipeline pressure.

The opposite of column separation is the collapse of the vapor pocket. The upstream column will be accelerated and the downstream column decelerated if the back-pressure increases, and the upstream column overtakes the downstream column. As a result, the column can collapse if this process occurs quickly. If the difference in velocity at the instant of collapse of the column is ΔV, the expected pressure increase is $\Delta P = a^*(V_u - V_d) = a^*\Delta V$, where V_u = velocity of the upstream column and V_d = velocity of the downstream column. When a separated column collapses, it can be destructive enough to rupture the pipe if the velocity difference is high and the resulting pressure increase is sufficient. Hence, care needs to be taken to control this phenomenon (see Section 5.1.3).

Since the liquid columns are unstable, they will eventually return to a stable condition. After a vapor pocket is formed it may continue to increase in size until the upstream liquid column starts to move faster than that downstream. When the back-pressure increases, the flow velocity of the upstream column would remain the same while the downstream column movement will slow down. Then, the upstream column catches up to the downstream column, resulting in collapse of the liquid columns. The pressure increase due to the collapse can be so large that the pipeline can be ruptured. Practically speaking, by taking into account various pressure drops caused by surge and minor pressure losses, a pressure higher than the vapor pressure must be maintained at pump stations, delivery locations, and at high elevation points along the pipeline.

For pipeline system design and particularly operation, transient simulations are required. A transient model calculates time-dependent flow, pressure, temperature, and density behaviors by solving the time-dependent flow equations. Therefore, a transient model generates more realistic hydraulic results than a steady-state model and is capable of performing not only all time independent functions performed by the steady-state model but also time-dependent functions such as the effect of changes in injection or delivery, system response to changes in operation, and line pack movement.

Many pipeline failures, particularly for liquid pipelines, occur because improper provisions are made to manage transient related problems such as pump trip, etc. In order to manage these adequately, the following operating conditions should be properly taken into account in design and operational analysis:

- Changes to pump operations,
- Power failures,
- Valve operation,
- Line fills.

Transient simulation offers the following advantages over steady-state simulation for analyzing pipeline system operations:

- The study of normal pipeline operations — pipeline operation changes are simulated to find a cost effective way of operating the pipeline system. The transient model allows the operation staff to determine an efficient control strategy for operating the pipeline system and analyzing operational stability.
- Analyzing startup or shutdown procedures — different combinations of startup or shutdown procedures are simulated to determine how they accomplish operation objectives. The transient model can model a station, including the pump or compressor unit and associated equipment.
- Determining delivery rate schedules — a transient model can be used to determine delivery rate schedules that maintain critical system requirements for normal operations or even upset conditions.
- Studying system response after upsets — a pipeline system can be upset by equipment failure, pipe over pressuring, or supply stoppage. The transient model is used to evaluate corrective strategies by modeling various upset responses.
- Studying blow-down on a HVP line or pipe rupture — the transient model allows the operation engineers to study the effects of blow-down on a compressor station and piping or to develop a corrective action when a leak or rupture occurs.
- Predictive modeling — starting with current or initial pipeline states, future pipeline states can be determined by changing one or more boundary conditions.

In summary, a transient analysis is required for short-term operational study because pipeline states, in all operations, change with time. When an operation change takes place, the flow rate and pressure change immediately, and the subsequent change will have an impact on the pipeline system. With a transient analysis, the following problems can be addressed:

- Over or under pressuring along the pipeline,
- Equipment operations such as pump tripping,
- Potential column separation.

5.1.3 Surge Control

The main purpose of transient control or surge control is to protect the pipeline system by reducing the magnitude of surges to the allowable strength limits of pipe, valves, and pumping and other equipments. Controlling surge pressure is critical in pipeline system control, because it is directly related to safe operation of liquid pipeline systems while maintaining maximum throughput. Broadly, there are two ways of managing surges:

- Direct control of the surges
- Extra protection of the pipeline and/or equipment

Surge control is particularly important during the following operations, because they generate the largest magnitude of pressure surges:

- Start-up and shut-down operations
- Valve operations including pressure or flow control
- Injection and delivery condition changes

The implementation of an automated control system begins with the development of control strategies. Since surge control is more or less specific to a facility and control points along the pipeline, the control strategies discussed in this section are generic in terms of control devices, timing and magnitude of surges. Broadly, the following three levels of control strategies may be required:

- Train the pipeline system operation and maintenance staff in operating and maintaining their pipeline system. They have to be familiar with their system including its instrumentation and control system, and its maintenance because the system including the host SCADA can help identify over-pressuring or under-pressuring areas, so that unsafe operation can be prevented before any undesirable consequences occur. The training issue of operation and maintenance staff is discussed in ASME B31Q Standard for Pipeline Operator Qualification.
- Install basic control devices such as valves. Control devices mainly include various types of valves, and control timing includes opening and closing speeds of valves and timing of pump control in terms of speed (if applicable), pump start and shutdown.
- If surges are expected to be very large at certain locations, the piping systems have to be reinforced and/or a special surge relief device implemented.

Pipeline operating personnel are responsible for protecting the pipeline system from failure. The protection of pipeline and its related equipment is required to maintain the integrity of the pipeline system and to prevent potential system failure due to events that are beyond the control of pipeline operators. Even during normal operations, operators occasionally encounter undesirable operational problems, which can damage the pipeline system and whose consequences can be serious. A partial list of these problems is:

- Power failure
- Equipment failure
- Valve failure
- Accidents
- Human error

Experienced pipeline operators may be able to mitigate the consequences of these problems, but it is not possible to always respond to and control surges manually in a consistent and timely manner. Therefore, an automated surge control system needs to be implemented to operate the system safely and at the same time economically.

During the normal operation phase, operation and maintenance staff have to try to:

- Open and close pump station control valves gradually for fixed speed pumps, or ramp speed up and down slowly for variable speed pumps.
- Open and close line or station valves slowly.
- Allow an extra margin between the design pressure and pressure gradient, because it can reduce potential damage to pipe caused by large surges.
- Avoid any vapor in the pipeline and pump (particularly in the discharge region of a pump), because substantial transient pressures can be developed when the vapor collapses. Filling the vapor region can produce velocities that are above the expected steady-state velocities.
- Minimize power or engine failures to avoid pump trips by understanding the operating conditions and maintaining the pumping equipment adequately.

Certain factors need to be considered as part of control strategies for design and operation phases. During the pipeline design phase, particular attention has to be paid to the locations where frequent and severe surges are expected and their consequences can be significant. Typical design considerations for controlling surges and reducing potential risks and consequences are:

- Install thicker pipes at the locations where severe surges are expected during operations.
- Install block and check valves along the pipeline and at each pump station.
- Install automated control system such as a PID controller at each pump station.
- Avoid high flow rate, because low flow velocities ensure that changes in velocity cannot be too large. Note, potential surge is proportional to the flow velocity.
- Select longer station spacing, because it ensures minimum effect of surge due to large attenuation while the surge wave propagates upstream and downstream along the pipeline.
- Install multiple pumping units at each pump station, minimizing the chance for a complete station failure, hence the surge effect can be reduced.
- Install variable speed pumps instead of fixed speed pumps because variable speed pumps reduce surge pressure by ramping pump speed.
- Install special control devices such as surge relief tanks to relieve surge pressures, if the surge pressure increase is too large to control without them.
- Avoid sharp bends of pipe, where surges are likely to hit the sharp bends hard.
- Avoid, if possible, sudden changes in slope, where vapor pockets may be formed.

The application of some of the above issues/control methods are illustrated in the example of section 5.1.4 of this chapter. For further details of surge control techniques and equipment refer to Addendum to this chapter.

5.1.3.1 Control Devices

All transmission pipelines are installed with various types of valves; control valves, check valves, and block valves. However, valve movements can create a surge, and the magnitude of the surge depends on the type of valve and the valve movement in terms of the position and timing, in addition to the liquid compressibility and the elastic properties of the pipe. Different types of valve have different responses to their opening and closing operations; as well as flow characteristic behaviors and valve coefficients that are dependent on valve opening or closing position. Therefore, the responses of pipeline pressures to valve movement need to be fully evaluated for proper design of the control system. In general, it is safe to close the valves slowly, i.e., closure times longer than 2L/*a*. However, detailed computer simulation studies are necessary to understand the hydraulic behavior of the pipeline system in response to various valve operations and to implement an efficient valve control system. Reference [1] details valve types and controls.

Check valves allow fluid to flow in the normal flowing direction only and are installed to stop any reverse flows, thereby protecting pipe and equipment such as meters and pumps. Check valves are usually installed usually downstream of river crossings to protect rivers from contamination by preventing backflow in the event of a pipe rupture. However, they can cause large surge pressure if the reverse flow passes through them before they are closed completely. Certain check valves can slam almost instantly before a reverse flow can become large, creating a large surge pressure on the check valves. On the other hand, most modern check valves close slowly, allowing the reverse flow to ebb gradually while reducing the magnitude of the surge. For slow reacting check valves, one way to limit the amount of reverse flow is to install two check valves in series. It is not easy to analyze the check valve problem due to check valve response time delay and repeated flapper closing and opening actions.

5.1.3.2 Pump Unit and Pump Station Operations

Potentially, pump start-up and shut-down operations can be most disruptive, because they add or remove energy to the pipeline system. Upon start-up, the pump operates against a closed valve. As the valve opens, the flow into the pipeline gradually increases to the full pump capacity. Pump start-up operations can cause a rapid increase in fluid velocity that may result in an undesirable surge, but they seldom cause a problem in actual operations. As the pump starts up at a pump station, the flowing liquid forces open the check valves downstream of the pump and the liquid in the line begins to move. The flowing liquid develops an upsurge in the downstream section of the pipeline and a down-surge on the upstream side. The magnitude of the surge pressure depends on the starting speed of the pump and density and bulk modulus of the liquid. The pressure increase is a manageable size because generally no vapor is present in the piping system on the discharge side.

However, tremendous surge pressure can be created if there is vapor in the discharge piping system and the discharge pressure increases quickly. This is due to the collapse of the vapor pocket under the increasing discharge pressure and consequently the collapsing vapor pocket can create a huge local pressure instantaneously on the discharge side. This sudden pressure increase will disrupt the normal pump start-up process and potentially affect upstream and downstream pump station operation, potentially resulting in pump trip and pipeline system shut-down.

During a pump start-up period, surge control is most critical, requiring effective surge controlling methods. These may include, but not be limited to, the following:

- For fixed speed pumps, open a control valve slowly after the motor starts, reducing transients by interlocking the pump with control valves;
- If multiple pumps need to be brought online, start them one at a time at an interval of two times the critical period;
- For variable-speed pumps, ramp up the pump speed so slowly that large surges can be avoided;
- Install a PID (proportional-integral-derivative) controller to balance the flow movements and pressure behaviors at the pump station.

Normal pump shut-downs also cause surges; resulting in an increase in suction pressure and decrease in discharge pressure. The surges have to be controlled to keep the pipeline system at a pressure level between maximum and minimum pressure limits so that vapor can be avoided in the pipeline system, including pump stations and at the same time the pump stations and pipeline system can be restarted smoothly. Figure 5-4 shows the pressure changes during a pump station shutdown process [3]. The following approaches can be adopted in the scheduled shut-down of a pump station to minimize surge pressure:

- Turn pumps off one at a time at intervals at least two times the critical period, which is the time obtained by dividing the pipeline length between the adjacent operating pump stations by the acoustic speed of the fluid.
- For fixed speed drivers of pumps, close a control valve slowly (at least two times the critical period) before the driver is stopped. Upon shut-down, close slowly the control valve to decelerate the flow, after which power to the pump is shut off, but not until the valve is fully closed.
- If pumps are equipped with variable-speed drives, ramp down the pump speed so slowly that the surge pressure associated with a station shutdown can be minimized.

Figure 5-4 shows an initial pressure profile and the profile after the PS_2 station shutdown. When the station is shut down, its suction pressure increases and discharge pressure drops, resulting in low flow rate in such a way that the PS_1 and PS_3 stations can pump the liquid flow. Right after the station shutdown, an up-surge occurs upstream of PS_2 and a down-surge downstream of the station. The up-surge and down-surge pressures can exceed the controllable station discharge or suction pressure limit of the neighboring station. If the limit is set tight, either or both stations can be tripped. Otherwise, the flow rate reaches a steady state and the final pressure profile is established as shown in the plot. Then, the pressure gradient downstream of PS_3 would be similar to the pressure gradient upstream of the station.

Power failure or other non-scheduled events at a pumping station can cause pump tripping, which occurs almost instantaneously and cannot be avoided. A pump trip results in an initial rapid down-surge on the discharge side and pipe section close to the pump station. Control valves cannot prevent down-surge on power

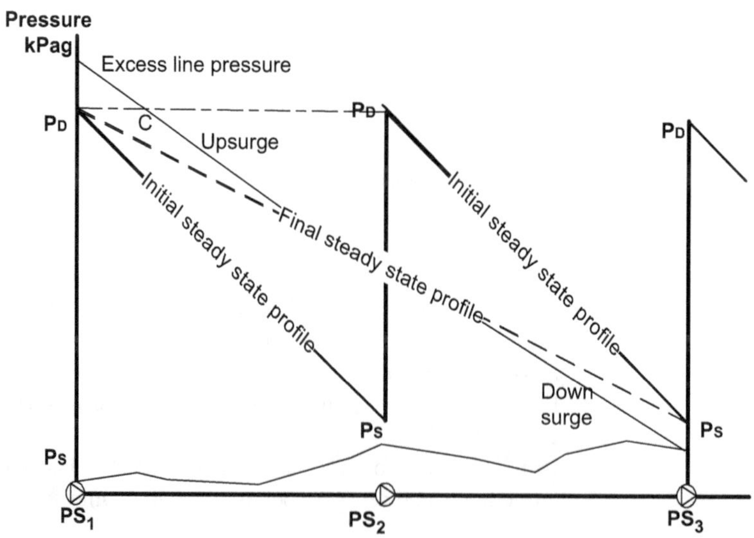

Figure 5-4. Pump station shut-down and pressure profile change

failure. In addition, the pump trip can cause the adjacent pumps to be tripped and eventually the whole pipeline system can be shut down. The most severe transients can occur upon power failure at a pump station where the elevation profile rises steeply downstream of the pump station. If the power or fuel source is cut off, the pressure just downstream of the pumps drops quickly, and this pressure drop propagates downstream at the acoustic speed. This sudden drop in pressure can cause a column separation and lead to subsequent column collapse of large magnitude in the downstream segment. Also, a flow reversal in the system can occur and eventually lead to significant overpressures in the system, generally in the vicinity of the pumps, if the surge pressure is not properly relieved. As identified earlier, various mitigating factors should be considered if such a severe surge problem is expected in the design phase:

- Install thicker pipes at the locations where severe surges are expected,
- Avoid high flow rate,
- Install multiple variable pumping units at the pump station,
- Install special control devices such as surge relief tanks to relieve surge pressures,
- Avoid sudden changes in slope, where vapor pockets may be formed.

Refer to Section 5.1.4 for the operation of a pipeline running in a mountainous region with a severe elevation profile.

Most modern pump stations are equipped with a PID controller to control the pump station including surge; PID is the acronym for proportional, integral and differential. A PID controller is a commonly used feedback controller used to maintain the stable operation of pump units and stations. PID controllers operate control valves for pressure, flow or other parameters. They are primarily used to control pump stations and subsequently the pipeline system, by continually monitoring the actual pump station conditions, comparing them with the expected or set conditions, and then adjusting the control valve position or driver speed. Initially, the PID controller actions can effect quick changes in pressure, flow rate, and possibly temperature for certain liquids at the controlled pump station, and the changes in pressure and flow rate will result in changes in the upstream and downstream parts of the pipeline. It is pointed out that the rate of pressure change determines the required controller characteristics for limiting overshoots from a set point to acceptable values [4]. This continuous monitoring and controlling capability at the pump stations provides the pipeline system with fast and accurate control capability initially at the pump station. Such an automated controlling capability allows the operators to reduce their system operating responsibilities by focusing on monitoring the whole pipeline system instead of controlling each station. A typical PID controller for a fixed speed driver is shown in Figure 5-5.

As shown in Figure 5-5, the PID controller receives the pump suction and discharge pressures. If a flow transmitter is available, it may use the flow data for control. The controller determines the difference between the measured pressures and the pump station set point to generate a controller output to adjust the position of the control valve. The position change alters the suction and discharge pressures, which will be fed back to the controller to come up with another controller output. This control process is repeated continuously to maintain stable pump unit and station operation.

ASME B31.4 requires that pipeline sections downstream of the pump stations should be protected by the maximum station discharge pressure control system and the independent maximum pump pressure shutdown system. As an additional protection, surge relief devices need to be installed along the pipeline.

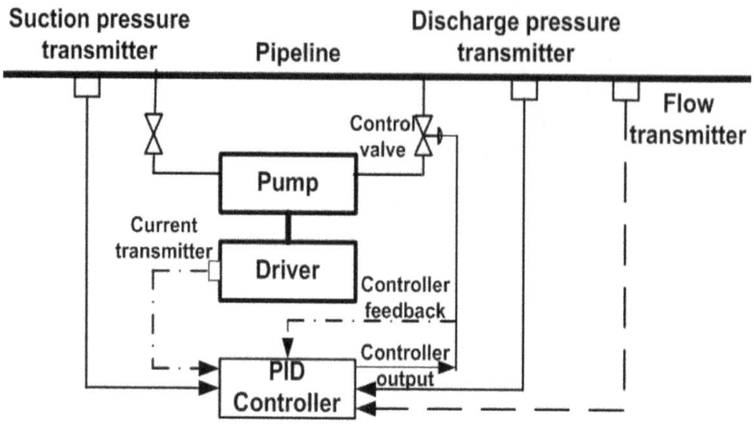

Figure 5-5. PID controller

5.1.3.3 Special Surge Relief Devices

In addition to the above two measures, surge relief devices may be needed to safeguard the pipeline system further. Various pressure relief devices are available, among which only two are suitable for use in petroleum transmission pipelines; pressure relief valves and surge tanks. Occasionally, rupture disks can be used to arrest surge, but they are destroyed when activated and hence need to be replaced.

Pressure relief valves are used to protect the pipeline system from excessive pressure, opening the valve when the pressure exceeds a specified pre-set pressure. When the pressure increases above the set value, the valve opens and the liquid flows out the attached tank through the valve, dampening the surge pressure significantly. The valve closes when the pressure in the line decreases below the set value. When the tank is full, the liquid is pumped back into the pipeline if the pipeline pressure is in a safe operating pressure range.

Pressure relief valves are designed to reduce large upsurges mainly in the pipeline, but not to control the down-surge that may occur on pump shut-down or power failure. Also, they are intended to reduce short and steep pressure increase along the pipeline where significant flow reversal occurs after a pump trip, being followed by an upsurge. Relief valve selection is most critical because the relief valve may not open quickly enough to prevent a very short surge of high pressure and an undersized valve may not be able to discharge the liquid into the tank fast enough to reduce the pipeline pressure.

The pressure relief valves are installed on points along the pipeline, usually closer to the discharge side, where surge pressure can be high but maintenance is easy. Sometimes, the pressure relief valve can be installed just downstream of the pump within the station yard to prevent the pump station and pipeline from operating near shut-off pressure, especially during pump start-up. If the relief valve is installed in the pump discharge line, it can work as a bypass valve during pump startup.

A typical pressure relief valve and tank assembly is illustrated in Figure 5-6. If the main pressure exceeds the relief valve pressure set point, the valve opens and the liquid flows from the main line to the tank to relieve the main line pressure. The pipe size from the main line to the tank has to be large to let the liquid flow quickly. If the liquid in the tank may be contaminated, strainers are installed before the pumps. The injection pump is activated to inject the liquid back into the main pipeline when the tank is nearly full. Usually, the pipe size downstream of the pumps is smaller than the pipe size upstream of the tank.

A surge tank is pressurized with a gas that absorbs the pressure surges. It is a closed container filled with the system liquid in the lower chamber and with a pres-

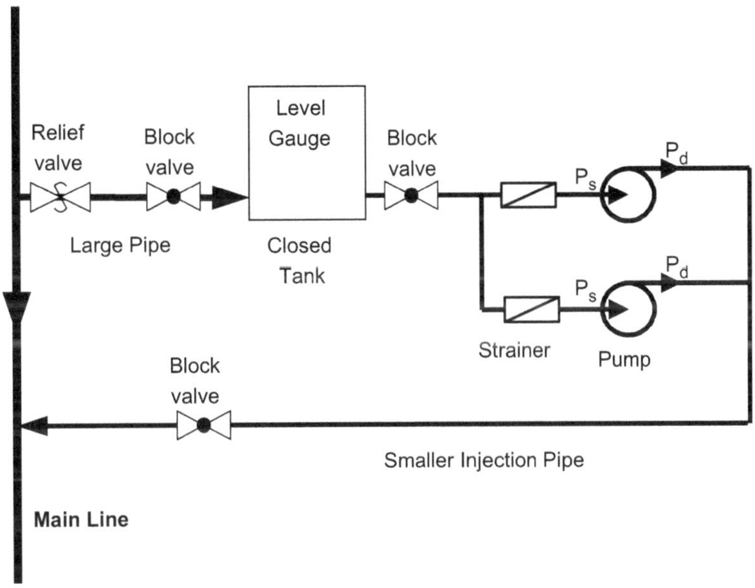

Figure 5-6. Pressure relief valve assembly

surized gas in the upper chamber. The lower chamber is attached to the pipeline and separated from the upper chamber by a piston or a flexible membrane. The pressurized surge tanks are designed to keep the local pressure from exceeding a pre-set pressure. If the local pressure exceeds the pre-set limit, the liquid in the pipeline enters the lower chamber, relieving the peak pressure in the pipeline, and the gas in the chamber is compressed through the separator. If the pipeline pressure is reduced to the level below the chamber pressure, then the liquid in the lower chamber is pushed by the gas chamber pressure. The surge tanks also prevent the local surge pressure from propagating to another section of the pipeline. Figure 5-7 shows a pressurized surge tank attached to the pipeline.

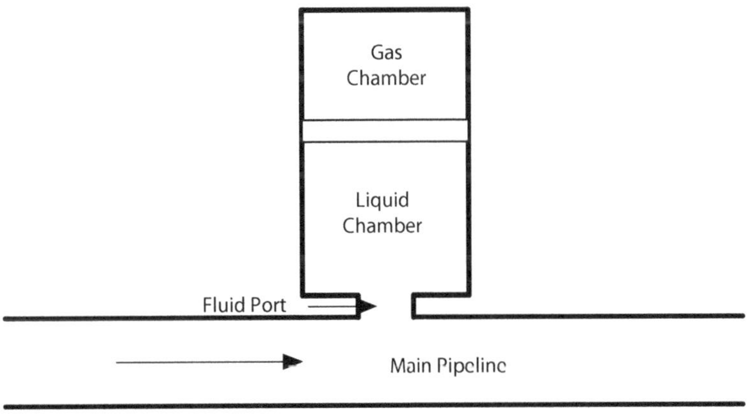

Figure 5-7. Pressurized surge tank attached to pipeline

Pressurized surge tanks must be adequately sized so that the lower chamber can contain sufficient liquid volume to reduce the local pressure. If the tank sizes are adequately designed, the tanks can be effective in smoothing out short but steep surge pressures. They do not require repairs because there are no moving parts; however, they are expensive to install. Regular maintenance is required to maintain the volume of gas in the tank.

Rupture disks are non-mechanical pressure surge control devices which consist of a bursting membrane designed to rupture at pre-set pressures. They are less expensive than the other two pressure surge control devices. Like pressure relief valves, a tank is attached to rupture disks to accept relief flow. Disks must be replaced after being ruptured. They are seldom installed for transmission pipelines.

Surge and its control can be analyzed and best understood by means of computer simulations. During the design phase, various normal and abnormal operating scenarios can be simulated and reviewed, and then the most reliable and responsive control system can be selected from among the simulated results. The following steps are suggested to design a surge control scheme:

Step 1: Identify the key points where potential hydraulic transient problems may occur by simply reviewing the pipeline system and its elevation profile; peak points, deep valleys, pump station locations relative to steep elevation changes, etc.
Step 2: Determine the realistic worst case operating scenarios for both normal and abnormal operating conditions.
Step 3: Simulate the pipeline system operations for the operating scenarios using a transient software package. The transient software must have the capability to realistically simulate not only the pipe flow but also timing and functions of control equipment.
Step 4: Select the locations where surge control devices are installed and the most effective control devices and timing.

The results of the simulation can reveal the surge behaviors and critical points where surges must be controlled and control devices are required. If control devices are not able to control surges within an economic limit, either pipe segments are strengthened or a combination of strengthening pipe and installing control devices must be implemented to provide proper protection of the pipeline system.

5.1.4 Example of Pipeline Operation and Surge Control

This section describes the pipeline system operations of the OCP Ecuador Pipeline (OCP Ecuador has kindly provided certain parts of the information contained in this section). The pipeline system and operations are greatly simplified to describe the concepts only. The following operations are discussed:

- Scheduled pipeline system start-up
- Scheduled pipeline system shutdown
- Emergency pipeline system shutdown
- Batch movement in mountainous regions

This pipeline is selected as an example because it has a unique design and challenging operation requirements due to severe elevation changes, as shown in Figure 5-8. The unique design aspects will be briefly described while discussing the above operations.

Figure 5-8 shows the OCP pipeline configuration and elevation profile. The OCP pipeline system transports heavy crude (18 °API) over 485 km across a mountainous area. The pipe grade is X70 and sizes range from 24" to 36". The pipe wall thicknesses also vary with elevation and other factors. The MAOP changes significantly due to se-

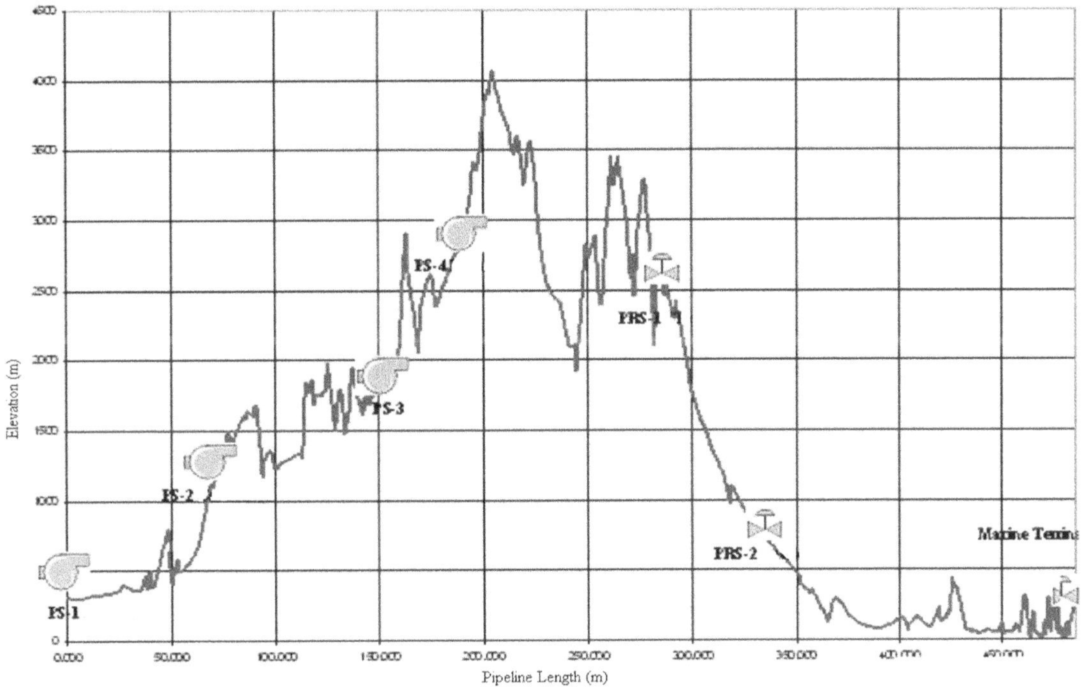

Figure 5-8. Configuration and elevation profile of OCP Ecuador pipeline

vere elevation changes along the pipeline. For example, the elevation drops more than 2100 m downstream of the peak point, resulting in a significant static pressure gain, while the static pressure gain is a small upstream of the Marine Terminal. Therefore, the MAOP requirements vary significantly along the pipeline. In order to satisfy such varying MAOP requirements, the OCP pipeline is constructed of significantly different pipe sizes and wall thicknesses but with the same pipe grade throughout the pipeline. Operating and design pressures change along the route because of large elevation differences. Each pipeline section between stations is designed to withstand full static pressure. For example, the pipe wall is thicker than an inch in the valley between the peak point and PRS-1.

There are four pump stations; one initiating station and three intermediate stations. The discharge pressures may be greater than 11,900 kPag, and the minimum suction pressure is 400 kPag in order to operate the pipeline in safe pressure ranges. All pump stations are equipped with five variable speed pump units with one unit as a spare. As pointed out earlier, such a combination of multiple units with variable speed pump unit control capability can reduce the surge effects significantly. Since the PS-4 station discharges at a pressure head higher than the peak point, no additional pump station is required because of the pipeline pressure gains due to elevation drop. Instead, two pressure reducing stations (PRS), PRS-1 and PRS-2, are installed to reduce the static pressures. Two PRS stations are needed to control the static pressure gain of approximately 30,000 kPa. Their locations are determined through a surge analysis to lower the operating pressure limit while effectively addressing the control requirements such as slack flow conditions around high elevation points.

For viscosity higher than a certain limit, the heavy crude needs to be heated to reduce friction losses in the uphill sections of the pipeline. Heaters are installed at each station, but heating requirements are different at different pump stations because friction loss in one section is different from that in other sections.

5.1.4.1 Scheduled Pipeline System Start-Up

Since this pipeline transports heavy crudes, the main factors to be considered for a pipeline start-up are the oil viscosity with heating requirements and the inlet pressures to pump stations. For this pipeline, the maximum viscosity is 100 cP or 105 cSt to protect the mechanical seals of the main pumps from damage and to satisfy the pressure requirements for the design flow rate. Beyond this limit, the crude has to be heated before a startup. In addition, the inlet pressure to the pump stations must be high enough to prevent pump cavitation.

In general, the pipeline operator checks the following before starting the pipeline system:

- Check if all the main line valves are open along the pipeline and the valve positions including pressure control valves (PCVs) at the stations,
- Check if one tank is connected to the inlet booster header at the lifting station,
- Check if one tank is connected to the inlet header at the delivery terminal.

At the completion of this checking procedure, the operator performs the following tasks to start the pipeline system:

- Set the flow rate at the lifting station and the inlet pressures at the main line pump stations,
- Start booster pumps at the lifting station after checking that the suction and discharge valves are open,
- Start pumps at the lifting pump station after checking that suction and discharge valves are open,
- Start pumps at the next station when its suction pressure is above the minimum pressure level and keeps increasing, after checking that suction and discharge valves are open,
- Start pumps at the next stations sequentially for some pipelines or in different orders for other pipelines. In general, pump start sequence depends on the elevation profile and pressure conditions,
- Start controlling the pressure at the PRS station according to the valve opening sequence, assuming that the PRS is installed,
- At the delivery terminal, adjust the delivery pressure set point in multiple steps, until reaching the desired flow rate.

The criterion for starting a station is to let the inlet pressure increase to above a certain minimum pressure (700 kPag for the OCP). Even though the pressure of each station is high at the time of station shutdown, the station pressures would drop if the crude oil in the line cools down. If the pressure at each station inlet reaches the minimum pressure at the start-up time, each station can be started as soon as the pressure wave arrives from the upstream station. Wave speed is about 1.1 km/s, and the wave travelling time between pump stations can be estimated by dividing the intervening distance by the wave speed. However, the presence of closed intermediate check valves can hinder the wave propagation and thus reduce the wave speed until the upstream pressure exceeds the downstream pressure. As a result, it can take a longer time for the pressure wave to reach the next operating station.

Once the suction pressure is set at a pressure above the minimum station pressure, the discharge pressure at the upstream station will be adjusted to deliver the same flow entering the station, unless there is an obstruction in the section. The minimum station pressure is determined in such a way that the suction pressure is above the net positive suction head required by the pump and at the same time no vapor pocket is in the upstream segment of the station. If the discharge pressure exceeds the maximum allowable pressure limit, the suction pressure control will be overridden by the discharge pressure control.

While a station is shut down, back flow of high pressure crude oil from the discharge pipeline is prevented by the closure of a check valve in the station discharge header, main pump discharge valves, and even outlet emergency shutdown (ESD) valves. Station startup begins by opening the main outlet ESD and pump discharge valves.

As the pipeline is shut down, the pressure control valves (PCVs) and block valves at each PRS are closed. After receiving the open signal, the block valves are opened first, while PCVs remain closed until the pressure begins to increase at the inlet of PRS-1 (refer to Figure 5-8). The inlet pressure starts to increase about 90 seconds after ramping up the pressure at PS4. The time may vary somewhat with the number and size of vapor pockets around the peak point. After confirming that the pressure reducing stations are in an automatic mode, the set pressure at the station is gradually changed by opening the PCVs to the final value to be reached in a steady-state flowing condition. The outlet pressure is controlled to protect against overpressure of the downstream pipeline. Approximately 50 seconds after opening PCVs at PRS-1, the flow arrives at PRS-2 and its inlet pressure starts to increase. The PRS-2 operation proceeds in the same way as for PRS-1.

Intermediate block valves should remain open even after the pipeline is shutdown. In the case where some valves are closed, they are to be opened sequentially, starting from the valve closest to the delivery station, except at PRS-1 and PRS-2 to avoid static pressure build-up.

The Marine Terminal is the delivery station of the OCP pipeline. When the pipeline system is shut down, the block valves at the delivery terminal and pump stations are closed. After confirming that the Marine Terminal is ready to receive the incoming crude, the main inlet valve is opened first and then the valve connected to the tank selected to receive the crude is opened, while keeping the PCVs closed. The PCVs, assuming that they are in automatic mode, are set at a pressure above actual inlet pressure. The pressure set point is slowly changed until reaching the required steady-state inlet pressure. The flow rate must be kept lower than the flow through PRS-2 until the upstream segments are in full flow. The pressure wave reaches the terminal approximately 160 seconds after opening PRS-2, the inlet pressure to the terminal starts to increase and the PCVs start to automatically regulate the pressure to the set point. It is critical to adjust the set points gradually in order to remove the vapor pockets slowly so that the vapor can be absorbed into the liquid without creating spikes in pressure. Such a sequence of gradual operation is intended to minimize transients in the line so that flow and pressure spikes can be avoided through the station piping and at the meter station.

When the pipeline system starts, vapor pockets may exist on the suction side of PS-2, PS-3 and/or PS-4 stations, around high points, and also downstream of pressure reduction stations PRS-1 and PRS-2. Since the pump stations are equipped with multiple variable speed pumps, the flow and pressure can be increased very slowly and thus vapor pockets can slowly disappear. One pump unit is started at a time, first ramping up the pump speed, and then the next. The vapor pockets downstream of PRS-1 and PRS-2 can be eliminated by increasing back pressure slowly, but can be present for a longer period of time depending on the initial volume of vapors and the difference between the flow rate entering and leaving the sections between stations. The gradual adjustment of the set points of the inlet pressure controller at PRS-1, PRS-2 and the Marine Terminal are the only control variables that the operator can use to minimize this period.

Note that a leak detection system based on mass balance does not work in the presence of vapors in the pipeline until the vapors are reabsorbed into the liquid due to the increase of pressure after starting the pipeline system.

5.1.4.2 Scheduled Pipeline System Shutdown

In mountainous sections of a pipeline, the pressures at high points or on the suction side of a pump station can drop below vapor pressure. While the pipeline is shut in, the pressure can drop further as the ambient temperature drops. In order to avoid cavitation, the pipeline pressure is usually kept high when the pipeline is shut down. For example to avoid violating the vapor pressure limit at the peak point, the minimum required static pressure on the discharge side of PS-4 should be kept above 11,500 kPag. Here, the example is focused on the pipeline section between PS-4 and PRS-2, because it is the most challenging to operate this pipeline section.

When a pipeline is to be shutdown, some control actions are initiated simultaneously and other actions follow in sequence. The operator initiates the simultaneous closing actions at the following facilities:

Lifting station: Shut down all heaters including heat exchangers, close ESD valves and main pump discharge valves, open main pump recirculation valves, and then shut down all booster pumps but one which is shut down after shutting down all main pumps one after the other. It is intended to keep the suction pressure high in order to avoid potential cavitation when the pump is restarted.

Pump stations: Shut down all heaters including heat exchangers, close outlet ESD valves and main pump discharge valves, open main pump recirculation valves, and then ramp down pump speed to the minimum before shutting it down. Since multiple pump units may be operating at the time of shutting down, a single unit is shut down one after the other to minimize transient effects in pressure and flow rate.

Pressure Reduction Stations (PRS): To maintain the upstream pressure, the inlet ESD valves should be closed and then PCVs.

Marine Terminal: Normal inlet pressure control remains in operation. The PCVs close gradually and the last section of the pipeline is depressurized immediately downstream of PRS-2. The control system will automatically change the set pressure to 525 psig, so that only the section about 10 km downstream of PRS-2 may remain below atmospheric pressure.

An automatic block valve station is installed upstream of the Marine Terminal. The station is closed only if the shutdown procedure is initiated automatically by the closure of the Marine Terminal.

A scheduled shutdown is based on an operational strategy of maintaining the entire pipeline system above the liquid vapor pressure, while keeping pressure transients to a minimum. This strategy should result in high discharge pressure and a suction pressure greater than the vapor pressure due to static pressure in uphill segments.

Figure 5-9 shows the MAOP and elevation profile as well as pressure profile from PS-4 to PRS-2, when the pipeline sections are shut-in. Note that the PS4 downstream pressure is set at 12,000 kPag to keep the peak point pressure about 750 kPag and that downstream pressure at PRS-1 is kept low to avoid any MAOP violation in the downstream section.

The pressure profiles for very low and high flow rates are shown in Figure 5-10, in which the PS4 discharge pressure range is very narrow, less than 3000 kPa, and the upstream pressure at PRS-2 for a low flow rate is much higher than that for a high flow rate. These are the consequences of the fact that the static pressure changes are significant for low flow rates, while the frictional pressure drops are small.

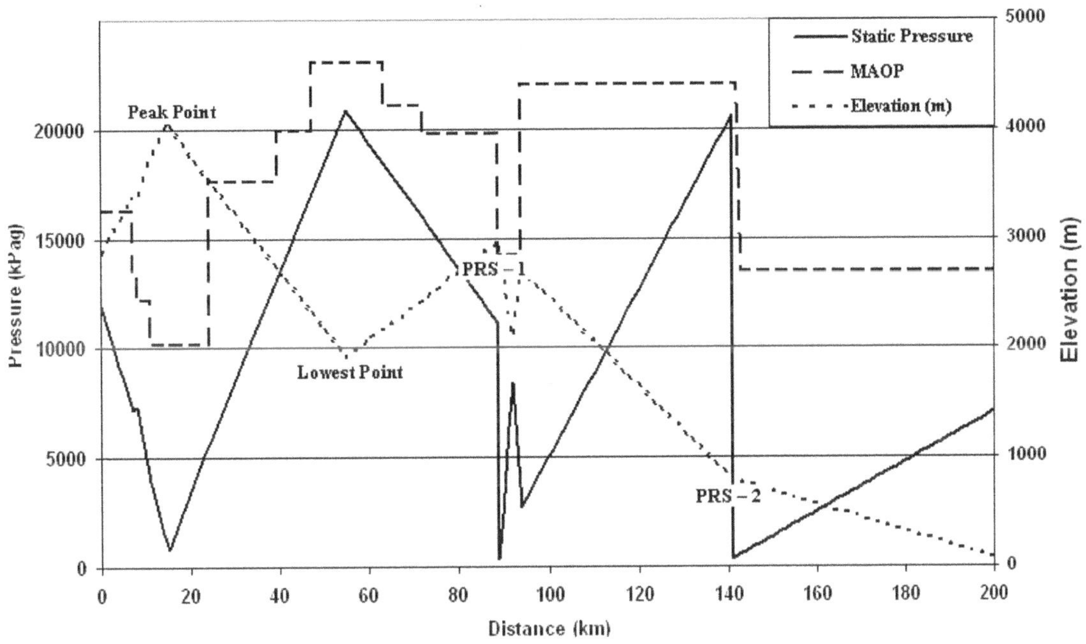

Figure 5-9. Static pressure profile

Pipeline sections downstream of the pump stations are protected by the maximum station discharge pressure control system such as PSVs and the independent maximum pump pressure shutdown system, thus meeting ASME B31.4 requirements. Main line block valves are seldom closed in a scheduled shutdown because the MAOP is increased with larger pipe wall thicknesses where static pressure increase is high.

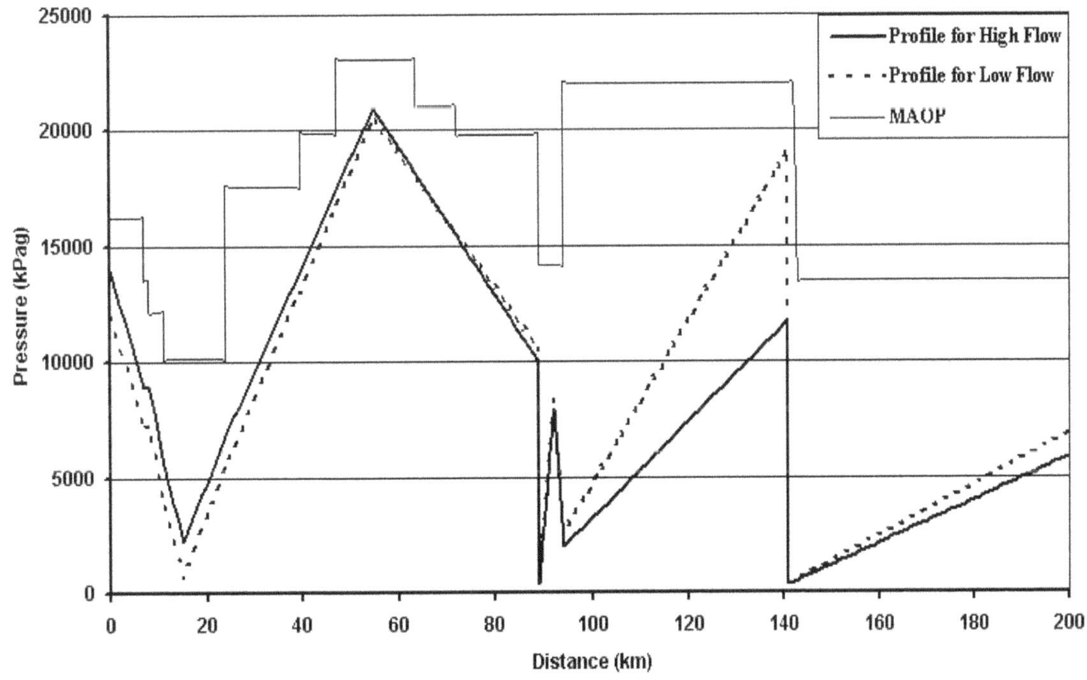

Figure 5-10. Pressure profiles for low and high flow rates

Therefore, each station is shut down in a specific order with a certain time delay instead of shutting down all stations at once.

A normal shutdown sequence can vary with the pressure levels at the time of shutdown; start shutting down pump stations from upstream, but not necessarily in a sequential order, in order to confine pressure to a specified pressure range in each section. For example, the shutdown sequence can be as follows:

- Shutdown PS-1 pump station at the lifting point
- Shutdown PS-3 after PS-1 is shut down, and then PS-2 and PS-4, PRS-1, PRS-2, and then marine terminal.

However, if the pressure level in the pipeline is relatively low, the shutdown sequence can start from the marine terminal in order to build up the pressure level. If the pressure level in a certain section is too low at the time of shutdown, the downstream pump station or PRS can be shutdown first and then the rest follow.

5.1.4.3 Emergency Shutdown of the Pipeline System

Emergency shutdown capability and facility isolation is required at a specific station or in a section between two stations. The causes of emergencies may include a fire at the station, line rupture, terrorist attack, earthquake, etc. Normally, the shutdown procedure is fully automated and the operator initiates an emergency shutdown by pushing a dedicated ESD button on the operator console of the host SCADA system.

The emergency shutdown procedure is usually similar to the scheduled shutdown at the stations that are not directly affected by the emergency. However, an additional action is required for emergency shutdown; in addition to ESD valves, all valves connected to fuel and crude sources into the station must be closed.

If a pipeline rupture occurs, the operator follows the procedure specified in the emergency response manual. The first step may include the identification of the leaking section. When the leaking section is identified, the station upstream of the section is shutdown first and the nearest block valves are closed, and then the downstream station is shutdown. The rest of the stations are shutdown in a manner similar to a scheduled shutdown. This step is intended to reduce the overall spillage.

In terrain where the elevation profile changes significantly, it is very difficult to keep the pipeline pressure above the vapor pressure in the event that a pump station, especially PS-4, is tripped. A station trip would not likely occur in the OCP pipeline, because the pump stations are designed with multiple variable speed pump units and the station spacing is sufficiently long. These pump units are relatively small and their drivers are of variable speed. Therefore, the magnitude of pressure surge would be limited when pump speed changes or a unit is brought on-line in normal operations. Unless the fuel supply is abruptly disconnected, a pump station is not likely to be tripped, because multiple pumps would not drop all at once. If one unit is operating, it can be tripped but the flow rate is too small to cause an emergency shutdown.

5.1.4.4 Batch Operation

Only the hydraulic behaviors due to batch movements are discussed in this section, but general aspects of batch design and operations are discussed in Section 5.2.

It is challenging to operate a batch pipeline in such a mountainous terrain as that between PS-4 and PRS-2, because pressure varies significantly with the changes in elevation and product properties. The pressure behaviors differ at different points along the pipeline if the points are located at different elevations. However, these changes take place over a long period of time. Assuming that the batches have different density and viscosity, static pressures change as these batches move along the changing terrain.

Pressure changes with respect to movement of the batches are illustrated in Figure 5-11, plotting the pressures and batch movement along the pipeline. Even though

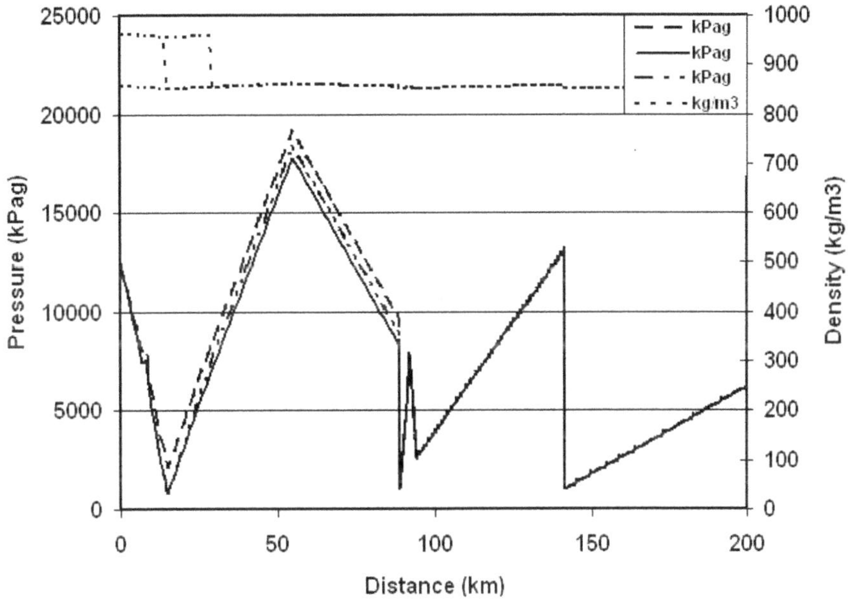

Figure 5-11. Batch movement and pressure behaviors

OCP Pipeline is not a batch pipeline, a section between PS-4 and PRS-2 is used for this illustrative example. In this simulation, the discharge pressure at PS-4 and flow rate in the pipeline section is kept constant, while transporting two batches; light oil (35 °API = specific gravity of 0.850) being followed by heavy oil (18 °API = specific gravity of 0.9465).

The dashed line (first in the legend above) represents the pressure profile for the light crude in the pipeline, only. The solid line represents the pressure profile for the heavy crude moving to the first 15 km point, and the dash and dotted line (third in the legend above) for the heavy crude moving to another 15 km position. As the heavy crude flows in the uphill segment, the pressure drop is higher due to the elevation gain and larger friction loss than the drop with the light crude only. However, as the heavy crude flows past the peak point, the static pressure increases due to elevation drop and higher density of the heavy crude. This increasing trend will continue until the heavy crude reaches the lowest point. The pressure profile downstream of PRS-1 and PRS-2 remains the same until the head of the heavy crude batch passes the respective points.

Figure 5-12 shows the trends of pressure changes recorded at the peak point, lowest point downstream of the peak point, and the upstream of PRS-1. The batch movements over 20 hours were simulated. The figure shows that the peak point pressure dropped until the head of the heavy crude batch arrives at the peak point and that the pressures downstream of the peak point also dropped along with the peak point pressure. However, the peak point pressure remains the same after the arrival of the heavy crude, while the downstream pressures increased due to elevation gains as the heavy crude moved downhill. In this example, the peak point pressure drop due to the batch movement was significant, so pressure control is required to keep the pressure above the vapor pressure.

5.1.5 Transient or Surge Analysis

As mentioned earlier, a steady-state assumption is not valid for analyzing operations and for certain design problems, instead their analysis, particularly of the following

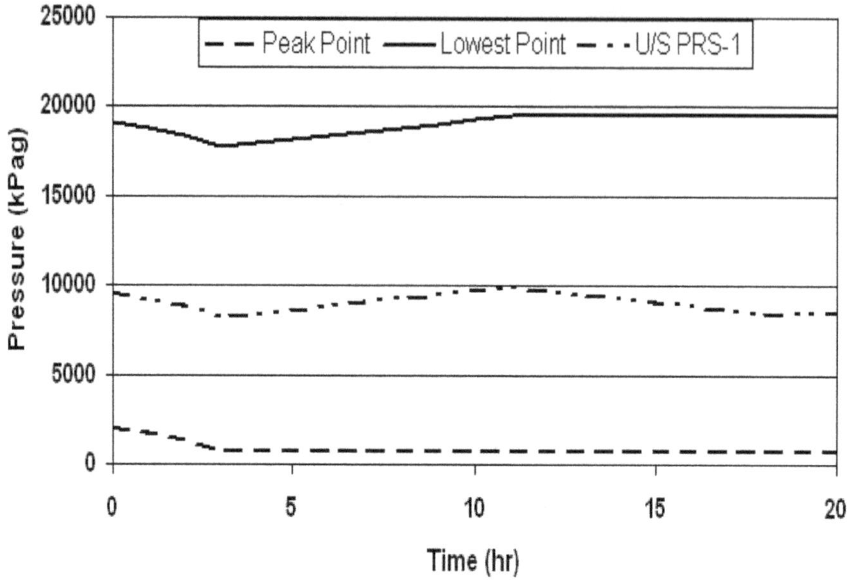

Figure 5-12. Trends of pressure changes

operation problems, should be based on the time-dependent transient flow, because these create significant pressure and flow changes over a short period of time:

- Determination of pipe wall thicknesses of pipe segments along the pipeline where significant elevations change;
- Locating and sizing surge control facilities such as pressure relief stations (PRS), surge relief valves, etc.;
- Design of station control systems such as PID controllers, and determination of control parameters;
- Analysis of pump station operations including starting or stopping of pumps, changes in control valve setting or pump speed, etc.;
- Analysis of a change in valve settings including open or close status change;
- Analysis of batch movement along the pipeline including batch interface change at a pump station;
- Analysis of pump tripping, caused by power or pump station equipment failure;
- Analysis of pipeline rupture or large leak;
- Analysis of column separation and its subsequent collapse.

Batch movement may not cause a significant pressure change, but other operational problems can potentially cause over- or under-pressuring along the pipeline or line pack changes. Even though these operational problems are properly taken into account during the design phase, the abnormal operating conditions such as pump trips and pipeline rupture cannot be avoided. These surge causing problems can be directly related to pipeline system safety and extra provision has to be made to manage them adequately.

Surge analysis is performed using a transient model provided by Energy Solution International. A transient model calculates time-dependent flow, pressure, temperature and density behaviors by solving the time-dependent flow equations discussed in the first section of this chapter. The model should be able to perform all time independent functions produced by a steady-state model, as well as time-dependent functions such as the effect of changes in injection or delivery, system response to changes in operation, and batch movement.

An effective transient model needs to be used as a pipeline system design and operation tool. As a design tool, it is used to determine pipe wall thickness by locating high pressure points in normal and abnormal operating conditions, the valve opening/closing time, the location and size of a pressure relief valve and its tank capacity, the minimum and maximum allowable transient pressures, the surge control system, the pump station spacing, and the leak detection system.

As an operation tool, both normal and abnormal pipeline operation changes can be simulated to find a cost effective way of operating the pipeline system. The transient model allows the operations staff to determine an efficient control strategy for operating the pipeline system and analyzing operational stability. It also helps study the effects of supply and demand changes on the pipeline and equipment and the line purge and load during pipeline commissioning. Different combinations of startup or shutdown procedures are simulated to determine how they accomplish operation objectives. The transient model must be able to model pump stations, including the pump, associated equipment and control strategy of the pump station.

A transient model should be able to realistically simulate system response to upsets. A pipeline system can be upset by pump trips, pipe rupture, or supply stoppage. The transient model is used to evaluate corrective strategies by modeling various upset responses. It allows the operation engineers to study the effects of blow-down on a compressor station and piping or to develop a corrective action when a leak or rupture occurs.

In addition, the model can be used as a predictive tool and a hydraulic training simulator for operation staff. As a predictive tool, it can be used to predict future pipeline states from the current states by changing one or more boundary conditions. Note that a transient model is more complex to use and its computer execution time is generally longer than that of a steady-state model. It requires extensive data, particularly equipment and control data.

5.2 LIQUID BATCHING TRANSPORTATION

Liquid pipelines are either designed to carry single products such as crude oil, refined products, high vapor pressure fluids, water, bitumen, condensate, or a number of products in a batch form. Others include slurry (mixed product pipelines, which are in essence single products pipeline that are designed to transport a heavier fluid/solids using a carrier fluid such as water, carbon dioxide, air, etc. [5, 6]).

5.2.1 Types of Liquid Pipelines

Liquid transportation in a batch form allows multiple products to be shipped in the same pipeline (Figure 5-13). Sequential movement of liquids in a batch form is commonly exercised by refineries and liquid pipeline companies to transport multitudes of products through a single pipeline. This form of transportation includes batch transportation of low as well as high vapor pressure fluids.

Liquid pipeline companies can transport and deliver a multiple of petroleum liquid products to many customers. For example, in year 2000, Colonial Pipeline delivered 90 different products for 85 customers to 270 terminals and into more than 1000 storage tanks [7].

5.2.2 Liquid Hydrocarbon Batching

The most reliable, safe, and economical way of transporting and delivering large volumes of a wide range of crudes and refined products from refineries to distant depots

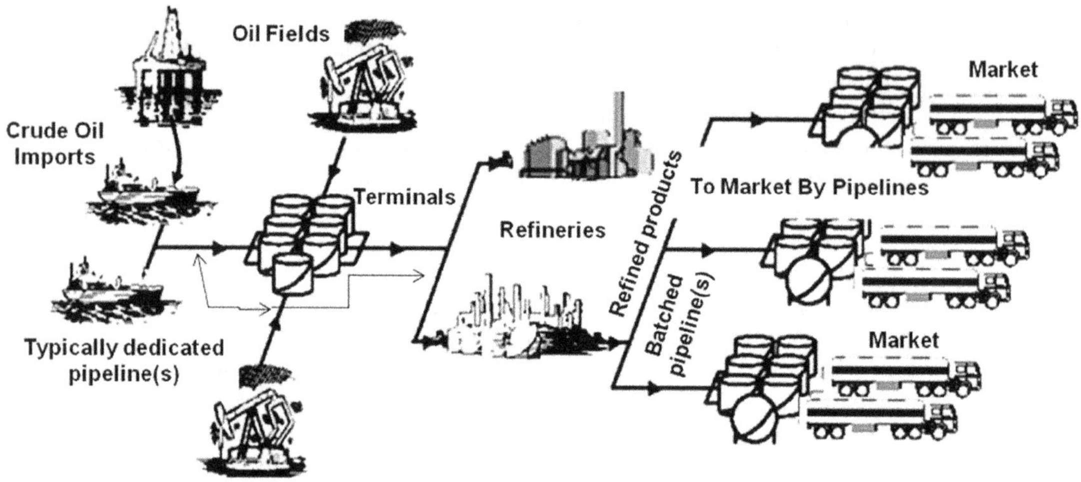

Figure 5-13. Single products and/or batched products pipelines applications [8]

is by means of batched pipelines. Batches of different crudes, grades, and products are pumped back-to-back in the line most often without any separating device. Batches move forward in the line and products are transferred to terminals whenever a new batch is injected at the head terminal (Figure 5-14).

Typical products commonly transported as single or batched products are listed below:

- Crudes
  - Heavy to Extra heavy, Heavy, Medium heavy, and light heavy
  - Intermediate
  - Light
  - Synthetic
- Diesel
  - Low sulfur
  - Ultra low sulfur
  - Low Pour
- Gasoline
  - Leaded
  - Unleaded
  - Super
  - Jet fuels — Jet A and Jet B
- Distillate and kerosene including
  - Jet kerosene
  - Jet naphtha

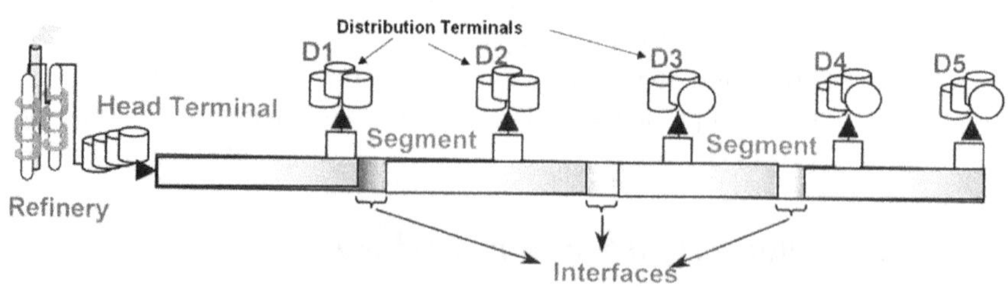

Figure 5-14. Batched products pipeline system [8]

    - Iso-octane
    - Alkylate
- Condensate
    - Raw
    - Sweet
    - Residuals
- Propane
- Butane — Iso-butane and Normal butane

For commodity properties transported and stored, refer to Chapter 8, Table 8-1.

5.2.3 Batched Product Pipeline Growth and Technique

The growth of transporting different products through transmission pipelines has increased substantially in the past several decades (Figure 5-15). This form of transportation includes batch transportation of low vapor pressure as well as high vapor pressure fluids.

Batching in the early days was achieved by injecting a liquid into the pipeline followed by a separation pig (usually a sphere) and then the second batch of another product/fluid. Batching of a multitude of products without a separation pig is more common today. Different product batches are "pushed" through the system abutting each other (Figure 5-16).

In order to ensure operational efficiency and safety, pipeline operation is generally composed of maintaining a constant steady-state flow through the understanding of pipeline hydraulics, pumping and flow controls. In a batched pipeline, this understanding also considers the characteristics of the different batches and how to adjust facilities operations to accommodate the unique characteristics of each batch.

A batch interface is the region where two batches abut in a pipeline and where some mixing of the products occurs. Batches with different densities (specific gravities) and viscosities, and the interfaces between them can cause significant changes in the flow rate and hence in the pipeline hydraulics. Fluid density and viscosity have a great impact on pipeline operations. Density affects the differential pressure, as well as pressure due to elevation head. Viscosity is a major cause of friction losses in pipeline

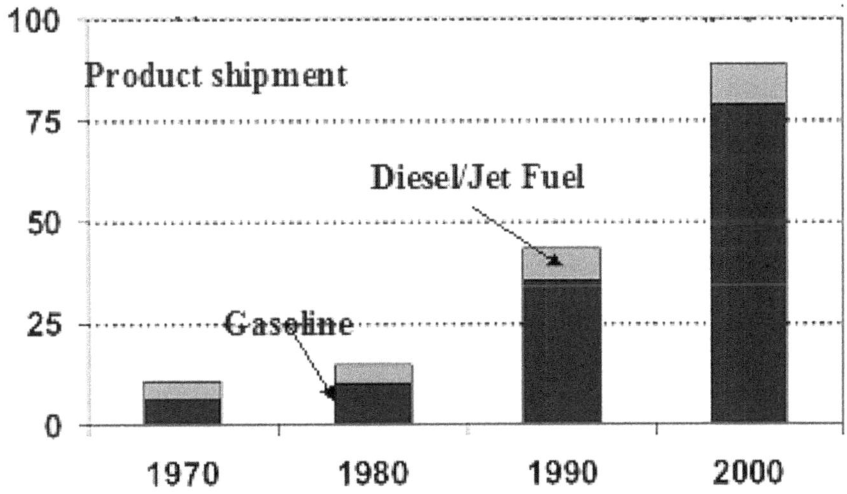

Figure 5-15. Growth in the number of product transportation [7]

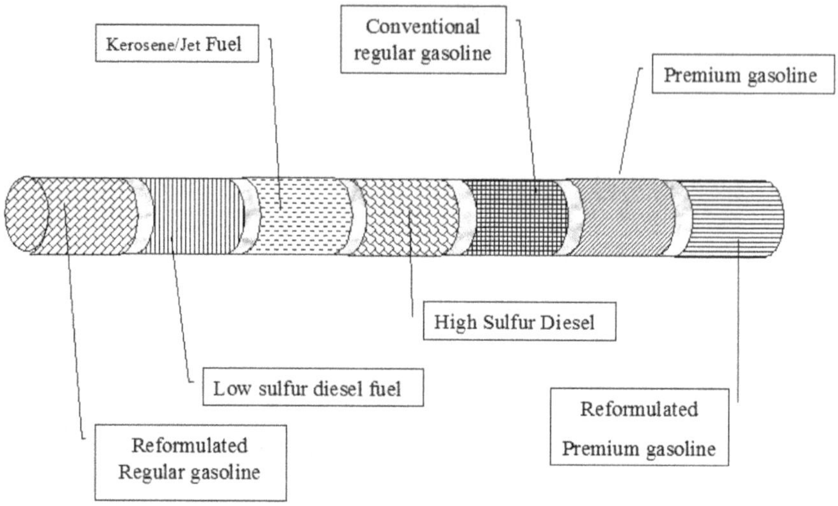

Figure 5-16. Typical batched products pipeline

operations. Operators must be aware that batch changes are particularly important when batch interfaces pass through pump stations, and thereby impact on the pump head pressure developed and pipeline frictional losses.

In batched product pipelining, increasing the number of distinct product types and delivery locations can complicate pipeline system control. Hence, the product scheduling for distribution systems as can be seen from Figure 5-17 (indicating different products transportation for different shippers A, B and C) can become complex [7].

A typical batched quality specification is shown in Table 5-2. Some of the standards (for test parameters) are also listed in Table 5-2 [9].

5.2.4 Products Batching Definitions and Terms

The following terms are commonly referred to in batching of multi products pipelines and in batched pipeline operations and reporting:

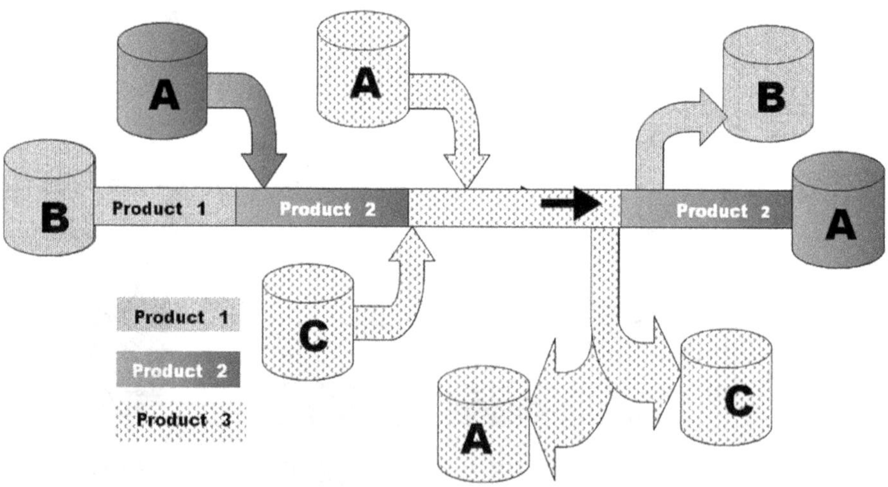

Figure 5-17. Batched product pipeline scheduling

TABLE 5-2. Typical batched liquid quality specification

Test Parameter	Refined Products	Synthetic Buffer	Natural Gas Liquids
Sediment and water or particulates	< 5 mg/L. particulate	< 10 mg/L. particulate	< 5 mg/L. particulate
density @ 15°C @STP, IP 216	(< 5%)	(< 10%)	(< 5%)
Viscosity @ lower of –Receipt temp. –Reference temp.		0.4–2 cs	< 20 cs
Vapor Pressure, kPa	< 103 kPa Reid vapor Pressure	< 103 kPa Reid vapor Pressure	< 1100 kPa absolute @ 37.8°C
ATM copper strip corrosion test (ASTM D130)	< 1b (expect change)	< 2a	< 2a

Batch: Batches are the means by which product movements can be tracked. A batch always starts out as a defined continuous volume of product. It may be split, partially delivered, or stored in one or more location product pools before it is finally delivered to its nominated destination.

Batches also exist for accounting and shipper/transport purposes. A single batch must be unique with respect to a shipper, product, origin, "carrier from," and ratability.

Fungible Batches: A "fungible batch" is defined as a batch of petroleum product meeting carrier's established specifications, which may be commingled with, or delivered in substitution for, other quantities of petroleum product meeting the same specifications. (i.e., they are interchangeable)

Fungible product specifications are established based on industry standards, federal and state requirements, and the pipeline operator's ability to handle various products. Fungible products usually provide shippers with a significant degree of flexibility for scheduling lifting and delivery times.

Segregated Batches: A "segregated batch" is defined as a batch of petroleum product meeting carrier's established specifications, which may not be commingled with other quantities. A batch may be segregated because it has properties that differ from the fungible specifications.

5.2.4.1 Batch Sequencing

Liquid pipeline operators transport various liquid petroleum products or grades of the same product in sequence through a pipeline, with each product or "batch" distinct from that preceding or following. Once a refined product or crude oil grade is injected and begins its journey, subsequent products may be injected and shipped. It can be fungible, in which case same products from different shippers can be allowed to be shipped in as one batch. As a consequence of batching without a separation pig, interface development occurs between batches. This interface is a petroleum mixture which occurs during normal pipeline operations between adjacent batches of petroleum products having different specifications. It may also be called "slop" or "transmix."

Factors that affect the order of batches include:

- Compatibility with adjacent batches
- Batch cycle requirements
- Buffering requirements
- Transmix or contamination levels
- Capacity and power availability and scheduling and operational requirements
- Ratability
- Product availability

A typical batch sequencing is shown in Figure 5-18.

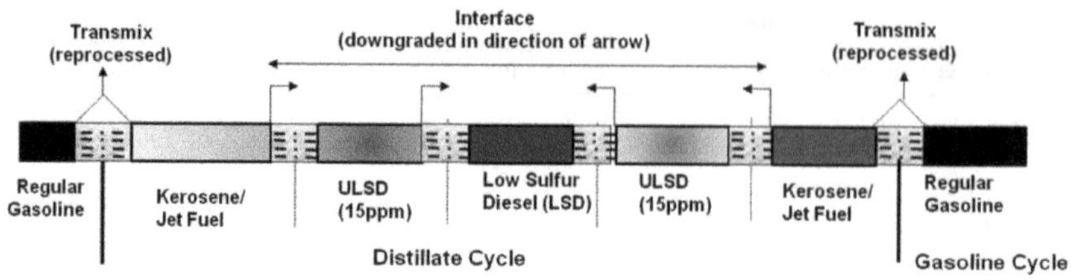

Figure 5-18. Typical batch sequencing [10]

Figure 5-18 is representative of a typical distillate and gasoline cycle; however, it may be noted that in observing environmental concerns, low-sulfur and high-sulfur diesels (LSD/HSD) are no longer produced in North American refineries.

Also, it may be noted that another factor that impacts batch sequencing is the product availability. Sometimes, a batch may be scheduled to be shipped, but it is not available in sufficient volume when a shipment is required. Thus, the batching sequence must be modified to account for this factor.

5.2.4.2 Batch Cycle/Slug

A batch cycle is a set pattern of batches of similar commodity type that takes into account:

- Interface contamination of batches
- Injection and delivery patterns of batches
- Repeated usage for production of batch sequences

Incompatible batch cycles are buffered to minimize contamination. A cycle is a defined period (generally, 1 to 10 days, short term or 10 to 45 days long term) where a prescribed set of products is transported, generally in a particular order and in particular batch sizes. Cycles simply repeat one after the other throughout the month.

A typical batch cycle for transportation is shown in Figure 5-19 and consists of the following products and their sizes, butane (352 m^3), gasoline (regular, 1500 m^3; premium 5200 m^3), kerosene (4200 m^3) and diesel (8200 m^3). The sizes given are examples only.

A slug/batch train identifies a continuous stream of a single homogenous product within a pipeline. A slug may contain one or more batches, so long as the batches consist of the same product. Slugs are operational movement tools, and are defined purely for the convenience of the scheduler. At any time, one or more batches, or even one or more partial batches, may be contained within a slug.

5.2.4.3 Buffers

A buffer is a petroleum commodity that physically separates dissimilar or different commodity types so as to minimize contamination. Some liquid pipeline companies use a synthetic crude oil, a semi-refined, "clean" crude product as a buffer between liquids such as propane gas (LPG) and refined products. Use of buffers ensures that the

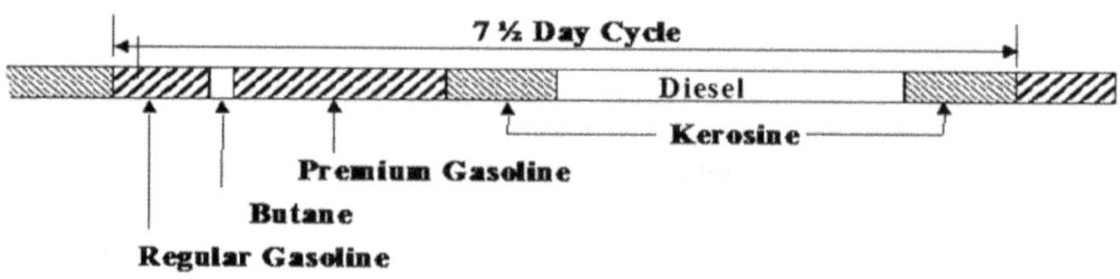

Figure 5-19. Typical batch cycle

lighter product does not migrate into heavier products and affect their vapor pressure and hence flash point.

When used, buffers are delivered to separate facilities (interface or "slop" or "transmix" tanks) or are mixed in with the contaminated batch when it is delivered.

5.2.4.4 Batching Travel Time

The time that it takes for a batch to travel depends on factors such as product flow rate, the number and type of preceding and sequential batches, compressibility differences between batches, pipeline elevation, etc. However, knowing the flow rate, a rough estimate can be made as to the time for batch travel from the point of injection to delivery location.

Long-distance liquid pipeline companies usually create a batch transit table or a chart that shows the approximate time required for a batch to move from one terminal location in a system to another.

The charts are most useful when batch transit times are long and batches are pumped in one month and delivered in the following month. Shippers utilize the transit table to estimate batch delivery times, even though the carrier may have yet to issue a new monthly schedule.

For pipeline shipment of varied products, a seasonal or other upheaval in delivery or intermediate terminals logistics can be disruptive. The shipment depends on delivering each product to each terminal in sequence. A batch travel chart together with delivery terminal restriction will assist in scheduling deliveries into facilities, diverting a portion of flow to an available terminal and sending the remainder to more distant delivery locations.

5.2.4.5 Batch Interface Marking and Detection

Batch interface detection is necessary to send the incoming batch to the proper tank or to perform other necessary operations. Batch interfaces can be detected by a densitometer if batches have different densities or by other types of detector if the batch densities are similar. Densitometers are most frequently used for detecting batch interfaces. Interface detectors are normally installed upstream of the delivery locations. If batches are separated by a sphere, then the sphere has to be inserted at the time of batch launch, bypassed when it passes a pump station, and removed when it arrives at the delivery point.

Some pipeline companies use a fluorescent dye to mark refined products' batch interfaces and use fluorometers to detect the arrival of batch interfaces. Others use optical interface detectors (OID) [11].

For interface detection with fluorescent dye, a single shot of dye is injected into the interfaces between refined products. To indicate the beginning and end of a new batch cycle, two shots of dye may be injected into the leading synthetic crude buffer as well as the last refined products batch. One injection is at the refined product (RP)/buffer interface, and the other is upstream of the interface, to provide early warning of the arrival of the following RP/buffer batch.

Advantages and disadvantages of such marking are:

- Advantages:
  - Allows the use of one pipeline for different products
  - Lower capital and operating costs
- Disadvantages:
  - Complex dye injection design and operation
  - Potential for missing an interfacial cut due to a misplaced or missing dye shot
  - Possible contamination and degradation
  - Possibility of lower-quality product impacting higher-quality product

With OID, the change in product is determined by calibrating and measuring the OID output as a function of product gravity. In this way, interfaces can be detected easily and accurately [11]. One commercially available dye product that is used in pipeline batch interface marking is Fluorescent Yellow 131 SC, produced by the Rohm and Hass company [12]. This is a liquid, concentrated fluorescent solvent soluble dye. It contains Solvent Red 175 in a high flash, severely hydro-treated naphthenic solvent system with the following characteristic/properties (Table 5-3):

The disadvantage is that it is completely miscible in all petroleum fuels and many other related hydrocarbon solvents. Also, a disadvantage of interface dye marking is the increased risk to batch cuts due to the potential for failures in the marker injection/detection.

The dye emits an intense white-yellow fluorescence when energized with a high intensity black light, making the smallest of leaks detectable. Because of this property, it makes for a good leak detection application for a variety of process fluids.

Another method that may be used in interface detection involves the use of a sensitive and precise viscometer. Typically, the same product coming from two different refineries will have slightly differing viscosities. Therefore, such a viscometer can be used effectively in determining and differentiating interfaces between two batches.

Calculating batch interface positions requires the following information:

- Pipeline system configuration
- Current storage levels
- Injection/receipt and delivery volumes
- Batch sequences

5.2.4.6 Batch Injection, Transportation, and Delivery

Liquid batches are injected and transported through the pipeline for delivery into terminals or refineries in the following fashion:

- Full stream injection
- Side stream injection
- Straight injection and delivery

In full stream injections the fixed portion of the main line is completely filled with the fluid, displacing its current contents.

Once batch sequences are determined, the pipeline is then scheduled from the first injection into the system through to the final delivery out of the system. Midline breakout tankage facilities are usually provided to be able to schedule start and end times for deliveries and injections. Breakout tankage is also used to manage differing flow rates in pipeline segments of differing diameters. This, in turn, allows for proper tankage/storage facility management.

In side stream injections, the liquid pumped into the main line shares the line with the same type of liquid already flowing in the line (Figure 5-20), side stream injection operations are useful when:

TABLE 5-3. Characteristic/properties of batch interface marking, fluorescent yellow 131 SC (manufactured by ref. [12])

Physical Form	Dark-Colored Liquid
Specific Gravity	0.92
Flash Point, ASTM D-3278	> 85 °C
Excitation wavelength	494 nM
Emission wavelength	535 nM

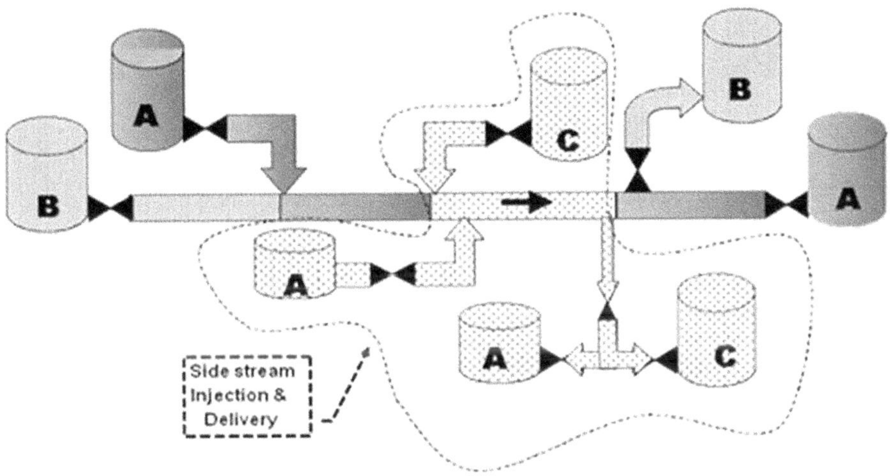

Figure 5-20. Side stream injection

- The allowable flow rate downstream of the injection supply/terminal is greater than the allowable flow rate upstream of a terminal.
- The maximum injection flow rate at a supply terminal is lower than the required pipeline flow rate.

With this method, batch injections can be achieved with minimum impact on overall pipeline flow rates. If there are flow rate or pressure restrictions, then the upstream flow rate can be reduced in order to allow the side stream injection.

Side stream delivery operations allow the pipeline to be used more efficiently, by increasing throughput or capacity, and are therefore advantageous when

- the upstream allowable flow rate is higher than the downstream flow rate and
- the delivery rate at a terminal/delivery location is low.

With straight injection/delivery, one liquid is delivered to a terminal from the main line, simultaneous with a second liquid being injected from that same location into the main line. The injection and withdrawal may be of same volume and product or may be of different volumes and products. This method allows for optimizing and increasing pipeline throughput, through tank/storage usage optimization when the product batching cycle is created.

A big advantage of side stream injection is that it facilitates the increase of the size of a fungible batch, thereby reducing the numbers of interfaces required, and any associated contamination or transmix costs.

5.2.4.7 Batch Reporting

Batching reports are used for the following purposes:

- Assist control center operators in operating the pipeline, and
- Help pipeline schedulers to schedule and coordinate batch movement
- Help maintenance staff schedule pipeline maintenance
- As an indicator of pipeline capability/integrity

5.2.5 Minimum Batch Size

Minimum batch size can be determined from a knowledge of the allowable contamination of one product in another. Most product pipeline companies allow themselves a

TABLE 5-4. Industry accepted contamination levels

Contaminant	Product	% of Contamination in Product
Butane	Gasoline	*
Premium gasoline	Regular gasoline	3
Regular gasoline	Premium gasoline	1
Jet fuel (kerosene)	Regular gasoline	1
Jet fuel (kerosene)	Premium gasoline	2
Jet fuel (kerosene)	Diesel	2
Any	Jet fuel	Nil

Note: *Depends on butane already added

reduction in the quality of specification of the product. However, such a degradation allowance depends on contractual and/or legal requirements rather than practical necessity. From a legal point of view, a product (e.g., unleaded gasoline) must be delivered to the market with quality specifications that meet government specified limits. Contamination levels in batch product pipelines are related to the size and type of batches or tenders required for entrance to the pipeline and the interfacial contamination that occurs during transport.

Tolerable levels of contamination usually varies from company to company and, as previously mentioned, is dependent on the specification of the final product and the quality of the products as produced by refineries. Typical allowable contamination levels for refined products accepted by the industry are shown in Table 5-4.

5.2.6 Crude Oil Contamination

5.2.6.1 Natural Crude

Acceptable contamination in a crude oil is difficult to define. Because it is a naturally produced substance, it contains natural materials, some of which are undesirable for a refiner, such as sulfur, nitrogen, carboxylic acids, metals, sand, clay, etc. The levels of these contaminants vary according to the source of the crude oil and its treatment history. A refiner will carry out an assay on a crude oil stream prior to purchasing it, in order to establish its value to that refinery. The value of the crude oil is based on the value and amounts of products that the refiner can make from that crude oil, as well as the cost to the refiner of removing the contaminants that are present in the feed stock. Mixing a different crude oil with the originally assayed crude oil will "contaminate" that oil, and change its value to the refiner, positively or negatively. Typically, a refiner is willing and prepared to accept interfacial levels of contamination, but not levels that are associated with tank bottoms service changes.

5.2.6.2 Synthetic Crude

Synthetic crude oils are significantly different from natural crude oils, both because of their source as well as the fact that they have been processed in an upgrader. Synthetic crude oils (e.g., Alberta based), generally have high aromatics content (considered to be problematic to a refiner) and very low contaminant concentrations (beneficial to refiners). They also usually have no bottoms or residual content because they have been distilled.

Refiners sometimes look at the contamination level in synthetic crude oil based on the residual content. Most synthetic crude oils are "bottomless" when they leave the upgrader site, meaning that they contain no residue. However, during transportation, they can acquire a measurable amount of residue as a result of contamination with other crude oils. This contamination occurs usually from being interfaced with those oils. Sulfur content is also a good indicator of contamination in synthetic crude oils because they usually have very low sulfur contents, of the order of 0.10% by weight. Contamination with heavy crude oil will result in a marked increase in the sulfur content of synthetic crude oil due to the high sulfur content of heavy crude oil (2 to 4 wt.%).

It is well understood that contamination by heavy crude oil of lighter grades of crude, such as synthetic, is problematic. It is a common misperception however that contamination of heavy crude oil with a lighter grade of crude oil improves the heavy crude oil. This is unlikely to be correct. The refiner has selected a heavy crude oil for his feedstock based on the properties and distillation yield of that crude oil. Altering that yield and properties may well result in the refiner suffering a loss, perhaps even a serious loss. In one example, a refiner received a heavy crude oil batch, intended for asphalt shingle production, which had been seriously contaminated with a light sweet crude oil. The produced shingles did not meet the requirements for durability, and the asphalt simply melted off the shingles under summer sun.

Contamination can occur from many sources as the crude oil stream moves from its production site to the refiners' tankage. Common sources include tank farm manifolds, tank bottoms from tank service changes or common tank usage, leaking valves, interfacial contamination, dead legs, etc. The impact of each of these sources varies according to the volume of the contaminating crude oil. Tank farm manifolds can be a significant contributor, depending on the design and operation of a tank farm. Simple tank farms are easy to analyze, and operating procedures can be developed for minimization of their impacts, including operational sequences and tank service restrictions. Complex tank farms should utilize a manifold specifically designed for minimization of contamination. The use of dead legs should be minimized in the design process, through tight placement of valves, no more than three pipe diameters from a pipe tee. Leaking valves can be identified through careful inventory tracking and control. The largest contributor to contamination as a result of operations (assuming full turbulent flow) is tank bottoms.

Tanks have a minimum working volume (working bottoms). At levels below that minimum working volume, special precautions must be taken for filling and taking suction from that tank, so normally a tank is not operated below its working bottom. When changing service from one crude oil to another, the working bottoms are not considered if the crude oil types are the same, e.g., from one heavy crude oil to another heavy crude oil. However, when planning to change service of a tank between two dissimilar types, the working bottoms must be taken into account. The contamination caused if a tank is simply swapped from one service to another would be unacceptable. There are two alternatives for handling this "incompatible" type of crude oil service change. One is to reduce the volume of bottoms below the working level; by pumping the tank out to minimum suction levels, which could be followed by using vacuum trucks to remove the non-usable volume. The second alternative would be to negotiate with a refiner to accept one or two batches known to be heavily contaminated. Both have disadvantages, the former results in operational restrictions on tank fill rates to restore the working bottoms of the new crude oil in the tank, the latter will probably result in a financial penalty on the contaminated crude oil batch(es).

5.2.6.3 Contamination Level

As a rule of thumb, contamination from interfaces will amount to a few percent of between batches, while contamination from tank bottoms can be as high as 20% to 40% (first batch after service change), see Chapter 8 for details. However, the successful batching of differing grades of crude oil depends on maintaining turbulent flow. Reducing the pipeline flow rates such that Reynolds numbers drop below 5000 tends to cause stretching of interfaces, with higher associated contamination costs as a result. The critical parameter in maintaining full segregation is the Reynolds number of the midpoint of the interface between the two batches. Acceptable midpoint Reynolds number (*Re*), and the associated contamination, depend on the type of product being transported. Refined products systems typically use a minimum Reynolds number of 20,000, while crude oil lines typically accept lower values of *Re*, in the range 3000 to 5000. It is generally accepted that Reynolds numbers in the range 2000 to 3500 or so represent transition flow. That is a flow regime that is partially laminar and partially turbulent. Exactly which flow regime is present depends on the nature of the flow history. If the flow rate is increasing and passing from laminar to turbulent, it is more likely that the transition from laminar to turbulent will occur at higher Reynolds numbers, as long as no external source of turbulence is present. Examples of such an external source of turbulence might be a pump impellor, a set of short diameter pipe bends, a sudden change in pipe diameter, or any significant pipe roughness. Experiments with water flow have shown that, given optimal flow conditions, laminar flow can be made to continue as high as *Re* 10,000. Similarly, when decreasing flow rates, turbulent flow tends to drop to transition and laminar flow regions at lower Reynolds numbers. If a pipeline system must operate at marginal flow rates for turbulent flow, it could be advantageous to bring flow rates initially to a rate that corresponds to a Reynolds number of 5000 or so and then reduce the flow rate to the target operating zone.

It is also possible, although rarely done, to operate with heavy crudes in transitional or even laminar flow, separated by light crude oils that are in turbulent flow. The midpoint interface must be in turbulent flow for batch segregation to be maintained. This is a sub-optimal operation, and the interface will be slowly shifted as the laminar flow zone captures the heavy tail of the interface. The interface will increase in size from turbulent model predictions. It will also become asymmetric, with a reasonably short interface length from the light crude to the midpoint of the interface (by density), and a longer interface from the midpoint to the heavy crude end of the interface. Little data has been published on this type of operation, but heavy crude has been successfully batched at Reynolds numbers less than 2000 between light crude batches having Reynolds numbers over 10,000.

An example of this is the line-fill operation of a 2000-km NPS 30 line at 200,000 BBLSD with heavy crude and light synthetic oil. The heavy crude oil batches will flow from the KMP 0 to 300 segment at a Reynolds number of about 840, while the synthetic crude would be at a Reynolds number of about 71,000. The Reynolds number of the interface will be about 12,400, which is expected to provide good batch integrity, as described above.

5.2.7 Interface-Volume Estimations

A method for estimating the interface length and volume is described by Austin and Palfrey [13]. The following describes the steps involved in arriving at interface length and volumes and is described further by Mohitpour et al. [14].

Step 1

The blended viscosity for a 50/50 mix can be derived from the following equation:

$$\upsilon_B = 0.5 \log \upsilon_1 + 0.5 \log \upsilon_2 \tag{5-3}$$

where

υ_B = blended viscosity, CS
υ_1 = product 1 viscosity, CS
υ_2 = product 2 viscosity, CS

Step 2

The Reynolds Number for the blended product is then:

$$Re = \frac{V \cdot d}{\upsilon_B} \times 1{,}000{,}000 \tag{5-4}$$

where
Re = Reynolds number
V = fluid velocity, m/s
d = inside pipe diameter, m

Step 3
The critical Reynolds number is found from

$$Re_c = 10{,}000 * \text{Exp}\left(2.75 \times \sqrt{d}\right) \tag{5-5}$$

where
Re_c = critical Reynolds Number

Step 4
The interface length (LC), if $Re > Re_c$, is

$$\text{LC} = \frac{0.3716\sqrt{L \cdot d}}{Re^{0.1}} \tag{5-6}$$

and if $Re < Re_c$, then

$$\text{LC} = \frac{582.49\sqrt{L \cdot d}\,\text{Exp}\left(2.19\sqrt{d}\right)}{Re^{0.9}} \tag{5-7}$$

where
LC = length of interface, km
L = length of travel, km

Step 5
Finally the interface volume can be found from

$$\text{VC} = 250\,\pi d^2 \cdot \text{LC} \tag{5-8}$$

where
VC = interface volume, m^3
d = inside pipe diameter, m.

The above computation assumes that the product velocity and pipeline diameter remain constant throughout the pipeline. However, situations arise where products are delivered to different delivery (drop off) points, and where the pipe diameter changes, subsequent to drop off. In this case, a new interface does not begin at each of the velocity change locations. Rather, the interface length continues to grow at a rate dependent on the new velocities. The following steps are thus followed for calculation of interface lengths in situations where velocity changes in the pipeline as a function of diameter.

At each velocity change point, a new Reynolds Number and a critical Reynolds Number must be calculated. If $Re > Re_c$, then from Eq. (5–7) $\text{LC} = \frac{0.3716\sqrt{\text{L} \cdot d}}{Re^{0.1}}$

Step 6

The next step is to calculate LC_1 for the end of the first section (with inside diameter d_1). The length LC_1 must be converted to an equivalent length that the interface will occupy in the second section with an inside diameter d_2.

$$LC_{(1-2)} = LC_1 \times \left(\frac{d_1}{d_2} \right)^2 \qquad (5-9)$$

Step 7

The length of second section pipeline that would produce the equivalent length of interface received from Section 1 is then calculated by manipulating the Austin and Palfrey equation.

$$L_{2\,\text{equiv}} = \frac{\left(\dfrac{LC_{(1-2)} \cdot Re_2^{0.1}}{0.3716} \right)^2}{d_2} \qquad (5-10)$$

Step 8

The total cumulative length of interface at the end of the second section is now.

$$LC_{2\,(\text{Total Length})} = \frac{0.3716\sqrt{\left(L_{2\,\text{equiv}} + L_2\right) \cdot d_2}}{Re_2^{0.1}} \qquad (5-11)$$

L_2 = Length of second segment

Step 9

Steps 6 through 9 are repeated for all sections of the pipeline system where either velocity or pipe diameter changes occur.

5.2.7.1 Batch Calculation and Tracking Example

The following example provides a result of interface calculations for a pipeline transporting multi-products from two refineries to market locations.

Design Data

- Pipeline Schematic (refer to Figure 5-21)
- Pipeline Inlet Pressure:
  - Refinery 1: 300 kPa gauge
  - Refinery 2: 300 kPa gauge
- System inlet temperature: 30 °C
- Minimum operating pressure: 300 kPa gauge
- Maximum design pressure: 10205 kPa gauge
- Maximum operating pressure: 9184 kPa gauge
- Maximum design temperature: 50 °C
- Minimum delivery pressure: 300 kPa gauge
- Pump station operating efficiency: 75%
- Pipeline depth of burial: 1 m
- Pipeline roughness: 0.045 mm
- Ground temperature: 27 °C
- Ground conductivity: 1.14 W/m°C
- Pipeline route elevation: refer to Table 5-6
- Corrosion allowance: 1.6 mm

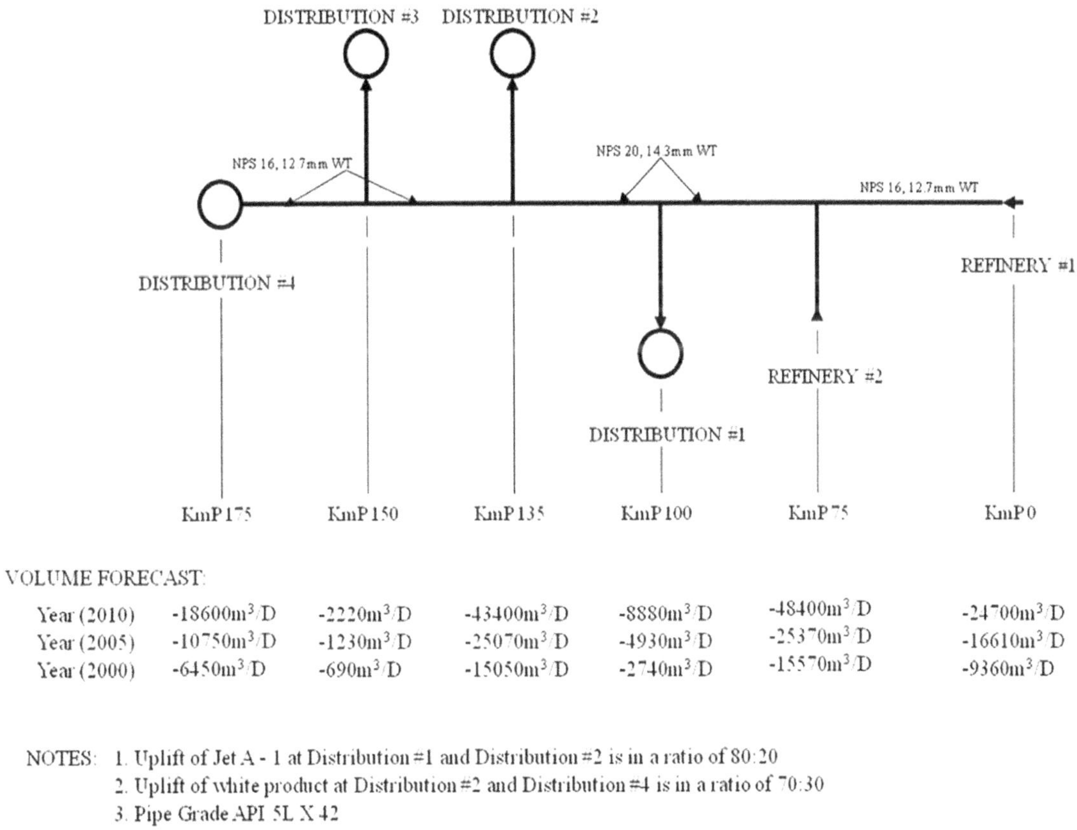

Figure 5-21. Multi-product pipeline system schematic

Table 5-5 below lists the properties of the petroleum products transported for the given example. For properties of other petroleum products refer to Chapters 1 and 2.

Table 5-6 and Figure 5-22 below provide an elevation profile for the multi-products pipeline.

Contamination criteria from Table 5-4 above was utilized in the prediction of interface lengths and volumes, as shown in Tables 5-7 through 5-8 below.

5.2.7.2 Results

The pipeline was designed for transporting the heaviest of the products (i.e., diesel). Pumping facilities were selected such that under the worst scenario, a single batch of diesel can be transported for the design conditions stipulated previously (refer to design data and Figure 5-21).

Maximum interface volumes and minimum batch size meeting the contamination criteria of Table 5-4 are provided in Tables 5-7 and 5-8, respectively. A sample calculation of interface volume and length is provided in Table 5-9 for regular and unleaded gasoline.

Typical interface volume accumulation tracking is shown in Figure 5-23 for the interface between diesel and jet A-1. From this figure, it can be inferred that as velocity or flow rate increases, interface volume decreases.

5.2.8 Batched Products Pipeline Design and Operational Issues

In the transportation of any batched products (LVP and HVP), there are a number of issues that will require specific attention to ensure that the products are transported and

TABLE 5-5. Product properties information

Hydrocarbon Liquid	Viscosity cS	Temperature (°C)	Density (kg/m3)*
Diesel	6.86	5	847
	5.10	15	820
Gasoline (leaded)	0.68	5	711.3
	0.61	15	–
Gasoline (unleaded)	0.7	5	699
	0.63	15	690 (assumed)
Jet A fuel	8	0	774
	1.5	29	–

Note * Density at STP (101.325 kPa and 15 °C)

delivered with the appropriate quality/integrity, safety, and least environmental impact in the event of a leak or pipeline failure.

5.2.8.1 Design and Operational Issues

Design and operational issues that specifically affect batched products pipelines are summarized below:

- Technical design of facilities (pipeline, pumps, measurements) including piping arrangements
- Pipeline valve spacing (for high vapor pressure product transportation)
- Dead leg impacts and remediation
- Minimal active taps/valves
- Manifold piping design
- Minimal dead legs
- Valves
- Multi-service tankage/tank residuals
- Parallel flow paths in piping systems
- Start/stop operations vs. continuous flow

TABLE 5-6. Pipeline route elevation profile

Kilometer Position (kmp)	Elevation (m)
0	72
1	72
21	72
27	49
30	47
42	76
57	76
64	59
68	99
82	60
92	43
101	119
109	8
128	60
132	60
136	17
146	6
150	28
155	7
170	7
175	7

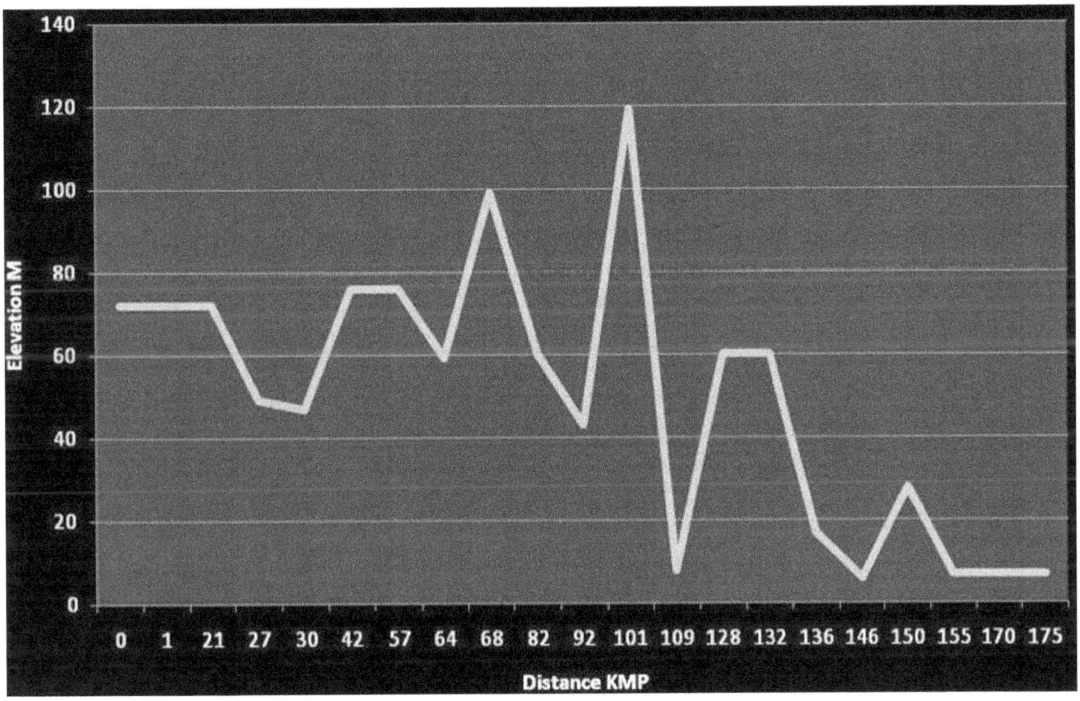

Figure 5-22. Pipeline elevation profile

Batch interfacial management including:

- Batch sequencing
- Buffering
- Contamination impacts
- Cut-points
- Degradation cost impacts

Operational controls and procedures:

- Quality testing
- Pump control procedures
- Impacts of supply forecasting
- Impacts of shared load pumps/meters — quality and degradation
- Post pipeline treatment for interfaces/contaminants

Delivery considerations — filtration, interfacial cuts, contamination, and degradation impacts, tankage residence times, transmix segregation, and disposal

- Safety, environment, and risk
- Emergency response

TABLE 5-7. Predicted interface mixture volume (m^3)

	Year 2000	Year 2005	Year 2010
Diesel/Jet A-1	259.37	245.71	234.08
Diesel/regular gasoline	250.28	237.95	226.69
Regular/unleaded gasoline	225.43	214.33	204.18
Unleaded/premium gasoline	219.93	209.09	199.20
Premium gasoline/diesel	244.17	232.14	221.15

TABLE 5-8. Minimum batch size

Fuel	Minimum batch, m^3		
	Year 2000	Year 2005	Year 2010
Diesel/Jet A-1	12968.50	12285.50	11704.00
Diesel/regular gasoline	25028.00	23795.00	22669.00
Regular/unleaded gasoline	22543.00	21433.00	20418.00
Unleaded/premium gasoline	7331.00	6969.67	6640.00
Premium gasoline/diesel	24417.00	23214.00	22115.000

5.2.8.2 Operation and Control

Pipeline control in a batched products pipeline involves scheduling to transport different shipper nominated volumes of crude, products, and NGLs from designated injection locations to designated delivery points, while maintaining the quality and quantity of the product.

Batching information is critical to the success of scheduling because it influences pipeline operation (Figure 5-24). Batch information provides a means for:

- Determining the accuracy of the schedule
- Identifying the baseline, or starting point, for scheduling new operations
- Pipeline system operational requirements

Three major components of liquid pipeline operation are:

- Nominations (supply control, pipeline supply scheduling)
- System operation (operations scheduling, pipeline control [products receipt, pumps and tankage/storage operation, product delivery])
- Inventory accounting.

A high level scheduling process cycle for a pipeline network transporting multiple products is shown in Figure 5-25. The process involves nominations, data review and input from schedulers to develop a cycle plan, a batch plan, and system schedule. This scheduling information is then distributed to shippers, terminals, schedulers, third parties, and pipeline controllers to keep everyone with an interest in product movements abreast of the schedule. The process usually allows schedulers to schedule up to 45 to 60 days in advance, and to modify that schedule as required.

Nominations: Shipper nominations involves receiving and processing all Notice of Shipments (typically volume, type of commodity/product, injection and delivery points), performing system line balance calculations on all originating lines and determining the system capacity adequacy to transport all products.

Generally, if adequate capacity exists, pipeline companies commence the transportation scheduling process. However, if adequate capacity does not exist, the following apportionment procedure is performed:

TABLE 5-9. Interface between regular and unleaded gasoline (year 2000 volume)

Item	km 0–75	From Dist#2 (km 26)	km 75–100	km 100–135	km 135–150	km 150–175
Blended Viscosity cs	0.794	0.794	0.794	0.794	0.794	0.794
Reynolds Number (Re)	456,206	456,206	965,715	965,715	481,561	481,534
Critical Reynolds # (Rec)	54,701	54,701	67,271	67,271	54,701	54,701
Interface Length, LC, km	0.5397	0.31777	0.32427	0.38368	0.24006	0.30992
Interface Volume m^3	61.50	36.21	58.50	69.22	27.36	35.32
Cumulative Interface Vol m^3	61.50	97.71	156.21	255.43	27.36	62.67

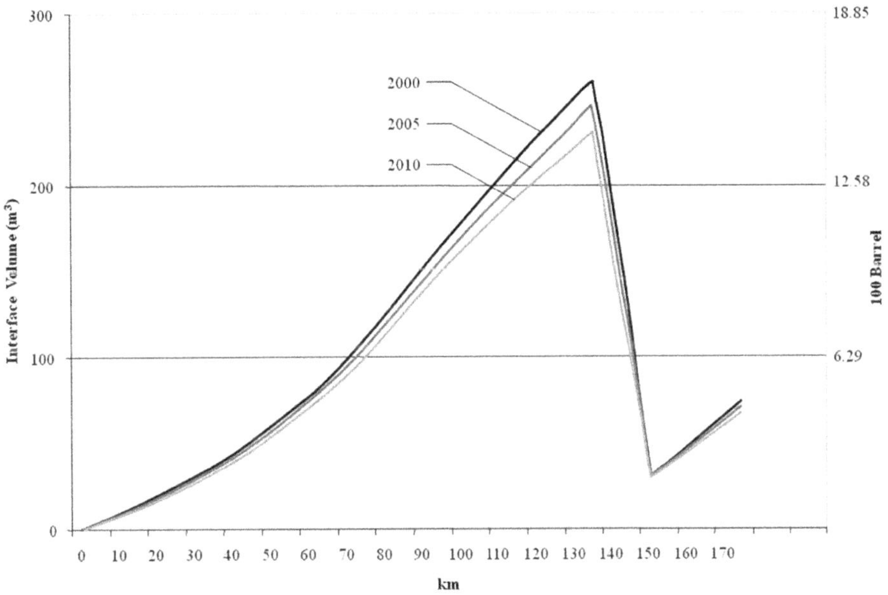

Figure 5-23. Interface volume growth and tracking between diesel and Jet A-1

- Perform feeder pipeline nominations verification
- Apportionment calculations to determine the percentage of apportionment by line
- Notify each shipper of the maximum allowable by shipper and by line
- Receive from the shipper the adjusted/apportioned nominated volumes, i.e., revised Notice of Shipments
- Commence scheduling process

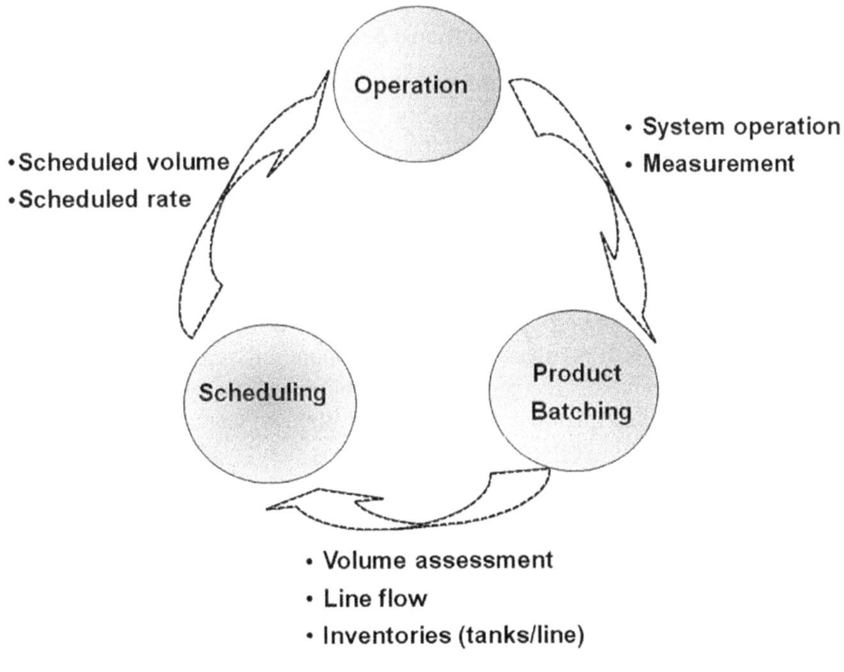

Figure 5-24. Pipeline operation

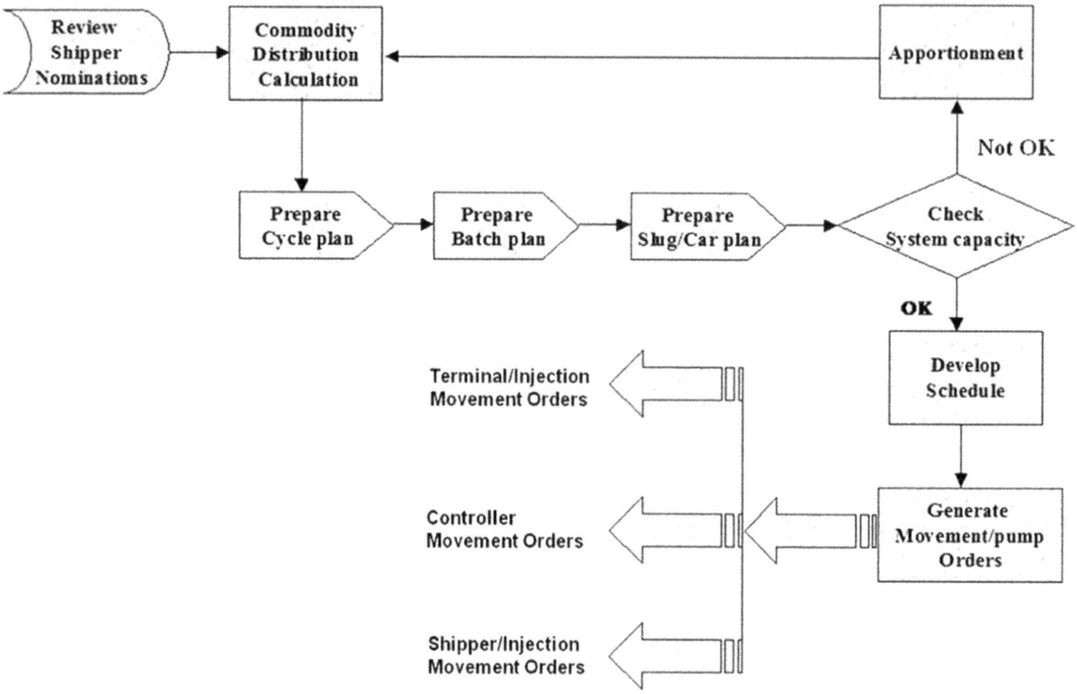

Figure 5-25. high level overview of scheduling process

Ratability: Defined as the even distribution of batches carried by the pipeline according to the shipper, the commodity type, the volume and the destination. The purpose is to help ensure that all shippers have equal access to the pipeline. It is also considered to be a measure of the ability of a pipeline to deliver a scheduled batch to its delivery point within a defined window of its originally scheduled date and time.

Capacity calculations: Several factors determine how pipelines are scheduled

- Design capacity
- Sustainable capacity
- Operating capacity
- Expected capacity

Proration/capacity matching/apportionment: If nominated quantities exceed available capacity within any pipeline segment, the industry uses proration rules/apportionment, to determine maximum total nomination by segment.

The purpose is to pinpoint areas of the pipeline where potential bottlenecks will occur. It involves calculating net pumping requirement, which is the total volume of commodity that a pipeline must pump every day of the month through each section of pipe in order to meet the product shipment volumes. Usually, all product shipments are totaled to determine capacity requirement/availability.

At the same time, schedulers determine each shipper's minimum inventory level and maximum inventory level by product and location. These levels are necessary to ensure that all shippers share the responsibility for maintaining the working volume in the bottoms of the tanks based on each shipper share of the total nominated movements for that product. At the same time, each shipper is able to utilize the working storage capacity in proportion to their nominations. When overcapacity volume is determined, proration/apportionment is the process used to reduce the net pumping volume tendered for a section of pipeline. This apportionment or proration is calculated as follows:

$$Apportionment = (1 - ECV/NTV) \times 100$$

where
NTV = net tendered volume
ECV = expected capacity volume

Overcapacity occurs when the net pumping requirement exceeds the volume that a pipeline section is capable of moving for a given nomination period.

Pipeline supply scheduling: involves the following:

- Receiving information for products to be shipped
- Schedule batch movements downstream
- Follow-through scheduling
  - Issue all schedules as segment fills are completed
  - Issue injection and delivery schedules

For all shippers, connecting carriers and delivery locations:

- Schedule updates and communication for current and subsequent periods
- Maintain system throughputs by pipeline segments
- Direct contact with all shippers and connecting carriers

A typical planning schedule involving batch nominations and scheduling events is shown in Figure 5-26. The following provides a general sequential list of scheduling process involving nominations, batch planning and movements.

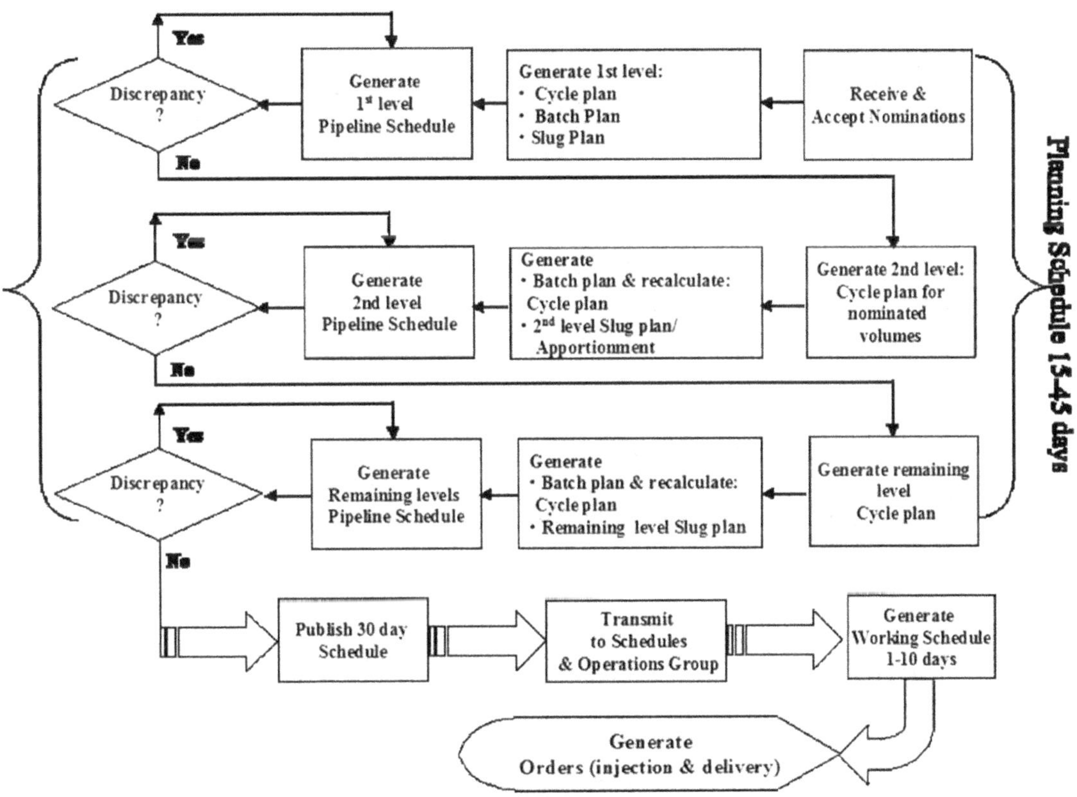

Figure 5-26. Typical batch planning schedule

Notification of shippers for monthly nominations

- Proration or apportionment if nominated quantities exceeds available capacity within any pipeline section, to determine shipper's maximum total nomination by segment.
- Determination (by schedulers) of each shipper's minimum inventory level and maximum inventory level by product and location.
- Communication of prorated deliverability and inventory to Shipper.
- Designation by the Shipper which products are required to be cut back to meet the prorated deliverability and inventory requirements by resubmitting new nominations.
- Generation of a preliminary cycle plan for, and communications of plan to all parties
- Provision of comments by Shippers
- Revision of cycle plan and communication to Shippers
- Completion of the cycle and batch plans by pipeline schedulers.
- Completion of slug/batch train plan.
- Communication of the Planning Schedule to the shippers and all interested parties
- Obtaining the connecting carrier schedules and communicating the Working Schedule to the pipeline operations group.
- Generating modified schedules and issue short-term movement orders as required.
- Determining and communicating the status of Shippers nominations/batches

Factors that influence batch scheduling are:

- Reductions in supply
- Increases in supply
- Delivery pattern changes
  - Third-party impacts
  - Line upsets
- Scheduled and unscheduled maintenance
  - Tankage/storage requirements
  - Facility limitations
  - Refinery/delivery upsets

It is normal practice that once times have been assigned to batch injections and deliveries, schedules are sent to the

- Shippers
- Connecting pipelines/carriers
- Injection and delivery sites on the pipeline
- Refineries/delivery locations

Shipper schedules include information about

- All batches that the shipper owns
- All injections of volumes tendered
- All deliveries for both the current tender period and previous tender periods

In today's technological environment shippers can electronically submit via the Internet:

- Monthly forecasts
- Monthly nominations

- Batch changes
- Batch trades and swaps

Further details on the batch scheduling and shipper information system including a web-based system such as T4 are detailed by Yoon et al. [3].

5.2.8.3 Pipeline System Operation/Control

Pipeline system operational control, particularly for a batched system, involves operations scheduling, pipeline control (products receipt, pumps and tankage/storage operation, product delivery) and oil/product transportation accounting.

Generating Operational Schedule: Pipeline operational scheduling generally involves the following, depending on the complexity of the batched product network:

- Availability of information to generate the schedule such as previous days/period pipeline activities
- Availability of tankage/storage facilities at all intermediate, breakout or delivery tankage locations
- Review of 1- to 10-day schedule and fine-tuning of all daily activities
- Review and consideration of maintenance/construction-related activities
- Review of shipper requests for changes
- Communicating with pipeline control and delivery terminal operations
- Revising and updating movement orders (as necessary)
- Generating the day's movement order

Once a schedule is generated, movement orders must be generated. There are several kinds of movement orders, tailored specifically to the recipient. Example types of product movements are shown in the following Table 5-10:

TABLE 5-10. Type of product movements

Name/Recipient	Characteristics
Detailed Movement Order/Scheduler	Contains virtually all movement information
Terminal Movement Order/Terminals	Contains virtually all scheduled pipeline activity for a single terminal
Shipper Movement Order	Contains all movement order information for a given shipper anywhere on the scheduling pipeline.
Connected Carrier Movement Order/ Connected Carrier (if applicable) or lines	Contains all movements anticipated to or from tankage or lines to the connected a carrier
Summary Movement Order/ Controllers and Schedulers	Includes most information and has an option to see the pass-by lineup—slugs that will pass by a location without affecting the local tank inventory.

Movement or pumping orders are the culmination of the scheduling process. These define the activities that pipeline control takes to direct product movements on the pipeline [15].

Operations and Maintenance: Pipeline operations are usually set up for approximately one week to 10-day periods; this includes:

- Full or side stream injections or deliveries
- Full or partial breakouts
- Land or replenish operations
- Constraints

Generally, pipeline and tankage maintenance events are built into the schedule, pump operations/orders, including today's and the future (e.g., next several days) pump operational orders, and any revisions that occur during operation and maintenance activities.

Pipeline/Terminal Control: Many factors affect liquid pipeline operation, including:

- Pipeline pumping operations, i.e., interruptions and/ or slow downs
- Number of types of product and specialty fuels such as MTBE, ethanol, ultra-low sulfur diesel, mid-sulfur diesel, which:
  - Limits substitution of products between markets
  - Limits effective pipeline capacity
- Lower inventory at pipeline and terminal tank farms; the number of days of inventory is usually reduced as:
  - Shippers try to improve return on capital.
  - Capacity demand increases without building additional tanks/facilities
- Transition of different seasonal grades of products, e.g., gasoline sold at a terminal is different on different days
- Many pipelines and/or segments nearing capacity
- Market supply becomes tight as demand increases above capacity
- Less spare capacity available to make up for interruptions

Generally, responsibility for safe and efficient operation of the pipeline (pipeline, pumps, custody transfer) and terminals (tankages) rests with the pipeline control center, which undertakes the following tasks:

- Follow daily pump/movement orders requirements
- Maintain required daily/hourly throughputs by segments
- Monitor station pressures
- Start/stop pumps and operate valves
- Diversion of products into breakout tanks
- Monitor receipts from feeder/supply pipelines
- Communication with feeder pipelines, connecting carriers and refiners
- Coordinate all movement injections, landings into tankage and deliveries
- Initiate emergency procedures (if required)
- Track batch positioning in pipelines lines/segments using a batch tracking system
- Specifications control using online instrumentation
- Communications and product movement management
- Shipper/customer communications
- Provision of information to accounting

Pump Operations/product movement: The Central Control room is responsible for remote operation of pumps, terminals/tankages, and custody transfer and deliveries (Figure 5-27). To be able to operate a multi-products pipeline with many batches, an understanding of batch interfaces, density and viscosity changes are required. Figure 5-27 shows the Koch modern control centre. Koch has about 22,000 km of pipelines throughout North America. In these two control centers, the lines carry a wide range of products including crude oil (3200 km), refined products (2500 km), natural gas liquids (9000 km) and anhydrous ammonia (3100 km) [16].

Similarly, TransCanada PipeLines Limited modern control centre provides operational control of the Keystone and Keystone XL liquid pipelines (Figure 5-28).

Figure 5-29 shows operational hydraulics of a typical batched products pipeline carrying NGLs, synthetic crude (SYN) and Refined Product (RP). The pipeline is designed to a maximum pressure of 1,440 psi (maximum pump discharge pressure).

Figure 5-27. Typical views of a hydrocarbon liquid pipeline/terminal control center

Figure 5-28. Keystone modern liquid pipelines operation center (Courtesy TransCanada)

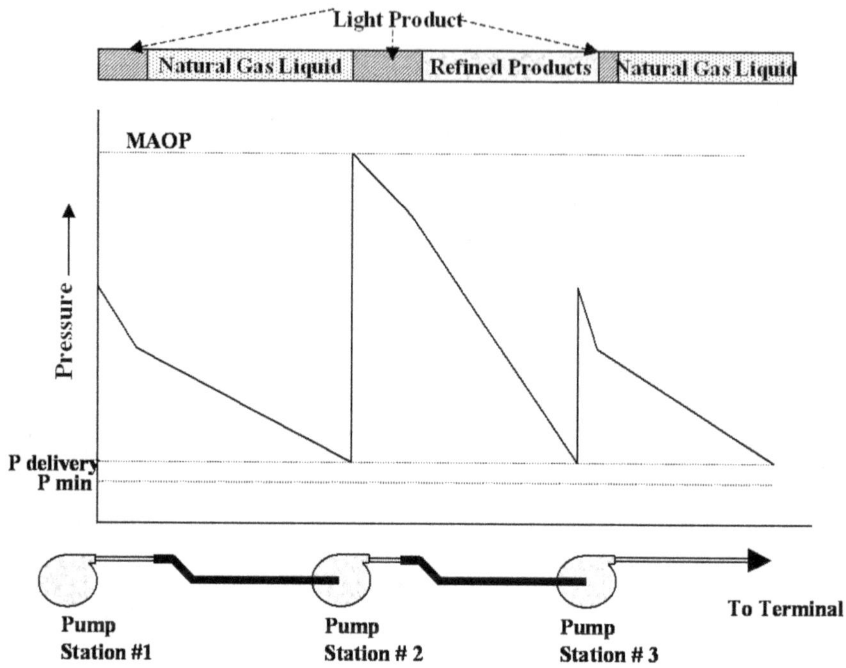

Figure 5-29. Batched product pipeline operational hydraulics

The suction pressure of pumps (Figure 5-29) are to be limited to at least the products' minimum vapor pressure, i.e., the pressure at which liquidity of the products are maintained. The vapor pressures of some commonly batched products are provided in Table 5-11.

The pipeline capacity for various products is represented by the pressure gradient as shown in Figure 5-30.

Density changes in the batch interface: Pump head remains constant for a given pump at a constant flow rate, but the pressure rise across the pump is dependent on density. If there is a change in the density of the fluid flowing through the pump, then the pressure rise across the pump also changes. Batch interfaces also cause changes in line pressure as elevation changes. Operators must be ready for these changes in density and pressure. A pipeline will generally have a densitometer at a short distance upstream of a pump station to give advance warning of changes in density.

Viscosity changes in batch interface: Friction head loss changes continually as the batch interface moves along the pipeline. The greater the difference in viscosities, the greater the change in friction head loss. The pump head downstream must

TABLE 5-11. Vapor pressure of common batched products (at standard temperature and pressure)

	Vapor Pressure	
Product	**kPa**	**Psi**
Ethane	5,536	802
Propane	1262	183
Butane	489	71
Synthetic	45	6.5
Distillate	172	25
Gasoline	172	25
Dilbit	50	7

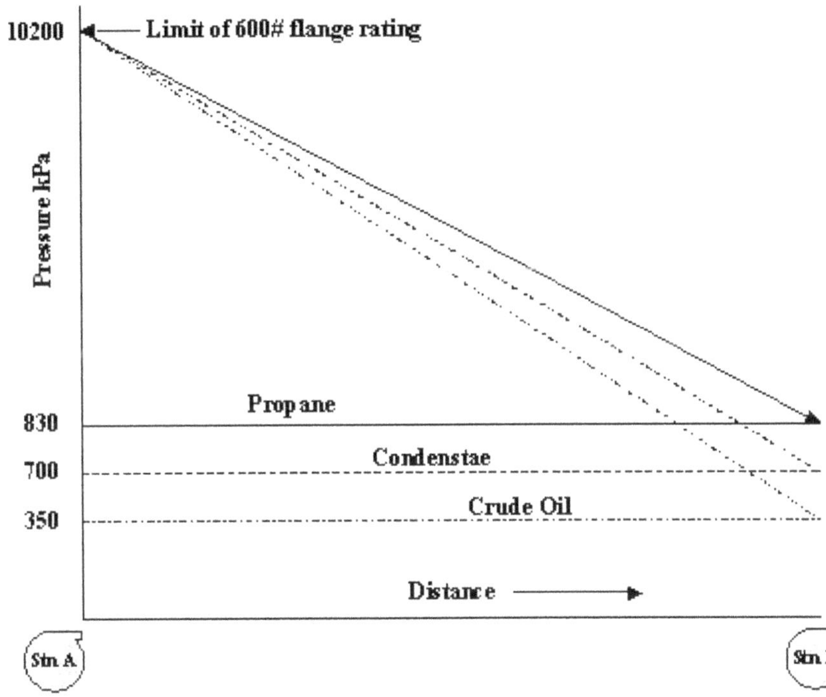

Figure 5-30. Line segment capacity for various products based on acceptable delivery suction pressures

be adjusted to compensate as head loss changes. Thus, to operate a batched products pipeline, the Central Control Room effectively controls pump and terminal valve operations to optimize products' movements based on the volume requirement while minimizing the interface and contaminations.

Quality Control and Product Movement: An important aspect of a batched pipeline operation is the control and maintenance of batch quality and batch movement. Quality Control considers contamination and degradation:

- Contamination: A change in quality due to commodity commingling in the system, e.g., crude into synthetic crude, the synthetic crude may be downgraded to a lower-value crude type if sufficiently contaminated.
- Degradation: A one-way exchange in volume between differently valued commodities, e.g., Commodity A to Commodity B—the interface lands at a tankage location. The commodity volume is under landed (or under delivered instead) say by 200 m^3 and the Commodity B volume is over landed (or over delivered) by the same 200 m^3. This 200 m^3 would be subject to the price differential value between the two commodity types. Degradation could result in a net dollar gain or loss.

Quality is maintained by:

- Designating commodity types by separate lines if possible
- Maintaining minimum batch size requirements
- Minimizing dissimilar commodity types adjacent to one another
- Cycle patterns
- Tankage/storage

- Maintaining line-pack
- Operating within predefined flow rate regimes

Deliveries are normally made to tankage, to connecting pipelines, or to a refinery; based on a mid-point density cut-point to maintain quality.

Product Movement: In a batched products, pipeline generally batches of different product are directly adjacent to each other. Thus, it is extremely important to keep the integrity of batched products in the transportation process by assuring that each product arrives at the termination point with as little contamination as possible. As noted previously, this is achieved through the following:

- Assuring turbulent flow minimizes interfaces (mixing zone) (Figure 5-31)
- Optimum batch sequencing reduces contamination/downgrade (Figure 5-32)

By keeping distillate and gasoline cycles separate as well as optimizing the sequence of each batch of products in each cycle (gasoline or distillate), it is possible to minimize interface contaminations.

An example is the batching method used for transporting high and low sulfur diesel. A typical batch sequence and appropriate interface cuts using a distillate (such as jet/kerosene) in between is shown in Figure 5-33.

However, when transporting Ultra Low Sulfur Diesel (ULSD) the question would be how best to arrange the batch sequencing and interface cut (Figure 5-34).

It has been reported [17] that ULSD quality will not be affected when a proper batch sequencing technique and optimal handling configurations are implemented in the pipeline transportation system (Figure 5-35).

Stasioski [17] concludes that each pipeline system is different and thus should be treated as such, but generally between the interface zones, content does not appear to increase during pipeline shipment. However, to avoid increasing contamina-

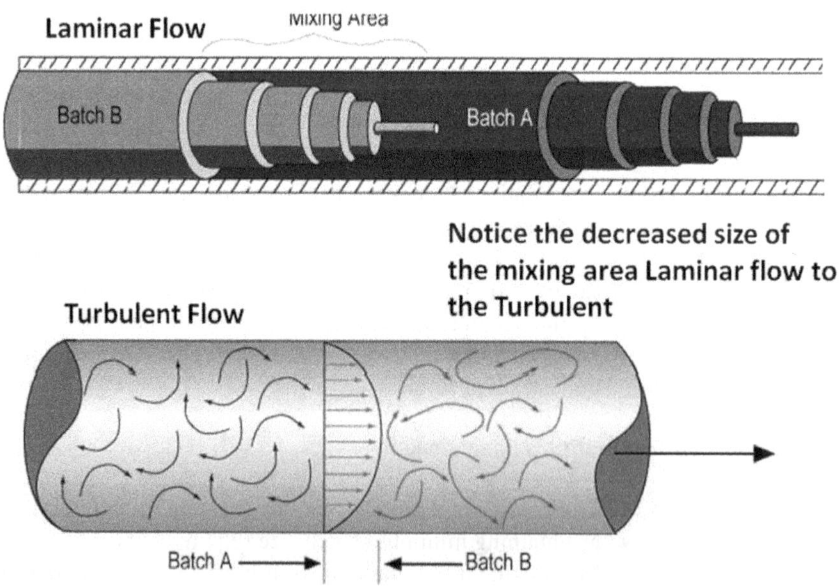

Figure 5-31. Minimizing interface in batched pipelines by assuring turbulent flow

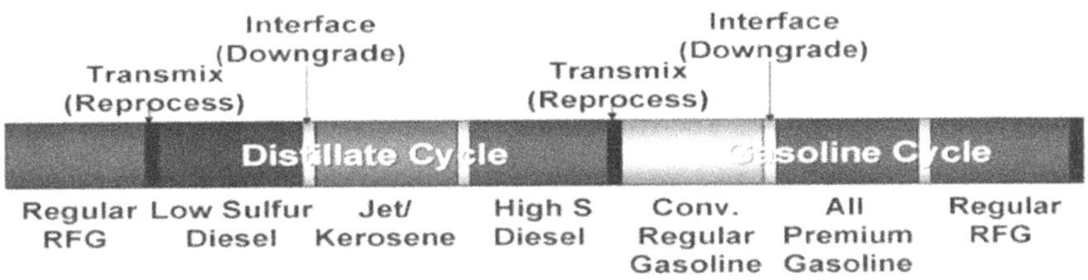

Figure 5-32. Optimum batch sequencing to minimize interface contamination

tion levels, optimal handling practice must be exercised which should include the following:

- Additional displacements and purges to protect ULSD (ultra low sulfur diesel)
- Automated sumps are locked out and do not pump into ULSD batch
- Pumps are not restarted during a batch of ULSD
- Minimal active taps and dead legs exist along the mainline
- In the interface, the sulfur content will begin to change before, and continue to change after, the detectible gravity change
- The ULSD batch must be protected by making more conservative cuts than cuts based on gravity
- The sulfur in the interface however generally will be 50% to 100% larger than the gravity interface.
- The larger interface should be additional down grade to LSD (low sulfur diesel), HSD (high sulfur diesel), or transmix depending on available tankage

Although the above comments are specific to ULSD, they could be viewed as applying to any extremely sensitive, quality critical batch type. Typically, extreme precautions such as those above are not used in crude oil movement.

From the point of view of tank farm contamination Stasioski, found batch sulfur contents increase within tank farms and contributing points of contamination including:

- Dead legs
- Valves
- Multi-service tankage/tank residuals

Communications and Oil Movement Management: This usually includes the following:

- Providing online assistance to industry in processing and monitoring petroleum transportation arrangements

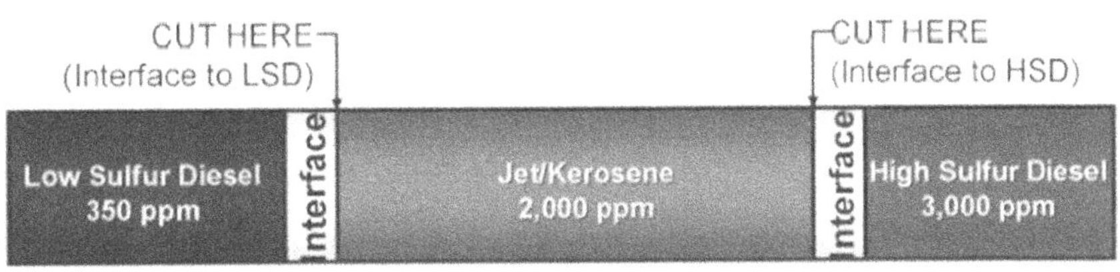

Figure 5-33. Batch sequence and cut in the transportation of high and low sulphur diesel with distillate

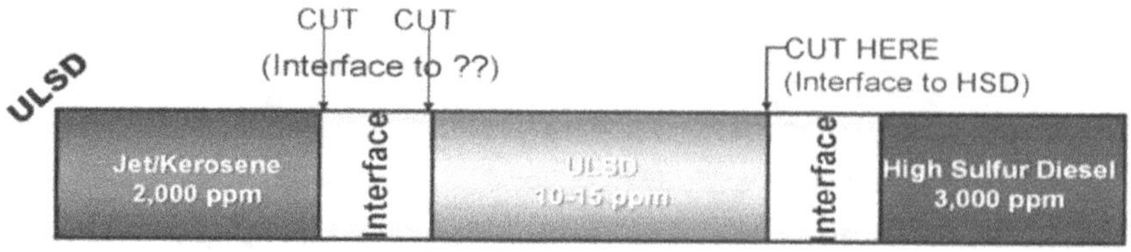

Figure 5-34. Options for batch sequencing and cut required in batch transportation of ULSD

- Providing customers with the ability to view their batches in the pipeline and their batch injection/delivery schedules
- Linking between the pipeline scheduling system and the customers scheduling system
- Supplies' pipeline and other system-related information as required announcements

Shipper/Customer Communications: This is achieved through:

- Telephone hot lines
- E-mail and fax documents
- Individual customer visits
- Refinery visits
- Industry shippers' meetings
- Supply and pipeline capacity management
- Association and task force meetings
- Industry business functions
- Orientation presentations

5.2.9 Practical Batch Operation in Real-time

As previous described, a hydrocarbon liquid batch is the volume of a product which is pumped into the pipeline in a continuous operation while keeping the same product properties. Once a batch is injected into the pipeline, the identity of this batch must be maintained until it is delivered to the shipper. The batches for each shipper are identified with product codes and numbers. A batch cycle is made up of several batches pumped in a continuous sequence. A buffer may be injected between certain batch cycles to prevent contamination of the expensive products.

Batch operations are based on the batch schedule and have three distinctive phases:

1. batch launch at the batch lifting and injection locations,
2. batch tracking while moving along the pipeline, and
3. batch delivery at the designated delivery locations.

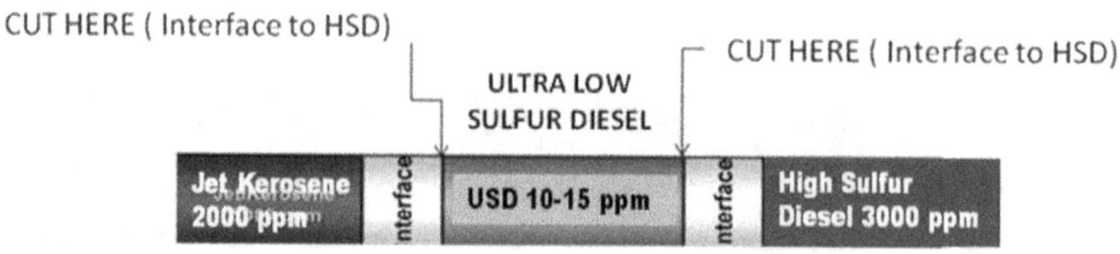

Figure 5-35. Options to minimize contamination in transportation of ULSD

For pipeline batch operations, the facilities (equipment and instruments) that need to be incorporated at the product lifting locations, upstream of the delivery or take-off points, and at the delivery point are described hereunder:

5.2.9.1 Batch Launch and Delivery

A batch is launched at the lifting tank designated to it in the tank farm and delivered to the receiving tank. The launch and receiving facilities are basically piping manifolds with valves and instruments. Instruments may include a densitometer, pressure and flow meter. A manifold schematic is shown in Figure 5-36.

The operation for launching and receiving the batches is normally designed to be controlled remotely from the pipeline system control center. To facilitate such a remote control, batch launching and receiving operations are fully automated.

1. On-line densitometers: These instruments provide continuous reading of density of batches at the lifting and delivery points in the pipeline. The instrument reading is sent to the control room by telemetry. As the batches are lifted or moved along the pipeline, this information is used to open and close appropriate valves, start and stop pumps, and direct the batches to the appropriate tanks. Additional densitometers are normally located a few kilometers upstream of pump stations and a delivery or take-off point to inform the operator of the impending arrival of a batch interface. They are required at the upstream of the final delivery tanks and downstream of the take-off points to confirm the arrival of batch interfaces. This is required to initiate certain actions needed for proper delivery of the batches.

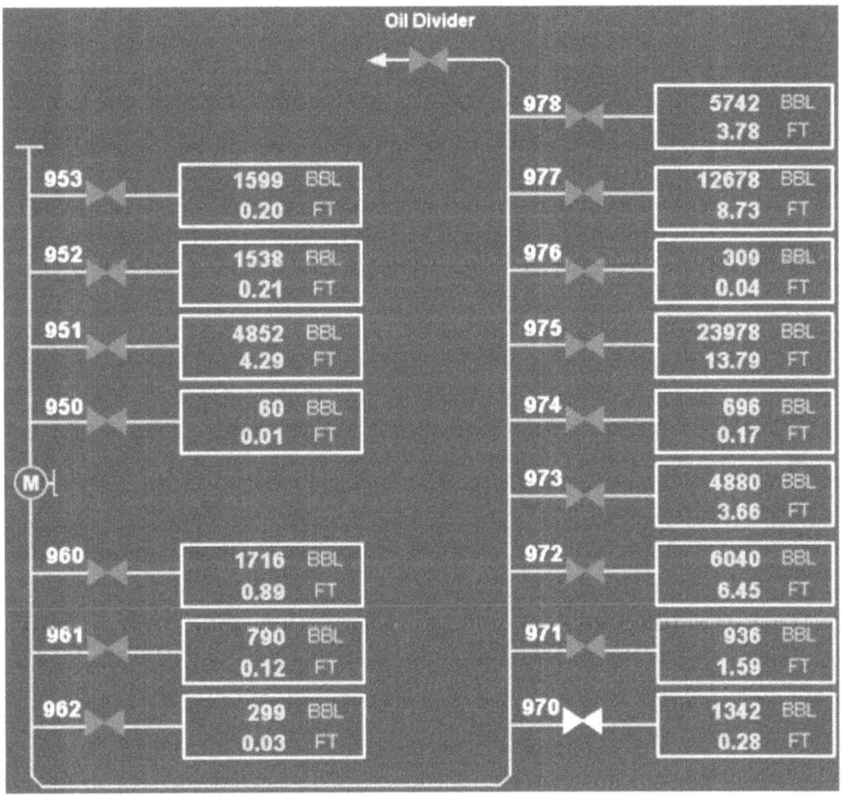

Figure 5-36. Tank farm manifold schematic (image courtesy of Telvent)

2. Flow meters and volume accumulators: The flow rates are measured and volumes are accumulated at the lifting and delivery locations in order to meet the nominated volume requirements during batch operation and to account for the total lifted and delivered volumes. Since tanks can be simultaneously receiving and discharging product batches, use of tank gauging cannot provide accurate volume data. The best approach to accurately measure batch volumes is to install flow meters with volume accumulators so that accurate information can be transmitted to the control center.

In addition, an on-line batch tracking capability is required to provide the pipeline and terminal operators with up-to-date and accurate information about batch locations, batch volumes and estimated arrival times. This capability reduces the possibility of errors and allows the operators to be much more effective and efficient in dealing with batch movement and operation.

5.2.9.2 Launching and Delivery Operation

Batch launching operations can be triggered by an indication from SCADA, a change in density, a change in valve status, and/or a schedule. The batch launcher is normally automated for the control center to remotely launch the batch at the batch interface. Conversely, the procedures for launching batches can be implemented in the tank farm controller or PLC, which performs the sequence of valve operations and/or checks density changes. After completing this sequence, the controller generates the batch launch signal, which is picked up by the SCADA system and used to start tracking batches along the pipeline.

When a batch is launched, the meter should be initialized with the batch ID and the actual start date and time. The meter factor for the new batch has to be determined based on the product and meter type. The meter factor is obtained from the meter proving records. The batch is injected until the metered batch size becomes the same size as the scheduled batch size. If the metered batch size is similar to the scheduled batch size within a defined tolerance, the batch lifting should be terminated and the actual batch size recorded. This actual batch size derived from an injection meter will be used to deliver the batch. The delivery will be active until the delivered batch size is the same as the actual batch size for the appropriate delivery meter.

The batch receiver at the delivery location is another set of manifold piping and valves designed to flow out of the line to the designated tank. Similar to the launch sequence, the sequence to operate the manifold has to be created to deliver the batch to the correct receiving tank.

5.2.9.3 Batch Tracking

Batch tracking monitors each batch for its volume, origin, current location, destination, and estimated time of arrival to the designated locations. A batch is defined as a contiguous entity of uniform fluid properties which moves through the pipeline system as a single entity. For example, a batch is assumed to have constant density, compressibility and viscosity.

Real-time batch tracking information helps the operators reduce unnecessary downgrading of product or contamination of product in tanks. In addition, up-to-date batch tracking information is useful in improving the accuracy of short-term batch schedules.

The batch tracking process (Figure 5-37) must be able to perform the following main functions:

- Determine and update the positions of the batch interfaces with each scan.
- Maintain batch volumes in the pipeline.
- Calculate batch overages and shortages in the pipeline.
- Calculate estimated time of arrival (ETA) of batch interfaces at designated locations.

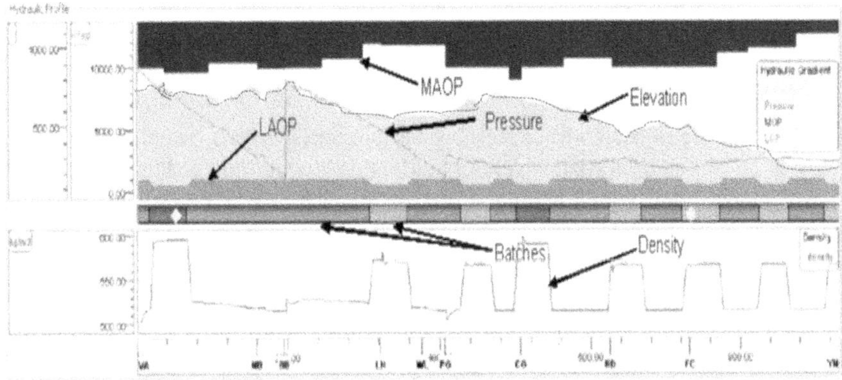

Figure 5-37. Typical batch tracking display

- Estimate interface mixing lengths and volumes.
- Detect an actual interface arrival automatically at a batch interface detector such as densitometer.
- Adjust batch volumes and interfaces automatically according to the specified rules in the event that a batch interface is detected, and provide the operator with the capability to modify batch volumes, batch positions, or batch ID manually.
- Alert the operator of batch arrivals.

Batch volumes are updated based on injection and delivery volumes obtained from metering locations along the pipeline. The interface positions can be determined, given the order and volume changes in and out of the pipeline. Given pipeline flow rates and interface positions, estimated times of arrival (ETAs) to the designated downstream locations can be determined. Upon completion of delivery and removal of the batch from the pipeline, an over/short volume is calculated and stored. The over/short volume reflects the difference between metered injections and deliveries along the pipeline as well as any manual adjustment that may have been made along the way.

If a side stream injection takes place, batch tracking is affected in two different ways: either the injected product is the same as the flowing product or the injected product is different from the flowing product. The former case maintains the same batch ID but the size is different and the flow rate downstream of the side stream injection point increases by the same amount as the injection rate. However, if a different product is injected into the flowing product, then the following changes take place:

- Two products are blended and the properties of the blended product should be determined for modeling;
- The batch size on the upstream side of the injection point reduces and eventually the batch disappears; and
- The blended product becomes a new batch downstream of the injection point, and its size grows.

The above figure displays the batch tracking information. The operator uses the batch tracking information for effective batch operations, which is normally made available to him/her through the SCADA system. The critical information for the operator includes the batch IDs, line fill volumes and flow rates of the lifting and delivering batches, batch interface positions along the pipeline, and ETAs for each batch to the next designated locations. If a drag reducing additive (DRA) is injected, the batches containing DRA together with their positions and the concentration of DRA and their ETAs to the next pump stations or significant facilities should be reported. Also, any contaminated batches and their positions should be tracked and reported to the operator.

Batch tracking may be integrated with a batch scheduling system, to determine an up-to-date batch schedule; this is accomplished by comparing actual batch tracking data with scheduled injection and delivery volumes and times. Current batch volumes and positions can also be used to update short-term batch schedules.

5.2.10 Multiproduct Pipeline Batch Optimization

Batch sizing and sequencing/cycling have been described previously. However, it is the industry's practice to decide the optimum batching cycle for a pipeline project based on contamination costs as a function of the cycle time as well on the basis of tankage costs for differing cycle sizes. The larger the cycle time, the larger the tankage that is required to store the cycle volume prior to injection and after delivery from the pipeline.

To determine the most economical batching cycle, the required analysis compares the costs resulting from various batching cycles by determining the net present value (NPV) of these costs, as indicated in Figure 5-38. The batch cycle with the lowest NPV is preferred as it will minimize the present value of the costs. In this example, the two-day cycle is shown as being optimum.

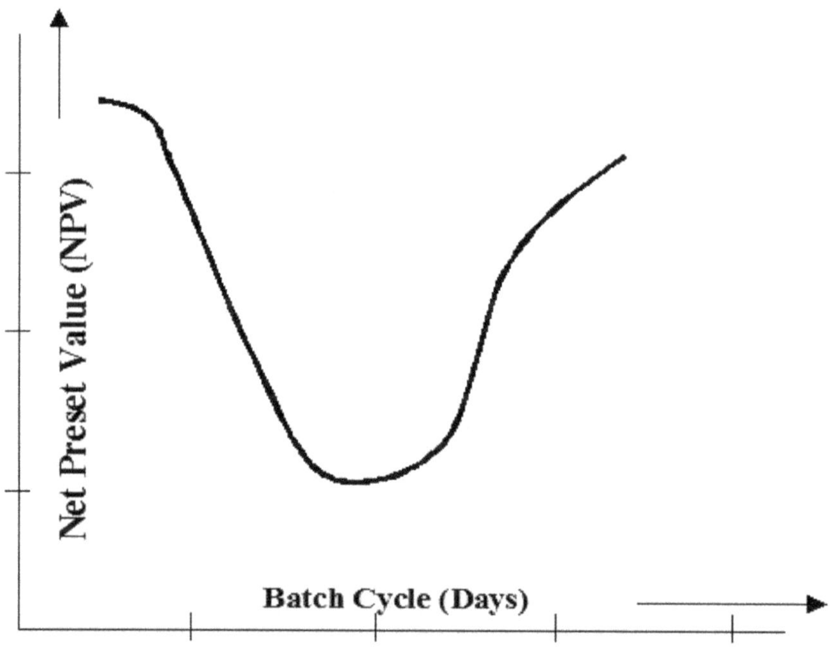

Figure 5-38. Batch sizing/cycling optimization based on NPV calculations of cost

ADDENDUM TO CHAPTER 5

Pipeline System Surge Mitigation Equipment

Surge control and mitigation/relief systems are widely used in hydrocarbon transmissions pipelines, marine terminals and tank farms. Generally, all systems where pressure is contained require some kind of pressure relief. Ignoring or dispensing with this requirement can endanger personnel and equipment and may lead to serious damage of valuable assets, affecting public safety and the environment. However, various preventive measures are made to avoid such damage.

As discussed in Section 5.1.3, surge pressure is a consequence of a sudden change of fluid velocity that can be caused by;

- Rapid valve closure, curtailment of supplies/deliveries
- Pump start-up, shutdown and emergency shutdown, including emergency shutdown of mainline pipeline or terminal facilities.

Long distance transmission pipelines, if not properly designed, can produce high transient/surge pressures that result in:

- Flanged connection failures;
- Pipe weld seam integrity damage;
- Pipeline leak or rupture
- Propagation of potential cracks;
- Misalignment of the pump outlet and discharge pipeline;
- Damage to pipe fittings
- Cavitation and erosional issues

Surge in a pipeline must be determined by engineering analysis usually through dynamic hydraulics assessments or by measurement. Such assessment techniques are highlighted by application in Section 5.1.3 and in Section 6.6 (refer also to Section A5.9 below).

There are a number of solution possibilities including those listed in Table A5-1 that the pipeline operator may use to mitigate any potentially harmful surge pressures. These solution possibilities depend on whether it is a new design or a retrofit to mitigate situations causing surge.

The best technique is to have a complete design and operation strategy/procedure devised to avoid surge in a pipeline system during the design process. However, the implication of various techniques must be assessed and equipment selected that meet the objective of surge mitigation and safety criteria throughout the life of a system. The following describes the surge mitigation devices that are most often used in pipeline system design or retrofits (as indicated in Table A5-1).

A5.1 FLOW CONTROL VALVES

A control valve (Figure A5-1) is a mechanical device installed in a pipeline to control the flow and/or pressure of the liquid. An automatic control valve opens or closes automatically in response to some signal to control the flow or pressure of oil in a pipeline. Technically speaking, an automatic control valve is one that closes automatically in response to a complete pressure loss or a flow rate increase, when either of which exceeds a predetermined set point.

In a pipeline with varying route elevation slopes that include ascending and descending gradients there is a great potential for down surge (i.e., column separation). In such situations, flow control valves can provide a means of changing the hydraulic gradient in a hydrocarbon liquid line to reduce the potential for column separation [18].

TABLE A5-1. Surge mitigation alternative summary (*indicates most often considered)

Surge Control Method	New Facilities	Existing System
• Pipe wall thickness increase	√	
• Pipe material higher grade	√	Replace sections
• Pressure reduction piping	√	Retrofit
• Pipeline reroute/alter elevation	√	
• Additional pipe supports to reduce resonance	√	√
• Flow/pressure control valve*	√	
• Air/Vacuum release valves	√	Retrofit
• Intermediate check valves/Non Slam check valves*	√	Retrofit
• Check valves*	√	Retrofit
• Pump/valve bypass*	√	
• Surge anticipation/relief valves	√	
• Liquid accumulators	√	
• Surge tanks (not for hydrocarbon liquid pipelines)	√	
• Pump station/pipeline valve relocation	√	
• Pump and valve location adjustment/relocation	√	No (expensive)
• Pressure bursting discs*	√	
• Replace weak pipe sections/higher grade pipe	√	Replace sections only
• Increase diameter of pipeline to reduce average velocity	√	Replace sections only
• Variable speed drives*	√	
• Soft start pumps	√	
• Valve closure and opening times*	√	√
• Increasing the inertia of pumps and motors (i.e. flywheels or by selection)	√	Retrofit only

Figure A5-1. Typical automatic control valve, [1], [18]

Advantages of such valves are:

- Multiple duties and scenarios
- Depending on the valve operator, power or instrumentation is not necessarily required
- Can be retrofitted on existing pipeline and facilities
- Moderate capital cost

However, such valves have to be maintained to ensure they remain effective.

If the pipeline hydraulic gradient intersects the actual pipe elevation profile during one or more of the operating scenarios, it can cause surge at those locations. It is a common solution to provide a flow control valve at the end of the pipeline to ensure that the hydraulics gradient level remains above the route profile. Typical flow/pressure control valve operation is indicated diagrammatically in Figure A5-2.

For typical pipeline application, pressure/flow control valves can provide the following control options:

- Upstream pressure/flow
- Downstream pressure/flow
- Upstream/downstream differential pressure

Control valve sizing is detailed in Chapter 4; however, the following equations can be utilized as applicable

$$Q = C_V \sqrt{\frac{\Delta P}{\gamma}} \qquad (A5-1)$$

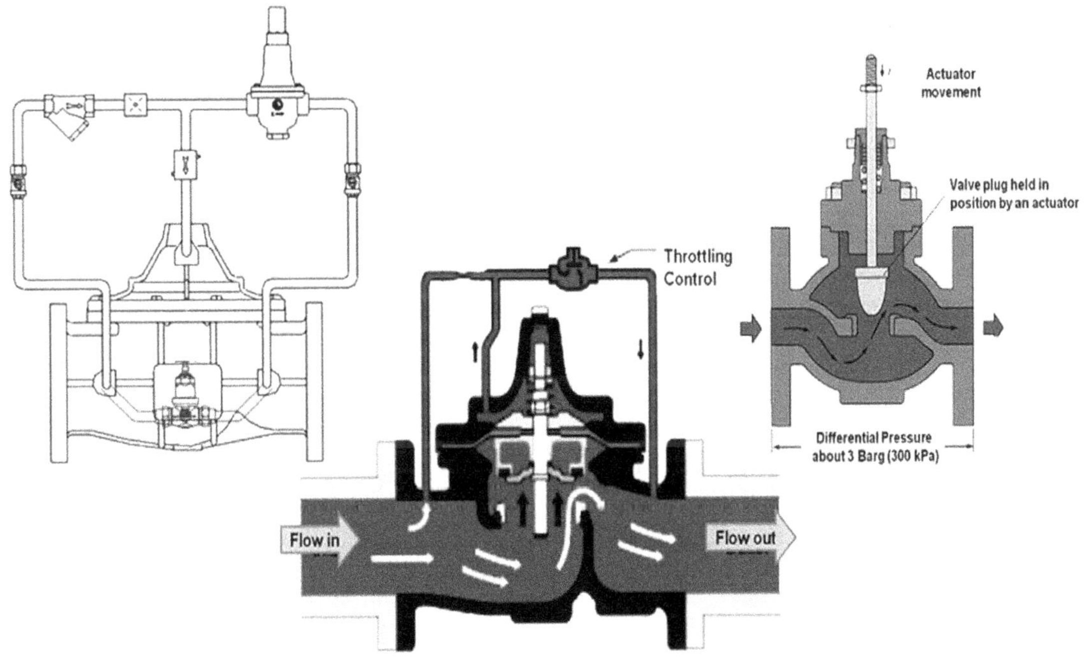

Figure A5-2. Typical Pressure Control Valve Operation (after ref. [18], Inset Spirax Sarco, http://www.spiraxsarco.com/resources/steam-engineering-tutorials/control-hardware-el-pn-actuation/control-valve-capacity.asp)

where Q is in US GPM, and ΔP in PSI. or

$$Q = 0.0856 * C_v \sqrt{\frac{\Delta P}{\gamma}} \tag{A5 – 2}$$

where Q is in m^3/hr, ΔP in kPa and γ is specific gravity. Control valve sizing is thus a function of the flow coefficient C_v which is obtained experimentally by individual valve manufacturers. The flow coefficient C_v is determined for the required flow rate and related pressure drop using a relationship such as that developed by Paracol [19]:

$$Cv = Q * \left[\left(\Delta P_{(C_v)} * \rho_o\right) / \left(\Delta P * \rho\right)\right]^{1/2} \tag{A5 – 3}$$

where:

Q = volumetric flow rate, GPM or m^3/hr
$\Delta P_{(Cv)}$ is the static pressure drop of 1 psi or 1 kPa
Δp is the static pressure drop from upstream to downstream expressed in psi or kPa
ρ is the density of the fluid expressed in lb/ft^3 or kg/m^3
ρ_o is the density of the water expressed in lb/ft^3 or kg/m^3

The above equation is valid at base pressure and temperature. Once the appropriate flow coefficient C_v has been calculated, the size of control valve can be selected, or a number of control valves from different manufacturers can be compared in terms of flow capacity for certain pressure drop and the same control valve size.

A5.2 CHECK VALVES

A check valve (Figure A5-3), prevents flow reversal, thus preventing damage caused by the reverse flow velocity at which the fluid reaches a pump station, or an upstream location along a pipeline route at a lower elevation, where a pipeline segment could be vulnerable to surge. It effectively reduces the pressure surge. However, check valves may end up being located in remote locations where road access can be challenging. There are several types of check valves available including:

- Swing clapper (top swing)
- Swing ball
- Spring assisted split disk /wafer double disk
- Nozzle/annulus type

The pipeline industry mostly uses the swing clapper type as these are through bore type of valves, thus facilitating the passage of pipeline pigs.

A check valve has a clapper type moving part (Figure A5-3, inset). During normal flow, the clapper swings up (in the direction of the flow, Figure A5-4), allowing free forward flow. When the pressure is lost or when the downstream pressure is higher than the upstream pressure, the clapper closes. The clapper falls and creates a barrier to back flow usually in milliseconds (Figure A5-4, inset B). The higher the downstream/backflow pressure, the tighter will be the seal. It may be noted that some clapper types (on side hinged check valves) are known to vibrate and sometimes stick.

Advantages of using a check valve are:

- Effective in preventing the surge pressure damaging vulnerable pipeline segments

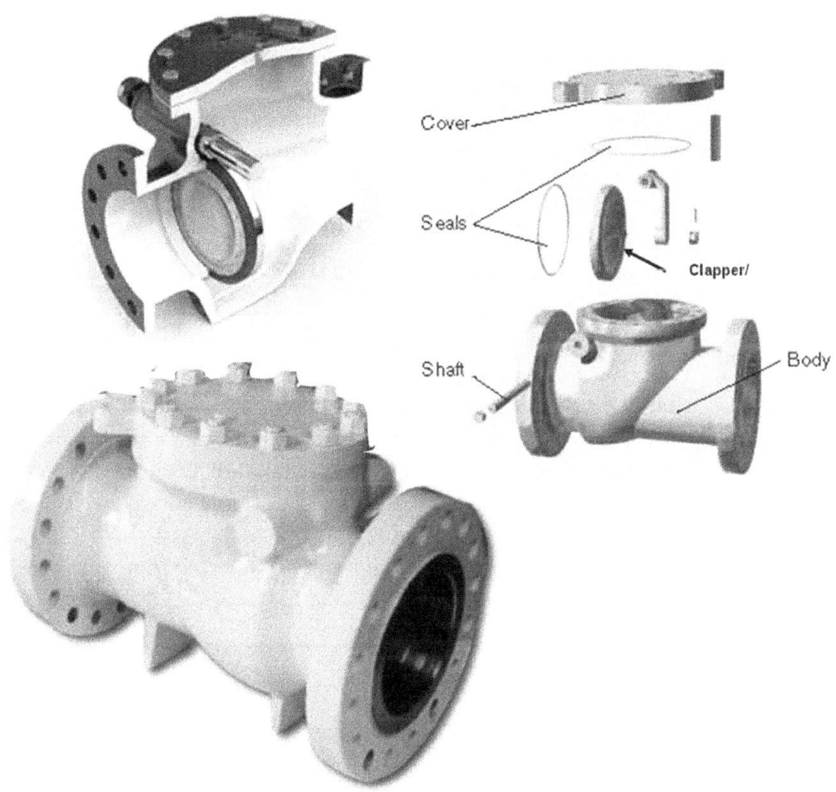

Figure A5-3. Typical pipeline check valve [20] — top inset air cushioned swing check valve)

- Minimal maintenanceProtecting pipeline and facilities from highest peak pressure

However, it may be noted that check valves are not considered an adequate form of isolation and they do require maintenance. Hence, check valves should be installed in conjunction with upstream and downstream isolation valves as recommended by the industry.

Check valve seats are subject to cavitation/erosion where elevation changes in a pipeline result in liquid column separation due to pressures falling below the vapor pressure of the liquid. In these situations, non-slam check valves are preferred.

Figure A5-4. Check valve operation (A — clapper open, B — clapper closed) [20]

The ideal check valve is one that closes immediately when the flow velocity at the valve reaches zero and/or when downstream pressure is higher than upstream pressure. This velocity reduction to zero is likely to control valve slam but not necessarily maintain pressures at acceptable levels. Potential for a high pressure surge downstream of the valve must be analyzed and the adjoining piping checked for suitability. The decision as to the best type of valve in a particular installation depends on the characteristics of the valve and the pipeline system (Figure A5-5).

Valve dynamic characteristics should be checked against pressure transients in pipeline system design. The deceleration flow rate is the most important parameter and it can be determined by first analyzing the system without the check valve. When the fluid deceleration rate has been determined, then the maximum backflow/reverse velocity V_r can be determined from the dynamic characteristic of the check valve and thus the suitability of the check valve for the application [18].

From Figure A5-5 above, it is evident that a nozzle/annulus type check valve with strong springs allows the lowest maximum reverse velocity to develop and hence this type of check valve (Figure A5-6) is closest to the ideal check. This particular non-slam check valve has a rapid closing time of less than 0.4 s. When flow reverses, the valve is already closed and thus the pump can be protected from any reverse high pressures [2].

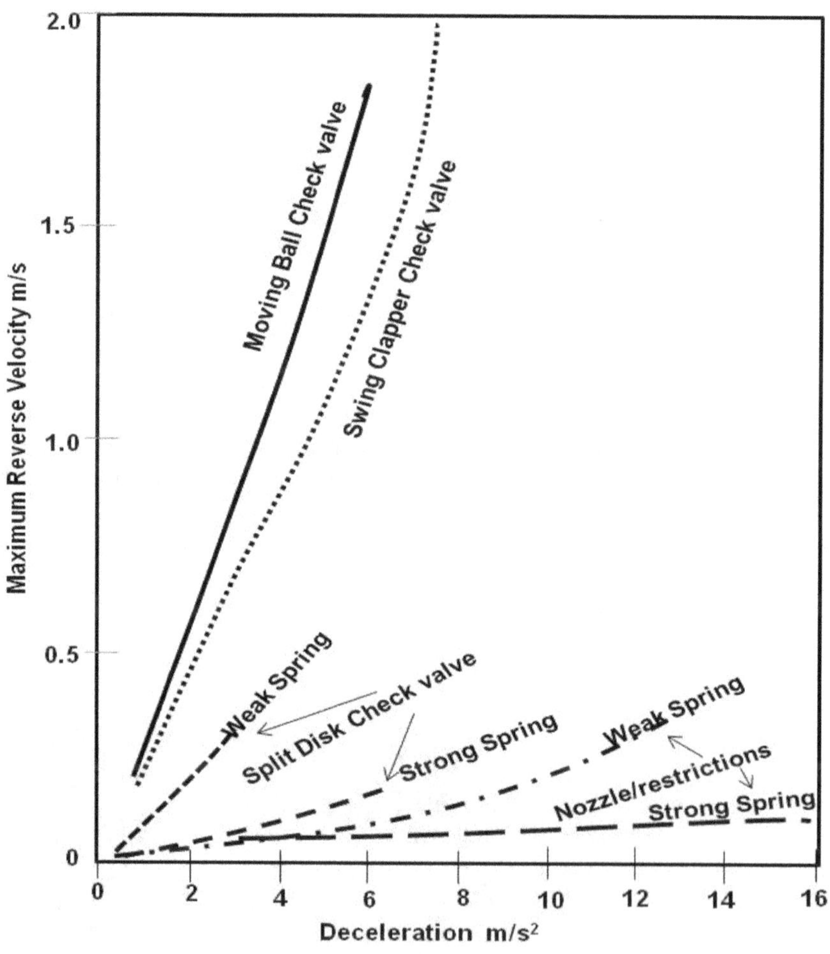

Figure A5-5. Comparison of dynamic characteristics of check valves (reproduced from [18]) (from [21, 22])

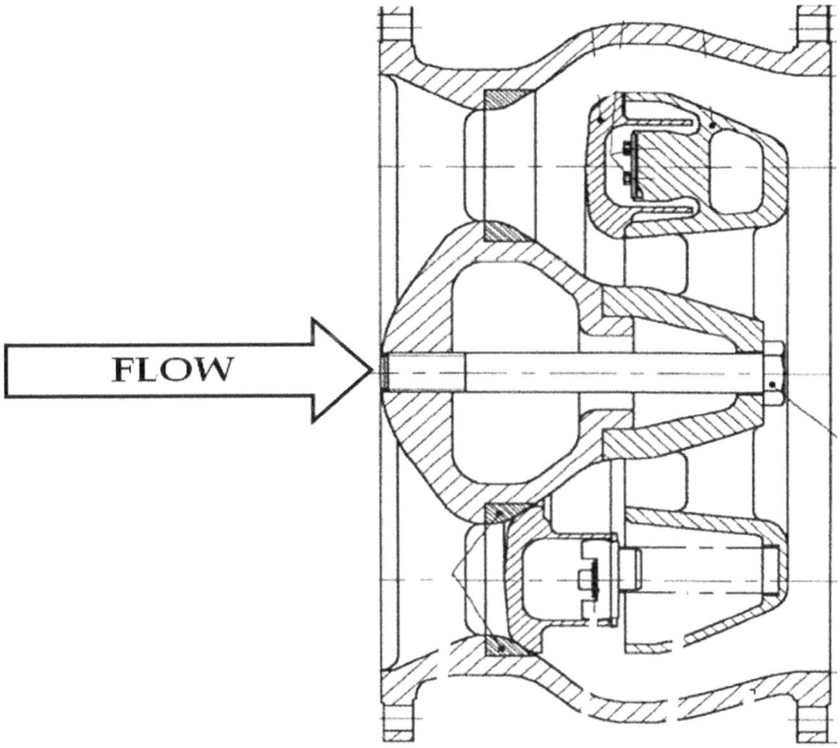

Figure A5-6. Nozzle/annulus type check valve (Noreva)

Many facilities exposed to check valve slam with conventional swing check valves have been made silent by the use of the nozzle type check valve. There are two types of nozzle check valves; single spindle single spring or annulus multiple spring type.

As an example, a plot of dynamic pressure/head transients downstream of such a check valve (located downstream of a pump that has been tripped) is shown in Figure A5-7. As the check valve is able to respond quickly to changing flow condition, the maximum back flow velocity V_r is small. However, if the valve motion was slow to respond, the associated pressure rise will be much larger due to a much higher back flow velocity V_r.

However, while this type of check valve is suitable for pump station protection, its use is inappropriate in transmission pipeline application where a full bore application is required for pigging. Mainline swing check valves can be fitted with a device called a slam retarder. This is a device designed to prevent the clapper of a check valve from slamming as it closes upon flow reversal.

Note on check valves for mainlines: the US DOT Office of Pipeline Safety (OPS) now known as the Pipelines and Hazardous Materials Safety Administration (PHMSA), alerted the operators of hazardous liquid pipelines to test check valves located in critical areas to assure the proper closure during a pipeline failure. The failure of such valves to close during an incident could increase the risk to public safety or damage to the environment.

This alert was due to a pipeline incident that has caused PHMSA to reevaluate the safety of pipeline check valves. The clapper in these valves had hinges on the side rather than at the top. As a consequence a top hinged clapper is preferred along with a surge analysis to simulate the functionality of check valves in pipeline transient flow situations.

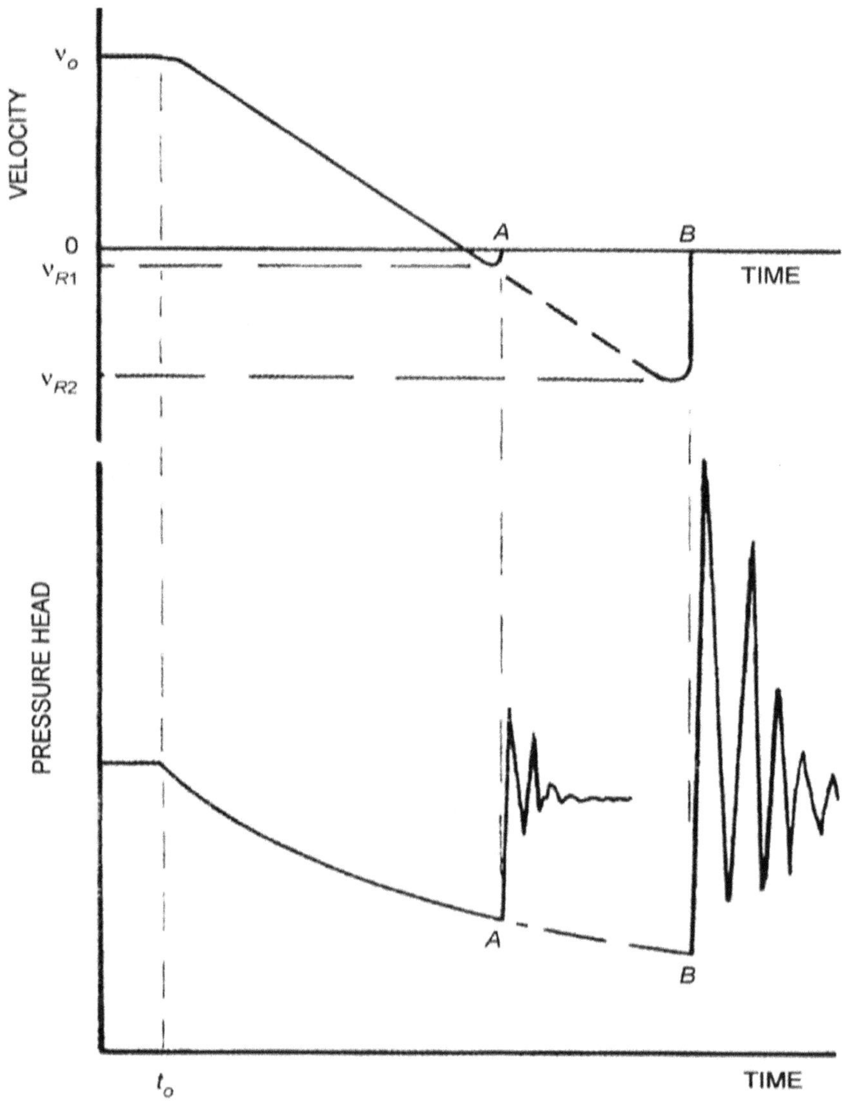

Figure A5-7. Pressure and velocity transients downstream of a check valve after a pump trip/ shutdown [18]

A5.3 RELIEF VALVES

Relief valves come in a variety of designs. A simple conventional spring loaded relief valve is most unlikely to operate sufficiently fast to relieve a pressure wave as it passes the relief valve nozzle. To be effective against pressure surges, a pressure relief valve must be placed as close as practicable to the pipeline segments being protected. If a valve is located on a lateral (branch) pipe, the shock wave will have passed the branch by a distance of about twice the branch length before the reflected wave from the relief valve returns to the pipe junction as a reduced pressure wave. The use of pressure relief valves is addressed in ANSI B31.4 Liquid Petroleum Transportation Piping standard [23] and is also reflected in other standards

A5.4 BURSTING/RUPTURE DISC

The bursting disc is another form of relief device and is used mostly in connection with offshore loading facilities where pressure surges due to ship loading (from oil tank farms/terminals) is more prevalent. Bursting Discs are non-reclosing devices that are designed to burst or rupture at a pre-determined pressure in order to relieve dangerous levels of pressure or vacuum (Figure A5-8). The drawbacks from the use of a bursting disc are similar to thc rclicf valvc.

If deployed, when they have served their purpose, they have to be replaced before the process can be restarted. Additionally, holding tanks will be required to collect the liquid discharged through the relief piping. Some design factor has to be employed to ensure that protection will occur at the anticipated set design pressure. Additionally, a statistical burst pressure determination should be considered by consulting different manufacturers. Metal bursting discs have been known to suffer from fatigue and fail prematurely and thereby requiring replacement of the rupture disk before the facility can be placed back in service.

A5.5 SURGE DIVERSION VALVE

A surge diversion valve provides a diversionary flow of fluid in the event of a transient pressure occurring. The concept is designed to release energy in a system before damaging pressures can occur. Similar to relief valves, such valves need to be placed as close to the point where a full pressure transient event effect is expected and/or close to the initiating surge point. The device is usually designed to allow fluid in or out of the system. In the hydrocarbon liquid pipeline industry high pressure nitrogen gas operated valves have been used to rapidly open such valves to dissipate energy. Such valves can also have electric, hydraulic, or pneumatic operators.

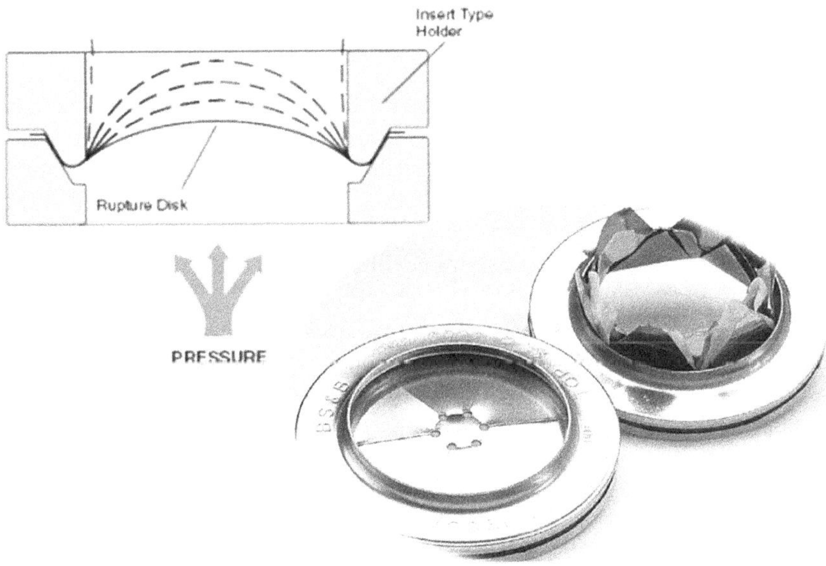

Figure A5-8. Typical burst/rupture disk used for pipeline over-pressure protection

These devices are complex and rely upon a high degree of maintenance to ensure that they work effectively. Control systems need to be backed up by an uninterruptible power supply to ensure that the valves continue to operate.

A5.6 INCREASING PIPELINE DIAMETER AND/OR WALL THICKNESS

If it is possible to increase the diameter of the pipeline, the immediate effect is to reduce the surge pressure. This occurs as the velocity is reduced. When the fluid is brought to rest the change in momentum is reduced in direct proportion to the maximum velocity. However, the increased pipeline size reduces the frictional losses so the damping of any pressure fluctuations is reduced though the transient may also be occurring longer. This can add to the fatigue loading of components in the system.

Advantages of increasing the pipe diameter are:

- Reduction in celerity and surge pressure
- Increased pipeline facilities' life
- Lower maintenance
- Reduction in pumping power

However, disadvantages are

- Settling of solids more likely in case of increase in diameter
- Increase in capital expenditure

Another way of ensuring that the pipeline system can withstand pressure surges is to increase the pipe wall thickness in areas along the pipeline that could be susceptible to surge. Pipeline design must ensure that any surge pressures remain within the Maximum Allowable Operating Pressure (MAOP) for that section of pipe.

A5.7 VARIABLE SPEED DRIVES AND SOFT STARTERS

Variable speed drives (VSD) for pumps provide a reliable means of prevention of damage from surge events (see Chapter 6 for an example of the application). Variable speed drives (VSD) provide the best method of reducing the impact of pressure surges and fatigue damage to pipeline components. Check valve slams are also avoided as the liquid column decelerates slowly. The VSD allows the pump speed to increase slowly to achieve slow line filling and thus any air entrainment can be removed without damaging the pipeline. They also provide flexibility of operation for batched products pipeline where density change between batches occurs and also for a process where flows can be increased for future needs without changing the equipment.

Variable speed drives have a serious disadvantage when there is power event such as a loss of power. This can result in causing the highest positive and negative pressures in a system. In most applications, this type of event is rare. However, power backups are generally provided for vulnerable pipeline pumping facilities. Other disadvantages include increased cost and more frequent replacements and upgrades.

Soft starters are widely deployed in many pumping stations to reduce the electrical load on the power supply to a facility, particularly where the pump station is at the end of a long power transmission line. Soft starters have an economic advantage over variable speed drives. They have some but not all of the features of a VSD. They are able to control ramp speed up and down to some extent. Therefore, they do provide benefits in reducing surge in some applications.

A5.8 VALVE OPENING AND CLOSURE TIMES

Closing and opening valves can generate pressure surges in a pipeline system depending upon the time it takes to close or open a valve. Generally, the industry strives for a valve closure timing of about 5 seconds per inch diameter to avoid undue surge/fatigue effects. The worst type of valves for causing undue surges are the pneumatically operated butterfly, gate, globe, and gate valves as these can be rapidly closed or opened. If the time of closure of a valve is less than the time taken for a pressure wave to travel from its point of initiation to the end of a pipeline and return then the valve is described as having a *rapid closure*. This results in the maximum head predicted by the Joukowsky equation or column separation or pressure transients [24, 25].

Extending the closure times is often restricted to short pipelines. Some facilities employ the two stage closing process whereby the valve is closed to a 15% to 20% open position rapidly and then the last closure occurs over an extended period. Similarly the valve opening is a two stage process. In high pressure systems dual valves (of different sizes) are installed in parallel. The smallest valve opens first. It subsequently closes while the bigger valve closes and then reopens to obtain maximum flow rates. Varying the closure time may however be dictated by some other process requirement such as an Emergency Shutdown Valve (ESD).

The advantages of varying valve closure times are:

- Low capital cost solution
- Effective in reducing surge pressures
- Can be modified during commissioning or operation if valves are automated and fitted with adjustable opening/closing devices

Disadvantages include:

- Requires power supply in the form of hydraulic, pneumatic or electrical energy to be totally reliable and effective
- Needs uninterruptible power supply for secure operation
- Requires extensive modeling to cover all operational scenarios
- Requires routine testing to be effective

A5.9 SURGE TANKS

Surge tanks are not common for liquid hydrocarbon pipeline applications to mitigate pressure surge transients. They are principally used on water transmission pipelines. The principal demand on a surge tank is to compensate the mass oscillation of the flow or load changes of turbines and/or pumps. In connection with a suitable throttling device, the surge tank provides a most powerful damping of the amplitude usually seen in the very first period of oscillation [26].

Surge tanks can only function when the local hydraulic gradient falls below the liquid level in the tank. Under transient conditions, the places in the pipeline where this is most likely to occur will be at significant reductions in upward slope and in the vicinity of peaks along the pipeline route.

Surge tanks do not provide protection against positive pressures and as such are not commonly used for liquid hydrocarbon pipeline applications.

A5.10 PUMP BYPASS CHECK VALVES

A bypass check valve is usually used in a pump station, and it takes the form of a valving arrangement in parallel to the pump(s) (Figure A5-9). The concept is that on loss of power there is still a reduced flow into the pipeline via this valve. This prevents the column separation occurring immediately downstream of the pump discharge check valve. This valve is used where there is sufficient pressure/head to drive liquid into the pipeline with a low static head. The second application is where there is a booster pump in the pipeline. The bypass valve opens and thus protects the pipeline when the booster pump fails [18, 27].

The bypass valve is generally a reduced size check valve. However a fail to open actuated valve may be used in lieu of check valve. Such a valve can be of the pilot actuated cylinder or diaphragm type valves. The selection of valve size, characteristics and location must be accurately analysed to fully evaluate the effectiveness of these devices. This should include varying flow, heads and pipeline roughness.

A5.11 APPLICATIONS

The application of surge mitigation technique using the above listed equipment/solutions are demonstrated in Section 5.1.3 and Section 6.6. The example in Chapter 6 specifically highlights the adaption of the following solution techniques to avoid damaging a pipeline system due to surge caused by operational scenarios and emergencies:

a) Higher wall thickness at selected locations
b) Valve closure timing

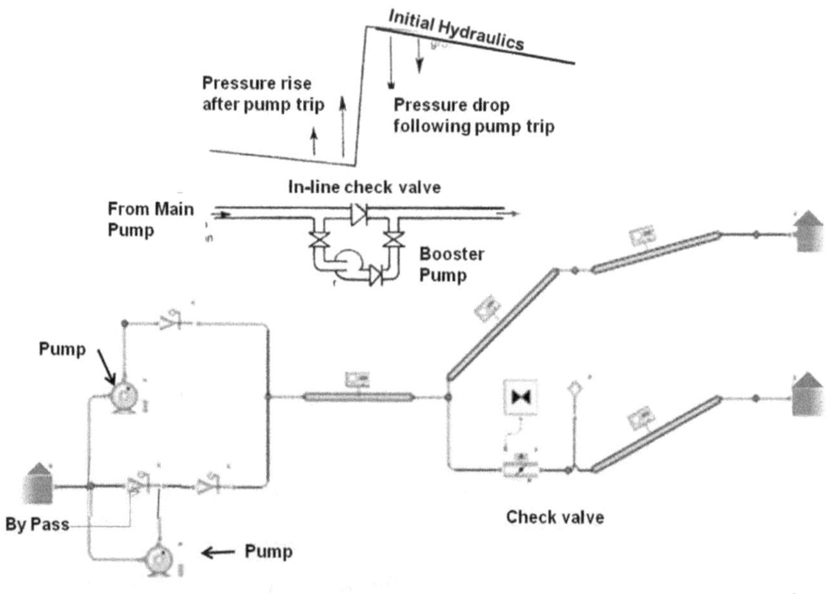

Figure A5-9. Bypass check valve

c) Optimized mainline valve location
d) Use of Variable Speed Drives for pumps
e) Use of pressure relief system
f) Use of check valves at strategic locations
g) Use of pressure control valves d/s of pump stations
h) Use of effective operational procedures

The above solution technique was applied to the example shown in Figure A5-10 using detailed surge analysis for a range of scenarios to ensure that the predicted performance for the selected surge control/reduction option was effective.

Figure A5-11 below indicates the pressure build-up process caused by the closure of the ESD valve inside a pipeline equipped with conventional spring type pressure relief valves. The rate of a surge pressure rise in transmission pipelines is extremely fast and requires relief valves with high relief volume capacity and very rapid response times in the order of 100 milliseconds. The conventional spring type relief valve can respond (open) to fast rising pressures but tends to immediately slam closed, leading to rapid opening and closing cycling which seriously restricts the valves capacity to relieve and causes the upstream pressure to rapidly climb to unacceptable levels.

The analysis indicated that basic characteristics required for liquid pipeline surge pressure relief must have:

- very fast opening speed check valve, Such a valve should react fast enough to be able to equalize rapidly rising pressure;
- valve with non-slam capability;
- valve which requires to return to the normal (closed) state quickly but without causing additional pressure surge during closure;
- high capacity valve which should be capable of passing the entire flowing stream if it is required.

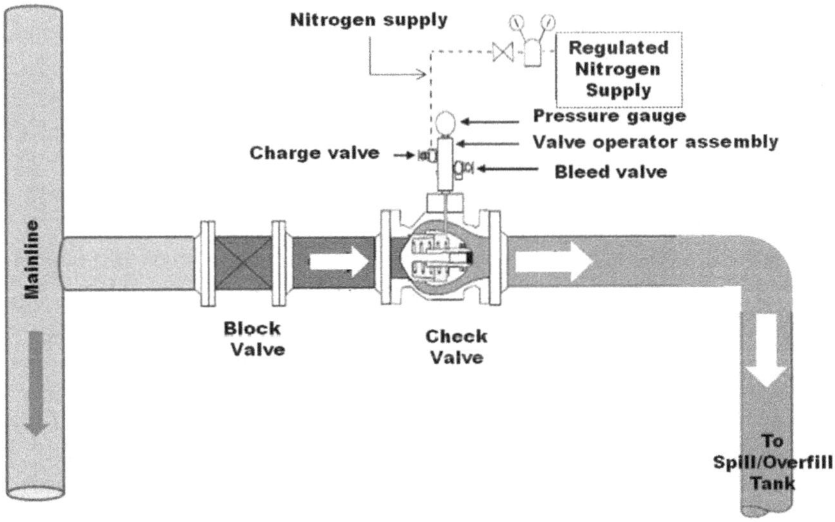

Figure A5-10. A liquid pipeline subjected to pressure surge (after ref. [25])

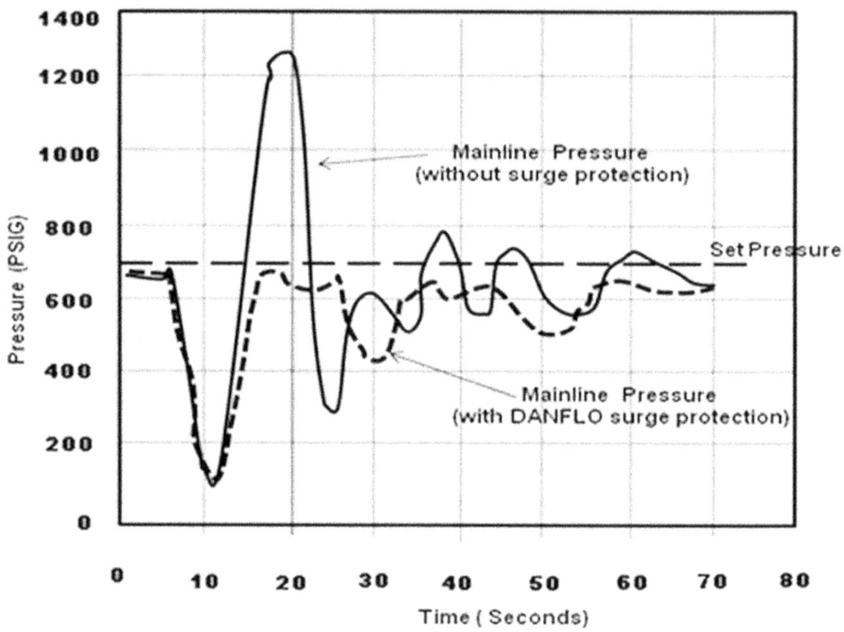

Figure A5-11. Transient pressure profile in pipeline system with and without surge protection facility (reproduced from ref. [28])

REFERENCES

[1] Fisher Controls, 2005. *Control Valve Handbook*, 4th ed., Emerson Process Management, http://www.documentation.emersonprocess.com/groups/public/documents/book/cvh99.pdf.

[2] Wylie, E. B., and Streeter, V. L., 1983, *Fluid Transients*, FEB Press, Ann Arbor, MI, USA.

[3] Yoon, M., Warren, B., and Adam, S., 2007, *Pipeline System Automation and Control*, ASME Press, New York, N.Y.

[4] Brainerd, H. A., 1982, "Good surge control can help pipeline throughput," Oil & Gas J., pp. 126–137.

[5] Derammelaere, R. H., and Shou, G., 2002, "Antamina's Copper and Zinc Concentrate Pipeline Incorporates Advanced Technologies," *Proc of HydrotransportEnergy Information Administration (EIA), 2001* U.S. Regions for Distribution of Petroleum and Their Key Pipelines—Appendix C, http://www.eia.doe.gov/oiaf/servicerpt/ulsd/appendix_c.htm.

[6] Heywood, N. L, 2003, "Developments in Slurry Pipeline Technologies," *Chemical Engineering Progress*, **72**(3), pp. 180-220.

[7] Jacobs, S., 2002, "Pipeline Factors Affecting Gasoline Prices," *Paper, U.S. Federal Trade Commission (FTC) Conf.*, May.

[8] Cerda, J., 2008, "Oil Pipeline Logistics," Pan American Study Institute on Emerging Trends in Process Systems Engineering, August 11-21, Mar del Plata, Argentina, http://cepac.cheme.cmu.edu/pasi2008/slides/cerda/library/slides/jcerda-pasi-2008-1page.pdf.

[9] Fuel Technology Pty, Ltd., 1997, "FTC Addition- No Change in Fuel Specification," Technical bulletin 104-97, http://www.fpc1.com/tests/ftc/ftpl/tb104-97.htm.

[10] EIA (Energy Information Administration), 2001, "The Transition to Ultra-Low-Sulfur Diesel Fuel: Effects on Prices and Supply," May www.walshcarlines.com/pdf/ulsd.pdf.

[11] Yarborough, V., 2001, "Colonial Pipeline Tests Interface-Detector Methods," *Oil & Gas Journal*, pp. 54-56, Aug.

[12] Rohm, and Hass, 2010, "Fluorescent Yellow 131SC Liquid, Concentrated Solvent Soluble Fluorescent Dye," http://www.rohmhaas.com/wcm/products/product_detail.page?product=1120557.

[13] Austin, J. E. and Palfrey, J. R., 1964, "Mixing of Miscible but Dissimilar Liquids in Serial Flow in a Pipeline," Proc. Instn Mech Engrs., **178**(I), pp. 377–396.

[14] Mohitpour, M., Golshan, H., Murray, A., 2007, *Pipeline Design & Construction—A Practical Approach*, 3rd Ed., ASME Press, New York, NY.

[15] Koenig, S., Youngbery, E. D., and Wright, D. L., 1999, " Liquid Pipeline Nominations Processing and Batch Scheduling," *Pipeline Simulation Interest Group Conf.*, St Louis, http://www.psig.org/papers/1007/9912.pdf-125k.

[16] Bullion, I., 2002, "Koch Pipeline Reduces Leaks With Strict Monitoring and Maintenance," *Pipeline & Gas Journal*, March.

[17] Stasioski, W., 2004, "Ultra-Low Sulfur Diesel Fuel Testing Results," *Industry/EPA Ultra Low Sulfur Diesel Workshop*, Astor Crown Plaza, New Orleans, LA.

[18] Stone, G. D., 2005, "Avoiding Pressure Surge Damage in Pipeline Systems," *Presented at Australian Inst. Chem Eng-Sydney Division*, http://www.pipingdesign.com/articles/solutions_to_pressure_surge_in_piping_systems.pdf.

[19] Parcol, 2003, " Handbook for Control Valve Sizing," PARCOL S.p.A. Via Isonzo, 2—20010 Cannegrate (MI), Italy, http://www.parcol.com/docs/1-i_gb.pdf.

[20] SPX, 2009, "M&J Valve," www.spx.com.

[21] Thorley A.R.D, 2004, "Fluid Transients in Pipeline Systems" (2nd Edition), Professional Engineering Publishing, I.Mech.Eng, UK.

[22] BHRA, 2004, "Dynamic performance of Air Valves," International conference on Pressure Surges, Cranfield, UK

[23] ASME (American Society of Mechanical Engineers), 2009, "ASME B31.4: Pipeline Transportation Systems for Liquid Hydrocarbons and Other Liquids," ASME Press, New York.

[24] Tijsseling, A. S., and Anderson, A., 2004, "A Precursor in Waterhammer Analysis — Rediscovering," *Proc., 9th Int. Conf. on Pressure Surges*, J. von Kries, S.J. Murray, ed. BHR Group, Cranfield, UK. pp. 739–751. Also: TUE-RANA 04-02.

[25] Tijsseling, A. S., and Anderson, A., 2005, "The Joukowsky Equation for Fluids and Solids," http://www.win.tue.nl/analysis/reports/rana06-08.pdf.

[26] Steyrer, P.,1999, "Economic Surge Tank Design by Sophisticated Hydraulic Throttling," http://www.iahr.org/membersonly/grazproceedings99/pdf/B126.pdf.

[27] Nakayama, Y., and Boucher, R. F., 1998, *Introduction to Fluid Mechanics*, Butterworth-Heinemann, ISBN: 0340676493, 9780340676493.

[28] Bahar Sanat, N., 2010, "Pipeline Surge Protection," http://www.baharsanat.com/?lng=en&cid=cms&gid=294&content=185.

Chapter 6

NON-CONVENTIONAL HYDROCARBON LIQUIDS, PRODUCTION, AND TRANSPORTATION

Non-Conventional hydrocarbons are:

- Oil
  - Heavy Oil
  - Extra Heavy Oil and Bitumen
- Gas
  - Coal Gas
  - Tight Gas
  - Gas Hydrates
  - Aquifer Gas

The world contains large quantities of non-conventional oil and gas and various oil substitutes. However, the rapidity of the decline in the production of conventional oil and gas has attracted interest and investment in non-conventional fluids.

Conventional oil is defined fairly generally as oil produced by primary or secondary recovery methods (specifically: under own pressure, physical lift, water flood, and water or natural gas pressure maintenance). This definition is not universal, but is further clarified in the ADDENDUM at the end of this chapter.

This chapter provides concepts and details for modes of transportation of non-conventional oil and specifically provides details on pipeline modes of transportation. The technology of heavy oil transportation by pipeline is further exemplified through design requirements and operational controls of a long distance pipeline "TCPL Keystone-XL Heavy Oil Pipeline" to highlight design aspects as well as operational control opportunities.

To complement the transportation techniques, the ADDENDUM to this chapter further highlights world heavy oil/extra heavy oil resources their recovery techniques, rheology, and characteristics.

6.1 HEAVY OIL TECHNOLOGY AND TRANSPORTATION

6.1.1 Background

Heavy oil is sometimes interchangeably referred to as "Bitumen". The two types of heavy oils are natural bitumen and extra heavy oil which are the remnants of very large volumes of conventional oils (extracted from deep reserves) that have been generated and subsequently degraded, principally by bacterial action. Chemically and texturally, they resemble the residuum produced by refinery distillation of light oil.

Oil sands or, more technically, bituminous sands, are a type of unconventional petroleum deposit. The oil sands are loose sand or partially consolidated sandstone containing naturally occurring mixtures of sand, clay, and water, saturated with a dense and extremely viscous form of petroleum referred to as bitumen.

There are significant resources of heavy oil, extra-heavy oil, and bitumen primarily in North and South Americas and in Russia, and in smaller deposits in many other countries. These viscous oils are more costly to extract, transport, and refine than conventional oils. However, their production levels have increased to more than 1.6 million barrels per day (MMBLLSD), or just under 2% of world crude oil production. While these resources only provide a small percentage of current oil production existing commercial technologies could allow for significantly increased production and transportation. These unconventional oils can be economically exploited, produced, and transported (Figure 6-1).

Canada, Venezuela, and the United States are leading producers of these unconventional oils. In Canada, open-pit mining of shallow oil sands provides approximately 50% of the country's heavy oil production. In situ production of heavy oil with sand and thermal production using injected steam provides the remainder of the heavy oil/bitumen production in Canada. Besides Venezuela, there are also large extraction, production, and transportation of heavy crudes in other countries of South America particularly in Colombia, Peru, Ecuador, and Cuba.

However, there are several barriers to the rapid growth of heavy oil, extra-heavy oil, and bitumen productions and transportation. These are mostly due to production methods and transportation options that affect costs involved in getting such heavy oils to the market, often long distances away. In reservoirs with heavy or extra heavy oils, it is generally not possible to employ conventional recovery methods due to the fact that high oil viscosity hinders its movement within the porous medium unless surpassed by light/heated fluid injections rendering low seep efficiencies.

Figure 6-1. Heavy oil, from extraction, pipeline transportation to storage

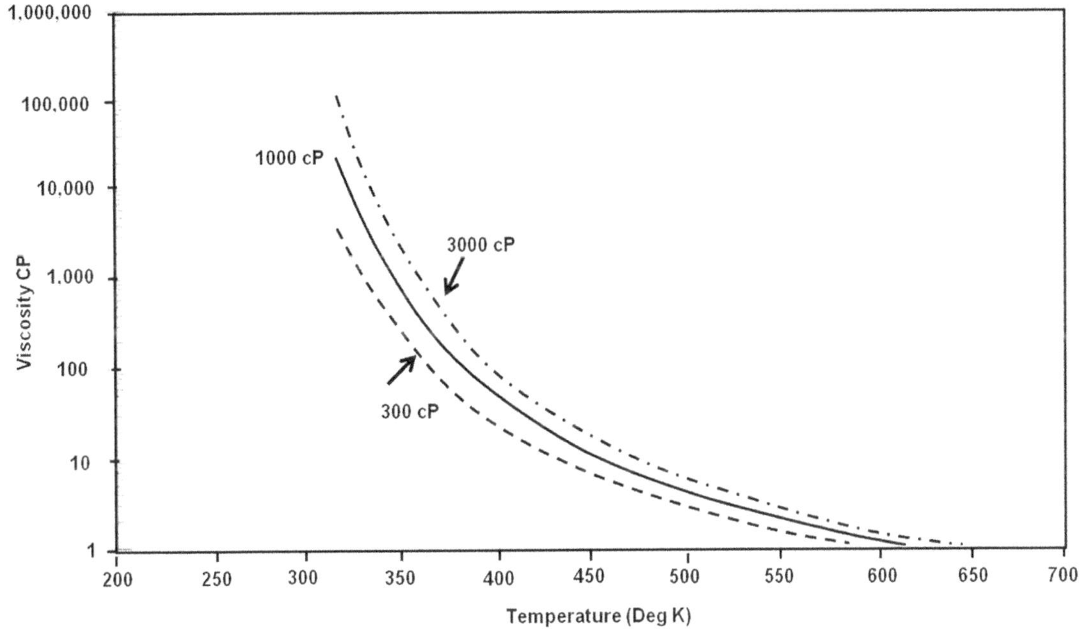

Figure 6-2. Viscosity-temperature for heavy oil at various initial viscosities

Extra heavy oil and heavy oil typically have a viscosity between 100 and 10,000 cP and therefore do not flow easily unless diluted with a solvent or heated. Figure 6-2 shows the reduction in heavy oil viscosity with temperature for different initial viscosities [1]. However, it may be noted that US Geological Survey states that natural bitumen is oil having a viscosity greater than 10,000 cP at reservoir conditions. Extra-heavy oil is commonly defined as oil having a gravity of less than 10 °API and a reservoir viscosity of no more than 10 000 cP.

In this chapter, a rationalized approach is provided highlighting world heavy oil resources their recovery techniques, rheology, and characteristics together with their modes of transportation. The technology of heavy oil transportation by pipeline is indicated through design requirements and operational controls. An example of a long distance pipeline "TCPL Keystone-XL Heavy Oil Pipeline" is provided to highlight design aspects as well as operational control opportunities.

6.2 HEAVY OIL TYPES AND GLOBAL DISTRIBUTION

As indicated previously, heavy crudes, which often result from a bacterial oxidation of conventional oils inside the reservoir rock, have different physical and chemical properties, generally degraded. They have much higher viscosity, higher heavy metals, and higher sulfur and nitrogen contents.

Heavy oil is a type of crude oil which is very viscous, making it difficult to flow easily. These oils are characterized by their high viscosity, high density/specific gravity (low API gravity), and high concentrations of nitrogen, oxygen, sulphur, carbon residues, asphaltenes, and heavy metals.

The American Petroleum Institute gravity, or API gravity, is a measure of how heavy or light a petroleum liquid is compared to water. Fluids with an API gravity greater than 10 are lighter and can float on water. API gravity is thus a measure of the relative density of a hydrocarbon liquid to the density of water, but it is used to compare the relative densities of petroleum liquids and hence it is important for the assessment of pumps and transportation facilities design. For example, if one petroleum liquid floats on another and is therefore less dense, it has a greater API gravity.

Figure 6-3 below indicates the viscosity and range of densities of as produced natural bitumen, extra heavy oil as well as upgraded heavy oil and light crude.

The three different categories of heavy crude (bitumen, extra heavy and heavy oils) are usually defined according to their density (API Gravity) (Figure 6-3, inset):

- the **heavy oils**, the °API of which is between 10 and 20;
- the **extra-heavy oils** and **bitumen**, the °API of which is less than 10 (the *in situ* level of viscosity makes the distinction between extra-heavy oils and bitumen).

Generally, 40 to 45 API gravity degree oils have the greatest commercial value for refining purposes. Above 45 °API gravity, the molecular chains become shorter and less valuable to a refinery.

Crude oil is classified as light, medium, or heavy, according to its measured API gravity. Light crude oil is defined as having an API gravity higher than 31.1 °API.

Medium oil is defined as having an API gravity between 22.3 °API and 31.1 °API.

Bitumen derived from the oil sands deposits generally has an API gravity of about 8 °API. It is 'upgraded' to an API gravity of 31 °API to 33 °API as synthetic crude.

It must be noted that a fourth category also exists, oil shale. In this category, reservoir rock and source rock are the same because the oil has not migrated. These specific

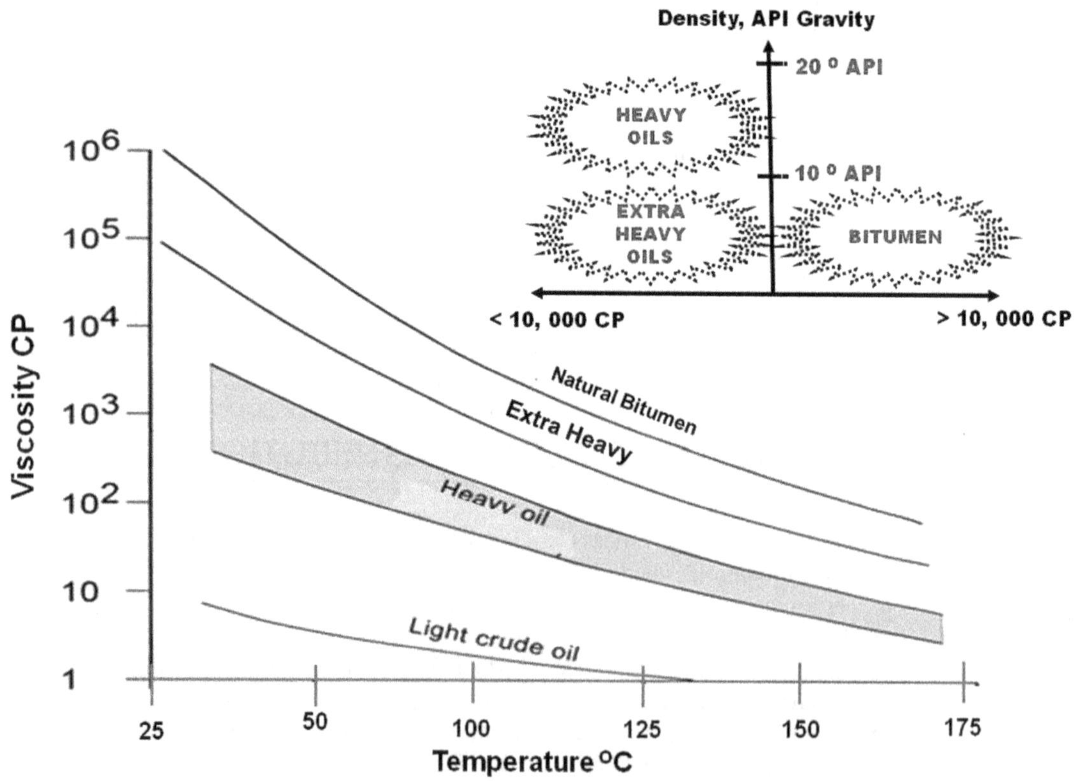

Figure 6-3. Typical viscosities of as-produced bitumen and crudes (inset: viscosities for pipeline transportation)

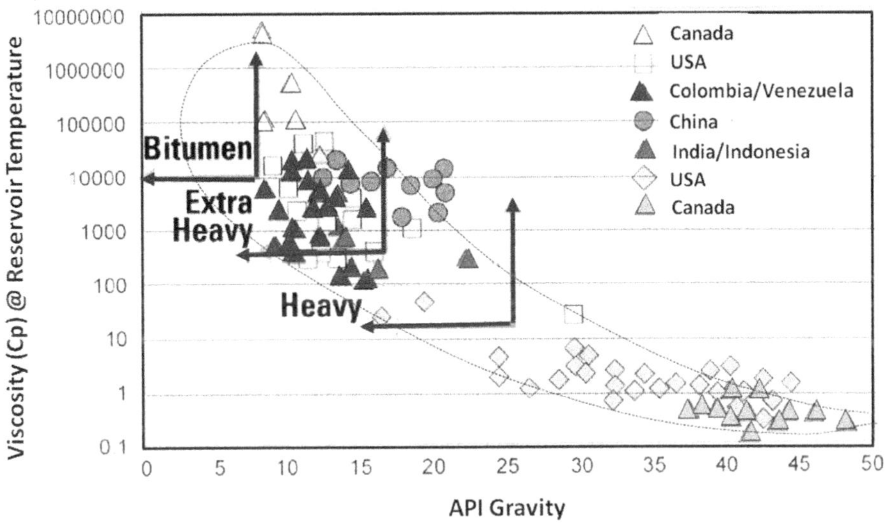

Figure 6-4. Regional distribution of bitumen, extra heavy oil and heavy oil based on API gravity (data reproduced, [2])

properties imply specific advanced technical solutions throughout their exploitation, from production, to transport, and refining.

The heavy oils are transported in large volumes to processing plants. Specialized techniques are then used at in such processing plants to take heavy oil and bitumen feedstock and process it into high quality, low sulfur, synthetic crude oil. Further processing is required to produce more useful fractions, such as: naphtha, kerosene, and gas oil.

It is the synthetic crude that is transported to market areas often long distances away.

The API grading of bitumen, extra heavy oil and heavy oil based regional distribution are shown in Figure 6-4 [2].

However, most reserves are located outside the Middle East, refer to ADDENDUM to CHAPTER 6, Figure A6.2. Thus, with ever increasing importance of the Middle East region in terms of oil reserves, production, and supply constraints, heavy crude has become more and more attractive for western countries: exploitation of heavy crude will reduce the potential pricing power of the leading conventional oil producers.

6.3 HEAVY OIL PROPERTIES AND TYPE

Properties of oil that are in place have a significant impact on reservoir extraction/recovery methods, yield/production, transportation, and marketing. The properties thus provide information for:

- Estimating hydrocarbon reserves in pace
- Understanding production reservoir processes
- Predicting reservoir behavior
- Assessing well-flow performance
- Designing appropriate facilities
- Transportation and
- Marketing

6.3.1 Types/Grouping

Crude oils (heavy or otherwise) cover a wide range in physical properties and chemical compositions, and it is often important to be able to group them into broad categories of related oils to associate the appropriate properties and behavior (Newtonian versus non-Newtonian) for pipeline transportation purposes.

Figure 6-5 signifies properties of heavy oil, in terms of pressure-temperature diagrams in the context of types of oil.

In general, crude oils are commonly classified into the following classifications:

- **Ordinary black oil** (light/medium crude oil or heavy/extra heavy crude oils), initial gas-oil ratios < 2000 scf/stb, API < 40° (usually between 15° and 40°)
- **Low-shrinkage** crude oil (often referred to as black oil), API < 35°
- **High-shrinkage** (volatile) crude oil (initial gas-oil ratios in the 2000 to 3300 scf/Bbl range), API > 40° (typically 45–55° API)
- **Near-critical** crude oil (condensate/Retrograde gas), gas-oil ratio > 3000 scf/STB

Black oils are made up of a variety of components including large, heavy, and non-volatile hydrocarbons. The stock tank that contains such oil is usually brown to dark green in color.

Black oil is often called low shrinkage crude oil or ordinary oil [4]. It is black or deeply colored.

High shrinkage (volatile) oils contain fewer heavy molecules and more intermediate components (ethane through hexane) than black oils. The color is generally lighter than black oil — brown, orange, or green. This type of crude oil is commonly characterized by a high liquid shrinkage immediately below the bubble-point. Such oil is greenish to orange in color.

The phase envelope for a volatile oil tends to cover a much narrower temperature range when compared to a black oil; but like a black oil, the reservoir temperature is always lower than the critical temperature for the fluid. As the reservoir temperature approaches the critical temperature a volatile oil will become more gas-like such that

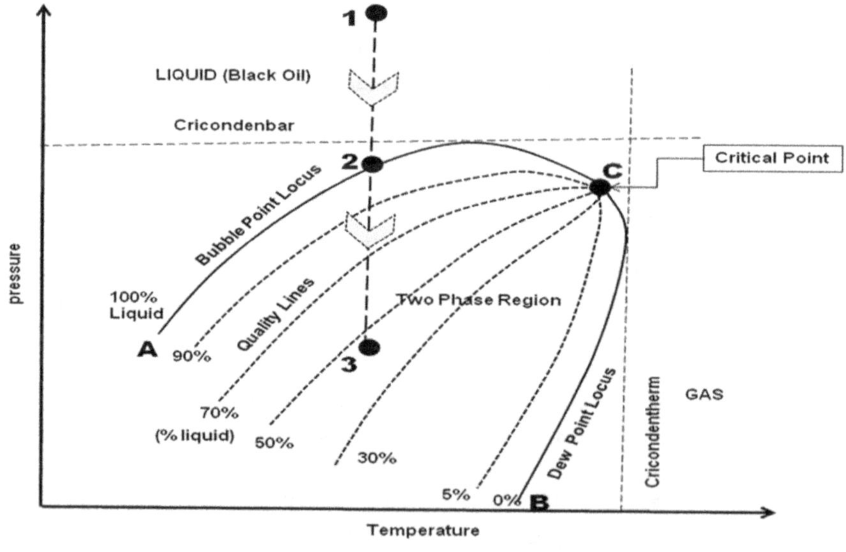

Figure 6-5. Crude oil phase diagram defining the type of oil [3]

with even moderate depletion, a volatile oil reservoir can flash mainly to gas and have a relatively low liquid content.

The above classifications are essentially based upon the properties exhibited by the crude oil, including physical properties, composition, gas-oil ratio, appearance, and pressure-temperature phase diagrams.

6.3.2 Oil Viscosity Prediction

Viscosity is the single most important transport property necessary for the calculation of flow through reservoirs porous media and pipelines. Such calculation will require accurate viscosity data and predictions tools.

The basis for oil-viscosity calculations using a traditional black-oil approach is the determination of dead- or gas-free-oil viscosity.

It may be noted that dead oil is defined as black oil in pressure-temperature of phase diagram (Figure 6-5). Dead oil is also defined as oil at sufficiently low pressure that it contains no dissolved gas or a relatively thick oil or residue that has lost its volatile components.

If the oil is under-saturated, there is no free gas in the reservoir pores and thus it is one phase flow (in absence of water). Under-saturated oil, the oil pressure as per definition (refer to ADDENDUM to CHAPTER 6, Figure A6-9) is higher than the bubble point pressure.

In absence of viscosity measurement data, the viscosity is determined through various correlations that were previously determined through empirical data. Thus, the viscosity correlations are empirically derived equations for estimating oil viscosity and are a method of estimating oil viscosity when laboratory data does not exist.

There are a number of correlations for the estimation of fluid viscosity based on measured fluid properties. These correlations can be divided into three categories:

TABLE 6-1. List of some viscosity correlations [5, 6]

Correlation	Type/Use
1. Al-Marhoun 1985	Saudi Arabian oil
2. Beal	(Chart)
3. Beal	(Equation)
4. Beggs & Robinson	
5. Chew & Connally	
6. De Ghetto et al	Heavy oil (10–22.3 API)
7. De Ghetto et al	Extra heavy oil (API < 10)
8. Egbogah-Jacks	(Without pour point)
9. Hanafy et al	Egyptian oil
10. Modified Egbogah-Jacks	(Extra heavy oils)
11. Modified Egbogah-Jacks	(Heavy oils)
12. Glaso	North Sea oil
13. Labedi	
14. Labcdi (modificd)	(Extra heavy oils)
15. Labedi (modified)	(Medium oils)
16. Kahn et al	(Saudi Arabian crude oils)
17. Kartoatmodjo	
18. Kartoatmodjo (modified)	(Extra heavy oils)
19. Kartoatmodjo (modified)	(Heavy oils)
20. Kartoatmodjo (modified)	(Medium oils)
21. Petrosky & Farshad	Gulf of Mexico oil
22. PVTsim	
23. Standing	California oil
24. Vazquez & Beggs	

- Dead oil viscosity (μ_{od}),
- Bubble-point viscosity (μ_{ob}) and
- Undersaturated oil viscosity (μ_o).

Some of the available Viscosity correlations are listed in Table 6-1 [5, 6]. Benisson [5] compared all the predictions listed in Table 6-1 above with the measured dead oil viscosity data, and indicates that Beal's equation {#2 in Table 6-1} is perhaps the best showing the smallest mean difference, though none of the correlations provide a reliable estimate of the dead oil viscosity (μ_{od}).

$$\mu_{od} = \{0.32 + (1.8 \times 10^7/°API^{4.53})\} \times \{360/(T + 200)\}^{\alpha} \qquad (6-1)$$

where $\alpha = \text{antilog}\{0.43 + 833/°API\}$

$$\mu_{ob} = \alpha \times \{\mu_{od}\}^{\beta} \qquad (6-2)$$

where $\alpha = 10.175 \times (R_s + 100)^{-0.515}$

$$\beta = 5.44\ (R_s + 150)^{-0.338}$$

$$\mu_o = \mu_{ob} + 0.001 \times (P\text{-}Pb) \times \{0.024 \times (\mu_{ob})^{1.6} + 0.038 \times (\mu_{ob})^{0.56}\} \qquad (6-3)$$

In the above equations, the following units of measure are applicable

°API	Gravity of oil at 60°F	°API
GG	Gas gravity	Air = 1.000
P	(Reservoir) Pressure	psia
Pb	Bubble-point pressure	psia
Rs	Solution gas-oil ratio	scf/STB
T	(Reservoir) Temperature	°F
μ_o	Undersaturated oil viscosity	cP
μ_{ob}	Bubble point oil viscosity	cP
μ_{od}	Dead oil viscosity	cP

Generally, heavy crude oils and bitumen exhibit non-Newtonian shear thinning behavior; that is, viscosity (affected by shear mechanical energy) decreases with increasing oscillation frequency or shear rate over temperature ranges of industrial interest for production, storage, and transportation. For such behavior, viscosity correlation can be generally expressed as [7]:

$$\mu = -CT^{-A}\,\text{Ln}\,\dot{\varepsilon} + K\,T^{-B} \qquad (6-4)$$

where A, B, C and K = Fluid Constants, T = Temperature and $\dot{\varepsilon}$ = Shear Rate, μ = absolute Viscosity

6.4 HEAVY OIL TRANSPORTATION TECHNOLOGIES

As indicated previously, there are significant accumulations of heavy oils and bitumen throughout the world, some of which are "stranded" or "economically constrained". However, in order to reach the market in situ heavy oil must be extracted and appropriately upgraded or modified to allow for transportation to market. Consequently, heavy oil upgrading is receiving considerable attention as much of the remaining unexploited petroleum reserves in the world are heavy and extra heavy (bitumen).

Transporting heavy, high viscosity oils requires additional pumping power and may require a technique to lower its viscosity such as heated transport systems or diluent to allow transportation at acceptable volumes/rates.

There are a number of different methods to assure effective and efficient transportation of heavy oils. These methods include:

- Dilution
- Partial upgrading
- Heating/steam injection/Inline injection
- Water Emulsion
- Core annular flow
- Surfactants/flow improvers (i.e., use of additives as pour point depressant)
- Slurry transportation (oil-solid slurry)
- Viscosity reducers

The first two methods are often associated with the same project or process. However, it may be indicated that currently dilution and partial upgrading are primarily used for transportation of heavy oils to the market.

The methods are described below:

6.4.1 Dilution

A method for enhancing heavy crude oils transportation is by blending the oils with less viscous hydrocarbons such as condensate, naphtha, kerosene, light crudes. Typically, there is an exponential relationship between the resulting viscosity of the mixture and the volume fraction of the diluents (Figure 6-6).

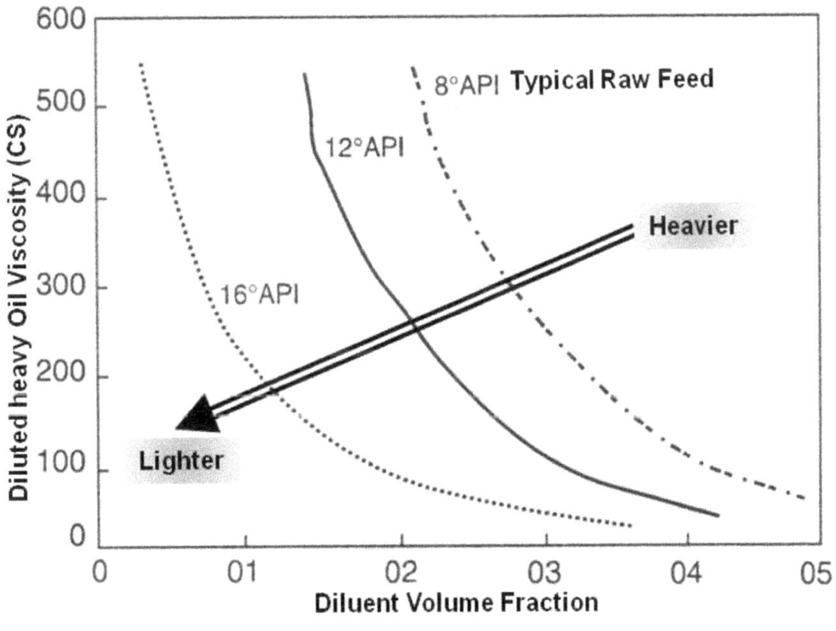

Figure 6-6. Dilution of heavy oils with condensate for different API (from [8])

Dilution can be performed in two different ways, depending on whether the diluent is recycled or not. Due to the relative difficulty in sourcing diluents for bitumen and heavy oil, efforts are made to recover the diluent at the delivery refinery and transport it back to the bitumen/heavy oil production facilities.

In many cases, the blended bitumen is exported to distant markets, and the diluents cannot be economically recycled.

The most common diluent currently used is a very light natural gas condensate (C5+ or "Pentane Plus") which is a by-product of natural gas processing, or other light hydrocarbons. For transportation purposes, a diluent typically constitutes 20% to 50% (refer to Figure 6-7) of the bitumen blend.

6.4.2 Upgrading/Partial Upgrading

In the upgrading/partial upgrading method, the heavy crude is upgraded by modifying its composition to make it less viscous. Upgrading technologies such as hydro-treating processes which are traditionally used in refineries are typically used for this application. As an example, this method is applied in Canada where large volumes of synthetic crude are produced for export. In such a case, the upgrading unit is located on the production site: the produced coke is stored in the open mine and the synthetic crude is transported to refineries.

Other upgrading techniques include:

- Chemical upgrading such as vapor reforming
- Use of Viscosity Reducers- dispersion of asphaltene (see item F below) and
- Bioconversion

A partial upgrading process is shown in Figure 6-7 [9].

Typical relationship between viscosity and °API for four heavy oils (extra heavy to heavy and light crude from four different fields) and the resulting products using an upgrading technology (Rapid Thermal Process, [10]) is presented in Figure 6-8. The heavy oils are all from heavy oil fields in USA (Blerige Heavy and Medway-Sunset, Kern County, and San Ardo Oil Field in Monterey County, California in California, and Canada (Athabasca northern Alberta).

From Figure 6-8, it is apparent that there is a dramatic reduction in the viscosity using the High Yield configuration by this technology touted as HTL (Heavy to Light) [11].

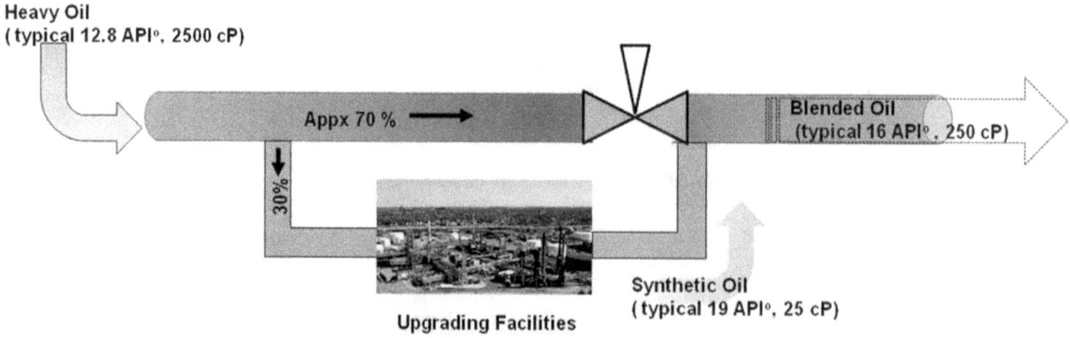

Figure 6-7. A partial upgrading scheme [9]

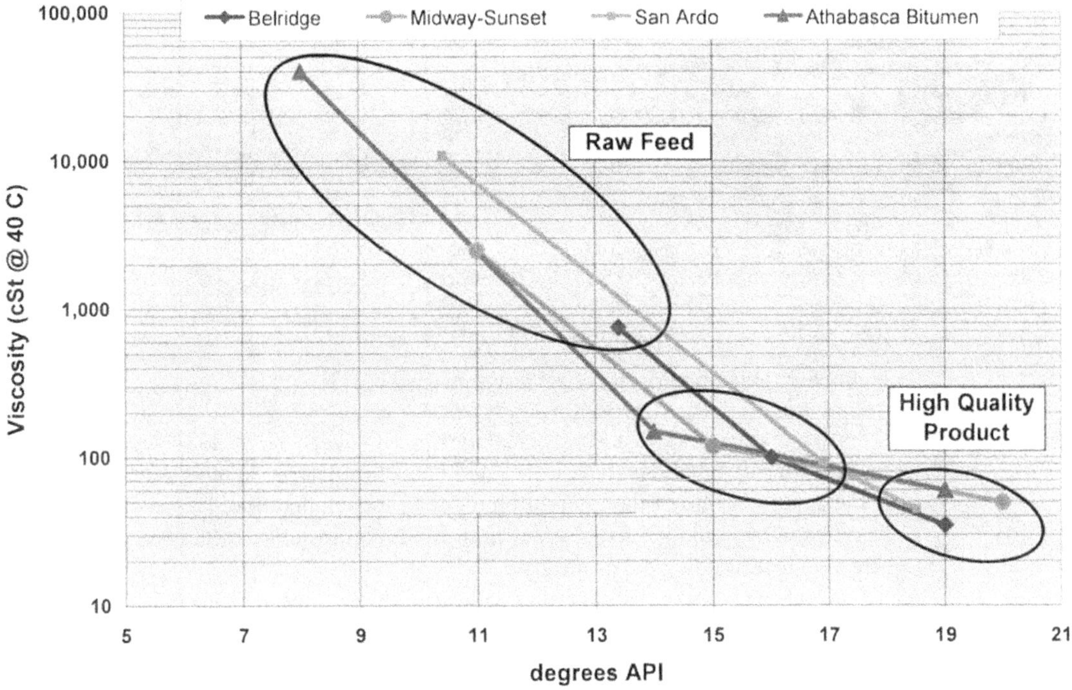

Figure 6-8. Viscosity versus °AP for typical raw feed and upgraded heavy oil/bitumen [10]

Upgrading processes generally involve heavy investments with associated high operating costs. Additionally, some technologies require infrastructure and skilled personnel not readily available in remote location where most heavy oil reserves are available.

6.4.3 Heating/Thermal Upgrading

A way to transport heavy oils is to heat the oil as the viscosity decreases very rapidly with increasing temperature (refer to Figure 6-2). This is well indicated by review of ASTM equation that

$$\log \log (\upsilon + 0.7) = m \log T + C \qquad (6-5)$$

where

υ = kinematic viscosity, cSt and T = temperature absolute (°K)

m = slope of temperature–viscosity curve and C = constant depending on type of diluents.

Generally, heavy oils that are required to be shipped to a regional upgrader or major pipeline terminal without diluents will require a heated/insulated pipeline. An example is the **Enbridge Pipeline** system (NPS 12, 35 KM, Operating Temperature 120°C) for transporting bitumen (for PetroCanada) from MacKay River production site to Fort MacMurray (both in Alberta) where it is blended before being exported through the extensive network of pipelines to the North American market (Figure 6-9) [11].

It may be noted that the design and construction of a heated pipeline involves many considerations: including pipelines expanding/extending due to heat, number of

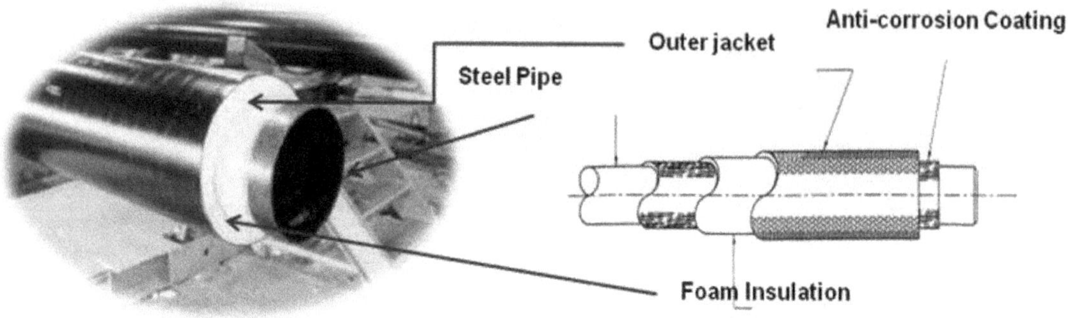

Figure 6-9. Enbridge MacKay river heavy oil pipeline [11]

pumping/heating stations, heat losses and hence insulation. Another significant issue is the corrosion rate of the internal pipe due to the temperature. It is also shown that heat treatment could induce changes in the colloidal structure of the heavy crude oils affecting their rheological properties.

Another technique is generally to heat trace the pipe (Figure 6-10). Heating is provided by a capacitive power-generating facility to maintain the carrier pipe temperature to about 65°C. In most cases, two heat tracing cables are installed at 3 and 9 o'clock positions on the pipe and then together with the pipe are wrapped in insulation

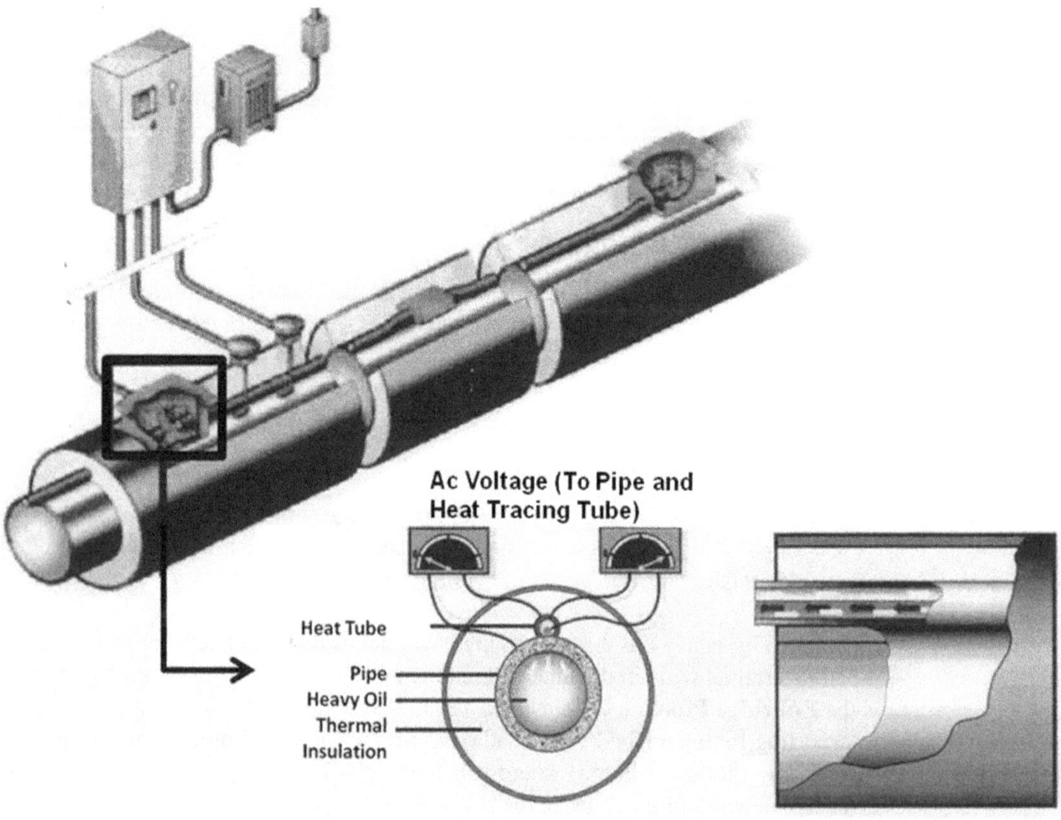

Figure 6-10. Typical insulated heat-traced (to 60°C) heavy oil pipeline [12]

(typically 90 mm PU foam). Suitable systems are also provided for emergency power usually, DC power systems, and uninterrupted power supply. The system is suitable for Class, Division 1 Hazardous Area Classification and other standards as applicable. Heat output : 40 W/m (at 65°C) with overall effective heat transfer coefficient : 0.654 W/m-K. Control of the system is accomplished by solid-state temperature controls with RTDs as sensors.

Generally, heating is an expensive way of transporting heavy oil, as for long distance line and in cold climates may affect pipeline hydraulics [13], requiring several heating stations along the pipeline. Furthermore, during pipeline stoppage (e.g., pump station shutdown), cooling could occur with a resultant requirement for a high-pressure restart.

6.4.4 Water Emulsion

This method consists of dispersing the heavy crude oil in water in the form of droplets stabilized by surfactants (surface- active agents), which would lead to a reduction in viscosity. Oil in water emulsion is significantly less viscous than oil itself no matter how viscous and dense the oil is. A typical emulsion is composed of 70% crude oil, 30% aqueous phase, and about 500 to 2000 ppm of chemical additives as surfactant (Surface Active Agents). The resulting emulsion has a viscosity in the 50 to 200 cP range at pipeline operating conditions and is particularly stable. Typical oil in water emulsion viscosity is shown in Figure 6-11.

Oil in water emulsion is used in Venezuela to produce ORIMULSION R [14]. In this process, the emulsion is not broken and used as such to feed power generation plants [15].

Emulsified transportation is a naturally lubricated flow (of oil in water mix), and it is indicated that if such a product pumped at higher rates, then there is likelihood of lip flow occurrence that would reduce pressure losses (Figure 6-12).

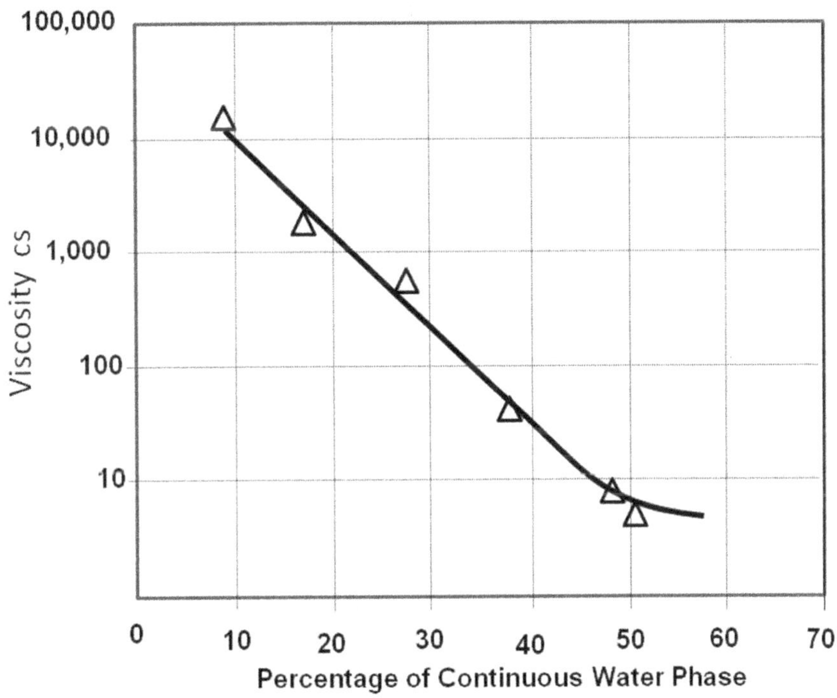

Figure 6-11. Oil in water emulsion viscosity [14]

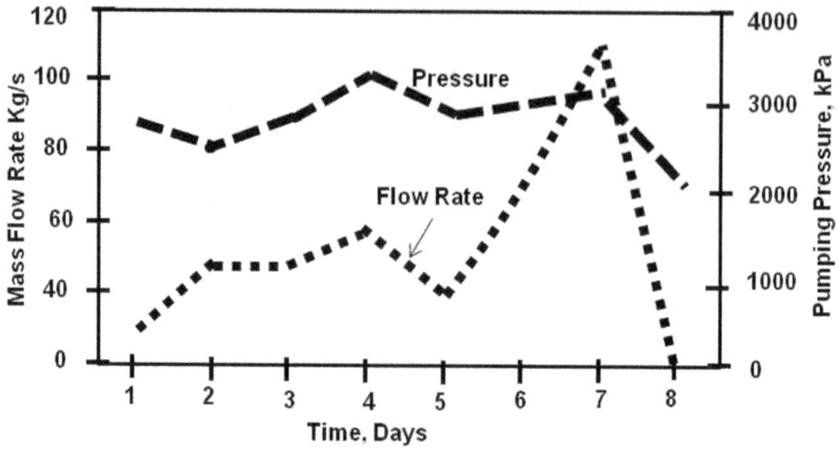

Figure 6-12. Effect of flow rate on pipeline pumping pressure [14]

The technique is stable and the emulsion is relatively easy to prepare. However, it is costly (for additives and surfactants) and requires skilled personnel for operation.

6.4.5 Core Annular Flow (CAF)

This technique was developed by Syncrude in Canada in early 1970 and is used for oil sand hydro-transport of extra heavy bitumen over very short distance. In this method of transportation, a water film surrounds the oil core (Figure 6-13) and acts as a lubricant so that the pumping pressure necessary for the lubricated flow is comparable to the one for water alone. The water fractions are typically in the range of 10% to 30%. An example of CAF is the Aurora pipeline operated by Syncrude. It is 36 km long from the Aurora bitumen extraction site to the Syncrude upgrader site located in Northern Alberta, Canada. In this system, water is heated to about 55°C, depending on the temperature of bitumen, ground temperature, and flow rate.

In this technique, the flow regimes of two-phase flow (oil in water) configurations are dependent on the fluid properties such as density, surface tension and on the shear

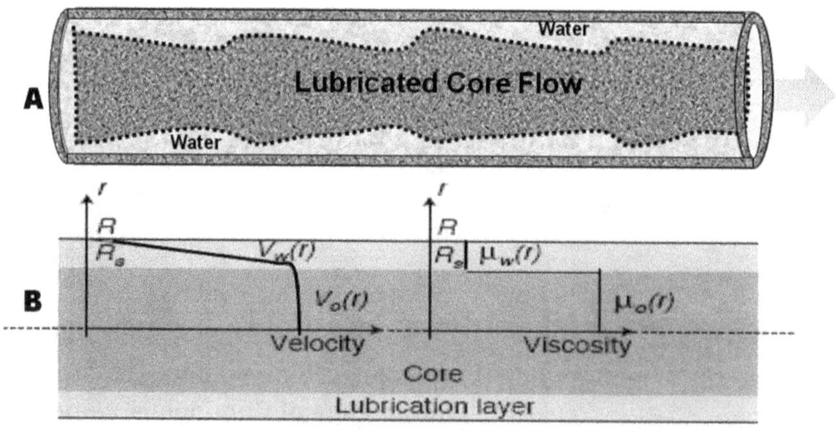

Figure 6-13. Core Annular Flow (CAF) [16]

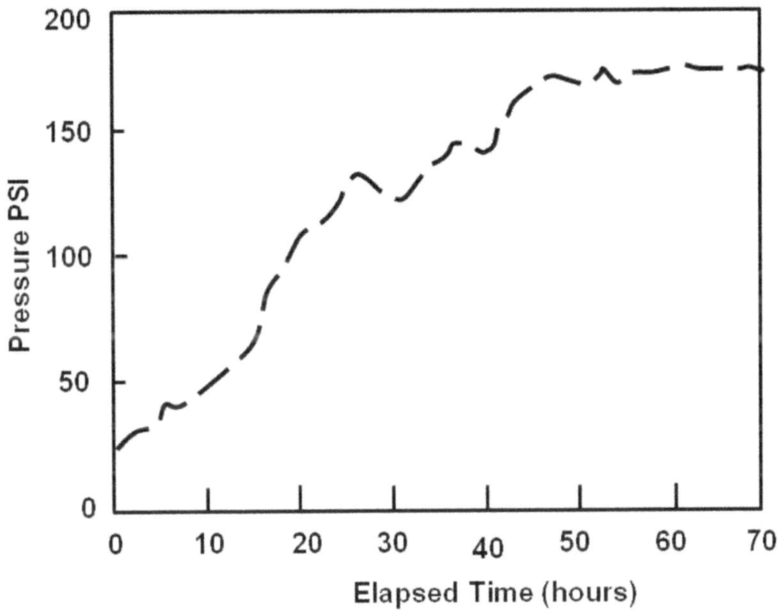

Figure 6-14. Typical fouling in Core Annular Flow (CAF) technique, 1 km long test circuit transporting heavy Venezuelan crude (ZUATA) [14]

rate in the flow. Mean injection velocities are also key parameters for the flow regime determination.

In core annular flow water surrounds the lubricated oil core Figure 6-13A. A perfect CAF is shown in Figure 6-13B, small water layer is sheared and the velocity field is approximately linear if the difference of viscosity between oil and water is large. In that case, the oil core is nearly like a plug flow. However pressure losses are that of water transported through a pipeline.

Limitations are fouling, Figure 6-14 and will require high-pressure pumping start up. It is however indicated that Core Annular Flow (CAF) technique is well established and that to avoid fouling a chemical type additive (made up of silicate) along with internal coating can prevent severe fouling.

6.4.6 Surfactants/Flow Improvers

The transportation of heavy crude oils as emulsions is an alternative to blending the crude oil with natural gas condensate or other diluents. In this technique, aqueous surfactant solutions are employed to convert high viscous heavy crude oils into low viscosity oil in water emulsion.

Such a surfactant is a chemical that stabilizes mixtures of oil and water by reducing the surface tension at the interface between the oil and water molecules. Because water and oil do not dissolve in each other a surfactant has to be added to the mixture to keep it from separating into layers.

With this technique emulsion viscosity decreases considerably with the increase in volume concentrations of the dispersed phase, thus giving a possible alternative is to use surfactants to emulsify the heavy crude oil in water and to transport the low viscosity emulsion in a pipeline [17].

Flow improvers/drag reducers are chemicals that are injected in the flowing heavy oils to reduce the pressure loss as these are transported through a pipeline. These pressure reductions are through reducing turbulence in the pipeline (Figure 6-15). Lowering these internal fluid pressure losses decreases the operating pressure (Figure 6-16) or increases the bulk throughput of the pipeline, for a given pumping energy thus reducing the operating costs.

The energy saving using DRAs may entail any of the following:

- Bypass intermediate pump stations
- Shutdown incremental pumps and
- Increase flow
- Reduce pumping power (for a given flow rate) to fuel cost and reduce emission

Flow improvers, also known as "Drag Reducers", only improve the flow efficiency when added to the upgraded bitumen/heavy oils for transportation through pipelines [18]. This is because, generally, oil pipelines operate in laminar region (Figure 6-17). Injection of DRAs may also reduce the friction factors by forcing the pipeline operation to move into the region with higher Reynolds numbers, but with smoother flow pattern (Figure 6-18).

Different types of additives can be used in these systems and include surfactants, fibers, aluminum disoaps, and high polymers. Drag reducing additives are effective because they reduce the turbulent friction of a solution. This result is a decrease in the pressure drop across a length of a pipeline or conduit and likewise reduces the energy required to transport the liquid.

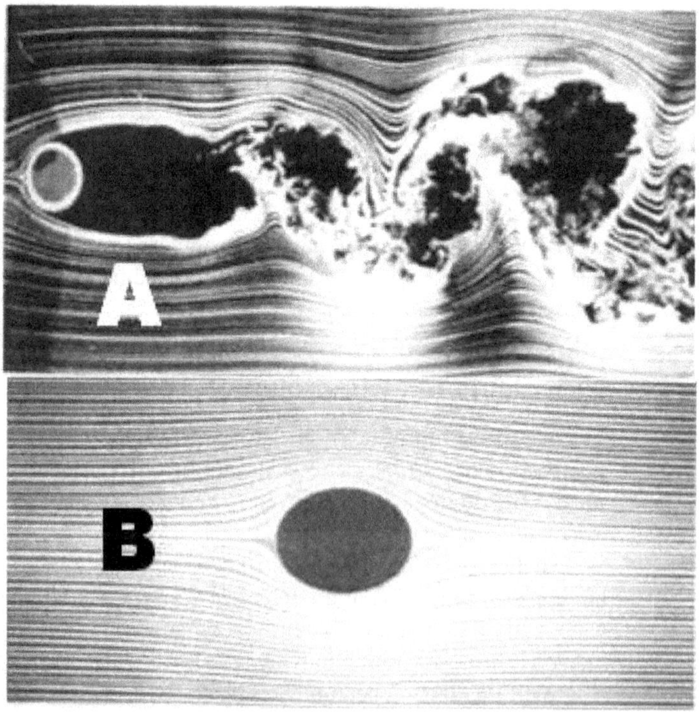

Figure 6-15. DRA/Flow improvers in action: A — turbulence before DRA injection and B — linear flow pattern after DRA injection (Turboflo™, courtesy of FLOWECHEM)

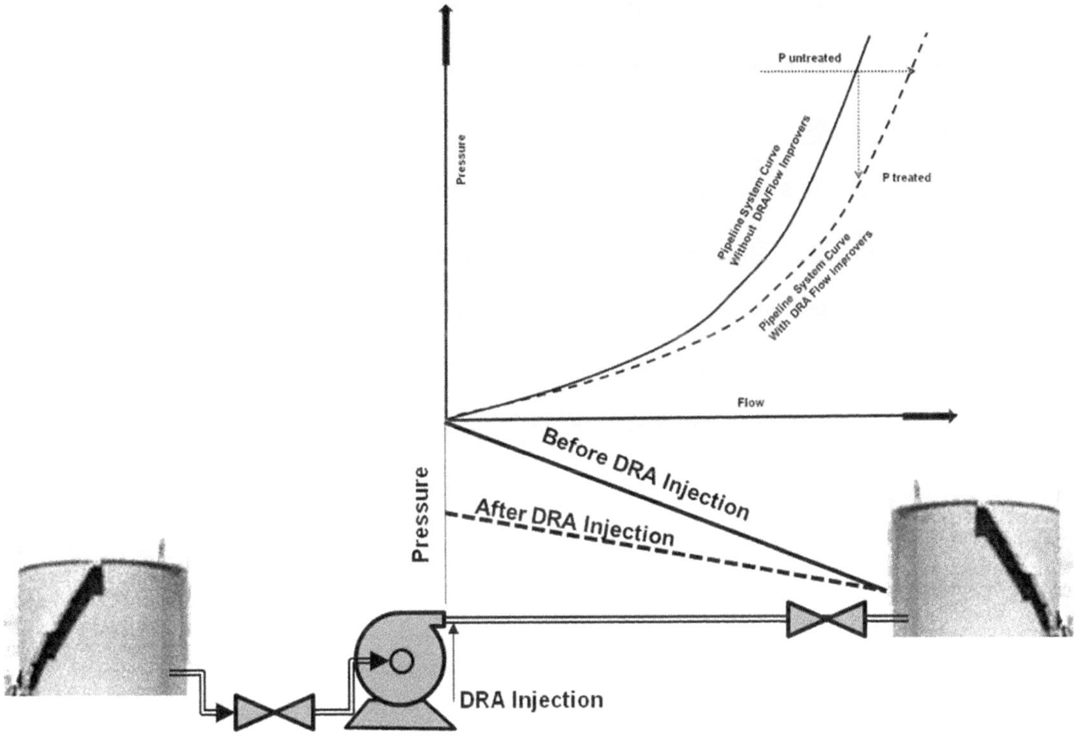

Figure 6-16. Effect of chemical drag reducers (CDR) on pipeline pump pressure and flow rate

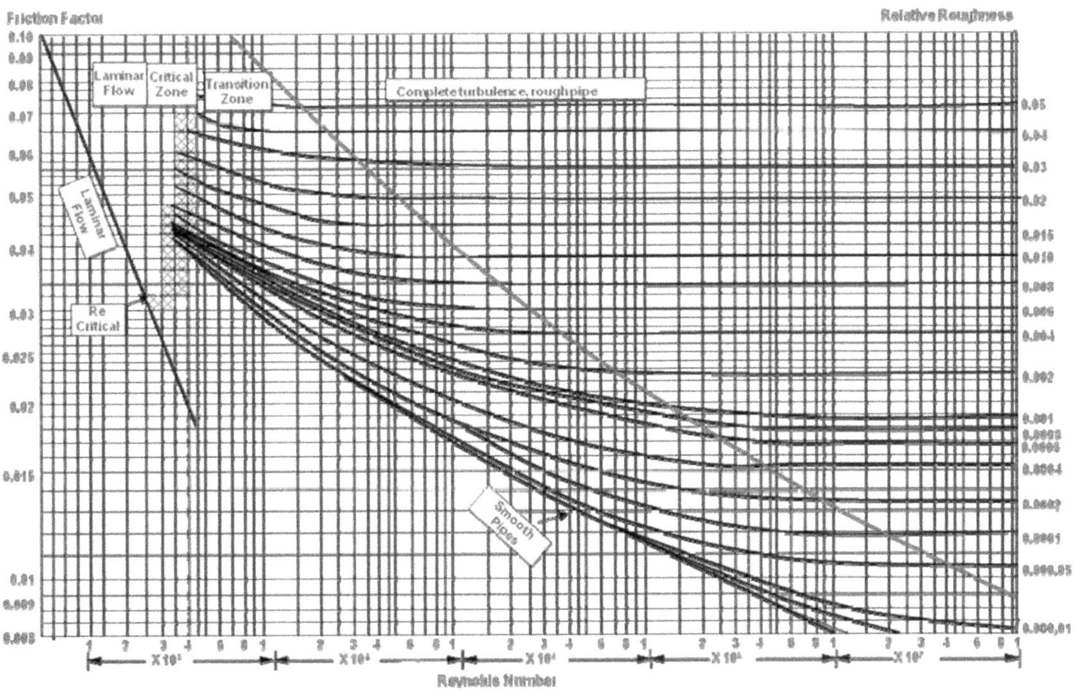

Figure 6-17. Moody diagram friction factor for flow of fluids in pipelines [19]

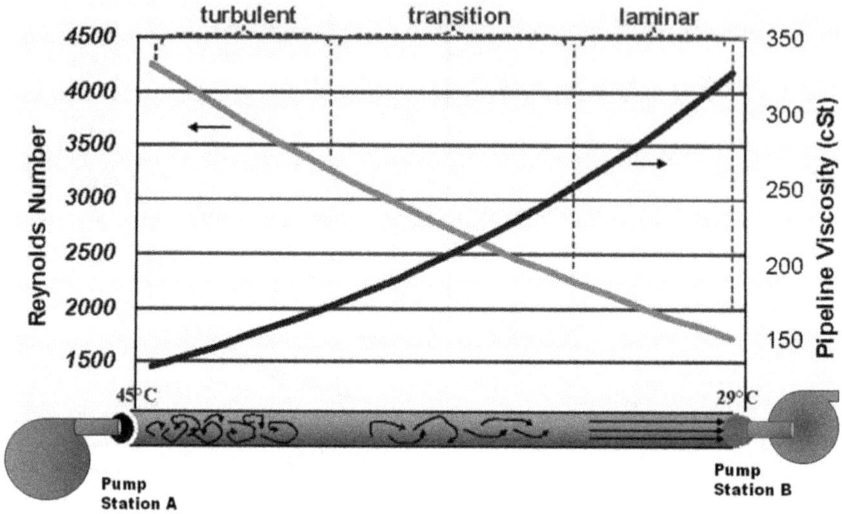

Figure 6-18. Drag reducing agent /flow improvers; pipeline regime profile (based on Castilla blend ("heavy", 18 °API))

6.4.7 Slurry Transportation Method

This is also known as the ***Asphaltene Dispersion*** method. Heavy crude oils are described as a colloidal suspension composed of a solute (asphaltenes) and a liquid phase (maltenes). These are overlapped and this asphaltene thus contributes to high viscosity of heavy oils. By suppressing the effect of asphaltene (i.e., by making it into solids particles in suspension), less viscous fluid would result. It appears that one way to lower the viscosity of heavy oils is to diminish the apparent volume occupied by their asphaltenes [15].

Viscosity reducers are molecules (resins, surfactants etc.) that disrupt the asphaltene/maltene colloidal interactions, thus reducing the heavy oil viscosity.

6.4.8 Comparison of Transportation Techniques

The above techniques together with their advantages and limitations are summarized in Table 6-2.

Comparison of major transportation techniques for bitumen, extra heavy, and heavy oils indicate that the adoption of each technique depends on the following:

- Bitumen/heavy facilities requirement and cost
- Environmental condition and requirements, topography and locations of supply and market
- Availability of water, diluents (type and property), chemical additives/surfactants
- Source of power

Figures 6-19 and 6-20 provide a comparison between heating, emulsification, dilution, and core annular flow (CAF) for transportation of bitumen and heavy oil.

TABLE 6-2. Heavy oils transportation technologies comparison (from ref. [8])

Transportation Preparation Method	Dilution Technique	Upgrading/Partial Upgrading	Heat Application (use of inline heaters)/Steam	Water Emulsion	Core Annular Flow	Surfactants	Flow improvers	Slurry
Issues								
Heavy Oil Pipeline Size	Large	Normal	Normal	Large	Normal	....	As designed	Normal
Limitations Concerns	Availability of diluent	Applied at Plant	Corrosion potential and High Costs	Water recycling to be resolved	Oil adhers to pipe walls & eventual blockage	Viscosity depends on type & ppm added	Expensive, third party supply	....
Additional Facilities Requirement	Parallel Diluent or Injection Systems	Field Refinery	Heaters	Water source and pumps	Decloging-Large restart pressure	....	Injection facilities	....
Potential Field Concern			Electricity Sources for Heating	Electricity Sources for	Electricity Sources for Heating	....	Re-injection facilities after pumps	....
Advantages	Advanced technique	Used in Refineries/plants	Good process to reduce oil viscosity	Comercialized Extensively	Industrial application limited	Easy & cost effective	Improves capacity	Asphaltene reduction
Experience	Extensive	Extensive	Extensive	Extensive	Only two applications reported	New	Pioneed in 1970, Conoco Phillips	New upgrading

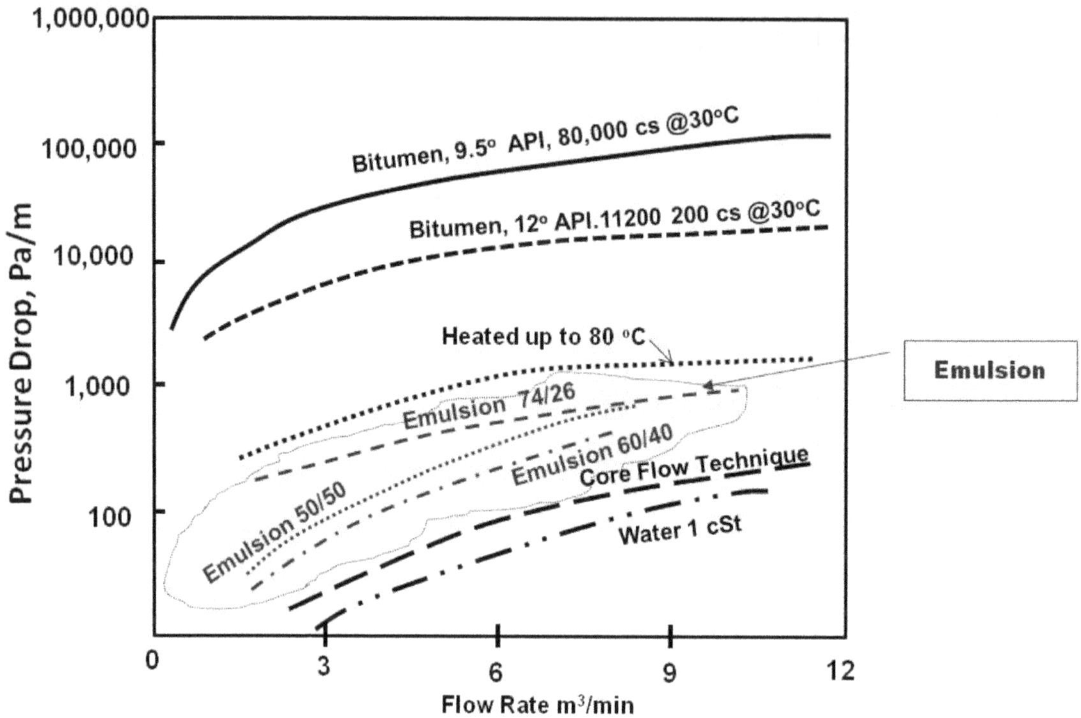

Figure 6-19. Comparison of pressure drop versus flow rate for different transport methods, reproduced from [20]

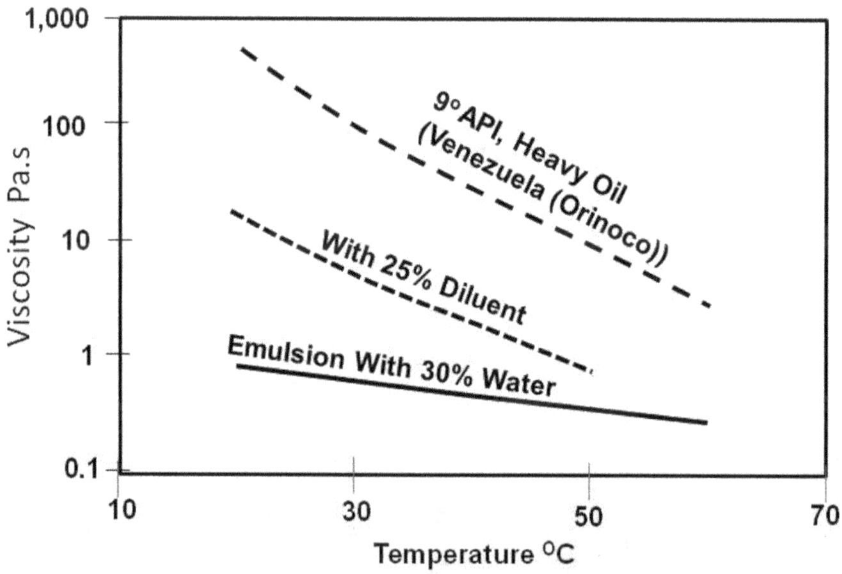

Figure 6-20. Comparison of heating, dilution and emulsification (oil and water) for transportation of extra heavy crude [14]

6.5 HEAVY CRUDES PROPERTIES FOR PIPELINE TRANSPORTATION

6.5.1 Grouping of Crudes and Designations

A — Grouping: For the purpose of market use and pipeline transportation, the heavy oil industry uses labels, and in some cases, generalizes crude grades into light, medium, heavy, sweet, and sour groups [21].

Not all parties follow the same conventions on what constitutes light, medium, and heavy. Some sub-groupings are present such as high and low TAN (total acid number, [22]) variants of heavy sour crudes. Proximity to common carrier pipelines, equalization systems (similar in operation to the API gravity and sulphur banks), and other business factors are used to determine the ultimate location of wellhead production. In general, various grades are described as follows;

1) **Condensate (CRW)** (density ~725 kg/m3, 63 °API, sulphur ~0.2 wt.%)
2) **Synthetic crude (OSA, SYN, SSB, HSB)** (density ~860 to 870 kg/m^3, 31 to 33 °API, sulphur typically < 0.2 wt.%)
3) **Light sweet crude** (density ~830 kg/m^3, 39 °API, sulphur < 0.5 wt.%)
4) **Light sour crude (Light Sour Blend also known as LSB)** (Crudes density ~850 to 860 kg/m^3, ~34 °API, sulphur ~1.0 to 1.5 wt.%)
5) **Medium sweet crude** (density ~880 to 890 kg/m^3, ~30 °API, sulphur < 0.5 wt.%)
6) **Medium sour crude** (density 885 to 890 kg/m^3 (~ 30 °API) and 2.0 wt.%)
7) **Heavy sour crude** (density 925 to 940 kg/m^3, ~20 °API, sulphur 2.9 to 3.6 wt.%)

Table 6-3 below indicates commodity classification for a typical pipeline transportation [23]:

B — Designation: The industry, depending on where heavy oils are recovered and processed utilizes certain commodity designations, some of which are list below:

CRW = Condensate
OSA, OSC = Suncor Oil Sands Blend A or C
SYN, SSB = Synthetic, Suncor Sweet (Blend)
LSC = Light Sour Crude
MSC = Medium Sour Crude
NSA = Newgrade Premium Synthetic
HSB = Husky Synthetic Blend
HSC = Heavy Sour Crude
PAS = Premium Albian Synthetic
SYNBIT = Synthetic Blend
SSB = Sycrude Sweet Blend
DILBIT = Diluted Blend (with condensate)
SSX = Shell Sweet Blend

TABLE 6-3. Pipeline transportation commodity classification

Viscosity (mm^2/s)	density (kg/m^3)	Classification
100–350	904–940	Heavy
20–99	876–903	Medium
2–19	800–875	Light
0.4–1	600–799	Prod. & Condensate.
to 0.3	to 599	NGL

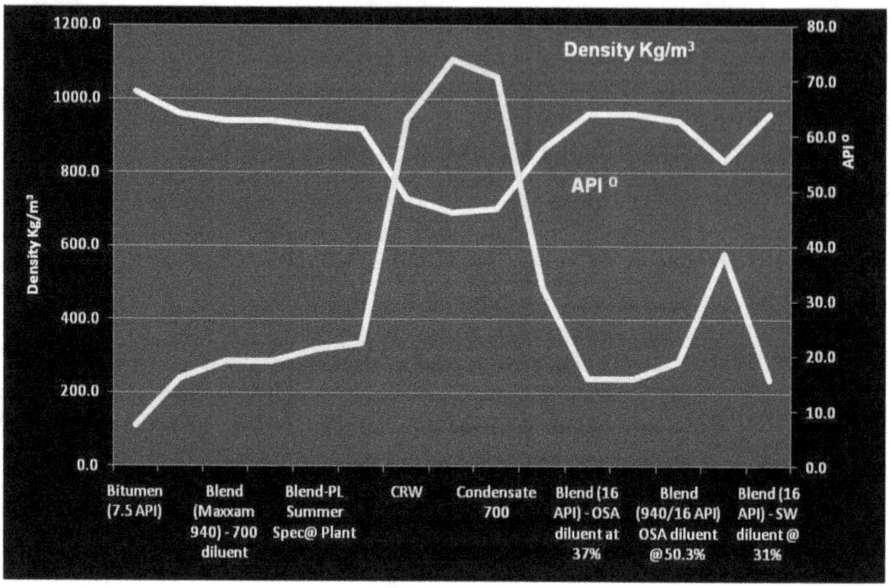

Figure 6-21. Density/API gravity of bitumen and typical treated heavy and light oil blends

6.5.2 Typical Properties

Crucial to the success of heavy oil transportation and addressing design and operational issues is having realistic properties for bitumen, crudes, and diluents so that acceptable bitumen blending and trimming scenarios can be established for the delivery of the final products to the required specifications.

The properties of several possible diluents examined by industry are shown in Figure 6-21 through to 6-23.

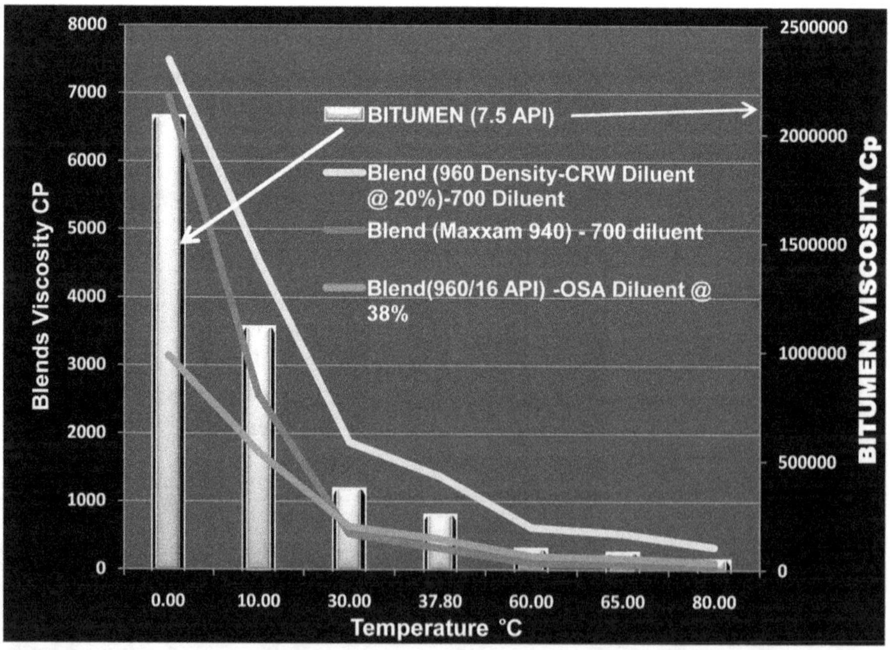

Figure 6-22. Viscosity of bitumen and typical treated heavy oil blends

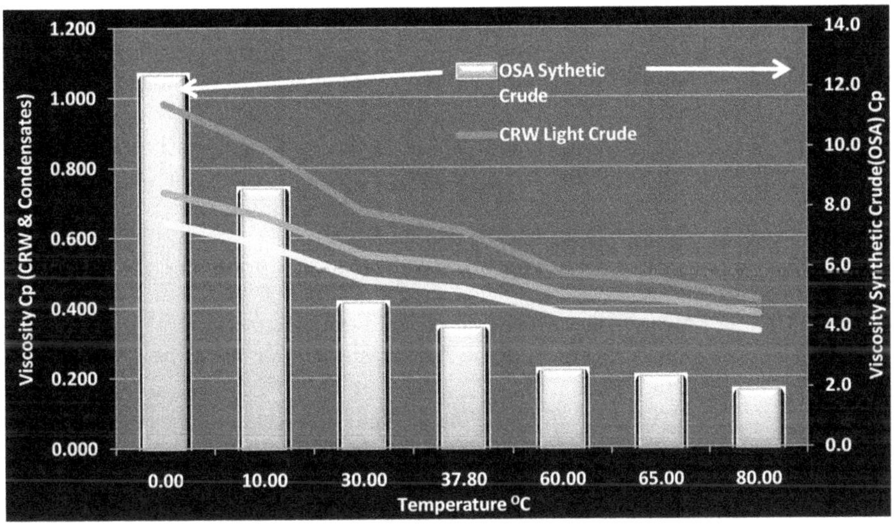

Figure 6-23. Viscosity of typical condensate and light crude

6.6 HEAVY OIL PIPELINE TRANSPORTATION EXAMPLE — ROLE OF DESIGN FOR OPERATIONAL CONTROL

Based on Paper by Victor Cabrejo and Mo Mohitpour, ASME IPC 2010-31650 (with permission) [24].

6.6.1 Summary on Role of Design

Most liquid pipelines design and operational control is based on steady-state flow analysis. This neglects dynamic effects that occur as a result of occurrence of surges in a pipeline caused by rapid changes in pressure as a consequence of changes in the flow rate.

A transient analysis of liquid pipelines on the other hand assures pipeline performance under all conditions (steady state and dynamic situations) including evaluating the following:

- Impact from pump station start up, delivery restriction or shutdown (zero delivery)
- Pump unit trip/failure
- Rapid mainline valve closures including slam shut of a non-return (check) valve
- Effect of running the pipeline with minimum flow and maximum pump discharge pressure operating condition
- Variation in demand including rapid reduction/curtailment of delivery volumes
- Bubble formation/collapse (the transition from slack-line to tight-line flow)
- Tight-line/slack-line operation
- Unintentional changes in operational position of control valves
- Fluid property delivery conditions
- Liquid injection assessment
- Surge protection including pressure relief/control system evaluation
- Restart requirement to avoid slack-line conditions prevalent in hilly/mountainous parts right of way (ROW)

Such a dynamic analysis would indicate whether liquid surges are of concern from design, as well as system operational conditions. It also would provide an evaluation of an automated control or potential automated strategies for overpressure protection.

In this paper, the dynamic analysis of liquid pipelines resulting in design and operational benefits will be described. Finally, their benefits in application to a heavy oil pipeline facilities "Keystone" will be highlighted.

6.6.2 Need for Transient Analysis

Pressure surges occur in all fluid pipeline systems. This phenomenon can cause two types of damage fatigue and catastrophic failure as well as incorrect system response/operation [25, 26].

Traditionally, pipeline transmission systems have been designed using steady-state simulations. Steady-state simulations provide the designer with information on system capability and a reasonable level of confidence when the system is not subject to radical changes such as mass flow rates or operating conditions, and for various mixes of fluids. However, a conventional steady-state analysis has limitations in dealing with surges in mass flow rates, the loss of facilities and facility operation, etc. In these and other instances, the designer will want to perform a dynamic (or transient) analysis to test the capability of the system for various fluids including determining delivered products properties, choose the system components, setting operational conditions, and maintaining the appropriate level of safety.

Steady-state hydraulics process generally involves reviewing flows and pressure drops and determining capacity, pipeline diameters, pipeline loop lengths, and overall pump station power requirements. System capability, limitations, and some operational conditions can be determined.

In liquid lines pressure surges occur. Pressure surge is a term used to describe a relatively rapid process that occurs with "almost" incompressible fluids. With an incompressible fluid such as oil or water, there is no storage capability obtained by pressurizing the fluid. In addition, when an attempt is made to compress an incompressible fluid, the pressure of the fluid will rise rapidly throughout the system. This phenomenon is very important during the design of liquid pipeline systems as it affects not only the system design but also the operational controls.

Fundamental correlations and transient solution techniques for compressible and non-compressible fluids have been described previously [27]. Examples have also been cited for a long distance gas pipeline for several systems [28, 29]. Similar fundamentals also apply to liquid pipelines [26]. Therefore, the following addresses the dynamic phenomenon from the viewpoint of the available solutions rather than the mathematics and modelling involved in determining the magnitude of the surge pressure and method for controlling it.

6.6.2.1 Information Required for Pipeline Dynamic Assessment

Operation of various pipeline components including, pump and motor, control valve, and other parameters affect the magnitude of the transients in the pipeline system. Information and parameters required for a detailed assessment of these pressure transients and their impact on systems design and operational controls are:

Pipeline Systems Characteristics and Data Including

- Pipeline/pipe element diameters, lengths, wall thickness, grade, class location, maximum operating pressure, test pressure, allowance
- Pipe inside roughness
- Elevations (pipeline and appurtenances)

- Valve and fitting locations
- Pump location and arrangements
- Tankage facilities location and information

Fluid Property Information

- Type and description of the fluid being transported
- Specific gravity
- Bulk modulus of elasticity
- Viscosity at various temperatures
- Vapor pressure

Valve Characteristics

- Size and flow characteristics at various openings (C_v versus percentage (%) of opening)
- Valve operator speed and characteristics
- Type of check valves, damped or un-damped
- Description of pump station discharge or suction control valves for normal operation and rapid operation (emergency situation)

Pump and Driver Information

- Pump performance data (head, efficiency, horsepower, or torque versus flow), pump type
- Number of stages (for specific speed calculation)
- Changes expected for increased throughput
- Rated conditions (conditions at the best efficiency point for head, flow, speed, and torque)
- Efficiencies, adiabatic, and mechanical
- Pump characteristics diagram or synoptic chart (if not available, curves from a pump of similar speed)
- Driver type: variable/constant speed (induction motor, synchronous motor, turbine, etc.)
- Driver torque versus speed curve (for pump start-up cases)
- Safe current versus time data for electric motors if start-up analysis is to be performed
- Special devices on pump/driver and any auxiliary facilities and appertaining loads
- Pump station controls description (minimum flow shutdown, flow discharge pressure shutdown, suction/discharge pressure control, etc.)

Operational Data

- Normal start-up and shutdown procedures
- Emergency operational procedures
- Unplanned operations (inadvertent closures, pump trip/shutdowns, etc.)
- Constraints on pipeline and equipment operation

Information/Preferences on Surge Pressure Controls

- Surge tanks (tank area and height)
- Accumulators (tank volume, initial gas volume, other parameters)

- Relief devices (set pressure, relief devices performance data)
- Specific surge control devices or schemes preferred

Note: Parameters for surge suppression devices is usually determined by systems hydraulic assessment.

6.6.3 Surge Mitigation Methods

There are numerous techniques for controlling the harmful effects of pressure transients and surges, some involving design considerations and others the consideration of surge protection devices. Some of the mitigation techniques include:

- Pipe Material of Higher Grade/Higher Wall Thickness
- Rerouting Pipelines
- Flow/Pressure Control Valve
- Adding Check Valves
- Adding Bypass Valves
- Liquid Accumulators
- Adding Surge Relief Tanks
- Surge Relief Valves
- Pressure Bursting Discs
- Increased Diameter of Pipeline
- Relocation of Facilities (Pumps Stations, Valves, etc.)
- Pump and/or Valve Bypass
- Pumps With Variable Speed Drives
- Pumps With Soft Starters
- Valve Opening and Closing Times
- Increase the Moment of Inertia of Pumps
- Minimizing Mechanical Resonance Hazards by Additional Supports (where applicable)

However, there must be a complete design and operational strategy devised and assessments made to combat potential pressure surges and adverse pressure transients in a system. The transient event may either initiate a low-pressure event (*down-surge*) as in the case of a pump power failure, or a high-pressure event (*up-surge*) caused by the closure of a downstream valve.

Down-surge can lead to undesirable situations such as the unwanted occurrence of fluid column separation, which itself can result in severe pressure rises following the collapse of a vapor cavity. In some systems negative pressures are not even allowed because of (1) possible pipe collapse or (2) ingress of outside fluid (such as water in offshore systems) or air.

The means of controlling the transient will in general vary, depending upon whether the initiating event results in an up-surge or down-surge. For pumping plants, the major cause of unwanted transients is typically the complete outage of pumps due to loss of electricity to the motor. For full pipelines, pump start up, usually against a closed pump discharge valve for centrifugal pumps, does not normally result in significant pressure transients.

The majority of transient problems in pumping installations are associated with the potential (or realized) occurrence of *fluid-column separation* and *vapor-pocket formation and collapse*, resulting from the failure of one or more pumps, with or without valve action. The pump discharge valve, if actuated too suddenly, can even aggravate the down-surge problem.

Generally, the pipeline industry's method to control adverse effects from high surge pressures is through effective automation systems (telecommunication and SCADA) and strict operating procedures for implementing a timely, sequential pump station shutdown upstream of an offending happening/situation (e.g., rapid valve closure, rapid reduction in delivery/flow, high station discharge pressures, ruptures, etc.) which cause undue high pressures threatening pipeline physical and operational integrity.

6.6.4 Code Requirement

Codes and Standards have requirements to design pipeline systems to take account of the effects of occasional loads such as pressure transients in systems. This not only concerns positive pressures but also negative pressures. Full vacuum can occur when there is column separation in a pipeline. This commonly occurs when there is a loss of power or rapid closure of an upstream valve or pipelines traversing down slope in a steep mountainous environment.

North American codes that apply include:

- ANSI-ASME B31.4 — 1998 "Pipeline Transportation Systems For Liquid Hydrocarbons and Other Liquid Systems"
- CSA Z662 - 07 "Oil & Gas Pipeline Systems"

In addition, the US Department of Transport (DOT) Office of Pipeline Safety's (OPS) Code of Federal Regulation (CFR) Title 195 "TRANSPORTATION OF HAZARDOUS LIQUIDS BY PIPELINE" is also followed for design (including overpressure protection) and operational purposes.

Other relevant international codes that consider transients are:

North American codes indicate the following:

- For normal operation the maximum steady-state operating pressure must not exceed the internal design pressure of the pipeline and pressure ratings for the components.
- Surge calculations must be made, and adequate controls and protective equipment shall be provided (if required), so that the level of pressure rise due to surges and other variations from normal operations do not exceed the internal design pressure at any point in the piping system and equipment by more than 10%.

It may be noted that there also other international codes that relate to dynamic situations as a consequence of pipeline operations. It is imperative to review such a codes for the appropriate design and control conditions to avoid the harmful effects of occasional loads. Some of these codes are listed below; however, they have not been reviewed for the purpose of this case study

- UK PD 8010 — Parts 1 and 2
- BS EN 14161 — Petroleum and Natural Gas Industries, Pipeline Transportation Systems
- Institute of Petroleum Pipeline Code IP6
- DNV OS-F101 Submarine Pipeline Systems 2000
- Australian Gas and Liquid Pipeline code AS 2885
- Submarine Pipeline Code AS 1958
- Buried Flexible Pipelines Design AS 2566

- Pipes Design for Dynamic Stresses
- Pressure Vessel Code AS 1210

6.6.5 Case Study — Application to A Heavy Oil Pipeline Projects

The benefits accrued from the aforementioned dynamic analysis for heavy oil pipeline projects are demonstrated by its application to the Keystone Pipeline Project.

TransCanada's proposed Keystone XL pipeline system consists of NPS 36 (API X70, 9930 kPa MOP) crude oil pipeline that would begin at Hardisty, Alberta and extend southeast through Saskatchewan, Montana, South Dakota, and Nebraska. It would incorporate a portion of the current Keystone Pipeline (between Steele City and Cushing) that is being constructed through Kansas (Steele City) to Cushing, Oklahoma, before continuing through to a delivery point near existing terminals in Nederland, Texas to serve the Port Arthur, Texas refiners as well as through 80-km pipeline extension to the Houston, Texas marketplace (Figure 6-24).

The portion of the Keystone XL Pipeline between Steele City and Cushing (the Cushing Extension) is being constructed with the current Keystone project. Once built, this segment will be operating with the Keystone pipeline system and will serve the Cushing market as required. Thus, it is envisioned that both systems will have the same operational philosophy and pipeline control as well as overpressure protection strategy.

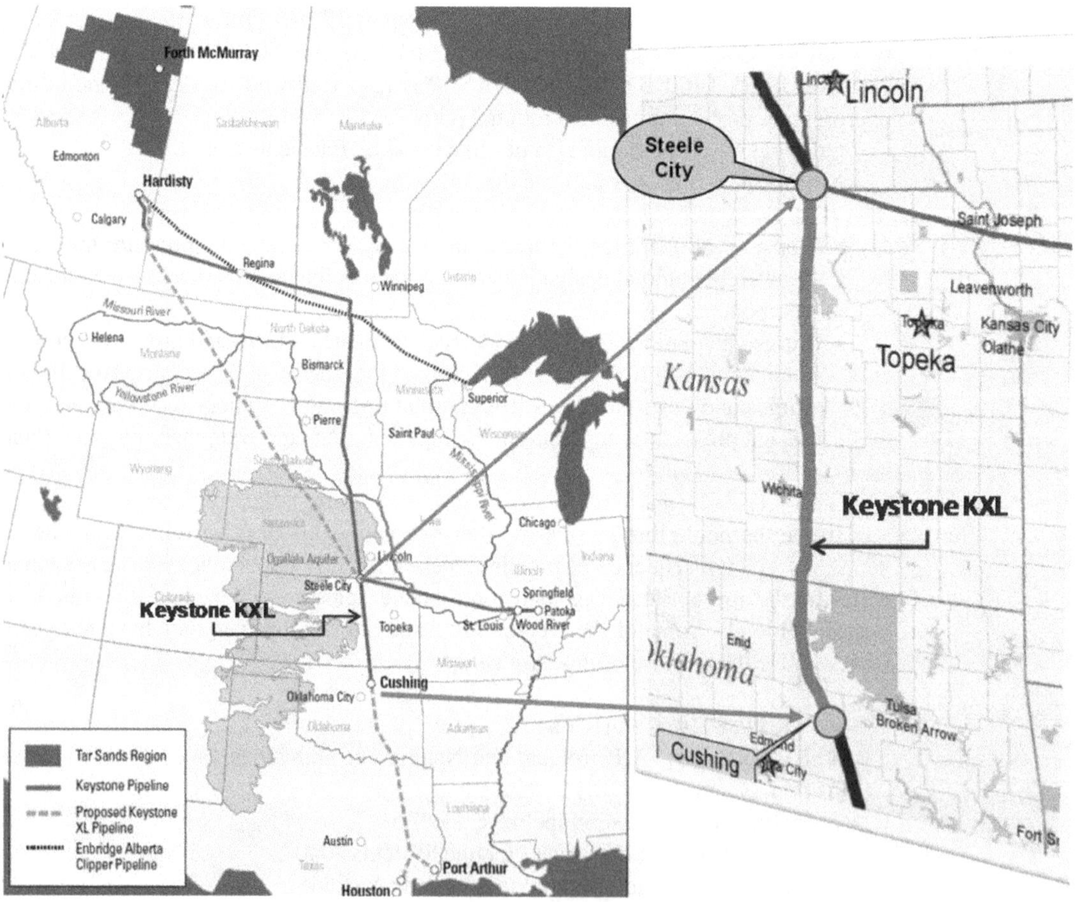

Figure 6-24. Keystone pipeline — Cushing extension

To assure the consistency of approach, the impact of operation of key pipeline facilities (mainline valves closures, pump unit or pump station trips/shutdown and delivery closures) as well as other operational issues are to be addressed for this segment first. This will assist in determination of operational control systems under steady and dynamic situations.

The purpose of the study was first to determine the entire system capacity and data (with predetermined pipeline route, size and operating pressure as well as predetermined facilities location for pump station and mainline check and block valves) for transportation of Heavy DilBit, and implementing batch transportation of a volume of Synthetic Crude Oil (SCO) through the pipeline from Hardisty in Alberta Canada through to delivery points in Texas, USA. The focus was then to specifically model the Keystone XL segment of the pipeline between Steele City and Cushing to address dynamic effects of dynamic responses due to the following:

- Effects of mainline valve closures,
- Impact of pump station shutdowns,
- Impact of Cushing delivery restriction (zero delivery)/shutdown, and
- Minimum flow operation (with maximum pump station discharge pressure), steady state only.

Modeling of the Cushing Extension was however based on the modeling of the entire Keystone XL pipeline system (from Hardisty to Nederland, Texas) for determining the pipeline capacity and other operating parameters/data (under given conditions) and then focusing only on the Cushing Extension for the required simulations.

6.6.5.1 Fluid Properties

Fluid properties important to hydraulic simulation are viscosity and density (API gravity) at given reference temperatures.

Generally, the Keystone XL pipeline will transport Heavy DilBit with batches of Synthetic Crude Oil (SCO).

DilBit is abbreviated for "diluted bitumen", meaning bitumen blended with naphtha or condensate or light crude oils. The diluents are added to create a mixture that can be transported by pipeline (i.e., reduce viscosity).

Synthetic crude oil (SCO) is the processed product from a bitumen/heavy oil upgrader facility.

Figure 6-25 illustrates kinematic viscosities for Dilbit. Typical densities of DilBit are as follows:

$$952 \text{ kg/m}^3 \text{ at } 0^\circ\text{C}$$

$$922 \text{ kg/m}^3 \text{ at } 37^\circ\text{C}$$

The properties for SCO depend on the processes used in the upgrading. Typical values are:

- Viscosity

10°C	4.29 cSt
20°C	3.30 cSt
30°C	2.83 cSt
40°C	2.29 cSt
45°C	2.10 cSt

- Density

812.9 kg/m^3 at 15°C

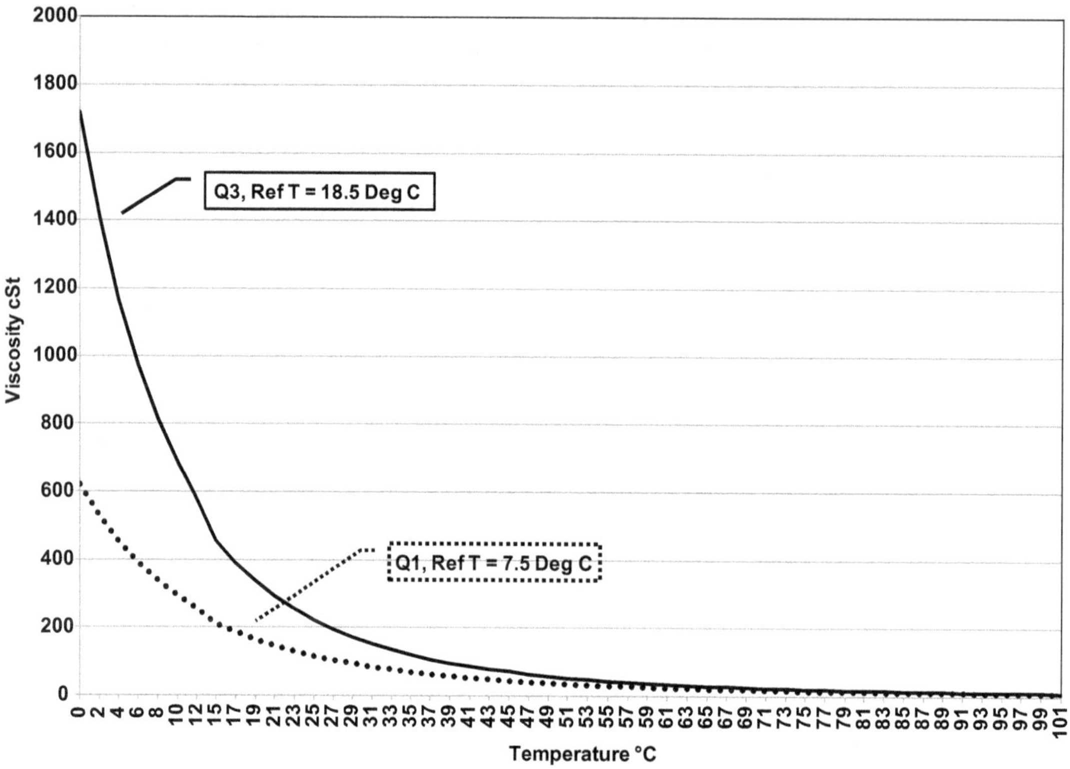

Figure 6-25. Viscosity of heavy Dilbit (typical 1st quarter and 3rd quarter)

6.6.5.2 Simulation Model and Data

Hydraulic models for pipeline systems must be such that steady state and dynamic simulation of the pipeline system for transportation of Heavy DilBit as well as a batched volume of Synthetic Crude Oil (SCO), for comparing alternatives and dynamic effects due to equipment operation (mainline block valve, pump stations and pump units) would be possible.

The simulation model, which considered the stipulated characteristics and facilities (supply and delivery, pump stations (PS), valves (BV), regulators, etc.), was developed and implemented using Energy Solutions International Pipeline Studio, "TLNET" and is based on the route profile and facilities location information shown in Figure 6-26.

For the purpose of simulation typical M&J Series 303 slab gate valves were used as typical mainline block valves. Valve characteristics (Cv 20,000) are shown in Figure 6-27.

Centrifugal pumps (designed for heavy crude oil service operating at 16°C, specific gravity 0.94) were incorporated in the model:

- Pump (Nuovo Pignone), Model DVS, single stage, refer to Figure 6-28 for pump curve.
- Size — 30 × 29
- Rated flow —7250 m^3/hr
- Flow at BEP — 6290 m^3/h
- Differential head /pump — 216.5 m
- Rated speed — 1790 RPM
- Power — 4.885 MW (at Sg =1)

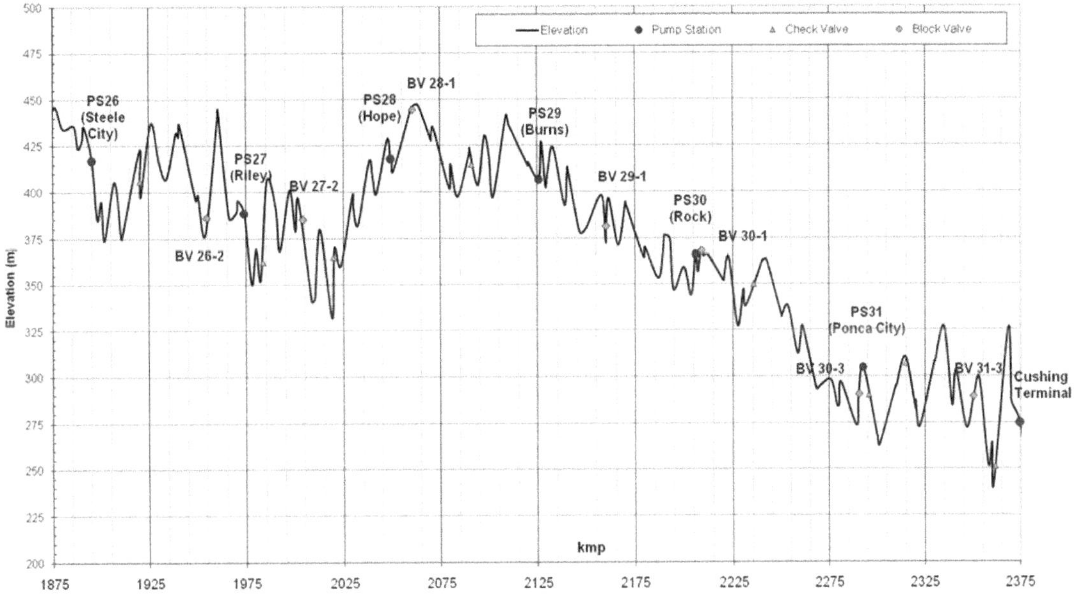

Figure 6-26. Steele City–Cushing pipeline segment elevation profile and facilities locations

- NPSH R (3%) — 23.8 m
- Efficiency at rated flow — 87.6%
- Best efficiency Point (BEP) — 89.1%
- MCSF — 3089 m^3/h

Ground temperature profile (at pipeline burial depth) utilized for hydraulics modeling is shown in Figure 6-29.

Pipeline Facilities Location Along Steele City Cushing Segment of KXL Pipeline and TLNET Model of the pipeline are, respectively, shown in Figures 6-30 and 6-31.

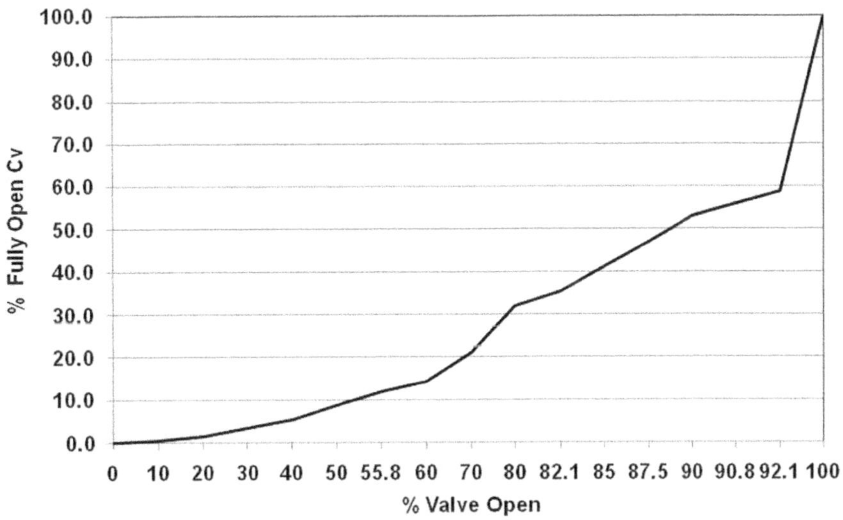

Figure 6-27. C-2 M&J valve characteristic used as input into hydraulics modeling NPS 36, (CV 20,000)

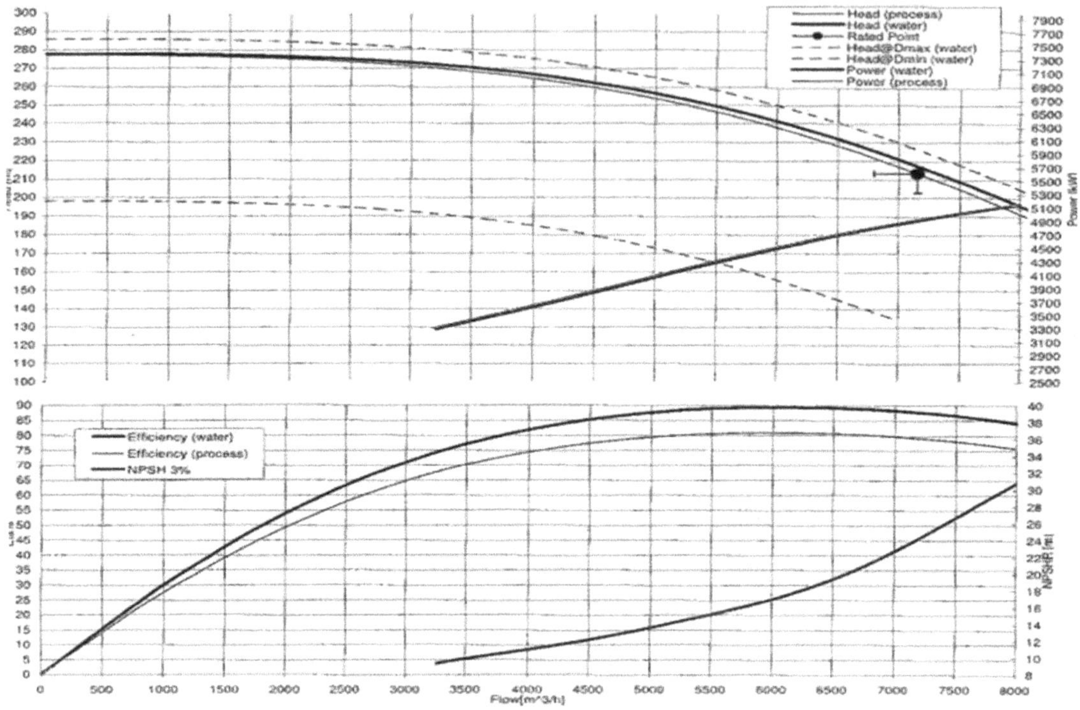

Figure 6-28. Pump curves for Nuovo Pignone Model DVS

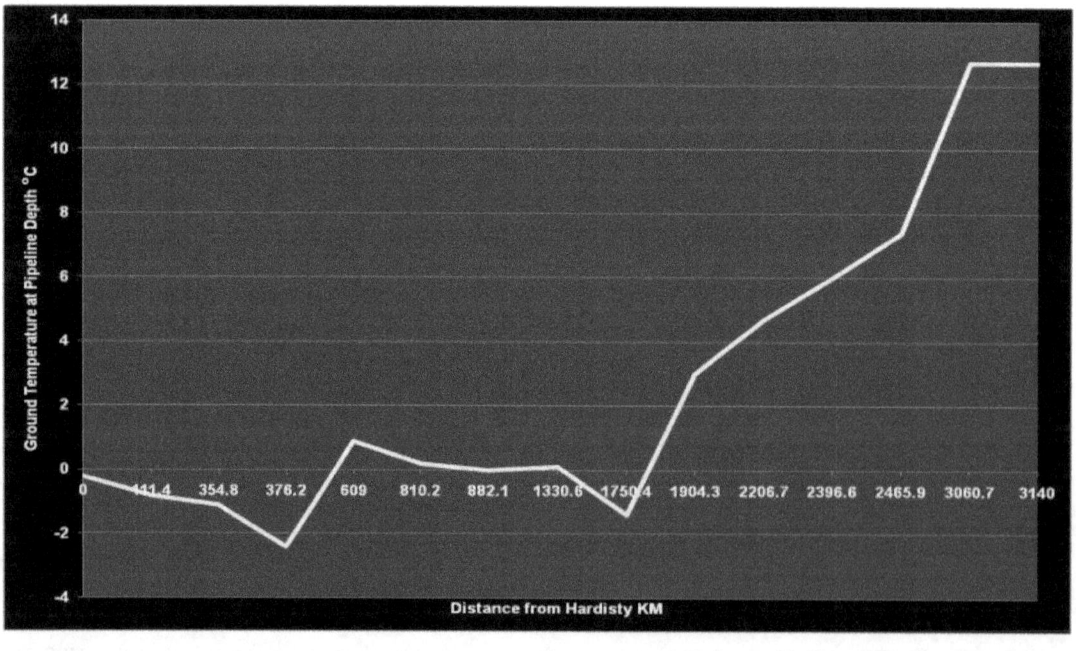

Figure 6-29. Pipeline ground temperature profile (winter, at 1 m pipeline depth)

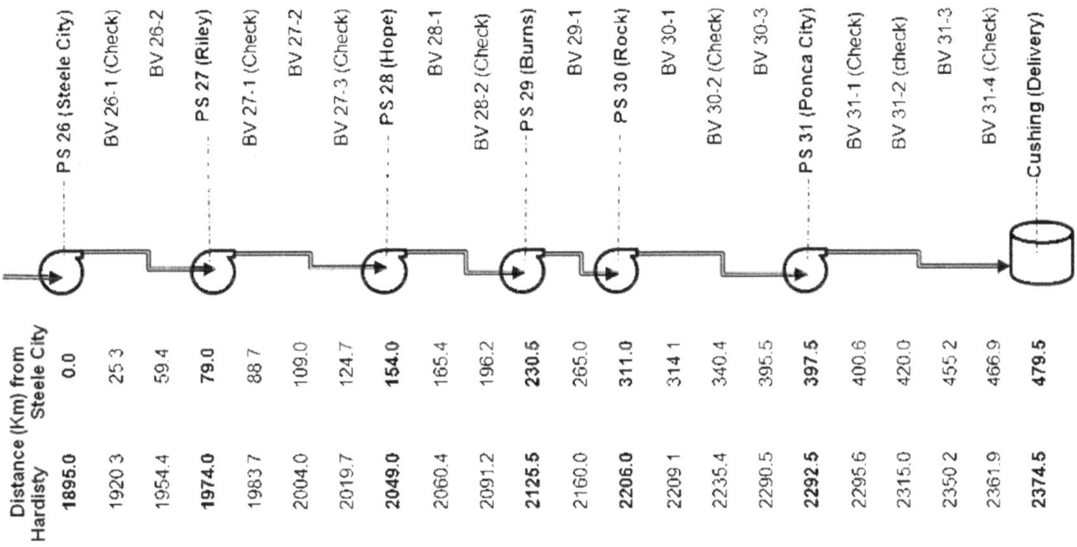

Figure 6-30. Pipeline facilities location

6.6.6 Batch Movement/Transient Simulation Time

For various transient analyses, emphasis was to be made from the effect of a volume of Synthetic Crude Oil batch (32,000 m^3) transported between batches of DilBit. As the Synthetic Crude Oil (SCO) batch is lighter than Dilbit, it is possible that the pipeline would experience lower frictional losses (or lower pressure drops) when such a batch is transported. Dynamic effects (due to pressure surges) caused by mainline block valve closures, pump, or pump station shutdowns could thus be more onerous.

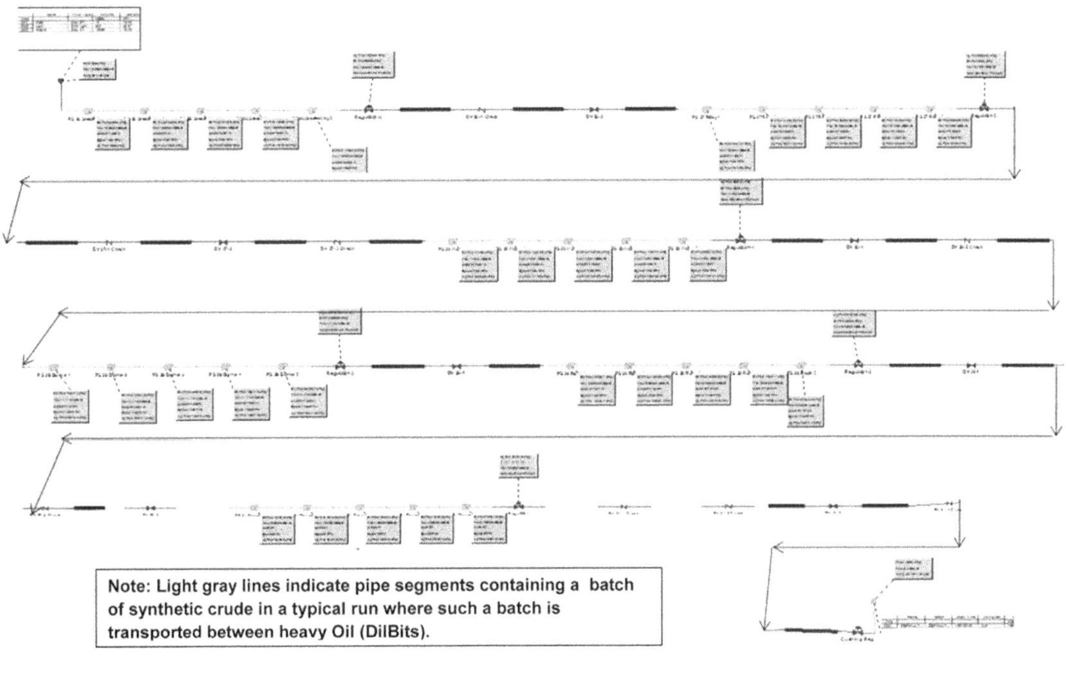

Figure 6-31. TLNET model of keystone pipeline (Steele City–Cushing segment)

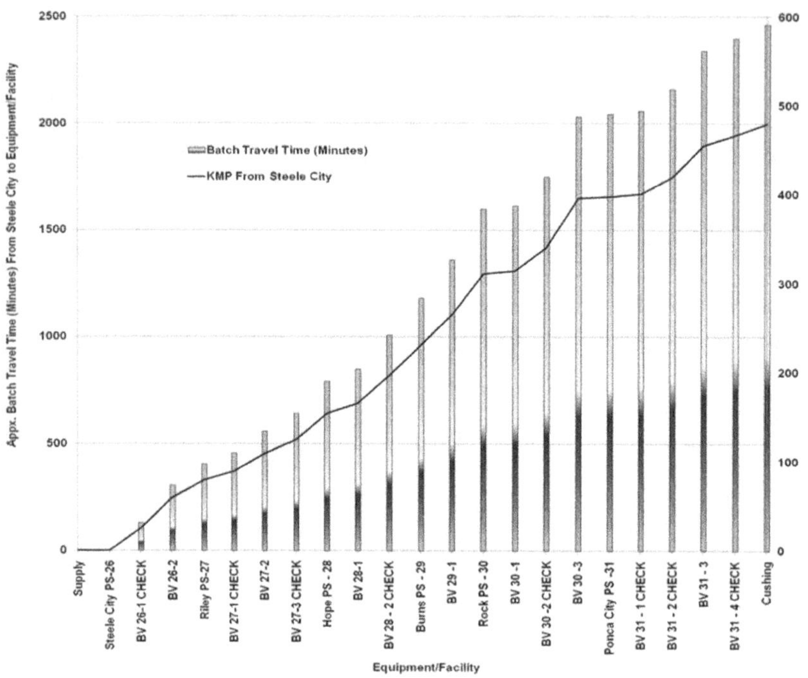

Equipment	KMP
Supply	1895
Steele City PS-26	1895
BV 26-1 CHECK	1920
BV 26-2	1954
Riley PS-27	1974
BV 27-1 CHECK	1984
BV 27-2	2004
BV 27-3 CHECK	2020
Hope PS - 28	2049
BV 28-1	2060
BV 28 - 2 CHECK	2091
Burns PS - 29	2126
BV 29 -1	2160
Rock PS - 30	2206
BV 30 -1	2209
BV 30 -2 CHECK	2235
BV 30 -3	2291
Ponca City PS -31	2293
BV 31 - 1 CHECK	2296
BV 31 - 2 CHECK	2315
BV 31 - 3	2350
BV 31 - 4 CHECK	2362
Cushing	2375

Figure 6-32. Heavy DilBit/synthetic crude oil (SCO) batch travel time cycle

While TLNET can monitor movements of batched products, it is unable to pinpoint the time where a high-pressure event has occurred along the length of the pipeline. However, this can be achieved interactively by observing the pipeline pressure profile over time. Thus, it was decided to simulate the required dynamic situation (e.g., valve closure) when SCO is just arrived before the selected equipment (Figure 6-32).

6.6.7 Simulations Scenarios and Techniques

Dynamic simulation of the pipeline involved first determining optimum time steps for the simulation and subsequently performing the simulation for various scenarios for assessing the dynamic effects caused by pressure surges due to equipment operation (mainline block valve closures, pump stations shutdown/pump unit trips, delivery curtailment, etc.) in a given/stipulated sequence.

6.6.7.1 Time steps and Pipe Segment "Knot Spacing"

Transient (or dynamic) runs in TLNET involves first running a steady-state model followed by dynamic simulation of selected device/equipment, valve, pump, PCV, etc. or transient scenario for the required time. Stability of transient simulation is in selecting small time intervals (as surge pressure waves travel at sonic velocity or speed of sound). Pipeline segments should also further be divided by smaller length sections (knots) to capture the intermediate hydraulic calculations along the pipeline. The smaller the time increment and knot spacing, the more accurate the results. However, as the time increment and knot spacing are decreased, CPU and overall computation time are increased. For a single run with small time steps and knot spacing, the computing time could be in the order of hours.

Based on various run experimentation, a knot spacing of 0.1 km and time interval of 1 second (0.0166667 minutes) was found to provide optimum results (accuracy and least overall run time).

However, where instabilities were observed or results indicated very high-pressure anomalies, the time increment was further reduced to 0.1 second (0.00166667 minutes) and smaller knot spacing implemented as required.

6.6.7.2 Valve Closure and Station Shutdown Timing Sequence

For transient simulations, the following were assumed

- Assumed pump station shutdown time — 1 minute
  (i.e., signal time to complete shutdown 0 to 1 minute)
- Assumed valve closure timing — 3 minutes
- Cushing PCV closure time — <10 seconds

Pump station/unit operations are configured according to shippers nominations and operators delivery schedules. Additional pumps can be brought online or shut down as the flow volume and pressure are required. Pump operation can also be interrupted or tripped due to an unforeseen event.

Mainline block valves are generally kept open except to isolate or direct flow, unless emergencies occur.

The industry rule of thumb for closing mainline valves is in the order of 10 seconds per OD inch valve size. The modeling of 36-in. valves thus would generally require a closure of up to 6 minutes. A more aggressive valve closure time of 3 minutes is also considered, as generally valves could be set to close faster.

With respect to the timing for upstream pump shut downs, the timing interval of 1 minute is on the conservative side. This is because the controls and automation are generally configured such that the upstream pump will trip off once the downstream valve actuator starts to close (in an abnormal situation). One could also trip off the upstream pump station once the valve was completed closing or based on a "sensed" pressure limit on the upstream side of the valve. The final time to pump trip is a matter of choice, and ranges from how much control to give to the pipeline operator versus how much automation are desired. A conservative figure of 1 minute after the valve was closed to begin sequentially tripping off the upstream pumps was selected to be consistent with allowing an operator 1 minute to acknowledge the abnormal closing of a valve and either accept the action or take other measures before the pump trips automatically [18, 30].

6.6.8 Simulation Results

6.6.8.1 Effect of Valve Closures

For a selected valve typical scenario runs included the following:

- Run steady-state simulation
- Determine SCO batch arrival time for the selected valve location
- Run transient simulation. Transient timing simulation started from the time of the arrival of the batch, incremented to the valve closure time (of 3 minutes) followed by closing upstream pump stations sequentially in 1 minute intervals to the final/desired overall simulation time.

Up to 2 days (i.e., 2880 minutes) was simulated as it takes up to two days by a batch cycle to traverse the pipeline between initiating supply and delivery locations, i.e., the Cushing Extension.

For the scenarios simulated, it was evident that generally the pipeline can be protected from overpressurization by effective automation systems (telecommunication and SCADA) and strict operating procedure for implementing timely sequential pump shutdowns upstream of a valve experiencing closure. However, an anomaly was

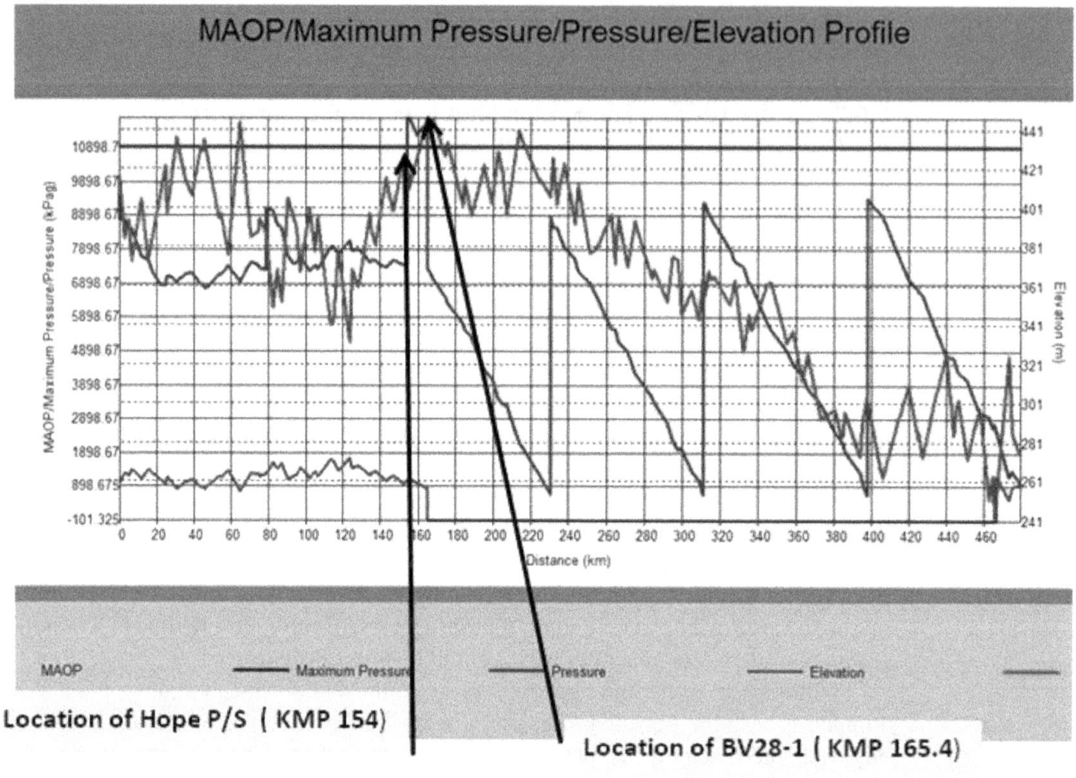

Figure 6-33. Areas and extent of overpressurization as a consequence of closure of valve BV 28-1

observed when closing mainline block valve BV 28-1 situated just 11.4 KM downstream of a Cushing Extension pump station (PS 28, KMP 154) (Figure 6-33).

Fluctuation in surge pressure to the maximum of 11789 kPa can be seen in Figure 6-33 as mainline block valve BV 28-1 is closed and upstream pump stations are shutdown sequentially in 1 minute intervals (Figure 6-34).

It may be noted that the area (i.e., the length 11.4 KM of pipeline affected) is in the highest elevation location with the Cushing Extension pump station (PS 28) at lower elevation than mainline block valve BV 28-1. Also, it may be noted that this segment is filled with the lighter Synthetic Crude Oil (SCO) and is thus more vulnerable to surge pressure fluctuations as anticipated.

To avoid the pressure surges affecting pipeline integrity in this area, any of the following can be adapted.

- Heavier wall pipe (between pump station to BV 28-1);
- Alternative site for BV-2 further downstream;
- Replacement of BV28-1 with a check valve; or
- Hard wiring the BV-28 communication with the upstream pump station (PS 28) PCV for pressure reductions.

6.6.8.2 Effects Due to Pump Stations Shutdown

Each pump station contained five centrifugal pumps in series. Simulation was attempted by shutting down a selected pump station in 1 minute followed by shutdown of all upstream pump stations sequentially in 1 minute intervals.

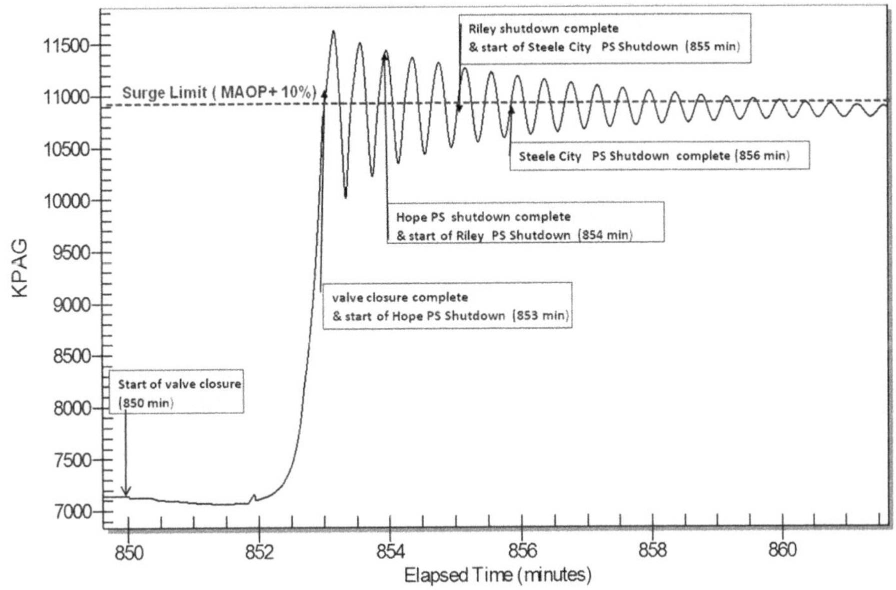

Figure 6-34. Pressure surge fluctuations due to BV 28-1 closure

From many simulation runs, it was noted that under the given/assumed conditions, the pressure surges caused by shutting down any of the pump stations are within the allowable permitted by the code. Highest pressure recorded (10142 kPa) was noted for shutting down Pump Station 28 with the batch just upstream of mainline block valve BV 27-2 (KMP 109) (Figure 6-35). The area corresponds to the location where highest surge pressure was seen as the case for shutting down mainline block valve BV 28-1.

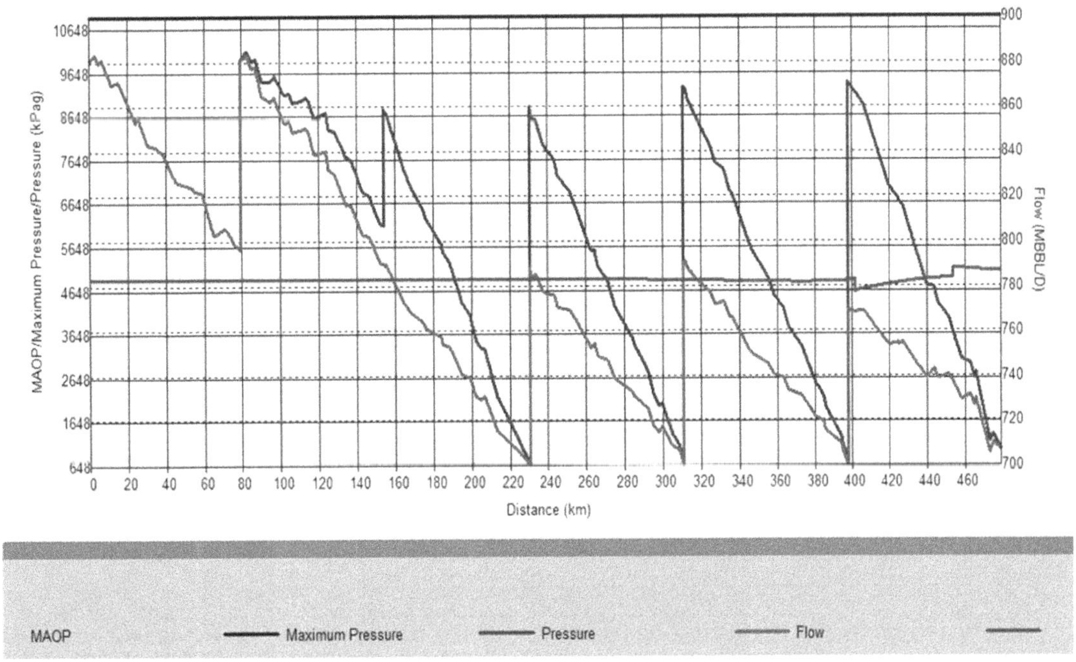

Figure 6-35. Typical example: pressure surges as a consequence of shutting down pump station 28 only (maximum surge pressure 10142 kPa vs. 10923 kPa allowable)

6.6.8.3 Delivery Restriction (Zero Delivery)

Two scenarios were simulated with the Synthetic Crude Oil (SCO) batch just upstream of mainline block valve B31-3 Cushing delivery was reduced to zero flow in 10 seconds.

Scenario 1 covered the simulation run for closing down Cushing Pressure Control Valve (PCV) in 10 seconds followed first by shutting down the 1st upstream pump station in 1 minute and subsequently shutting down of all remaining upstream pump stations sequentially in 1 minute intervals. Results indicate that for the sequence of pump stations shutdown, no undue pressure surges will occur. The maximum pressure that is experienced by the pipeline is 9788 kPa at KMP 397.5.

Scenario 2 considered curtailing the Cushing delivery to zero (PCV closed in 10 seconds) and subsequent closure only of the 1st pump station. The runs indicated the violation of pipeline allowable surge limits (MAOP+10% = 10923 kPa) at a number of locations (Figure 6-36).

6.6.8.4 Terminal PCV Closure

This clearly demonstrates that under such a circumstance, closure of all upstream station must be implemented through an effective SCADA/communication system and strict operating procedures. This procedure (i.e., shutdown of upstream pump stations) can be effectively implemented because an analysis on the timing of the first surge pressure upstream pump station (PS 31) indicates a lapse of about 10 minutes prior to reaching the highest surge pressure of 11218 kPa at KMP 394 after initiation of PCV closure and shutting down the upstream pump station (PS 31).

6.6.8.5 Effect of Minimum Flow Delivery at Maximum Pump Stations Discharge Pressure — Line Packing Conditions

The pipeline segment was simulated for the minimum flow (DilBit only) under line packing condition, i.e., all pump stations discharging to MAOP and their suction pressure fluctuating. This simulation was to observe pressures exceeding MAOP (9930 kPa) along the pipeline and for an effective solution for not exceeding the MAOP under such a condition.

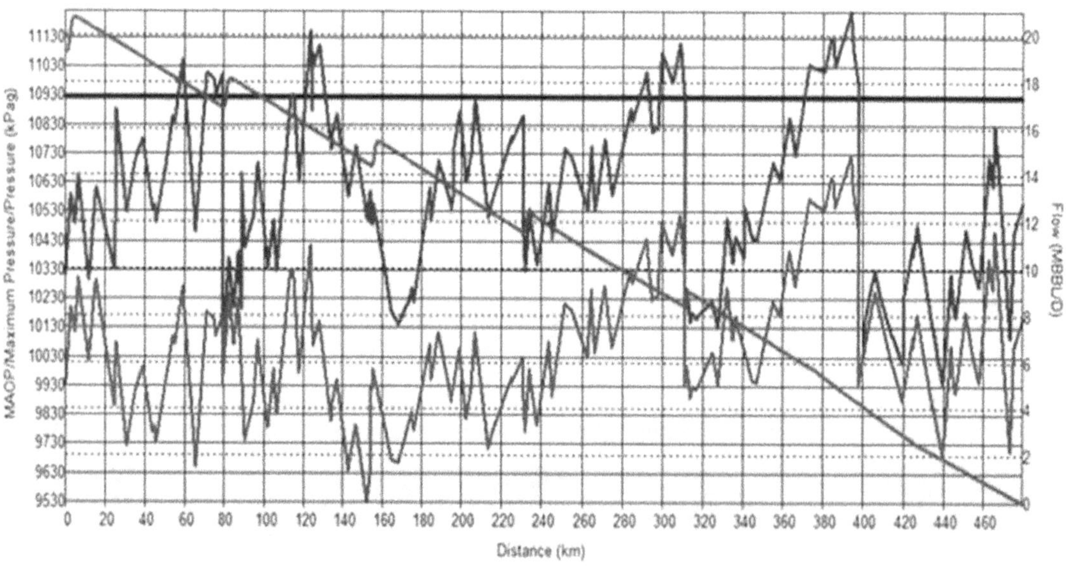

Figure 6-36. Pipeline surge pressure limit violations due to the delivery

High pump station suction pressures were noted when pipeline is line packing under minimum delivery conditions. To avoid such a situation, since the control mechanism can be set to rely on telecommunications, any SCADA system can be set to include a safe failure mode which could be achieved by forcing a discharge pressure setting such that only a desirable line packing pressure gradient can be achieved under minimum flow conditions. Alternatively, and in order not to rely on telecommunication/SCADA, protective controls can be local, such that undesirable high suction pressure would not be reached. In reality, however, when flow is reduced, either there will be a reduction in the number of pump stations or a number of operating units to match the flow.

6.6.9 Conclusion

Steady-state modeling is limited in providing a realistic system response for adequately establishing the system design, capability, and for the determination of pipeline control to avoid undesirable situations caused by undue pressures. The use of transient modeling in the design and operational assessment of a liquid pipeline however provide assurance of the system capability, control, safety, and integrity as well as setting of pipeline controls through effective SCADA systems for avoiding undue pressure surges.

The strategies that were derived from hydraulic transient modeling for the Keystone XL included the following:

- Local Flow-based Station Discharge Pressure Control
- Station Overpressure Protection
- Use of Heavy-wall Pipe at Selective Locations
- Pre-determined Intermediate Mainline Valve Closure Time
- Remote Pressure-based Station Discharge Pressure Control
- SCADA-based Shut-down Commands
- Safe Operating Parameters under Communications Failure Status

ADDENDUM TO CHAPTER 6

Heavy Oil Resources and Recovery Techniques

A6.1 HEAVY OIL RESOURCE BASE

Natural bitumen occurs in clastic (i.e., cemented rocks such as sandstone and shale) and carbonate (i.e., such as limestone and dolomite) reservoir rocks/formations and commonly in small deposits at, or near, the earth's surface. In Canada, specifically in Alberta the formation is mostly of loose sand and sandstone.

The resource base of natural bitumen and extra-heavy oil is immense and can easily support a substantial expansion in production. This resource base can make a major contribution to oil supply, if it can be extracted and transformed into useable refinery feedstock at sufficiently high rates and at costs that are competitive with alternative resources.

Natural bitumen accumulations have been mined since antiquity for use as paving materials and sealants. In some places, such deposits are extremely large, both in areal extent and in the resources they contain, most notably those in northern Alberta,

Canada, in the Western Canada Sedimentary Basin. The three Alberta oil sands areas, Athabasca, Peace River, and Cold Lake, together contain at least two thirds of the world's discovered bitumen in place (1.75 trillion barrels, WEC 2007 estimate [31]) (Figure A6-1).

The U.S. heavy oil resource approaches 100 billion barrels of original oil in-place (OOIP). These resources are concentrated in 248 large reservoirs, holding 80 billion barrels of OOIP, primarily located in California, Alaska, and Wyoming. Numerous other states, such as Arkansas, Louisiana, Mississippi and Texas, also contain significant volumes [32].

Heavy oils, extra heavy oils, and bitumen are of a high strategic importance for OECD (Organization for Economic Co-operation and Development countries) in terms of diversification of oil supply and enhancement of security of supply. General distribution of world heavy oil resources is shown in Figure A6-2. Geographically, major bitumen, extra heavy oil and resources are located in North and Latin Americas as well as the former Soviet Union (Transcaucasia) (Figure A6-3).

As indicated in Figure A6-3, the identified volumes in place of heavy oil, extra heavy oil, and bitumen are estimated at about 4800 GBBLS. About 87% of those resources are represented by oil sands and bitumen in Canada (see Figure A6-3), extra heavy oil in Venezuela, and heavy oil in Russia.

The distribution of the resources is given in Table A6-1. The western hemisphere accounts for about 52% of the world's heavy oil and more than 85% of its natural bitumen. The Middle East and South America have the largest in-place volumes of heavy oil, followed by North America. North and South America have, by far, the largest in-place volumes of natural bitumen. Very large resource deposits (un-estimated) are also known in eastern Siberia [33, 34].

Largest field of crude oil (in place) by type are indicated in Table A6-2. This table indicates the importance of heavy oil and oil sands production, and it may be noted that as much as 70% of oil in place (OIP) is recovered through stream flood [35].

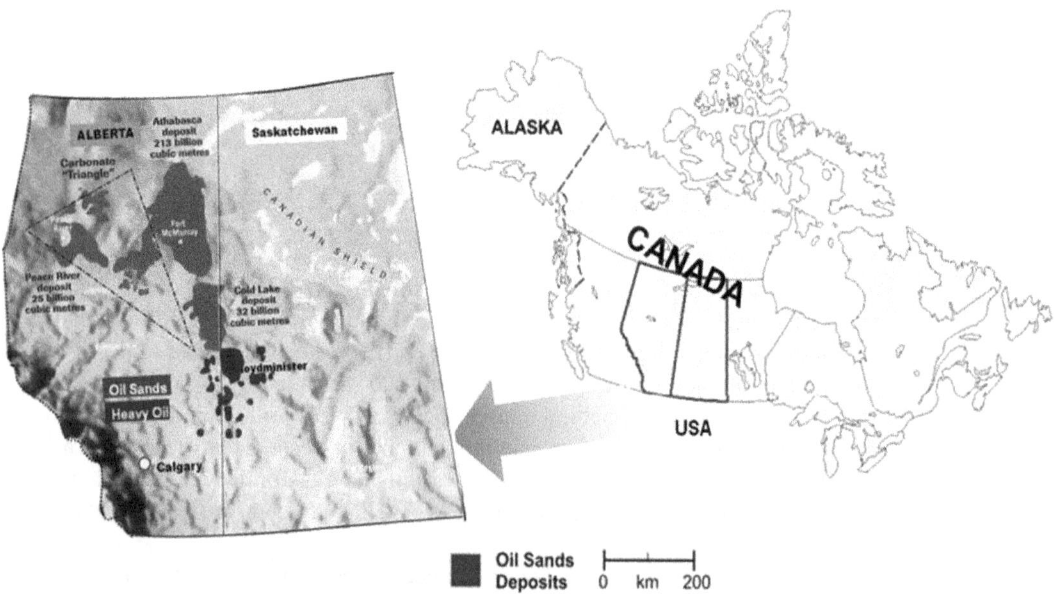

Figure A6-1. Location of major Canadian oil sand and heavy oil deposits [31]

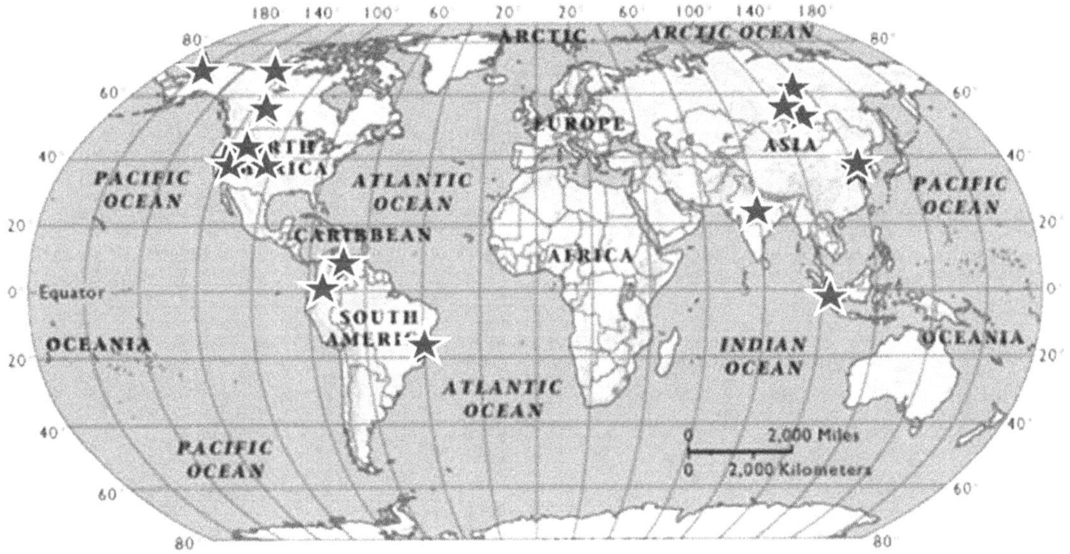

Figure A6-2. General global distribution of heavy oil resources (indicated by stars)

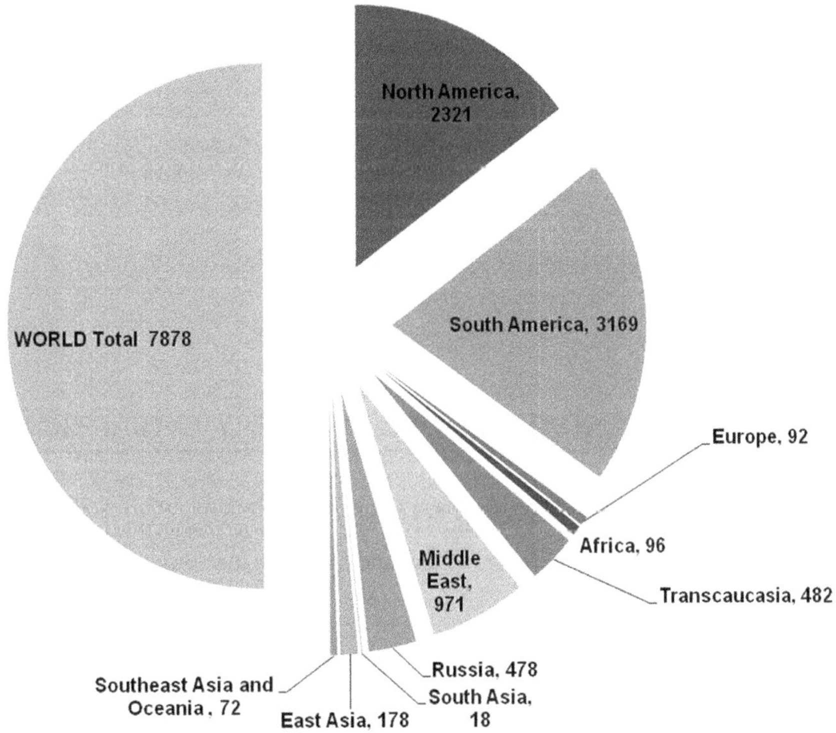

Figure A6-3. Distribution of bitumen, extra heavy and heavy oils resources (GBBLS) [33]

TABLE A6-1. Regional distribution of heavy oil and natural bitumen (billion barrels). Volumes may not add to totals due to independent rounding [33]

Region	Discovered original	Prospective additional	Total
HEAVY OIL			
North America	650	2	651
South America	1099	28	1127
Europe	75	0	75
Africa	83	0	83
Transcaucasia	52	0	52
Middle East	971	0	971
Russia	182	0	182
South Asia	18	0	18
East Asia	168	0	168
Southeast Asia and Oceania	68	0	68
Total	3366	29	3396
BITUMEN			
North America	1671	720	2391
South America	2070	190	2260
Europe	17	0	17
Africa	13	33	46
Transcaucasia	430	0	430
Middle East	0	0	0
Russia	296	51	347
South Asia	0	0	0
East Asia	10	0	10
Southeast Asia and Oceania	4	0	4
Total	4512	993	5505

TABLE A6-2. World's largest reserves of crude oil in place by type (Recreated from ref. [36])

Field	Crude Type	Country	Oil in Place (Billion BBLS)
Orinoco	X-Heavy Oil	Venezuela	1,200
Athabasca	Oil Sand	Canada	869
Cold Lake	Oil Sand	Canada	271
Ghawar	Oil Field	Saudi Arabia	190
Burgan	Oil Field	Kuwait	190
Bolivar Coast	Oil Field	Venezuela	160
Melekess	Oil Sand	Russia	123
Wabasca	Oil Sand	Canada	119
Chicontepec	Unconventional	Mexico	100
Peace River	Oil Sand	Canada	92

A6.2 BITUMEN AND HEAVY OILS RECOVERY/EXTRACTION TECHNIQUES

A6.2.1 Extraction/Recovery Techniques

There are a number of ways and processes that bitumen and heavy oils can be extracted [37, 38].

Generally, the oil recovery/production technologies are divided as follows:

- **Primary Recovery**: through natural pressure of reservoir
- **Secondary Recovery**: Through water and natural gas flooding/injection after subsequent decrease in the reservoir natural pressure (to act as driving force)
- **Tertiary Recovery**: through enhanced technique such as horizontal drilling or injection of thermal (through application of heat (such as heated air/steam), and

in-situ combustion), chemicals (hydrocarbon solvents), carbon dioxide gas all to restore reservoir pressure as well as optimum levels of rock and fluid properties.

- **Other Recovery Techniques** including:
  - Mining
  - Immiscible carbon dioxide and nitrogen
  - Polymer and microbial injection
  - Extraction from unconventional reservoirs (considered as Non-Conventional)
- Tar-sands and Oil Shales (considered as Non-Conventional)

The recovery of bitumen and heavy oils falls through two main methods (Figure A6-4) [39]:

- Surface mining (open pit) method
- In situ processes such as steam injection

However, the recovery of heavy oil, extra-heavy oil, and bitumen resources is categorized based on their field/reservoir, location, environment, and characteristics as follows:

- Shallowest resources (< 50 m)
- Shallow resources (50 to 100 m (average 75 m), too deep for mining but no caprock seal)
- Medium/Intermediate-depth resources (100 to 300 m (average 200 m), caprock seals pressures < 200 psi)
- Intermediate-depth resources (300 to 1000 m, seal for pressure > 200 psi)
- Deep resources (> 1000 m)
- Arctic resources (permafrost)
- Other resources (offshore, carbonate, thinly bedded, highly laminated shales, and hydrates)

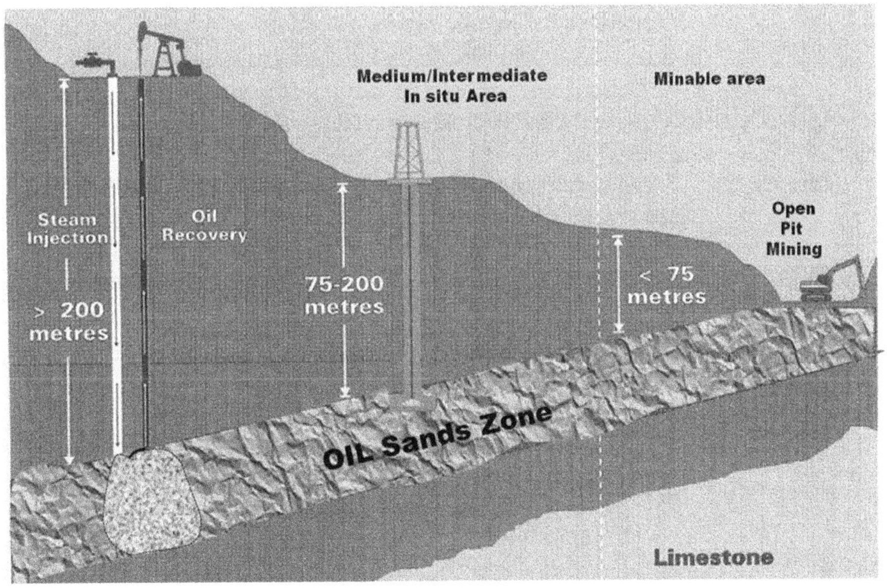

Figure A6-4. Bitumen and extra heavy oil recovery technologies

Surface mining is mostly for oil sands deposits. However, depths up to 120 m are considered to be surface mineable, and deposits from 120 m to 750 m are amenable to *in situ* processes [37]. Originally, the oil sands were mined with draglines and bucket-wheel excavators and moved to the processing plants by conveyor belts. In recent years, companies have switched to lower cost shovel-and-truck operations using the largest power shovels (100 or more tons) and dump trucks (400 tons) in the world.

In situ processes include the following [40, 41]:

- Cold-production horizontal wells and multilaterals
- Water-flood
- Cold heavy oil production with sand (CHOPS)
- Cyclic steam stimulation (CSS)
- Steam flood
- Steam-assisted Gravity Drainage (SAGD)
- Solvent without heat or steam (e.g., Vapex)
- Hybrid: Solvent with heat or steam (SAGD, CSS and steam-flood wells or Expanding Solvent SAGD (ES-SAGD))
- Toe to heel air injection (THAI)- hot air
- Fire flood with vertical wells (~20 °API oil only)
- In-situ combustion or fire flood with vertical and horizontal wells
- Down-hole steam generation (CSS, flood, SAGD)
- Electric, induction or RF heating
- Gas Injection I including supercritical fluids (e.g., CO_2 from flue gases), Nitrogen (N_2) and, NGL.
- Biotechnology (such as polymer and Microbiological Enhanced Oil Recovery (MEOR) techniques) [42].

Cyclic steam stimulation (CSS) is a well-developed process; however, its major limitation is that typically less than 20% of the initial oil-in-place (IOIP) can be recovered. With steam-assisted gravity drainage (SAGD) technique more than 50% of the oil in-place can be recovered [43].

The CSS is predominantly a vertical well process, with each well alternately injecting steam and producing bitumen and steam condensate. It can also be applied through horizontal wells. The heat injected warms the bitumen and lowers its viscosity. A heated zone is created through which the warmed bitumen can flow back into the well (Figure A6-5).

The development of horizontal and multi-lateral well drilling techniques has enhanced the in situ processes since such drilling techniques provide both greater reservoir access and the development of novel recovery processes based on gravity drainage mechanisms. Also, improvements in reservoir characterization tools (for example, 3-D Seismic) have further enhanced horizontal well technology by allowing for accurate placement and location of wells [44].

The challenges of horizontal well drilling include:

- Limitations to drill in shallow oil sands deposits
- Limitation in formation evaluation particularly in bitumen and heavy oil fields due to higher logging, coring, and seismic costs
- Complication in operations of horizontal wells because there is less control of fluid entry over the length of the well. Remedial action is more complex and costly

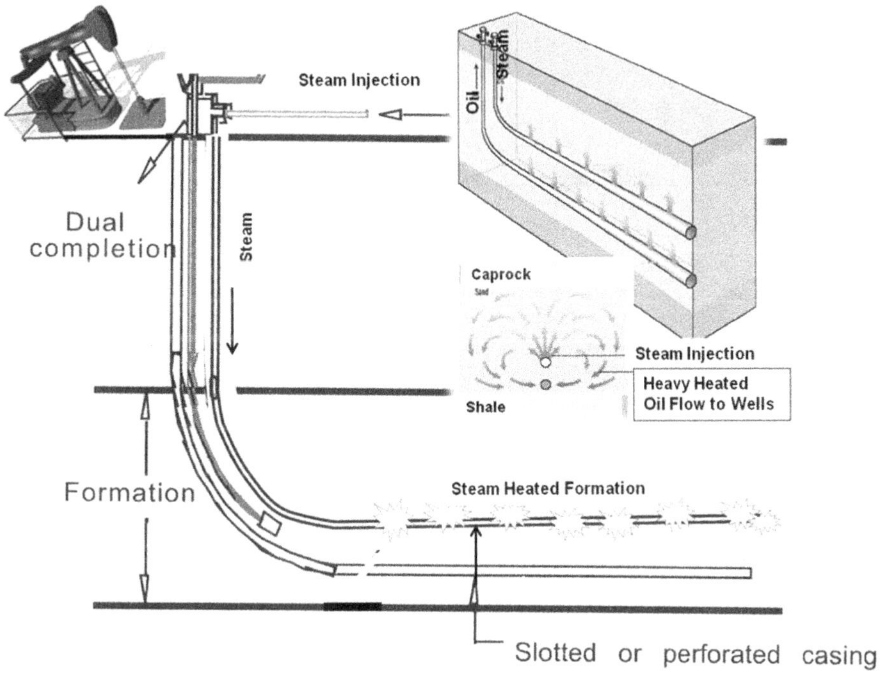

Figure A6-5. Heavy oil recovery technique through steam injection (from refs. [37], inset from refs. [44, 45])

A6.2.2 Production Techniques Scope

As indicated previously, heavy oil recovery can generally be through:

1. Cold Production
2. Warn Cold Production (including Cyclic Steam Simulation (CSS) and SAGD)
3. Warm Vapor Extraction (VAPEX), solvent dilution – N-Solv
4. Top-Down Combustion
5. Thermal Gravity Process (steam-assisted gravity drainage (SAGD))
6. In situ hydrogenation, pyrosys (Fireflood)
7. Other EOR (Enhance Oil Recovery) Techniques

Cold Production involves the co-production of heavy oil and sand and is a commercial process whereby oil and sand are pumped to the surface. This only works well in areas where the oil is fluid enough to be pumped. In cold production, production rates are often low, with marginal recovery factors and high water cuts. Allowing sand to be produced along with the oil helps and is known as CHOPS (Cold Heavy Oil Production with Sand).

The process is mostly applied to lower viscosity (higher gravity) heavy oils with some mobility and results in the development of high-permeability channels (called "wormholes") in the adjacent low cohesive strength sands. Sand transport is facilitated by the flow of "foamy oil" caused by solution gas drive (Figure A6-6). The main mechanism involves foamy oil flow, sand failure, and sand transport from wormhole tip to the well. This technique allows a higher recovery of oil from the sand (up to 10%) but results in disposal issues with the residual sand [37].

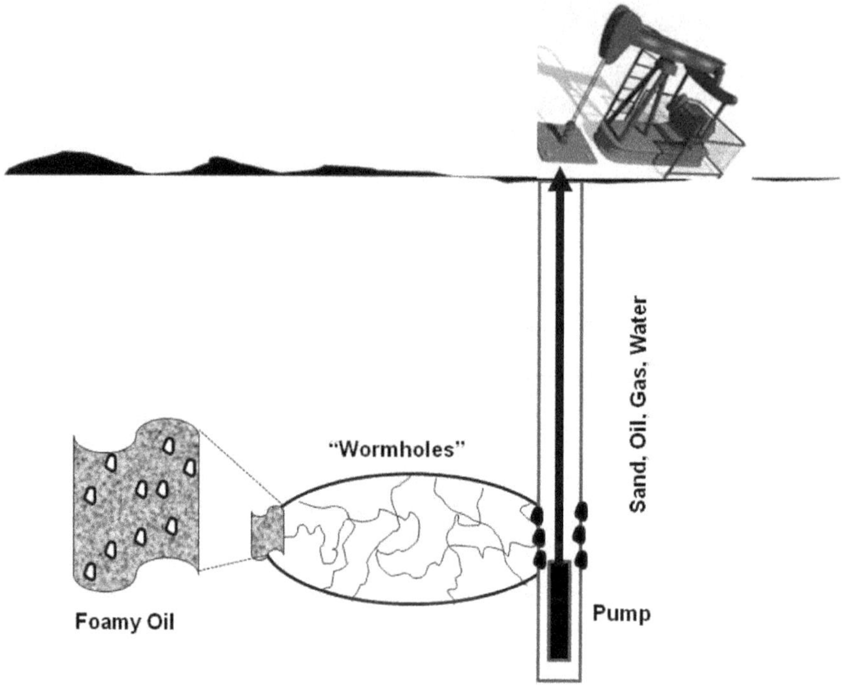

Figure A6-6. Cold simultaneous production process of sand and oil using solution gas (from ref. [37])

Viscosity reduction can be achieved by thermally reducing the viscosity (heating), by mass transfer (solvent dilution), or by molecular scission (pyrolysis, *in-situ* hydrogenation, etc.). Thermal methods cover warm/cold production and all the thermal process (items 2 through 5 listed above) and include steam-assisted gravity drainage (SAGD) [46, 47], cyclic steam stimulation (CSS), and in-situ top-down combustion. The in situ combustion method begins with the injection of heated air into heavy oil reservoir. Heat is generated as a result of oil oxidation, thus increasing the temperature. With the continuation of the oxidation process, the temperature reaches the "ignition point" when combustion occurs. Cold air is then injected to continue the process. The combustion front thus displaces any trapped fluids in the reservoir up through the producing well.

The viscosity of heavy oil that is in the reservoir can be reduced by dilution or the mass transfer of a light hydrocarbon solvent into the heavy oil as in the case of vapor extraction (VAPEX) or NSolv. VAPEX is considered the solvent analog to SAGD). Both NSolv and VAPEX involve injecting a solvent into a heavy oil reservoir to reduce the viscosity of the heavy oil via mass transfer. The solvent-enhanced live oil then drains via gravity drainage and is produced through a lower horizontal production well [44].

A typical representation of a thermal and solvent recovery method using horizontal wells is indicated in Figures A6-7 and A6-8.

Solvent or steam is injected into the upper well where the following actions can assist the recovery of heavy oil/bitumen:

> ***A. Steam Injection***: the steam condenses on the cold bitumen surface, reducing the viscosity of the bitumen through heat conduction (steam assisted gravity drainage — SAGD, or other thermal steam processes). Steam-Assisted

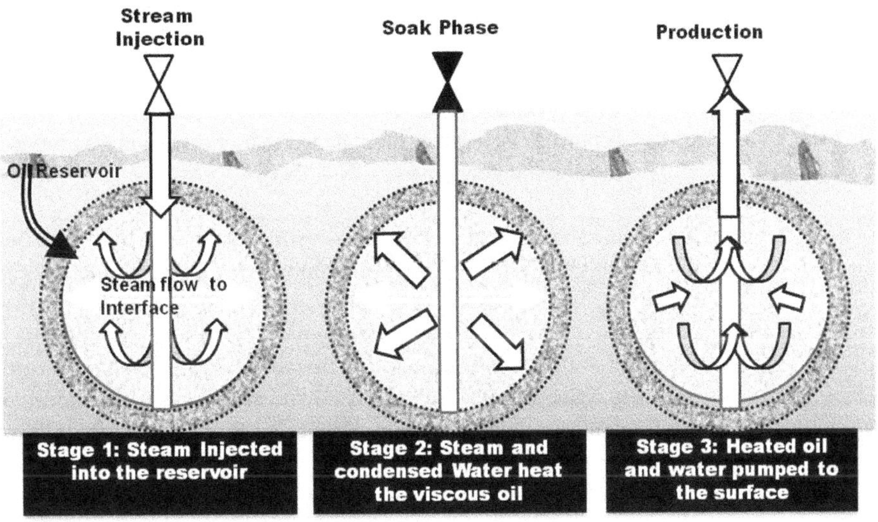

Figure A6-7. Thermal recovery process through steam injection (known as huff n'puff [48]

Gravity Drainage (SAGD) generally uses paired horizontal wells. Steam, which is continuously injected through the upper well, creates a steam chamber along the walls of which the heated bitumen flows and is produced in the lower well (Figure A6-8). Several variations of this process have been developed. One variation uses a single horizontal well, with steam injection through a central pipe and production along the annulus. Another variation involves steam injection through existing vertical wells and production through an underlying horizontal well.

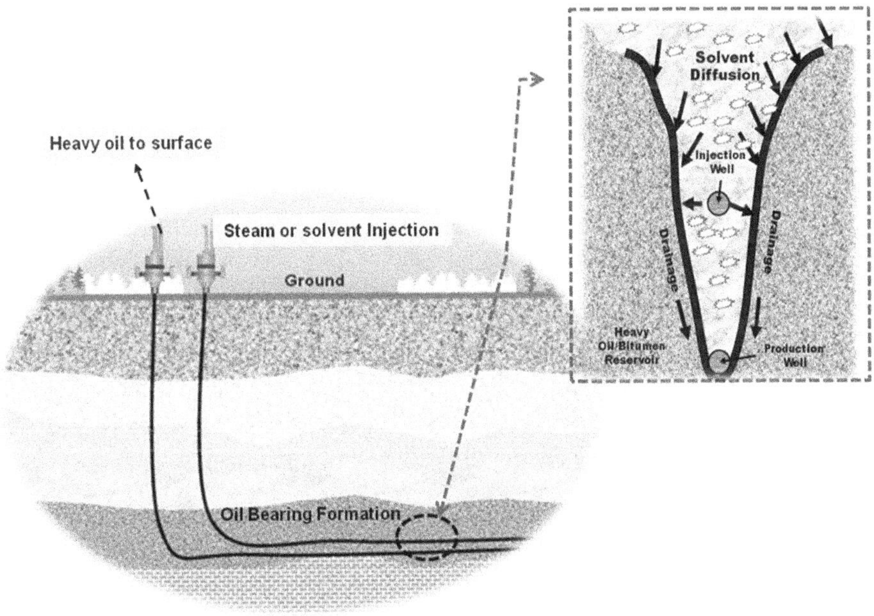

Figure A6-8. Typical solvent based method for the recovery of heavy oil through two horizontal wells (from ref. [44])

B. Vapor Assisted Extraction: if solvent is used, solvent vapor diffuses into the bitumen and reduces the viscosity (solvent processes operated in the vapor phase such as Vapor Extraction – VAPEX) or
C. Solvent dilution: solvent condenses on the bitumen interface and reduces the bitumen viscosity through heat conduction and mass transfer (when operated so the solvent condenses — N-Solv)

In all above situations, the viscosity reduced mobile "live heavy oil" is gravity drained to the bottom of the formation zone into the production pipe. It may be noted that the growth of steam or the solvent chamber is generally upwards first to the top of the heavy oil "pay zone" during what's called the chamber rising phase. Then, the chamber starts to spread laterally outwards sweeping the oil bearing formation as the viscosity reduced heavy oil drains downwards. The pores drained of oil become filled with steam or solvent (steam or solvent chamber) which grows laterally in time (chamber spreading phase). Finally, when the solvent/steam chamber reaches the extent of the oil formation, the height of the oil filled pores decreases during the chamber falling phase, as indicated in Figure A6-8 inset, [44].

A6.2.3 Recovery Techniques Summary

Table A6-3 provides a summary of production techniques for recovery of heavy, extra heavy and bitumen [40].

A6.2.4 Oil Reservoir Classifications

Oil reservoirs are classified on the basis of the location of the point representing the initial reservoir pressure $\mathbf{P_i}$ and temperature $\boldsymbol{T}$ with respect to the pressure-temperature diagram of the reservoir fluid. This diagram is also known as phase diagram shows equilibrium temperature-pressure relationships for different phases of a substance (Figure A6-9).

Depending upon initial reservoir pressure $\mathbf{P_i}$, oil reservoirs are generally subclassified as [3]:

1. ***Under-saturated oil reservoir.*** If the initial reservoir pressure $\mathbf{P_i}$ is greater than the bubble-point pressure $\mathbf{P_b}$ of the reservoir fluid (Point 1, Figure A6-9), the reservoir is classified as under-saturated oil reservoir. This is also known as dead oil-black oil.
2. ***Saturated oil reservoir.*** When the initial reservoir pressure is equal to the bubble-point pressure of the reservoir fluid, (Point 2, Figure A6-9), the reservoir is called a saturated oil reservoir.
3. ***Gas-cap reservoir***. When the initial reservoir pressure is below the bubble point pressure of the reservoir fluid (Point 3, Figure A6-9), the reservoir is termed a gas-cap or two-phase reservoir, in which the gas or vapor phase is underlain by an oil phase. The appropriate quality line gives the ratio of the gas-cap volume to reservoir oil volume.

When the reservoir pressure lies anywhere along line 1 → 2 (Figure A6-9), the oil could dissolve more gas if more gas were present. If the pressure is at 2, the oil contains the maximum amount of dissolved gas and can't hold any more gas. A reduction in pressure at this point will release gas to form a free gas phase inside the reservoir. Additional gas evolves from the oil as it moves from the reservoir to the surface. This causes some shrinkage of the oil.

TABLE A6-3. Heavy oil production techniques versus location resources summary [40]

Production Technique	Open-pit maining	Cold Production(Horizontal and Multilateral wells)	Water Flood	Cold Production with Sand	Cyclic Steam Simulation (CCS)	Steanflood	SAGD	Solvent Only	Solvent With heat/steam	Fire Flood(Vertical wells(20° API oil))
Usage Stage	Commercial	Commercial	Commercial	Commercial	Commercial	Commercial	Commercial	Pilot test	Pilot test	Commercial
Shallowest (<50 m)	Only solution	No	No	No	No	No	No	No	No	No
Shallow (50 to 100 m)	Limited economically	No	No	No	No	No	No	Possible, but unproven	Unknown	No
Medium depth (100 to 300 m)	No	Only for very low viscosity or high solution gas& high permeability	Only for very low viscosity and high permeability	Unlikely, may require solution gas, but may be possible	No, unless good sealing caprock	No, unless good sealing caprock	For good vertical and horizontal permeability,payzone > 10m	Unproven, needs good vertical and horizontal permeability	Unproven, needs good vertical and horizontal permeability	Possible
Intermediate depth (300 to 1,000 m)	No	Only for very low viscosity or high solution gas& high permeability	Requires low viscosity and/or high formation temperature	Requires unconsolidated formation and solution gas	Yes, deep zones require high temperature steam & are less economic	Yes, but deep zones need higher temperature steam & are less economic	Yes, deep zones require high temperature steam & are less economic	Unproven, needs good vertical and horizontal permeability	Unproven, needs good vertical and horizontal permeability	Yes
Deep (>1,000 m)	No	Requires low viscosity with solution gas or high formation temperature and high permeability	Requires low viscosity and/or high formation temperature	Unlikely because requires unconsolidated formations	No, needs high temperature and high-pressure steam and too much heat losses to overburden through injection wellbore	No, needs high temperature and high-pressure steam and too much heat losses to overburden through injection wellbore	No, needs high temperature and high-pressure steam(too much heat losses to overburden through injection wellbore)	Possible, but unproven	Unknown	Possible
Arctic	No	Maybe	Maybe	Disposal of sand and water an issue	Maybe if permafrost can be managed	Maybe if permafrost can be managed	Maybe if permafrost can be managed	Possible, but unproven	Unproven, must manage permafrost issue	Possible, but unproven
Offshore	No	Maybe	Yes, North Sea	Disposal of sand and water an issue	No, too much heat loss in riser to ocean water	No, too much heat loss in riser to ocean water	No, too much heat loss in riser to ocean water	Possible, but unproven	Unlikely	Possible, but unproven
Carbonates	No	No	No	No	No	No	No	Unknown	Unknown	Unknown
Thin beds (<10 m thick)	Can be mined if near surface and thin overburden	Maybe	Maybe	Yes	Possible with horizontal wells	No, needs at least 10 m bed, heat losses to overburden are too great	No, need at least 10 m bed	Possible, but unproven	Possible, but unproven	Unknown
Highly laminated	Mined if near surface, thin overburden	Yes, if multilaterals can penetrate multiple layers	Maybe for vertical wells	Yes	Possible with horizontal wells	May be possible with horizontal wells, but unproven	No, need at least 10 m bed	Unlikely	Unknown	Unknown

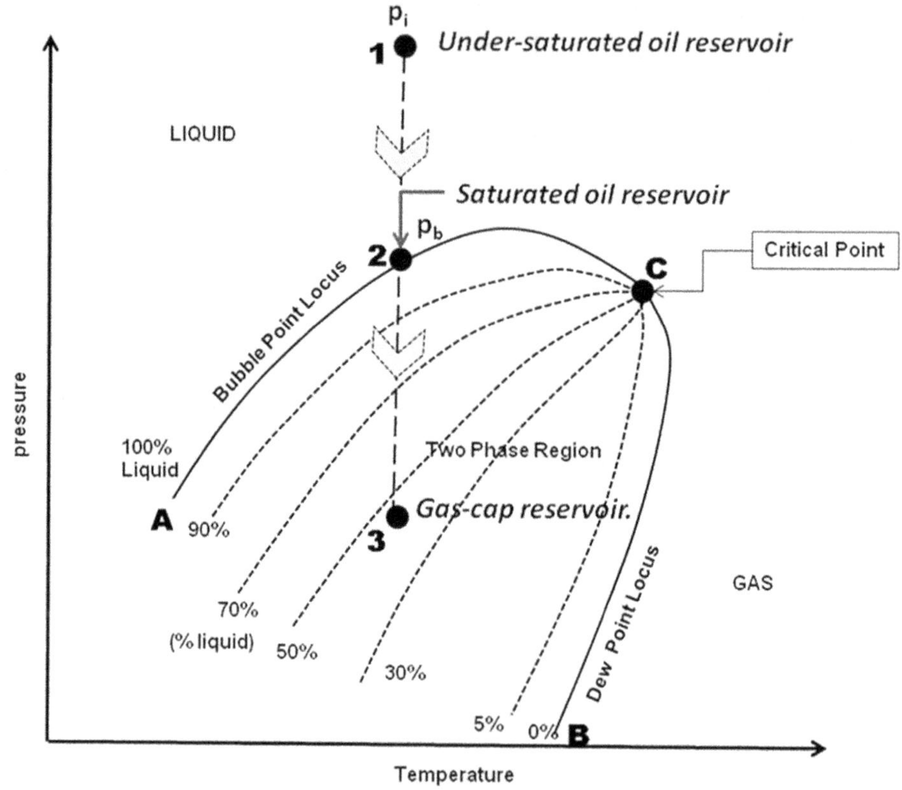

Figure A6-9. A typical phase (p-T) diagram for a crude oil (from [3])

REFERENCES

[1] Barillas, J. l. M., Dutra, T. V., and Mata, W., 2008, "Improved Oil Recovery Process For Heavy Oil: A Review," *Brazilian Journal of Petroleum and Gas*, **2**(1), pp. 45–54.

[2] Alboudwarej, H., 2006, "Fluid Property Measurement: A Challenge for Heavy Oil," *Heavy Crude Oil Symposium*, Galveston, TX, USA, Oct. http://webdelprofesor.ula.ve/ingenieria/mabel/HeavyCrudeTransport.pdf.

[3] Ahmed, T., 2000, "Reservoir Engineering Handbook," 2nd edition, Gulf Professional Publishing, Houston TX, USA.

[4] Fekete Harmony™, 2011, "Reservoir Fluid Types," http://www.fekete.com/software/feketeharmony/media/webhelp/Harmony/Reservoir_Fluid_Types.htm.

[5] Bennison, T., 1998, 'Prediction of Heavy Oil Viscosity," *Presented at IBC Heavy Oil Development Conference*, London, UK, Dec 2–4. http://www.ecltechnology.com/subsur/reports/pvt_tgb.pdf.

[6] Bergman, D. F., and Sutton, R. P., 2009, "A Consistent and Accurate Dead-Oil-Viscosity Method," *Society of Petroleum Engineers Reservoir Evaluation & Engineering* (*SPE Res Eval & Eng*), 12(6), pp. 815–840. SPE-110194-PA. doi:10.2118/110194-PA.

[7] Tovar, J., Salazar, A., and Salzar, N., 2006, "The Impact of Non-Newtonian Fluid Behavior on Well Performance For the Orinoco Belt Reservoir," *SPE/IBP Workshop on Artificial Heavy Oil Offshore*, Armaca de Buzios, Brazil, May 28.

[8] Saniere, A., Henaut, I., and Argillier, J. F., 2004, "Pipeline Transportation of Heavy Oils, a Strategic, Economic and Technological Challenge," *Oil & Gas Science and Technology, IFP (Institut France du Petrole)*, **59**(5), pp. 455–466.

[9] Galvin, J., 2006, "A New Approach to Heavy Oil and Bitumen Upgrading," *Heavy Oil Symposium*, Galveston, TX, Oct.

[10] Veith, E., 2006, "Releasing the Value of Heavy Oil and Bitumen: HTL Upgrading of Heavy to Light Oil," *Proceedings of 1st World Oil Conference (WHOC)*, Paper # 2006 727.
[11] Winter, C. H, and Glowach, A. M., 2002, "High Temperature Insulated Coating and Construction Methodology for the Mackay River Pipeline," *Proceedings of ASME, IPC (International Pipeline Conference)*, September 29–October 3, 2002, Calgary, Alberta, Paper # 27318.
[12] NEERI (National Environmental Engineering Research Institute), 2010, "Mangala terminal to Salaya Terminal Heavy Oil Pipeline Project," http://pipelinesinternational.com/news/heating_up_in_india_the_mangala_to_salaya_oil_pipeline/040184/.
[13] Mohitpour, M., 1991, "Temperature Computation in Fluid Transmission Pipelines," *ASME, PD*, **34**, Pipeline Engineering Boo # G00587.
[14] Briceno, M. I., 2006, "Heavy Crude Oil Pipeline Transportation," http://webdelprofesor.ula.ve/ingenieria/mabel/HeavyCrudeTransport.pdf.
[15] Layrisse, R., 1998, "Viscous Hydrocarbon-In-Water Emulsions," U.S. Patent 4,795,478.
[16] Bensakhria, A., Peysson, Y., and Antonini, G., 2004, "Experimental Study of the Pipeline Lubrication for Heavy Oil Transport," *Oil & Gas Science and Technology – Rev. IFP*, **59**(5), pp. 523–533.
[17] Al-Roomi, Y., George, R., Elgibaly, A., and Elkame, A., 2004, "Use of a Novel Surfactant for Improving the Transportability/Transportation of Heavy/Viscous Crude Oils," *Elsevier, Journal of Petroleum Science and Engineering*, **42**, pp. 235–243.
[18] Mohitpour, M., Van Hardeveld, T., Peterson, W., and Szabo, J., 2010, *Pipeline Opeartion and Maintenance – A Practical Approach*, ASME Press, New York.
[19] Moody, L.F., 1944, "Friction Factors for Pipe Flow," Transaction of ASME, 66, p. 671.
[20] Bensakhria, A., Peysson Y., and Antonini G., 2004, "Experimental Study of the Pipeline Lubrication for Heavy Oil Transport," Oil & Gas Science and Technology - Rev. IFP, Vol. 59, No. 5, pp. 523–533.
[21] Crude Monitor, 2005, "Grades and Types of Crudes from the Western Canadian Sedimentary Basin," www.CrudeMonitor.ca, Dec 16.
[22] HPC (Hydrocarbon, Publishing Company), 2008, "Opportunity Crudes Report II: Technology and Strategies for Meeting Evolving Market and Environmental Challenges," http://www.hydrocarbonpublishing.com/ReportP/Prospectus-Opportunity%20Crudes%20II_2011.pdf.
[23] Anand, A., 2004, "Synthetic Crude Logistics in the Enbridge System," http://www.coqa-inc.org/20040129Enbridge.pdf.
[24] Cabrejo, V., and Mohitpour M., 2010, "Transient Flow Assurance for Determination of Operational Control of Heavy," Proc. 8th ASME International Pipeline Conference, IPC 2010," Calgary Alberta, Sept 27–Oct 1.
[25] Mays, L. W., 2000, *Hydraulic Transient Design for Pipeline Systems*, McGraw Hill Companies Inc, http://www.digitalengineeringlibrary.com/dxreader/opendxreader.asp?chapterid=p2000aed999706_1001.
[26] Van Vuuren, S. J., 2001, "Theoretical Overview of Surge Analysis," University of Pretoria, South Africa, http://www.up.ac.za/academic/civil/divisions/water.html.
[27] Mohitpour, M., Thompson, W., and Asante, B., 1997, "The Importance of Dynamic Simulation on the Design and Optimization of Pipeline Transmission Systems," *Proc. of ASME 1st Int. Pipeline Conf.*, Vol. 2, p.1183.
[28] Mohitpour, M., Kazakoff, J., and Brittin, R., 1998, "Gas Pipeline Design for Operational Reliability," *Proc. Pemex 3rd Congreso y Expo Internacional de Ductos*, Monterrey, Mexico, Dec 7–9.
[29] Mohitpour, M., Golshan H., and Murray, A., "Pipeline Design and Construction – A Practical Approach," 3rd Edition, ASME Press, New York.
[30] Mohitpour, M., Trefanenko, B., Tolmasquim, S. T., and Kossatz, H., 2004, "Valve Automation to Increase Oil Pipeline Safety," *ASME 5th International Pipeline Conf.*, Hyatt Regency, Calgary, AB, Canada, Oct 4–8.
[31] WEC (World Energy Council), 2007, "4- Natural Bitumen and Extra-Heavy Oil," *Survey of Energy Resources*, pp. 119–143. http://www.worldenergy.org/.
[32] US DOE (Office of Petroleum Reserves), 2007, "U.S. Heavy Oil Resources," http://fossil.energy.gov/programs/reserves/npr/Heavy_Oil_Fact_Sheet.pdf.

[33] Meyer, R. F., Attanasi, E. D., and Freeman, P. A., 2007, "Heavy Oil and Natural Bitumen Resources in Geological Basins of the World," USGS Open File Report 1084, http://pubs.usgs.gov/of/2007/1084/OF2007-1084v1.pdf.

[34] WEC (World Energy Council), 2010, "Survey of Energy Resources," World Energy Council Regency House, 1-4 Warwick Street, London, W1B 5LT, United Kingdom, ISBN: 978 0 946121 021, http://www.worldenergy.org/documents/ser_2010_report_1.pdf.

[35] Croft, G., and Feder, T., 2007, "Impact of Enhanced Oil Recovery and Unconventional Reservoirs on Oil Supply," Report ER291, Transportation Energy Seminar, U.C. Energy Institute, Berkeley, California, pp. 16.

[36] Roadifer, R. E., 1986, "How Heavy Oil Occurs Worldwide," *Oil & Gas Journal*, March 3.

[37] Isaacs, E., Cyr, T., Chu, H., and Singh, S., 1998, "Recovery Methods for Heavy Oil and Bitumen in the 21st Century," *7th Unitar Int. Conf. on Heavy Crudes and Tar Sands*, Beijing, China, Vol. 1. www.gbv.de/dms/tib-ub-hannover/306749947.pdf.

[38] Clark, B., 2007, "Heavy Oil, Extra-Heavy Oil and Bitumen- Unconventional Oil," *Working Document of the NPC Global Oil and Gas Study*, July 18 Meeting.

[39] Isaacs, E., 2011, "Advances in extra heavy oil development technologies," 20th World Petroleum Congress, Doha, Forum BPK04: Advances in Extra-Heavy Oil Development, http://www.ai-ees.ca/media/39655/presentation-advances-heavy-oil-development-technologies.pdf

[40] NPC (The National Petroleum Council), 2007, "Heavy Oil," Topical Paper # 22, Working Document of the NPC Global Oil & Gas Study, www.npc.org.

[41] OGJ (Oil and Gas Journal), 2008, "Worldwide EOR Survey," April 21.

[42] Lazar, I., Petrisor, I. G., and Yen, T. F., 2007, "Microbial Enhanced Oil Recovery (MEOR)," *Petroleum Science and Technology*, **25**, pp. 1353–1366.

[43] Mago, A. L., 2006, "Adequate Description of Heavy Oil Viscosities and a Method to Assess Optimal Steam Cyclic Periods for Thermal reservoir Simulation," MSC. thesis, Texas A&M University, Petroleum Engineering.

[44] James, L. A., 2009, "Mass Transfer Mechanisms During the Solvent Recovery of Heavy Oil," Ph.D. thesis, University of Waterloo, Chemical Eng Department, Ontario, Canada. http://uwspace.uwaterloo.ca/bitstream/10012/4478/1/James_Lesley.pdf.

[45] Curtis, C., Kopper, R., Decoster, E., Guzman-Garcia, A., Huggins, C., Knawer, L., Minner, M., Kupsch, N., Lineares, L. M., Rough, H., and Waite, M., 2002, "Heavy-Oil Reservoirs," *Oilfield Review,* **14**(3), pp. 30–52.

[46] Butler R. M., 1991, *Thermal Recovery of Oil and Bitumen*, Department of Chem and Petroleum Eng., Prentice Hall, New Jersey, # 7, pp. 285–358.

[47] Butler, R. M., 2001, "Application of SAGD, Related Process Growing in Canada," *Oil and Gas Journal*, pp. 74–78, May 14.

[48] Nasr, T. N., and Ayodele, O. R., 2005, "Thermal Techniques for Thermal Recovery of Heavy Oil and Bitumen," *SPE 79488, Int. Improved Oil Conf.*, Asia Pacific, Kula Lumpur, Malaysia, Dec. 5–6.

Chapter 7

LIQUID MEASUREMENT

7.1 INTRODUCTION

Liquid hydrocarbons are produced, transported, bought and sold throughout the world in a large network of pipelines, tankers and storage systems. Ownership of the liquids may change many times from wellhead to final market. Each time the product changes ownership en-route to market or there is a transfer of product into another pipeline system or terminal, there is measurement of the product and custody of the product is transferred to another party. Due to the high value of liquid hydrocarbons, accurate measurement is fundamental for custody transfer facilities. Contractual arrangements between buyer and seller specify that certain measurement standards be applied to the custody transfer facilities' design, construction and operation. Typically, these standards include the *American Petroleum Institute Manual of Petroleum Measurement Standards (API MPMS)* and, internationally, the *International Organization for Standardization (ISO)*.

The purpose of a measurement system is to determine a numerical value that corresponds to the variable being measured. Measurements are required for producers, customers/shippers and transportation companies for custody transfer or change in ownership or responsibility as well as pipeline system monitoring and control. Also, measurements may be required if petroleum products move across national boundaries. Transportation companies include pipeline, tanker, and other transportation media. Pipeline companies charge their shippers for the transportation services based on the measured quantities of the products they have transported, assuming that they satisfy other transportation requirements such as the product quality. Therefore, a custody transfer quality measurement system should generate accurate measured values.

The quantities typically measured for custody transfer and monitoring or controlling facilities are as follows:

- Volume flow rate, accumulated volume or mass at all custody transfer points,
- Pressure at the custody transfer and pressure control points such as pump stations and peak points
- Temperature at the custody transfer and temperature control points such as injection/delivery and heater locations
- Density at the custody transfer and batch lifting/delivery points
- Quality at the custody transfer and batch lifting/delivery points

The measurement systems used to establish custody transfer are dependent on the fluids being measured and/or the different regulations or contractual conditions applicable. For certain products such as ethylene, custody transfer is based on mass, but for the majority of petroleum liquids, custody transfer is based on volume measurement. Most pipeline systems are monitored and controlled for flow and pressure, while liquid density is required for batch control.

These quantities are measured with various instruments using many different techniques. The actual value of the measurement is the value of the process quantity, obtained by an instrument. An instrument is a device that measures and regulates the process variables. Instruments may include devices that can change a process variable such as pressure and provide control capabilities. A pipeline system requires instrumentation in order to gather measurement data from the field and change the process variables.

There are many types of measurement systems in use in the liquid hydrocarbon pipeline industry. Measurement systems are installed for a number of purposes beyond custody transfer including inventory control, batch size determination, and leak detection line balancing purposes. Depending on the type of commodities being transported, there are several options for establishing the most appropriate measurement system.

Hydrocarbon liquids can be measured either statically or dynamically. Static measurement is achieved by measuring tank volume and its changes over time (Section 7-2). API and ISO Standards dictate procedures for verifying tank volumes and measuring tank levels, product temperatures, and product densities/specific gravities, all necessary to determine the volume of product referenced to some standard condition.

Dynamic quantity determination occurs when the hydrocarbon is measured by under flowing conditions by meter (Section 7-4). Meters frequently used by liquid pipeline industry or recognized by API and ISO for custody transfer purposes include positive displacement (PD) meters, turbine meters, Coriolis meters, and ultrasonic meters. These types of meters can be manufactured and calibrated to the high measurement accuracy levels necessary for custody transfer applications. Other types of meters in use in the pipeline industry include vortex shedding meters and, in some cases, orifice meters. These types of meters do not exhibit the accuracy levels necessary for custody transfer but may be suited to certain other applications such as process control or internal product transfer measurement.

7.2 STATIC MEASUREMENT

The calculation of petroleum quantities by tank measurement requires exacting attention to detail and precision not only in the calculation process but also in the underlying supporting processes. The supporting processes include tank calibration, calibration of temperature and pressure transmitters, densitometer calibration and tank gauging equipment verification or calibration.

7.2.1 Tank Calibration

Calibration of storage tanks is the process of accurately determining the capacity of a tank and expressing this capacity as a volume for a given linear increment or height of liquid in the tank [1]. There are several measurement methods that are in use today to calibrate above-ground cone and floating roof steel storage tanks. Procedures are outlined in the *API MPMS Chapter 2* and *ISO/TR 7507*.

7.2.1.1 Manual Tank Strapping Method (MTSM)

Tanks must only be calibrated after they have been filled at least once with a liquid of density equal to or greater than that of the liquid which they will hold when in use. The hydrostatic test applied to new tanks will satisfy this requirement in most cases. This requirement is common to all tank calibration methods described below.

The following dimensional measurements and determination of weights are necessary inputs to determining accurate tank capacities. *API MPMS Chapter 2.2A* addresses necessary measurement procedures to determine total and incremental tank volumes and procedures for computing volumes.

Circumferential measurement — This method of calibration uses a calibrated steel tape for a procedure known as tank strapping. Procedures for strapping the circumference of the tank at predetermined levels are provided in *API MPMS Chapter 2.2A*. The strapping tape is of a length great enough to encompass the circumference of the tank. The tape will have a calibration report that accounts for the thermal coefficient of the steel used. All strapping tapes used in the calibration process should be checked against a master tape (calibrated by the *National Institute for Standards and Technology (NIST)*). *API MPMS Chapter 2.2A* details the location and number of strappings to be taken on each tank ring. At each strapping location, the tape is read after sliding to distribute surface tension and then applying a predetermined tension to the tape ends.

Height — Total tank height must be measured as well as the height of plates in each ring. If the tank contains liquid, then the product temperature and its density must be recorded. In addition, the ambient air temperature must be recorded.

Tilt — The tank should be measured to determine if it is tilted from the vertical. If the tank is tilted less than one in seventy the correction is negligible and can be disregarded.

Deadwood — It is necessary to determine the volume of space taken internally in the tank by mixers, manholes, ladders, floating roof stands, and other facilities that affect the volume of liquid in the tank. Accurate measurement of the deadwood is required to arrive at the net tank volume.

Floating roof — Floating roofs displace a volume equal in weight to the weight of the roof. It is necessary to determine the weight of the roof material used in construction as well as any appurtenances such as ladders, drain lines, tank wall seals, and support legs.

Tank bottom survey — Tank bottoms are measured for variances from a datum floor plate. This is usually determined by means of a surveyor's level or transit measuring the height differences between the datum plate and various selected points on the tank bottom. Starting at the datum plate, a series of survey points are taken into the center of the tank. Then, from the datum plate, level readings are taken around the circumference and from these points into the center of the tank. It follows that the more level readings that are taken, the more accurate will be the bottom calculations. Another method of determining the bottom volume of the tank is by metering quantities of water into the tank and recording the relative heights to volumes above and below the datum plate.

Other measurements — To complete calculation of the internal volume of the tank, it is necessary to obtain the wall thickness of the circumferential plates and also the thickness of the paint used on the tank walls.

The above measurements are required input into the tank capacity calculations detailed in *API MPMS Chapter 2.2A*.

7.2.1.2 Optical Reference Line Method (ORLM)

The Optical Reference Line Method is an alternative method for determining tank diameter using an optical device. This method still requires manual strapping of a reference circumferential course. This reference strapping of the first ring, 20% below the top horizontal weld seam, forms the datum. Deviations in diameter are then measured using the optical equipment. The equipment consists of an optical device consisting of a theodolite and a precision level mounted on a tripod. A traversing magnetic trolley with graduated slide is used to measure offsets at different vertical stations, as shown in Figure 7-2 below. The trolley is equipped with a horizontal graduated scale in 0.01 ft or 1 mm increments. The combined resolution of the scale and the optical devices must allow the operator to read the offset measurement to the nearest 0.005 ft or 1 mm at any given station (Figure 7-1).

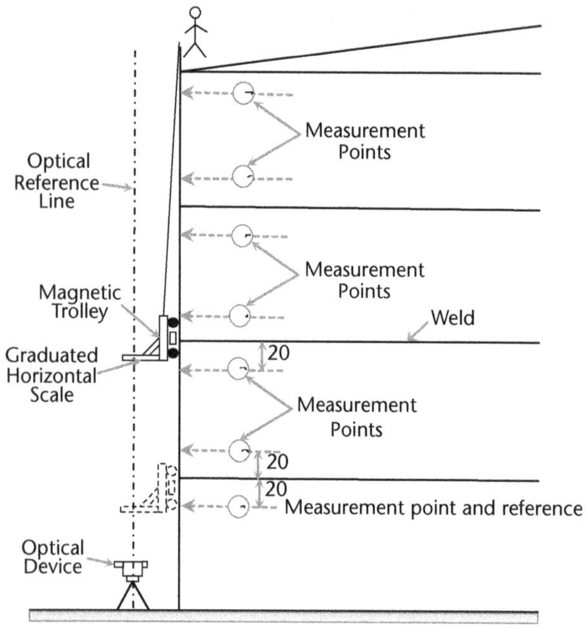

Tank Diameter (ft.)	Minimum Number of Stations
50	8
100	12
150	16
200	20
250	26
300	32
350	36

Figure 7-1. ORLM measurement process

The number of stations required for the optical measurements is based on the tank diameter. Obviously, the more stations used, the higher the accuracy of measurement. API standards state the minimum number of stations to be used.

The location of each horizontal station must allow the vertical traverse of the trolley to not be impeded by vertical weld seams. Two vertical measurement points per tank course must be established approximately 20% from the horizontal weld seam of each course.

Before starting the vertical scale readings, the optical device must be leveled along the three axes and the perpendicularity of the device must be verified. Verification is accomplished by raising the trolley toe the uppermost level. A reading of the scale is recorded. The optical device is then rotated 180° vertically and the reading is again noted. The difference of the two readings should not be greater than 0.005 ft. or 1 mm.

CALIBRATION PROCEDURES

The reference circumference must be measured using a master tape at or very near to the first vertical station. The optical device must then be stationed and aligned correctly with the perpendicularity verified. This establishes the optical reference line and allows measurement of the offset distance ***a***.

The trolley is then moved upwards to the next vertical station and the offset ***m*** is read. This process is repeated at each vertical station. After reading the uppermost offset, the trolley is lowered to the first vertical station and the reference offset measurement is repeated. The readings must be within 0.005 ft. or 1 mm. If not, the procedure must be repeated until the accuracy of measurement is confirmed.

The above procedures are then repeated at each horizontal station on the tank.

The determination of ring circumferences/radii are determined as outlined below. Once these measurements are completed, tank capacity tables can be developed using procedures provided in *API Standard 2550*.

As the distance from the tank center to the vertical reference line is constant for each given horizontal station, the following is true:

$$(r' + m) = (r + a)$$

$$r' = r + (a - m)$$

$$r' = \frac{C}{2\pi} + (a - m)$$

$$r' = \frac{C}{2\pi} + \frac{\sum (a-m)}{n}$$

where

r = reference radius based on reference circumference
r' = radius at given vertical station
a = reference offset
m = offset at given vertical station
C = reference circumference
N = number of horizontal stations

Wall thicknesses of each tank ring and paint coating thickness are required to determine internal tank radii. *API MPMS Chapter 2.2B* (*ISO/TR 7507-2*) provides detailed procedures for this method of tank calibration.

7.2.1.3 Optical Triangulation Method (OTM)

With this method of tank shell calibration, a theodolite or laser theodolite is used to externally calibrate vertical cylindrical tanks by measurement of angles. This method has an advantage where there is no access to tank roofs for operating a magnetic offset trolley. As with ORLM, this method requires that a measured reference circumference be determined by manual strapping at a location on the bottom ring.

The OTM calibration method requires a minimum number of stations be established to provide the level of accuracy required. *API MPMS Chapter 2.2C (ISO 7507-3)* provides the following minimum number of theodolite stations for external OTM procedures (Table 7-1).

The horizontal stations should be spaced approximately equal distances along a circle concentric to the tank.

A reference measurement is determined by manually strapping around the circumference of the bottom ring of the tank measured 20% below the first horizontal weld seam.

The number of vertical stations is the same as in the ORLM method (two per ring) and is established at 20% of the distance from the upper and lower horizontal weld seams for each tank ring.

TABLE 7-1. Minimum number of theodolite stations for external OTM

Tank Circumference, *C* (m)	Tank Circumference, *C* (ft)	Minimum Number of Points
$C \leq 50$	$C \leq 164$	4
$50 < C \leq 100$	$164 < C \leq 328$	6
$100 < C \leq 150$	$328 < C \leq 492$	8
$150 < C \leq 200$	$492 < C \leq 656$	10
$200 < C \leq 250$	$656 < C \leq 820$	12
$250 < C \leq 300$	$820 < C \leq 894$	15
$300 \leq C$	$894 \leq C$	18

The tank is then sighted from the first horizontal station (T_1) using a theodolite (see Figure 7-2). Two sightings must be made tangentially to the tank, on the left and right from each station, recording the angle subtended between the two sightings (2θ). The first vertical sighting should be made at the same height as the reference circumference was taken. This measurement will determine the reference angle. The theodolite is then angled upwards to sight at the next vertical statin. In order to prevent any correction for tilt in the tank, the vertical angle for each pair of sightings should not be changed during the measurement procedure.

After the angle between each pair of sightings has been recorded for all vertical stations at T_1, the theodolite is relocated to the next horizontal station. All measurements and procedures are then repeated for all remaining horizontal stations (Figure 7-3).

The tank is strapped at the reference circumference, 20% below the horizontal weld seam at the top of the first course. Two sets of target points are to be established for each course, each set being located 20% below and above the horizontal welds of the particular course being measured.

Calculations for optical triangulation follow:

T is the horizontal station site of the theodolite. The siting T ⇒ B and T ⇒ B′ at the exact location of the manual strapping level determines the reference horizontal angle 2θ.

Therefore:

$$TZ = r \times \frac{1}{\sin\theta} \times \frac{C}{2\pi} \times \frac{1}{\sin\theta}$$

Sightings T ⇒ A and T ⇒ A' to any vertical station give the horizontal angle 2θ′.

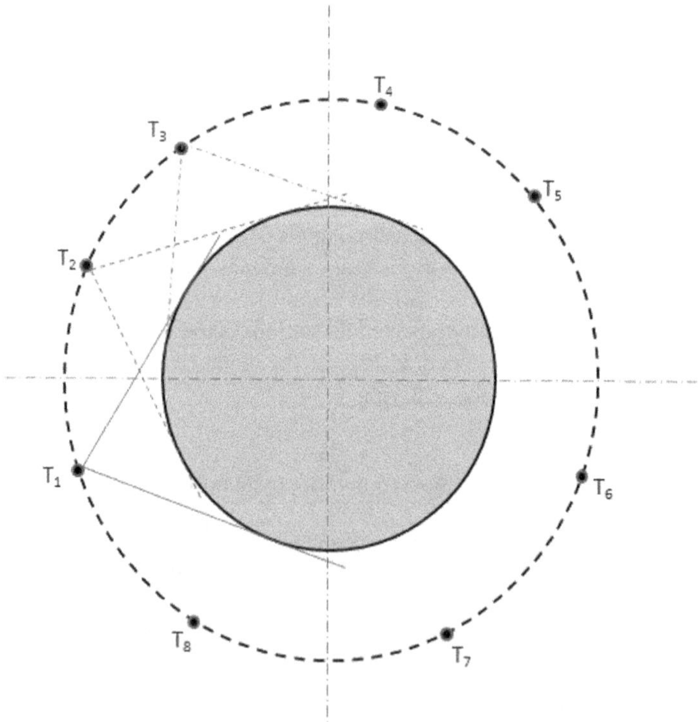

Figure 7-2. ORLM horizontal station sightings

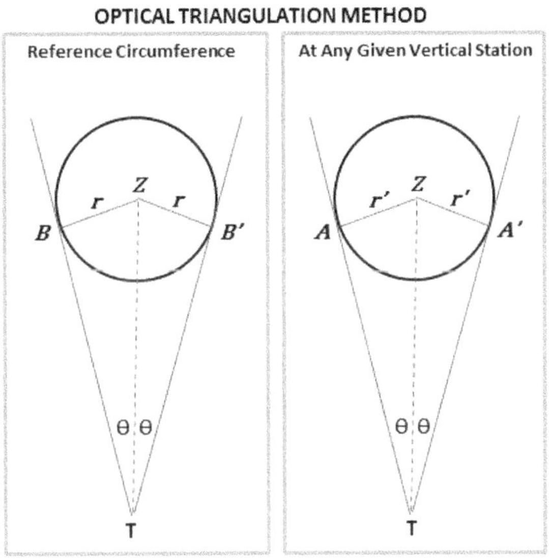

Figure 7-3. Angle between sightings

Therefore:

$$r' = TZ \times \sin\theta' = \frac{C}{2\pi} \times \frac{\sin\theta}{\sin\theta}$$

As there will be two average radii per ring, the mean value of the two will be the average radius to the external tank surface for that course.

API MPMS Chapter 2.2C (ISO/TR 7507-3) provides detailed procedures for this method of tank calibration. Ring wall thicknesses and paint thickness will be required input to preparation of the capacity tables outlined in *API MPMS Chapter 2.2A (ISO/TR 7507-1)*.

7.2.1.4 Electro-Optical Distance Ranging Method (EODRM)

This method of calibration of tank shell dimensions uses internal measurement procedures. Procedures are outlined in *API MPMS Chapter 2.2D (ISO 7507-4)*.

The equipment utilized by this method of calibration consists of an electro-optical distance-ranging instrument capable of precision angular measurement and a laser beam emitter that may be part of the instrument or a separate device. Distances and slope angles are measured at a predetermined number of target points around the internal circumference of the tank (Table 7-2).

TABLE 7-2. Minimum number of target points per set

Tank Circumference, C (m)	Tank Circumference, C (ft)	Minimum Number of Target Points
$C \leq 50$	$C \leq 164$	8
$50 < C \leq 100$	$164 < C \leq 328$	12
$100 < C \leq 150$	$328 < C \leq 492$	16
$150 < C \leq 200$	$492 < C \leq 656$	20
$200 < C \leq 250$	$656 < C \leq 820$	24
$250 < C \leq 300$	$820 < C \leq 894$	30
$300 \leq C$	$894 \leq C$	36

Two sets of target points are to be established for each course, each set being located 20% below and above the horizontal welds of the particular course being measured.

The distance-measuring part of the instrument must have accuracy limits to within limits described in *Chapter 2.2D (ISO/TR 7507-4)*. Equipment required for this procedure also includes stadia which are used for field equipment verification purposes. The stadia are usually 2 m in length with graduations between the two stadia marks.

The EODR instrument is placed on the tank floor and stabilized to prevent movement. The instrument should be located at or near the center of the tank. All target points are sighted along the horizontal plane at each course location and the slope distance, horizontal angle and vertical angle for each point are measured as shown in Figure 7-4 below.

The dimensional coordinates obtained by the calibration procedure can be converted to Cartesian coordinates using the following equations:

$$X = D \times cos\,\theta \times cos\,\phi$$

$$Y = D \times sin\,\theta \times sin\,\phi$$

$$Z = D \times sin\,\phi$$

where:

D = measured slope distance;
θ = measured horizontal angle
ϕ = measured vertical angle

Once the internal radii of the tank are established, the development of the capacity table can be conducted in accordance with *API MPMS Chapter 2.2A (ISO/TR 7507-1)*.

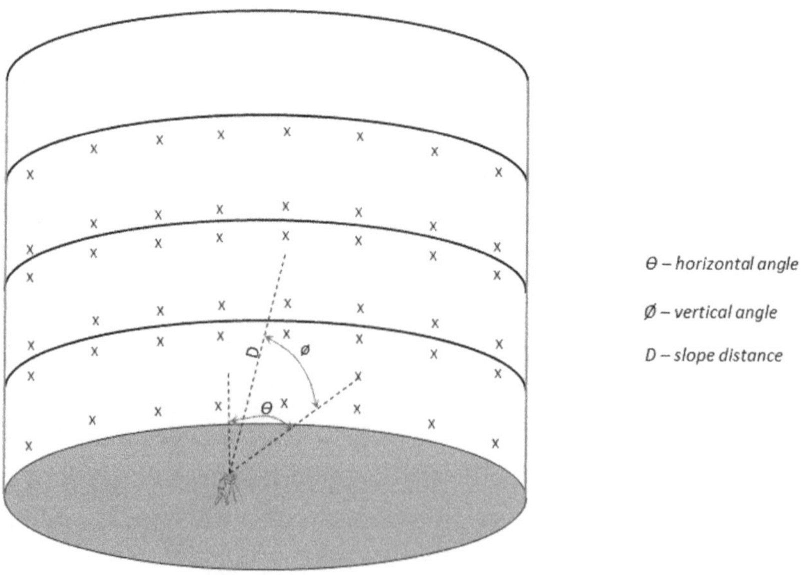

Figure 7-4. Illustration of calibration procedure

The tank bottom shall be calibrated by the liquid method in accordance with *API Standard 2555*, or by use of the electro-optical ranging instrument operated as a surveyor's level, or by other methods outlined in *API MPMS Chapter 2.2A (ISO/TR 7507-1)*.

The overall height of the tank is measured by a dip tape and weight. The following data will be required to complete the calculations for tank capacity:

- the density and working temperature of the liquid to be stored in the tank;
- the height of each course;
- the plate thickness of each course;
- the safe filling height and maximum filling height;
- deadwood dimensions
- the tilt of the tank as shown by the deviation from a vertical line.

7.2.2 Tank Capacity Tables

Tank capacity tables are determined from the dimensional measurements obtained through the above calibration procedures. This is normally a procedure that is performed by the tank manufacturer following construction. Recalibration of the tank will require establishing revised tank capacity tables. For tanks in custody transfer service, verification of the bottom course diameter, bottom course plate thickness and tank tilt is suggested once every five years. Should variation in any of these three parameters exceed the criteria provided in *Chapter 2.2A*, a total recalibration of the tank should be considered. For tanks not in custody transfer service, verification of diameter, thickness of bottom course and tilt may be considered once every five to fifteen years.

Because tank volumes do change with time and service, *API MPMS* Standards state that it is justifiable practice to recalibrate tanks on a periodic basis to reassure good measurement accuracy. A total recalibration at 15-year intervals for tanks in custody transfer service and at 15 to 20 years for others is reasonable according to *API*.

Throughout the pipeline industry, many operators have installed tank capacity tables into their computer systems. Tank Inventory programs and terminal supervisory control and data acquisition systems require the input of capacity tables. Exact replication of capacity tables in computer systems is required for custody transfer transactions.

7.2.3 Liquid Calibration of Tanks

If tank shapes are irregular, unknown or inaccessible, liquid calibration may be required. The procedure used for liquid calibration depends upon the equipment available and the size of tank to be calibrated. The most common procedure for liquid calibration is the use of a positive displacement meter. The meter should be proved before and after the calibration procedure. If the tank is large, additional proving should be considered.

See *API Standard 2555 — Method of Liquid Calibration of Tanks, September 1966, Reaffirmed, March 2009* or *ISO 4269:2001 Petroleum and liquid petroleum products — Tank calibration by liquid measurement; Incremental method using volumetric meters* for detailed procedures on this method of tank calibration.

7.3 TANK GAUGING

7.3.1 Manual Tank Gauging

This system still plays a significant role in the custody transfer of crude oil and refined products [2]. To obtain accurate, reliable, and repeatable product levels with manual gauging, there are guidelines and procedures outlined in *API MPMS Chapter 3.1A — Standard Practice for the Manual Gauging of Petroleum and Petroleum Products*.

For custody transfer operations, the manual gauge tape must be calibrated and certified to a national metrological traceable standard such as National Institute of Standards and Technology (NIST). In performing a manual gauge, an operator or gauger must take all safety and precautionary measures to assure personal safety and avoid any possible incidents. Some operations do not allow open access to tank contents because of hazardous vapors or for emission control purposes. In these cases, it is desirable to implement manual closed tank gauging systems (Figure 7-5).

7.3.2 Servo Tank Gauge

This method of automatic tank gauge (ATG) level measurement involves a physical connection of the Servo instrumentation to a level sensor or float or by physical connection to a floating roof. The servo tank gauge can provide product level indication as well as product density and thus can be used to determine the Net Standard Volume (NSV) of the tank contents (Figure 7-6).

Floating roof tanks are broadly divided into internal (IFR) and external (EFR) floating roof tanks. IFR tanks are used for liquids with low flashpoints (e.g., gasoline, ethanol, etc.). These tanks are cone roof tanks with a floating roof inside the tank. EFRs are open at the top and do not have a fixed roof. They are suitable for medium flash point liquids (naphtha, kerosene, diesel, crude oil, etc.). When mounting a servo gauge on a floating roof tank, a gauging platform is required in order to mount the gauge over the product. This may be the tank roof on covered internal floating roof tanks or an actual platform that extends out from the tank wall over an open roof tank. Many platforms incorporate a stilling well. The stilling well provides stability for the floating roof movement and a "calm" surface on the product to enable an accurate measurement. Servo gauges require a stilling well.

Accurate temperature information is required for custody transfer purposes. This is accomplished by installation of multiple resistance temperature detector (RTD) elements each of a different length in a probe that traverses the entire height of the tank.

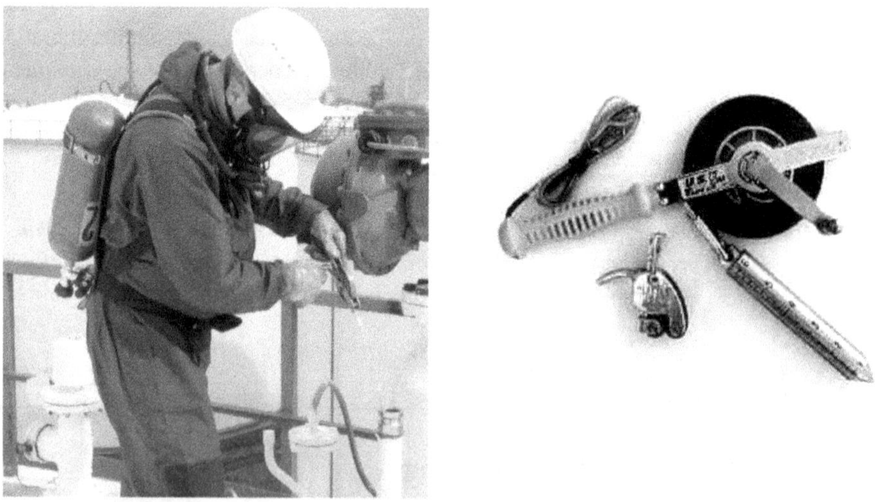

Figure 7-5. Manual gauging procedure

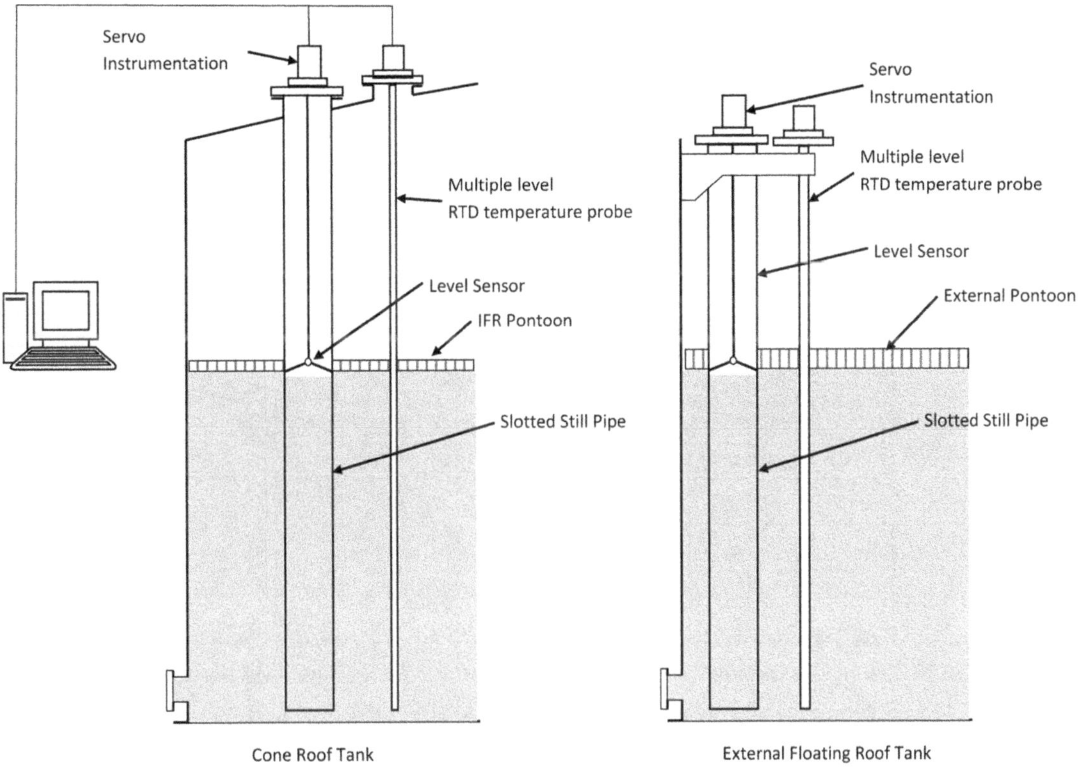

Figure 7-6. Servo tank measurement system

Microprocessors determine the RTDs that are submerged in the liquid by the level indication and then average the temperature indicated by the immersed RTDs.

7.3.3 Radar Tank Gauge

This system of measuring tank liquid levels has become very reliable and accurate and is suitable for custody transfer purposes in tanks with open visibility to the liquid. A radar tank gauge is a non-contact device that accurately measures the product level. This type of automatic tank level measurement is the most accurate level measurement device in the industry today. With its capability to deliver product level measurements in the accuracy range of less than a millimeter and fraction of an inch, this device is ideal for providing the most accurate calculation of tank product volumes corrected to base temperature and pressure. The radar tank gauge needs to be installed with instruments of similar accuracy in the measurement of product temperature and density. For increased accuracy, a stilling well is recommended for radar level measurement (Figure 7-7).

Detailed information on the selection, installation, commissioning, calibration, and verification of an ATG system comprised of either Servo or Radar level measurement is available in the following standards:

- *API MPMS Chapter 3.1B — Standard Practice for Level Measurement of Liquid Hydrocarbons in Stationary Tanks by Automatic Tank Gauging*
- *ISO 4266 — Part 1 — Petroleum and liquid petroleum products — Measurement of level and temperature in storage tanks by automatic methods*

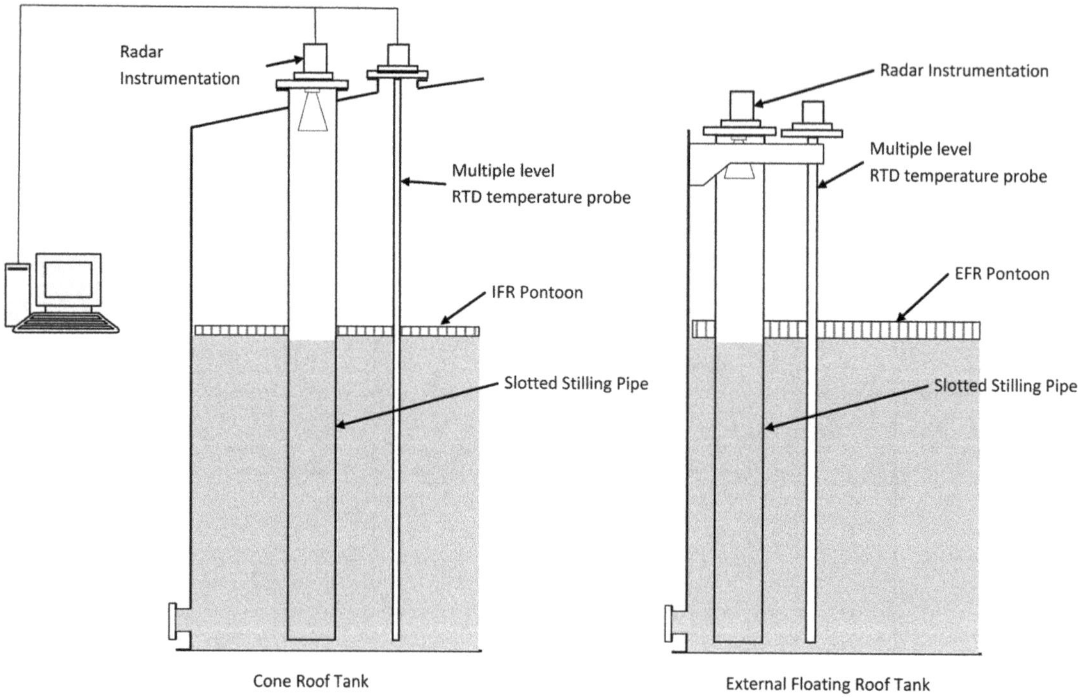

Figure 7-7. Radar tank measurement system

7.3.4 Hybrid Tank Measurement Systems

Hybrid tank gauging combines accurate level gauge instrumentation such as servo or radar systems with temperature sensors and pressure transmitters. By utilizing the best of both level-based and mass-based systems, hybrid tank gauging obtains tank level, temperature compensated volumes, mass and density measurements (Figure 7-8).

A Hybrid Tank Measurement System (HTMS) consists of four major components:

- An automatic tank gauge (ATG)
- Tank thermometer system that provides accurate temperature measurement of tank volumes
- Tank bottom pressure sensor and transmitter that provides accurate pressure
- A processor that provides an accurate conversion of pressure to density

These four components of the system are required to calculate the tank Net Standard Volume (NSV). The product level is directly measured by either Servo or Radar tank gauging systems. The product temperature is directly measured by the tank RTD system. The true (observed) density is determined from hydrostatic pressure measured by the pressure sensor at tank bottom and the product height above the pressure sensor, as measured by the ATG. Total static mass is computed by a hybrid processor from the true density and the tank capacity table. Gross observed volume, standard volume, and reference density are computed using industry practice for static calculations (see *API MPMS Chapter 12.1*).

Detailed information on the selection, installation, commissioning, calibration, and verification of an HTMS is provided in the following standards:

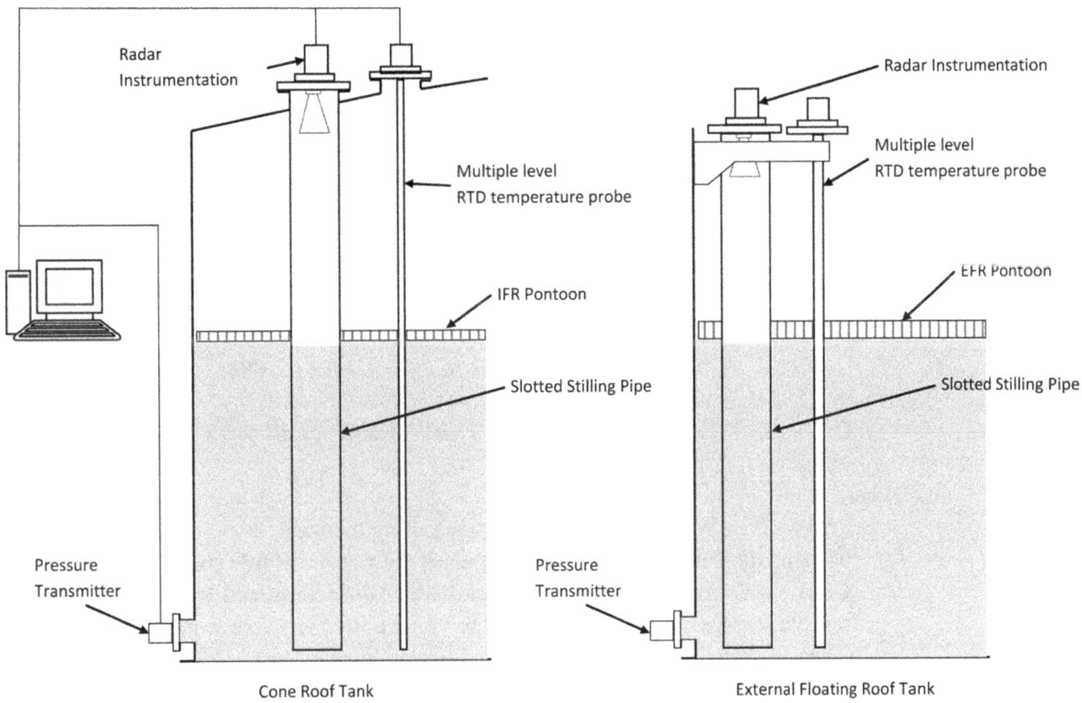

Figure 7-8. Hybrid tank measurement system

- *API MPMS Chapter 3.6 — Measurement of Liquid Hydrocarbons by Hybrid Tank Measurement Systems*
- *ISO 15169 — Petroleum and liquid petroleum products — Determination of volume, density, and mass of the hydrocarbon content of vertical cylindrical tanks by Hybrid tank measurement systems*

7.3.5 Calculation of Tankage Volumes

The petroleum industry has developed standardized calculation methods for determining the volumes in tankage [3]. These methods are expressed in API and ISO standards as follows:

- *API MPMS Chapter 12.1.1 — Calculation of Static Petroleum Quantities, Part 1* — Upright Cylindrical Tanks and Marine Vessels
- *ISO 4267-1:1988 — Petroleum and liquid petroleum products — Calculation of oil quantities — Part 1:* Static measurement

Net Standard Volume (NSV) is the primary unit of measurement for custody transfer and product inventory control. NSV is the equivalent volume of a liquid at its base temperature and pressure conditions that does not include non-merchantable items such as sediment and water. NSV can be determined by establishing the following known values:

The determination of NSV is accomplished by the following process.

$$TOV \rightarrow GOV \rightarrow GSV \rightarrow NSV$$

Total Observed Volume (TOV) — is determined by measuring the liquid height of the tank. TOV consists of all liquids and sediment contained in the storage tank including free water and water in suspension (S&W). TOV is derived by referencing the liquid height in the tank to the calculated volume from the tank capacity table.

Manual tank gauging is accomplished by gauge tape and performed in accordance with *API MPMS Chapter 3.1A*. Automatic tank gauging is performed in accordance with *API MPMS Chapter 3.1B*. Temperature measurements are performed in accordance with *API MPMS Chapter 7*.

The mathematical formulae for the various required values can be expressed as follows:

$$\boldsymbol{GSV = \{[(TOV - FW) \times CTSh] \pm FRA\} \times CTL}$$

$$\boldsymbol{NSV = \{[(TOV - FW) \times CTSh] \pm FRA\} \times CTL \times CSW}$$

where

FW — Free Water is determined by measuring the water level in the tank and then utilizing the tank capacity table to establish the volume of free water.

CSW — Correction for sediment and water — determined by obtaining a representative sample of product from the tank above any free water. Percentage of S&W is determined by one of the methods outlined in **API MPMS Chapter 10 — Sediment and Water**.

CTL — Correction for temperature of the liquid — corrects a volume at an observed temperature to a standard temperature (60 °F or 15 °C). This factor is obtained from API MPMS Chapter 11.1 — Temperature and Pressure Volume Correction Factors for Generalized Crude Oils, Refined Products, and Lubricating Oils. This standard provides the algorithm and implementation procedure for the correction of temperature and pressure effects on density and volume of liquid hydrocarbons which fall within the categories of crude oil, refined products, or lubricating oils.

CTSh — Correction for temperature of the shell — the correction factor for the effect of the temperature, both ambient and liquid, on the shell of the tank. CTSh may be calculated by the following:

$$CTSh = 1 + 2a\Delta T + a^2 \Delta T^2$$

where:

$\boldsymbol{a}$ = linear coefficient of expansion of the tank shell material (for mild carbon steel $\boldsymbol{a}$ = 0.00000620/°F or 0.0000112/°C)

ΔT = Tank shell temperature (T_{Sh})–base temperature (T_b)

FRA = Floating roof adjustment — the adjustment made to offset the effect of the displacement of the floating roof. If the tank capacity table has been prepared as a table of open tank or shell capacity only, the roof correction is calculated as follows:

FRA = (Weight(apparent mass) of roof)/(Density of product × CTL)

FW = Free water quantity deduction (may include bottom sediments) — the water present in the tank that is not suspended in the liquid hydrocarbon. This is derived from a manual tape gauge with use of a water detection paste or an automatic tank gauge with the capability of measuring the interface between product and water.

GOV= Gross observed volume — the total volume of all petroleum liquids and sediment and water, excluding free water, at observed temperature and pressure.

GSV = Gross standard volume — the total volume of all petroleum liquids and sediment and water, excluding free water, corrected by the appropriate volume correction factor (CTL) for the observed temperature and API gravity, relative density, or density to a standard temperature (60 °F or 15 °C).

NSV = Net standard volume — the total volume of all petroleum liquids, excluding sediment and water and free water, corrected by the appropriate volume correction factor (CTL) for the observed temperature and API gravity, relative density, or density to a standard temperature (60 °F or 15 °C).

TCV = Total calculated volume — the total volume of all petroleum liquids and sediment and water corrected by the appropriate volume correction factor (CTL) for the observed temperature and API gravity, relative density, or density to a standard temperature (60 °F or 15 °C) and all free water measured at observed temperature (gross standard volume plus free water).

TOV = Total observed volume — total measurement volume of all petroleum liquids, sediment and water, free water, and bottom sediments at observed temperature. TOV is the volume obtained from the tank capacity table prior to any corrections, such as those for floating roof and the temperature of the tank shell.

7.4 DYNAMIC MEASUREMENT

A flow meter is a device that measures the rate of flow or quantity of a moving fluid in an open or closed conduit. It usually consists of primary and secondary devices. The secondary devices for flow measurement may include not only pressure, differential pressure, and temperature transducers but also other associated devices such as chart recorders and volume totalizers.

Since volume and flow rates vary with pressure and temperature, the measured volume of a fluid at measured conditions will change with differing pressures and temperatures. Normally, base pressure and temperature conditions are defined for custody transfer in the contract between the parties involved. The correction of measured quantities to base conditions depends on the fluid's properties, particularly density, and thus requires the comparison of pressures and temperatures in order to be calculated. This relationship can be obtained from experimental data or an equation of state, and its accuracy influences the accuracy of the measured value at the base conditions. In North America, API 1101 Volume Correction Factor is often used for hydrocarbon liquids.

7.4.1 Measurement Systems and Characteristics

When selecting instruments, the following aspects of measurement requirements should be fully understood:

- Characteristics of and requirements for the pipeline system operation in terms of custody transfer, monitoring, or control
- Requirements for the accuracy, precision, etc. of the measured values

A dynamic measurement system consists of four elements:

- sensing element or transducer (primary device) that is mounted internally or externally to the fluid conduit which produces a signal with a defined relationship to the fluid flow in accordance with known physical laws relating the interaction of the fluid to the presence of the primary device

- signal conditioning element (secondary device) that takes the output of the sensing element and converts it into a form more suitable for further processing, such as ampere to voltage conversion and amplification
- signal processing element (secondary device) that converts the output of the signal conditioning element into a form suitable for presentation such as analog to digital conversion
- measured data presentation element (secondary device) that presents the measured value in a form that is easily usable such as on a visual display

A sensing element has certain characteristics that have an effect on overall measurement performance. Measurement performance characteristics that apply to sensing devices, particularly custody transfer meters, include:

- Linearity is a measure of how close the flow meter output maintains a linear relationship with actual flow rate, to which the following definitions are applied;
    - Accuracy - accuracy can be defined as how close to the true or actual flow the instrument is indicating. When this is applied over a flow range, it can then be described as linearity.
    - Repeatability - is the ability of the meter to indicate the same reading each time the same flow conditions exist.
- Resolution - is a measure of the smallest increment of total flow that can be individually recognized.
- Rangeability - is the ratio of the maximum flow to the minimum flow over which the specified accuracy will be maintained. This is sometimes referred to as the meter turndown ratio. Most meter manufacturers offer a normal flow range with specified linearity and extended minimum and maximum ranges.
- Pressure loss - is the measure of pressure loss at specified flow rates and specified product viscosities.
- Back pressure - is the minimum pressure required immediately downstream of the meter that will prevent cavitation.

A sensing element is considered to be linear if measured values establish a linear relationship between the minimum and maximum values. If the measured values deviate from a linear relationship, then the sensor is said to be non-linear. Non-linearity, hysteresis, and resolution effects in modern sensors and transducers are so small that it is difficult to exactly quantify each individual error effect. Often, the sensor performance is expressed in terms of error and response to changes. Maintaining measurement operations with a small error is the most important factor in custody transfer, while response characteristics are more important for system control.

7.4.2 Measurement Uncertainty

Measurement uncertainty or errors are inherent in all measurement systems. The measured numerical value will not be equal to the true value of the variable due to measurement errors. From a custody transfer point of view, measurement uncertainty is critical because it is directly associated with the transaction cost. The pipeline industry deals with measurement uncertainty problems by implementing a technical standard acceptable to all stakeholders.

Measurement uncertainty can be biased and/or random, and change with time and environmental factors such as humidity and temperature. An error bias is the difference between the average and true values. It is directional and must be added or subtracted from the instrument reading. Bias error, if known, can be eliminated by a bias correction process. In practice, it is difficult to determine a true bias error, unless standard equipment such as the equipment at the National Institute of Standard and Technology (NIST) in the U.S. is used.

A random error is called a precision error in the *ANSI/ASME PTC 19.1* document. Precision can be improved only by selecting a different measuring device than the one in which the error occurred. Three cases regarding accuracy are illustrated in Figure 7.9 and are discussed below.

In the example of Figure 7-9(a), bias error is negligible; however, the measured data is widely scattered around the true value, so the precision is poor. While the average may be close to the true value, implying that there may be no significant bias; this device is not considered accurate due to large precision error.

In Figure 7-9(b), bias error is not negligible, but precision is good. The measured data is tightly clustered about an average value but offset from the center. The difference between the average value and true value is the bias error. This device is not considered accurate, because it is precise but largely biased.

In Figure 7-9(c), bias error is small and precision is good; this is an accurate device. The measured data is tightly clustered and close to the true value. This device is considered accurate, because it is precise and unbiased.

Measurement errors are expressed in terms of accuracy, systematic error, bias, repeatability, resolution, and precision. In the pipeline industry, accuracy and repeatability are more widely used. Repeatability or precision error is the ability of a sensor or transducer to generate the same output for the same input when it is applied repeatedly. Poor repeatability is caused by random effects in the sensor or transducer and its environment. Accuracy is the combination of bias and repeatability.

To determine the accuracy of a variable measurement, the accuracy of the primary measuring device must be combined with the individual accuracies of other measuring devices and then properly weighted in the accuracy calculation. The final accuracy figure is arrived at by taking account of both the primary and secondary device errors, which include their respective electronic errors. (The electronic errors come from current/voltage conversion error, amplification error and analog/digital conversion error.) These errors are combined by statistical methods to obtain the total errors for the measured quantity. Refer to *ANSI/ASME PTC 19.1 — Test Uncertainty* for detailed error analysis.

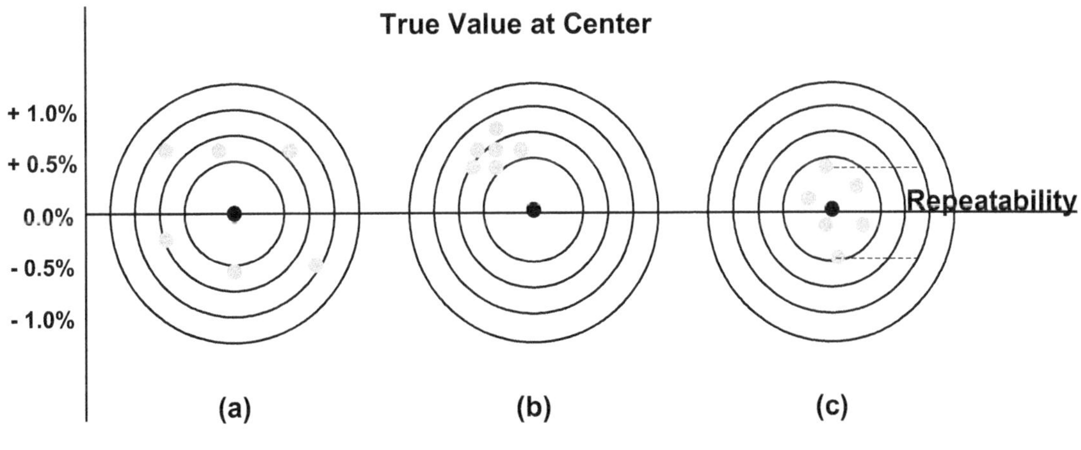

Figure 7-9. Bias vs. precision

Fluid properties and other factors affect measurement accuracy. Various factors need to be taken into account to achieve overall flow measurement accuracy. The measurement of flow rate requires instruments to measure temperature, pressure and/or differential pressure and density. The sensitivity of a flow meter is dependent on the sensitivity of each instrument. The accuracy of a flow meter depends on the steady flow of a homogeneous, single-phase Newtonian fluid, and thus departure from these quantities, known as influence quantities, can significantly affect the measurement accuracy. The influence quantities include velocity profile deviation, non-homogeneous flow, pulsating flow, non-Newtonian flow, and cavitation. These quantities are discussed in the *Flow Measurement Engineering Handbook* [5], and the quantities related to product properties in Section 2.11. The total error is obtained by the square root of the sum of the square of individual errors (known as the "RMS" value or root-mean square).

7.4.2.1 Quality of Liquids

The quality of liquid is defined differently for different liquids. For example, gasoline is specified for its octane value and diesel for its sulfur contents and cetane number. Contaminants for certain pure products like ethylene are strictly limited to very small amounts of impurities.

The following are important factors for most petroleum liquids:

- Sediment and water (S & W) — The amount of S & W should be limited within the contractual specified percentage.
- Air content — Air has to be removed to avoid cavitation problems and measurement error.
- Transmix — A transmix occurs as a result of the mixing of two adjacent products in a batch operation. Transmixes have to be handled as off-spec products and may be collected in a slop tank or refined again to meet the required specifications.

Other process or operating conditions and preventive maintenance programs to consider in the design and operation of meter stations are [4]:

- Pseudo Fully Developed Flow (for inferential flow-meters)
- Cleanliness of Stream-Quality
- Pigging Frequency and Operation
- Possibility of Multiphase (in case of retrograde condensate)
- Wax Formation
- Hydrate Formation
- Hydrochloric Acid Formation
- Sulfuric Acid Formation
- Elastomer Compatibility
- Presence of Drag Reducing Agent
- Auto-decomposition and Polymerization (for ethylene)
- Inhibitor Program
    - Anti-flocculation agents
    - Oxygen scavengers
    - Fungicides (to control bacteria)
    - Internal corrosion inhibitors

7.4.2.2 Device Degradation

As the primary and secondary devices age and operating environments change, the performance of the transducers, including sensors, degrades. The primary devices degrade more frequently than the secondary devices. Recalibration process can restore the performance of the primary device.

7.4.2.3 Operational Problems

In practice, various operational problems are associated with measuring devices and facilities. The typical operational problems associated with liquid measurement are caused by factors such as gas entrapment or solid particles in the liquid. The capacity of the measuring devices used and of facilities to cope with such operational problems must be taken into account in their design, selection, and operation. The *Flow Measurement Engineering Handbook* [5] addresses these problems in detail.

7.4.2.4 Calibration

Calibrating is the process of ensuring that a measuring instrument is accurate and in good operating condition, by adjusting the sensor and/or transducer to improve accuracy and response (e.g., zero level, span, alarm and range). The need for and frequency of calibration depends on the application and accuracy requirements, and is usually specified in a custody transfer contract if applicable. Both the primary and secondary devices need to be calibrated.

7.4.2.5 Transducer/Transmitter

The terms "transducer" and "transmitter" are used interchangeably in connection with instrumentation and measurement but they are not the same. All measuring instruments involve energy transfer and a transducer is an energy conversion device. A transducer is defined as a sensing element capable of transforming values of physical variables into either equivalent electrical signals or a packaged system which includes both sensing and signal conditioning elements. At a minimum, a transducer gives an output voltage corresponding to an input variable such as flow rate, temperature, and pressure.

A transmitter is a general term for a device that takes the output from a transducer and generates a standardized transmission signal on a transmission medium and is a function only of the measured variable. Like a packaged transducer, a transmitter in a pipeline system amplifies the signal from the sensor and converts it into a more convenient form for transmission.

Certain types of transducers are classified as smart sensors. They contain a dedicated computer which digitizes and linearizes a standardized 4 to 20 mA signal in order to minimize sensor errors. Smart flow transducers combine all of the measured values such as pressure and temperature to correct the flow rate to a reference condition as a way to improve flow measurement accuracy.

The following types of primary flow measuring devices are discussed. For the last thirty years, several linear flow meters have been widely accepted by the pipeline industry. Linear flow meters such as turbine, positive displacement, ultrasonic flow meters, Coriolis mass meters and vortex shedding meters are in common use. Due to technical advances, they have become more reliable and produce more accurate measurements than when they were first developed. All linear flow meters measure flow volumes directly, based on the principle that the measured volume increases linearly with flow velocity.

7.4.3 Custody Transfer Requirements

Liquid petroleum custody transfer metering systems must meet requirements set by industry bodies such as API or ISO [6, 7], and national metrology standards such as:

- International Organization of Legal Metrology (OIML) (International),
- National Institute of Standards and Technology (NIST) (U.S.),
- National Metrology Institute of Germany (PTB),
- Certification Management Committee (CMC) (China), and
- GOST Standards (Russia)

These requirements can be of two types:

- Legal — The weights and measures codes and regulations according to the country or jurisdiction in which the sale is conducted controls the wholesale and retail trade requirements to facilitate fair trade. The regulations and accuracy requirements vary widely between countries and commodities, but they all have one common characteristic — traceability. There is always a procedure that defines the validation process where the duty meter is compared to a standard that is traceable to the legal metrology agency of the respective region.
- Contract — A contract is a written agreement between buyers and sellers that defines the measurement requirements. These are large-volume sales between operating companies where refined products and crude oils are transported by marine, pipeline or rail. Custody transfer measurement must be at the highest level of accuracy possible because a small error in measurement can amount to a large financial difference. Due to these critical natures of measurements, petroleum companies around the world have developed and adopted standards to meet the industry's needs.

In Canada, for instance, all measurement of a custody transfer nature falls under the purview of Measurement Canada. In the USA, the Federal Energy Regulatory Commission (FERC) controls the standards which must be met for interstate trade.

7.4.4 Types of Meters

7.4.4.1 Positive Displacement Meters

A positive displacement (PD) meter measures flow by isolating segments of the liquid while it flows through the meter and then counting the segments. PD flow measurement consists of a class of devices which measure a specific amount of fluid volume for each cycle. Meters of this design divide the fluid stream into unit volumes and totalize these unit volumes by means of a mechanical counter. The volume displaced during the revolution is multiplied by the number of revolutions to give the accumulated volume passed by the meter [8, 9].

This type of meter is described as a capillary seal PD meter as the capillary action of the metered product forms a liquid seal between moving and stationary parts. Product slippage across clearances between moving and stationary parts [4] of the meter affects the accuracy of a capillary seal PD meter. The viscosity of the liquid is a major factor in determining the suitability of PD meters to certain applications. PD meters are best suited to liquids of medium to high viscosity.

The measurement parameters required for the PD meters are pressure, temperature, and density. If the fluid is a homogeneous single product, a proper equation of state, together with the measured pressure and temperature, is used to correct the measured volume to the base conditions (refer to Section 4.8.2).

There are a number of types of PD meters that have been developed and used in the process industry. These include the bi-rotor, oscillating piston, nutating disc, sliding vane, oval, rotating paddle amongst others. However, two types of PD meters have predominated for custody transfer applications in liquid petroleum pipeline transportation. These are rotating vane and bi-rotor meters.

A PD meter consists of three basic components; the external housing, the measuring component, and the counter drive train. The external housing is the pressure vessel of the meter. It can be a single or double case design. The single case design has the housing and the measuring chamber as one integral unit. Double case design has an external housing separate from the measuring chamber. The double case design has two major advantages. First, the measuring chamber tolerances are not affected by changes in operating pressures and thus have more repeatable measurement if product pressures vary during the meter operation. Secondly, system piping stresses are absorbed by the external casing and do not affect the measurement chamber dimensions.

7.4.4.1.1 Rotating Vane Meter This type of positive displacement meter measures flow by isolating segments of known volume and counting them as the liquid moves through the meter.

The measuring function is accomplished in a chamber of precise volume created by the moving blades. Product flow causes the blades to rotate around a fixed cam. The blades move out to the inner unit housing and create a precise volume that is measured by counting the revolutions of the rotor. Close tolerances in the blade clearances ensure volumetric accuracy.

The following diagram below demonstrates the mechanics of this type of meter (Figure 7-10).

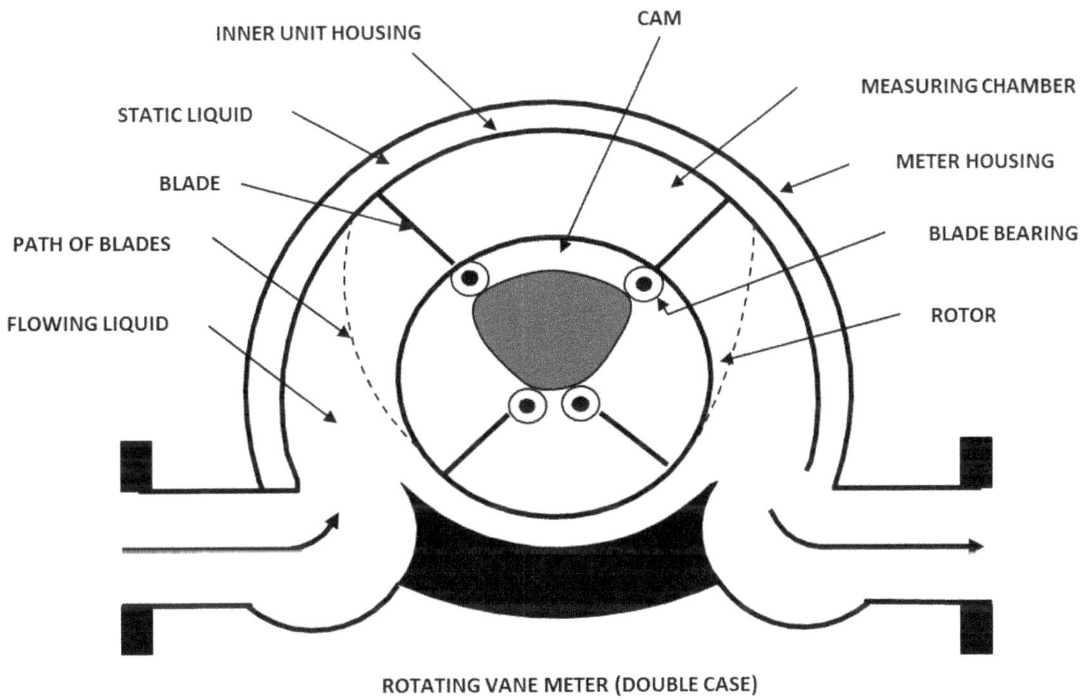

Figure 7-10. Rotating vane meter (double case)

7.4.4.1.2 Bi-Rotor Meter

Principle of Operation

Two spiral fluted rotors within the measuring unit are dynamically balanced to minimize bearing wear. As the product enters the intake of the measuring unit (see Figure 7-10), the two rotors divide the product into precise segments of volume momentarily and then return these segments to the outlet of the measuring unit. During this "liquid transition," the rotation of the two rotors is directly proportional to the flow rate of the liquid throughput. A gear train located outside the measuring unit chamber conveys mechanical rotation of the rotors to a mechanical or electronic register for totalization of liquid throughput (source — Brodie International) (Figures 7-11 and 7-12).

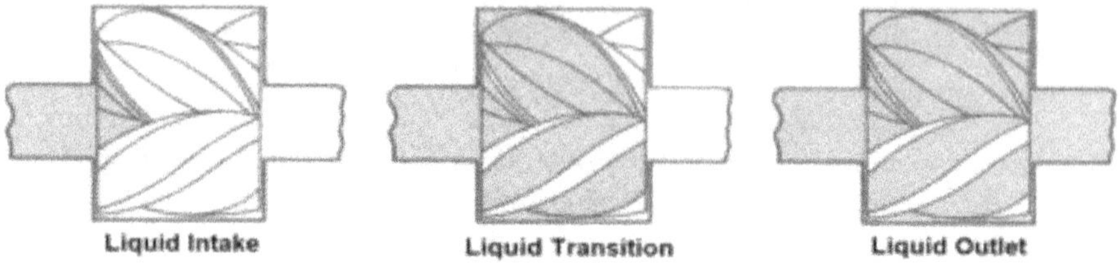

Figure 7-11. Bi-rotor flow pattern — courtesy of Brodie Meter

The main advantages of PD meters are:

- Wide applicable range (about 10:1)
- High accuracy (0.5% error)
- Minimum viscosity effects (accurate even for heavy crude measurement)
- Good for low flow rates
- Simple calibration
- No special piping requirement

However, a PD meter can only be used for clean fluids and is expensive to maintain because of its many moving parts. Also, a PD meter with large sizes is relatively expensive.

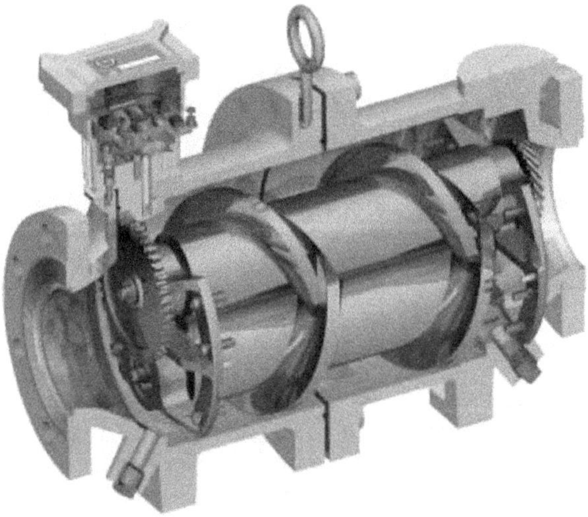

Figure 7-12. Bi-Rotor Meter — courtesy of Brodie International

7.4.4.2 Turbine Meters

Turbine meters have been used for custody transfer of petroleum liquids such as liquefied petroleum gases (LPG), refined products, and light crude oil since the early 1970s [10, 11]. With the introduction of helical turbine meters in the 1990s, turbine meter applications were expanded to higher viscosity crude oils, waxy crude oil and other products that were troublesome for the conventional bladed turbine meters (Figure 7-13).

Operating Principle

A turbine meter measures volume directly based on the principle that when a fluid passes over a turbine the fluid makes it rotate proportional to the amount of fluid passing over the turbine at a speed that is proportional to fluid velocity. Turbine rotation is a measure of velocity, which is detected by a non-contacting magnetic detector or by other means (Figure 7-14).

Figure 7-13. Helical turbine meter — reproduced courtesy of FMC Technologies, Inc

A turbine metering system consists of a meter run, turbine wheel and housing, bearings, pulse detector, straightening vanes, and pressure and temperature measurement devices. The turbine wheel rotates in the direction of fluid flow. Figure 4-15 shows the basic construction of a turbine meter. The axis of the turbine coincides with the longitudinal axis of the meter run, which is supported by bearings on both sides of the turbine wheel. These bearings are lubricated by the metered fluid. A permanent magnet embedded in the wheel generates pulses and a small coil mounted on the housing picks them up. Each pulse represents a distinct unit of volume. The total number of pulses integrated for a period of time represents the total volume metered. The straightening vanes provide flow straightening, eliminating the need for long piping upstream and downstream of the turbine meter. A uniform velocity profile is recommended for accurate measurement, but no strict requirements for fully developed flow profiles are required. The flow rate through a turbine meter is determined using the following equation:

$$Q = V/t$$

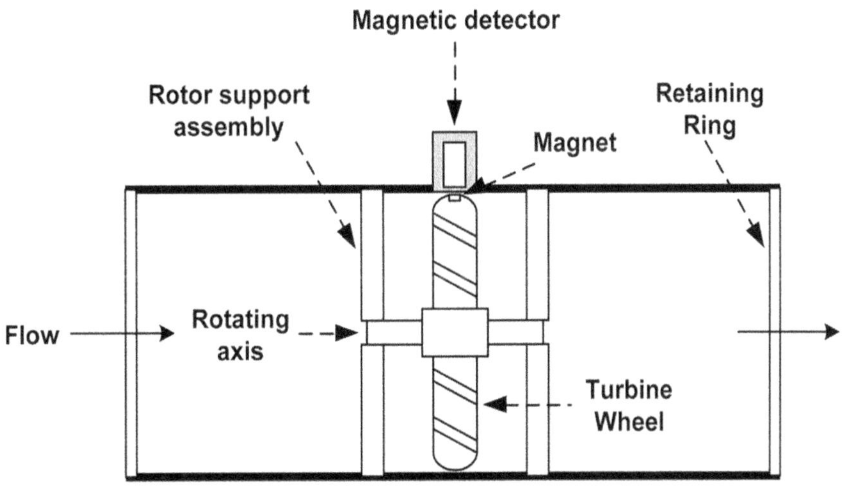

Figure 7-14. Conventional turbine meter

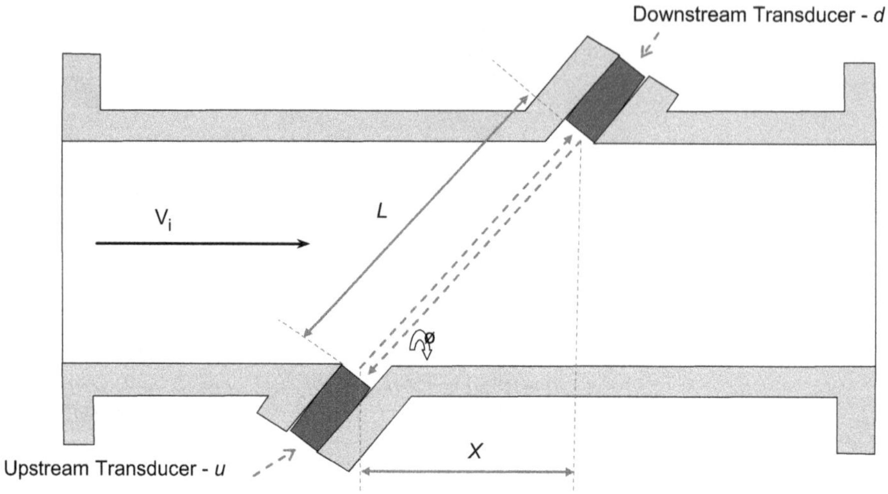

Figure 7-15. Time transit ultrasonic meter variables

where

Q = flow rate at flowing conditions
t = time
V = volumetric output over time period t

The volumetric output of the turbine meter is recorded by a revolution counter on the turbine wheel. It is expressed as:

$$V = C/k$$

where

C = pulse counts
k = meter factor

The meter factor is expressed as pulses per unit volume. It is unique for each turbine meter and used by a flow computer to calculate the totalized volume through the meter over a given time. The meter factor is a mechanical meter correction factor which accounts for effects such as bearing friction, fluid drag, and many other mechanical and electrical conditions. It is determined by a meter calibration process, using a meter prover under normal flowing conditions.

The liquid volume can be corrected to base conditions using the procedures outlined in Section 4.8.2 and detailed in *API MPMS 11.1*. Further, there is a minimum operating backpressure level that will prevent cavitation, depending on the characteristic of the specific fluid. A conservative statement of sufficient back pressure necessary when utilizing a turbine meter is given in *API Standard 2534 — Measurement of Liquid Hydrocarbon by Turbine Meter Systems, 1970.*

The liquid volume flowing through a turbine meter is calculated by correcting the raw meter pulses to base pressure and temperature conditions and taking into account the effects of flowing pressure and temperature on the fluid and the meter. The net volume at base conditions is expressed as:

$$\text{Net volume} = (\text{Number of pulses}/K\text{-factor})\ C \times C_p\ C_t\ M_p\ M_t$$

where:

K-factor is a meter factor obtained from meter proving, pulses/m^3
C_p is pressure correction factor for liquid to base conditions
C_t is temperature correction factor for liquid to base conditions
M_p is pressure correction factor for steel to base conditions
M_t is temperature correction factor for steel to base conditions

C_p and C_t can be determined from the procedures described in *API 2534*, while M_p and M_t may be obtained from a steel reference manual or the meter manufacturer.

The accuracy of a turbine meter is based on two assumptions. First the flow area remains constant and; secondly, the rotor velocity accurately represents the stream velocity. A turbine meter has advantages of high accuracy in the order of 0.25% over the flow range, has a large range of up to 100:1 at high pressure and high flow conditions, of negligible pressure loss across the metering system, and is easy to calibrate and maintain. Turbine meters are most suitable for flow control because of their fast response time to changes. Since turbine meters measure fluid volumes directly, they are known to provide accurate totalized volumes. Because of these qualities, turbine meters are widely accepted for use in custody transfer in pipeline industry.

They do however have certain limitations; they are sensitive to viscosity and their performance is adversely affected by solids or liquids in the gas stream and solid debris in the liquid stream. Therefore, the turbine metering system requires a strainer on the upstream side of the meter run. The effective rotor flow area can change for a number of reasons:

- Erosion, corrosion, deposits — small buildup or erosion of the bladed rotor can have a significant effect on meter performance.
- Boundary layer thickness — is relatively constant and insignificant for liquids of low viscosity such as refined products or light crude oils. As viscosity increases, the boundary layer increases and reduces the effective flow area.
- Cavitation — due to local vaporization of liquid in the meter will increase rotor velocity and dramatically affect the meter's accuracy.
- Obstructions — product impurities can impinge on rotor blades and decrease the effective flow area through the rotor and cause a shift in the fluid velocity profile.

Turbine meters can provide cost-effective measurement and long service life. To maximize meter accuracy and performance, flow conditioning is required. This is typically provided by upstream and downstream straightening sections of piping and, in certain cases, flow straightening vanes.

7.4.4.3 Ultrasonic Meters

Ultrasonic meters provide volumetric flow rates by measuring flow velocity. They operate either on transit time/frequency or on the Doppler effect. The transit time method is more frequently used for determining flow where sound waves transmitted in the direction of fluid flow travel faster than those traveling against the fluid flow. The transit time difference is proportional to fluid velocity. Ultrasonic flow meters have negligible pressure drop, have high turndown capability, and can handle a wide range of applications. Crude oil and refined product custody transfer are typical applications for this technology [12].

Recently, multiple beams have been used to increase accuracy and repeatability. Multipath ultrasonic flow meters use more than one pair of sending and receiving transducers to determine flow rates. The transducers send and receive a signal alternately through the same path. Flow rate is determined by averaging the values obtained by the different paths, resulting in greater accuracy and reliability than provided by single-path meters. The applicable standard for liquid measurements in North America is ASME MFC-5M.

Ultrasonic flow meters can be classified in terms of the mounting options:

- Insertion flow meters are inserted perpendicular to the flow path, with ultrasonic transducers being in direct contact with the flowing fluid.
- Clamp-on flow meters are clamped on existing pipes. Clamp-on flow meters tend to be less accurate than insertion types, but installation cost is low.

They can also be inserted between two pieces of flanged pipes or threaded into pipes. The ultrasonic transducers can be mounted in one of two modes. The upstream and downstream ultrasonic transducers can be installed on opposite sides of the pipe (diagonal mode) or on the same side (reflect mode).

Transit time ultrasonic flow meters have ultrasonic transducers facing each other as shown in

To determine the fluid velocity, the transducers transmit ultrasonic pulses with the flow and against the flow to the corresponding transducer. Each transducer alternates as a transmitter and a receiver. A pulse traveling with the flow arrives sooner than the one traveling against the flow and this time difference is related to the product velocity in the meter.

At zero flow, there is no difference in the time it takes for signals to transmit from one transducer to another. When flow is introduced, the time for the transmission of a signal from the downstream transducer to the upstream transducer will take longer than the upstream to downstream transmission. The time differential forms a relationship with the velocity of the fluid as follows:[2]

$$t_{ud} = \frac{L}{C + V_i \cos\phi}$$

$$t_{du} = \frac{L}{C - V_i \cos\phi}$$

$$Vi = \frac{L}{2\cos\phi} - \frac{t_{du} - t_{ud}}{(t_{ud})(t_{du})} = \frac{L^2}{2X} \frac{t_{du} - t_{ud}}{(t_{ud})(t_{du})}$$

$$C = \frac{L}{2} \frac{t_{du} + t_{ud}}{(t_{ud})(t_{du})}$$

where:

T_{ud} = transit time from transducer *u* to *d*
T_{du} = transit time from transducer *d* to *u*
L = path length between transducer faces *u* and *d*
C = velocity of sound in the liquid in still conditions
V_i = mean chord velocity of the flowing liquid
ϕ = acoustic transmission angle

The principle of measurement is simple but determining the true average velocity is difficult, especially to obtain custody transfer measurement accuracy. Manufacturers often provide multiple sets of transducers to enhance measurement accuracy.

A key difference between ultrasonic meters and other meters is inertia. In PD meters, turbine meters and Coriolis mass meters, there is an inertia transfer from the flowing stream to the measuring element. The ultrasonic meter measures the flow stream directly without imposing any constraints. Without inertia, an ultrasonic meter detects a small change in stream velocity. For this reason, ultrasonic meters are far more sensitive to systematic error than conventional meters.

Like turbine meters, ultrasonic meters are affected by boundary layer thickness. With medium to high viscosity liquids, this effect must be compensated to achieve accurate measurement. Multi-path ultrasonic meters have methods to minimize this effect.

As with turbine meters, flow straightening piping is required to provide maximum ultrasonic meter accuracy. Field proving of liquid ultrasonic meters is difficult for two reasons:

1. Ultrasonic meters have output pulses that are not related in real time to the meter throughput. There is a time delay that exists between what is being measured and the pulse output. Reducing the meter's response time and/or increasing the prove volume are recommended.
2. The inertia-free principle makes the ultrasonic meter far more sensitive to systematic error than conventional meters. Measurement accuracy is improved by taking more samples.

A Doppler ultrasonic flow meter uses the fact that fluid flow causes sound frequency shifts which are proportional to the fluid velocity. Doppler meters also send an ultrasonic signal across a pipe, but the signal is reflected off moving particles in the flow, instead of being sent to a receiver on the other side. The moving particles are assumed to be travelling at the same speed as the flow. A receiver measures the frequency of the reflected signal, and the meter calculates flow by determining the frequency shift of the detected frequency from the generated frequency. Doppler ultrasonic flow meters require the presence of particles in the flow which deflect the ultrasonic signal. Because of this, they are used mainly for slurries and liquids with impurities but their accuracy is poor and only applicable to liquids.

Ultrasonic transit time flow meters offer the promise of high accuracy, low cost, wide flow range, low pressure drop, and low maintenance because of the lack of moving parts. The range of the flow meter is 20:1 while its accuracy for a multi-path system is better than ±1.0%. However, they do not work well for liquids with suspended solid particles or air gaps. Doppler ultrasonic flow meters can be used for liquids with bubbles or suspended solid particles.

7.4.4.4 Coriolis Meter

Coriolis mass meters are relatively new compared to other flow measuring devices [13]. They were introduced to industry in the early 1980s and have gained acceptance as accurate and reliable flow measuring devices. An advantage of the Coriolis meter is that it measures the mass flow rate directly, eliminating the need to compensate for pressure and temperature. Coriolis meters are available in a number of different designs. A popular configuration consists of one or two U-shaped, horseshoe-shaped, or tennis-racket-shaped flow tubes with inlet on one side and outlet on the other. The tubes are enclosed in sensor housing and connected to a flow computer unit. A more recent single straight or slightly bent tube design is available to measure some dirty and/or abrasive liquids that may clog the U-shaped design.

API first published *MPMS Chapter 5.6 — Measurement of Liquid Hydrocarbons by Coriolis Meters* in October 2002. This standard describes methods to achieve custody transfer levels of accuracy when a Coriolis meter is sued to measure liquid hydrocarbons. Also, *ISO 10790 Standards* cover Coriolis mass flow meters for liquid applications (Figures 7-16 and 7-17).

Operating Principle

When an oscillating excitation force is applied to the tube causing it to vibrate, the fluid flowing through the tube will induce a rotation or twist to the tube because of the Coriolis force acting in the opposite direction on either side of the applied force. For example, when the tube is moving upward during the first half cycle created by the vibration source, the

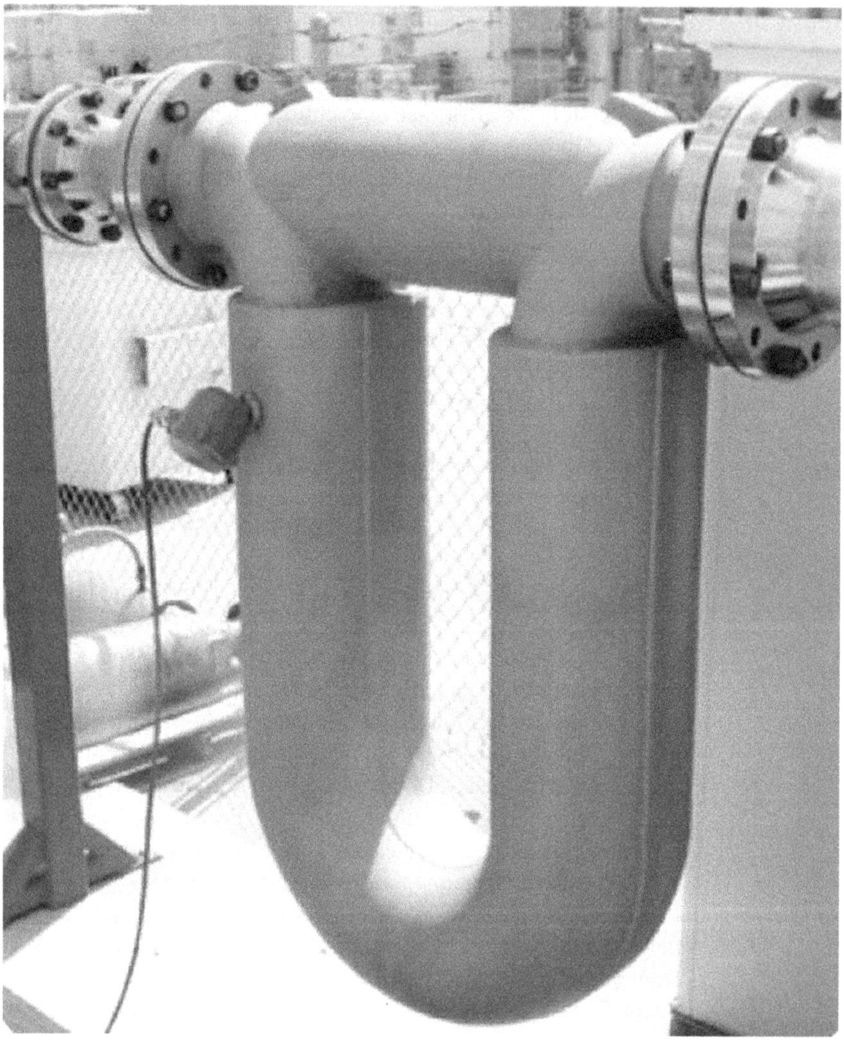

Figure 7-16. Micro Motion ELITE Coriolis meter courtesy of Emerson Process Management

Figure 7-17. High capacity Promass X 4 tube Coriolis meter courtesy of Endress+Hauser

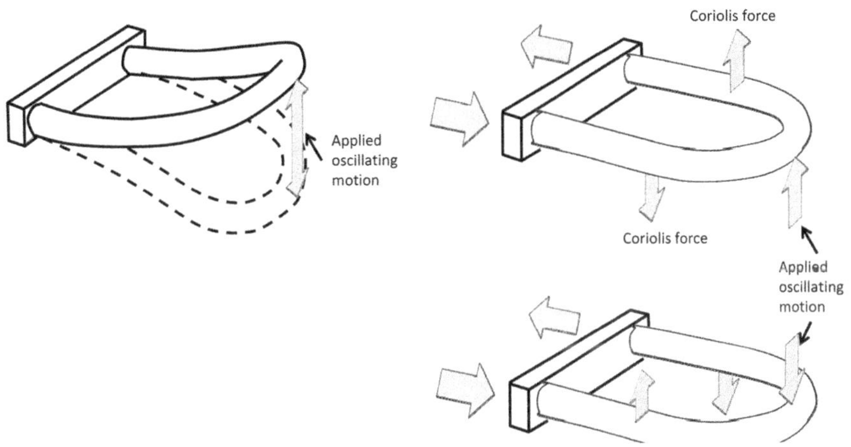

Figure 7-18. Coriolis operating principle

fluid flowing into the meter resists being forced up by creating an opposing force down on the tube (Figure 7-18). On the opposite side, the liquid flowing out of the meter resists having its vertical motion decreased by creating an opposing force up on the tube. This action causes the tube to twist. When the tube is moving downward during the second half of the vibration cycle, it twists in the opposite direction. This twist is recorded by sensors on either side of the tube between the source of the oscillation force. At zero flow, both sensors are in phase. Under flowing conditions, the phase difference (time lag) between the sensors on the inlet side and the outlet side directly correlates with the mass flow through the tube.

The amplitude of the Coriolis force depends on the moving mass and its velocity in the system. Manufacturers use different technologies to manipulate the phase shift and frequency to determine mass flow and density of the product. The frequency output is expressed as a pulse scaling factor (PSF), representing the number of pulses for a given mass flow rate. The factor defines the relationship between the mass flow rate and frequency.

Coriolis mass meters are known to be very accurate and their measurements repeatable and have "out of the box" accuracies of the order of 0.5%, independent of flow profile and composition, and do not require ancillary measurement equipment such as pressure and temperature to determine the mass flow rate. The meters have low maintenance requirements. They are ideal for relatively low flow rate and for custody transfer in mass rather than volume such as for ethylene. They are even suitable for liquid flow measurements with a small amount of gas. Even though the meter generally costs more than other types of meters, it does not require flow conditioners and thus overall cost is comparable to others. The pressure drop can be high however and thus it may not be suitable for measuring large mass flow rate without its use resulting in excessive pressure drop; this limitation is a factor of the size of Coriolis mass meters. Recently, certain manufacturers have offered higher capacity meters such as the versions seen in Figure 7-16 and Figure 7-17.

Density measurement

Density is determined at zero flow by measuring the natural frequency of the tube containing the particular fluid. The measuring tubes are excited at their resonance frequency by electromagnetic sensors. A change in the mass and thus the density of the oscillating system (consisting of both measuring tubes and fluid) results in a change in the resonance frequency. Resonance frequency is thus a function of fluid density. A microprocessor flow computer utilizes this relationship to obtain a density signal.

Temperature measurement

The temperature of the measuring tubes is determined in order to calculate the compensation factor due to temperature effects. This signal corresponds to the product temperature and together with the mass flow determination and density of the fluid; the flow computer is able to accurately determine volumetric flow.

Volume Measurement

Although the Coriolis meter directly measures mass, it is not common in the petroleum industry to use mass measurement for custody transfer of petroleum and its products. Most Coriolis meters can measure the density of the fluid in addition to the mass flow rate. Since volume is equal to mass flow divided by density, a flow computer can convert output to volume. At this point, Coriolis meters become volume meters and can provide output similar to other types of meters. To determine overall accuracy capabilities of Coriolis meters, it is necessary to determine their accuracy of both density measurement as well as mass measurement. Coriolis meters can differ significantly in their specification of density accuracy and therefore would differ significantly in their volume accuracy.

Advantages

- Direct mass flow measurement — this feature is particularly advantageous for liquid measurement in the petrochemical industry where product is measured by mass rather than by volume.
- High accuracy — Coriolis meters are highly accurate with accuracies typically at or better than +/– 0.1 % of range.
- High reliability — no moving parts other than the vibrating flow tube means high reliability and low maintenance.
- High flow rate turndown ratio — Coriolis meters have a large turndown ratio, more than twice the turndown of a turbine meter

Disadvantages

- Cost — Coriolis meters are expensive in terms of purchase price. However, their low maintenance and high reliability provide advantage on life cycle cost. Some manufacturers have responded by making available lower cost Coriolis flow meters that are comparable in cost and accuracy with other types of flow measurement such as ultrasonic, positive displacement and turbine meters.
- Pressure drop — until recently, Coriolis meters were limited in size and configuration of tubes that resulted in relatively high pressure losses through the meter. However, in the past several years, a number of companies have introduced Coriolis flow meters in line sizes above NPS6 and recently as large as NPS14. As well, there have been developments in straight tube Coriolis meters that will reduce pressure losses through the meter.

7.4.5 Meter Selection

Factors that should be considered in the selection of meter type, size and quantity include:

- Accuracy — meters are designed to operate within a specified accuracy or linearity range
- Fluid properties including viscosity, density and contaminants
- Pressure losses through meter and piping
- Dimensional requirements for meter and necessary flow conditioning piping
- Back pressure requirements

The most important operating conditions that affect the accuracy of liquid measurement are flow range and viscosity range [14, 15].

Flow Range — is the minimum/maximum flow rate at which a meter/measurement system can operate within the stated accuracy. Custody transfer meters are normally specified to operate over a 10:1 (10% to 100% of maximum flow) flow range with a linearity of +/–0.15%. The measurement accuracy can be improved by reducing the range of flows or establishing a separate meter factor for each flow rate. Today's flow computers allow multiple meter factors to be input and automatically linearize the meter factor over the flow range.

Viscosity Range — can vary for petroleum liquids from less than 0.1 cP for LPG to over 1000 cP for heavy oils. All meters are sensitive to viscosity (with the exception of Coriolis meters) but each metering technology is affected differently. Because of slippage through internal clearances, PD meters are affected by low viscosity liquids whereas turbine and ultrasonic meters are sensitive to high viscosity crude oils.

With current meter proving methodology, the total uncertainty (UT) of a liquid custody transfer measurement can be in the range of UT = +/–0.1% at a 95% confidence level. This is industry norm for most large volume liquid transactions.

For optimum performance, the selected meters must be capable of covering a wide flow range over which the meter maintains a linear pulse output with respect to flow rate; typically 0.25% for a 10:1 flow range or turndown ratio. Meters with this capability include turbine meters, positive displacement (PD) meters, ultrasonic meters and Coriolis meters. Turbine and ultrasonic meters offer the advantages of high flow capacity and reduced weight, space and maintenance. Turbine and ultrasonic meters require flow conditioning piping upstream and downstream of the meter to prevent fluid swirl and non-uniform velocity profiles. Flow straightening tube bundles or conditioning plates reduce the amount of upstream piping required for flow conditioning. PD meters and Coriolis meters do not require upstream or downstream flow conditioning.

Positive displacement meters and turbine meters are the most commonly used meters in custody transfer applications. However, ultrasonic and Coriolis meters have been making inroads as they have definite advantages in some applications.

Turbine meters are preferred for high flow rates and low viscosity applications. Turbine meters are susceptible to deposition of wax in certain crude oils. However, a limited amount of fine abrasives have less effect on the life and performance of a turbine meter because solids in suspension continue to flow uninterrupted through the meter. Positive displacement meters are more affected by fine abrasives because of the close tolerances of the moving parts.

Ultrasonic meters have the advantage of low pressure losses and capability of handling corrosive liquids or liquids with contaminants.

Coriolis meters have relatively high pressure losses through the tubing assemblies and can be susceptible to deposition of certain products such as wax.

Positive displacement and Coriolis meters are only slightly affected by installation conditions whereas velocity meters such as turbine and ultrasonic meters can be highly affected. Turbine and ultrasonic meters require flow conditioning piping assemblies.

7.4.5.1 Meter Sizing

Meters should have capacity to handle the minimum and maximum expected flow rate for the meter run. PD meters are normally selected for continuous operation at about 75% of the manufacturer's nameplate capacity if the liquid has reasonable lubricity. The capacity of PD meters is reduced to as low as 40% of nameplate capacity for liquids with poor lubricity. Turbine, ultrasonic, and Coriolis meters may be operated at full nameplate capacity with any liquid. However, pressure losses through the meter and piping at full rated capacity may be a factor in choosing the most appropriate meter size for the particular application.

7.4.5.2 Instrumentation and Accessories

Strainers and Filters — Strainers and filters should be designed to remove only solids that could damage a meter or create uncertainty of measurement. Meters can be protected individually or as a bank. With turbine and ultrasonic meters, the strainer should be placed well ahead of the meter runs to prevent the problem of liquid swirl from affecting meter performance. This is not an issue with PD or Coriolis meters. The strainer should be equipped with a pressure differential monitoring system to warn of accumulation of material in the strainer.

Sediment and Water (S&W) Determination — S&W determination procedure including the frequency of sampling must be representative of the entire volume transaction as well as the subsequent S&W sample analysis. There are two methods to obtain the measurement; sampling or on-line analysis using a suitable product analyzer. Sampling can be categorized by two methods; spot or grab sampling or continuous proportional sampling. It is important that the sample location be carefully selected such that the flowing stream is adequately mixed. Manual sampling procedures and equipment are addressed in *ASTM D4057 (API MPMS Chapter 8.1) — Manual Sampling of Petroleum and Petroleum Products.*

Most pipeline metering systems employ automatic sampling. Procedures are covered in *ASTM D4177 (API MPMS Chapter 8.2) — Automatic Sampling of Petroleum and Petroleum Products.* An accurate analysis of any sample depends on the appropriate handling and mixing of that sample from its sourcing through to its analysis. Procedures are covered in *ASTM D5854 (API MPMS Chapter 8.3) — Practice for Mixing and Handling of Liquid Samples of Petroleum and Petroleum Products.*

Back-pressure valves — a back-pressure valve should be installed downstream from the meter station if the line resistance downstream is insufficient to prevent vaporization at the meter assemblies under any flow conditions.

Flow control valves — Flow control valves may be placed on individual meter runs or act collectively for a number of meters. A flow control valve placed downstream of the meter may also act as a back-pressure valve by the application of control logic to the valve actuator.

Electronic Flow Measurement (EFM) — An EFM is any flow measurement and related system that collects data and performs flow calculations electronically. This may be part of a Distributed Control System (DCS), supervisory control and data acquisition (SCADA) system, a Programmable Logic Control (PLC) system, or a specialized flow computer.

The table below summarizes the meter sizes and applicable liquids for the selection of meters.

Flow meter	Pipe size in (mm)	Clean	Viscous	Dirty	Corrosive
Turbine	0.25–24 (6–600)	D/A	N/A	N/A	A
PD	< 12 (300)	D/A	D/A	N/A	A
Ultrasonic	> 0.5 (12)	D/A	A	N/A	D/A
Coriolis	0.1–4 (2.5–100)	D/A	D/A	D/A	A

D/A: Designed for this application; A: Normally applicable; N/A: Not applicable.

The table below summarizes the meter accuracy without smart transmitter and applicable maximum pressure, temperature, and Reynolds number [2]. The accuracy is over the upper range value of the flow rate.

Flow meter	Accuracy (+/– %)	Pressure, Psig (kPag)	Temperature, °F (°C)	Reynolds Number
Turbine	0.25	3,000 (21,000)	–400 to 500 (–268 to 260)	<15 cSt
PD	0.5	3,000 (21,000)	600 (315)	<8000 cSt
Ultrasonic	1.0	Pipe rating	–300 to 500 (–180 to 260)	No limit
Coriolis	0.5	2,000 (14,000)	570 (300)	No limit

7.4.5.3 Flow Computers

Flow computers are widely used in the pipeline industry. Flow computers not only collect measured flow and other data, calculate volumes, correct flow rates to base conditions, and store all measured and calculated data, but also provide the flow information rapidly on a real-time basis.

The hardware structure of flow computers is similar to personal computers (PC). They should be rugged due to severe or even hazardous environments in which they operate. Flow computers are interfaced with flow meters and other measuring devices through their transducers and have programmable capabilities necessary for various applications. Also, they have the capabilities to upload the flow computer data to a host SCADA system and provide measurement data security by setting the security code and authorization. Unlike PCs, flow computers work in real-time and are dedicated to applications related to flow measurements.

Flow computers do not have flexible display capabilities. Their screens are small and a keypad is used to set up and view parameters. Typically, flow computers require the following displays, mostly menu-driven, to enter and access the necessary parameters and data:

- The flow computer configuration data such as unit ID, location, base pressure and temperature, etc.
- Meter specific parameters such as meter type, meter factor, etc.
- Liquid product property parameters such as API gravity, volume correction factor, etc.
- Communication parameters
- Security
- Calibration
- Alarm parameters such as analog and digital input alarm, rate/volume alarm, etc.
- Diagnostic messages

In general, flow computers perform flow measurement and process calculations, monitor transducer inputs (both analog and digital inputs) in real-time, produce and store multiple measured and calculated output including reports, and can serve as a remote terminal unit (RTU).

Flow computers are able to provide all measurement related functions. They not only read and monitor all inputs of flow and/or volume, temperature, and pressure for most flow meters but also differential pressure from differential pressure meters and pulse input from turbine and positive displacement meters. Input also includes fluid properties such as liquid density and viscosity.

For custody transfer, flow computers need to be able to process the flow and other measurement data as specified in the standards appropriate to the measured fluids. They should be able to correct volumes to base conditions and totalize volumes from meter run totals and station totals for each product. Flow computers monitor and store batch operation data, which include batch ID, volume, and batch lifting and delivery times.

When a flow computer is used for proving meters, it not only controls the meter prover and calculates the meter factor during the proving time, but also uses the meter factor and *K*-factor to determine accurate volume. The *K*-factor is the number of pulses per unit volume and the meter factor a correction applied multiplicatively to the *K*-factor.

Most flow computers are able to display limited outputs and produce various reports. The minimum required reports generated by a flow computer may include volume totals and quality, batch, alarm, and audit trail reports. This information can be directly accessed from the flow computer or uploaded to the host and accessed from the SCADA database.

7.4.6 Meter Station Design

A meter station is typically located where there are injections or deliveries from a pipeline. Flow metering can be provided for control purposes to supply the operator with operational flow data. Alternately, the meter station may be part of a pipeline leak detection system. Metering stations may also contain a custody transfer meters, measuring the amount of liquid for commercial purposes [16–18].

These meter stations are designed for a high degree of accuracy and a wide range of flow rates. In order to meet both of these requirements, a meter station is usually installed with one or more parallel meter runs, each containing metering devices. The number of parallel meter runs will determine the flow range measurable by the meter station. A multi-run station needs to be remotely controllable, with meter station control logic that will automatically put the required number of meters into operation to meet the demand. Flow or pressure regulators may be required to control flow or pressure at the injection site or the delivery station.

Dynamic liquid measurement systems typically consist of a primary measurement element, such as a meter, secondary measurement devices such as temperature and pressure transmitters and, in some cases, level transmitters and densitometers. Tertiary devices for electronic flow measurement include distributed control systems (DCS), supervisory control and data acquisition systems (SCADA), remote terminal units (RTU) and flow computers (Figure 7-19).

Figure 7-19. A custody transfer meter station — Source [27]

7.4.6.1 Meter Station Components

Meter stations should be designed as per the requirements specified in the appropriate standards. A typical meter station has the following components as shown in Figure 7-9.

Headers: Headers are the upstream and downstream pieces of pipe, to which the meter runs and the yard piping in a meter station are connected. Headers should be designed to allow for the addition or deletion of meter runs to or from the meter stations as the flow requirements change. At a multi-run meter station, it is recommended that all headers be the same size in order to split the flow equally among the meter runs under varying flow conditions (Figure 7-20).

7.4.6.1.1 Meter Station Valves Meter station valves include check valves to avoid back flow, block and bypass valves to regulate flow direction, and a flow control valve to regulate the flow rate. Valves for isolation of each meter are installed so that individual meter elements can be removed for repair without shutting down the meter station. The control valve in each meter run is used by the meter station controller to balance flows between each meter run.

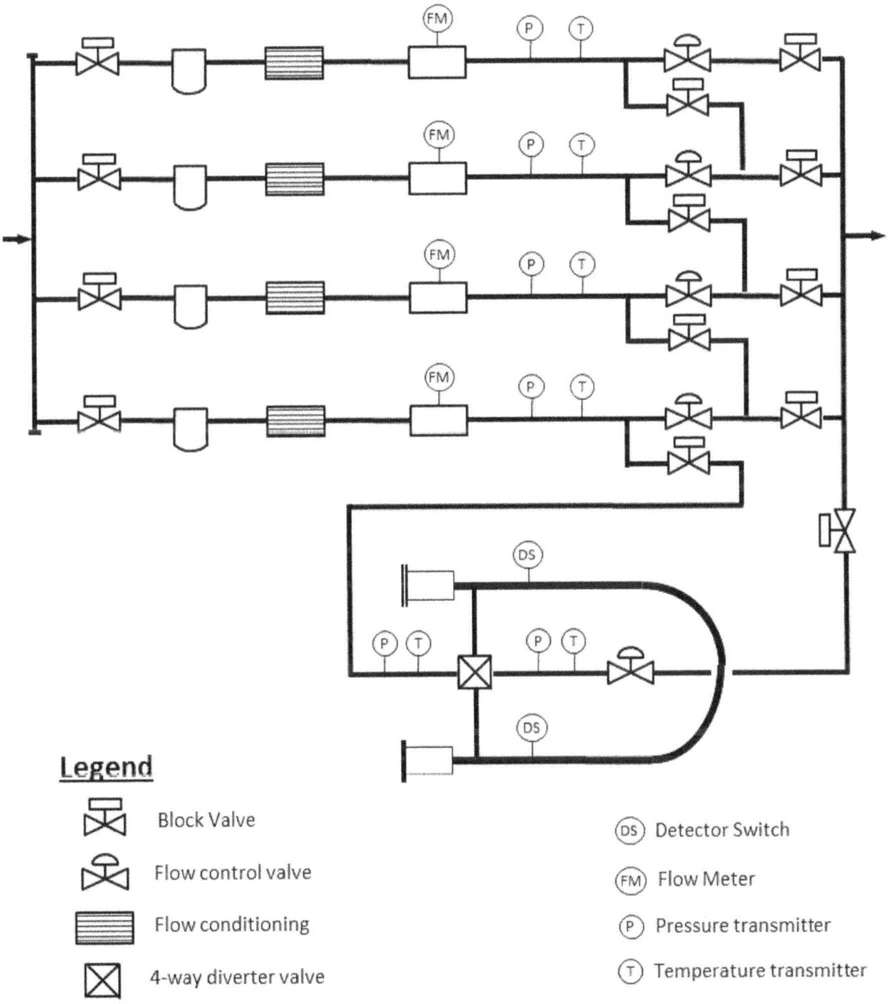

Figure 7-20. Simplified meter station with prover

Many valves are required at a meter station to control flow and protect flow meters. Depending on their function and location, valve types can be different as described below:

- Control valve to regulate the flow rate: globe or ball valve
- Check valve to stop back flow for protecting the meter
- Upstream or downstream valve assembly: gate or ball valve
- Upstream or downstream meter run: ball valve

7.4.6.1.2 Meter Prover A meter prover is used with turbine and PD meters to establish a relationship between the number of counts or revolutions of the meter and the volume flowing through the meter. The number of counts on the meter being proved is related to the volume passing the detectors on the prover.

7.4.6.1.3 Other The pressure and temperature transmitters are installed together with the flow transmitters for flow or volume correction. A pressure regulator maintains a constant downstream pressure regardless of the flow rate in the line.

7.4.6.2 Meter Run

A meter run consists of straight pipes, valves, straightening vanes, and other equipment before and after a flow meter. A strainer or filter is installed at the turbine and the PD meter station, since the metering is susceptible to damage from solid particles. A strainer is installed on each meter run.

Fluid enters a meter through a meter run. A meter run is defined as the length of straight pipe of the same diameter between the primary measuring element and the nearest upstream and downstream pipe fittings. It is required to reduce flow turbulence so that the measurement accuracy and consistency can be maintained in any operating conditions.

A meter run consists of a meter and pipes, pressure and temperature measuring devices, valves including a check valve, a strainer and straightening vanes for turbine and orifice meters. The flow range of the meters installed and the volume of fluid flowing through the meter station primarily determine the size and number of meter runs.

Pipe lengths and fittings of the upstream and downstream headers are different depending on the pipe size and flow meter type. These requirements are specified in the standards of the selected flow meter type. These standards are listed at the end of this chapter. The meter tube diameter used in a meter run should be consistent with the size of the meters such as turbine and PD meters. The meter tube is used to maintain the accuracy of flow measurements. Straightening vanes may be installed in the upstream section of the meter tube to minimize turbulence ahead of the meter. Where in-line proving is to be performed, proving taps and a double block and bleed diverter valve must be installed.

Components of a liquid hydrocarbon measurement system are shown in Figure 7-21.

The number of meters for a particular station is determined by a number of factors. A large number of small meters would result in a smaller prover, wider rangeability of flow rates and standby capacity. However, a station with too many meters has high costs for all associated auxiliary equipment and instrumentation.

Most system designs include a spare meter run. This ensures that system capacity is not compromised should one meter or critical component become inoperable. A normal multi-meter station usually has a minimum of two meter runs and a maximum of 5 or 6 m runs.[3]

API MPMS Chapter 5.3 — Measurement of Liquid Hydrocarbons by Turbine Meter — Appendix A presents an empirical method for computing the length of upstream straight pipe required for various installation configurations and operating conditions. Experience has shown that a nominal length of 20 diameters of meter-bore piping

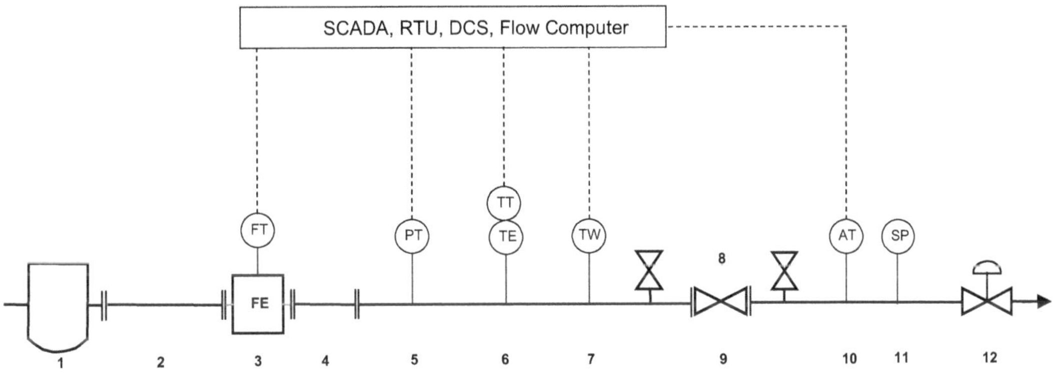

Components

1 - Strainer
2 - Upstream straight lengths
3 - Meter
4 - Downstream straight lengths
5 - Pressure transmitter
6 - Temperature transmitter
7 - Thermo well
8 - Prover valves
9 - Double block and bleed diverter valve
10 - Analyzer (e.g. S&W monitor, densitometer)
11 - Sample point (manual or on-line)
12 - Flow control valve

Notes

- Schematic is generic in nature. All elements may not be required for a specific application
- Air eliminator is not necessarily required for pipeline applications
- Strainer is required for most but not all meter types
- Upstream and downstream flow conditioning straight lengths vary depending upon meter type and upstream piping disturbances
- Flow transmitter (FT) may be close coupled to flow sensor (FE) or remote mounted
- No components such as analyzer loops or pressure relief valves should be between meter and prover taps.
- Pressure relief valves should be located in such a manner as to preclude unmeasured product being diverted via a leaking relief valve

Figure 7-21. Metering assembly schematic

upstream of the meter and 5 diameters of meter-bore piping downstream of the meter provide effective conditioning in many installations; however, the required length of upstream piping should be verified for each installation, using the method presented in *Appendix A*. This technique does not predict the length of straight pipe required downstream of the meter. A minimum of 5 diameters of meter-bore piping should be provided downstream of the meter unless a different length is supported by the manufacturer's recommendations or tests.

Where there is limited space for upstream piping, flow straightening vanes can be installed. *API MPMS Chapter 5.3* describes the methodology for determining the upstream piping requirements as shown in Figure 7-22.

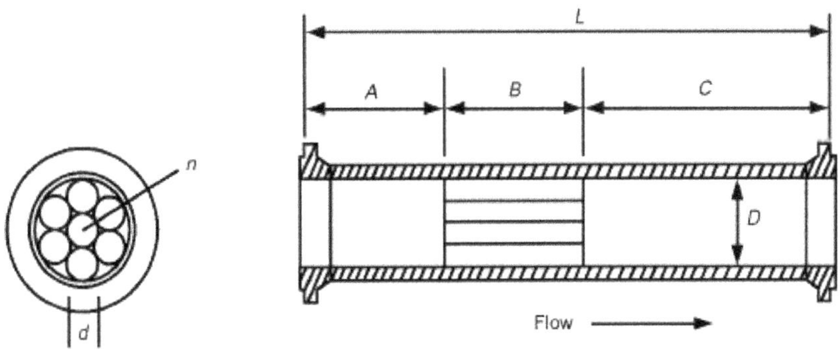

Figure 7-22. Upstream piping requirements — source API MPMS 5.3

7.4.6.3 Meter Provers

In dynamic measurement systems used for custody transfer, the proving system and process are essential to validate the accuracy of volume measured. There are various worldwide standards and regulations that define the measurement system requirements but each are fundamentally the same. Each is traceable to a national standard which, in turn, is traceable to an international standard. In dynamic measurement systems, this link is the prover volume. There are gravimetric and volumetric methods where the prover volume is traceable to the International Bureau of Legal Metrology (BIML). The technical arm of the bureau is the International Organization for Legal Metrology (OIML). This is an intergovernmental organization that deals with all aspects of legal metrology. It has 59 member states and 50 corresponding members. In this way, all the countries are tied to the same weights and measurement standards [19–22].

Proving is a method of checking a measuring device against an accepted standard to determine the accuracy and repeatability of that measuring device. Meters are proven immediately after repair, removal from service for any reason, when changing fluid products being measured, when product viscosity changes, or on demand if meter history indicates that it is required. On liquid pipelines that carry multiple products or products of varying properties, it is important to be able to perform meter proving on demand in order to test the accuracy of the meter station.

When proving or determining the accuracy of meters, proper practices and procedures must be followed. The API Manual of Petroleum Measurement Standards (MPMS) Chapters 4 and 5 as well as ISO 7278 provide guidelines on meter proving techniques and standards. A meter prover is used to verify the accuracy of the liquid meters. Essentially, the prover determines the meter factor that is representative of the volume being put through the meter. The *API MPMS* defines the meter factor as a number obtained by dividing the actual volume of liquid passed through a meter during proving by the volume registered by that meter. The meter factor accounts for non-ideal effects such as bearing friction, fluid drag, and mechanical or electrical readout drag. In addition, turbine and PD meters are subject to accuracy variations as a result of temperature, pressure, viscosity, and gravity changes. Therefore, the meter should be proved under the same operating conditions as those that the meter experiences during normal operation.

During proving, the meter outlet is diverted through the prover and the measurement of the prover is compared to the meter's measurement. If the difference between the two readings exceeds the allowable tolerance, a new meter factor is calculated and used to correct the meter until it is proven again. This correction factor is called the meter factor and will be one of the data points sent to the station control and SCADA system. A history of meter factor changes is required to monitor the meter's performance and for billing audit purposes. It is also used to adjust previous meter tickets according to the metering policy of the pipeline company.

The method of determining the meter factor is to put a known volume through the meter and count the number of pulses generated from the test meter. Temperature and pressure need to be stable before running the prover and should be measured and recorded during the proving to correct their effects on volume. The meter factor or the meter K-factor is determined by

$$\text{Meter } K\text{-factor} = \text{number of pulses from meter/actual volume}$$

If direct volume readout is obtained from the meter, the meter factor is determined by

$$\text{Meter factor} = \text{corrected prover volume/corrected meter volume}$$

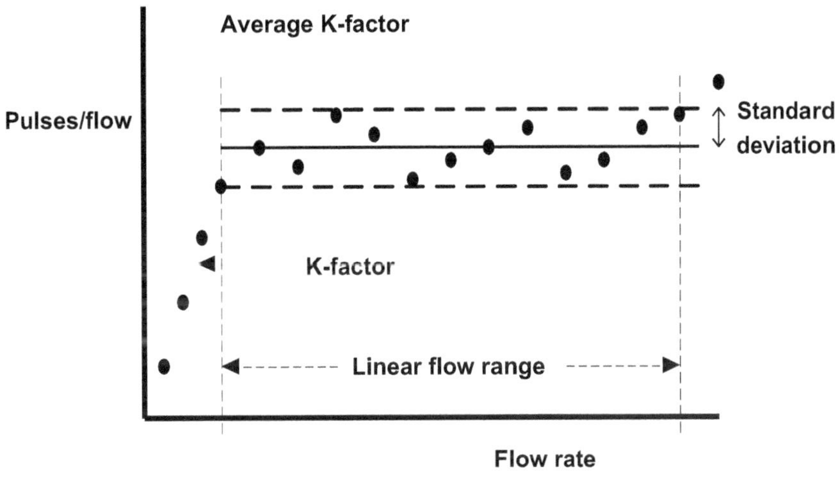

Figure 7-23. Meter factor

Since the meter factor depends on the fluid properties, different meter factors are required for different fluids. The following data are required for meter proving (Figure 7-23):

- Data on the test meters and meter prover
- Properties of all the products that are put through the test meter
- Temperature, pressure and density during a proving run
- Meter factor for each product and flow range including their variables such as pulse counts and actual volumes used to calculate the meter factor
- Status information of the prover and the proving progress, pump starts/stops, ball launch/detection, etc.

K-factors are determined over various flow rates by conducting several proving runs and a representative *K*-factor is selected for an applicable linear flow range. The *K*-factor is acceptable if the average from multiple consecutive proving runs is within 0.05%. A typical *K*-factor linearity curve is shown above.

The meter factors are maintained in a table illustrated below. The meter factor table includes the meter factors with the flow rate and density ranges over which the meter factors are applicable. In addition, it includes the pressure and temperature at which the meter is proven, and the date and time when the meter is proven, and prove report number.

Meter proving can be done using a flow computer or the SCADA system, depending on the telemetry equipment availability. If meter proving is done locally using the flow computer or the SCADA, either the local facility is instructed to perform meter proving or it is automatically initiated upon the detection of a batch interface for a batched pipeline. The host SCADA may be able to control the entire proving sequence, from initiating each prover run, gathering all data at the completion of each run, to calculating the meter factor (Figure 7-24).

A custody transfer meter station must have either a fixed prover on site or have connections for a portable prover or a master meter proving system.

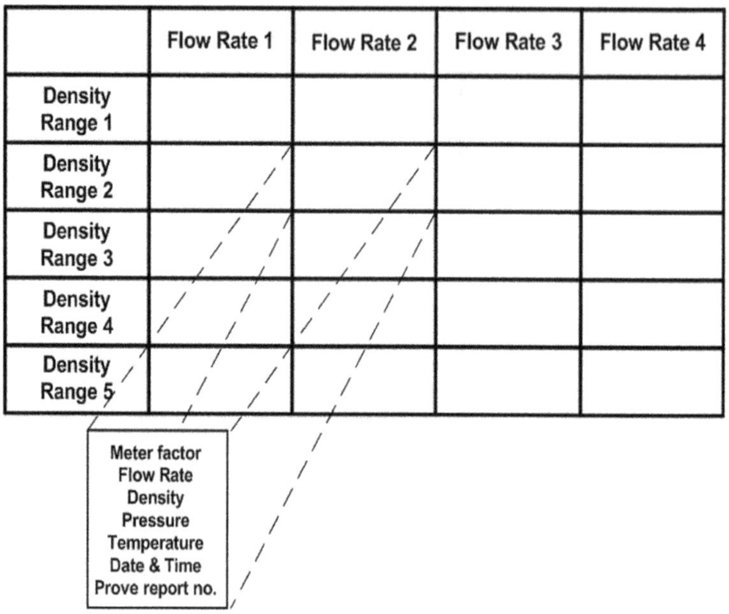

Figure 7-24. Meter factor

7.4.7 Prover Types

7.4.7.1 Tank Provers

Tank provers have a capital cost advantage over pipe provers in fixed installations. They are best suited to low capacity or remote locations without electric power. Tank provers are not suitable for high vapor pressure liquids or for viscous liquids that may not completely drain from the tank prover inner surfaces.

Operation of static fill type provers, such as tank provers, requires that the flow be started and stopped and thus they do not provide for a meter correction factor under actual measurement conditions of flow, pressure and temperature.

7.4.7.2 Conventional Pipe Provers

Conventional pipe provers operate by displacing a known volume of liquid within a calibrated section of pipe. Repeatable displacement of fluids is achieved by an oversized sphere or piston travelling through the pipe between detectors. The first conventional pipe provers standardized by API achieved the required measurement resolution of 0.01% pulse resolution by generating no less than 10,000 m pulses during a proving pass. The pipe prover design is such that the full flow through the metering stream being proved will pass through the pipe prover.

Pipe provers are an important part of a custody transfer metering station and are used to calibrate the meter *K*-factor periodically when flow rates or conditions change. The pipe prover is used to prove the accuracy and repeatability of flow meters on actual operating fluids by measuring the volume of fluids passing through the meter in relation to the number of pulses generated by the meter. The basic operating principle of the pipe prover is to meter the fluids swept by the displacer through a calibrated volume of pipe by counting the number of meter pulses between the start and stop detectors at either end of the calibrated volume. The displacer, either a piston or sphere, actuates the start and stop detectors and is designed to form a sliding seal which moves at the same rate as the flowing liquid. Temperature and pressure corrections are required to convert the calibrated volume at standard conditions to process conditions. The volume indicated by the meter is compared to the calibrated volume to determine a meter

K-factor. Generally, meter *K*-factor calibration must achieve repeatability over five successive runs to within a band of 0.1% to meet the generally accepted requirement for overall measurement uncertainty of ±0.25% at the meter station.

API MPMS Chapter 4 sections 1 to 9 and ISO 7278 parts 1 to 4 are the current publications standardizing pipe prover design and operation.

The conventional positive displacement pipe prover generates no fewer than 10,000 pulses for each proving pass to achieve a measurement resolution of 0.01% as defined in the *API MPMS* and *ISO* standards. Conventional pipe provers can be constructed in a number of configurations such as unidirectional or bi-directional pipe provers with piston or sphere displacers.

7.4.7.2.1 Uni-directional Provers The uni-directional prover utilizes an oversized sphere which is launched from a sphere handling valve during the proving cycle. The sphere enters the prover piping and passes the first detector switch actuating the meter pulse counter. The sphere travels with the flow through the calibrated section of pipe past the second detector which stops the pulse count and the sphere then drops into the receiver piping and back to the entrance of the sphere handling valve. The system is then ready to begin the proving cycle again (Figure 7-25).

The uni-directional provers are typically used for proving large meters with high flow rates. The unidirectional prover assembly does not need the 4-way diverter valve required for the bi-directional provers and this reduces the system cost. Unidirectional provers are more suitable for those applications where the product always flows through the prover such as products with high viscosity or high pour point at ambient temperatures.

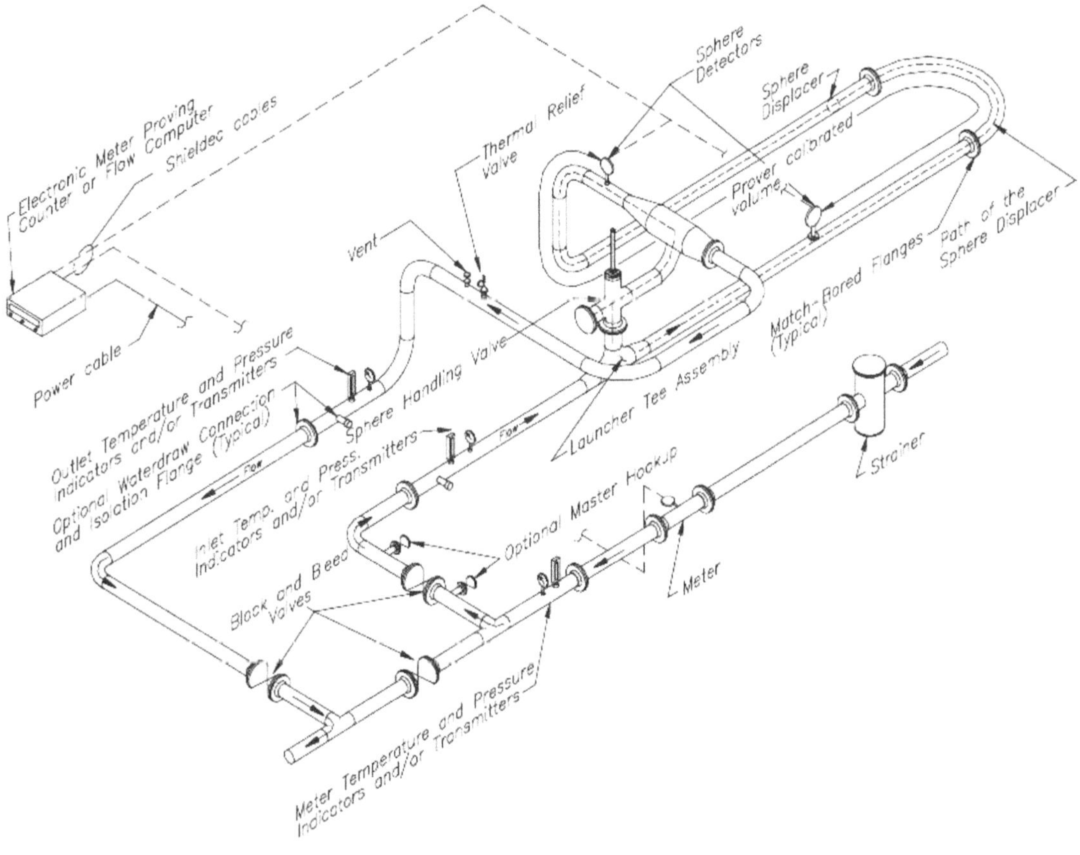

Figure 7-25. Uni-directional sphere prover — MPMS 4.2 (reproduced courtesy of the American Petroleum Institute)

7.4.7.2.2 Bi-directional Provers With this type of prover, a sphere is moved by the product flow from the launch piping into the proving loop. The sphere then continues past the first detector switch into the calibrated section of pipe, then past the second detector and out into the receiver chamber. When the sphere passes the first detector switch, the pulse counter is triggered to count meter pulses until the second detector switch is triggered. The number of pulses accumulated on the pulse counter while the sphere moves between detector switches is compared to the calibrated volume of the prover section to obtain a meter factor.

The proving cycle of a bi-directional prover is one round trip of the sphere. The travel of the sphere is reversed by activating the four-way flow reversal valve and the pulse counter accumulates pulses for the two runs of the sphere. Bi-directional pipe provers require 20,000 pulses for each prover round trip 10,000 pulses for each pass (Figure 7-26).

A bi-directional piston prover (Figure 7-27) works on the same principle as the sphere prover with the only difference being the way that the displacement piston is stopped at the end of each proving run. In this case, the piston comes to rest once it is past the diversion piping at the end of the prover pipe section. When the four-way diversion valve is actuated, flow in the prover pipe is reversed and the check valve in the diversion piping closes and flow is directed to the end of the prover pipe. This actuates the piston in the opposite direction.

7.4.7.2.3 Small Volume Displacement Provers Compact or small volume provers are similar to standard pipe provers with the noticeable difference that the displacer is not free; it is in fact a piston connected to a piston rod. The rod extends outside the barrel of the prover and is usually fitted with an indicator either of the micrometer type, or is fully electronic using a displacement transducer and flags/proximity switches (Figure 7-28).

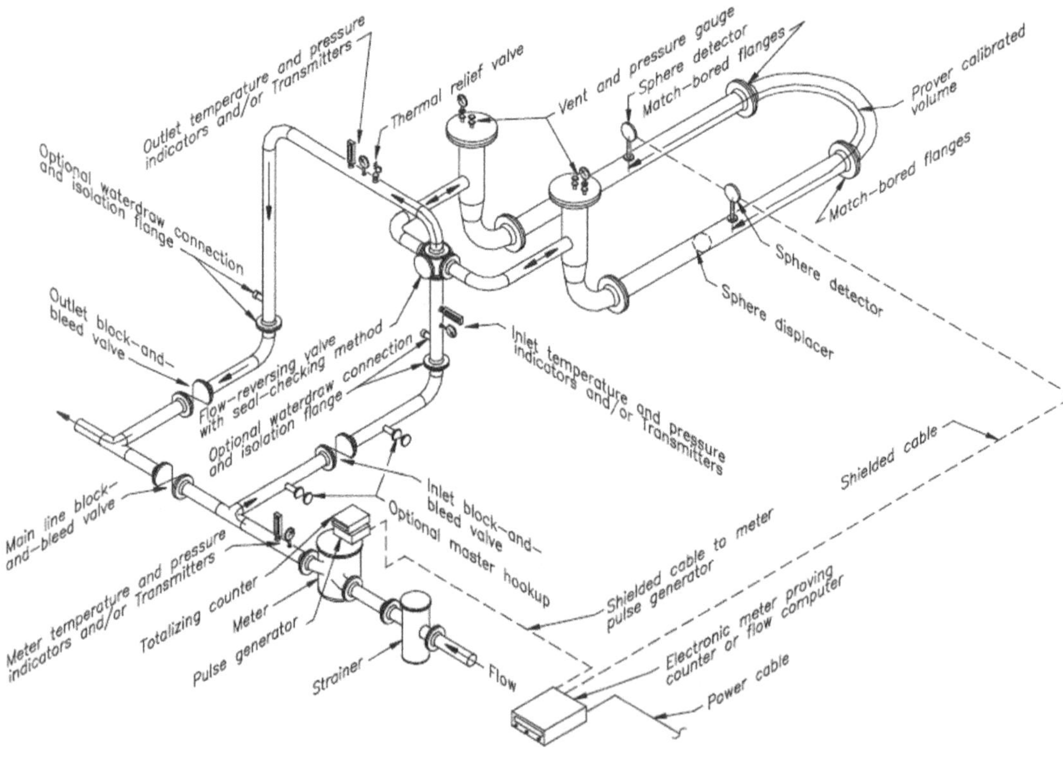

Figure 7-26. Bi-directional sphere prover — source MPMS 4.2 (reproduced courtesy of the American Petroleum Institute)

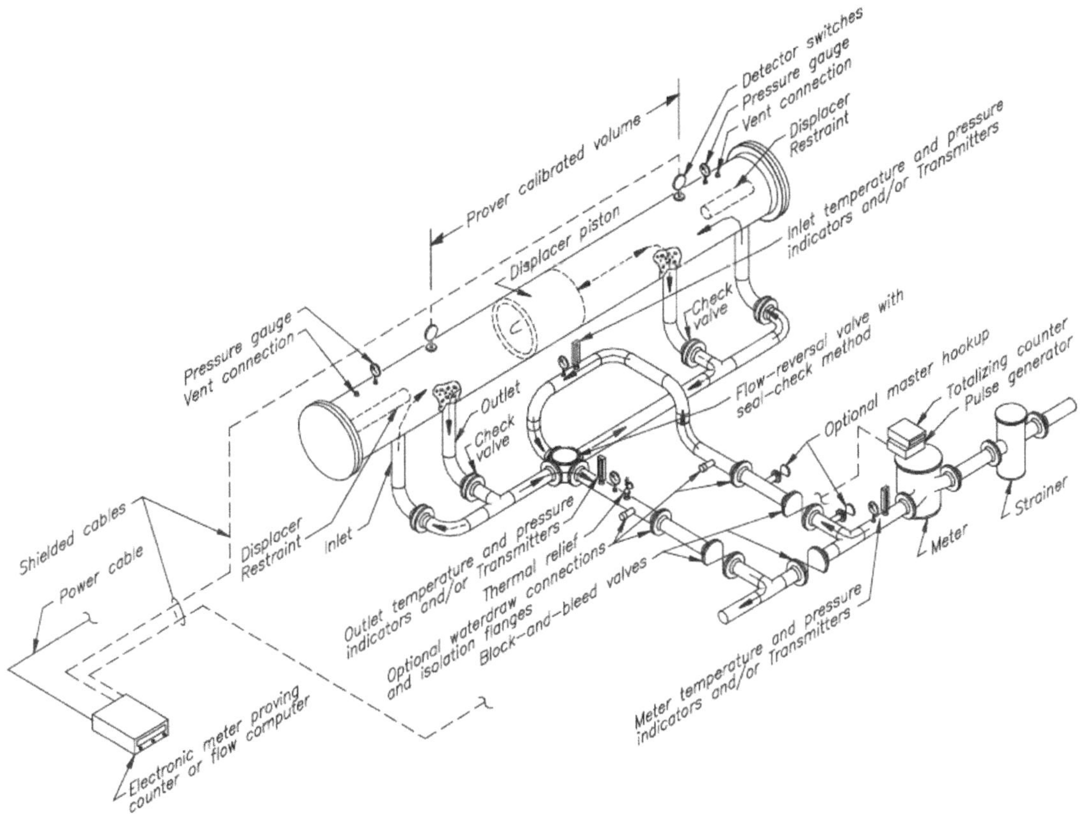

Figure 7-27. Bi-directional piston prover — MPMS 4.2 (reproduced courtesy of the American Petroleum Institue)

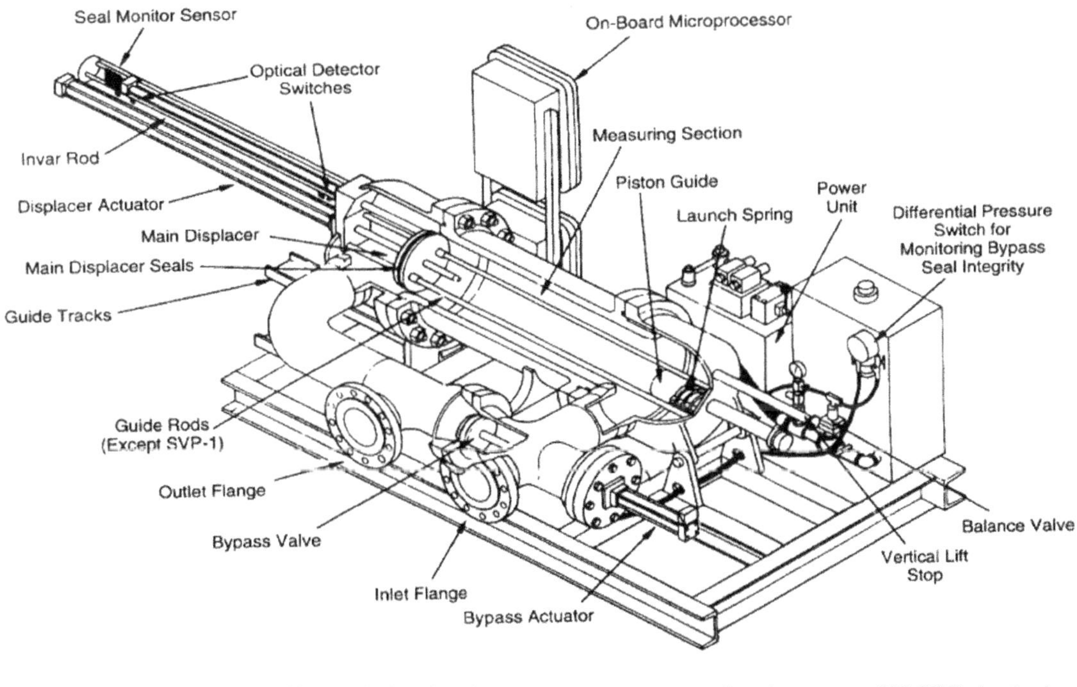

Figure 7-28. Uni-directional compact prover — reproduced courtesy of FMC Technologies, Inc

Due to the relatively small volume of product displaced by the piston's travel or stroke, a means had to be found to ensure compliance with the API's recommended standard minimum of 10,000 pulses per run. The means employed was pulse interpolation.

There are several methods of employing pulse interpolation, but by far the most used is the dual chronometry (double clock) method (see *API MPMS Chapter 6. Proving Systems — Pulse Interpolation*). Essentially, this method involves the use of two timers (T1 & T2) both driven by the same high-speed clock.

The sequence employed is:

- Start Timer 1 (T1) when first detector switch is activaed.
- Start Timer 2 (T2) at the leading edge of the next flow meter pulse following T1 start. (This is also where the pulse counter will start counting.)
- Stop Timer 1 (T1) when the final detector switch is activated.
- Stop Timer 2 (T2) at the leading edge of the next flow meter pulse following T1 stop. (This is also where the pulse counter stops counting.)

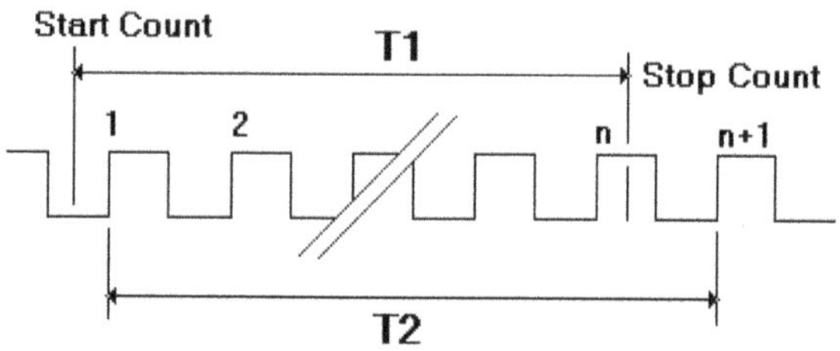

The formula to determine the number of full and partial pulses is then:

$$K = \frac{T1}{\text{Vol}} \times \frac{n+1}{T2}$$

where

K = proving K factor
$T1$ = time elapsed for Timer 1 over duration of run
$T2$ = time elapsed for Timer 2 over duration of run
$n+1$ = number of pulses counted during T1 timing
Vol = base volume of the prover

7.4.7.2.4 Master Meter Provers Master meter proving is typically used when other prover methods are not practical. Master meters add another layer of calculations that provide additional uncertainty to the measurement. The master meter must first be proven to establish its meter factor and then this factor is used to determine the field operating meter's meter factor. In the master meter method, a meter is first proved by a prover that has been calibrated by the water draw method.

7.4.8 Prover Calibration

The purpose of the calibration of a liquid prover is to determine its certified base volume with traceability to recognized standards and accepted practices. The base volume of a prover is the gross operating volume corrected to standard conditions (60 °F and 14.696 psia in US customary units or 15 °C and 101.325 kPa in SI units). The commonly used methods of calibrating provers are waterdraw by either volumetric or

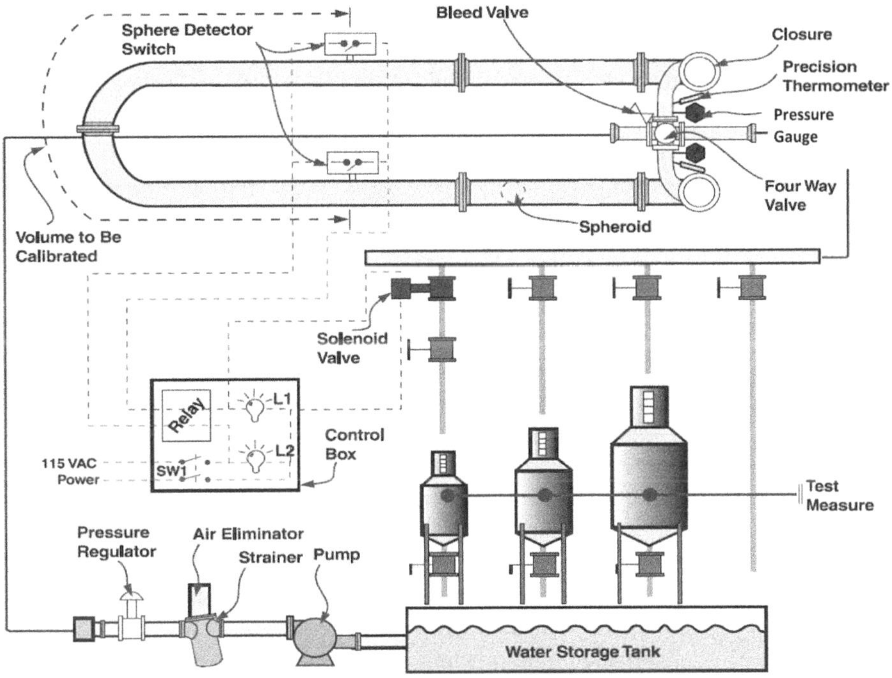

Figure 7-29. Conventional Prover Calibration — reproduced courtesy of FMC Technologies, Inc

gravimetric measurement or by master meter. In the waterdraw method, water is measured from the prover into certified field standard measures (see Figure 7-29).

Calibrations are performed initially at the manufacturer's plant usually by water draw. Stationary provers are typically calibrated again once on site to confirm prover volume. Periodically, provers are recalibrated to confirm that no changes have occurred during repeated operations while in service. A method for determining the frequency of recalibrating pipe provers is outlined in *API MPMS 4.9.1 Appendix B*. API states that under no circumstances should the projected time to the next calibration exceed 60 months. If mechanical repairs, alterations or changes which affect the certified volume are made to the prover, then the prover should be immediately recalibrated after completion of this work. The prover should then be scheduled for recalibration again in 12 months.

When the pipe prover volume is too great to perform a volumetric waterdraw, calibration is generally carried out by the master meter/master pipe prover method [23]. Using a compact prover as the master pipe prover, the calibrated volume of the compact prover is used to calibrate the master meter which is then used as a transfer standard for the volume of the permanently installed pipe prover (Figure 7-30).

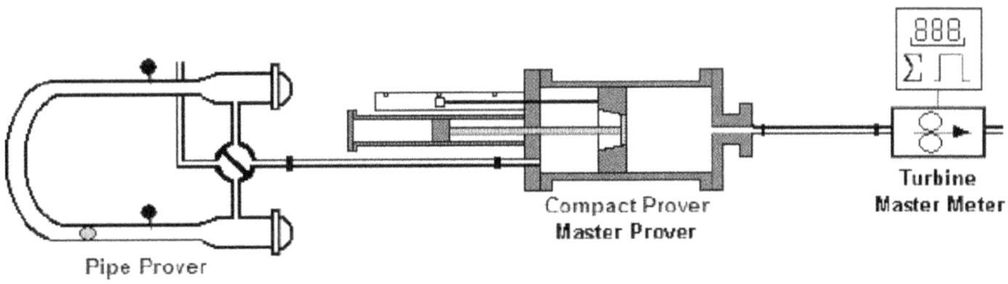

Figure 7-30. Master meter/master pipe prover

This calibration method is undertaken with the compact prover, master meter and pipe prover connected in series and allows both the master meter and pipe prover to be calibrated on operating fluids at the same pressure, temperature, and flow rate.

7.5 VOLUME ACCOUNTING SYSTEM

The previous sections describe various flow or volume measurement methods and issues. The data supplied by each instrument may not be sufficient to provide the information useful for transportation services. A volume accounting system addresses various measurement issues, which make the measured values useful for the transportation services and to provide the required information and reports on tickets [24].

The volumetric accounting system performs various functions such as ticketing and volume balancing functions. It also determines net metered volumes for each meter in the pipeline system. Since the accounting system collects and maintains a large amount of data, it should be able to manage the large amount of data efficiently and reliably. This requires not only efficient data storage and retrieval management capability but also data entry and editing functions with auditing capability. In addition, the system should have the reporting functions for shippers and internal customers such as marketing and management.

Depending on pipeline operations, flow meters are classified as batch meter or continuous meter. The batch meter is applicable to batch operation and each batch is measured and represented by its corresponding ticket for the batch meter, while the continuous meter is applicable to a single product operation and a ticket is issued on a daily basis. A ticket is a record of metered batch volume for batch operation issued in the event of batch receipt/delivery or of metered receipt/delivery volume for single product issued daily or at a specified interval.

A typical batch meter summary is shown in Figure 7-31. The summary includes the batch ID and product name, status in the pipeline, total volume, batch lifting status and schedule, and other data for batch operation.

File Edit Options Help

Batch Id.	Product	Stat.	Tot. Vol.	Man. Cor.	Location	Comp.	Ship.	Stat.	Sch. Start	Act. Start	Est. Compl.	Act. Compl.	Sch. Vol.	Curr. Vol.	O/S
PROP 030	PROP	actv	12812	13806											
					00EM	PTC	PTC	activ	Jul/15 10:37	Jul/22 18:00			12580	994	
					05SV	PTC	PTC	sched	Jul/16 02:19				314	0	
					30RE	PTC	PTC	sched	Jul/16 00:12				2139	0	
IBUT 029	IBUT	actv	6321	0											
PROP 028	PROP	actv	12089	12089											
					05SV	PTC	PTC	activ	Jul/15 18:54	Jul/22 18:00			314	0	
					15MJ	PTC	PTC	sched	Jul/17 05:37				566	0	
					30RE	PTC	PTC	sched	Jul/13 05:11				5346	0	
NBUT 027	NBUT	actv	6321	0											
PROP 026	PROP	actv	11804	11922											
					15MJ	PTC	PTC	activ	Jul/15 10:37	Jul/22 18:00			629	118	
					30RE	PTC	PTC	activ	Jul/10 15:04	Jul/22 18:00	Jul/23 00:17		2516	0	
IBUT 025	IBUT	actv	6290	0											

Figure 7-31. SCADA batch summary

7.5.1 Ticketing Functions

A measurement ticket is a written acknowledgement of a receipt or delivery of petroleum and petroleum products. If a change in ownership or custody occurs during the transfer, the measurement ticket or report serves as documentation between the parties involved as to the measured quantities and tested qualities of the liquids transferred.

The first step in volume accounting is to perform the ticketing functions. The host SCADA system collects the data from the measurement points throughout the pipeline system including tank farms, and the volume accounting system consolidates all the required data for further processing. The ticketing functions include the following:

- Capture tickets from field instruments either automatically and/or manually. Ticket types are meter and tank tickets.
- Validate meter tickets by applying validation rules to remove errors, verify scheduled volumes to actual tickets, and convert gross to net volumes for the customers at all measurement points on the pipeline system.
- Collect pipeline and tank inventory data.

Normally, the following input data are required to perform the ticketing functions:

- Meter configuration for each meter includes the name, size, location, type of meter, meter factors, applicable standard, etc. The meter configuration data is not frequently changed, unless a new meter is added, an existing meter is replaced or modified, or a new facility such as storage tank is added.
- Product definition includes the density and/or composition, volume correction factors for temperature and pressure, etc. The product data is not frequently changed, unless a new product is included in the transportation service or an existing product is removed from the service.
- Tickets are produced at the beginning and end of the batch for batching meters, while tickets are produced on a daily basis for continuous meters.
- A gross volume is corrected to net volume either by the flow computer or the host SCADA. An incrementing volume of a rolling accumulator rolls over to zero at a specified maximum value, and that of a batching accumulator resets to zero at the onset of batch change flowing through the meter.
- Tank gauge input requires the tank level change and direction of flow, i.e., in or out of the tank to accumulate tank volume and calculate its net volume.
- Pressure, temperature, and density measurements are required to correct volumes to base conditions.

In addition, pipeline inventory and tank inventory data are required for daily scheduling, volume balancing and gain/loss analysis. These data are available through the SCADA and batch tracking application.

- Pipeline inventory data includes dates and times, line fill volumes, products and shippers for the pipeline.
- Tank inventory data includes dates and times, tank volumes, products, and shippers for each tank.

All these data items need to be integrated into a single ticket database, because a single integrated database allows the system users to easily access the required data, improve data quality, and reduce redundant efforts.

7.5.2 Meter Ticket

The meter and tank ticketing work according to the daily schedule, from which a pending queue is created, listing pending tickets. A new ticket is started when a ticket is cut for the current batch and the status of the next batch in the pending queue turns to active ticket. If a single product is transported, a new ticket is cut at the designated time of the day. When a ticket is started, the volume of the ticket is initialized, its product is identified, and start time and ticket number with the corresponding schedule are created.

As the ticket meter is accumulating the ticket volume, the scheduled volume is compared to the ticket in order to ensure that the ticket volume does not exceed the scheduled volume beyond the specified tolerance. When the receipt or delivery of the batch is completed, a ticket for the batch is cut and the ticket status changes. The ticket for the current batch may be split, if the meter factor is changed or if the contract day ends for continuous flow metering. If a ticket for an unscheduled batch is created, it has to be reconciled. When an active ticket is completed, the ticket becomes a completed ticket. The completed ticket needs to be reviewed and edited, if necessary, for validation and verification before it is stored in a completed ticket database.

Figure 7-32 exhibits an example of a meter ticket display. To sum up, the ticket status can be one of the following:

- The pending ticket is a ticket that is not yet started.
- The active ticket is a ticket whose product is flowing and thus volume is collected.
- A ticket becomes the completed ticket when receipt or delivery is completed.

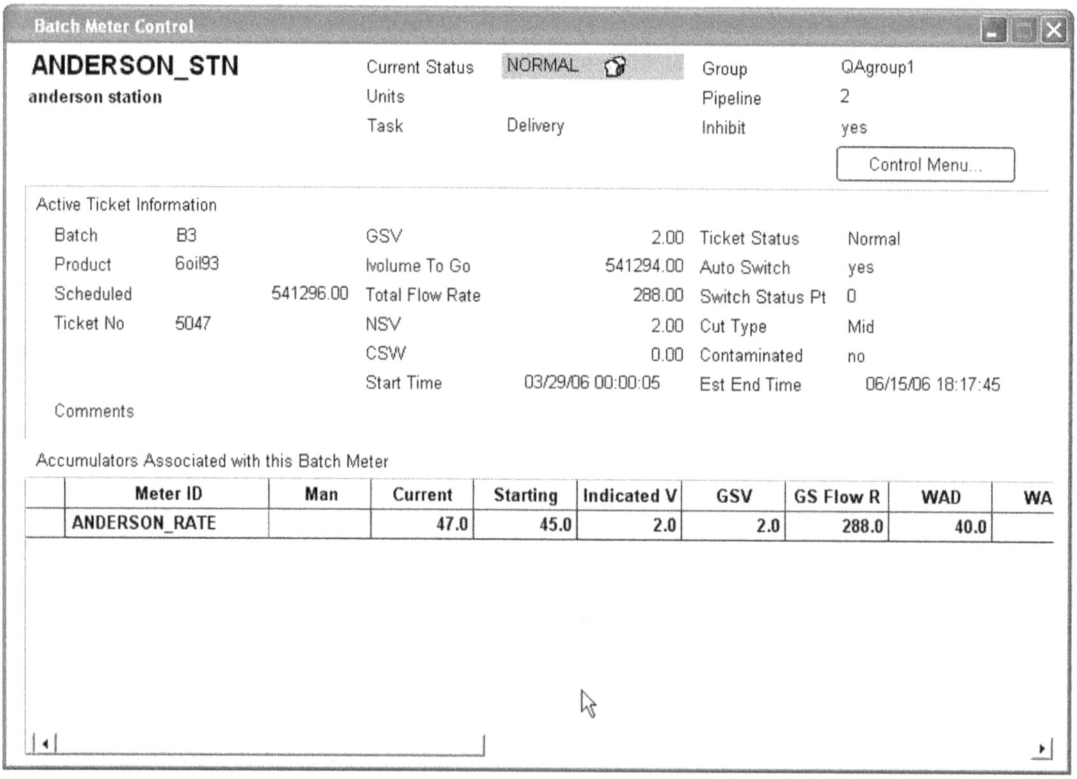

Figure 7-32. Meter ticket display — reproduced courtesy of Telvent

Depending on the ticket status, different data is allowed to be modified, entered or deleted. For example, the pending ticket allows the batch ID and scheduled data to be edited, while the active ticket allows the batch ID and current measured quantities such as product and volume to be edited. After the active ticket is changed to the completed ticket, the batch ID, shipper, measured or manually entered quantities such as meter factor and product gravity, and product quality such as Sediment and Water (S&W) content can be edited. Any manual data editing requires the editor to record the reasons for future auditing.

A meter ticket may include the ticket number, shipper, ticket start and stop times and location, ticket status, meter ID and meter factor, average temperature and pressure, scheduled and ticket gross and net volumes, batch ID, product properties such as density, volume correction factor, S&W content, etc. When the ticket is completed, the ticket data is stored in the ticket database after the data is validated and verified. They can be validated automatically with specified rules or manually verified before they are stored.

7.5.3 Tank Ticket

The tank inventories are mostly monitored and their operations are controlled by the common carrier. However, some tanks belong to its customers so that their inventories cannot be used for scheduling and not accounted for. The common carrier is still responsible for the lifting from and delivery to these tanks and thus responsible for producing tickets.

The data for a tank ticket is similar to the meter ticket data, except that the tank ticket includes the tank ID, flow direction, tank type with roof adjustment, tank gauges associated with this ticket, and tank strapping table to calculate tank volume. A strapping table is a set of calibrations that mathematically relate tank level to tank volume. The flow rate or volume in and out of a tank is measured using a tank gauge or flow meter connected to the tank. The flow rate, determined from the tank gauge, is less accurate than the flow rate measured by a flow meter. A tank ticket can be started or stopped automatically when receipt or delivery is completed or manually on dispatcher's or field operator's command. A floating roof tank with heavy snow or rain can affect the volume measurement reading significantly, so it must be taken into account when taking tickets based on tank levels.

A tank ticket can be generated either automatically or manually. Automatic ticket generation is done based on a sequence of the valve status changes. For manual operation, either the dispatcher or field operator issues the ticket with a unique tank ticket number and sends it to the SCADA system. To operate a tank safely, the dispatcher or operator requires the following data:

- Percentage of full value of the tank
- Volume to fill the tank to the maximum tank level
- Time to fill the tank to the maximum tank level (at a given flow rate)
- Volume to pump out from the tank to the minimum tank level
- Time to empty the tank to the minimum tank level (at a given flow rate)

Tank alarms are generated if certain conditions are violated. Some of the tank alarms are:

- Tank level alarms for the violation of the maximum or minimum level
- Time to fill or empty alarms if the calculated time-to-fill/empty passes a specified time limit
- Rate of change alarms for exceeding the highest rate of change

7.5.4 Volume Tracking

The dispatcher often requires the volume remaining in a batch to control the batch volume to be lifted or delivered. The volume tracking can be performed either automatically or initiated by the dispatcher. The warning message is generated when the volume remaining is less than a preset warning volume or when the time remaining for completion is less than a preset limit.

The volume to be tracked is taken from the batch schedule and then reset with each batch. The remaining batch volume is calculated by subtracting the lifted or delivered volume from the scheduled volume. In addition, the expected completion time of the receipt or delivery is determined using the current flow rate. The volume tracking allows the dispatcher to see the remaining volumes and the expected time to complete the scheduled volume.

7.5.5 Volume Calculation and Balancing

7.5.5.1 Volume Calculation

Transportation services are charged based on net volumes or mass. Since mass is independent of flowing pressure and temperature, volume correction due to the pressure and temperature is not necessary. As described in the previous sections, the metering devices measure raw flow or volume and also require regular calibration of meter and its associated measurements. Therefore, several steps with their parameters are needed to determine the net volume [25, 26].

7.5.5.2 Meter Factor and Calibration

Calibration is required to maintain the meter accuracy. The meter factor is determined as a result of meter proving. The meter factor is a meter correction factor, varying about 1.0. It accounts for fluid and mechanical effects on measurement, depending on product density and flow rate. Since meter factors can vary with product density and flow rate changes, the factors are either input into a local flow computer for computation of the appropriate meter factor applicable to the product and its flow rate or this information is transmitted to a SCADA system. The SCADA system then would apply the appropriate meter factor in a similar manner. Meter factors are tracked over time to identify any drift in meter factors that would indicate need for meter maintenance.

When a meter measures a flow rate of a product, the meter factor corresponding to the product and flow rate is selected from the meter factor table and applied to the flow rate in order to adjust the value. The selected meter factor is checked against the flow range, and if the flow limit is violated an alarm is generated to indicate invalid meter factor.

7.5.6 Determination of Liquid Volume

The volume of a product depends on pressure and temperature. The volume correction factor is the ratio of a unit volume at base pressure and temperature to a unit volume at the flowing pressure and temperature at the time of measurement. There are several volume correction factors depending on the product density, because the volume correction factor varies with the product density and there is no universal equation to determine the factor. The volume correction method and applicable standards are for different products are described further in this section.

The petroleum industry has developed standardized calculation methods for determining petroleum volumes at reference conditions which are expressed in the following API and ISO standards:

- API MPMS Chapter 12.2.1 — Calculation of Petroleum Quantities Using Dynamic Measurement Methods and Volume Correction Factors, Part 1 — Introduction
- ISO 4267-2:1988 — Petroleum and liquid petroleum products — Calculation of oil quantities — Part 2: Dynamic measurement

The reference (base) conditions for the measurement of liquids having a vapor pressure equal to or less than atmospheric at base temperature are as follows:

United States Customary (USC) Units:
Temperature — 60.0°F
Pressure —14.696 psia

International System (SI) Units:
Temperature — 15.0°C
Pressure — 101.325 kPa

For liquids having a vapor pressure greater than atmospheric at base temperature, the base pressure shall be the equilibrium vapor pressure at base temperature.

Base pressure and temperature can be different for different countries.

7.5.7 General Equations for Determining Liquid Volumes at Base Conditions

Indicated Volume

$$IV = \text{closing meter reading} - \text{opening meter reading}$$

or

$$IV = (\text{closing pulses} - \text{opening pulses})/KF$$

Gross Standard Volume

$$GSV = IV \times CTL \times CPL \times MF \text{ or;}$$

$$GSV = IV \times MF \times DENobs/DENb \text{ or;}$$

$$GSV = Mass/DENb$$

Net Standard Volume

$$CSW = 1 — (\%S\&W/100)$$

$$NSV = GSV \times CSW \times SF$$

where

***CPL** — Correction for the effect of Pressure on Liquid* — Correction for compressibility of liquid at normal operating conditions. This factor is obtained from *API MPMS or ISO Standards* identified below.

***CTL** — Correction for the effect of Temperature on Liquid* — Correction for effect of temperature on liquid at normal operating conditions. This factor is obtained from *API MPMS or ISO Standards* identified below.

CSW — *Correction for Sediment and Water* — Correction for sediment and water to adjust the gross standard volume of the liquid for these non-merchantable items.
***DENb*-Base Density** — Liquid density at base pressure and temperature.
DENobs — *Observed Density* — Liquid density at observed pressure and temperature.
GSV — *Gross Standard Volume* — The volume at base conditions corrected also for the meter's performance (MF or CMF).
IV — *Indicated Volume* — The change in meter reading that occurs during a receipt or delivery.
KF — *K-factor* — A term in pulses per unit volume determined during a factory or field proving. The number of pulses generated by a linear meter divided by the *K*-factor will determine the indicated volume.
MF — *Meter Factor* — A dimensionless term obtained by dividing the volume of the liquid passed through the prover corrected to standard conditions during proving.
NSV — *Net Standard Volume* — The gross standard volume corrected for non-merchantable quantities such as sediment and water.

Pressure and Temperature Compensation

CTL and ***CPL*** must be calculated in accordance with the current Standards shown below for the applicable density and temperature range.

- *API MPMS Chapter 11.1 — Temperature and Pressure Volume Correction Factors for Generalized Crude Oils, Refined Products, and Lubricating Oils* — provides the algorithm and implementation procedure for the correction of temperature and pressure effects on density and volume of liquid hydrocarbons which fall within the categories of crude oil, refined products, or lubricating oils; NGLs and LPGs are excluded from consideration in this Standard.
- *API MPMS Chapter 11.2.2 — Compressibility Factors for Hydrocarbons: 0.350–0.637 Relative Density (60 °F/60 °F) and –50 °F to 140 °F Metering Temperature* — provides tables to correct hydrocarbon volumes metered under pressure for the metered temperature. Contains compressibility factors related to the meter temperature and relative density (60 °F/60 °F) of the metered material. 2nd Edition October 1986 Reaffirmed: December 1, 2007
- *API MPMS Chapter 11.2.2M — Compressibility Factors for Hydrocarbons: 350–637 Kilograms per Cubic Meter Density (15 °C) and –46 °C to 60 °C Metering Temperature* — Provides tables to correct hydrocarbon volumes metered under pressure to corresponding volumes at equilibrium pressure for the metered temperature. The standard contains compressibility factors related to the meter temperature and density (15 °C) of the metered material. 1st Edition October 1986 Reaffirmed: December 1, 2007
- *API MPMS Chapter 11.2.4 — Temperature Correction for the Volume of NGL and LPG Tables 23E, 24E, 53E, 54E, 59E, 60E* — The Standard represented by this report consists of the explicit implementation procedures. Sample tables, flow charts, and specific examples created from a computerized version of these implementation procedures are included. The examples are to provide guides and checkpoints for those who wish to implement a computerized procedure to represent the Standard, however these are not part of the actual Standard. This standard covers a 60°F relative Density range of 0.3500 to 0.6880 which nominally equates to a density at 15°C of 351.7 to 687.8 kg/m^3

and a density at 20 °C of 331.7 to 686.6 kg/m^3. The temperature range of this Standard is –50.8 to 199.4 °F (–46 to 93 °C). At all conditions, the pressure is assumed to be at saturation conditions (also known as bubble point or saturation vapor pressure).

Or:

ASTM D1250 — Volume Correction Factors (Joint Standard with API MPMS Chapter 11.1)
ASTM D1250-04 — Standard Guide for Use of the Petroleum Measurement Tables; 2000
Adjunct to D1250 — Temperature and pressure volume correction factors for generalized crude oils, refined products, and lubricating oils. 01-May-2004, D1250 ADJ CD-2-10

7.5.8 Volume Balancing

The volume balancing functions are necessary to verify that the batch operation and ticketing is consistent with daily schedule. When the active ticket is completed and the volume is corrected to base conditions, the ticketed volumes should be reconciled to scheduled volumes and the tickets are adjusted for the product by shipper. The volume balancing functions are performed on a regular basis, hourly, daily and/or monthly, to identify any overages or shortages.

The ticketed volumes are reconciled to scheduled volumes based on daily scheduling information, determining the balance between scheduled and actual volumes. When a ticket is completed for the scheduled batch, the actual ticket is compared to the scheduled volume. It is intended to catch errors up-front by reviewing the ticket earlier in the accounting process and to verify that the scheduled batch is entered and the actual ticket is correct with respect to the schedule of the batch.

In addition, batch movements should be balanced as part of the verification process. Balancing of batch movements is required to check the correctness of both inventories and batch movements with respect to the daily schedule. It results in the comparison of all batch movements of actual volumes to scheduled volumes and eventually gain/loss for the specified periods such as daily and monthly.

Gain/loss can be calculated for shipper, product, or their groupings, adding receipts to the beginning inventory of the specified period and then subtracting deliveries and ending inventory. The percentage of gain/loss is the percentage of gain/loss volume with respect to the delivered volume. As shown in the calculation method, the data sources of the gain/loss calculation include both the ticket and inventory data.

The completed ticket is closed after the verification process, confirming that the ticket is accounted for and that the scheduled volume vs. ticket volume and the gain/loss percentage are within the respective specified tolerances. This verification can be performed automatically or manually. If violations are detected during an automatic verification process, manual review and correction of the violations is required before it is closed. Otherwise, the ticket may be closed automatically.

As a result of the volume balancing including gain/loss, ticket volumes are adjusted and allocated to shippers. When ticket volumes are allocated at the end of the month, the actual volumes delivered to a shipper or customer is known. The allocated delivered volume is the basis for revenue accounting and used for transportation charges.

ADDENDUM: STANDARDS RELEVANT TO LIQUID PETROLEUM MEASUREMENT

A7.1 American Petroleum Institute (API) — www.api.org

Manual of Petroleum Measurement Standards

Chapter 1 — Vocabulary
Chapter 2 — Tank Calibration
Chapter 2.2A — Measurement and Calibration of Upright Cylindrical Tanks by the Manual Strapping Method; 01-Feb-1995
Chapter 2.2B — Calibration of Upright Cylindrical Tanks Using the Optical Reference Line Method; 1st Edition, 01-Mar-1989 Reaffirmed December 2007
Chapter 2.2C — Calibration of Upright Cylindrical Tanks Using the Internal Electro-optical Distance Ranging Method; 1st Edition, January 2002 Reaffirmed May 2008
Chapter 2.2D — Calibration of Upright Cylindrical Tanks Using the Optical Reference Line Method; 1st Edition, August 2003 Reaffirmed February 2009
Std 2555 — Liquid Calibration of Tanks; 1st Edition Sept 1966 Reaffirmed March 2009
Chapter 3 — Tank Gauging
Chapter 3.1A — Standard Practice for Manual Gauging of Petroleum and Petroleum Products in Stationary Tanks; First Edition August 2005; Reaffirmed Sept 2010
Chapter 3.1B —Standard Practice for Level Measurement of Liquid Hydrocarbons in Stationary Tanks by Automatic Tank Gauging; Second Edition, 01-Jun-2001; Revision 1, Reaffirmed 10/06
Chapter 3.3 — Standard Practice for Level Measurement of Liquid Hydrocarbons in Stationary Pressurized Storage Tanks by Automatic Tank Gauging, Revision 1996; Reaffirmed 10/06
Chapter 3.6 — Measurement of Liquid Hydrocarbons by Hybrid Tank Measurement Systems — First Edition, April 2001; Reaffirmed 10/06
Chapter 4 — Proving Systems
Chapter 4.1 — Introduction — 3rd Edition February 2005 Reaffirmed: September 10, 2009
Chapter 4.2 — Displacement Provers 3rd Edition September 2003
Chapter 4.4 — Tank Provers — 2nd Edition May 1998 Reaffirmed: September 21, 2010
Chapter 4.5 — Master-Meter Provers — 2nd Edition | May 2000 | Reaffirmed: August 1, 2005
Chapter 4.6 — Pulse Interpolation — 2nd Edition | May 1999 | Reaffirmed: May 1, 2008
Chapter 4.7 — Field Standard Test Measures — 3rd Edition April 2009
Chapter 4.8 — Operation of Proving Systems — 1st Edition November 1995 Reaffirmed: March 1, 2007
Chapter 4.9.1 — Methods of Calibration for Displacement and Volumetric Tank Provers — Part 1 —Introduction to the Determination of the Volume of Displacement and Tank Provers — 1st Edition October 2005 Reaffirmed: September 21, 2010
Chapter 4.9.2 — Methods of Calibration for Displacement and Volumetric Tank Provers — Part 2 — Determination of the Volume of Displacement and Tank Provers by the Waterdraw Method of Calibration — 1st Edition December 2005

Reaffirmed: September 21, 2010
Chapter 4.9.3 — Methods of Calibration for Displacement and Volumetric Tank Provers — Part 3 — Determination of the Volume of Displacement Provers by the Master Meter Method of Calibration — 1st Edition April 2010 Product Number: H409031
Chapter 4.9.4 — Methods of Calibration for Displacement and Volumetric Tank Provers — Part 4 — Determination of the Volume of Displacement and Tank Provers by the Gravimetric Method of Calibration — 1st Edition | October 2010 | Product Number: H4090401 | Price: $80.00
Chapter 5 — Metering — Covers the dynamic measurement of liquid hydrocarbons, by means of meters and accessory equipment.
Chapter 5.1 — General Considerations for Measurement by Meters — 4th Edition October 2005
Chapter 5.2 — Measurement of Liquid Hydrocarbons by Displacement Meters — 3rd Edition October 2005 Reaffirmed: September 21, 2010
Chapter 5.3 — Measurement of Liquid Hydrocarbons by Turbine Meters — 5th Edition September 2005
Chapter 5.4 — Accessory Equipment for Liquid Meters — 4th Edition September 2005 Reaffirmed: September 2010
Chapter 5.5 — Fidelity and Security of Flow Measurement Pulsed-Data Transmission Systems — 2nd Edition July 2005 Reaffirmed: September 2010
Chapter 5.6 — Measurement of Liquid Hydrocarbons by Coriolis Meters — 1st Edition October 2002 Reaffirmed: March 20, 2008
Chapter 5.8 — Measurement of Liquid Hydrocarbons by Ultrasonic Flowmeters Using Transit Time Technology — 1st Edition February 2005
Chapter 6 — Metering Assemblies
Chapter 6.1 — Lease Automatic Custody Transfer (LACT) Systems — 2nd Edition May 1991 Reaffirmed: April 2007
Chapter 6.6 — Pipeline Metering Systems — 2nd Edition May 1991 Reaffirmed: February 2007
Chapter 6.7 — Metering Viscous Hydrocarbons — 2nd Edition May 1991 Reaffirmed: April 1, 2007
Chapter 7 — Temperature Determination
Chapter 8 — Sampling — Covers standardized procedures for sampling crude oil or its products.
Chapter 8.1 — Manual Sampling of Petroleum and Petroleum Products — (ANSI/ASTM D4057) — 3rd Edition October 1995 Reaffirmed March 2006
Chapter 8.2 — Automatic Sampling of Petroleum and Petroleum Products — (ANSI/ASTM D4177) — 2nd Edition October 1995 Reaffirmed: March 2010
Chapter 8.3 — Standard Practice for Mixing and Handling of Liquid Samples of Petroleum and Petroleum Products (ANSI/ASTM D5854) — 1st Edition October 1995 Reaffirmed: March 2010
Chapter 9 — Density Determination
Chapter 9.1 — Hydrometer Test Method for Density, Relative Density (Specific Gravity), or API Gravity of Crude Petroleum and Liquid Petroleum Products; 01-Dec-2002
Chapter 9.2 — Standard Test Method for Density or Relative Density of Light Hydrocarbons by Pressure Hydrometer, Second Edition; 01-Mar-2003
Chapter 9.3 — Standard Test Method for Density, Relative Density, and API Gravity of Crude Petroleum and Liquid Petroleum Products by Thermohydrometer Method; 01-Nov-2002
Chapter 10 — Sediment and Water

Chapter 10.1 — Standard Test Method for Sediment in Crude Oils and Fuel Oils by the Extraction Method; Second Edition, 01-Oct-2002
Chapter 10.2 — Determination of Water in Crude Oil by Distillation; 1st Edition, 01-Apr-1981; Revision 2005.
Chapter 10.3 — Standard Test Method for Water and Sediment in Crude Oil by the Centrifuge Method (Laboratory Procedure), Second Edition: 01-May-2003
Chapter 10.4 — Determination of Sediment and Water in Crude Oil by the Centrifuge Method (Field Procedure); Third Edition, 01-Jan-1999
Chapter 10.5 — Manual of Petroleum Measurement Standards, Chapter 10-Sediment and Water, Section 5-Standard Test Method for Water in Petroleum Products and Bituminous Materials by Distillation; Fourth Edition, 01-Aug-2005
Chapter 10.6 — Standard Test Method for Water and Sediment in Fuel Oils by the Centrifuge Method (Laboratory Procedure); Fourth Edition, 01-Dec-2004
Chapter 10.7 — Sediment and Water — Section 7 — Standard Test Method for Water in Crude Oils by Potentiometric Karl Fischer Titration; 01-Dec-2002
Chapter 10.8 — Standard Test Method for Sediment in Crude Oil by Membrane Filtration; Second Edition, 01-Nov-2005
Chapter 11 — Physical Properties Data (Volume Correction Factors)
Chapter 11.1 — Temperature and Pressure Volume Correction Factors for Generalized Crude Oils, Refined Products, and Lubricating Oils — September 2004
Chapter 11.2.2 — Compressibility Factors for Hydrocarbons: 0.350–0.637 Relative Density (60 °F/60 °F) and –50 °F to 140 °F Metering Temperature — 2nd Edition October 1986 Reaffirmed: December 2007
Chapter 11.2.2M — Compressibility Factors for Hydrocarbons: 350–637 Kilograms per Cubic Meter Density (15 °C) and –46 °C to 60 °C Metering Temperature — 1st Edition October 1986 Reaffirmed: December 2007
Chapter 11.2.4 — Temperature Correction for the Volume of NGL and LPG Tables 23E, 24E, 53E, 54E, 59E, 60E — 1st Edition September 2007
Chapter 12 — Calculation of Petroleum Quantities
Chapter 12.1.1 — Calculation of Static Petroleum Quantities, Part 1—Upright Cylindrical Tanks and Marine Vessels — 2nd Edition November 2001 Reaffirmed: February 2008
Chapter 12.2.1 — Calculation of Petroleum Quantities Using Dynamic Measurement Methods and Volume Correction Factors, Part 1—Introduction — 2nd Edition May 1995 Reaffirmed: February 2009
Chapter 12.2.2 — Calculation of Petroleum Quantities Using Dynamic Measurement Methods and Volumetric Correction Factors, Part 2—Measurement Tickets — 3rd Edition June 2003 Reaffirmed: September 2010
Chapter 12.2.3 — Calculation of Petroleum Quantities Using Dynamic Measurement Methods and Volumetric Correction Factors, Part 3—Proving Reports — 1st Edition October 1998 Reaffirmed: March 2009
Chapter 12.2.4 — Calculation of Petroleum Quantities Using Dynamic Measurement Methods and Volume Correction Factors, Part 4—Calculation of Base Prover Volumes by Waterdraw Method — 1st Edition December 1997 Reaffirmed: March 2009
Chapter 12.2.5 — Calculation of Petroleum Quantities Using Dynamic Measurement Methods and Volumetric Correction Factors, Part 5—Base Prover Volume Using Master Meter Method — 1st Edition September 2001 Reaffirmed: November 2006
Chapter 14 — Natural Gas Fluids Measurement
Chapter 14.4 — Converting Mass of Natural Gas Liquids and Vapors to Equivalent Liquid Volumes. First Edition, April 1991

Chapter 14.7 — Natural Gas Fluids Measurement: Standard for Mass Measurement of Natural Gas Liquids; 01-Jan-1995, Revision 2006
Chapter 15 — Guidelines for the Use of the International System of Units (SI) in the Petroleum and Allied Industries. API Publication 2564. Second Edition, December 1983, Reaffirmed, August 1987
Chapter 16 — Measurement of Hydrocarbon Fluids by Weight or Mass
Chapter 16.2 — Mass Measurement of Liquid Hydrocarbons in Vertical Cylindrical Storage Tanks by Hydrostatic Tank Gauging. First Edition, November 1994 Reaffirmed 3/02

A7.2 ASTM International (American Standard for Testing Materials) — http://www.astm.org/

D287-92e1 — Test Method for API Gravity of Crude Petroleum and Petroleum Products (Hydrometer Method) 2000
D1250 — Volume Correction Factors (Joint Standard with API MPMS Chapter 11.1)
D1250-04 — Standard Guide for Use of the Petroleum Measurement Tables; 2000
Adjunct to D1250 — Temperature and pressure volume correction factors for generalized crude oils, refined products, and lubricating oils. 01-May-2004, D1250 ADJ CD-2-10
D1298-99e2 — Test Method for Density, Relative Density (Specific Gravity), or API Gravity of Crude Petroleum and Liquid Petroleum Products by Hydrometer Method; 10-Jun-1999
D2161-05 — Conversion of Kinematic Viscosity to Saybolt Universal Viscosity or to Saybolt Furol Viscosity; 01-Apr-2005
D4052-96 — Density and Relative Density of Liquids by Digital Density Meter; 10-Apr-1996
D4052-96 (2002) e1 — Standard Test Method for Density and Relative Density of Liquids by Digital Density Meter; Revised 2002
D4007-11 — Standard Test Method for Water and Sediment in Crude Oil by the Centrifuge Method (Laboratory Procedure)
D4057-95(2010) — Manual Sampling of Petroleum and Petroleum Products
D4177-95(2010) — Automatic Sampling of Petroleum and Petroleum Products
D5854-95(2010) — Practice for Mixing and Handling of Liquid Samples of Petroleum and Petroleum Products
D5002-99 — Standard Test Method for Density and Relative Density of Crude Oils by Digital Density Analyzer; Revised 01-Nov-2005

A7.3 American National Standards Institute/American Society of Mechanical Engineers

ANSI/ASME PTC 19.1-2005 — Test Uncertainty

A7.4 International Organization for Standardization (ISO) — www.iso.org

ISO 91-1:1992 Petroleum measurement tables — Part 1: Tables based on reference temperatures of 15 degrees C and 60 degrees F
ISO 91-2:1991 Petroleum measurement tables — Part 2: Tables based on a reference temperature of 20 degrees C

ISO/CD 3170 Petroleum liquids — Manual sampling
ISO 3170 AMD 1 Petroleum liquids — Manual sampling; Amendment 1, 01-Feb-1998
ISO 3171:1998 Petroleum liquids — Automatic pipeline sampling
ISO 3675:1998 Crude petroleum and liquid petroleum products — Laboratory determination of density, Hydrometer method
ISO 3838:2004 Crude Petroleum and liquid petroleum products — Determination of density or relative density, Capillary-stoppered pyknometer and graduated bicapillary methods.
ISO 3993:1984 Liquefied petroleum gas and light hydrocarbons — Determination of density or relative density—Pressure hydrometer method
ISO 4257:2001Liquefied petroleum gases; Method of sampling
ISO 4266:1994 Petroleum and liquid petroleum products — Measurement of temperature and level in storage tanks — Automatic methods
SO 4266 — Part 1: 2002 Petroleum and liquid petroleum products — Measurement of level and temperature in storage tanks by automatic methods — Part 1: Measurement of level in atmospheric tanks
ISO 4266 — Part 3:2002 Petroleum and liquid petroleum products — Measurement of level and temperature in storage tanks by automatic methods — Part 3: Measurement of level in pressurized storage tanks (non-refrigerated)
ISO 4266 — Part 4: 2002 Petroleum and liquid petroleum products — Measurement of level and temperature in storage tanks by automatic methods — Part 4: Measurement of temperature in atmospheric tanks
SO 4266 — Part 6: 2002 Petroleum and liquid petroleum products — Measurement of level and temperature in storage tanks by automatic methods — Part 6: Measurement of temperature in pressurized storage tanks (non-refrigerated)
ISO 4267-2:1988 Petroleum and liquid petroleum products — Calculation of oil quantities — Part 2: Dynamic measurement
ISO 4268:2000 Petroleum and liquid petroleum products — Temperature measurements — Manual methods
ISO 4269:2001 Petroleum and liquid petroleum products — Tank calibration by liquid measurement; Incremental method using volumetric meters
ISO 4512:2000 Petroleum and liquid petroleum products — Equipment for measurement of liquid levels in storage tanks; Manual method.
ISO 5024:1999 Petroleum liquids and liquefied petroleum gases — Measurement; Standard reference conditions
ISO/TR 7507-1:2003 Petroleum and liquid petroleum products — Calibration of vertical cylindrical tanks; Part 1: Strapping method
ISO/TR 7507-2:2005 Petroleum and liquid petroleum products — Calibration of vertical cylindrical tanks; Part 2: Optical-reference-line method
ISO/TR 7507-3:2006 Petroleum and liquid petroleum products — Calibration of vertical cylindrical tanks; Part 3: Optical-triangulation method
ISO/TR 7507-4:1995 Petroleum and liquid petroleum products — Calibration of vertical cylindrical tanks; Part 4: Internal electro-optical distance-ranging method
ISO/TR 7507-5:2000 Petroleum and liquid petroleum products — Calibration of vertical cylindrical tanks; Part 5: External electro-optical distance-ranging method
ISO/TR 7507-6:1997 Petroleum and liquid petroleum products— Calibration of vertical cylindrical tanks; Part 6: Recommendations for monitoring, checking and verification of tank calibration and capacity table
ISO 8222:2002 Petroleum measurement systems — Calibration–Temperature corrections for use when calibrating volumetric proving tanks

ISO 9770:1989 Crude petroleum and petroleum products — Compressibility factors for hydrocarbons in the range 638 kg/m^3 to 1,074 kg/m^3
ISO 11223:2004 Petroleum and liquid petroleum products — Direct static measurements — Measurement of contents of vertical storage tanks by hydrostatic tank gauging
ISO 15169:2003 Petroleum and liquid petroleum products — Determination of volume, density and mass of the hydrocarbon content of vertical cylindrical tanks by Hybrid tank measurement systems.

REFERENCES

[1] Yeandle, M., 2008, "The Calibration of Storage Tanks," *83rd International School of Hydrocarbon Measurement*, Oklahoma City, OK.

[2] Arias, R., 2009, "Measurement Methods for Liquid Storage Tanks," *84th International School of Hydrocarbon Measurement*, Oklahoma City, OK.

[3] Sivaraman, S., 1990, "Guidelines Help Select Storage Tank Calibration Method," *Oil and Gas Journal*, February 5th http://www.ogj.com/articles/print/volume-88/issue-6/in-this-issue/refining/guidelines-help-select-storage-tank-calibration-method.html

[4] Gallagher, J., 2008, "Effects of Petroleum Properties on Pipeline Measurement," *83rd International School of Petroleum Measurement*, Oklahoma City, OK.

[5] Miller, R., 1996, *Flow Measurement Engineering Handbook,* 3rd Ed., New York, N.Y.

[6] Hwang, E. D. A. G., 2010, "Custody Transfer: Flowmeter as Cash Register," Control Engineering - 57(9), November.

[7] Fiedler, J. C., 1991, "Liquid Measurement Station Design," FMC Measurement Solutions, Houston, TX.

[8] Barnes, R., 2008, "Displacement Meters for Liquid Measurement," *83rd International School of Hydrocarbon Measurement*, Oklahoma City, OK.

[9] Piskorski, T., 2009, "Displacement Meters for Liquid Measurement," *84th International School of Hydrocarbon Measurement*, Oklahoma City, OK.

[10] Seiler, D., 2009, "Application of Turbine Meters in Liquid Measurement," in 84th International School of Hydrocarbon Measurement, Oklahoma City, OK.

[11] Smith, J., 2008, "Helical Turbine Meters for Liquid Measurement," in 83rd International School of Hydrocarbon Measurement, Oklahoma City, OK.

[12] Seiler, P. S. A. D., 2007, "Use Liquid Ultrasonic Meters for Custody Transfer," *Hydrocarbon Processing*, pp. 91–96, August.

[13] Yon, M., 2008, "Coriolis Meters for Liquid Measurement," *83rd International School of Hydrocarbon Measurement*, Oklahoma City, OK.

[14] Baker, P. D., 1990, "Pipeline Meter Selection — Technical Paper 108," *FMC Energy Systems*, Houston, TX.

[15] "A Comparison of Liquid Petroleum Meters or Custody Transfer Measurement," FMC Technologies Technical Paper Bulletin TP0A014, Houston, TX.

[16] P. Jakubenas, 2002, "Liquid Measurement Station Design," in *77th International School of Hydrocarbon Measurement*, Oklahoma City, OK.

[17] Frey, M., 2008, "Liquid Measurement Station Design," *83rd International School of Hydrocarbon Measurement*, Oklahoma City, OK.

[18] Tansey, K., 2009, "Liquid Measurement Station Design," *84th International School of Hydrocarbon Measurement*, Oklahoma City, OK.

[19] Kalivoda, R., 2010, *Proving Liquid Ultrasonic Flow Meters for Custody Transfer Measurement*, FMC Technologies — Smith Meter, Houston, TX.

[20] Gradziel, J., 2007, "Liquid Flow Provers," *82nd International School of Hydrocarbon Measurement*, Oklahoma City, OK.

[21] Pon, M., 2009, "Proving Coriolis Meters," *84th International School of Hydrocarbon Measurement*, Oklahoma City, OK.

[22] Ryan, B., 1999, "Liquid Flow Provers," *74th International School of Hydrocarbon Measurement*, Oklahoma City, OK.

[23] Martin, P., 2009, "Realistic Pipe Prover Volume Uncertainty," *27th International North Sea Flow Measurement Workshop*, Tonsberg, Norway.

[24] Yoon, M., Warren, B., and Adam, S., 2007, *Pipeline System Automation and Control*, ASME Press, New York, NY.

[25] Kosewicz, P., 2008, "Calculation of Liquid Petroleum Quantities," *83rd International School of Hydrocarbon Measurement*, Oklahoma City, OK.

[26] Kosewicz, P., 2008, "Fundamentals of Liquid Measurement III — Dynamic," *83rd International School of Hydrocarbon Measurement*, Oklahoma City, OK.

[27] Pembina Pipeline Corporation, 2011, "ProjectTurbo-Early Project Bulletin," September, http://www.pembina.com/pembina/webcms.nsf/AllDoc/80E975E2DB852A3487257937005F8C96/$File/Turbo%20early%20Bulletin%20-%20singles%20-%20FINAL.pdf.

Chapter 8

HYDROCARBON PETROLEUM TANKAGE AND TERMINAL DESIGN

This chapter summarizes but brings together the history and use as well as salient features of design, operation and maintenance of hydrocarbon liquids storage facilities. Citations are provided in the list of reference to assist readers to source all documents necessary for full comprehension of issues related to components making up such systems. It is a reference chapter only on the subject matter and should not be used as turnkey. Readers are encouraged to refer to and consult all standards listed herein.

8.1 INTRODUCTION AND OVERVIEW

Oil terminals are widely used to store various liquids and gases such as chemicals, crude oil, and natural gas. Oil storage terminals play a vital role in oil transmission systems for temporary parking/storing liquid petroleum products. Such tankage provides the management of product inventory and shipment, side stream injection, pumping and product movement, pipeline maintenance, etc.

A **storage tank** is a container, usually for holding liquids (hydrocarbon, water, etc.), sometimes for compressed gases (gas tank). The term is also used for reservoirs (artificial lakes and ponds), and for manufactured containers. Tank farms cover a group of tanks for the commercial storage of oil and petroleum products, sited together.

Oil terminals are widely used to store various liquids and gases such as chemicals, crude oil, hydrocarbon liquids, natural gas, LNG/LPG, etc. Oil storage terminals play a vital role in oil transmission systems for temporary parking /storing liquid petroleum products. Such tankage provide the opportunity for the management of product inventory and shipment, side stream injection, pumping and product movement, pipeline maintenance, etc.

Figure 8-1 illustrates a typical liquid hydrocarbon pipeline system and storage and shipping facilities.

The tank farm is generally provided with a bund wall and contains pipe racks, drainage, and fire-suppressant piping (Figure 8-2). Such bunds are generally designed to contain any spills from the tank or tank piping.

From a regulatory point of view, Bunding is a legal requirement in many countries particularly around tanks, storage vessels and other plant that contain liquids which may be dangerous or hazardous to the environment.

As well, bunded hydrocarbon storage tanks are generally a requirement from most insurance companies as opposed to single skinned oil storage tanks. Due to the safer oil storage solutions brought about by bunded oil storage tanks, some environmental agencies require to install bunded oil tanks.

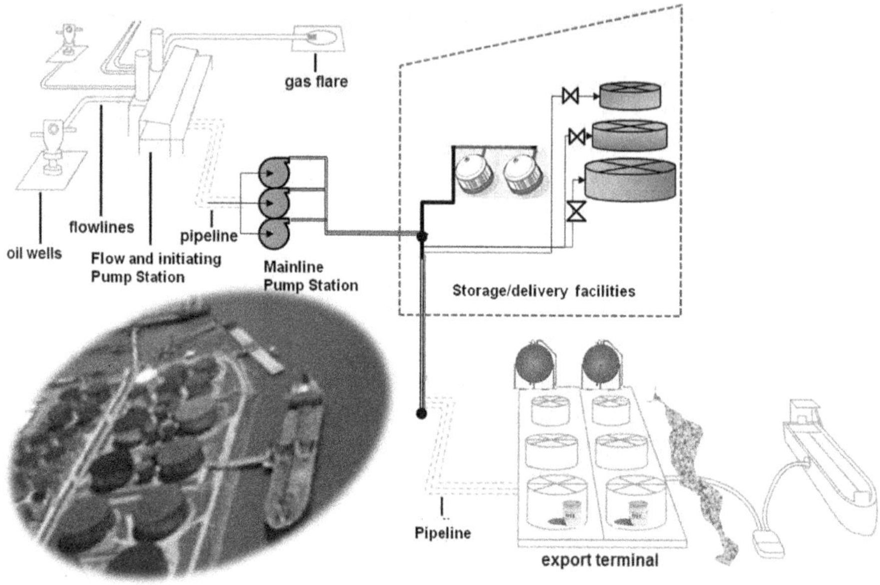

Figure 8-1. Liquid hydrocarbon pipeline transportation storage and shipping facilities

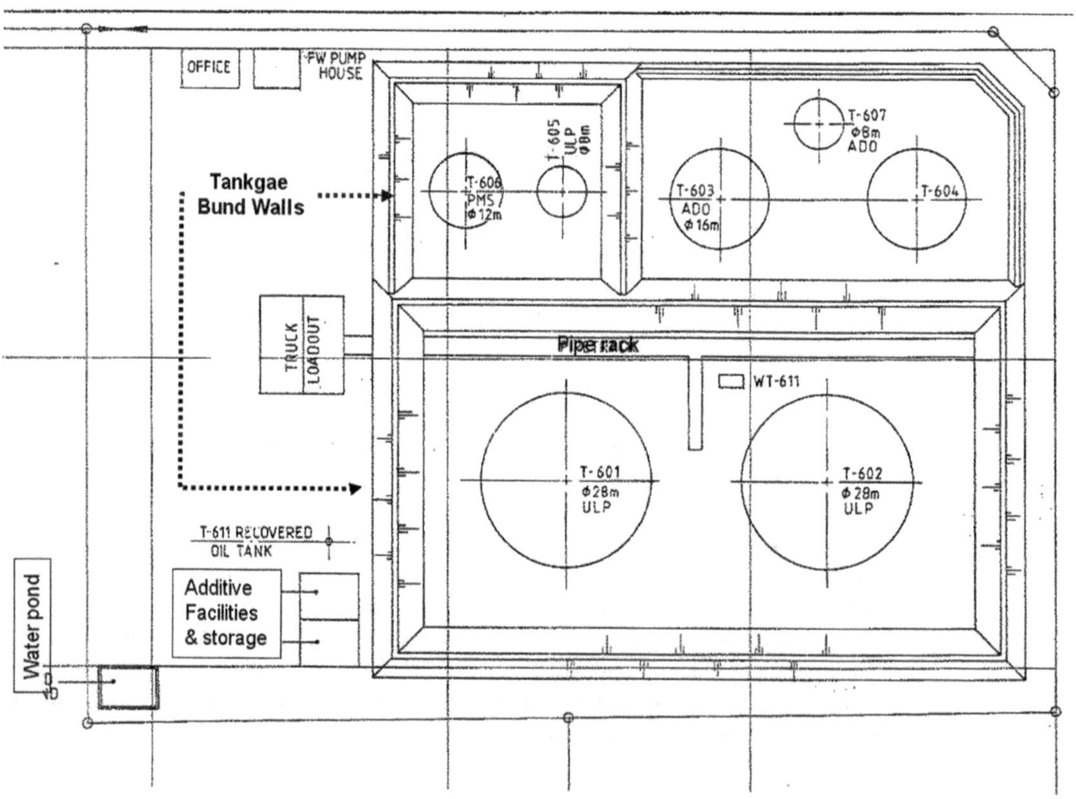

Figure 8-2. Typical tank farm with tankage bund walls

The height of the bund walls is typically designed to assure the retention of any fuel spillage from the tanks within the tank farm boundary. The industry requires that the bund containment generally exceed 110% of the storage capability in each bund (or 180% of the volume of the largest tank). Additionally, each tank is also separated by intermediate bund walls to hold minor spills.

One of the key imperative requirements for a tank farm is Health, Safety and Environment (HSE) and that the operators of a depot must ensure that all petroleum products are safely stored and handled and that there are no leakages (etc) which could damage the soil or the water table. Fire protection is a primary consideration, especially for the more flammable products such as gasoline and aviation fuel.

Fire prevention (fire protection/fire safety) often comes within the remit of health and safety professionals as well. The primary need is for safety measures must be in place to prevent fuel from exiting the tanks in which it is stored. Added safety measures are needed for when fuel does escape, mainly to prevent it forming a flammable vapor and stop pollutants from poisoning the environment.

Poor design, layout and lack of safety planning and execution can lead to disastrous consequences in an event of unforeseen incidents. An example is the fire incident at the Burchfield Oil Depot in the UK in 2005 (Figure 8-3).

The **Buncefield fire** was an inferno caused by a series of explosions on 11 December 2005 at the Hertfordshire Oil Storage Terminal (Total UK & Texaco facilities), an oil storage facility located near the M1 motorway by Hemel Hempstead in Hertfordshire, UK. The terminal was the fifth largest oil-products storage depot in the United Kingdom, with a capacity of about 60,000,000 imperial gallons (about 1.7×10^6 BBLS) of fuel. Figure 8-3 inset indicates the fire just ten minutes after the explosion.

Figure 8-3. Buncefield fire and explosion of Hertfordshire oil storage terminal [1, 2]

8.2 HISTORY AND REASONS FOR USE

Petroleum is not a substance new in the History of the World. This historical fact along with timelines indicating major discoveries/events can be illustrated as follows:

> 4000 years ago use of asphalt by Greeks (as per Greek historian Herodotus and confirmed by another Greek Historian Diodorus Siculus), asphalt was employed in the construction of the walls and towers of Babylon, and there were oil pits near Ardericca, a place located not far from Babylon

4000 BC Use of oil for ship caulking in Mesopotamia

Ancient Persian tablets Indicate the medicinal and lighting uses of petroleum in the upper levels of the Iranian society at the time

3rd Century AD The Romans began to use barrels

9th Century 1st distillation of petroleum by Iranian Chemist Zakariya Razi (865–925) producing chemicals such as kerosene

9th Century Exploitation of oil fields in the area around modern Baku, Azerbaijan to produce naphtha

13th Century Venetian trader and explorer Marco Polo (1254–1324) described the output of those oil wells in Baku area as hundreds of shiploads

1539 1st oil export, Venezuela [3]

1819 Discovery of oil in the USA

1846 Discovery of the process of refining kerosene from coal by Canadian Abraham Pineo Gesner

1854 First fractionation of petroleum by distillation by American Benjamin Silliman (1770 to 1864), a science professor at Yale University in New Haven (USA)

1854 1st recovery of by-product of oil from salt in the USA

1854 to 1856 Construction of world's 1st tank farm by Ignacy Lukasiewicz near Jasło, Austrian Empire (now in Poland)

1854 Kier's Refinery — First in Western Hemisphere, USA [3].

1856–57 Opening of world's first large refinery at Ploesti, Romania

1858 The birthplace of oil refining and kerosene distillation, Petrolia. South Africa

1859 1st application of well drilling technology by Edwin Drake

1859 Beginning of oil drilling in the USA

1861 Russian refinery in the mature oil fields at Baku built by Russian Meerzoeff. At that time Baku produced about 90 per cent of the world's oil

1861 1st commercial scale refinery in the USA by Lockhart, Frew and partners (at Brilliant Station on the south bank of the Allegheny River near Negley Run)

1870–80 Passage of the first oil tankers through the Suez Canal

1885 Discovery of oil in Sumatra (Dutch East Indies)

1889 Establishment of Chicago Bridge & Iron Company (CB&I)

1896 Formation of National Fire Protection Agency (NFPA)

1902 Invention of Fire fighting foam (FFF) by the Russian engineer and chemist Aleksandr Loran

1904 Formation of national Board of Fire Underwriters (NBFU), now NFPA

1908 Discovery of oil in Iran (Masjid Soleiman) (Figure 8-4)

1908 Discovery/production of oil in Peru, Venezuela, and Mexico, at an industrial level

Figure 8-4. Discovery of oil in Iran, Masjid Soleiman, 1908 (left photo [4]) and the Abadan Oil Refinery, 2008 (right photo)

1910 Discovery of significant oil fields in Canada (in the province of Ontario)
1912 Completion of 1st largest Oil Refinery, Abadan, Iran
1914 1st record of Storage of Oil Fuel [5]
1916 Establishment of National Tank Association (NSTA), now Steel Tank Institute (STI), USA
1916–17 Formation of Underwriters Laboratories (UL)
1919 Formation of American Petroleum Institute (API)
1922 First UL Standard UL142 "Steel Above Ground Tanks for Flammable and Combustible Liquids"
1923 First floating-roof tank for the oil industry By CB&I
1942 Destruction of oil tank farm by the bombing the US Army barracks at Fort Mears, Amaknak Island in Unalaska Bay
1950 Construction of Milford Haven oil port
1951 Formation of Petroleum Equipment Institute (PEI)
1964 The Texaco oil refinery Milford Haven
1964 Construction of 1st Largest Crude Oil Tank(160,000 BBLS, 109 m diameter, 18 m height), Kharg Island, Persian Gulf
1974 Construction of the largest Crude Oil Tank, Abu Zabi (182, 000 BBLS)

It is difficult to pin point the birth of hydrocarbon tankage and tank farms/depot and early years of professional design tank farms for storage of hydrocarbons fuels in large volumes. The only reference found is due to Lugoff and Camden [5], where in a book published in 1914, titled "Oil fuel for steam boilers", describe in a chapter the "Storage of Oil Fuel". There, they have a diagram of what they call a 'small or medium' size cylindrical steel tank. They also state that the common sizes of oil tanks. However, search of literature indicates the birth date for large tankage systems to be in the late 19th century.

Similar to liquid pipelines, storage tank systems have played an important role in the transportation industry, particularly in the post-World War II era.

Early in the twentieth century, the oil companies operated the pipelines and storage facilities as integrated subsidiaries and often along with refineries. They also often used them to control the oil industry.

Without storage facilities, pipelines are limited in the markets they can serve and would be limited in the commodities they can haul. Pipelines are the only mode of

Figure 8-5. General view of hydrocarbon tankages system and tank farm

transportation with no backhaul, i.e., they are unidirectional with products that only move in one direction through the line.

A tank farm includes many tankage systems and appurtenances (Figure 8-5). It is a facility where various petroleum products are stored prior to being transported further, shipped and disbursed to end consumers or retail facilities. They are also referred to as "oil depot."

Some tank farms are owned by a single company which uses the farm to meet its needs, while others are administered by a group. It is also possible for facilities to have their own tank farms for the purpose of storing fuel on site, with airports being a classic example of a facility which needs to have a large quantity of fuel (mostly jet fuel) on hand.

It is believed that the first permanent, professionally designed tank farm for storage hydrocarbon liquids was built at the turn of the century simultaneous with construction of major refineries.

8.3 PRODUCTS STORED AND PROPERTIES

Petroleum products that are stored in tanks, tank farms, breakout facilities depends on transportation company systems and the complexities of liquid hydrocarbon mixtures being transported and as well acceptable contamination levels.

It also depends where refineries/terminals are located along pipelines and in major cities and where there are only a few large-long refined products pipelines.

Products transported and/or stored generally include the following:

Crude Oil:

- Heavy
- Medium
- Light

Refined Product:

- motor gasoline
- diesel fuels
- aviation fuels

Natural Gas Liquids propane, butane and condensate mixture
Buffer (slops only) synthetic (semi-processed) oil
Synthetic Oil

Buffer is a synthetic crude oil, a semi-refined, "clean," crude product which is used as a buffer between the natural gas liquids and the refined products. This ensures that NGL does not migrate into refined products, affect the vapor pressure and thereby impact the flash point of the refined products. Therefore refineries and pipeline companies will have separate tanks for storing products used as buffer. Typical volume is about 2000 to 3000 m^3 in volume.

Products transported and stored have typical properties as indicated in Table 8-1 below.

Typically product storage depends on what, how, and rate and sequence various products are delivered. For example, crude storage depends on delivery of the crude to a refinery. While a pipeline is normally the preferred mode of transportation (as

TABLE 8-1. Typical properties product transported and stored

Commodity	Viscosity (CS)	Temperature (°C)	Density (kg/m^3)*
Diesel	6.86	5	847
	5.10	15	820
Gasoline (leaded)	0.68	5	711.3
	0.61	15	–
Gasoline (unleaded)	0.7	5	699
	0.63	15	690 (assumed)
Jet A Fuel	8	29	774
	1.5	0	–
Jet B Fuel	1.9	15	–
	1.5	35	–
Kerosene	3	15	–
	2.2	35	708.8
Condensate (sweet)	0.599	15	688.8
	0.548	25	572.3
	0.199	15	547.5
Condensate (raw)	0.171	30	500.6
	0.218	20	483.5
Propane (at 1000 kPa)	0.199	30	446.9
	0.166	52	560.8
	0.237	25	535.12
	0.235	44	529
Butane (at 470–520 kPa)	0.212	52	
	21.1	83	
	37.8	3.75	
Very Heavy Oil	40	3.28	
	50	2.44	
	21.1	37	
	37.8	19	
Heavy Crude	21.1	16.2	
	37.8	9.41	
Medium Crude	21.1	10.2	
Light Crude	37.8	6.25	

considered more reliable and less affected by weather), the crude tankage depends of various factors including:

Pipeline: Pipeline capacity, Reliability
Marine: Capacity of largest ship
Ship unloading rate
Ship travel/turn around frequency
Number of ships at one time

For intermediate storage, (the storages that are for transferring and importing), the following are typically applicable:

- Crude charge: 1 to 2 days supply per crude grade plus one identical tank
- Unit rundown: 1 to 4 days production capacity plus one identical tank per group
- Blending stock: 5 days storage at production (line blending) rates plus one identical tank per group

Similarly, in the refined product storage, the following applies:

- Pipeline shipment: for each product, frequency, rate, reliability
- Truck and Rail: truck or tank car size, Time for loading, loading capacity
- Marine Shipment: largest ship capacity and ship frequency, Number of ships at one time

The storage of liquid hydrocarbons depends on the quantity and its physical properties.

The physical properties include density/specific gravity and vapor pressure/boiling point; Their effects are described below:

- ***Density and specific gravity:*** the density of the liquid is its mass per unit volume. Water has a density of 1 gm/cm^3 at 4°C. The density of a liquid plays an important role in the design of a tank because larger densities require thicker shells.
- ***Specific gravity:*** another important physical property of the liquid stored. It is a measure of the relative weight of one liquid compared to water. Specifically, it is the ratio of the density of the liquid divided by the density of the water at 15.5°C. For example, petroleum oil, kerosene, and gasoline have a specific gravity of 0.82, 0.80, and 0.70, respectively. Care must be exercised if there is a significant increase in the specific gravity of the new liquid because the effective hydrostatic pressure acting on the tank walls will be greater if the design level is not reduced and could cause damage on the cylindrical shell.
- ***Vapor pressure and boiling point:*** the vapor pressure of a pure hydrocarbon liquid is the pressure of the vapor space above the liquid in a closed container, and increases with increasing temperature. It is an important consideration in order to select the type of tank and its roof and is crucial for the purpose of characterizing fire hazardousness.
- ***The boiling point:*** also important. It is necessary to know the temperatures at which some liquids should be stored, always below its boiling point. For example, some flammable and combustible liquids are prohibited by the fire codes to be stored at temperatures above their boiling point. A large number of tanks in oil refineries or petrochemical industries, store flammable liquids.

8.4 TYPES OF PETROLEUM STORAGE TANKS

8.4.1 Definition and Classifications

A storage tank is a container, usually for holding liquids, sometimes for compressed gases (such as LNG, NGL, etc.).

Generally (excluding spheres), there are six basic tank designs that are used for hydrocarbon liquid storage vessels: fixed roof (vertical and horizontal), external floating roof, domed external (or covered) floating roof, internal floating roof, variable vapor space, and pressure (low and high). These are described further herein.

There are many different forms to classify storage tanks. The most fundamental classification is based upon whether they are above or belowground. There are usually many environmental regulations applied to the design and operation of storage tanks, often depending on tank classification internal pressure (IP) or atmospheric pressure (AP) and on the nature of the fluid contained within. Aboveground storage tanks (AST) differ from underground storage tanks (UST) in the kinds of regulations that are applied. The aboveground tanks (AST) have almost all their structure exposed. The bottom part of these tanks is placed directly over soil or on a concrete foundation.

Classification based on the internal pressure: In the case that an internal pressure acts on the tank during storage, it is possible to classify these tanks based on this level of pressure. This pressure effect depends directly of the size of the tank. The larger the tank, the more severe effect of pressure is on the structure. This classification is commonly employed by codes, standards, and regulations.

Classification based on atmospheric tanks: These tanks are the most common. Although they are called atmospheric, they are usually operated at internal pressure slightly above atmospheric pressure. The fire codes define an atmospheric tank as operating from atmospheric up to 3.5 kN/m^2 above atmospheric pressure.

Low-pressure tanks: Within the context of tanks, low pressure means that tanks are designed for a pressure higher than atmospheric tanks. This also means that these tanks are relatively high-pressure tanks. Tanks of this type are designed to operate from atmospheric pressure up to about 100 kN/m^2.

Pressure vessels (high-pressure tanks): Since high-pressure tanks are really pressure vessels, the term high-pressure tank is not frequently used; instead they are called only vessels. Because these kinds of tanks are usually built underground, they are not included in this work and they are not covered in detail herein. However, they are treated separately from other tanks by all codes, standards, and regulations.

In the USA, storage tanks operate under no (or very little) pressure, distinguishing them from *pressure vessels*. Storage tanks are often cylindrical in shape, perpendicular to the ground with flat/sloping bottoms, and a fixed or floating roof.

Generally Tankage cost in refineries and terminals cost between 25% and 30% of total facilities capital investment. This includes utilities and design.

Tank capacities are determined based on the type of products, crude, intermediate and refined and the transportation system (pipeline, truck, and rail) as well as other requirements such as crude charge, unit rundown, blending stock, etc.

8.4.2 Types

Types of tanks includes the following:

- Atmospheric
  - Fixed roof, vertical and horizontal
  - Floating roof (external floating roof, domed external (or covered) floating roof, internal floating roof)
- Pressurized
  - Variable vapor space
  - High pressure
  - Low pressure
- Heated Tanks (Hot fluids)
  - High viscosity Oil (bitumen, asphalt, etc.)
- Refrigerated Tanks
  - Cryogenic
  - Refrigerated

The shape of the roof is an indicator of the type of a tank because it is self-explanatory to tank designer, fabricator and erector. The configuration of a vertical aboveground tank design can be either an open top with the roof floating on the stored liquid or a fixed roof. The safe design of floating roof tank offers a considerable level of fire safety over other vertical tank designs. As a result, fire codes allow closer spacing between floating roof tanks and for separation from adjacent properties or operations providing a cost advantage in tank farm layout and arrangement.

8.4.2.1 Fixed Roof Tanks

Fixed roof tanks are generally for hydrocarbon liquids with very high flash points (e.g., heavy crude oil, diesel, heavy kerosene, fuel oil, bitumen/asphalt etc.). Vapor pressure in such tanks is of the order of 3.5 kPa (0.5 Psia). Figure 8-6 illustrates vapor pressure of hydrocarbon liquids commonly transported by pipelines and stored in storage tanks [6]. Similar graphs were presented in Chapter 2, Figure 2-9.

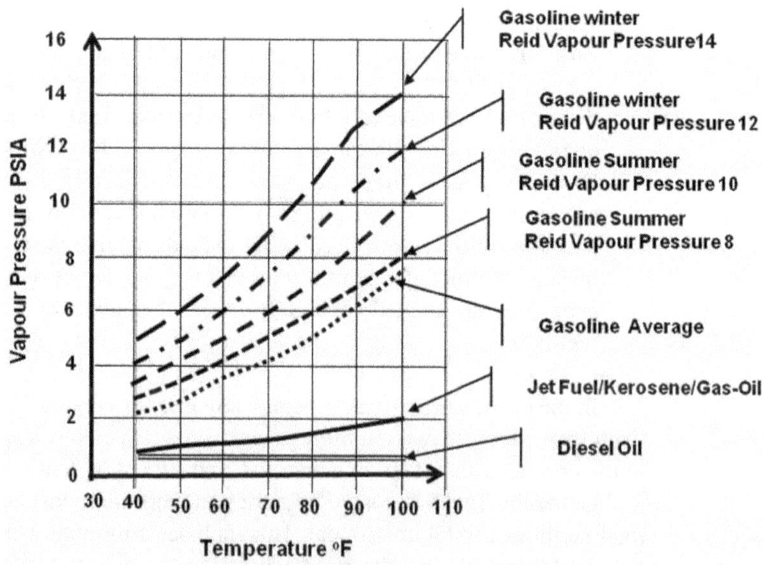

Figure 8-6. Vapor pressure of liquid hydrocarbons commonly transported by pipelines (reproduced from [6])

Losses from fixed roof tanks are caused by changes in temperature, pressure, and liquid level. Fixed roof tanks are either freely vented or equipped with a pressure/vacuum vent. The latter allows the tanks to operate at a slight internal pressure or vacuum to prevent the release of vapors during very small changes in temperature, pressure, or liquid level. Of current tank designs, the fixed roof tank is the least expensive to construct and is generally considered the minimum acceptable equipment for storing organic liquids.

Fixed roof tanks have several roof designs (Figure 8-7) including:

- cone roofs,
- dome roofs, and
- umbrella roofs

Cone-roof tanks have cylindrical shells in the lower part. These are the most widely used tanks for storage of relatively large quantities of fluid. They have a vertical axis of symmetry, the bottom is usually flat (or slightly sloped), and the top is made in the form of shallow cone as illustrated in Figure 8-7. A shallow cone roof deck on a cone roof tank approximates a flat surface and is typically built of a thick steel plate (approximately 4.76 mm thick).

Steel dome roof tanks have the same spherical shape as an umbrella roof. An umbrella roof is nothing more than a stiffened dome roof. If the umbrella roof framing is internal, visually there is very little difference in the appearance of a dome roof and umbrella roof. The dome roof shown in Figure 8-7 is an aluminum geodesic dome roof and looks very different than any other roof due to the triangular panels.

There are several ways to fabricate such tanks. One of these techniques is the tank airlift method, "in which the roof and the upper course of shell are fabricated first, then lifted by air that is blown into the tanks as the remaining lower courses of steel shell are

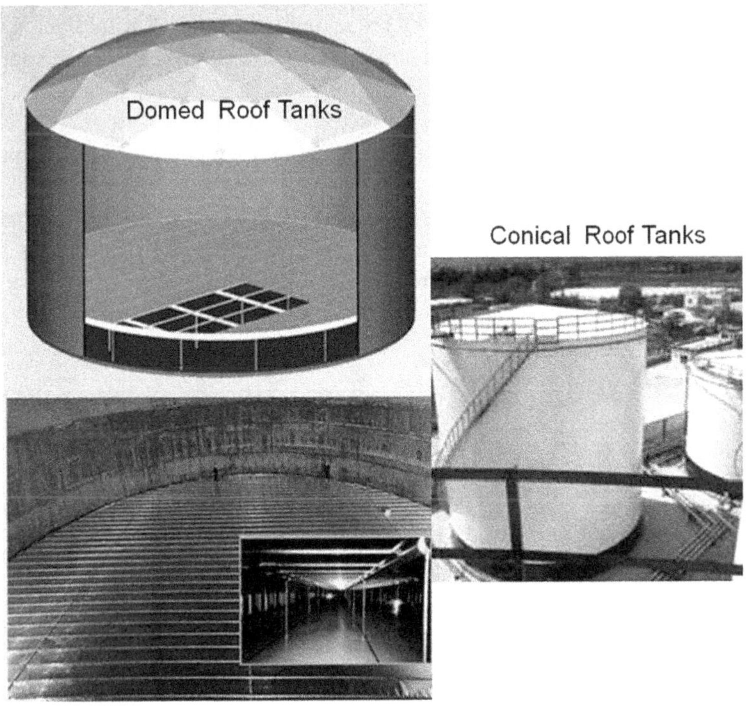

Figure 8-7. Fixed roof tanks (dome and conical roofs and float)

Figure 8-8. Fixed roof construction (airlift method) [7]

welded into place" (Figure 8-8). See Section 7 — Petroleum Storage/Terminal Design & Construction.

Cone and dome roofs can be either self-supported or column supported. Cone-roof tanks typically have roof rafters and support columns (Figure 8-9 (column support) and Figure 8-10 (Cone Roof and Roof Hob Details, [8])). Figure 8-9 is an indication of some of the older design.

Umbrella-roof tanks: They are very similar to cone-roof tanks, but the roof looks like an umbrella. They are usually constructed with diameters < 20 to 30 m. Another difference is that the umbrella-roof does not have to be supported by columns to the bottom of the tank, so that they can be a self-supporting structure. The umbrella types are spherical design.

Typical fixed roof storage tank together with appurtenances are shown in Figure 8-11.

Geodesic dome-roof tanks: Although most tanks are made of steel, some fixed-roof tanks have aluminum geodesic dome-roof. Some advantages include that they have a superior corrosion resistance for a wide range of conditions compared with steel tanks (Figure 8-12). Also they are often an economical choice and are clear-span

Figure 8-9. Cone roof tank with column support (under construction) [8]

Figure 8-10. Aluminum geodetic dome roof tank and roof top hub detail [8, 10]

structures that do not require internal supports. They can also be built to virtually any required diameter.

8.4.2.2 Floating Roof Tanks

There are two types of floating roof tanks. There are:

- External (EFRT = External floating roof tank)
- Internal (IFRT = Internal floating roof tank)

Products stored in such tanks have generally higher vapor pressure between 10 kPa and 77 kPa (1.5 to 11 psia). As such light and medium crudes, gasoline, reformate, naphtha and Jet fuels are stored in such tanks.

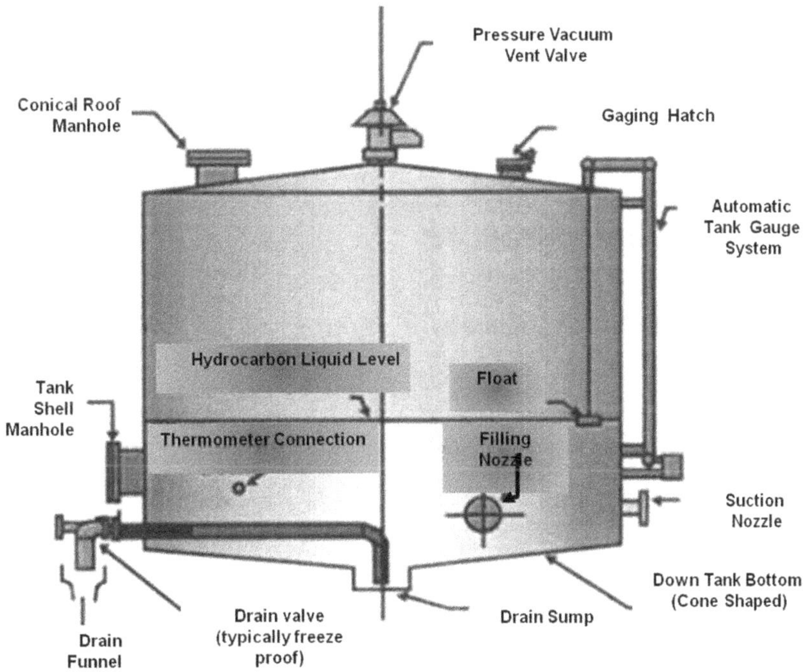

Figure 8-11. Typical fixed roof atmospheric tank and appurtenances (from ref. [11])

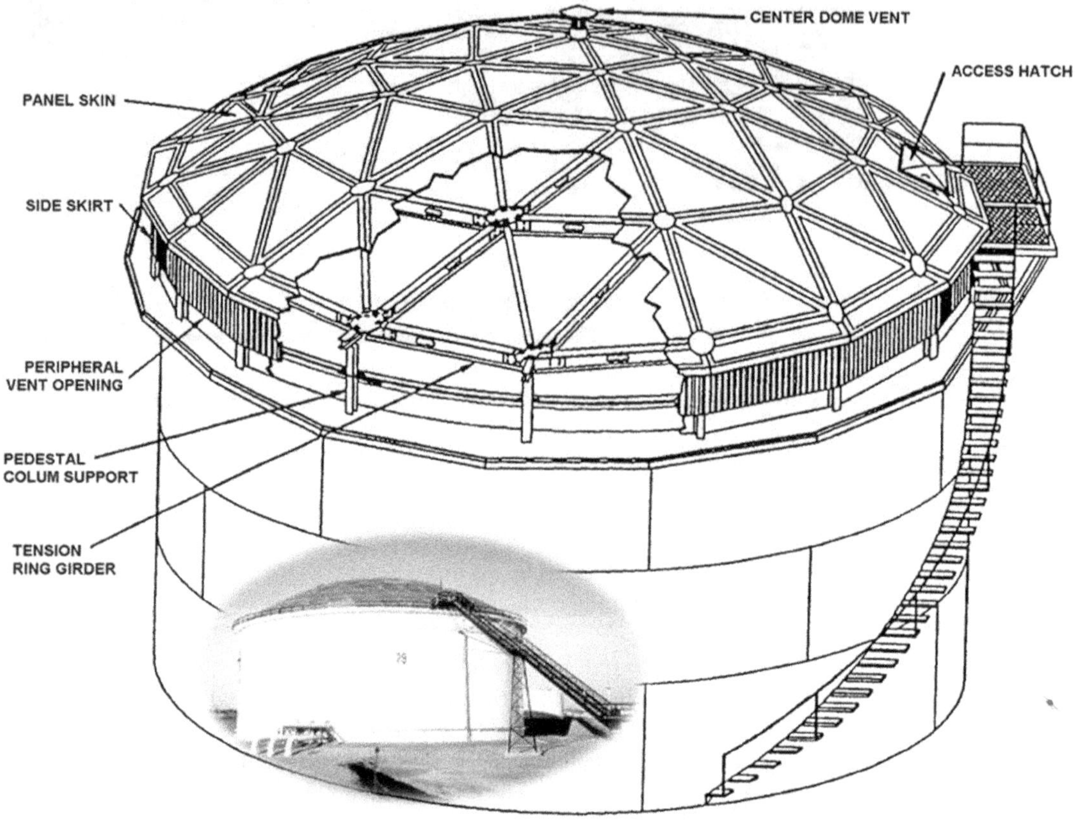

Figure 8-12. Typical external floating roof tank equipped with a geodesic aluminum dome [12]

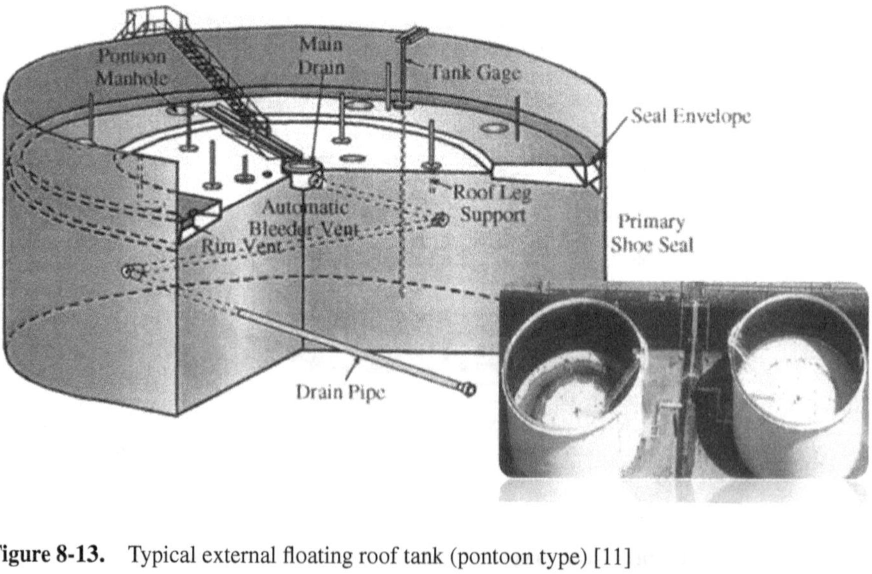

Figure 8-13. Typical external floating roof tank (pontoon type) [11]

These tanks have a roof that floats on the surface of the liquid (Figure 8-13). The floating cover or roof is a disk/pontoon structure that has sufficient buoyancy to ensure that the roof will float under all expected conditions, even if leaks develop in the roof.

External floating roof: An external floating roof tank is a storage tank commonly used to store large quantities of volatile petroleum products such as crude oil or gasoline (petrol). The external type floating roof tank is open on top (Figure 8-13) and the float can be of the following design types:

- Pan type
- Pontoon type
- Double deck type
- Buoy roof type

Most floating decks that are currently in use in the industry are constructed of welded steel plate and are of two general types: pontoon (Figure 8-13) or double-deck (Figure 8-14).

A typical external floating roof tank (EFRT) consists of an open-topped cylindrical steel shell equipped with a roof that floats on the surface of the stored liquid. The floating roof consists of a deck, fittings, and rim seal system. With all types of external floating roof tanks, the roof rises and falls with the liquid level in the tank.

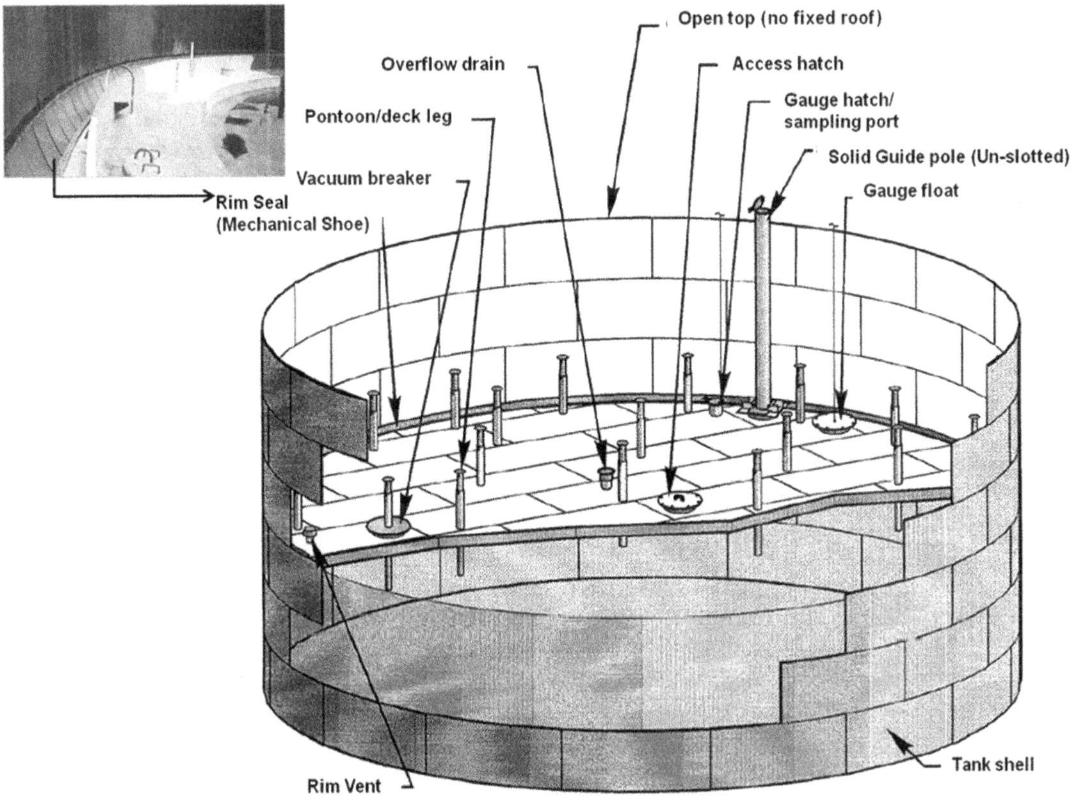

Figure 8-14. Typical floating roof tank component make up and appurtenances (double deck) [13,14]

As opposed to a fixed roof tank there is no vapor space (ullage) in the floating roof tank (except for very low liquid level situations). In principle, this eliminates breathing losses and greatly reduces the evaporative loss of the stored liquid.

External floating decks are equipped with a rim seal system, which is attached to the deck perimeter and contacts the tank wall. The purpose of the floating roof and rim seal system is to reduce evaporative loss of the stored liquid. Some annular space remains between the seal system and the tank wall. The seal system slides against the tank wall as the roof is raised and lowered. The floating deck is also equipped with fittings that penetrate the deck and serve operational functions. The external floating roof design is such that evaporative losses from the stored liquid are limited to losses from the rim seal system and deck fittings (standing storage loss) and any exposed liquid on the tank walls (withdrawal loss).

The roof has support legs hanging down into the liquid. At low liquid levels, the roof eventually lands and a vapor space forms between the liquid surface and the roof, similar to a fixed roof tank. The support legs are usually retractable to increase the working volume of the tank.

Components of external floating roof tanks are further detailed in Figure 8-15.

Different types of floating roofs design (as per API 650) for floating roof tanks are illustrated in Figure 8-16.

Advantages: External roof tanks are usually installed for environmental or economical reasons to limit product loss and reduce the emission of volatile organic compounds (VOC), an air pollutant.

Normally (roof not landed), there is little vapor space, and consequently a much smaller risk of internal tank explosion.

Disadvantages: Rain water and snow can accumulate on the roof, eventually the roof may sink. Water on the roof is usually drained from a flexible hose that runs from a drain-sump on the roof, through the stored liquid to a drain valve on the

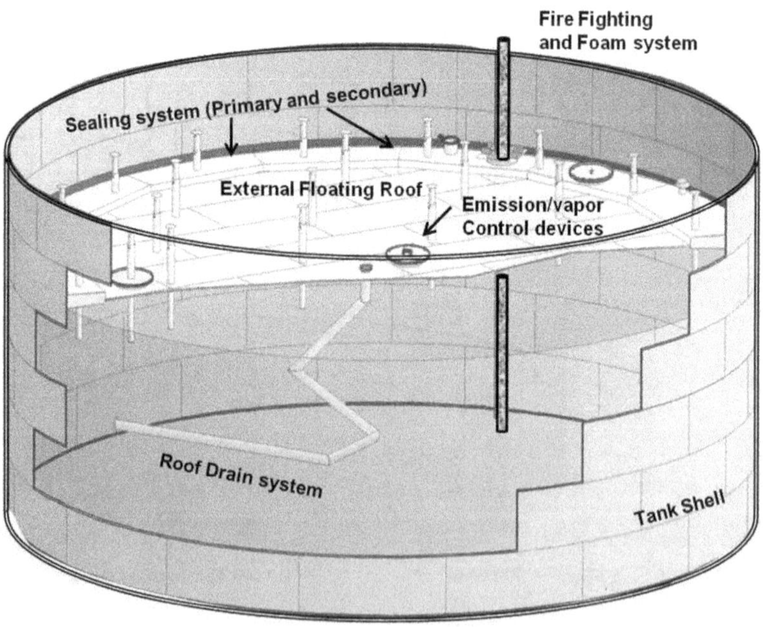

Figure 8-15. Components of external floating roof tank (EFRT) [15]

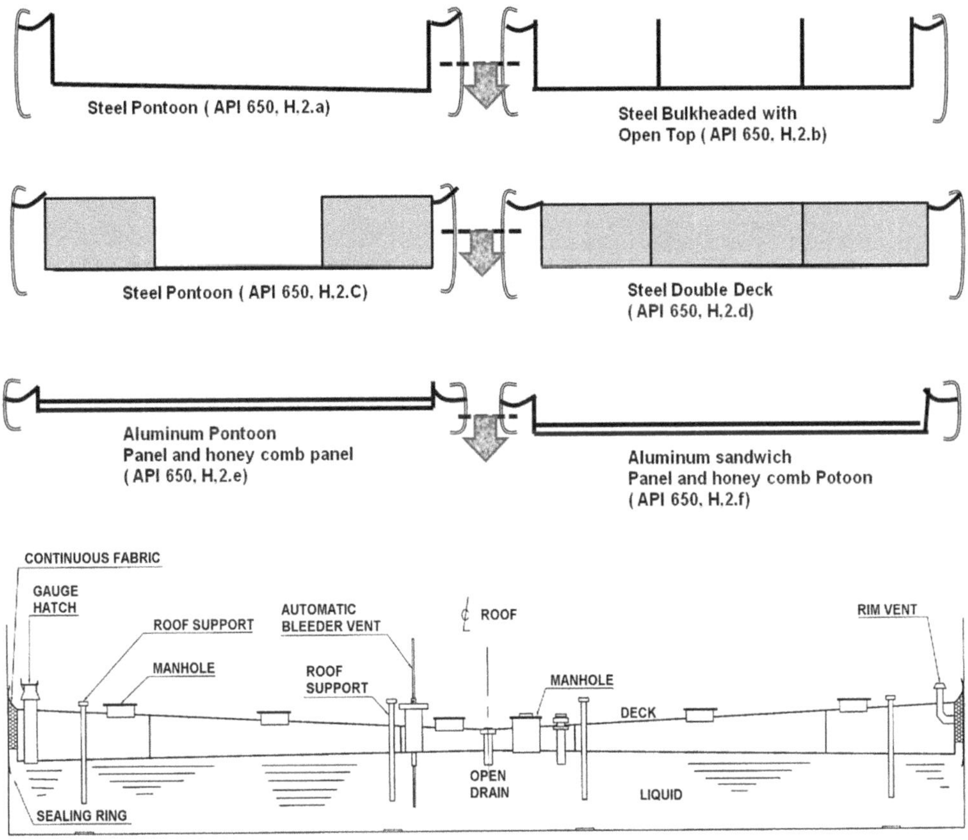

Figure 8-16. Sectional view of floating roofs (reproduced from ref. [11]) and a double design details [12]

shell at the base of the tank. The hose often develops leaks and drains both water and product.

Domed External Floating Roof Tanks: Domed external (or covered) floating roof tanks have the heavier type of deck used in external floating roof tanks (EFRT) described previously, as well as a fixed roof at the top of the shell like internal floating roof tanks. Aluminum floating roofs may also be used in a tank with an aluminum geodesic dome. In many cases the aluminum Internal Floating Roof (IFR) is supported from the dome rather than on legs.

Domed external floating roof tanks usually result from retrofitting an external floating roof tank with a fixed roof. This type of tank is very similar to an internal floating roof tank with a welded deck and a self-supporting fixed roof. A typical domed external floating roof tank is shown in Figure 8-17.

As with the internal floating roof tanks, the function of the fixed roof or geodesic dome is not to act as a vapor barrier, but to block the wind and as well keep rain and snow out of the tank and off the floating roof. The type of fixed roof most commonly used is a self-supporting aluminum dome roof, which is of bolted construction. Like the internal floating roof tanks, these tanks are freely vented by circulation vents at the top of the fixed roof. The deck fittings and rim seals, however, are identical to those on external floating roof tanks. In the event that the floating deck is replaced with the lighter IFRT-type deck, the tank would then be considered an internal floating roof tank.

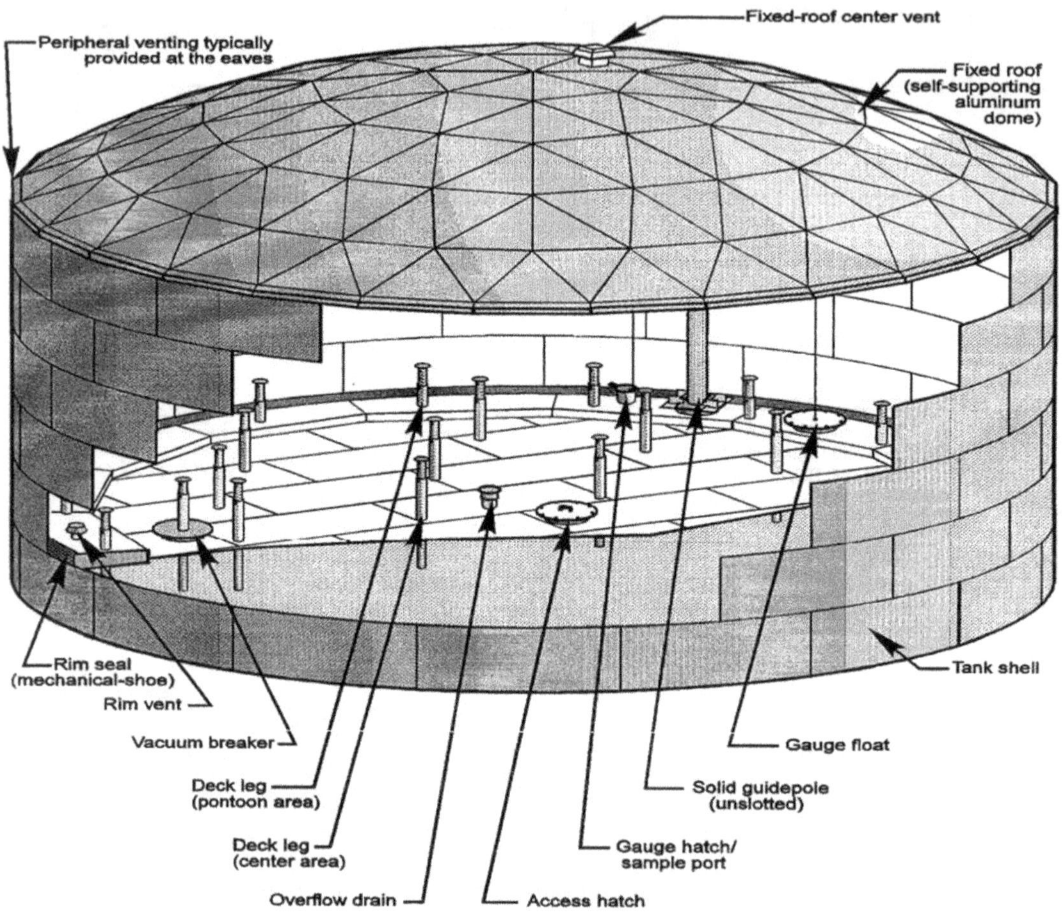

Figure 8-17. Domed external floating roof tank [13,14]

Internal Floating Roof: An internal floating roof tank (IFRT) has both a permanent fixed roof and a floating roof inside. There are two basic types of internal floating roof tanks: tanks in which the fixed roof is supported by vertical columns within the tank, and tanks with a self-supporting fixed roof and no internal support columns. Fixed roof tanks that have been retrofitted to use a floating roof are typically of the first type.

The advantages of using internal floating roofs are as follows:

- Decrease in the level of evaporation of stored product;
- Lower risk of fire — there almost are no cases of fire in the world practice;
- Aluminum internal floating roof. It has low height and storage capacity is increased;
- Protection from ambient climatic conditions therefore the tanks could be used in various earth regions;
- No requirement for mounting of roof drain.

External floating roof tanks that have been converted to internal floating roof tanks typically have a self-supporting roof. Newly constructed internal floating roof tanks may be of either type. The deck in internal floating roof tanks rises and falls with the liquid level and either floats directly on the liquid surface (contact deck) or rests on pontoons several inches above the liquid surface (noncontact deck).

The majority of aluminum internal floating roofs currently in service have non-contact decks (i.e. the floating roof is supported by pontoons on the liquid). A typical internal floating roof tank is shown in Figures 8-18 and 8-19.

Contact decks in IFRT can be:

- Aluminum sandwich panels that are bolted together, with a honeycomb aluminum core floating in contact with the liquid;
- Pan steel decks floating in contact with the liquid, with or without pontoons; and
- Resin-coated, fiberglass reinforced polyester (FRP), buoyant panels floating in contact with the liquid.

The majority of internal contact floating decks currently in service are aluminum sandwich panel-type or pan steel-type. The FRP decks are less common. The panels of pan steel decks are usually welded together.

Noncontact decks are the most common type currently in use. Typical noncontact decks are constructed of an aluminum deck and an aluminum grid framework supported above the liquid surface by tubular aluminum pontoons or some other buoyant structure. The noncontact decks usually have bolted deck seams.

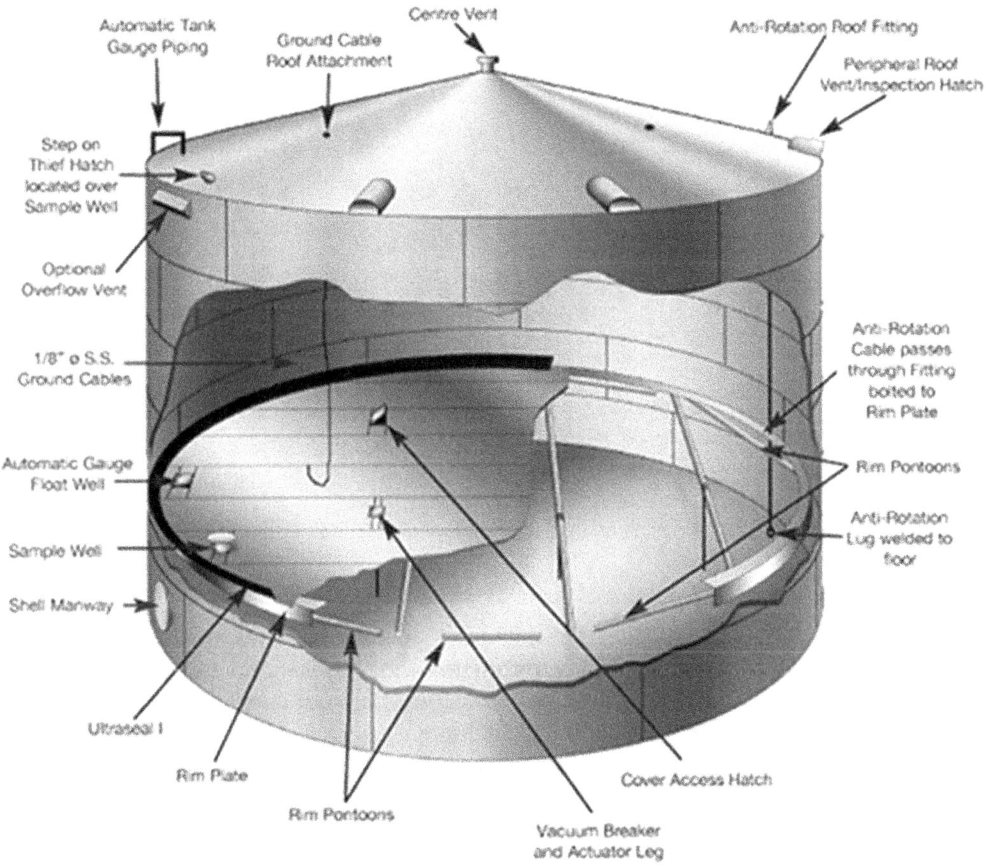

Figure 8-18. Major components of an internal floating roof tank (IFRT) [11]

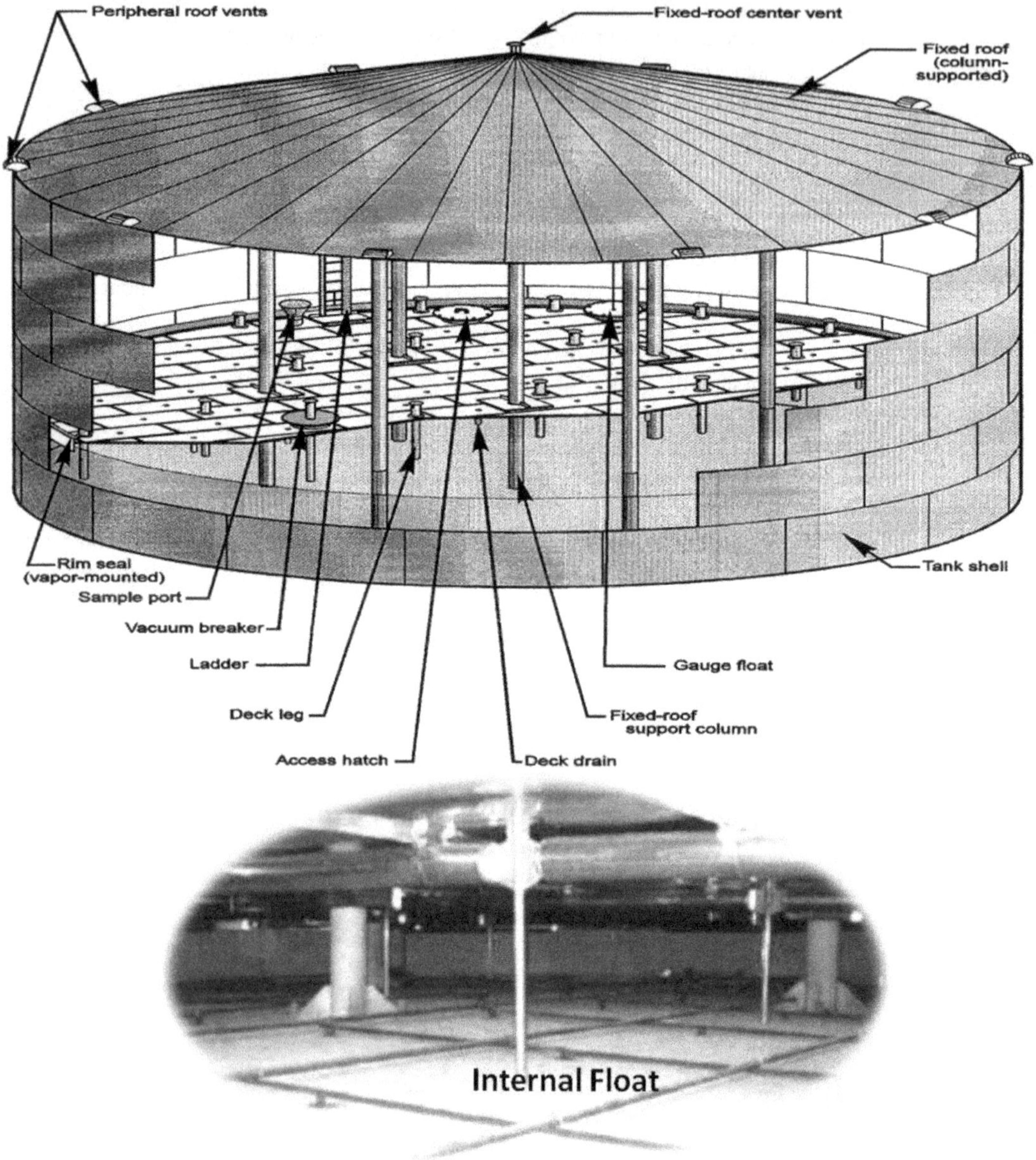

Figure 8-19. A typical internal floating roof tank (IFRT) details [13,14] (inset [11])

Installing a floating roof minimizes evaporative losses of the stored liquid. Both contact and noncontact decks incorporate rim seals and deck fittings for the same purposes previously described for external floating roof tanks. Evaporative losses from floating roofs may come from deck fittings, non-welded deck seams, and the annular space between the deck and tank wall. In addition, these tanks are freely vented by circulation vents at the top of the fixed roof. The vents minimize the possibility of organic vapor accumulation in the tank vapor space in concentrations approaching the flammable range.

It may be noted that an internal floating roof tank that is not freely vented is considered a pressure tank.

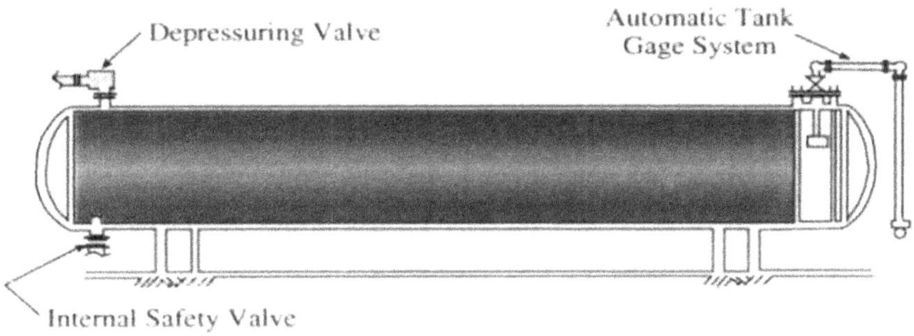

Figure 8-20. Typical high-pressure horizontal storage bullet [11]

Pressurized Storage Tanks: Pressure tanks generally are used for storing hydrocarbon liquids and gases with high vapor pressures and are available in many sizes and shapes, depending on the operating pressure of the tank. These depending on the pressure maintenance requirements will be either low pressure or high pressure as follows:

- Low-pressure tanks
  - Cylindrical (15 kPa to 100 kPa, 2.5 to 15 psig)
  - Spheroid (15 to 210 kPa, 2.5 to 30 PSI)
  - Noded Spheroid
- High-pressure storage tanks
  - Bullets (15–400 PSI), Figure 8-20
  - Spheres (15–3000 PSI), Figure 8-21

Pressure tanks are equipped with a pressure/vacuum vent that is set to prevent venting loss from boiling and breathing loss from daily temperature or barometric pressure changes.

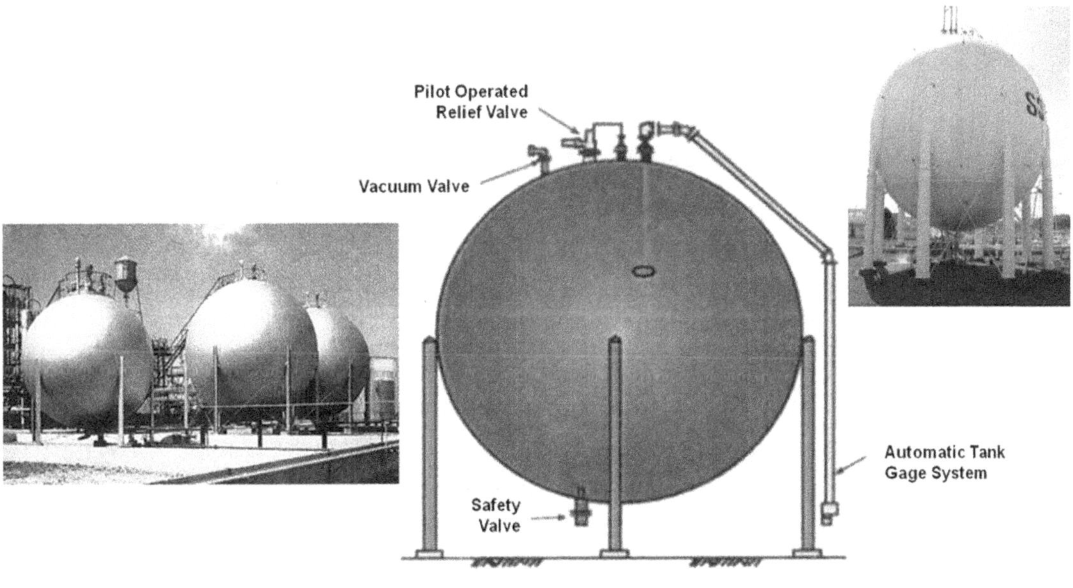

Figure 8-21. High-pressure storage sphere (Horton spheres) [11,16]

High-pressure storage tanks can be operated so that virtually no evaporative or working losses occur. In low-pressure tanks, working losses can occur with atmospheric venting of the tank during filling operations. No appropriate correlations are available to estimate vapor losses from pressure tanks.

Horton sphere was the name given to the storage vessels weld fabricated by the Chicago Bridge & Iron Co. These Horton spheres are at the Gulf Oil Corp. in Port Arthur, TX, completed in 1938, came in two sizes. One is 35 ft, 3 in., and the other 22 ft, 3 in., in diameter. Their respective storage capacities are 4000 and 1000 barrels [17].

Variable Vapor Space Tanks: Variable vapor space tanks are equipped with expandable vapor reservoirs to accommodate vapor volume fluctuations attributable to temperature and barometric pressure changes. Although variable vapor space tanks are sometimes used independently, they are normally connected to the vapor spaces of one or more fixed roof tanks.

The two most common types of variable vapor space tanks are lifter roof tanks and flexible diaphragm tanks. Lifter roof tanks have a telescoping roof that fits loosely around the outside of the main tank wall.

The space between the roof and the wall is closed by either a wet seal, which is a trough filled with liquid, or a dry seal, which uses a flexible coated fabric.

Flexible diaphragm tanks use flexible membranes to provide expandable volume. They may be either separate gasholder units or integral units mounted atop fixed roof tanks.

Variable vapor space tank losses occur during tank filling when vapor is displaced by liquid. Loss of vapor occurs only when the tank's vapor storage capacity is exceeded.

8.4.3 Emission Control In Storage Tanks

Storage vessels containing organic liquids are common in many industries, including

- Petroleum producing and refining,
- Petrochemical and chemical manufacturing,
- Bulk storage and transfer/distribution/terminal operations, and
- Other industries consuming or producing organic liquids.

Organic liquids/petroleum liquids, generally, are mixtures of hydrocarbons having dissimilar true vapor pressures (for example, gasoline and crude oil, etc.). Organic liquids in the chemical industry, usually called volatile organic liquids, are composed of pure chemicals or mixtures of chemicals with similar true vapor pressures (for example, condensate, benzene or a mixture of isopropyl and butyl alcohols).

Once stored, hydrocarbon products (unless they are under pressure) will evaporate. Total emission from storage tanks includes withdrawal as well as standing storage losses.

Therefore, storage tanks are usually equipped with seals and vapor controls and sometime leak detection systems to assure minimal emissions and as well increased safety.

8.4.3.1 Tank Rim Sealing Systems: Floating Roof Tanks

Emissions from floating roof tanks are the sum of withdrawal losses and standing storage losses. Withdrawal losses occur as the liquid level, and thus the floating roof, is lowered. Some liquid remains on the inner tank wall surface and evaporates. For an internal floating roof tank that has a column supported fixed roof, some liquid also clings to the columns and evaporates. Evaporative loss occurs until the tank is filled and the exposed surfaces are again covered.

Standing storage losses from floating roof tanks include rim seal and deck fitting losses, and for internal floating roof tanks also include deck seam losses for constructions other than welded decks. Other potential standing storage loss mechanisms include breathing losses as a result of temperature and pressure changes.

Sealing systems consist of primary and secondary rim sealing as follows:

- ***Primary seals are three types***
  - Vapor-mounted
    - highest emissions
    - fair service history
  - Liquid-mounted
    - Tube type: kerosene, foam filled
    - lowest emissions
    - but has poor service history
  - Mechanical-shoe type
    - relatively low emissions
    - has best service history
- ***Secondary seals***
  - Designed to reduce emissions
  - Additional safeguard against emissions
  - Types include:
    - Rim type
    - Mini-tube type
    - Shoe type: seal fabric
    - Wiper type: rubber
  - Mounted above the primary seal
    - may result in loss of tank capacity
  - Shoe-mounted secondary seal (not recommended)

The rim seal system is used to allow the floating roof to rise and fall within the tank as the liquid level changes. The rim seal system also helps to fill the annular space between the rim and the tank shell and therefore minimize evaporative losses from this area. A rim seal system may consist of just a primary seal or a primary and a secondary seal, which is mounted above the primary seal. Examples of primary and secondary seal configurations are shown in Figures 8-22 through 8-27 and described below.

The primary seal serves as a vapor conservation device by closing the annular space between the edge of the floating deck and the tank wall. As indicated previously, three basic types of primary seals are used on external floating roofs:

- Flexible wiper seals
- Resilient filled (nonmetallic), and
- Mechanical (metallic) shoe.

It may be noted that resilient foam filled primary seals have not been commonly used in the last 20 to 30 years.

Wiper seals generally consist of a continuous annular blade of flexible material fastened to a mounting bracket on the deck perimeter that spans the annular rim space and contacts the tank shell. This type of seal is depicted in Figure 8-22. New tanks with wiper seals may have dual wipers, one mounted above the other. The mounting is such

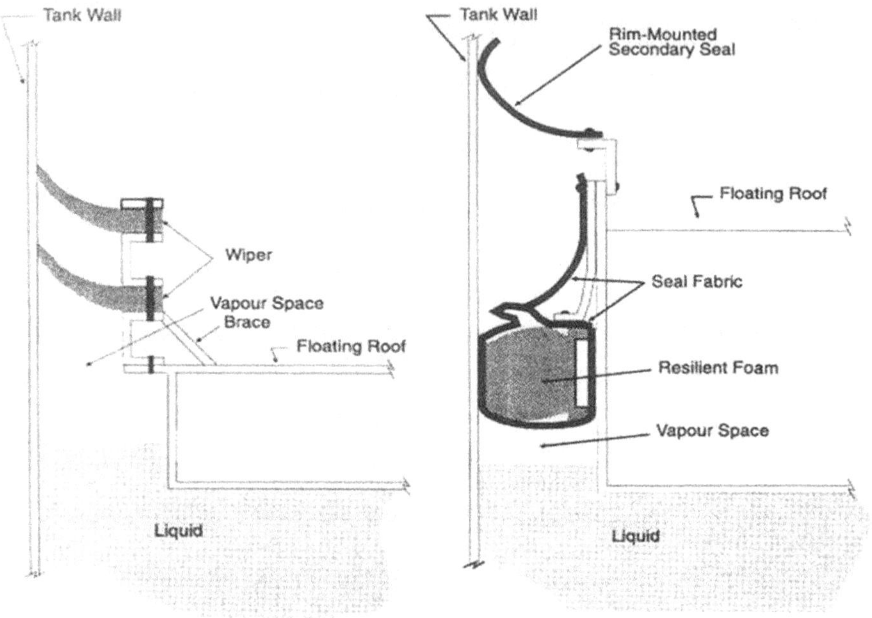

Figure 8-22. Vapor mounted primary rim seal with secondary seal (left — flexible wiper and right — resilient foam) [12]

that the blade is flexed, and its elasticity provides a sealing pressure against the tank shell.

Wiper seals are vapor mounted; a vapor space exists between the liquid stock and the bottom of the seal. For emission control, it is important that the mounting be vapor-tight, that the seal extend around the circumference of the deck and that the blade be in substantial contact with the tank shell. Two types of materials are commonly used to make the wipers. One type consists of a cellular, elastomeric material tapered in cross section with the thicker portion at the mounting. Rubber is a commonly used material; urethane and cellular plastic are also available. All radial joints in the blade are joined. The second type of material that can be used is a foam core wrapped with a coated fabric. Polyurethane on nylon fabric and polyurethane foam are common materials.

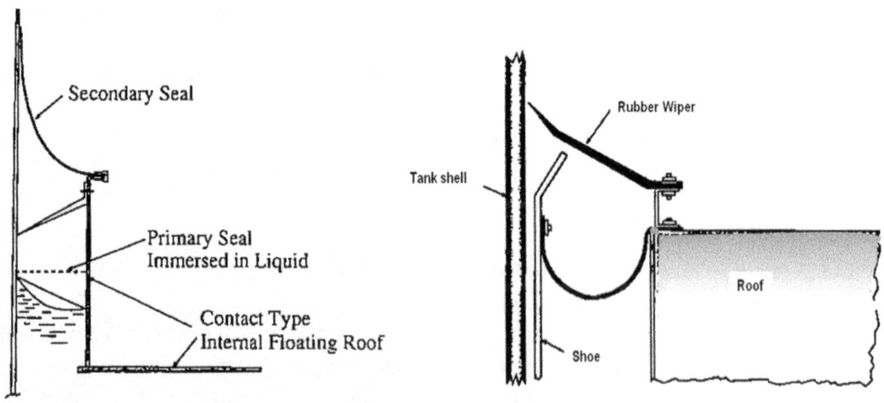

Figure 8-23. Typical secondary seal (A — rim mounted and B — wiper type) [11,12]

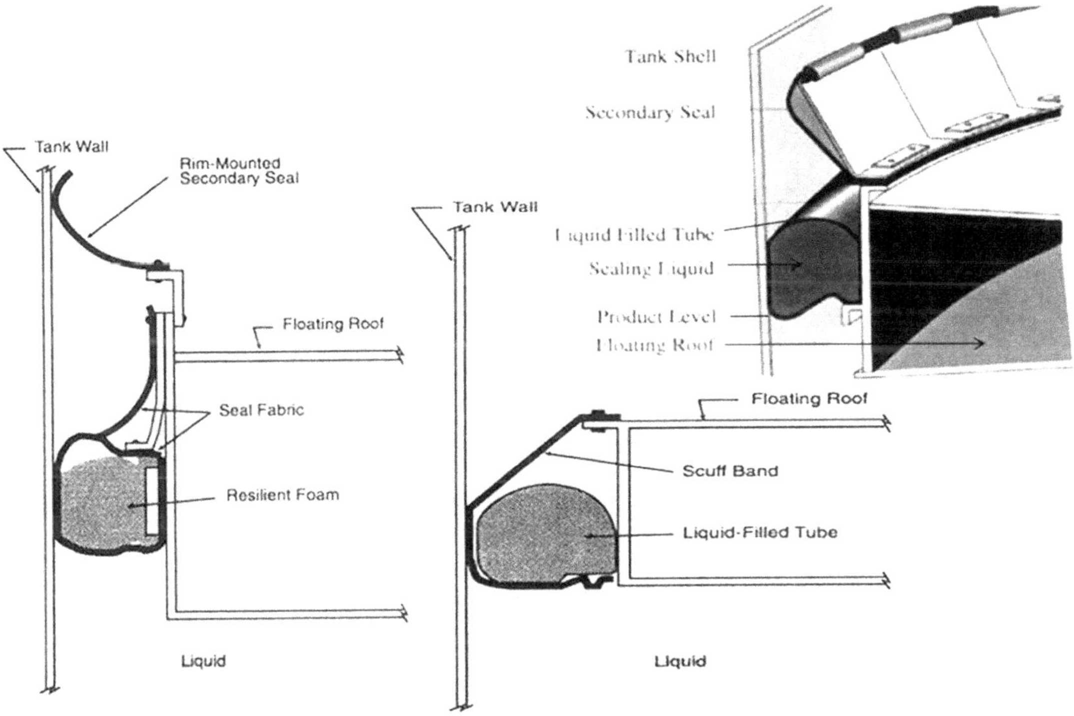

Figure 8-24. Resilient foam (left) or liquid filled (right) primary rim seal with secondary seal [11]

The core provides the flexibility and support, while the fabric provides the vapor barrier and wear surface.

A secondary seal (Figure 8-23) may be used to provide some additional evaporative loss control over that achieved by the primary seal. Secondary seals can be either flexible wiper seals or resilient filled seals.

A resilient foam-filled seal, can also be mounted to eliminate the vapor space between the rim seal and liquid surface (liquid mounted, Figure 8-24) or to allow a vapor space between the rim seal and the liquid surface (vapor mounted, Figure 8-25). Resilient filled seals work because of the expansion and contraction of a resilient material to maintain contact with the tank shell while accommodating varying annular rim space widths. These rim seals allow the roof to move up and down freely, without binding.

Resilient filled seals typically consist of a core of open-cell foam encapsulated in a coated fabric. The seals are attached to a mounting on the deck perimeter and extend around the deck circumference. Polyurethane-coated nylon fabric and polyurethane foam are commonly used materials. For emission control, it is important that the attachment of the seal to the deck and the radial seal joints be vapor-tight and that the seal be in substantial contact with the tank shell.

The scuff band as shown in Figure 8-24 is usually thin metal band that acts as protection against snags and tears in the foam bag fabric from the tank shell.

A mechanical/metallic rim mounted shoe seal (Figure 8-26) uses a light-gauge metallic band as the sliding contact with the shell of the tank. The band is formed as a series of sheets (shoes) which are joined together to form a ring, and are held against the tank shell by a mechanical device. The shoes are normally 1 to 1.5 m (3 to 5 ft) deep, providing a potentially large contact area with the tank shell. Expansion and

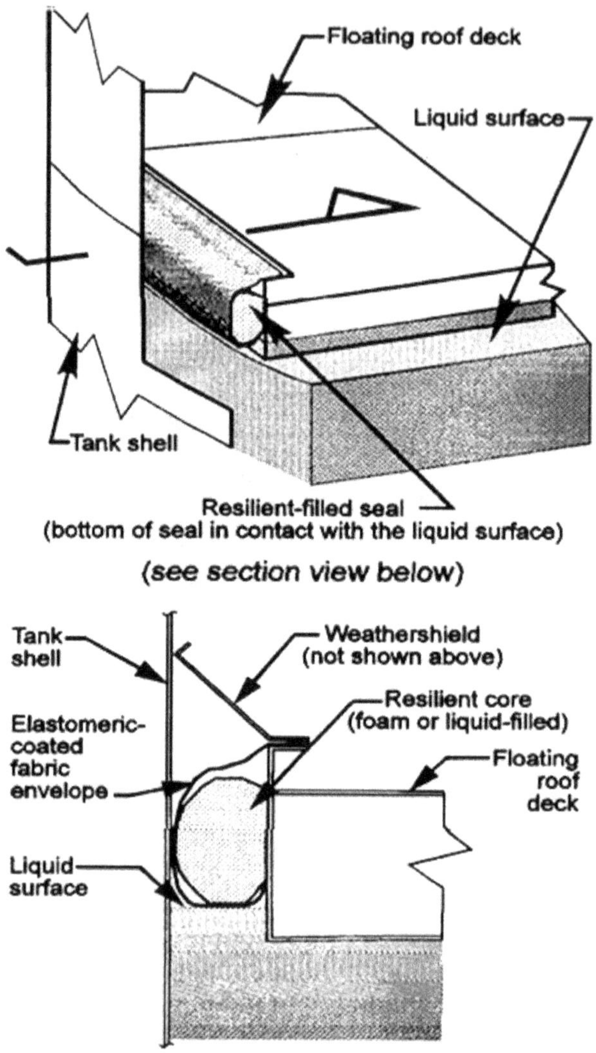

Figure 8-25. Schematic of resilient filled seal (vapor mounted) [13,14]

contraction of the ring can be provided for as the ring passes over shell irregularities or rivets by jointing narrow pieces of fabric into the ring or by crimping the shoes at intervals. The bottoms of the shoes extend below the liquid surface (as shown in Figure 8-25 above) to confine the rim vapor space between the shoe and the floating deck.

The rim vapor space, which is bounded by the shoe, the rim of the floating deck, and the liquid surface, is sealed from the atmosphere by bolting or clamping a coated fabric, called the primary seal fabric, which extends from the shoe to the rim to form an "envelope." Two locations are used for attaching the primary seal fabric. The fabric is most commonly attached to the top of the shoe and the rim of the floating deck. To reduce the rim vapor space, the fabric can be attached to the shoe and the floating deck rim near the liquid surface. Rim vents can be used to relieve any excess pressure or vacuum in the vapor space.

Some primary seals on external floating roof tanks are protected by a weather shield. Weather shields are usually simple metal pieces without fabric barriers but can be of elastomeric, or composite construction and provide the primary seal with longer

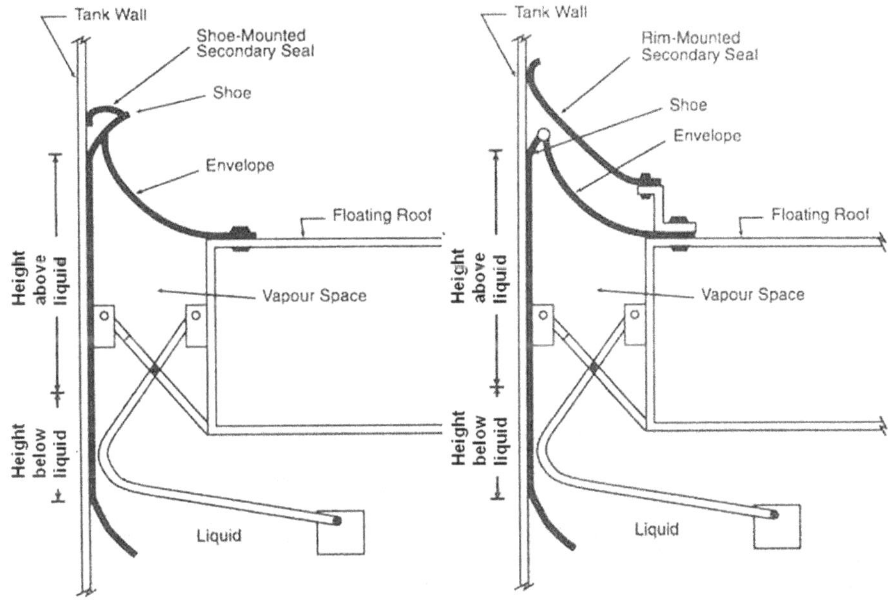

Figure 8-26. Typical mechanical shoe primary rim seal with a shoe mounted secondary seal system, from [11,18,19]

Figure 8-27. Flex-a-seal double wiper seal system including weather shield [11,20]

Figure 8-28. Weather guard secondary seal (from [21])

life by protecting the primary seal fabric from deterioration due to exposure to weather, debris, and sunlight. However, it may be noted that when a weather shield is combined with a continuous vapor fabric, it can also can act as a secondary seal. Without such a continuous vapor fabric, it cannot be considered as a secondary seal.

On external floating roofs, the most common secondary seal systems incorporate a compression/protection plate over a continuous fabric to provide additional vapor

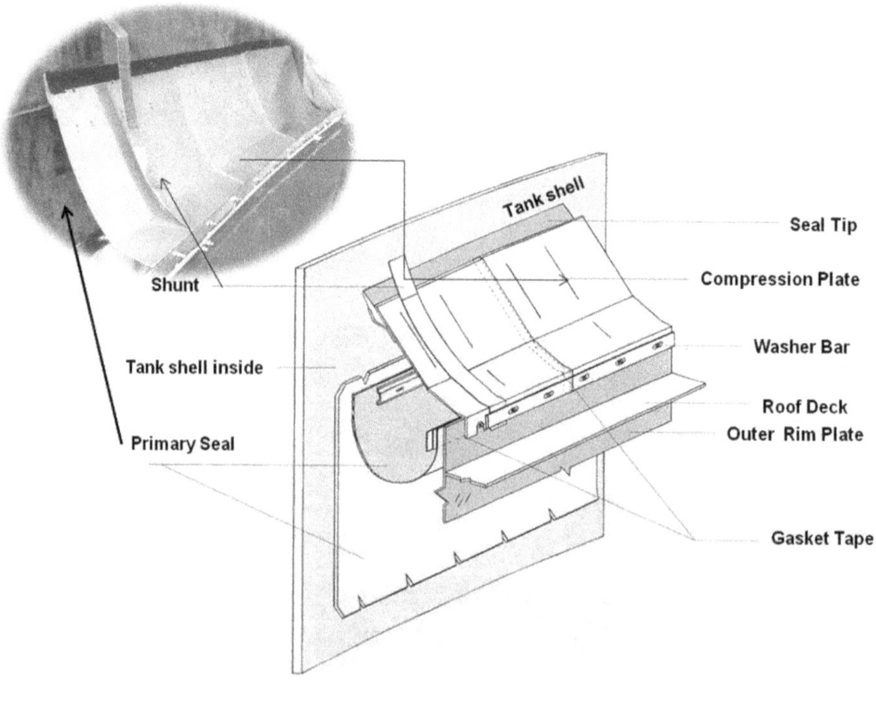

Figure 8-29. Weather shield and secondary seal system (Horton VS 200 [20])

preservations, besides protecting the primary seal. Some designs incorporate an exposed vapor barrier fabric.

Internal floating roofs typically do not have weather shields over the secondary seals, however, more recent designs incorporate steel compression plates and support to increase the longevity of the secondary seal over traditionally used wiper seals.

Internal floating roofs typically incorporate one of three types of flexible, product-resistant seals: Shoe seals, resilient foam-filled seals or wiper seals. It may be noted that rim mounted shoe seals are the most common type of primary seal used in the tanks and that Resilient Foam seals have not been common for many years. Wiper Seals are still used but they are not as common as they were years ago.

Some primary seals on external floating roof tanks are protected by a weather shield, Figures 8-27, 8-28, and 8-29, [20].

Other kinds of floating roof tank sealing systems are illustrated in Figures 8-30 through 8-34.

Consideration for Tank Seal Selection:

Tank floating roof rim seal are selected on the basis of the following:

- Tank shell wax or other scraping requirements
- Rain water shedding
- Sal tip life
- Fire protection
- Roof centering
- In-service installation (retro fitting)
- Life expectancy
- Cost (including maintenance)

8.4.4 Tank Fittings and Appurtenances

Floating Roof Tanks

Numerous fittings pass through or are attached to floating roof decks to accommodate structural support components or allow for operational functions. Internal floating roof deck fittings are typically of different configuration than those for external floating

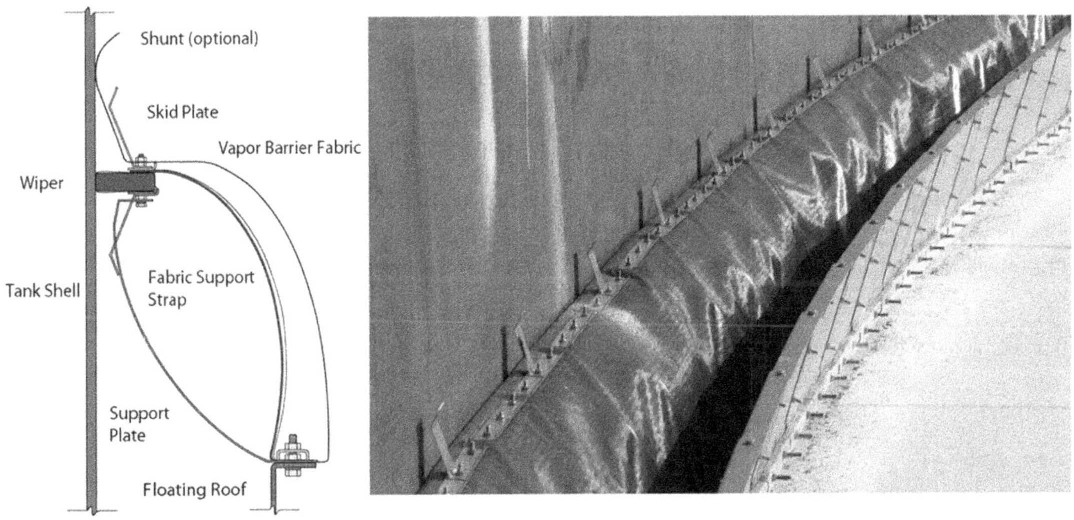

Figure 8-30. HMT seal-king (left) and HMT secondary low profile wiper (right) seals (courtesy of HMT Inc.)

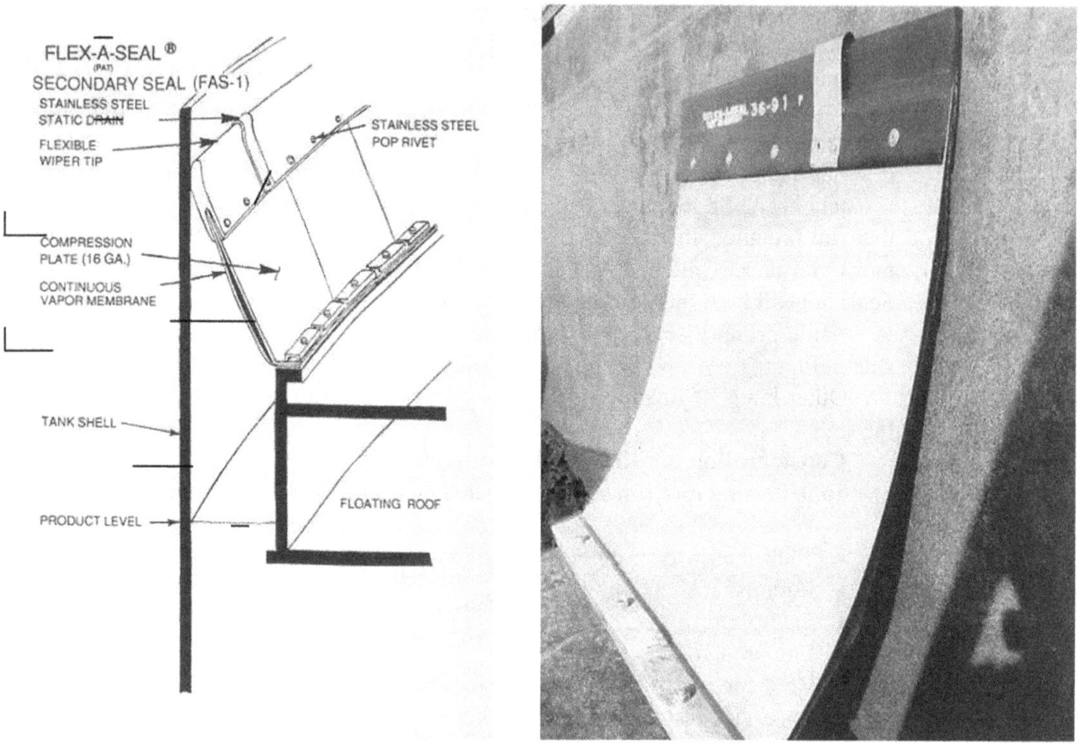

Figure 8-31. Flex-a-seal secondary seal

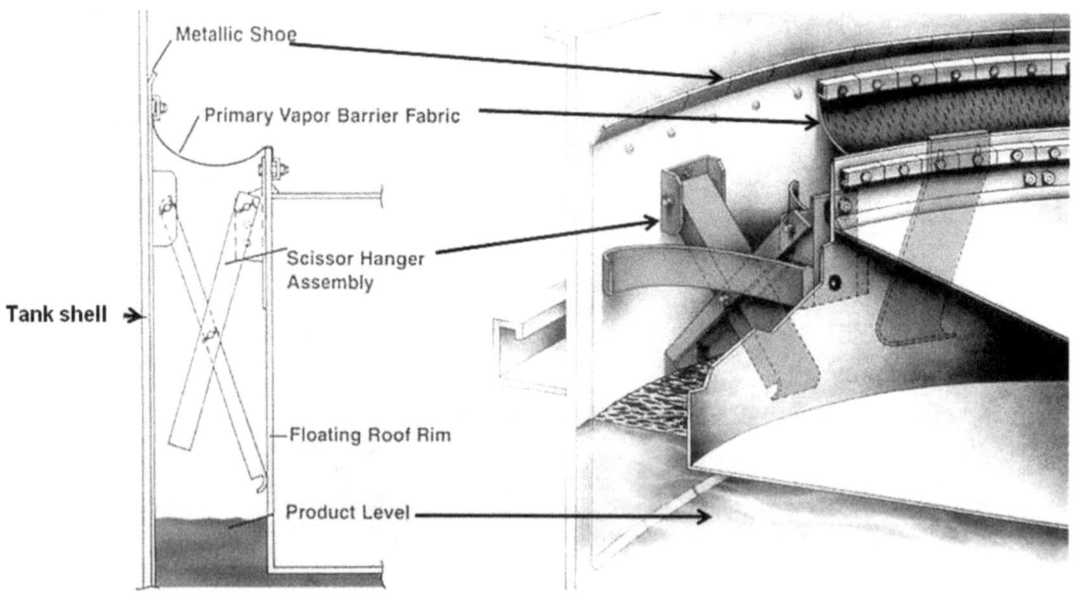

Figure 8-32. HMT scissor shoe primary mechanical seal [19]

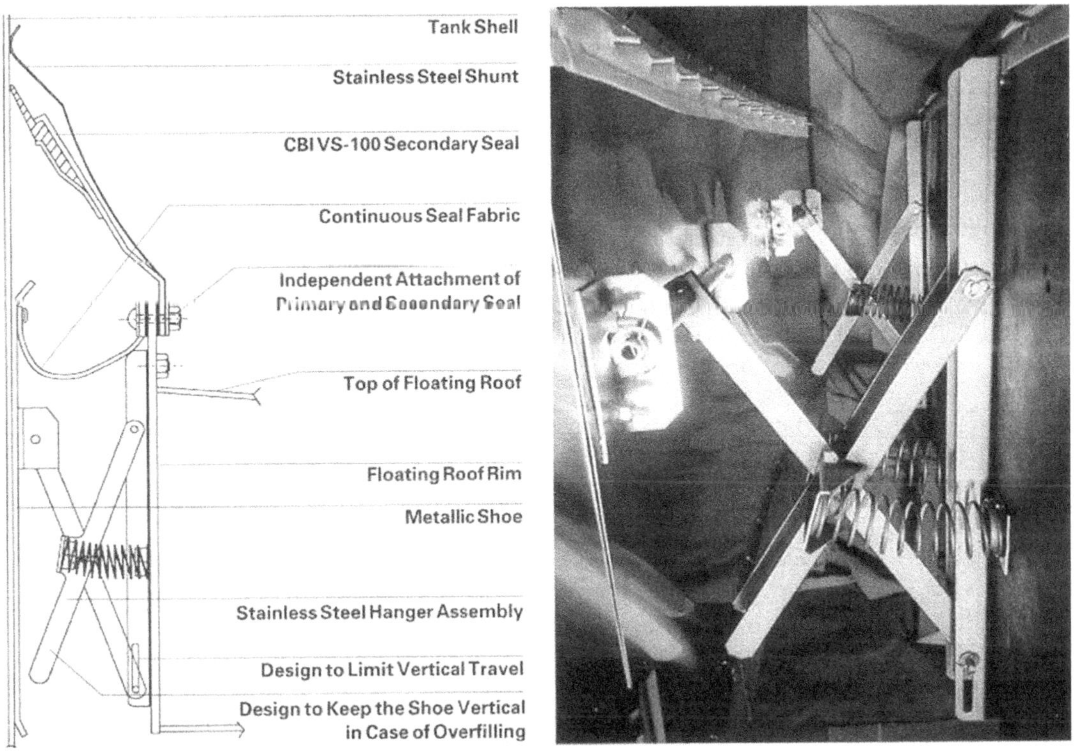

Figure 8-33. Horton SR-1A mechanical shoe seal [20]

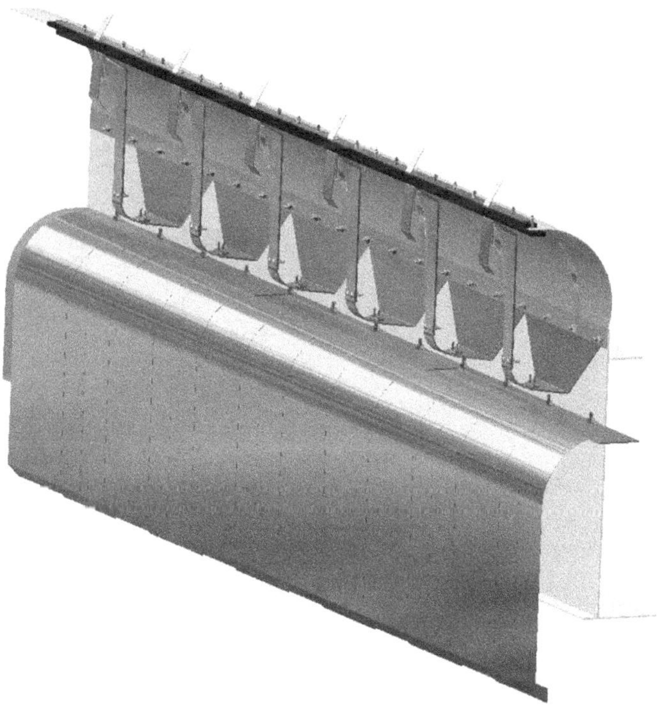

Figure 8-34. HMT seal-master primary and secondary seal (IFR) courtesy of HMT Inc.

roof decks. Rather than having tall housings to avoid rainwater entry, internal floating roof deck fittings tend to have lower profile housings to minimize the potential for the fitting to contact the fixed roof when the tank is filled. Deck fittings can be a source of evaporative loss when they require openings in the deck.

External Floating Roof: The most common components that require openings in the deck are listed below:

- Access hatches (Figure 8-35)
- Gauge-floats (Figure 8-36)
- Gauge-hatch/sample ports (Figure 8-37)
- Rim vents (Figure 8-38)
- Deck drains (Figure 8-39)
- Deck legs (Figure 8-40)
- Un-slotted guide-poles and wells and slotted (perforated) guide-poles and wells (Figure 8-41)
- Vacuum breakers (Figure 8-42)

Access hatches. An access hatch is an opening in the deck with a peripheral vertical well that is large enough to provide passage for maintenance personnel and materials through the deck for construction or servicing (Figure 8-35). Attached to the opening is a removable cover that may be bolted and/or gasketed to reduce evaporative loss. On internal floating roof tanks with non-contact (pontoon type) decks, the well usually extend down into the liquid to seal off the vapor space below the noncontact deck.

Gauge-floats. A gauge-float is used to indicate the level of liquid within the tank (Figure 8-36). The float rests on the liquid surface and is housed inside a well that is closed by a cover. The cover may be bolted and/or gasketed to reduce evaporation loss. As with other similar deck penetrations, the well extends down into the liquid on noncontact decks in internal floating roof tanks.

Gauge-hatch/sample ports. A gauge-hatch/sample port consists of a pipe sleeve equipped with a self-closing gasketed cover (to reduce evaporative losses) and allows

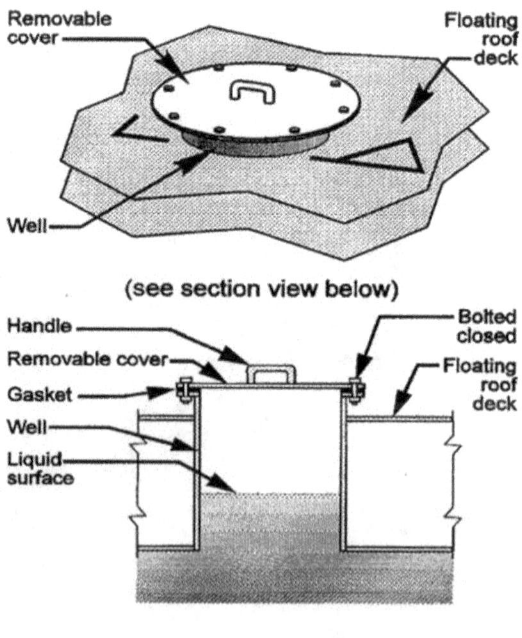

Figure 8-35. Floating roof deck fittings: access hatch [13,14]

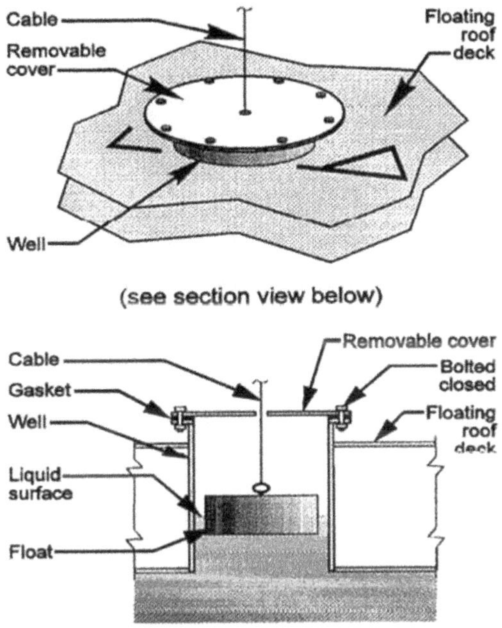

Figure 8-36. Floating roof deck fittings: gauge floats [13,14]

hand-gauging or sampling of the stored liquid (Figure 8-37). The gauge-hatch/sample port is usually located beneath the gauger's platform, which is mounted on top of the tank shell. A cord may be attached to the self-closing gasketed cover so that the cover can be opened from the platform if required.

Rim vents. Rim vents are used on tanks equipped with a seal design that creates a vapor pocket in the seal and rim area, such as a mechanical shoe seal. A typical rim vent

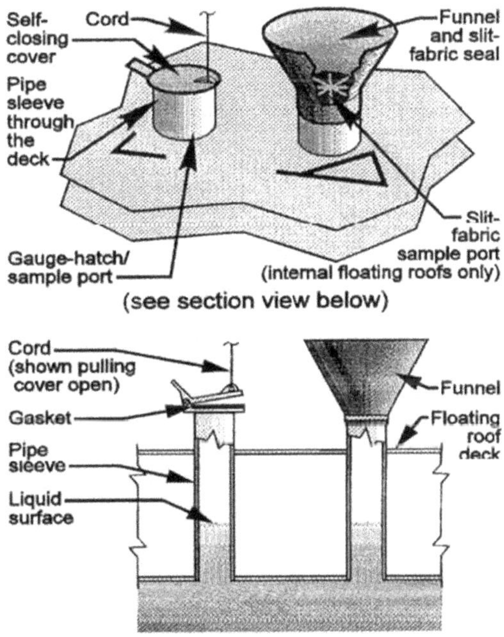

Figure 8-37. Floating roof deck fittings: gauge hatch/sample port [13,14]

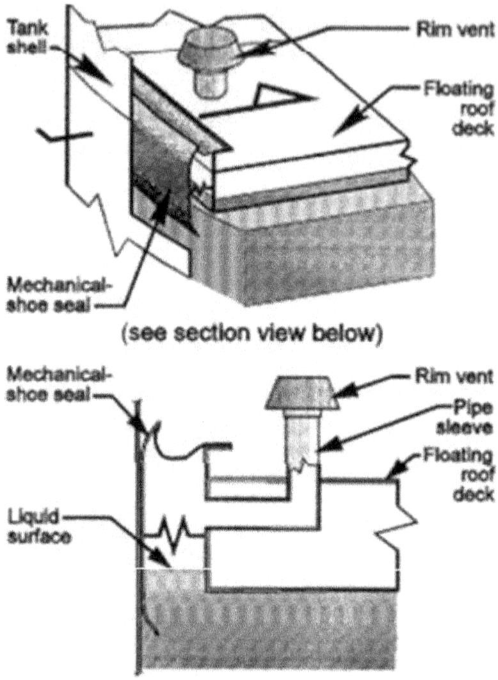

Figure 8-38. Floating roof deck fittings: rim vents [13,14]

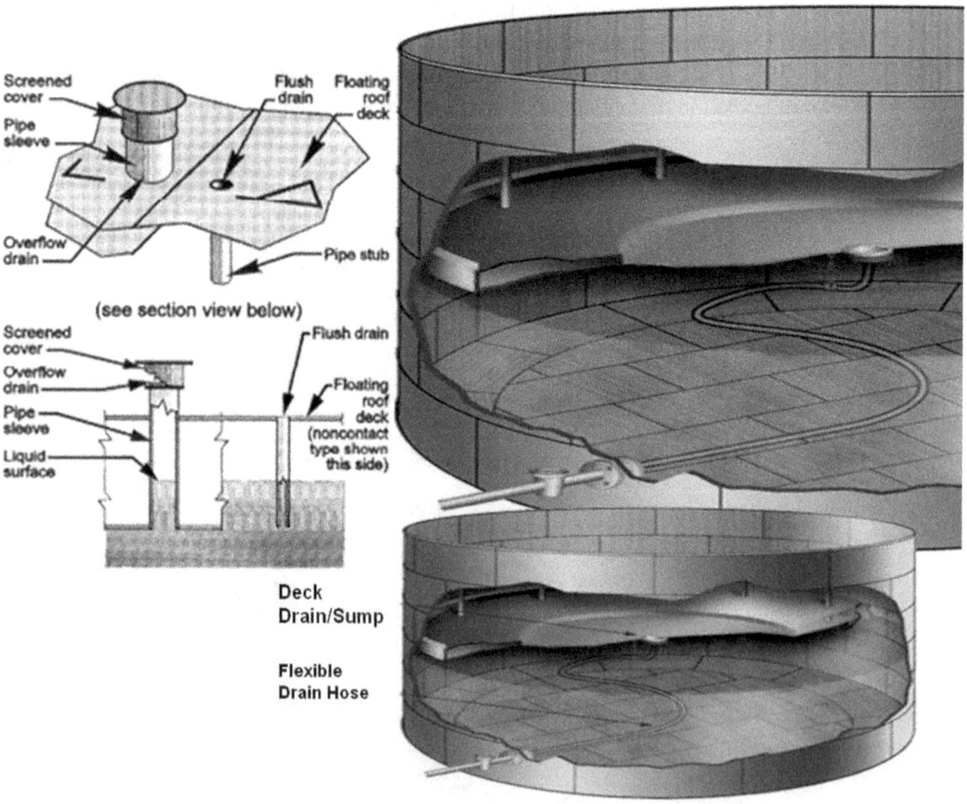

Figure 8-39. Floating roof deck fittings: deck drains/sumps [13,14]

is shown in (Figure 8-38). The vent is used to release any excess pressure or vacuum that is present in the vapor space bounded by the primary-seal shoe and the floating roof rim and the primary seal fabric and the liquid level. Rim vents usually consist of weighted pallets that rest on a gasketed cover.

Deck Drains. Currently, two types of deck drains are in use (closed and open deck drains) to remove rainwater from the floating deck (Figure 8-39). Open deck drains can be either flush or overflow drains (Figure 8-39, left illustration). Both types consist of a pipe that extends below the deck to allow the rainwater to drain into the stored liquid. Only open deck drains are subject to evaporative loss. Flush drains are flush with the deck surface. Overflow drains are elevated above the deck surface. Overflow drains are used to limit the maximum amount of rainwater that can accumulate on the floating deck, providing emergency drainage of rainwater if necessary.

Closed deck drains (Figure 8-39, right illustration) carry rainwater from the surface of the deck though a flexible hose or some other type of piping system that runs through the stored liquid prior to exiting the tank. The rainwater does not come in contact with the liquid, so no evaporative losses result. Overflow drains are usually used in conjunction with a closed drain system to carry rainwater outside the tank.

Float Roof Deck Legs. Deck legs are used to prevent damage to fittings underneath the deck and to allow for tank cleaning or repair, by holding the deck at a predetermined distance off the tank bottom. These supports consist of adjustable or fixed legs attached to the floating deck or hangers suspended from the fixed roof. For adjustable legs or hangers, the load-carrying element passes through a well or sleeve into the deck. With non-contact decks, the well should extend into the liquid. Evaporative losses may occur in the annulus between the deck leg and its sleeve. Typical deck legs are shown in Figure 8-19 and Figure 8-40 below.

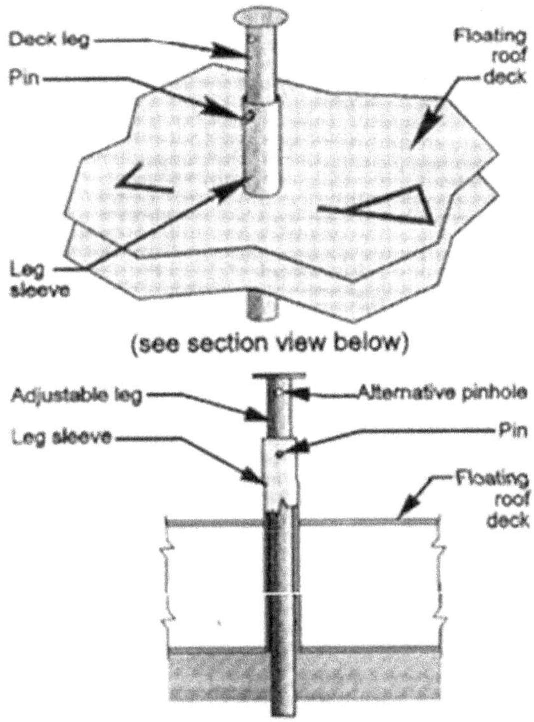

Figure 8-40. Floating roof deck fittings: deck legs [13,14]

Un-slotted guide-poles and wells. A guide-pole is an anti-rotational device that is fixed to the top and bottom of the tank, passing through a well in the floating roof. The guide-pole is used to prevent adverse movement of the roof and thus damage to deck fittings and the rim seal system. In some cases, an un-slotted (Figure 8-41A) guide-pole is used for gauging purposes, but there is a potential for differences in the pressure, level, and composition of the liquid inside and outside of the guide-pole.

Slotted (perforated) guide-poles and wells. The function of the slotted guide-pole is similar to the un-slotted guide-pole but also has additional features as shown in Figure 8-41B. Perforated guide-poles can be either slotted or drilled hole guide-poles. The guide pole is slotted to allow stored liquid to enter. The same can be accomplished with drilled holes. The liquid entering the guide-pole is well mixed, having the same

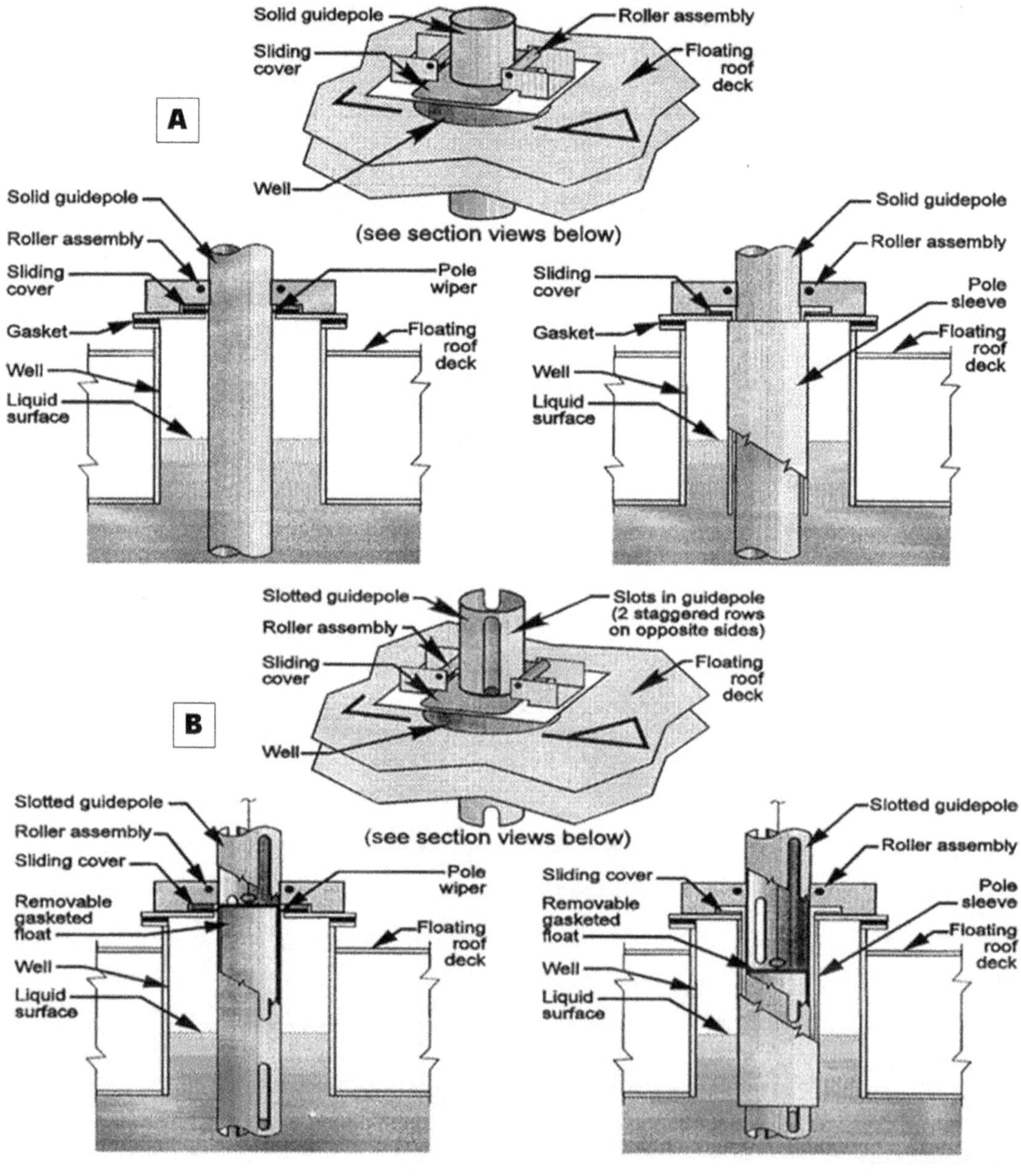

Figure 8-41. Floating roof deck fittings: un-slotted (A) and slotted (B) guide poles, [13,14]

composition as the remainder of the stored liquid, and is at the same liquid level as the liquid in the tank. Representative samples can therefore be collected from the slotted or drilled hole guide-pole. However, evaporative loss from the guide-pole can be reduced by modifying the guide-pole or well or by placing a float inside the guide-pole. Guide-poles are also referred to as gauge poles, gauge pipes, or stilling wells.

Vacuum breakers. A vacuum breaker equalizes the pressure of the vapor space across the deck as the deck is either being landed on or floated off its legs. A typical vacuum breaker is shown in Figure 8-42. As depicted in this figure, the vacuum breaker consists of a well with a cover. Attached to the underside of the cover is a guided leg long enough to contact the tank bottom as the floating deck approaches. When in contact with the tank bottom, the guided leg mechanically opens the breaker by lifting the cover off the well; otherwise, the cover closes the well. The closure may be gasketed or ungasketed. Because the purpose of the vacuum breaker is to allow the free exchange of air and/or vapor, the well does not extend appreciably below the deck.

Internal Floating Roof Tank (IFRT) Fittings

Fittings used only on internal floating roof tanks include

- Column wells (Figure 8-43),
- Ladder wells (Figure 8-44), and
- Stub drains.

Columns and wells. The most common fixed-roof designs are normally supported from inside the tank by means of vertical columns, which necessarily penetrate an internal floating deck (Figures 8-43 and 8-44). (Some fixed roofs are entirely self-supporting and, therefore, have no support columns.) Column wells are similar to unslotted guide pole wells on external floating roofs. Columns are made of pipe with circular cross sections or of structural shapes with irregular cross sections (built-up). The number of columns varies with tank diameter, from a minimum of 1 to over 50 for very large diameter tanks.

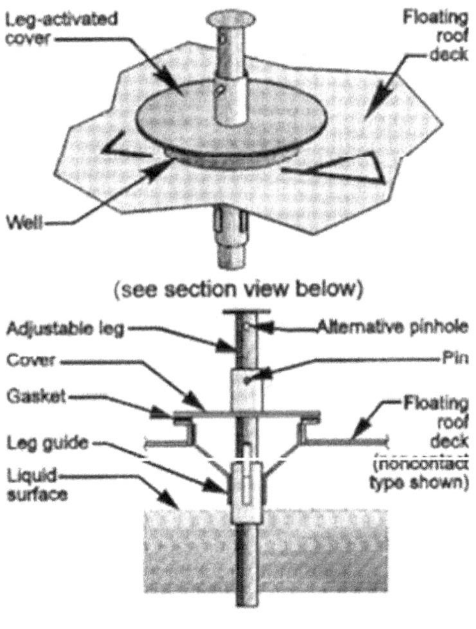

Figure 8-42. Floating roof deck fittings: vacuum breaker [13,14]

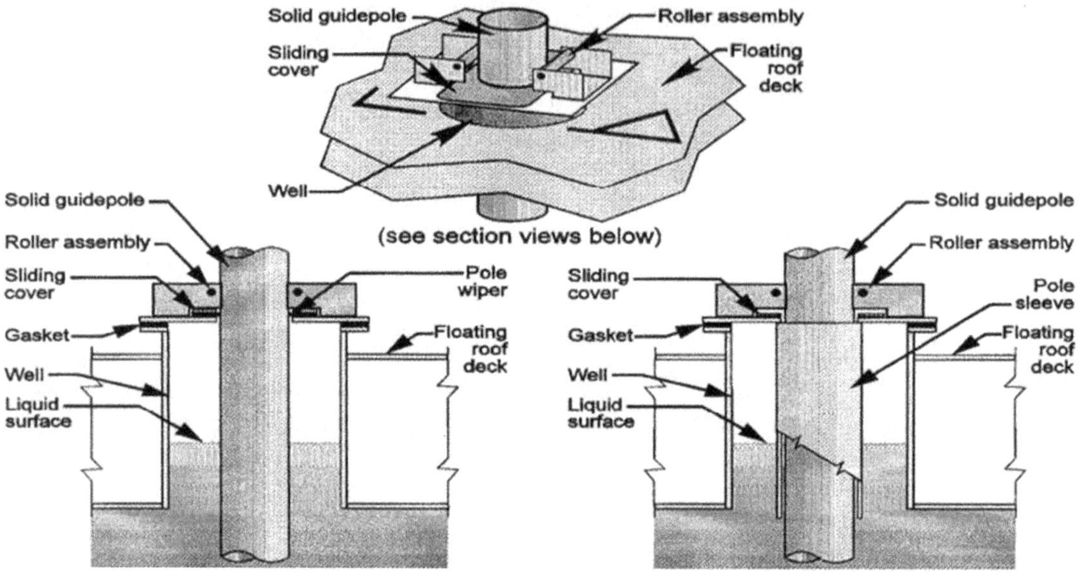

Figure 8-43. Internal floating roof tank (IFRT) — support (un-slotted), [13,14]

The columns pass through deck openings via peripheral vertical wells. With noncontact decks, the well should extend down into the liquid stock. Generally, a closure device exists between the top of the well and the column. Several proprietary designs exist for this closure, including sliding covers and fabric sleeves, which must accommodate the movements of the deck relative to the column as the liquid level changes. A sliding cover rests on the upper rim of the column well (which is normally fixed to

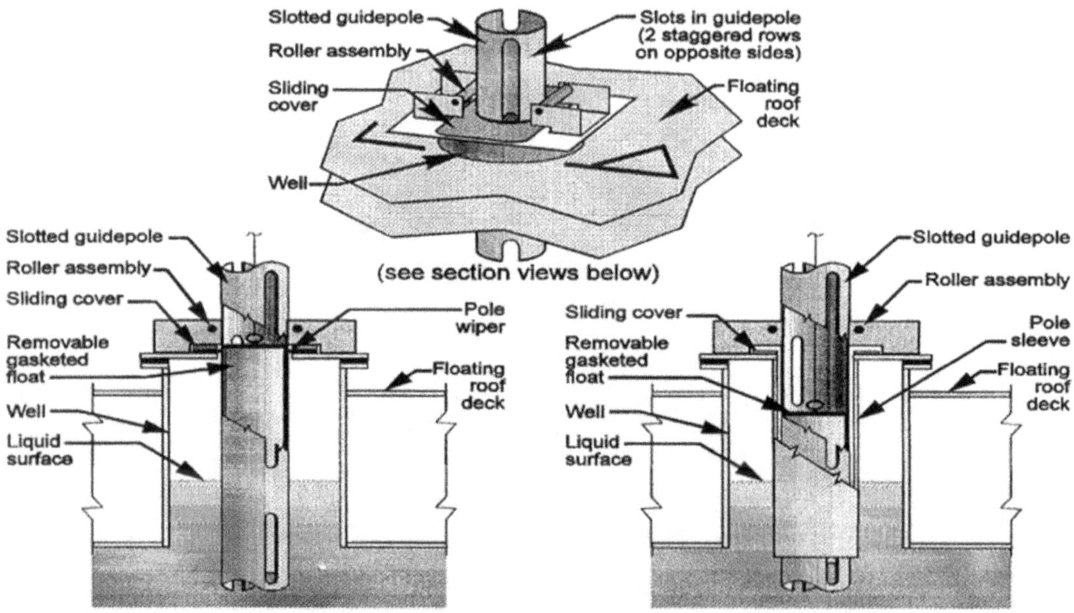

Figure 8-44. Internal floating roof tank (IFRT) - support (slotted) [13,14]

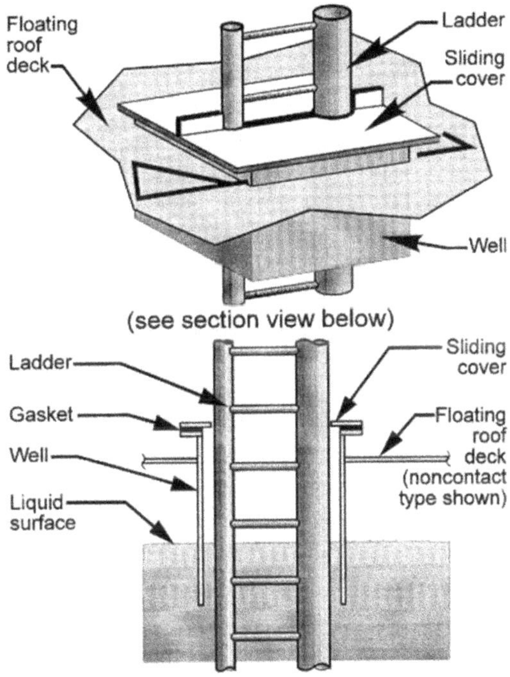

Figure 8-45. Internal floating roof tank (IFRT) — ladders and wells design [13,14]

the deck) and bridges the gap or space between the column well and the column. The cover, which has a cutout, or opening, around the column slides vertically relative to the column as the deck raises and lowers. At the same time, the cover slides horizontally relative to the rim of the well. A gasket around the rim of the well reduces emissions from this fitting. A flexible fabric sleeve seal between the rim of the well and the column (with a cutout or opening, to allow vertical motion of the seal relative to the columns) similarly accommodates limited horizontal motion of the deck relative to the column.

Ladders and wells. Some tanks are equipped with internal ladders that extend from a manhole in the fixed roof to the tank bottom. The deck opening through which the ladder passes is constructed with similar design details and considerations to deck openings for column wells, as previously discussed. A typical ladder well is shown in Figure 8-45.

8.5 PETROLEUM STORAGE TANKS STANDARDS (FOR DESIGN, OPERATION, AND PROTECTION)

Below is a list of frequently used storage tank standards and practices that may be referred to for the design, operation, maintenance, and protection of storage tanks [22–24]. There may be other applicable standards.

Note: The current (or most recent) edition/revision of a publication should be applicable.

A — American Petroleum Institute (API)

Standard/RP #	Title and/or Description
A-1 Construction Standards:	
API Spec 12D	Specifications for field welded tanks for storage of production liquids
API Spec 12F	Shop welded tanks for storage of production liquids
API Spec 12P	Specifications for fiberglass reinforced plastic tanks
API Std 620	Design and construction of large, welded, low-pressure storage tanks
API Std 650	Welded steel tanks for oil storage (replaced API 12 series Spec's)
API Std 2000	Venting atmospheric and low-pressure storage tanks
API Std 2610	Design, construction, operation, maintenance, and inspection of terminal and tank facilities
A-2 Inspection Standards: (including construction modification, and reconstruction standards)	
API Std 510	Pressure vessel inspection code (maintenance inspection, rating, repair and alteration)
API Std 570	Inspection, Repair, alteration, and rerating of in-service piping systems
API Std 653	Tank inspection, repair, alteration, and reconstruction
API Std 2015	Requirements for safe entry and cleaning of petroleum storage tanks
A-3 Recommended Practices (RP):	
API RP 12H	Installation of new bottoms in old storage tanks
API RP 12R	Setting, maintenance, inspection, operation, and repair of tanks in production service
API RP 574	Inspection practices for piping system components
API RP 575	Inspection of atmospheric and low-pressure storage tanks
API RP 580	Risk-based inspection
API RP 651	Cathodic protection of aboveground petroleum storage tanks
API RP 652	Lining of aboveground petroleum storage tank bottoms
API RP 1107	Pipeline maintenance welding practices
API RP 1110	Pressure testing of liquid petroleum pipelines
API RP 1604	Closure of underground petroleum storage tanks
API RP 1615	Installation of underground petroleum storage systems
API RP 1626	Storing and handling ethanol and gasoline-ethanol blends at distribution terminals and service stations
API RP 1627	Storing and handling of gasoline-methanol/cosolvent blends at distribution terminals and service stations
API RP 1631	Interior lining and periodic inspection of underground storage tanks
API RP 1632	Cathodic protection of underground petroleum storage tanks and piping systems
API RP 1637	Using the API color-symbol system to mark equipment and vehicles for product identification at gasoline dispensing facilities and distribution terminals
API RP 2003	Protection against ignitions arising out of static, lightning, and stray currents
API RP 2016	Guidelines for entering and cleaning petroleum storage tanks
API RP 2021	Management of atmospheric storage tank fires
API RP 2027	Ignition hazards involved in abrasive blasting of atmospheric storage tanks in hydrocarbon service
API RP 2350	Overfill protection for storage tanks in petroleum facilities

A-4 Other API Publications:

API-334 A	Guide to leak detection for aboveground storage tanks
API Pub 2009	Safe welding, cutting and hot work practices in the petroleum and petrochemical industries
API 2517	Evaporating losses from external floating roof tanks
API 2519	Evaporating losses from internal floating roof tanks
API Pub 2200	Repairing crude oil, liquefied petroleum gas, and product pipelines
API-2207	Preparing tank bottoms for hot work
API Pub 2217A	Guidelines for work in inert confined spaces in the petroleum industry
API 2550	Measurements and calibration of petroleum storage tanks

B — Petroleum Equipment Institute (PEI)

B-1 Recommended Practices:

PEI RP 100	Recommended practices for installation of underground liquid storage systems
PEI RP 200	Recommended practices for installation of aboveground storage systems for motor vehicle fueling

C — National Leak Prevention Association (NLPA)

C-1 Recommended Practices:

NLPA Std 631	Entry, cleaning, interior inspection, repair and lining of underground storage tanks

D — National Association of Corrosion Engineers (NACE International — The Corrosion Society)

D-1 Inspection Standards:

NACE TM 01-01	Measurement techniques related to criteria for cathodic protection on underground or submerged metallic tank systems
NACE TM 04-97	Measurement techniques related to criteria for cathodic protection on underground or submerged metallic piping systems

D-2 Recommended Practices:

NACE 1/SSPCSP5	Steel Structures Painting Council: "White Metal Blast Cleaning"
NACE 2/SSPCSP10	Steel Structures Painting Council: "Near White Metal Blast Cleaning"
NACE 3/SSPCSP6	Steel Structures Painting Council: "Commercial Blast Cleaning"
NACE 4/SSPCSP7	Steel Structures Painting Council: "Brush Off Cleaning"
NACE 10/SSPCPA6	Steel Structures Painting Council: "Fiberglass-Reinforced Plastic (FRP) Linings Applied to Bottoms of Carbon Steel Aboveground Storage Tanks"
NACE RP 0169	Control of External Corrosion on Underground or Submerged Metallic Piping Systems

NACE International — The Corrosion Society

NACE RP 0172	Surface preparation of steel and other hard materials by water blasting prior to coating or recoating
NACE SP 0177	Mitigation of alternating current and lightning effects on metallic structures and corrosion control systems
NACE RP 0178	Design, fabrication, and surface finish of metal tanks and vessels to be lined for chemical immersion service
NACE RP 0184	Repair of lining systems

NACE RP 0187	Design considerations for corrosion control of reinforcing steel in concrete
NACE SP 0188	Discontinuity (holiday) testing of new protective coatings on conductive substrates
NACE RP 0193	External cathodic protection of on-grade carbon steel storage tank bottoms
NACE RP 0275	Application of organic coatings to the external surface of steel pipe for underground service
NACE RP 0285	Corrosion control of underground storage tank systems by cathodic protection

E — National Fire Protection Association (NFPA) see also 37 PA Code Chapters 11 and 13, Flammable & Combustible Liquids Handbook

E-1 Construction Standards:

NFPA 70 (NEC)	National electric code®
NFPA 77	Static electricity
NFPA 30	Flammable and combustible liquids code
NFPA 30A	Motor fuel dispensing facilities and repair garages
NFPA 303	Marinas and boatyards
NFPA 326	Safeguarding tanks and containers for entry, cleaning, or repair

E-2 Recommended Practices:

NFPA 11	*Standard for low-, medium-, and high-expansion foam*
NFPA 15	*Standard for water spray fixed systems for fire protection*
NFPA 17	*Standard for dry chemical extinguishing systems*
NFPA 30	Flammable and Combustible Liquids Code
NFPA 72	National Fire Alarm Code
NFPA 1561	*Fire Department Incident Management System*

F — Underwriters Laboratories (UL)

Construction Standards:

UL Std 58	Standards for Steel Underground Tanks for Flammable and Combustible Liquids
UL Std 142	Standard for Steel Aboveground Tanks for Flammable and Combustible Liquids
UL Std 567	Standard for Emergency Breakaway Fittings, Swivel Connectors and Pipe-Connection Fittings for Petroleum Products and LP-Gas
UL Std 842	Standard for Valves for Flammable Fluids
UL Std 860	Standard for Pipe Unions for Flammable and Combustible Fluids and Fire Protection Service
UL Std 971	Standard for Nonmetallic Underground Piping for Flammable Liquids
UL Std 1316	Glass-Fiber-Reinforced Plastic Underground Storage Tanks for Petroleum Products, Alcohol and Alcohol-Gasoline Mixtures
UL Std 1746	Standard for External Corrosion Protection Systems for Steel Underground Storage Tanks
UL Std 2085	Standard for Protected Aboveground Tanks for Flammable and Combustible Liquids
UL Std 2245	Standard for Below-grade Vaults for Flammable Liquid Storage Tanks

American Society of Mechanical Engineers (ASME)-American National Standards Institute (ANSI)

Construction Standards:

ASME B31.3	American Society of Mechanical Engineers: "Process Piping"
ASME B31.4	American Society of Mechanical Engineers: "Liquid Transportation Systems for Hydrocarbons, Liquid Petroleum Gas, Anhydrous Ammonia and Alcohols"

Recommended Practices:

ASSE Z117.1	American Society of Safety Engineers: "Safety Requirements for Confined Spaces"

American Society for Testing and Materials (ASTM)

Construction Standards:

ASTM A182/A182M	Standard Specification for Forged or Rolled Alloy Stainless Steel Pipe Flanges, Forged Fittings and Valves and Parts for High-Temperature Service.
ASTM D2996	Standard Specification for Filament-Wound Fiberglass (Glass-Fiber-Reinforced Thermosetting-Resin) Pipe
ASTM D4097	Standard Specification for Contact-Molded Glass-Fiber-Reinforced Thermoset resin Corrosion Resistant Tanks
ASTM D5685	Standard Specification for Fiberglass (Glass-Fiber-Reinforced Thermosetting- Resin) Pressure Pipe Fittings

Recommended Practices:

ASTM E797	Standard Practice for Measuring Thickness by Manual Ultrasonic Pulse-Echo Contact Method
ASTM D2794	Standard Test Method for Resistance of Organic Coatings on the Effects of Rapid Deformation (Impact)

Steel Tank Institute

Construction Standards:

STI P3	Specification and Manual for External Corrosion Protection of Underground Steel Storage Tanks
STI F841	Standard for Dual Wall Underground Steel Storage Tanks
STI F894 Act-100®	Specification For External Corrosion Protection of FRP Composite Steel USTs (See also Association of Composite Tanks)
STI F921®	Standard for Aboveground Tanks with Integral Secondary Containment
STI F922	Specification for Permatank®
STI F941	Standards for Fireguard® Thermally Insulated Aboveground Storage Tanks
STI R951	Specification for Tanks Using Low Levels of Pressure in the Tanks Interstice
STI F961 ACT-100 U	Specification for External Corrosion Protection of Composite Steel Underground Storage Tanks

Inspection Standards:

STI SP001	Standard for Inspection of In-Service Shop Fabricated Aboveground Tanks for Storage of Combustible and Flammable Liquids

Recommended Practices:

STI SP031	Standard for Repair of In-Service Shop Fabricated Aboveground Tanks for Storage of Combustible & Flammable Liquids
STI R821 sti-P3	Installation Instructions
STI R891 RP	for Hold Down Strap Isolation
STI R892 RP	for Corrosion Protection of Underground Piping Networks Associated with Liquid Storage and Dispensing Systems
STI R912	Installation Instructions for Shop Fabricated Aboveground Storage Tanks for Flammable, Combustible Liquids
STI R913	Act-100® Installation Instructions
STI R923	Permatank® Installation Instructions
STI R931	F921® Installation Instructions
STI R942 F	Fireguard® Installation & Testing Instructions for Thermally Insulated, Lightweight, Double Wall Fireguard Aboveground Storage Tanks
STI R971	ACT-100-U® Installation Instructions
STI R972 RP	for the Addition of Supplemental Anodes to sti-P3® USTs

Steel Structures Painting Council (SSPC) see also NACE International

Recommended Practices:

SSPC Painting Manual volume I

SSPC Painting Manual volume II

Association of Composite Tanks

Construction Standards:

ACT 100	Specification for the Fabrication of FRP Clad Underground Storage Tanks

Fiberglass Petroleum Tank and Pipe Institute

Recommended Practices:

FPTP 1	Fiberglass Piping Systems Installation Check List for Underground Petroleum Pipe
FTPI RP T-95-02	Remanufacturing of Fiberglass Reinforced Plastic (FRP) Underground Storage Tanks

American Concrete Institute (ACI)

Recommended Practices:

ACI 350	Environmental Engineering Concrete Structures

8.6 REGULATIONS AFFECTING TERMINAL AND STORAGE FACILITIES

The Regulations Governing Aboveground Storage Tanks (AST) that exist are generally intended to address sources of pollution that may result from ASTs operation. To ensure the prevention and early detection of a Release of a Substance should one occur, new ASTs are required to meet acceptable design and installation criteria. An example of such a regulation is cited in ref. [25].

Regulations are purely for storage tanks and therefore generally the following types of aboveground facilities are not covered

- Pipelines regulated pursuant to 33 U.S.C. and 49 CFR 195;
- *Transportation of Hazardous Liquids by Pipeline*; or
- Pipelines regulated pursuant to 46 U.S.C. and 33 CFR 154 *Facilities transferring oil or hazardous material in bulk* and 33 CFR 156 *Oil and hazardous material transfer operations.*

Reference Standards: All regulatory bodies reference the standards listed in Section 8.5 and use them as the basis for the enacting various regulations.

Design: Regulations generally refer to the following Design and Construction Requirements for New AST:

- API Standard 620, *Design and Construction of Large, Welded, Low-Pressure Storage Tanks*;
- API Standard 650, *Welded Steel Tanks for Oil Storage*;
- API Standard 12D, *Specification for Field Welded Tanks for Storage of Production Liquids*;
- NFPA 30, *Flammable and Combustible Liquids Code*

Regulatory bodies require that all new metallic field-constructed ASTs be placed on a "Release Prevention Barrier" such as:

- An impervious soil layer or geo-synthetic clay liner with a permeability of 1×10^{-7} cm/sec or less
- An impervious geo-synthetic liner installed in accordance with manufacturer's recommendations such as a 60 mil unreinforced liner, 40 mil reinforced liner, or a material of similar or more stringent specifications
- A double bottom with Leak Detection Monitoring system
- An Impervious concrete slab foundation.

Regulations also require that all ASTs be equipped with normal and emergency venting in accordance with API 2000 and NFPA 30. Leak Detection methods other than visual must be installed, calibrated, tested, operated and maintained in accordance with the manufacturer's instructions, including routine maintenance checks for operability to ensure that the device is functioning as designed.

Protection: Generally, regulatory bodies require that new metallic ASTs installed on foundations consisting of sand, soil or other material that can allow moisture penetration and corrosion, and must install a Cathodic Protection System to mitigate external corrosion of the tank bottom as per following:

- NACE Standard RP0193 — *External Cathodic Protection of On-Grade Carbon Steel Storage Tank Bottoms*;
- API Recommended Practice 651 — *Cathodic Protection of Aboveground Petroleum Storage Tanks*;
- NACE Standard RP0169 — *Control of External Corrosion on Underground or Submerged Metallic Piping Systems.*

Additionally, regulatory bodies require that Cathodic Protection Systems be designed by individuals who have obtained a NACE Cathodic Protection Level 3 Certification and have relevant work experience in the design of Cathodic Protection Systems for ASTs.

Secondary Containment: Regulations usually require a Secondary Containment to collect spills or Leaks, which are required to be promptly removed with appropriate disposal and safety procedures.

Secondary Containment Options include diking or bund walls. Any diking must generally have a minimum capacity to contain 110% of the volume of the largest AST within the diked area or 100% of the volume of the largest AST plus six inches of freeboard for precipitation.

The extent of the diked area must be sufficient to capture overflows, splashing caused by overfilling and/or the trajectory of sidewall leaks.

- Sumps should be installed as part of all dikes and the dike floors sloped to the sumps to enhance material removal.
- Pumps to be permanently installed for the removal of collected material to areas outside the dike area as safely as required.
- The paving/curbing combination, to be considered Secondary Containment. Sumps should be installed as part of all curbs and the floors sloped to the sumps to enhance material removal.

Inert Requirement: Regulatory bodies usually require inerting for the Ullage Volumes of ASTs without a Floating Roof containing Flammable Substances as defined by NFPA 30.

Inspection and Monitoring Requirements for Leak Detection: Regulatory bodies require that a leak of a flammable substance be detected and contained to prevent an impact on surface water, groundwater, or soil outside the Secondary Containment.

A leak detection method other than visual is required to be calibrated, tested, operated, and maintained in accordance with the manufacturer's instructions, including routine maintenance checks for operability to ensure that the device is functioning as designed. The Leak detection method or combination of methods used, except for those ASTs equipped with a Release Prevention Barrier or a double bottom, are generally required to be inspected and monitored at least weekly to determine if a leak from the AST has occurred. A checklist for each Leak Detection monitoring point also required to be generated to document whether a Leak did or did not occur.

8.7 PETROLEUM STORAGE/TERMINAL DESIGN

8.7.1 Typical Layout and Spacing

Petroleum storage terminal layout generally will require the studies which must consider major Service requirements for the facilities. These include:

- Study of site facilities and utilities (where expansion is required for existing facilities)
- Coordination with the architects and other engineering consultants
- Space allocation for fuel tank farm, equipment rooms, and fire storage tank
- Equipment layout
- Equipment selection including pumps, air conditioning units, etc.
- Access

Considerations are:

- Land size
- Number of tanks
- Tank sizes (height and diameter)
- Types of tanks
- Consider the logical unit/equipment locations
- Look at overall site layout for process units
- Consider key separations on grounds of safety (e.g. Bund wall and distance from bund walls, tanks and fence line), wind directions, Security (Inner and outer walls/fences)
- Consider environmental factors (flood, spills …)
- Consider requirements for access/maintenance
- Locate key equipment into each module
- Produce initial plan & elevation to scale
- Prepare isometric sketches for pipe-runs, utilities, cabling
- Location of control centre and offices, parking
- Nearest community
- Future expansion
- Offices
- EMR facilities
- Emergency backup facilities
- Geotechnical considerations including
- Subsoil conditions
- Tank foundation design, monitoring of settlements, see Figure 8-46 for typical example (refer to Section 8.7.3 for typical settlement calculation)
- Hydrotesting (API 653)

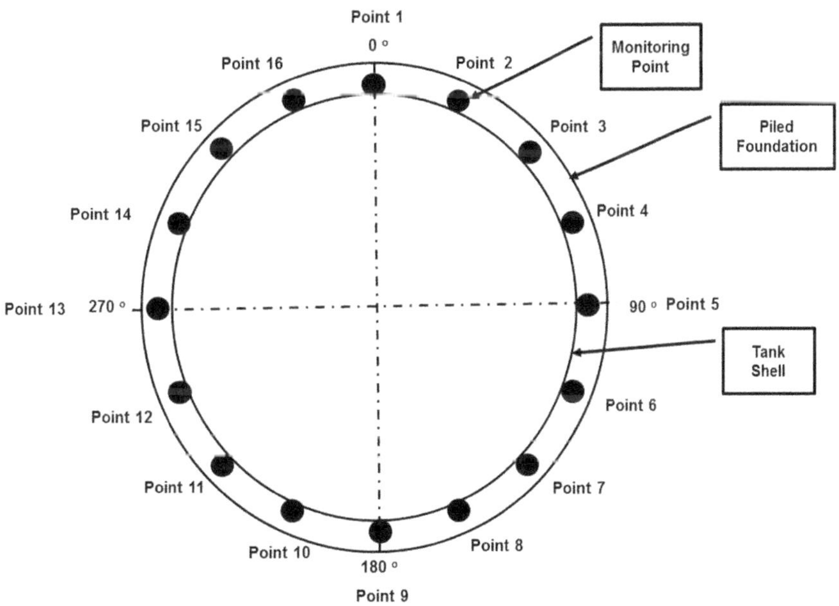

Figure 8-46. Layout of settlement points (refer to Section 8.7.3 [26])

Facilities Layout Review Hierarchy: In order of importance the reviews are indicated below:

(Overall site) Plot Plan • roads • access • storage • admin • utilities
Plant Process • roads • access • units • buildings
Unit/tankage location: Type, access • equipment
Tankage, Equipment • spacing • arrangement

Important issues in layout include:

- General terrain
- Safety and environment
- Regulations, land acquisition, native title, …
- Flammable/non-flammable products
- High-/low-pressure units
- Pumping facilities
- Maintenance, utilities

Important safety issues in plant layout:

- Accident containment and avoidance of "domino" effects
- High hazard operations
- Segregation of different risks
- Exposure to possible explosion overpressure
- Exposure to fire radiation
- Minimizing vulnerable piping
- Drainage and grade sloping
- Prevailing wind directions
- Provision for future expansion

Important equipment/appurtenances layout issues include:

- Pumps (NPSH, suction line, motor location)
- Instrumentation (CVs accessible)
- Heat exchangers (bundles, fin-fan vs water)
- Flares (radiation levels, alternative sites and types)
- Solids (use of gravity flow if possible, containment)
- Expensive piping (run lengths)
- Maintenance (access, removal)

Storage Tank Spacing for Chemical/Oil Plants: Codes and standards generally do not specify/recommend a safe distance between tankage facilities, however NFPA 30 "Flammable and Combustible Liquids Code" refers to minimum safe distances. American Institute of Chemical Engineers, Center for Chemical Process Safety indicate distances between tankages as given in Figure 8-47 [27].

Based on the layout consideration and the requirements for safe spacing between tanks a typical plot plan for a tank farm is indicated in Figures 8-48 and 8-49.

It may be noted that minimum standards of control (including tankage spacing, piping arrangement, access, etc.) must be in place at all locations/terminals storing large volumes of petroleum and similar products capable of giving rise to a large flammable vapor cloud in the event of a loss of primary containment [28,29].

	Floating & Cone Roof Tanks < 3000 BBLS	Floating & Cone Roof Tanks > 3000, < 10,000 BBLS	Floating & Cone Roof Tanks >10,000 < 300,000 BBLS	Large Floating Tanks > 300,000 BBLS	Pressurized Spheres	Pressurized Bullets	Refrigerated Dome Roof Storage Tanks
Floating & Cone Roof Tanks < 3000 BBLS	0.5 D						
Floating & Cone Roof Tanks > 3000, < 10,000 BBLS	0.5 D	0.5D					
Floating & Cone Roof Tanks >10,000 < 300,000 BBLS	1 D	1 D	1 D				
Large Floating Tanks > 300,000 BBLS	1D	1D	1D	1D			
Pressurized Spheres	1.5 D	1.5 D	1.5 D	1D	1D		
Pressurized Bullets	1.5 D, 15 m Min	1.5D, 30 m Min	1.5 D, 30 mMin	2D	1D, 30 m Min	1D, 30 m Min	
Refrigerated Dome Roof Storage Tanks	2D,60 m Min	2D, 60 m Min	2D, 60 m Min	2D	1D 30 m Min	1D 30 m Min	1D 30 m Min

Figure 8-47. Storage tank spacing for chemical/oil plants (adapted from ref. [27])

Figure 8-48. A typical well laid out tank farm layout

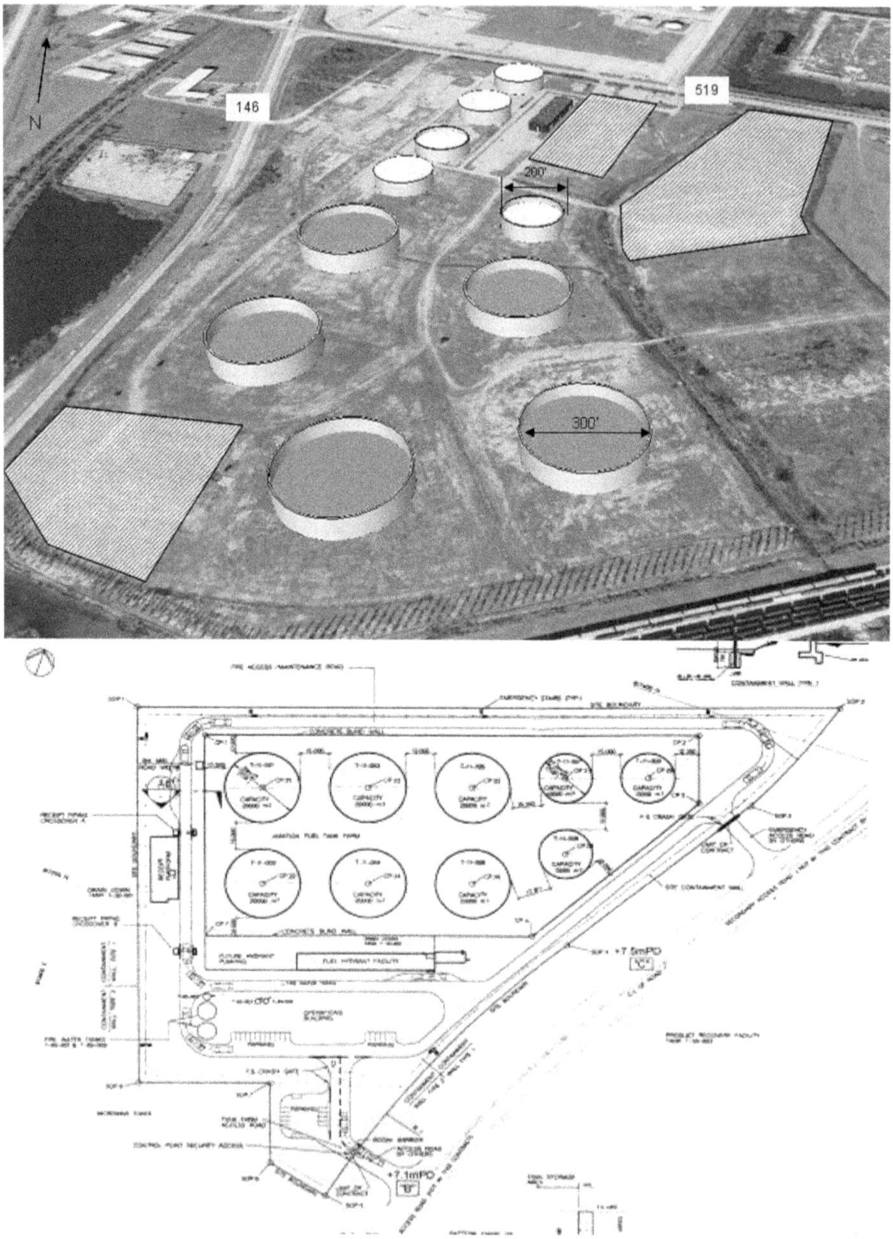

Figure 8-49. Typical tank farm with safe distances between tankages

8.7.2 Tank Design (Including Sizing, Materials, and Construction)

8.7.2.1 Design Data

Data required to undertake a tank design should include the following:

- Design Code
- Tank Location
- Nominal capacity of tank
- Tank diameter and/or height restrictions;
- Internal tank pressure

- Rate of filling and emptying tank
- Design pressure
- Operating Pressure
- Design Temperature
- Operating Temperature
- Density of Liquid
- Design Density
- Prevailing wind direction
- Maximum Wind Speed and Pressure
- Seismic Data
- Corrosion Allowance (shell/bottom and roof)
- Radiography requirements
- Stored Product and Class

Design of accessories should also be considered and includes the following:

- Access to Roof (Spiral stairway)
- Hand Rail at Roof
- Stairway, Platform and handrail
- Earth/foundation Connection
- Platform at Roof
- Primary and Secondary Seals
- Rolling Ladder
- Gauge Hatch/Well
- Primary Roof Drain
- Cooling & Foam System (including Fire water, fixed foam system, etc.)
- Foam riser/Dip Pipe
- Shell-Bottom
- Reinforcing pads
- Manhole Neck, flanges, covers
- Nozzle Necks
- Nozzle Flanges
- Gaskets (Spiral wound)
- Bolting for Structure
- Bolting for nozzles
- Structural components
- External Gussets
- Internals parts, pipe and other fittings
- Wind girder, support, and gussets
- Drains
- Annular plate
- Primer/Painting/external protection

8.7.2.2 Design Calculations

8.7.2.2.1 Tank Working Pressure Generally, there are regulatory requirements specifying the type of storage tank to be used, based on the storage tank. In addition, there are usually specific design requirements, for example in the type of seals to be used in a floating roof tank based on the tank design, the liquid type that is stored.

A design working pressure can be determined to prevent breathing, and thereby save standing storage losses. However, this should not be used in lieu of any environmental regulatory requirements regarding the design of storage tanks. The environmental regulatory requirements for the specific location should be consulted prior to the design of storage facilities.

Design pressure is usually the operating pressure plus a margin of safety.

Operating pressure of tank is the pressure at which the tank normally operates. It should not exceed the maximum allowable working pressure of the vessel. A suitable margin should be allowed between the pressure normally existing in the gas or vapor space and the pressure at which the relief valves are set, so as to allow for the increases in pressure caused by variations in the temperature or gravity of the liquid contents of the tank and other factors affecting the pressure in the gas or vapor space (API Standard 620).

The working pressure ϕ required to prevent breathing losses depends upon the vapor pressure of the product, the temperature variations of the liquid surface and the vapor space, and the setting of the vacuum vent.

$$\phi = B_{max} + (\Delta - B_{min})\left[(T_{max} + 273)/(T_{min} + 273)\right] - \text{Pa} \qquad (8-1)$$

where:

Δ = absolute internal tank pressure at which vacuum vent opens, kPa (abs)
ϕ = required storage pressure, kPa (abs)
B_{max} = vapor pressure of liquid at maximum surface temperature, kPa (abs)
B_{min} = vapor pressure of liquid at minimum surface temperature, kPa (abs)
P = absolute pressure, kPa (abs)
Pa = atmospheric pressure, kPa (abs)
T = temperature, K
T_a = ambient air temperature, °C
T_{max} = maximum average temperature, °C
T_{min} = minimum average temperature, °C

The above relation holds only when Bmin is less than Δ; that is, when the minimum vapor pressure is so low that air is admitted into the vapor space through the vacuum vent. When Bmin is equal to or greater than Δ, the required storage pressure is,

$$\phi = B_{max} - \text{Pa} \qquad (8-2)$$

Under this condition, air is kept out of the vapor space. Figure 8-50 is presented as a general guide to storage pressures for gasolines of various volatilities in uninsulated tanks.

These data for plotting the curves were computed from Eqs. (8-1) and (8-2) using the following assumptions:

- Minimum liquid surface temperature is 5°C less than the maximum liquid surface temperature.
- Maximum vapor space temperature is 22°C greater than the maximum liquid surface temperature.
- Minimum vapor space temperature is 8°C less than the maximum liquid surface temperature.
- Stable ambient conditions (ambient temperature 38°C).

These temperature variations are far greater than would be experienced from normal night to day changes. Therefore, the lower, nearly horizontal line, which shows a required storage pressure of 17 kPa (ga) for the less volatile gasolines is conservative and allows a wide operating margin.

Maximum liquid surface temperatures vary from 29 to 46°C. Sufficient accuracy will generally result from the assumption that it is 5°C higher than the maximum temperature of the body of the liquid in a tank at that location.

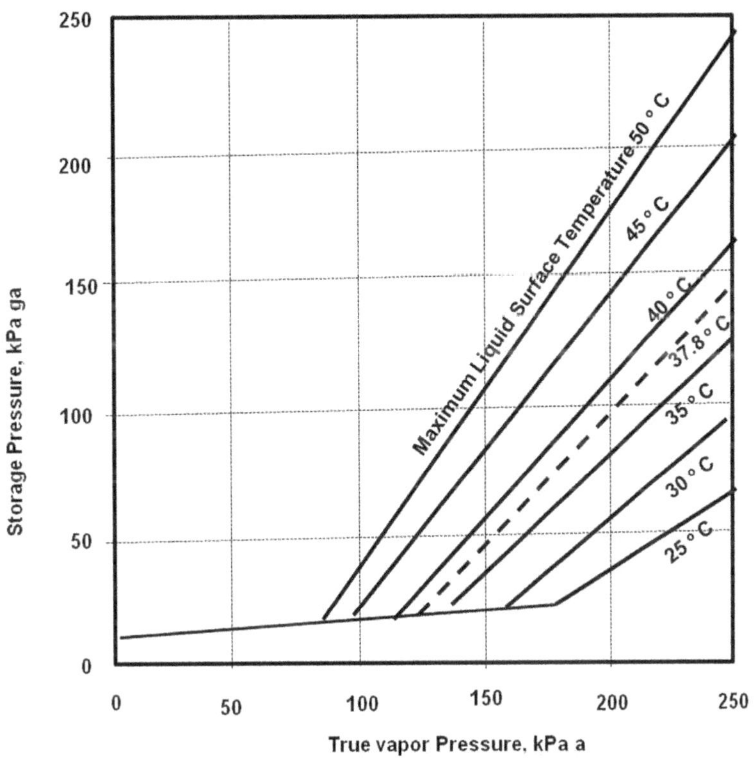

Figure 8-50. Storage pressure vs. true vapor pressure (TVP), redrawn from GPSA [30]

Example 8-1 — To illustrate the use of Figure 8-50, suppose a 165 kPa (abs) true vapor pressure (TVP) natural gasoline is to be stored where the liquid surface temperature may reach a maximum of 38°C. A vertical line extended upward from the 165 kPa (abs) mark at the bottom of the chart intersects the 38°C line at 64 kPa (abs). The design pressure of the tank should be a minimum of 70.5 kPa (abs) (64 kPa + 10%).

Figure 8 51 below can also be used for the determination of working pressure of storage tanks storing LPG and natural gasoline such as condensate. It is not to be used for crude oil.

In relation to Figure 8-51, the following definitions apply:

RVP = Reid Vapor Pressure is a vapor pressure for liquid products as determined by ASTM test procedure D-323.
The Reid vapor pressure is defined as pounds per sq in (kPa) at 37.8°C. The RVP is always less than the true vapor pressure at 37.8°C.
TVP = true vapor pressure is the pressure at which the gas and liquid in a closed container are in equilibrium at a given temperature.

Figure 8-51 can be used as follows:

- As quick reference to determine true vapor pressures of typical LPGs, natural gasolines, and motor fuel components at various temperatures.
- To estimate the operating pressure of a storage tank necessary to maintain the stored fluid in a liquid state at various temperatures.
- For converting from true vapor pressure to Reid Vapor Pressure (RVP).
- For simple evaluation of refrigerated storage versus ambient temperature storage for LPGs.

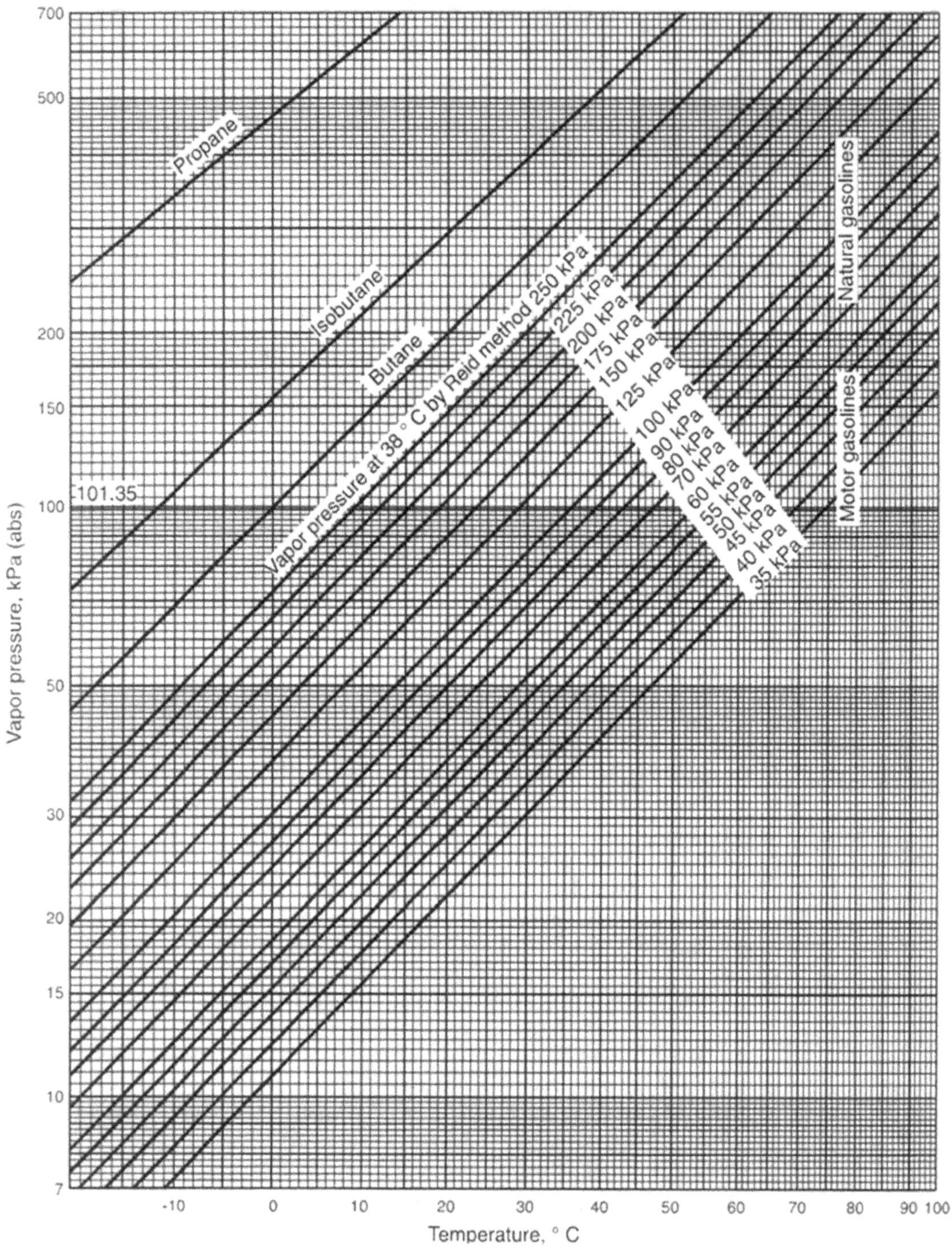

Figure 8-51. True vapor pressures vs. temperatures for typical LPG, motor, and natural gasolines (not to be used for crude oil vapor pressure), courtesy of GPSA [30]

Example 8-2 — Determine the TVP of an 83-kPa RVP gasoline. In addition, estimate the design pressure of a tank needed to store this same 83 RVP gasoline at a maximum temperature of 49°C.

Using Figure 8-51, a vertical line is extended upwards from the 38°C mark (38°C is used as the reference point for determining RVP) at the bottom of the chart to the intersection of the 83 kPa RVP line, read true vapor pressure of 91 kPa (abs). A vertical line is also extended from the 49°C mark to intersect the 83 RVP gasoline line. Now going horizontal, the true vapor pressure axis is crossed at approximately 125 kPa (abs).

The storage tank should therefore be designed to operate at 125 kPa [25 kPa (ga)] or above. The design pressure of the tank should be a minimum of 10% above the operating gauge pressure or approximately 137 kPa (abs).

Example 8-3 — Evaluate the options of refrigerated storage versus ambient temperature storage for normal butane.

From Figure 8-51, the vertical line is extended up from the 38°C (assumed maximum) mark to intersect the normal butane line at approximately 355 kPa (abs) [255 kPa (ga)]. The working pressure of the tank should be 255 kPa (ga) plus a 10% safety factor, or 280 kPa (abs). This same product could be stored in an atmospheric pressure tank if the product is chilled to 0°C.

This temperature is determined by following the normal butane line down until it intersects the 100 kPa (abs) horizontal vapor pressure line. Reading down to the bottom scale indicates the storage temperature at 0°C. The pressurized tank would require more investment due to the higher working pressure of 380 kPa (abs) [280 kPa (ga)] and the thicker shell requirement. The refrigerated tank would require less investment for the tank itself, but an additional investment would be necessary for insulation and for refrigeration equipment which requires additional operating expenses. The economics of each type of storage system can be evaluated to determine which will be the most attractive. The graphical method of converting from RVP to TVP is an approximation and is generally more accurate for lighter components.

Crude oils with very low RVPs could vary significantly from this graphical approach. This is due to the fact that during the Reid test the highest vapor pressure materials tend to evaporate leaving a residue which has a lower vapor pressure than the original sample. Equation 7-3 was developed by Kremser in 1930 [31] to relate the two vapor pressures at 37.8°C.

$$\text{TVP} = (1.07)(\text{RVP}) + 4.1 \qquad (8-3)$$

Using this formula for the 83 kPa, RVP gasoline example would calculate a 93 kPa (abs) TVP versus the 91 determined.

However, Figure 8-52 below can be used to determine the true vapor pressure of crude oils [32].

Similarly, Figure 8-53 below can be used for the determination of petroleum products.

8.7.2.2.2 Mechanical Design The following provides calculation methods for the determination of permissible stresses and of major tank elements:

Tank Internal Pressure: Internal pressure of tanks can be calculated as follows

$$P_{\max} = \frac{39W}{\pi D^2} + 77t, \text{ N/m}^2$$

or

$$= \frac{4W}{\pi D^2} + 7.85t, \text{ kgf/m}^2 \qquad (8-4)$$

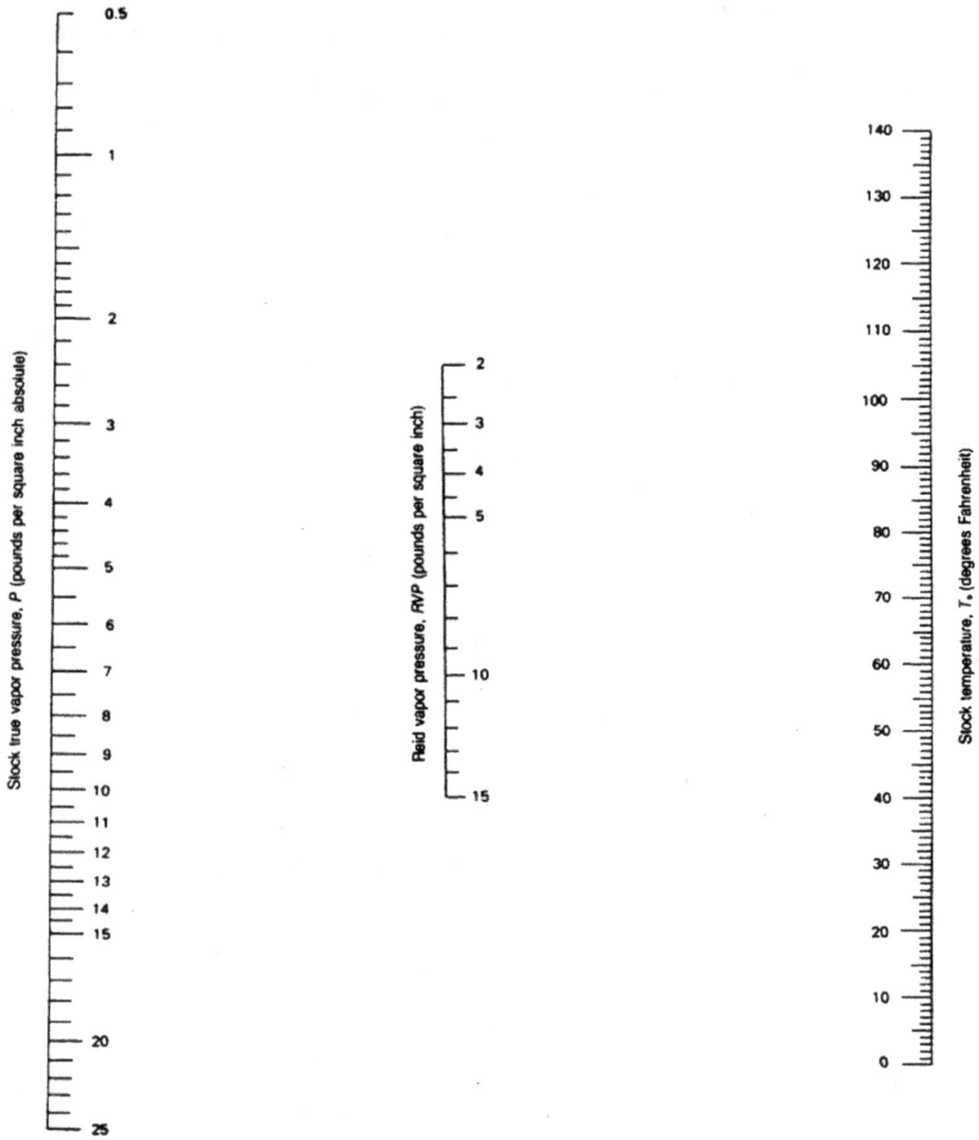

Figure 8-52. True vapor pressure of crude oils with a Reid vapor pressure of 2 to 15 pounds per square inch [32], courtesy of API

P_{max} = internal pressure,
W = total weight of shell and structure supported by shell in N (kgf),
D = diameter of tank in m, and
t = thickness of roof in mm.

Shell Thickness: The minimum thickness of shell plates shall not be less than that calculated from the following formula:

$$t = 4.9(H - 0.3)\ DG/SE \text{ if } S \text{ is in N/mm}^2 \qquad (8-5)$$

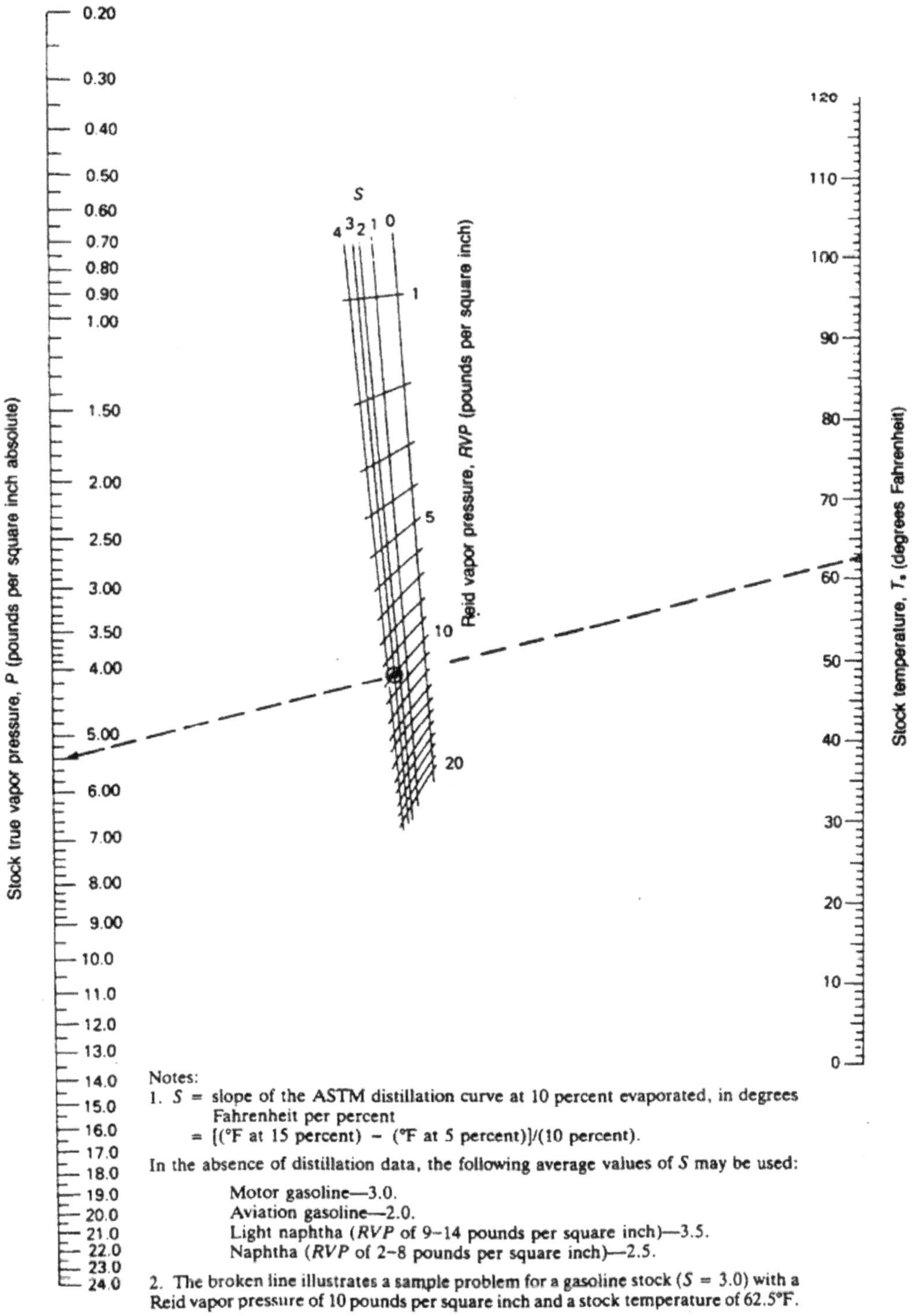

Notes:

1. S = slope of the ASTM distillation curve at 10 percent evaporated, in degrees Fahrenheit per percent
   = [(°F at 15 percent) − (°F at 5 percent)]/(10 percent).

In the absence of distillation data, the following average values of S may be used:

Motor gasoline—3.0.
Aviation gasoline—2.0.
Light naphtha (*RVP* of 9–14 pounds per square inch)—3.5.
Naphtha (*RVP* of 2–8 pounds per square inch)—2.5.

2. The broken line illustrates a sample problem for a gasoline stock (S = 3.0) with a Reid vapor pressure of 10 pounds per square inch and a stock temperature of 62.5°F.

Figure 8-53. True vapor pressure of refined petroleum stocks with a Reid vapor pressure of 1 to 20 pounds per square inch [32], courtesy of API

or

$$t = 50(H - 0.3)DG/SE \text{ if } S \text{ in kgf/cm}^2 \quad (8-6)$$

where

t = minimum thickness in mm;
D = nominal diameter of tank in m;
H = height from the bottom of the course under consideration to top of top curb angle or to bottom of any overflow which limits tank filling height in m;
G = specific gravity of liquid to be stored, but in no case less than 1.0;
S = allowable stress; and
E = joint efficiency factor

The following should be considered in the determination of shell plate thickness:

- *Loads:* Stresses in the tank shell can be computed on the assumption that the tank is filled with water of specific gravity 1.00 or the liquid to be stored, if heavier than water. The tension in each course shall be computed at 30 cm above the centre line of the lower horizontal joint of the course in question.
- Isolated radial loads on tank shells, such as that caused by heavy loads from platforms and elevated walkways between tanks may be distributed appropriately, preferably in a horizontal position.
- Wind and internal vacuum loads should be considered together to check the stability of tank shells. Wind loads may be as specified as per design data. Internal vacuum in the tank is usually taken as a minimum of 500 N/m^2 (50 kg/m^2, 0.5 kPa) vacuum by the industry. API 650 requires a minimum external pressure of 0.25 kPa
- *Joint Efficiency Factor* — This is usually taken as 0.85 for double welded butt joints, to determine the minimum thickness of shell plates computed from the stress on the vertical joints, subject to all vertical and horizontal butt welds being spot radiographed. Where welds are not to be so examined by radiography, the joint efficiency factor considered for design shall be 0.70. It may be note however that the base tank design procedure of API 650 requires a joint efficiency of 1.0 unless the tank is very small and has a maximum shell thickness of 13 mm (1/2").

The nominal thickness of shell plates refers to the tank shell as constructed and is based on stability rather than stress. Any required corrosion allowance for the shell plates must be added to the calculated thickness.

Stability of tank shells against external loads shall be checked by determining the maximum height of the shell from the tap curb angle or wind girder that does not buckle under this loading and providing stiffening to the shell if required.

The maximum height of un-stiffened shell in meters, is not to exceed H_1 as determined by the following equation:

$$H_1 = [14700t/p]\sqrt{\left(\frac{D}{t}\right)3} \quad \text{if } p \text{ is in N/m}^2 \quad (8-7)$$

$$H_1 = [1500\ t/p]\sqrt{\left(\frac{D}{t}\right)3} \quad \text{if } p \text{ is in kgf/m}^2 \quad (8-8)$$

where

H_1 = vertical distance between the intermediate wind girder and top angle of the shell or the top wind girder of an open top tank in m;

t = average shell thickness in height H_1 in mm determined from the actual thicknesses of plates used unless the it is specified that the net thickness (actual thickness used minus corrosion allowance specified) be considered.

8.7.2.3 Tank Material

Tanks are constructed from a number of different materials based upon cost of the material, ease of fabrication, resistance to corrosion, and compatibility with the fluid stored. Sometimes specialized composites and techniques are used in tank construction, but these are the exception.

Tanks made of metal materials such as carbon steel, are very prone to failure under natural hazards. Carbon steels used for storage tanks have specified minimum yield strengths of approximately 200 to 600 MPa. The principles of material selection are based on code requirements including brittle fracture and corrosion. In this respect it may be noted that while API 650 allows for a A285 Gr C and A83 Gr C material which has a yield strength of 200 MPa, most steel mill do not make this material any longer. A36 is the most common and widely available carbon steel. A36 has a minimum yield strength of 250 MPa.

Code requirements that govern tank designs often have very specific material selection requirements and limitations. Most codes include provisions in the material selection criteria that ensure materials with sufficient toughness under the service conditions to prevent brittle fracture.

- The susceptibility of the material to brittle fracture is also one of the most important material selection considerations. Brittle fracture is the tensile failure of a material showing little deformation or yielding. Brittle fractures typically start at a flaw and can propagate at high speeds, resulting in catastrophic failures.
- Corrosive effects in tanks may be divided into internal and external ones. The most common is the external corrosion that is usually minimized by the used of coatings for carbon steel tanks.
- The selection of a design metal temperature is important in ensuring that materials are selected which are tough enough to prevent brittle fractures under the service conditions.
- Ensuring a material with adequate toughness. One way to ensure that a selected steel has adequate toughness for the design metal temperature of the tank is to proof-test each plate by impact toughness testing samples at or below the design metal temperature (DMT).

It may be noted that the Minimum Design Metal Temperature (DMT) is usually defined as 8°C above the lowest one-day mean temperature.

Carbon steel is the most common material for tank construction. The material properties most commonly assumed for engineering calculations are a modulus of elasticity of 2.068×10^{11} N/m^2, Poisson's ratio of 0.3, mass density of 7849.7 kg/m^3, and yield strength of 2.156×10^8 N/m^2.

8.7.3 Civil Design

8.7.3.1 Tank Foundation

Tank foundation is the most import part of a tankage design, particularly for large volume tanks as they carried thousands of barrels of hydrocarbon fluid often very heavy. A

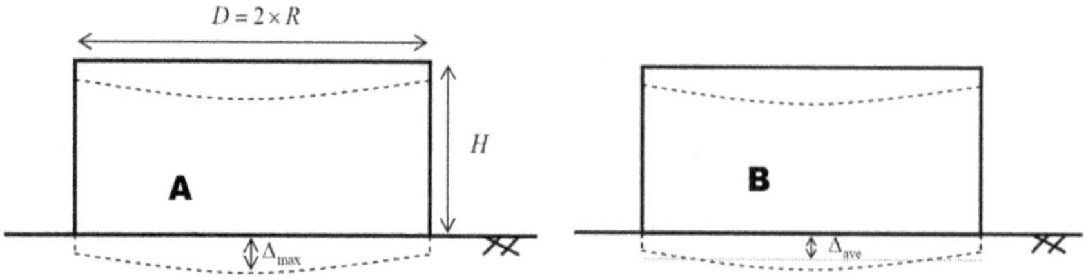

Figure 8-54. Settlement of steel tank (A — maximum and B — average)

geotechnical study of the site is thus required in the design of the foundation; however, in many cases (especially for tanks located in coastal areas), the soils are susceptible to have uniform or differential settlements. The buckling of the shell due to the foundation settlement however must be considered in details.

Various forms of settlements (Δ) could take place so it is important to define types of settlement (Figure 8-54) and all required variables for calculations:

D = Diameter of the tank.
R = Radius of the tank.
H = Height of the tank.
L = distance between two points with differential settlement.
Δ_{max} = max Total maximum settlement: This type of settlement illustrates in Figure 8-54A, *ave* = Average settlement: This type of settlement is an average of the settlement of all points of a tank Figure 8-54B. See also previous Figure 8-46.

Other parameters are:
w = Tilt: This component rotates the tank in a tilt plane (Figure 8-55A).
δ = Differential settlement between two points.
δ_{bottom} = Edge settlement occurs when the tank shell settles sharply around the periphery, resulting in deformation of the bottom plate near the shell-to-bottom corner junction (Figure 8-55B), or the depth of the depressed area of the bottom plate (Figure 8-56).
δ_{shell} = This component of settlement at the bottom edge leads to the lack of circularity and creates stresses in the shell. Shell δ is defined as differential outline settlement between settlement of one measurement point with respect to the average of settlements of its two adjacent points (Figure 8-54B).

$$\delta_i = U_i - \left(0.5 \times U_{i+1} - 0.5 \times U_{i-1}\right) \qquad (8-9)$$

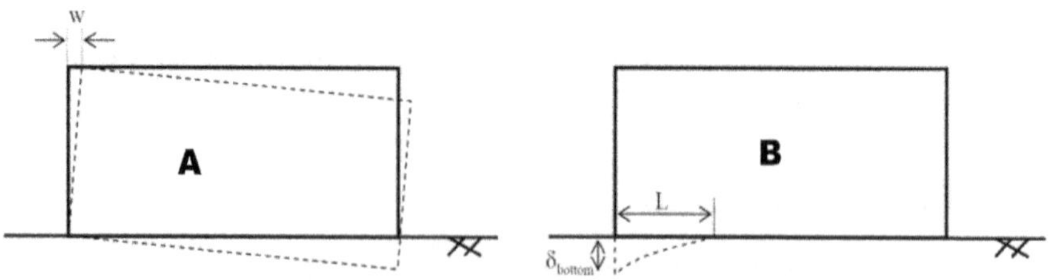

Figure 8-55. Steel tank settlement (A — tilt and B — bottom-edge differential settlement) [33]

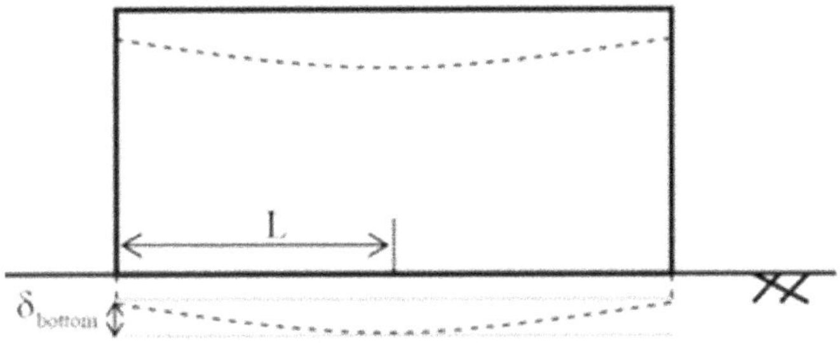

Figure 8-56. Steel tank settlement — bottom-center differential

where

δ_i = Differential settlement between one point and average settlement of its adjacent points

U_i = Settlement of each points in Figure 8-57 and previous Figure 8-46

API 653 is the related document for the calculation of tank settlements as follows [26, 33]:

- Uniform settlement: This component often can be predicted in advance, with sufficient accuracy from soil tests, and does not induce stresses in the tank structure. However, piping, tank nozzles, and attachments must be designed with adequate consideration to prevent problems caused by such settlement. Therefore, piping must be adequately supported to avoid undue stresses.
- Planer tilt: This type of settlement could affect tank nozzles which may have piping attached to them. The tilt will cause an increase in liquid level.
- Outline settlement of the shell: Use the following formula to calculate the maximum allowable outline settlement:

$$\delta_{shell} = \left(L^2 \times \varepsilon_y \times 5.5\right) / \left(H\right) \qquad (8-10)$$

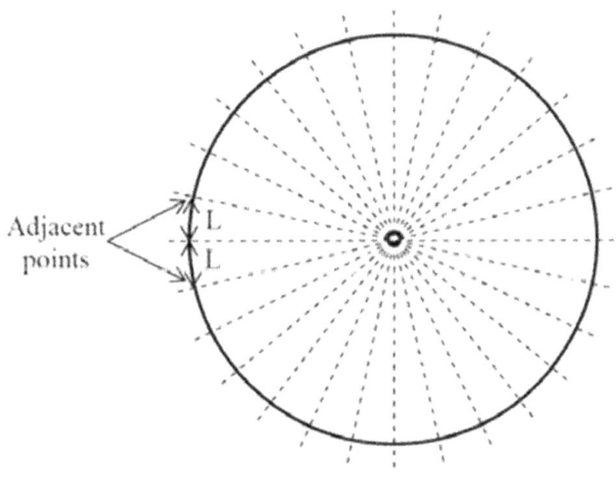

Figure 8-57. Shell differential settlement of steel tank [33]

where

δ_{shell} = shell deflection δ in meter,
L = arc length between measurement points in meter (Figure 8-57);
ε_y = yield strain, (conservatively equal to 0.001); and
H = tank height in meters.

Together with reference to API 653, and the above formula the allowable outline settlement between adjacent measurement points can be calculated. Thus:

- Bottom-Edge differential settlement: The maximum allowable bottom-edge settlement is shown in Figure 8-58. Conservatively $\delta_{bottom\text{-}edge} = B_{ew}$ could be assumed and it means areas with bottom lap welds approximately parallel to the shell.
- Bottom-Center differential settlement: Use the following formula to calculate the maximum allowable bottom plate settlement: $\delta_{bottom\text{-}center} = 0.031 \times R$.

Notes: With reference to Figure 8-58, it may be noted that tanks with larger edge settlements are to be repaired, or have detailed analysis of floor and floor-to-shell junction undertaken.

Welds in tanks with settlements ≥ 75% of B_{ew}, and larger than 2" are to be inspected with "Magnetic Particle" or "Liquid Penetrant Examination."

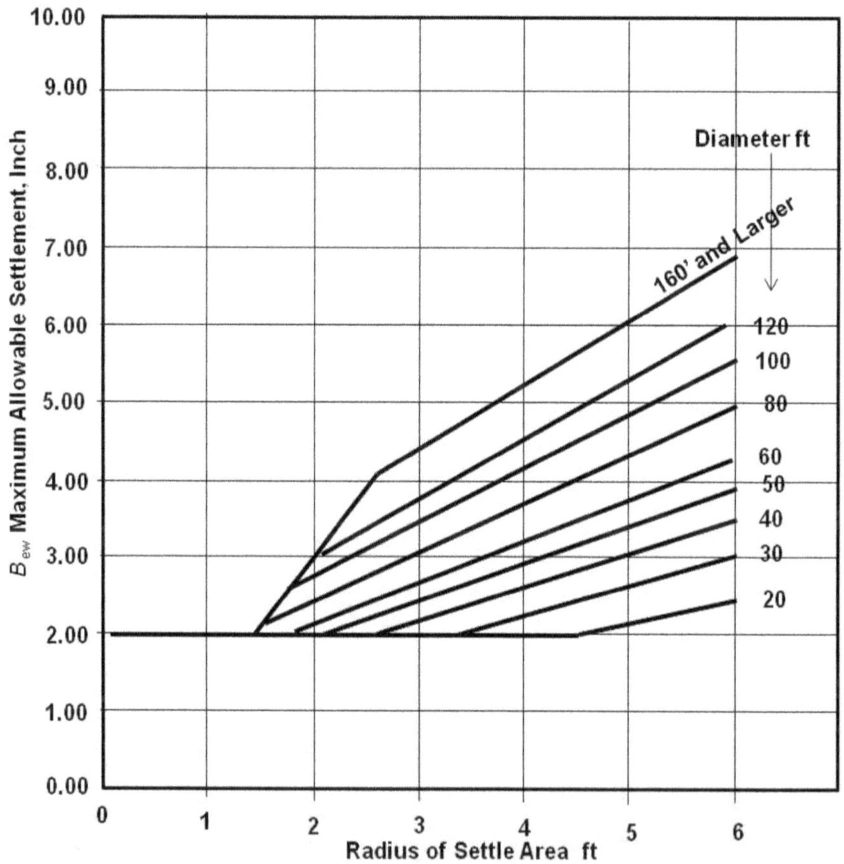

Figure 8-58. Allowable bottom-edge differential settlement, (Reproduced from API 653, Appendix-B)

8.7.3.2 Types of Foundations

There are a number foundation types for storage tanks, however, some of them are more common for specific applications. Foundation types may be broken into several classifications in generally increasing order of costs. These are:

- Compact soil foundation
- Crushed stone ring-wall foundation
- Concrete ring wall foundation
- Slab type foundation and
- Pile supported Foundation

Compact soil foundations: These foundations can be used where the soil quality and bearing capacity are good and for smaller tanks. Generally, the top 7 to 15 cm of soil is removed and replaced with a granular and sand backfill. These are often called sand pad foundations, laid directly on earth. The advantage of this type of foundation is the relatively low cost.

Crushed-stone foundations: This design generally must incorporate a leak detection system. The foundation has many of the advantage of the concrete ring-wall and can sustain large load if foundation is prepared properly with crushed stone and compacted to 90% proctor density. It provides uniform support of the tank bottom by dissipating concentrated loads in a granular pattern. Catastrophic failure of the bottom is possible if tank foundation is poorly prepared and or a leak starts and washes out the underlying support. Typical earth/gravel/crushed stone tank foundation for large tanks is shown in Figure 8-59.

Concrete ring-wall foundations: The concrete ringwall foundation is so called because of its appearance. It is used in foundations for tanks of a diameter of at least 10 m or more. In the large-diameter tanks, this is usually the most cost-effective reinforced concrete foundation, with many advantages such as reducing the probabilities of settlements failures in sandy areas and/or where differential settlements are possible. Typical concrete ring wall foundation is shown in Figure 8-60 which is the most common type of tank foundation.

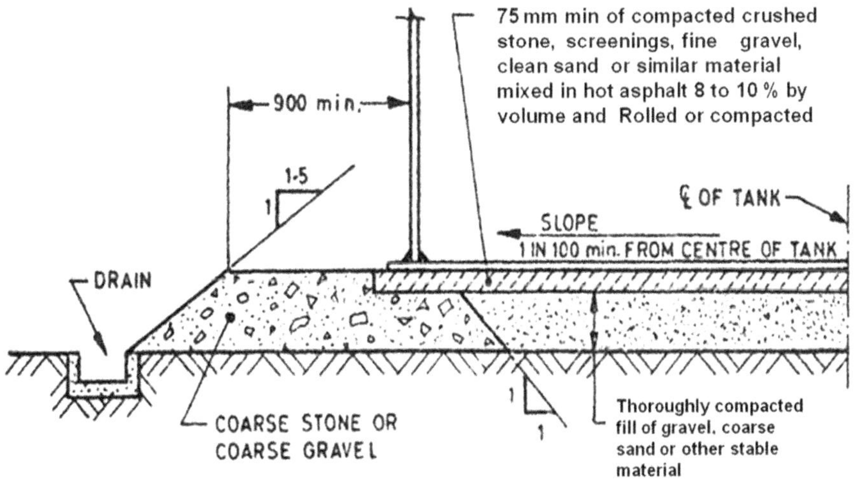

Figure 8-59. Typical Earth/gravel tank foundation (large tanks), all dimensions in mm

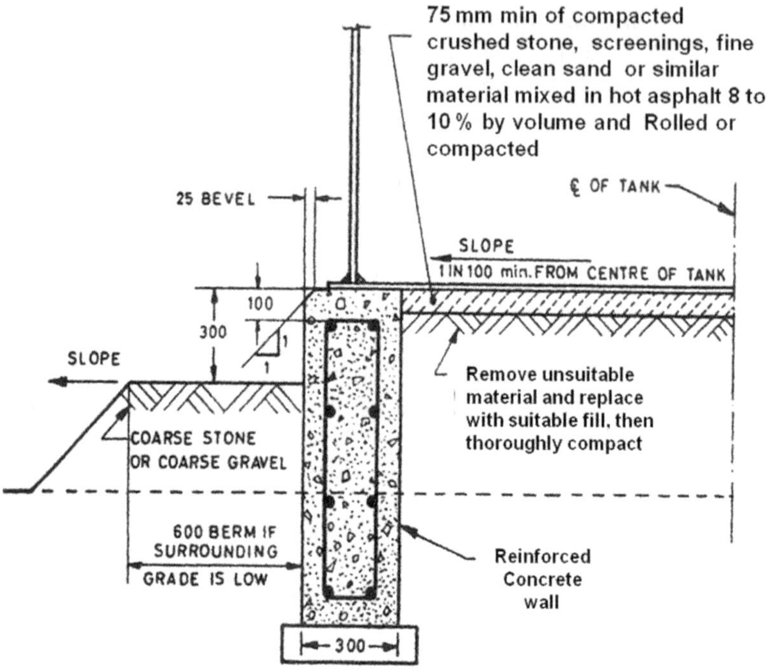

Figure 8-60. Typical concrete ring wall foundation (smaller diameter tanks), all dimensions in mm

Slab foundations: The concrete slab foundation has the advantages of the concrete ring-wall but is usually limited to tanks with diameters less than 10 m. Often the edge of the slab will be sufficiently thick to provide for anchorage. A slab foundation is very versatile, but its high cost limits it to use in small tanks. The slab provides a level and plane-working surface that facilitates rapid field erection.

Pile-supported foundations: The pile-supported foundation is usually found where the soil bearing pressures are very low. Examples might be river deltas and land adjacent to bays. They are also used where high foundation uplift forces are encountered resulting from internal pressure or seismic loading.

Foundation design must conform to welded tank bottom design. A tank bottom may be broadly classified as flat bottom or conical.

A Flat bottom: They are the most common end closures of tanks. These tanks appear flat but usually have a small designed slope and shape. These tanks are further categorized as follows:

- **Flat:** For tanks less than about 5 to 10 m in diameter, the flat-bottom tank is used. The inclusion of a small slope as described above does not provide any substantial benefit, so they are fabricated as close to flat as possible.
- **Cone up:** These bottoms are built with a high point in the center of the tank. Crowning the foundation and constructing the tank on the crown accomplish this. The slope is limited to about 25 to 50 mm per 3 m run. This is the design mostly adopted in large diameter above ground storage tanks.
- **Cone down:** The cone-down design slopes toward the center of the tank. Usually, there is a collection sump at the center. It is very effective for water removal from tanks. This design is inherently more complex because it requires a sump, underground piping, and an external sump outside the tank and therefore not used in design of very large tanks.

- **Single slope:** This design uses a planar bottom but it is tilted slightly to one side. This allows for drainage to be directed to the low point on the perimeter, where it may be effectively collected. Since there is a constant rise across the diameter of the tank, the difference in elevation from one side. Again, this design is not used for large diameter/volume storage tanks.

B Conical bottom: The second type is the conical bottom. The designers often use it to provide a complete drainage or even removal of solids. Since these types of tanks are more costly, they are limited to the smaller sizes and are often found in the chemical industry or in processing plants only.

8.7.3.3 Bund Walls/Dikes

Since most liquids can spill, evaporate, or seep through even the smallest opening, special consideration must made for their safe and secure handling. This usually involves building a bunding, or containment dike (Figure 8-61) around tanks, so that any potential leakage may be safely contained.

The term bunding also refer to dikes, but it is frequently used to describe liquid containment facilities that prevent leaks and spillage from tanks and pipes, though sometimes any barrier is referred to as bunding. Frequently, the liquids in tanks and pipes are toxic, but bunding is used to prevent the liquid from causing damage (either by force or its chemistry). If a large tank has a catastrophic failure, the liquid alone can cause extensive damage.

Bunded oil storage tanks are essentially a tank within a tank. The "bund" being the area surrounding the inner tank, hence bunded tank. Bunding is a legal requirement in many countries particularly around tanks, storage vessels and other plant that contain liquids which may be dangerous or hazardous to the environment.

Bunded oil tanks are generally a requirement from most insurance companies as opposed to single skinned oil storage tanks. Due to the safer oil storage solutions brought about by bunded oil storage tanks, most Environmental Agency require bunded oil tanks.

Figure 8-61. A typical dike or levee

Figure 8-62 shows a well-designed bundwall with an up-sloped dyke that leads to the perimeter fence line of tank farm. Prior to the fence line there is access road and perimeter lighting system typical of well secured facility. If there ever was to be a spill in the tank farm, the protection of underground waters is also ensured by the two concrete retention walls of 1.6 m and 2.0 m height, located at a distance of 15 to 30 m between them. Furthermore, in order to prevent the infiltration of oil-products into earth, there is a high density (2 mm) polyethylene underlay throughout the tank farm.

8.7.3.3.1 Bund-Wall Holding Capacity and Alarm Almost all regulations require a holding capacity of 110% of the maximum capacity of the biggest tank within the bund or 25% of the total capacity of all the tanks within the bund whichever is the greatest. In addition, further guidelines in some countries (e.g., the UK) recommend additional measures such as providing sufficient "freeboard" or height of wall above the maximum holding capacity to accommodate dynamic factors such as surge in situations of major tank failure or storm driven waves in larger bunds. As a rule (and unless specific local laws prevail) most operators work to the 110% capacity guide.

It may be noted that accumulation rain water in a well designed and sealed bund wall will reduce it s design capacity. Therefore regulations require implementation of a method for the removal of the rainwater. It is expected that a good system will work continuously and automatically. It should also provide alarms for conditions. These automatic pump systems are referred to as "BundGuard."

8.7.3.3.2 Bund Wall Construction It is reasonably easy to construct a "water-tight" bund around the base of a tank or vessel. A concrete base and a sealed wall of masonry, brickwork, concrete, or even prefabricated steel provides the holding capacity. There are two types of outer bund walls and their details are shown in Figure 8-63 below.

If built properly and bunding is large and strong enough, it will contain the contents of an entire tank, though regulations may require it to be up to a third larger. When multiple tanks share a bund, the capacity is based on the largest tank. One of the most common designs for large tanks is a concrete or masonry wall around the tank with a concrete floor. The outside of the wall may be reinforced with an earth berm.

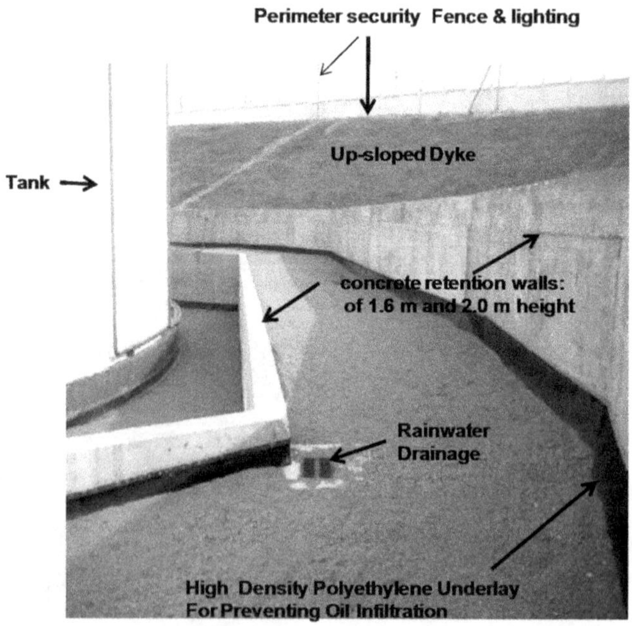

Figure 8-62. A well-designed bund wall system with an up-sloped dike

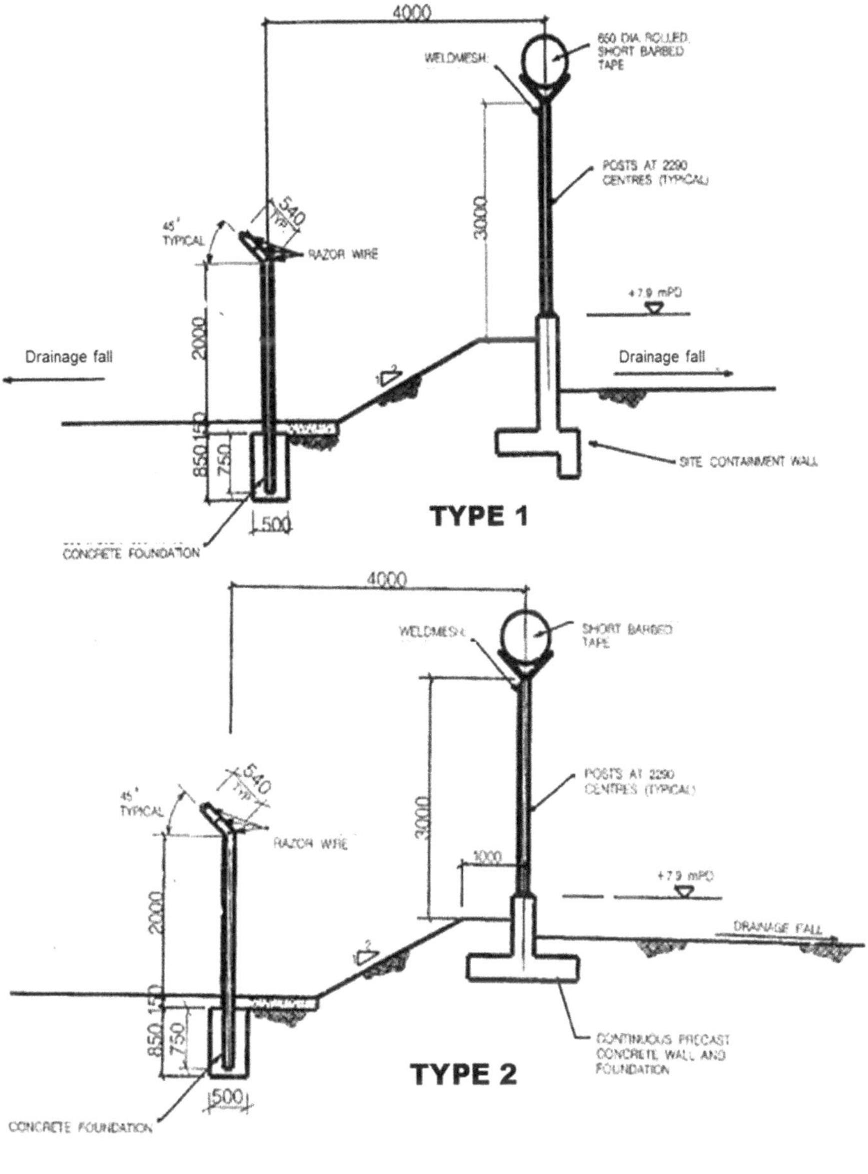

Figure 8-63. Typical oil storage tank bund wall and security fencing design

Concrete works very well for many liquids, but it is unsuitable for some applications like containing strong acids. Using earth berms for bunding is not recommended for most situations, though liners can be used to decrease permeability.

Large, exposed bunding will need a sump pump or some other system to remove precipitation, though it may also be used to transfer spilled liquid into another container. Rainwater must be treated if the liquid being stored is toxic because there may be small amounts of it surrounding the tank.

If the bund wall is over a meter high, it generally requires a ladder or steps to allow ease of escape from the bunded area. Another design uses a channel that drains the liquid to a secondary container.

8.7.3.3.3 Bunding Failures In 1919, a 15-m high molasses tank in Boston burst (Figure 8-64) killing 21 and injuring 150. A few other tanks have failed in a similar manner in the United States, but they have usually resulted in relatively few deaths.

Figure 8-64. Bund wall failure due to tank collapse, Boston 1919

However, a bunding's main purpose in unoccupied areas is to prevent environmental damage. Bund wall failures have occurred in the UK, such as at Warrington, Cheshire in 1994. A polypropylene tank holding about 30 tons of 40% aqueous caustic soda sprang a leak about halfway down the main wall. The corrosive fluid jetted out over the bund wall, causing great damage to the surrounding factory and adjacent premises. The bunding at Buncefield [2] also failed to contain gasoline and other hydrocarbon liquids.

8.7.4 Fabrication and Welding

8.7.4.1 Tank Construction — Fabrication and Welding

Figure 8-65 shows a three-dimensional structural view of a floating- roof tank, which consists of a flat roof, cylindrical shell, and flat bottom.

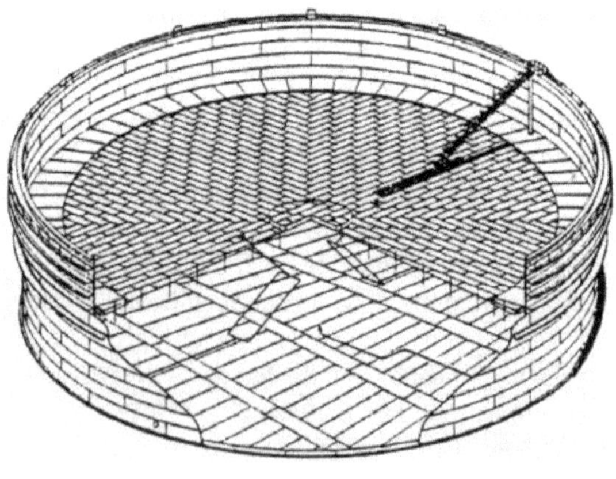

Figure 8-65. Schematic of constructing a typical floating-roof storage tank (for crude oil) [34]

Figure 8-66. Fabrication progress of a typical external floating roof tank

Typical fabrication of a floating roof tank is shown in Figure 8-66.

Depending on size and capacity, tank walls may have a constant thickness or a tapered wall with different values of the thickness at different elevations. The outer surface of the tank is smooth, and it is the surface where the thickness changes are observed. The maximum thickness is governed by the internal pressures and may be as high as 40 mm; however, in most tanks it is of the order of 10 mm.

Geometric parameters for typical tanks with floating roofs are summarized in Table 8-2 below.

Capacity of such larger tanks particularly for crude oil storage can be very high (about 200,000,000 l, 200,000 BBLS) (Figure 8-67).

The tank cylindrical shell, tank bottom and roof structure will have many plates that will require welding and several types of tanks may be involved.

The tank shell/cylinder itself is formed by curved plates that are welded. The quality of the welds is very important for the integrity of the shell, and several failures have been reported that initiated at welds in the lower part of the shell.

Processes and techniques must therefore be appropriate for welding of often very thick plates. Welding rods/wires must also be suitable for the technique such as sound welds are achieved.

TABLE 8-2. Floating roof tank, typical geometric parameters (* tapered shell thickness)

Tank Height, *H* [m]	Tank Diameter, *d* [m]	Shell Thickness, *t* [mm*]
10	20	6/6
	30	6/6
	40	8/8
	50	10/8
20	40	15/8
	60	23/8
	80	30/10
	100	40/10

Figure 8-67. Large storage tanks (Kharge Island, Persian Gulf)

8.7.4.2 Welding Techniques

The construction of large-scale cylindrical storage tanks requires automatic welding processes (Figure 8-68) such as submerged arc welding (SAW) and electrogas arc welding (EGW). In addition, manual shielded metal arc welding (SMAW) and semi-automatic gas metal arc welding (GMAW) may also be used according to the accessibility and applicability of the welding position and the targeted efficiency for individual welding joints, [34].

Table 8-3 below shows the typical welding joints for a crude oil storage tank, applicable welding positions and processes and well typical types of steel plates used for fabrication.

The welding Process (Notes: (1) HT steel: 550-610 MPa X80-90 HT steel (2) Dissimilar: Mild steel and 550-610 MPa, X80-90 HT steel), [34].

Figure 8-68. Automatic welding of large crude oil tanks (photo source: KHK, Safety & Tomorrow, Mar. 2000, [34])

TABLE 8-3. Main joints for cylindrical storage tank and suitable welding procedures (Reproduced from [34])

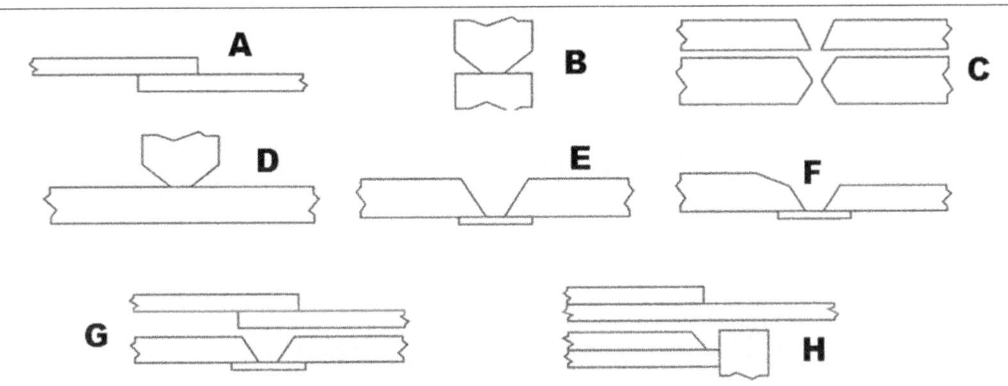

Configuration	Type of Joint	Weld Position	Welding Process	Steel Type(1)
A	Roof Plate	Horizintal Fillet	SMAW	Mild
B	Shell Plate	Horizontal	SAW-SMAW	Mild-HT
C	Shell Plate	Vertical	EGW-SMAW	Mild-HT
D	Shell Plate to Annular Plate	Horizintal Fillet	SAW-SMAW	Mild-HT
E	Annular Plate	Flat	SAW-SMAW	Mild-HT
F	Annular Plate to Bottom Plate	Flat	SAW-SMAW	Mild or Dissimilar (2)
G	Bottom Plate	Horizontal Fillet and Flat	SAW-SMAW	Mild-HT
H	Joint at Perifery of Nozzle and Manhole	Horizontal Fillet and Flat	SMAW-GMAW-SAW	Mild-HT

Some joint designs are indicated in Figure 8-69 below.

The following details the welding procedures for particular welding joints as referred to in the previous Table 8-3.

8.7.4.2.1 Shell Plate Horizontal Butt Joints: account for 90% of the total welding length of the shell, and the plate thickness of the joints is as high as 12 to 40 mm. Therefore, welding efficiency has a significant effect on the total construction cost.

To achieve high welding efficiency, the SAW process is usually carried out with special equipment for horizontal welding is generally used. Figure 8-70 outlines this process, in which a welding wire is fed at a certain angle into a granular flux that is sustained by conveyor tracking along the lower part of a double bevel groove. Figure 8-71 shows an application of this process at a construction site, in which the SAW equipment tracks along the shell plate.

This specific horizontal SAW process uses a particular flux and a thin solid wire (typically of about 3.2 mm Ø).

While SAW is the main welding process for shell plate horizontal welding in large-capacity cylindrical tanks, SMAW is equally indispensable. Because tack welding uses low hydrogen electrodes even for mild steel base metal, cold cracks are prevented by decreasing the diffusible hydrogen in the weld metal.

8.7.4.2.2 Shell Plate Vertical Butt: are typically welded by EGW, using portable equipment suitable for welding short lengths of large-capacity storage tanks. For this application, the SEGARC process (Figure 8-72) which features easy-to-handle equipment can be used.

As shown in Figure 8-72, the welding progresses while the weld pool is shielded with CO_2 gas and is dammed up by a water-cooled copper shoe on the front side and refractory backing (KL-4) or water-cooled copper backing on the backside of the welding joint. The welding head tracks during welding on a guide rail attached by a magnet on the surface of the base metal as shown in Figure 8-73.

- The SEGARC process is widely used in the construction of storage tanks due to the following outstanding characteristics of efficiency and operability:
- High deposition rates (e.g. 180 g/min at 380A) ensure high welding efficiency.
- Light-weight, compact equipment makes for easy set up.
- There is constant control of the wire extension in varied welding conditions.
- The welding line can be located either on the left side (Standard) or, by reassembling, the right side of the tracking rail.

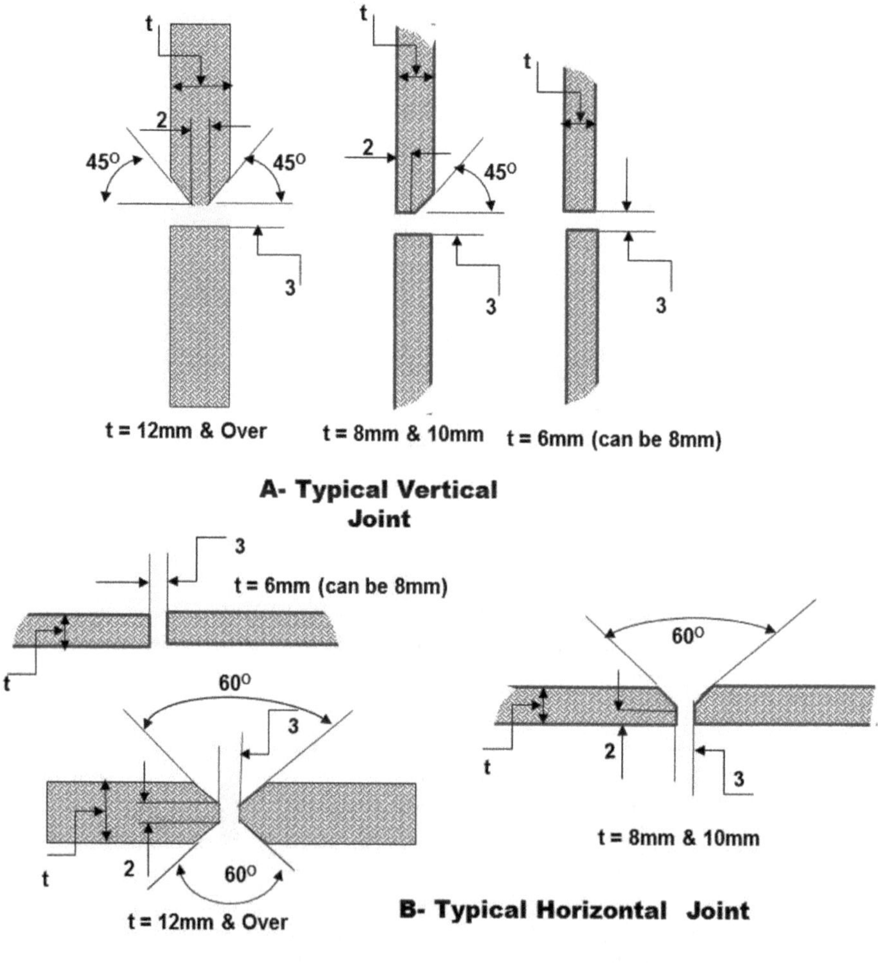

Figure 8-69. Typical vertical and horizontal weld joint design (all dimensions in mm) (Reproduced from [34])

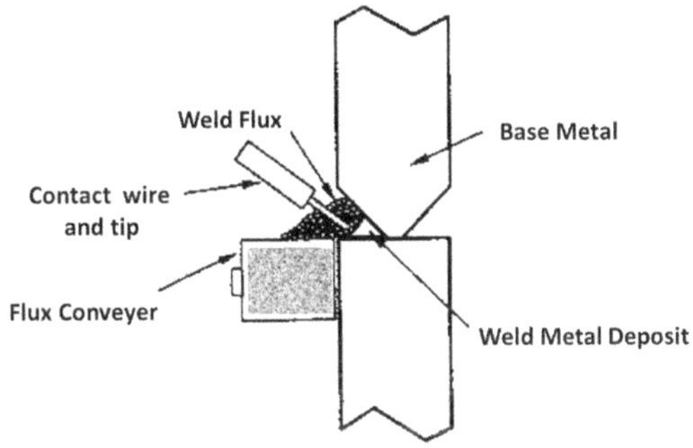

Figure 8-70. Outline of horizontal submerged arc welding (Reproduced from [34])

- With the oscillator (Optional), one-pass completion welding can be conducted for steel plates with a thickness of 32 mm max.

8.7.4.2.3 Shell to Annular Plate Tee Joints: can be subjected repetitively to severe bending stresses over the lifetime of the storage tank because of frequent loading and unloading of the liquid and uneven settling of the foundation under the tank. In addition, the welding of this joint is likely to be affected by sand, rust, dirt, oil, rain and dew because this joint is located close to the foundation at the construction site. Welding during fabrication must be conducted carefully in order to prevent welding defects and ensure the durability of the tank. In particular, root pass welds can easily contain porosity and cracks. To prevent these defects, the SMAW process is recommended for the root passes on the backing side and final side because it resists the difficulties in such a critical welding environment.

Figure 8-71. Application of horizontal submerged arc welding (photo source: KHK, Safety & Tomorrow, Mar. 2000 [34])

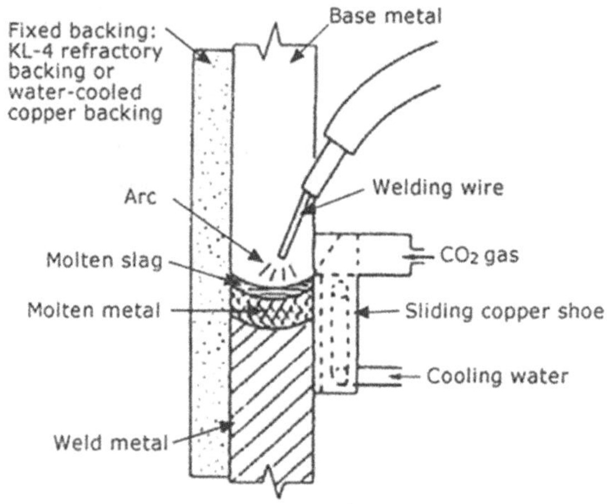

Figure 8-72. The SEGARC (submerged electro-gas arc welding) process for one-run vertical butt welding (KHK, Safety & Tomorrow, Mar. 2000 [34])

For the filling passes and the capping passes, SAW provides the highest efficiency. Due to the severe welding environment, the best flux-wire combination for this joint should emphasize porosity resistance.

With these flux-wire combinations, DCEP polarity will produce better bead appearance in single SAW than AC polarity. Figure 8-74 shows a typical weld pass sequence for this joint, combining SMAW for the root pass and single SAW for the filling and capping passes.

Figure 8-73. SEGARC (submerged electro-gas arc welding) process: a portable EGW process for vertical welding (Courtesy Kobe Steel [34])

8.7.4.2.4 Bottom Plate Joints (Figure 8-75): are typically made of mild steel lap joints in small-capacity storage tanks and butt joints with steel backing in 10,000 kl (10,000 BBLS) or larger storage tanks. As with the shell to annular plate joints, a similarly severe welding environment can lead to the inclusion of rust and dirt in the welding groove because the bottom plate rests directly upon the foundation of the tank. This is why SMAW is generally suggested for the root pass welds to minimize the occurrence of porosity. Because of its high efficiency, SAW is more beneficial for the filling and capping passes.

Bottoms are usually to either of the following two construction methods:

- Lap Welded Plates
- Butt welded Plates

The Lap welded plates must be reasonably rectangular and square edged. Three-plate laps are required not be closer than 300 mm from each other and also from the tank shell.

Plates are welded on top side only with a continuous fillet weld on all seams. Joints are lapped to 5 times the thickness of the thinner plate, but generally are not to exceed 25 mm (see Figure 8-75, Section BB).

Portion of the sketch plates shown in Figure 8-76A, coming under the bottom shell ring shall have the outer ends of the joints fitted and lap welded to form a smooth bearing for the shell plates, as shown in the same figure.

Bottom plate attachment with the shell plate may be made by an annular ring of segmental plates as shown in Figure 8-76B. Such annular rings (if used) must have their radial seams butt welded with a backing strip as shown in the same figure. Bottom sketch and rectangular plates can be lapped over the annular ring of segmental plates with the lap not less than five times the nominal thickness the thinner plates joined.

For bottoms butt welded construction, plates shall have the parallel edges prepared for butt welding with either square or V-grooves. If square grooves are employed, the root opening are generally required to be equal or over 6 mm. The butt welds may be

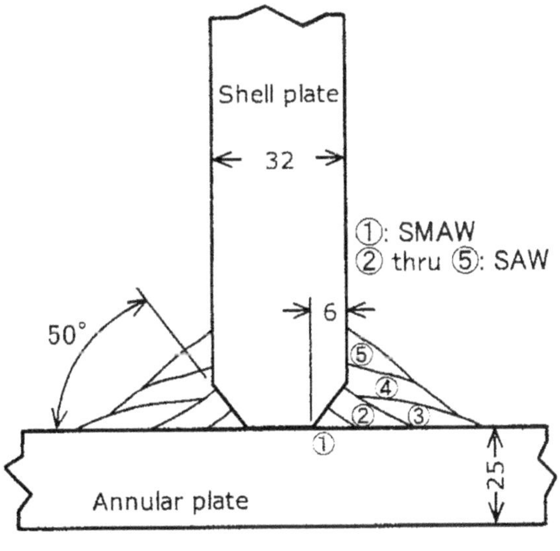

Figure 8-74. A typical pass sequence for shell plate to annular plate tee joint welded by using a combination of SMAW and single SAW processes [34]

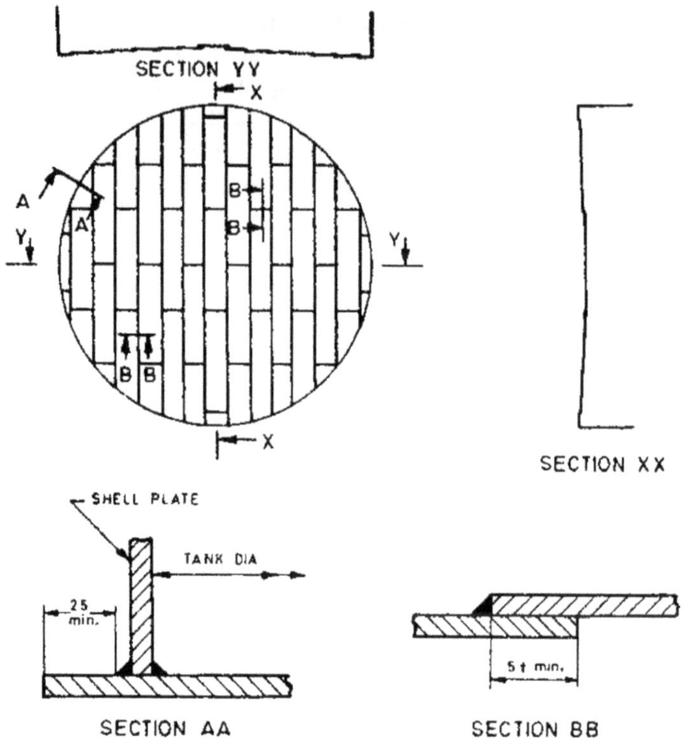

Figure 8-75. Typical layout of tank bottom and weld joint design (all dimensions in mm) [34]

made by applying a backing strip 3 mm thick or heavier by tack welding to the underside of the plate (see Figure 8-76B, Section XX). A metal spacer may be used, if necessary, to maintain the root opening between the adjacent plates.

In bottom plate butt welding three-plate joints are generally not be closer than 300 mm from each other and also from the tank shell.

Tank erectors may however propose other methods of butt welding the tank bottom.

8.7.4.2.5 Nozzle Peripheral and Manhole Joint (Figure 8-77): Some tanks are fitted with manholes. Such manholes are welded in the factory in advance. They can be welded by SAW using special automatic welding equipment that can track along the three-dimensional saddle-shaped welding line.

However, where such automatic welding equipment is not available, semi-automatic GMAW and SMAW are the second and third best processes respectively in terms of welding efficiency. These welds are postweld heat treated to remove the residual stresses; therefore, the filler metal should be selected taking into account the weld metal properties after the post-weld heat treatment at the specified temperature and soaking time.

8.7.4.3 Post Weld Heat Treatment of Welded Tanks Structures

Post Weld Heat Treatment (PWHT) is generally carried out to relieve residual stresses, remove diffusible hydrogen, and temper hard transformation microstructures of the weld, thereby preventing brittle fractures and obtaining the desired properties of the product. PWHT, however, can have some negative effects, such as stress relief cracking (SR Cracking) Rogantea et al. [35]. It may be noted that For High Strength Material with an Ultimate Tensile Strength > 415 MPa, shell openings ≥ 2" require a PWHT insert plate assembly in shells greater than 13 mm. For a material with a UTS < 415 MPa, A PWHT assembly is required for nozzles >12" in shell plates ≥ 25.4 mm.

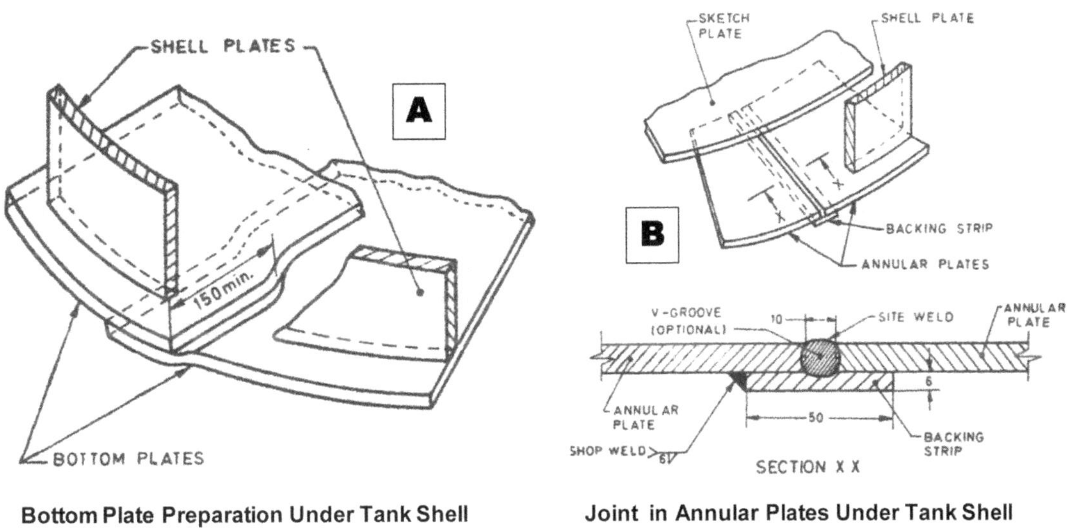

Figure 8-76. Tank bottom plate arrangement under tank shell [34]

SR cracking can become a problem particularly in the PWHT of high tensile strength steel, heat-resistant low alloy steel and stainless steel weldments. Figure 8-78 shows a typical example of SR Cracking that occurred in a 780-MPa high tensile strength steel weld that was heat treated at 600°C for 2 hours after welding. It is believed that this microscopic crack was initiated by the creep of the metal during relaxation of the residual stress at high temperatures particularly at the coarse grain area in the heat-affected zone (HAZ) at the toe

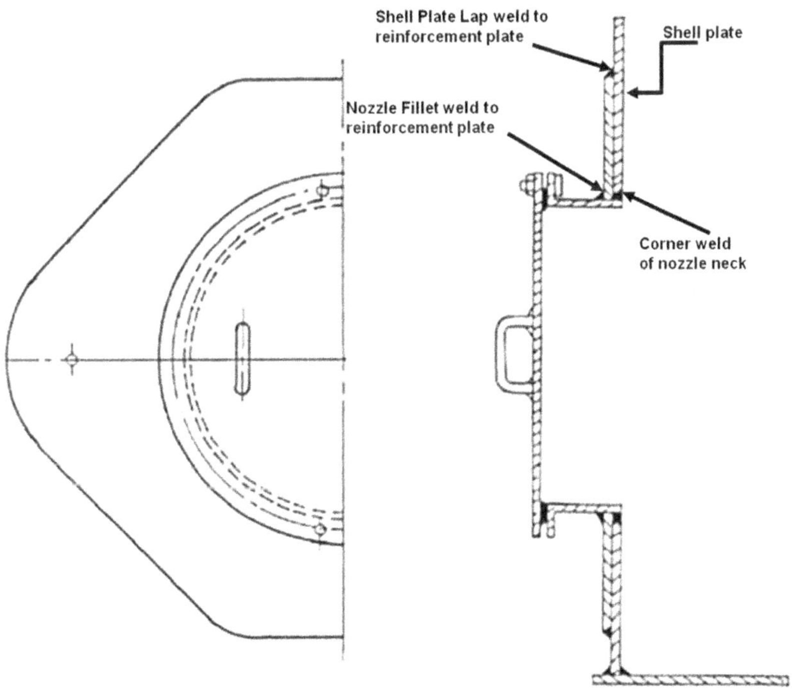

Figure 8-77. Welds at the periphery of a manhole in a tank shell plate [34]

of the weld. This is where the residual stresses are concentrated. And this crack propagates along the former austenite grain boundaries of the HAZ.

The SR-crack susceptibility of particular types of steel is governed by PWHT temperature and the alloying element. Figure 8-79 shows how SR crack susceptibility is affected by particular alloying elements contained in the testing steels and PWHT temperature. It clearly shows that the crack susceptibility becomes highest at 600°C. This is believed to be caused by the alloying element's carbide precipitation hardening of the crystal grains, thereby decreasing relatively the strength of the grain boundaries.

PSR cracking susceptibility index can be calculated as follows:

$$PSR(\%) = Cr + Cu + 2Mo + 10V + 7Nb + 5Ti - 2 \qquad (8-11)$$

where the applicable ranges of alloying elements are

- 1.5%Cr max,
- 0.10 to 0.25%C,
- 1.0%Cu max,
- 2.0%Mo max,
- 0.15%V max,
- 0.15%Nb max, and
- 0.15%Ti max.

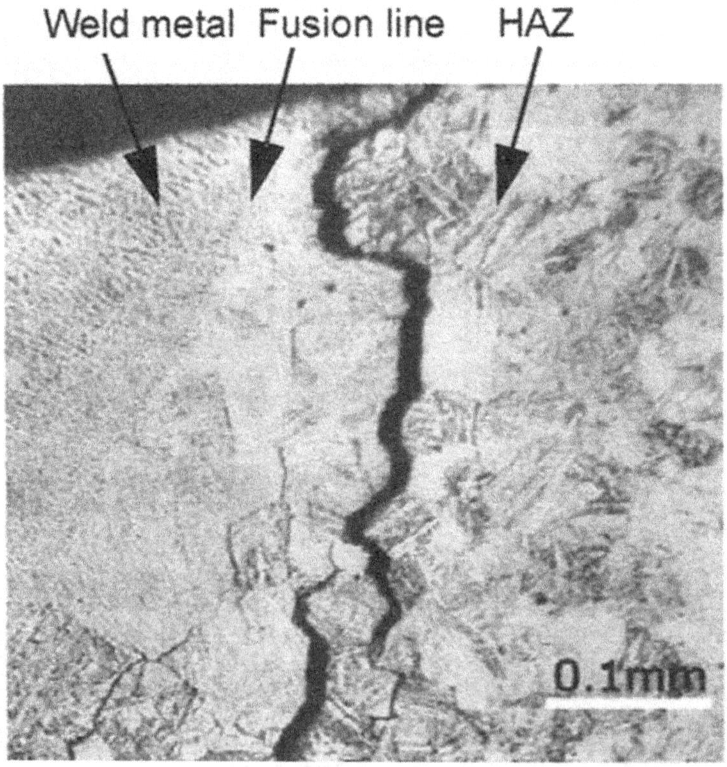

Figure 8-78. Typical SR cracks occurring in a 780-MPa high tensile strength steel weld (PWHT: 600°C × 2 h), KHK, Safety & Tomorrow, Mar. 2000 [34], http://www.kobelco.co.jp/english/welding/files/kwt2006-01.pdf

It is believed that where PSR is larger than zero SR crack can occur.
To avoid SR Cracking, the following measures may be taken:

1. Select a less susceptible steel taking into account, for instance, the SR cracking susceptibility index (PSR).
2. Refine the coarse grain HAZ at the toe of the weld by applying the temper bead technique.
3. Dress the weld metal to smoothen the transition to the surface of the base metal, or remove the reinforcement of the weld metal to be flush against the surface of the base metal to minimize or remove the site of stress concentration.
4. Avoid lapping a fillet weld onto a butt joint weld to prevent excessive residual stresses and stress concentration.
5. Avoid joining components of excessively dissimilar thicknesses to prevent the high concentration of stress.

8.7.4.4 Construction of Spheres

Figures 8-80 (A, B and C) through 8-81 [7], show the process of fabricating and welding spheres for the purpose of storing high vapor pressures hydrocarbons, such as LPG/Propane, etc. Typical LPG Sphere Fabrication and Completion is shown in Figure 8-82.

8.7.5 Mechanical/Piping Components and Instrumentation

8.7.5.1 Mechanical Appurtenances

Typical mechanical appurtenances and piping that make up a storage tank include the following:

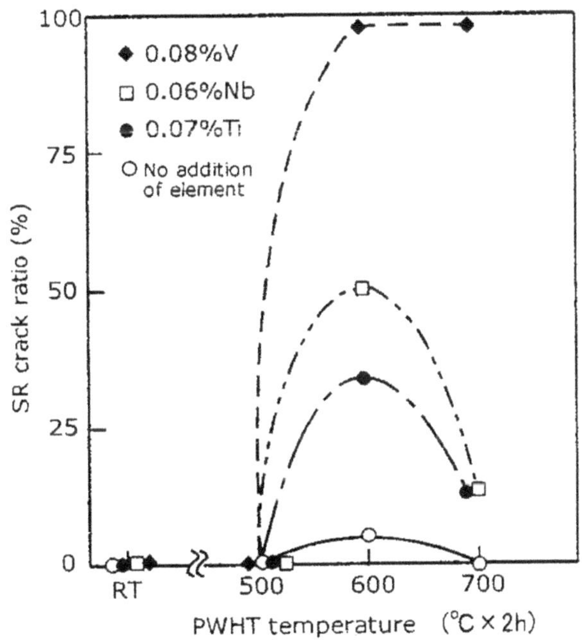

Figure 8-79. SR crack susceptibility of Cr-Mo steel (0.16%C, 0.30%Si, 0.60%Mn, 0.99%Cr, 0.46%Mo) as a function of PWHT temperature and additional alloying elements in groove restraint cracking test [31, 36]

- Gauge well pipe
- Gauge Well Sleeve
- Shell Nozzle pipe
- Water draw off nozzle
- Roof nozzle pipe
- Temperature hatch
- Pipe support sleeve
- Pipe support legs
- Auto bleeder vent pipe
- Auto bleeder vent sleeve
- Emergency drain pipe
- Roof drain pipe
- Fire water line
- Foam Water line
- Swivel for Primary Roof Drain
- Primary & Secondary Seal
- Foam & Water System- FR Tank
- Pressure Relief Lines
- Rolling ladder

Some of the above mechanical components along with other fitting arrangements are illustrated in previous Figures 8-33 to 8-43 (see Section 8.4.4).

8.7.5.2 Instrumentation and Controls

Storage tank instrumentation includes all gauges (sampling, temperature, pressure, densitometer, viscometer level/volume/mass measurements) and instruments (including rain and wind gauges) for hydrocarbon liquid handling and controls. It can also

Figure 8-80. A — Plate pressing [7]

Figure 8-80B. Plate cutting

Figure 8-80C. Bending/fabrication [7]

Figure 8-81. Bending, lifting, aligning, and welding [7]

Figure 8-82. Typical LPG sphere fabrication and completion [7]

include leak detection system with the appropriate software for data analysis and reporting and inventory management. Instrumentation will also include receiving alarms and the activation and control of fire and fire suppressant systems and for security requirements.

Inventory control of the hydrocarbon products in a pipeline tank farm is very important and must be carried out continuously to assist pipeline operation, particularly a batched pipeline system.

In a pipeline tank farm tank level measurement is integrated seamlessly into the automation system (Figure 8-83). Measurement hardware products could include

- Level Gauging through
  - Servo
  - Radar
- Tank Fittings
- Tank Field Control

Some of the other measurement and controls include:

- Measurement and control of received and dispatched products
- Vapor recovery
- Measurement of storage quantities for product inventory
- Loading technology, volume measurement and connection to automation system
- Access control and identification systems
- Special loading technology for railcars and barges (if applicable particularly at delivery ends of pipelines)
- Systems for inline and batch blending
- Additive Injection
- Pump units, systems and control systems for hydrocarbon product mainline transportation or terminal tank transfers and blending
- Valve opening and closure controls for product receipt and shipping as well as in terminal tank transfers
- Control of all product reception, inventory and delivery

8.7.6 Tank Venting Emission Calculations

Emissions from hydrocarbon liquids in storage occur because of evaporative loss of the liquid during its storage and as a result of changes in the liquid level. The emission sources vary with tank design, as does the relative contribution of each type of emission source.

Emissions from fixed roof tanks are a result of evaporative losses during storage (known as breathing losses or standing storage losses) and evaporative losses during filling and emptying operations (known as working losses).

External and internal floating roof tanks are emission sources because of evaporative losses that occur during standing storage and withdrawal of liquid from the tank. Standing storage losses are a result of evaporative losses through rim seals, deck fittings, and/or deck seams.

EPA document AP-42 [33], provides a detailed method for calculating emissions losses from single-wall storage tanks, both internal and external floating roof tanks. AP-42 can be viewed under section "Organic Liquid Storage Tanks" (Background

Figure 8-83. A typical level gauging system incorporated seamlessly into terminal automation systems [37]

Document) on the U.S. EPA's website at http://www.epa.gov/ttn/chief/ap42/ch07/index.html.

The following provide a stepwise approach of calculating emission factors for single-wall (Non-Insulated) and protected (Insulated) ASTs using AP-42. Only calculation of losses for fixed roof tanks are included in details. However, reference is also made attributed to floating roof tanks.

8.7.6.1 Total Losses from Fixed Roof Storage Tanks

The following equations apply to horizontal aboveground storage tanks (ASTs) that store organic liquids i.e. gasoline. These tanks must be substantially liquid and vapor-tight and must operate at atmospheric pressure. Total losses from ASTs are equal to the sum of the standing storage loss and working loss calculated for each month:

$$L_T = L_S + L_W \qquad (8-12)$$

where

L_T = total losses, lb/year
L_S = standing storage/breathing loss, lb/year
L_W = working loss, lb/year

Standing Storage or Breathing Loss is calculated as follows

$$L_S = nV_V W_V K_E K_S \qquad (8-13)$$

where

L_S = Standing storage losses, lb/month
n = number of days in the respective each year (usually 365, a constant)

V_v = vapor space volume of the ullage, ft^3
W_v = vapor density, lb/ft^3
K_E = vapor space expansion factor, dimensionless
K_S = vented vapor saturation factor, dimensionless

Tank vapor space volume, V_V, is considered to be equal to the ullage volume and is estimated as:

$$V_V = \left(\pi D^2/4\right) H_{VO} \quad (8-14)$$

where
V_V = vapor space volume, ft^3
D = tank diameter, ft,
H_{VO} = vapor space outage, ft, see Eq. (8-15)

The vapor space outage, H_{VO} is the height of a cylinder of tank of diameter D, whose volume is equivalent to the vapor space volume of a fixed roof tank, including the volume under the cone or dome roof.

The vapor space outage, H_{VO}, can be estimated from:

$$H_{VO} = H_S - H_L + H_{RO} \quad (8-15)$$

where
H_{VO} = vapor space outage, ft
H_S = tank shell height, ft
H_L = liquid height, ft
H_{RO} = roof outage, ft;

For a cone roof, the roof outage, H_{RO}, is calculated as follows:

$$H_{RO} = 1/3 H_R \quad (8-16)$$

where
H_{RO} = roof outage (or shell height equivalent to the volume contained under the roof), ft
H_R = tank roof height, ft

$$H_R = S_R R_S \quad (8-17)$$

where
S_R = tank cone roof slope, ft/ft; if unknown, a standard value of 0.0625 can be used
R_S = tank shell radius, ft

For a dome roof, the roof outage, H_{RO}, is calculated as follows

$$H_{RO} = H_R\left[\tfrac{1}{2} + 1/6\left(H_R/R_S\right)^2\right] \quad (8-18)$$

where
H_{RO} = roof outage, ft
R_S = tank shell radius, ft
H_R = tank roof height, ft

$$H_R = R_R - \left(R_R^2 - R_R^2\right)^{0.5} \quad (8-19)$$

H_R = tank roof height, ft
R_R = tank dome roof radius, ft
R_S = tank shell radius, ft

The value of R_R usually ranges from 0.8D to 1.2D, where $D = 2\ R_S$. If R_R is unknown, the tank diameter is used in its place. If the tank diameter is used as the value for R_R,

Equations (8-18) and (8-19) reduce to

$$H_{RO} = 0.137R_S \text{ and } H_R = 0.268R_S. \quad (8-20)$$

Alternatively the following approximate formulae can be used

$$V_V = ½ \text{ tank capacity (ft}^3\text{)}$$

Example: Tank capacity range = 751 to 1000 gallons
Average tank capacity = 875.5 gallons = 875.5/7.481 = 117 ft^3

$$V_v = 1/2*117 = 58.5 \text{ ft}^3$$

The standing storage loss L_S equation can be simplified by combining Eq. (8-12) with Eq. (8-13) and consider n = 365. The result is

$$L_S = 365\left(\pi D^2/4\right)H_{VO}K_S W_V \quad (8-21)$$

Vapor Space Expansion Factor, KE (of Eq. (8-12)

The calculation of the vapor space expansion factor, K_E, depends upon the properties of the liquid in the tank and the breather vent settings. If the liquid product has a true vapor pressure greater than 0.1 psia, or if the breather vent settings are higher than the typical range of ±0.03 psig. If the liquid stored in the fixed roof tank has a true vapor pressure less than 0.1 psia and the tank breather vent settings are ±0.03 psig, use either Eq. (8-22) or Eq. (8-23).

If the tank location and tank color and condition are known, K_E can be calculated using the following equation:

$$K_E = 0.0018\Delta T_V = 0.0018\left[0.72\left(T_{AX} - T_{AN}\right) + 0.028\alpha I\right] \quad (8-22)$$

where

K_E – vapor space expansion factor, dimensionless
ΔT_V = daily vapor temperature range, °R
T_{AX} = daily maximum ambient temperature, °R
T_{AN} = daily minimum ambient temperature, °R
α = tank paint solar absorptance, dimensionless
I = daily total solar insolation on a horizontal surface, Btu/(ft^2 day)
0.0018 = constant, (°R)$^{-1}$
0.72 = constant, dimensionless
0.028 = constant, (°R ft^2 day)/Btu

If the tank location is unknown, a value of K_E can be calculated using typical meteorological conditions for the area. For example, the typical value for daily solar insolation in the USA is 1370 Btu/(ft^2 day), the daily range of ambient temperature is 21°R, the daily minimum ambient temperature is 473.5°R, and the tank paint solar absorptance is 0.17 for white paint in good condition.

Substituting these values into Eq. (8-11) results in a value of 0.04, as shown in Eq. (7-12) below.

$$K_E = 0.04 \tag{8 – 23}$$

However, when the liquid stock has a true vapor pressure greater than 0.1 psia, a more accurate estimate of the vapor space expansion factor, K_E, is obtained by Eq. (8-24) below. As shown in the equation, K_E is greater than zero. If K_E is less than zero, standing storage losses will not occur.

$$K_E = \left[\left(\Delta T_V / T_{LA} \right) \right] + \left[\left(\Delta P_V - \Delta P_B \right) / \left(14.7 - P_{VA} \right) \right] \tag{8 – 24}$$

where

K_E = vapor space expansion factor, dimensionless
ΔT_V = daily vapor temperature range, °R
ΔP_V = daily vapor pressure range, psi
ΔP_B = breather vent pressure setting range, psi
14.7 = atmospheric pressure, psi
P_{VA} = vapor pressure at daily average liquid surface temperature, psi (derived earlier)
T_{LA} = daily average liquid (gasoline) surface temperature, °R (derived earlier)

- **The daily vapor temperature range, ΔT_V,** is calculated below using the daily maximum and daily minimum liquid (gasoline) surface temperatures (derived earlier):
  Assumption: The vapor temperature range is equal to the liquid (gasoline) surface temperature range.

$$\Delta T_V = T_{LX} - T_{LN} \tag{8 – 25}$$

- **The daily vapor pressure range, P_V,** is calculated using the following equation:

$$\Delta P_V = P_{VX} - P_{VN} \tag{8 – 26}$$

where

P_{VX} = vapor pressure P_{VA} at daily maximum liquid (gasoline) surface temperature, psi
P_{VN} = vapor pressure P_{VA} at daily minimum liquid (gasoline) surface temperature, psi

Using the daily maximum and daily minimum liquid (gasoline) surface temperatures, the respective vapor pressures can be calculated as:

$$P_{VX} = \exp\left[A - \left(B / T_{LX} \right) \right] \tag{8 – 27}$$

$$P_{VN} = \exp\left[A - \left(B/T_{LN}\right)\right] \tag{8-28}$$

where

RVP = 7 psi, A = 11.83 and B = 5500.90°R
RVP = 9 psi, A = 11.75 and B = 5314.31°R

- **The breather vent pressure setting range, delta P_B,** is calculated using the following equation:

$$\Delta P_B = P_{BP} - P_{BV} \tag{8-29}$$

where

ΔP_B = breather vent range (psi)
P_{BP} = breather vent pressure setting (psi)
P_{BV} = breather vent vacuum setting (psi)

For ASTs with a pressure/vacuum vent valve
P_{BP} = 2 in $H_2O \cong 0.0722$ psi
P_{BV} = –4 in $H_2O \cong -0.1444$ psi

For ASTs with no pressure/vacuum vent valve
P_{BP} = 0 in $H_2O \cong 0$ psi
P_{BV} = 0 in $H_2O \cong 0$ psi

- **The vapor pressure at daily average liquid (gasoline) surface temperature, P_{VA},** is calculated as shown earlier.
- **The daily average liquid (gasoline) surface temperature, T_{LA},** is calculated as shown earlier.

Vented Vapor Saturation Factor, K_S — The vented vapor saturation factor is calculated using the following equation:

$$K_S = \frac{1}{1 + \left(0.053 * P_{VA} * H_{VO}\right)} = 1 \tag{8-30}$$

where

K_S = dimensionless factor
P_{VA} = vapor pressure at daily average fuel surface temperature, psi
H_{VO} = vapor space outage = 0 ft
(as mentioned earlier, the ASTs being considered in this evaluation are horizontal with no roof outage or vapor space outage factor)

Calculation of Vapor Properties W_V

Vapor Density, W_V, is the density of the vapor and is calculated using the following equation:

$$W_V = M_V P_{VA} / R T_{LA} \tag{8-31}$$

where

W_V = vapor density, lb/ft^3
M_V = vapor molecular weight, lb/lb-mole
R = the ideal gas constant, 10.731 psia·ft^3/lb-mole·°R
P_{VA} = vapor pressure at daily average liquid-surface temperature, psia

T_{LA} = daily average liquid (gasoline/hydrocarbon) surface temperature, °R

- **Molecular weight of the vapor (M_V)** is obtained from the Table 8-4 below the physical properties of gasoline. The molecular weight of gasoline changes with the change in Reid Vapor Pressure (RVP).

Example: The RVP of gasoline for the summer months (April to October) is 7.0 psi and for the winter months (November to March) is 9.0 psi. Listed below would be the molecular weight of gasoline for each corresponding RVP:

April to October — RVP = 7 psi, $M_V = 68$
November to March — RVP = 9 psi, $M_V = 67$

- **True vapor pressure (P_{VA})** of gasoline products, at the daily average liquid surface temperature, can be determined using the following equation:

$$PV_A = \exp\left[A - \left(B / T_{LA}\right)\right] \qquad (8-32)$$

where

exp = exponential function
T_{LA} = daily average liquid (gasoline) surface temperature, °R
Figure 3-5 in AP-42 [38] shows the equations that can be used to determine vapor pressure constants, A (dimensionless) and B (°R) for each corresponding RVP of gasoline:

Example:
RVP = 7 psi, A = 11.83 and B = 5500.90,°R
RVP = 9 psi, A = 11.75 and B = 5314.31,°R

- **Daily average liquid (gasoline) surface temperature (T_{LA})** is calculated using the following equation:

$$T_{LA} = \left(T_{LN} + T_{LX}\right)/2 \qquad (8-33)$$

TABLE 8-4. Properties (M_V, P_{VA}, W_L) of typical petroleum liquids (reproduced from GPSA [30])

Hydrocarbon Liquid	Vapor Mole Weight (60°F) 1b/1b-mole	Liquid Density (60°F) lb/gal	True Vapor Pressure PVA (PSI)						
			40°F	50°F	60°F	70°F	80°F	90°F	100°F
Crude RVP 5	50	7.1	1.8	2.3	2.8	3.4	4	4.8	5.7
Fuel Oil #2, (Distillated)	130	7.1	0.0031	0.0045	0.0065	0.009	0.012	0.016	0.022
Gasoline RVP 7	68	5.6	2.3	2.9	3.5	4.3	5.2	62	7.4
Gasoline RVP 7.8	68	5.6	2.5929	3.2079	3.9363	4.793	5.7937	6.9552	8.2952
Gasoline RVP 8.3	68	5.6	2.7888	3.444	4.2188	5.1284	6.1891	7.4184	8.8344
Gasoline RVP 10	65	5.6	3.4	4.2	5.2	6.2	7.4	8.8	10.5
Gasoline RVP 11.5	62	5.6	4.087	4.9997	6.069	7.3132	8.7519	10.4053	12.2949
Gasoline RVP 13	62	5.6	4.7	5.7	6.9	8.3	9.9	11.7	13.8
Gasoline RVP 13.5	62	5.6	4.932	6.0054	7.2573	8.7076	10.3774	12.2888	14.4646
Gasoline RVP 15	60	5.6	5.58.02	6.774	8.1621	9.7656	11.6067	13.7085	16.0948
Jet Kerosene	130	7	0.0041	0.006	0.0085	0.011	0.015	0.021	0.029
Jet Naphta (JP -4)	80	6.4	0.8	1	1.3	1.6	1.9	2.4	2.7
Naphta Oil #6	190	7.9	0.00002	0.00003	0.00004	0.00006	0.00009	0.00013	0.00019

where

T_{LN} = daily minimum liquid (gasoline) surface temperature, °R
T_{LX} = daily maximum liquid (gasoline) surface temperature, °R

These values can be determined from ambient temperature data to calculate the average temperatures as follows:

$T_{amb.avg}$ = daily ambient average temperature, °F
$T_{amb.range}$ = daily ambient temperature range, °F or °R

Air Resources Board of Canada [39] conducted a study determining the effect of daily/diurnal ambient temperatures on gasoline surface temperatures (T_{LN} and T_{LX}) (Figure 8-84). The results indicate the following:

Single Wall (Non-Insulated) ASTs: Gasoline surface temperature range = Ambient temperature range attenuated by factor 0.17.
Protected (Insulated) ASTs: Gasoline surface temperature range = Ambient temperature range attenuated by factor 0.80.

The study indicated that diurnal changes in ambient temperature have a very small effect on the gasoline surface temperature.

Using the above attenuation factors and ambient temperature data, the daily minimum and maximum liquid (gasoline) surface temperatures can be estimated for both non-insulated and insulated ASTs as follows: Note that amb.avg temperature is converted from °F to °R by adding 460.

The daily minimum liquid (gasoline) surface temperature (T_{LN}) can be calculated as:

$$T_{LN} = \left[T_{amb.avg} + 460\right] - \left[\left(1 - \text{Attenuation factor}\right) \times \left(T_{ambrange}/2\right)\right] \quad (8-34)$$

The daily maximum liquid (gasoline) surface temperature (T_{LX}) can be calculated as:

$$T_{LX} = \left[T_{amb.avg} + 460\right] + \left[\left(1 - \text{Attenuation factor}\right) \times \left(T_{amb.range}/2\right)\right] \quad (8-35)$$

Working Loss (L_W), Eq. (7-12)

The working loss, L_W, refers to the loss of product vapors as a result of tank filling or emptying operations. Fixed roof tank working losses can be estimated from:

The general equation for working loss is as follows:

$$L_W = 0.0010 M_V \, P_{VA} \, Q \, K_N \, K_P \quad (8-36)$$

where

L_W = working loss, lb/yr
M_V = vapor molecular weight, lb/lb-mole; (M_V) can be obtained from the Table 8-4 [38]. It may be noted that the molecular weight of gasoline changes with the change in Reid Vapor Pressure (RVP).
P_{VA} = vapor pressure at daily average liquid surface temperature, psia; see Eq. (7-20)
Q = annual net throughput (tank capacity [bbl] times annual turnover rate), bbl/yr

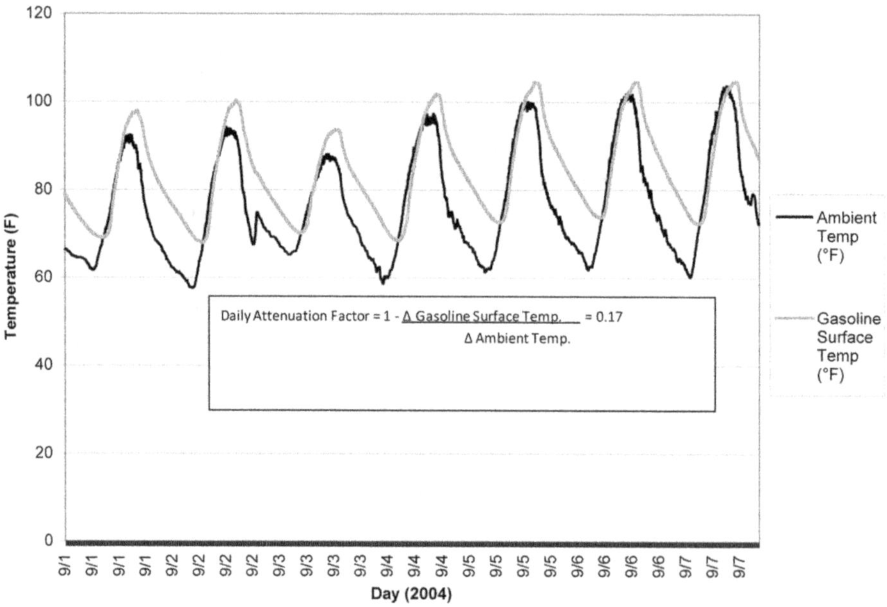

Figure 8-84. Ambient temperature compared to gasoline surface temperature in a single-wall (non-insulated) storage tank [39]

K_N = working loss turnover (saturation) factor, dimensionless
for turnovers > 36, $K_N = (180 + N)/6N$
for turnovers ≤ 36, $K_N = 1$
N = number of turnovers per year, dimensionless

$$N = 5.614\ Q/V_{LX} \qquad (8-37)$$

where
V_{LX} = tank maximum liquid volume, ft^3

$$V_{LX} = \pi/4D^2 H_{LX} \qquad (8-38)$$

D = diameter, ft
H_{LX} = maximum liquid height, ft
KP = working loss product factor, dimensionless for crude oils KP = 0.75
for all other organic liquids, KP = 1

Using the following steps, Eq. (7-36) can be simplified to combine all variables into one equation.

Using Eq. (8-31), the term "$M_V P_{VA}$" can be replaced with Eq. (8-38).

$$M_V P_{VA} = W_V\, R\, T_{LA} \qquad (8-39)$$

Using a combination of Eq. (7-37) and Eq. (7-38), the term "Q" can be replaced with the following Eq. (8-40).

$$Q = \left[N H_{LX}/5.614\right]\left[(\pi/4) D^2\right] \qquad (8-40)$$

Assuming a standard value of R to be 10.731 ft^3 psia/(lb-mole °R), the result is Eq. (7-40).

$$L_W = (0.001/6.614)(10.731) T_{LA} N H_{LX} (\pi/4) D^2 K_N K_P W_V K_B \qquad (8-41)$$

The above assumes that vent setting $K_B = 1$

where

- L_W = working loss, lb/yr
- N = number of turnovers per year, (year)$^{-1}$
- H_{LX} = maximum liquid height, ft
- D = diameter, ft
- K_N = working loss turnover (saturation) factor, dimensionless; see Figure 8-85 below
  for turnovers > 36, KN = (180 + N)/6N
  for turnovers ≤ 36, KN = 1
- K_P = working loss product factor, dimensionless for crude oils KP = 0.75 for all other organic liquids, KP = 1
- W_V = vapor density, lb/ft^3, see Eq. (8-31)
- K_B = vent setting correction factor, dimensionless for open vents and for a vent setting range up to ±0.03 psig, $K_B = 1$

8.7.6.2 Total Losses from Floating Roof Tanks

External Floating Roofs: For external floating roof tanks, the majority of rim seal vapor losses are wind induced. No dominant wind loss mechanism are reported by the industry for internal floating roof or domed external floating roof tank rim seals.

Rim losses can also occur due to permeation of the rim seal material by the vapor or via a wicking effect of the liquid, but permeation of the rim seal material generally does not occur if the correct seal fabric is used. Testing has indicated that breathing, solubility, and wicking loss mechanisms are small in comparison to the wind-induced loss.

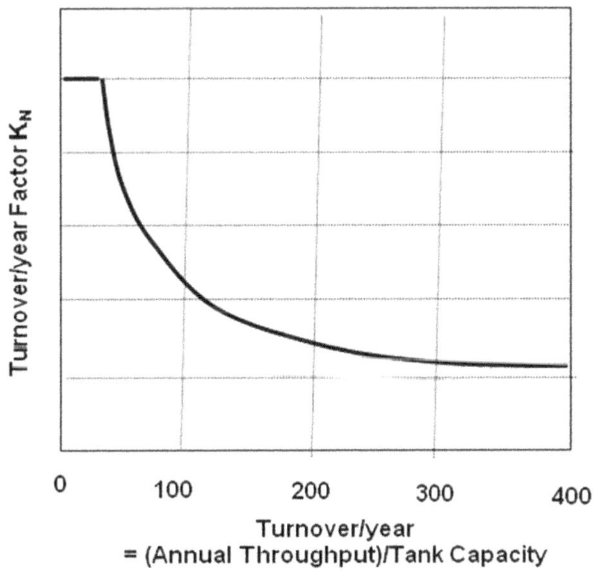

Figure 8-85. Turn over factors for fixed roof tanks (note: for 36 turnovers/year, or loss, K_N = 1.0)

A test method to estimates evaporation losses for floating rood tank turnover is describes by Sung [40].

Internal Floating Roofs: Deck seams in internal floating roof tanks are a source of emissions to the extent that these seams may not be completely vapor tight if the deck is not welded. Generally, the same loss mechanisms for fittings apply to deck seams. The predominant mechanism depends on whether or not the deck is in contact with the stored liquid. The deck seam loss equation accounts for the effects of all contributing loss mechanisms.

Total Losses Calculations: Total floating roof tank emissions are the sum of rim seal, withdrawal, deck fitting, and deck seam losses. The equations presented in this subsection apply only to floating roof tanks. The equations are not intended to be used in the following applications:

1. To estimate losses from unstable or boiling stocks or from mixtures of hydrocarbons or petrochemicals for which the vapor pressure is not known or cannot readily be predicted;
2. To estimate losses from closed internal or closed domed external floating roof tanks (tanks vented only through a pressure/vacuum vent); or
3. To estimate losses from tanks in which the materials used in the rim seal and/or deck fittings are either deteriorated or significantly permeated by the stored liquid.

Under normal operation, total losses from floating roof tanks may be written as:

$$L_T = L_R + L_{WD} + L_F + L_D \qquad (8-42)$$

where

L_T = total loss, lb/yr

L_R = rim seal loss, lb/yr;

L_{WD} = withdrawal loss, lb/yr;

L_F = deck fitting loss, lb/yr;

L_D = deck seam loss (internal floating roof tanks only), lb/yr.

Loss factors may be estimated for deck fitting configurations at the zero miles-per-hour wind speed condition (for internal floating roof tanks-IFRTs and coned floating roof tanks CFRTs), using standard AP-42 [38].

8.7.7 Operational Issues

There are number of issues related to the physical operation and maintenance of hydrocarbon storage tanks. Some of these issues are described below.

Storm Water on Floating Roof Tank - Drainage Operation: Storm water accumulated on a concave floating roof of an oil storage tank may affect its floatation, hence its operation, making it necessary to immediately drain the water. This is usually done through a flexible pipe, running from the floating roof down the tank, to an outlet above the ground near the bottom of the tank (Figure 8-86).

The following problems may occur in the operation of a floating roof drainage system:

- Small or large volumes of the product from the tank can penetrate the flexible pipe through pinholes or cracks that may appear in it. The product will then drain through the water drainage system unnoticed.

- Product from the tank may occasionally run over the floating roof through the roof's seal and exit through the water drainage pipe unnoticed.
- Sometimes the flexible pipe is bent or clogged preventing water from the roof to pass through it. In this case, water remains on the floating roof and this may disturb its floatation capability.

For safety and to avoid the above challenges the following are typically implemented:

1. **Installation of Oily Sheen/Water Detector:** An oil sheen leak detector (such as Leakwise ID-223 Oil Sheen Detector) can be installed in a settling tank (or in a sump or a separator) on the external water drainage pipeline. This settling tank will settle the liquid flowing in the pipeline from the roof drainage pipe outlet to allow detection by the detectors floating sensor. Normal indication of the such a sensor should be water on rainy days, or air on dry days. An alarm will be triggered if the detector detects oil or oil on water, indicating that oil is seeping through the roof drainage pipe (Figure 8-87).
2. **Detection of Clogs in the Roof Water Drainage System:** If air instead of water is indicated after a rainy day, it means that the flexible roof drainage pipe is bent or clogged and no water is running through it.
3. **Savings on Water Treatment Costs:** An oil sheen leak detector (such as the Leakwise ID-223 Oil Sheen Detector) can be used to control valves, pumps, and sump gates. By utilizing the control capabilities of detector sensors, the users can decide on the discharge of the storm water from the tank's roof directly to the sea, river, or public drainage system. Only the oily water will be diverted to treatment. This reduces the load from the local treatment system and brings substantial cost savings.

Figure 8-86. Problems in roof drainage operation — floating roof tank

Tank Overfill and Spill: Most tanks are equipped with high level alarm systems to reduce the risk of spills. However, the greatest concern with a tank is the possibility of an overflow spill due to instrumentation malfunction or operator error.

An ignited overflow spill will result in a major fire in the dike/bunded area around the tank as well as a fire at the surface level of the tank.

It is always important to control the spill fire before attempting to control the tank surface fire, since a spill fire would continue to re-ignite vapors coming from the tank. With the spill fire contained, efforts can be made to control the tank surface fire. Industry's practice is to secure the spill area with a foam blanket for the duration of the incident.

A floating roof, when in place, limits the amount of surface available to support a fire. If the floating roof is lost, due to an explosion or sinking, the fire can be expected to intensify rapidly.

Floating roof tanks may experience "rim seal" fires. These fires involve the seal area between the floating roof and the inside wall of the storage tank. The seals are typically made from synthetic rubber or plastic. The seal is about 200-300 mm (8 to 12 in.) wide, depending on the type of tank. These fires are generally controllable with small hand-lines or dry chemical extinguishers. In a cone/floating roof tank, the fire may be contained in the space between the floating roof and the fixed roof, out of the reach of hose streams.

When applying foam, care must be taken to avoid sinking or tilting the "floater." Therefore it is important that no plain water is applied onto a floating roof as it could cause the roof to sink.

Entry onto the roof of a floating roof tank can only occur after assessing the risk to the personnel and determining if the roof is constructed using an "inherently buoyant design." Floating roofs that are inherently buoyant are constructed of steel with multiple bulkheads that form liquid tight compartments.

The position of the floating roof in relation to the top of the tank shell should also be assessed. A floating roof that is more than 1.5 m (5 feet) below the top of the tank shell constitutes a confined space. If entry onto the roof is being considered as part of the incident strategy, the procedures for confined space operations must be applied.

The industry prefers the use of subsurface foam injection, topside fixed foam chambers, and hydro-foam monitor nozzles as a method to control a full surface fire.

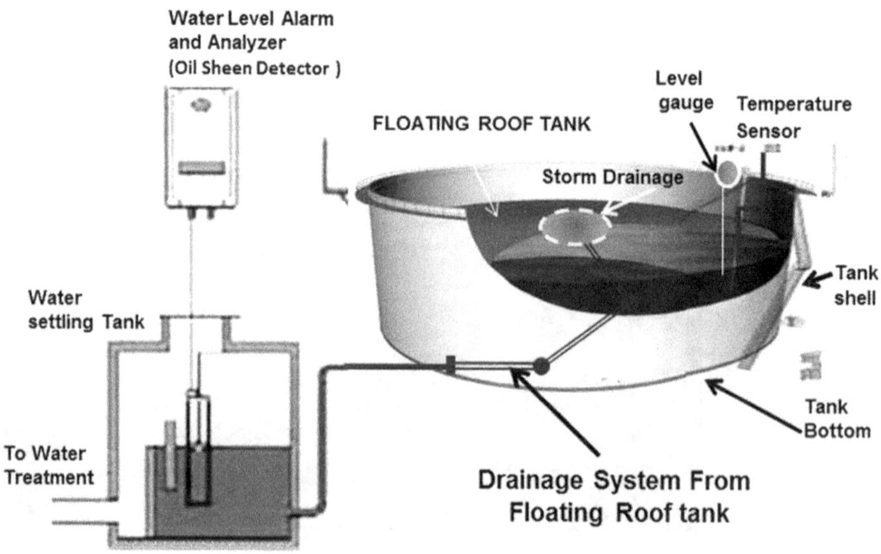

Figure 8-87. Treatment oil storm water from floating roof tank

When an entire tank top surface area is involved, hand-line streams will not be able to penetrate the thermal column to reach the burning liquid surface. Large volume Aqueous Film Forming Foams (AFFF) streams are the only alternative to subsurface injection or fixed system.

8.7.8 Cathodic Protection of Above Ground Hydrocarbon Storage Tanks

8.7.8.1 Definition of Corrosion

Corrosion is,

- The deterioration of a substance (usually a metal) or its properties because of a reaction with its environment.
- The result of interaction between a metal and environments which results in its gradual destruction.
- An aspect of the decay of materials by chemical or biological agents.
- An extractive metallurgy in reverse. For instance, iron is made from hematite by heating with carbon. Iron corrodes and reverts to rust, thus completing its life cycle. The hematite and rust have the same composition.
- Electrical energy added to metals when manufactured will leave when it is placed in a corrosive environment.

Generally Corrosion Can be Defined as Either: 1 — Practical Tendency of a Metal to Revert to its Native State or 2 — Scientific Electrochemical Degradation of Metal as a Result of a Reaction with its Environment [41].

8.7.8.2 Corrosive Environment

Corrosion cannot be defined without a reference to environment. All environments are corrosive to some degree. Following is the list of typical corrosive environment.

- Air and humidity
- Fresh, distilled, salt and marine water.
- Urban, marine and industrial atmospheres
- Steam and gases, like chlorine
- Ammonia and hydrogen sulfide
- Sulfur dioxide and oxides of nitrogen
- Fuel gases and liquid hydrocarbons
- Acids
- Alkalines
- Soils

8.7.8.3 Consequences of Corrosion

Some important consequences of corrosion can be summarized as follows:

- Facilities shutdown. Shutdown of refiner, power plants and pipelines may cause severe problems to industry and consumers.
- Loss of products, leaking containers, storage tanks, oil and gas transportation lines and hydrocarbon storage tanks cause significant loss of product and may generate severe accidents and hazards.
- Loss of capacity: efficiency loss in pipelines by corrosion reduces the transportation piping capacity.
- Contamination. Corrosion products may contaminate the stored and transported hydrocarbons and other products resulting in dire consequences.

The Importance of Corrosion Control are:

- Preserve Assets
- Reduce Maintenance Costs
- Reduce Inspection Cost
- Company/Government Requirement
- Preserve the Environment

For corrosion to take place, the formation of a *corrosion cell* is essential. A corrosion cell is comprised of the following four components:

- Anode
- Cathode
- Electrolyte
- Metallic path

These are described below (Figure 8-88).

1. Anode:
   - One of the two dissimilar metal electrodes in an electrolytic cell, represented as the negative terminal of the cell.
   - Electrons are released at the anode, which is the more reactive metal.
   - Electrons are insoluble in aqueous solutions and they only move, through the wire connection into the cathode.

$$Fe \rightarrow Fe^{+2} + 2e-$$

2. Cathode:
   - One of the two electrodes in an electrolytic cell represented as a positive terminal of a cell.
   - Reduction takes place at the cathode and electrons are consumed.

$$O_2 + 2H_2O + 4e- \ 4OH- \text{ (Figure 8-89)}$$

3. Electrolyte
   It is the electrically conductive solution (e.g. salt solution, soil) that must be present for corrosion to occur.

4. Metallic Path

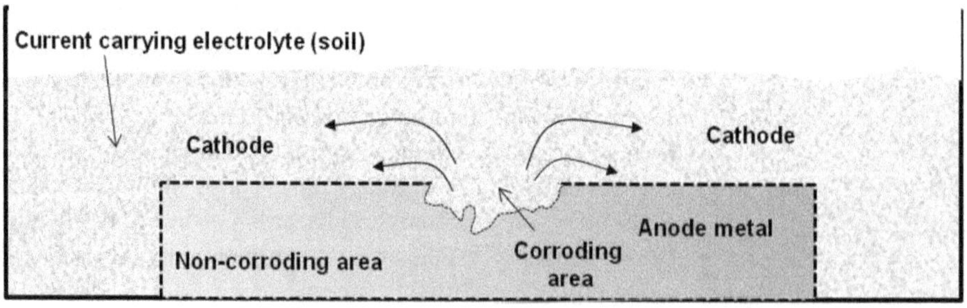

Figure 8-88. Corrosion cell [41]

The two electrodes are connected externally by a metallic conductor. Metals provide a path for the flow of conventional current which is actually passage of electrons in the opposite direction.

For corrosion to take place the four conditions states above must be present, previous Figure 8-89. The reaction characteristics can be anodic or cathodic.

Anodic Reaction Characteristics

(1) Oxidation of metal to an ion with a charge
(2) Release of electrons.
(3) Shift to a higher valence state.

Cathodic Reaction Characteristics

(1) Cathodic reactions are *reduction* reactions which occur at the cathode.
(2) Electrons released by the anodic reactions are consumed at the cathode surface.

The most common cathodic reactions in terms of electrons transfer are illustrated below:

(a) $2H^+ + 2e \longrightarrow H_2$ (in acid solution)
(b) $O_2 + 4H + 4e \rightarrow 2H_2O$ (in acid solution)
(c) $2H_2O + O_2 + 4e \longrightarrow 4OH^-$ (in neutral and alkaline solutions)
(d) $Fe^{3+} + e \rightarrow Fe^{2+}$ (metal ion reduction in ferric salt solutions)
(e) Metal deposition: $M^{2+} + 2e \rightarrow M$
$Ni^{++} + 2e \rightarrow Ni$
$Cu^{2+} + 2e \rightarrow Cu$
(f) Bacterial reduction of sulfate:

$$SO_2^- + 8H^+ + 8e \rightarrow S^- + 4H_2O$$

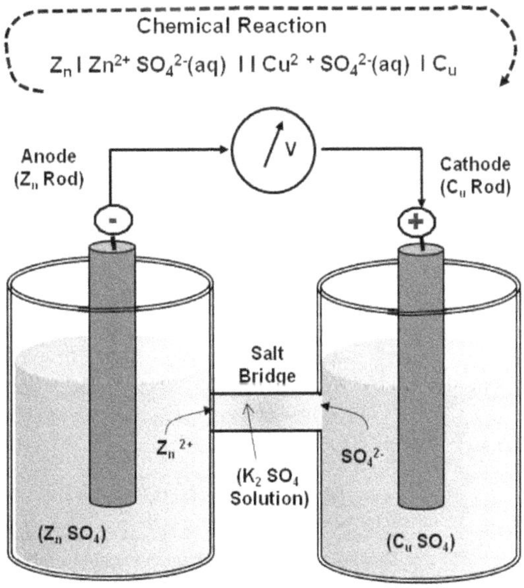

Figure 8-89. A galvanic cell (Daniel cell), (reproduced from [41])

Standard Electrode Potential of various material are indicated in Table 8-5. The electro Galvanic Series for Materials in Neutral Soils & Water.

The electro potential of some commercially available materials for anodes are provided in Table 8-6.

8.7.8.4 Types of Corrosion

1 Uniform Corrosion

It is the uniform thinning of a metal without any localized attack.

- Corrosion does not penetrate very deep inside. The most familiar example is the rusting of steel in air.
- Environment
  (1) Dry atmosphere.
  (2) Damp atmosphere.
  (3) Wet atmosphere.
  (4) Acids (HC_1, $HClO_4$, H_3PO_4).

TABLE 8-5. The electro galvanic series for materials [42]

Electrode Potential	Volt (Oxidation)
$Na = Na^+ + e^-$	2714 *AKT F*
$Mg = Mg^{+2} + 2e^-$	2363
$Al = Al^{+3} + 3e^-$	1662
$Mn = Mn^{+2} + 2e^-$	1180
$Zn = Zn^{+2} + 2e^-$	0.763
$Fe = Fe^{+2} + 2e^-$	0.440
$Cd = Cd^{+2} + 2e^-$	0.403
$H_2 = 2H^+ + 2e^-$	0
$Cu = Cu^{+2} + 2e^-$	–0.337
$4OH^- = O_2 + 2H_2O + 4e^-$	–0.401
$Ag = Ag^+ + e^-$	–0.799
$Au = Au^{+3} + 3e^-$	–1500

TABLE 8-6. Electro potential of some commercially available anodes

Material	Potential Volts (CSE)*
Carbon, graphite, coke	+0.3
Platinum	0 to –0.1
Mill scale on steel	–0.2
High silicon cast iron	–0.2
Copper, brass, bronze	–0.2
Mild steel in concrete	–0.2
Lead	–0.5
Cast iron (not graphitized)	–0.5
Mild steel (rusted)	–0.2 to –0.5
Mild steel (clean and shiny)	–0.2 to –0.8
Commercially pure aluminum	–0.8
Aluminum alloy (5% zinc)	–1.05
Zinc	–1.1
Magnesium Alloy (6% Al, 3% Zn, 0.15% Mn	–1.6
Commercially pure magnesium	–1.75

*Note: The typical potentials given in the above table are normally observed in neutral soils and water, measured with respect to copper sulfate electrode reference (CSE). It may be noted that material with more negative potential can protect the less electronegative potential. Thus for example copper can be protected by lead, but not by carbon graphite coke.

(5) Atmospheric contaminants.
(6) Process water containing hydrogen sulfide.
(7) Brines.
(8) Industrial atmosphere.
(9) Hydrocarbon containing wet hydrogen sulfide.

Examples of Uniform Corrosion

(1) Tarnishing of silver ware.
(2) Tarnishing of electrical contacts.
(3) Rusting of steels in open air.
(4) Corrosion of offshore drilling platforms.
(5) Corrosion of galvanized steel stairways.
(6) Failure of distillation columns.
(7) Corrosion of electronic components.
(8) Corrosion of underground pipes (composite asphalt coated).
(9) Corrosion of automobile bodies.
(10) Corrosion of heat exchanger tubes.
(11) Corrosion of structural steels.

2 Galvanic Corrosion

- Galvanic corrosion occurs when two metals with different electrochemical potentials or with different tendencies to corrode are in metal-to-metal contact in a corrosive electrolyte.
- When two metals with different potentials are joined, such as copper (+0.334 V) and iron (–0.440 V), a galvanic cell is formed. A cell in which the chemical change is the source of energy, is called a galvanic cell The corrosion which is caused due to the formation of the galvanic cell is, therefore, called galvanic corrosion.

3 Defect/Voids Corrosion

- This is a localized form of corrosion, caused by the deposition of dirt, dust, mud and deposits on a metallic surface or by the existence of voids, gaps and cavities between adjoining surfaces.
- An important condition is the formation of a differential aeration cell for crevice corrosion to occur. This phenomenon limits the use, particularly of steels, in marine environment, chemical and petrochemical industries.

Factors affecting crevice corrosion (Figure 8-90)

(a) Defect/Void type.
(b) Alloy composition.
(c) Passive film characteristics.
(d) Geometry of crevice.
(e) Bulk composition of media.
(f) Bulk environment.
(g) Mass transfer in and out of crevice.
(h) Oxygen

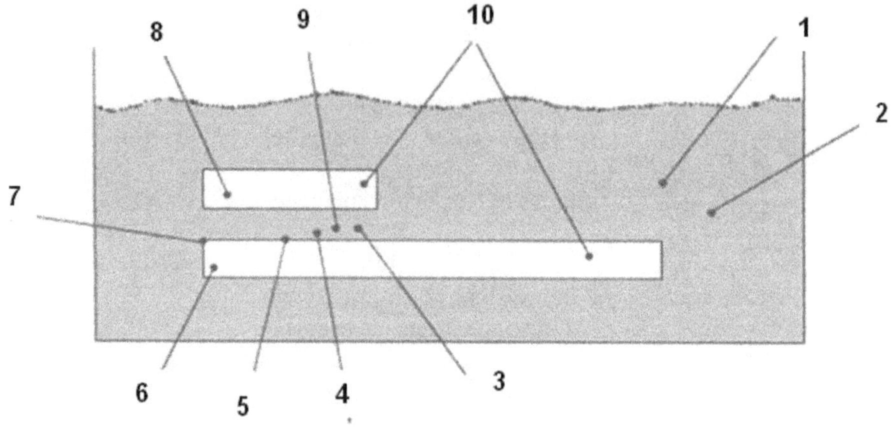

Figure 8-90. Factors affecting defect/crevice corrosion (reproduced after [41]), (Legend: 1–Electrolyte, 2–Electrolytic environment, 3–Mass convection (in/out of defect), 4–Defect solution, 5–Electrochemical reaction (metal solution, O_2 reduction, H_2 evolution), 6–Alloy composition (major/minor base elements, impurities), 7–Passive film characteristics (passive current or film stability), 8–Defect type (metal/metal, metal/non-metal, metal/marine growth), 9–Defect geometry (gap-width, depth), 10–Overall geometry (exterior to interior defect area ratio, number of defects), after [42])

4 Pitting Corrosion

- It is a form of localized corrosion of a metal surface where small areas corrode preferentially leading to the formation of cavities or pits, and the bulk of the surface remains un-attacked (Figure 8-91).
- Metals which form passive films, such as aluminum and steels, are more susceptible to this form of corrosion.
- It is the most insidious form of corrosion. It causes failure by penetration with only a small percent weight-loss of the entire structure.
- It is a major type of failure in chemical processing industry. The destructive nature of pitting is illustrated by the fact that usually the entire system must be replaced.
- Generally, the most conducive environment for pitting is the marine environment. Ions, such as Cl^-, Br^- and I^-, in appreciable concentrations tend to cause pitting of steel. The sulfate ions also induce pitting of steels.
- Aluminum also pits in an environment that cause the pitting of steel. If traces of Cu^{2+} are present in water, or Fe^{+3} ions are in water, copper or iron would be deposited on aluminum metal surface and pitting would be initiated.
- Oxidizing metal ions with chloride, such as cupric, ferric and mercuric, cause severe pitting.
- Presence of dust or dirt particles in water may also lead to pitting corrosion in copper pipes transporting seawater.

5 Stress Corrosion Cracking and Hydrogen Blistering/Damage

Stress corrosion is the failure of a metal resulting from the conjoint action of stress and chemical attack (Figure 8-92).

- It is a phenomenon associated with a combination of static tensile stress, environment and in some systems, a metallurgical condition which leads

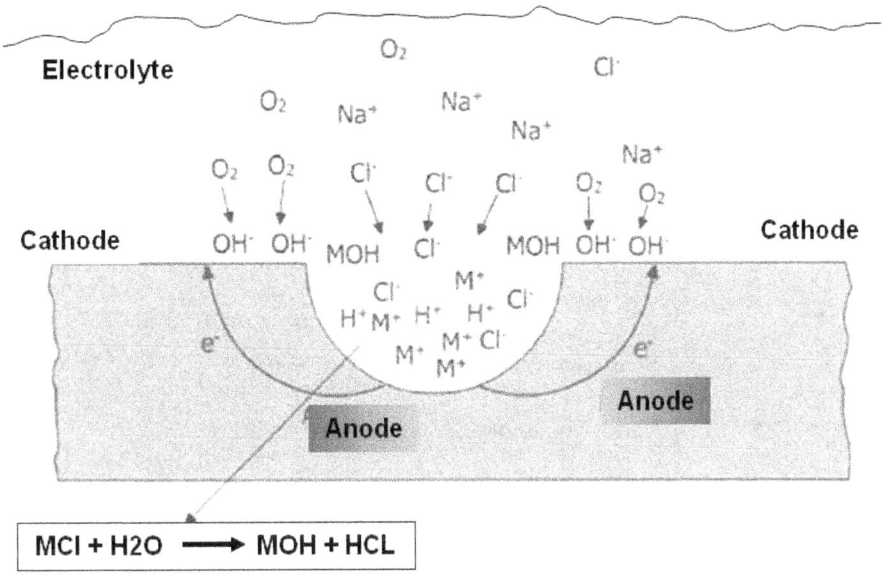

Figure 8-91. Pitting corrosion (reproduced after [41])

to component failure due to the initiation and propagation of a high aspect ratio crack.

- It is characterized by fine cracks which lead to failure of components are potentially the structure concerned. Stress corrosion cracking is abbreviated as SCC.

The following conditions are necessary for SCC to occur:

- o A susceptible metal.
- o A specific environment.
- o A tensile or residual stress.

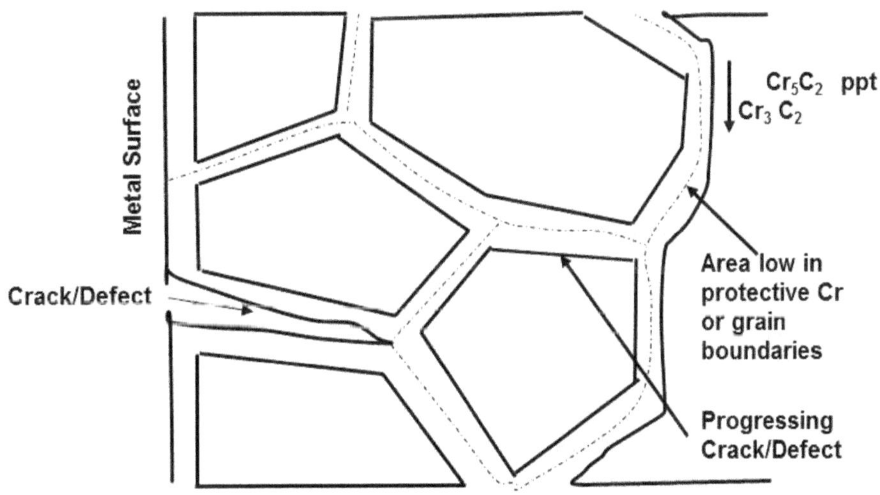

Figure 8-92. Causes of SCC (reproduced from [41])

TABLE 8-7. Sources of stress for SCC

Sources	Applications
Residual	Rapid heating and quenching
Welding	Thermal expansions
Shearing, punching, cutting	Vibrations
Bending, crimping, riveting	Rotation
Machining (lathe/mill/drill)	Bolting (flanged joints)
Heat treatment	Pressure (internal or external)
Straightening, breaking, deep drawing	Structural loading

Sources of SCC are indicated in Table 8-7.

Various types of SCC are distinguished as below:

a. Chloride stress corrosion cracking. It occurs in austenitic steels under tensile stress in the presence of oxygen, chloride ion and high temperature.
b. Caustic stress corrosion cracking. Cracking of steels in caustic environments where the hydrogen concentration is high, for instance, cracking of Inconel tubes in alkaline solutions.
c. Sulfide stress corrosion cracking. Cracking of steels in hydrogen sulfide environment as encountered in oil drilling industry.
d. Seasonal cracking. The term is now obsolete. It had a historical significance only. It refers only to SCC of brass in ammoniacal environment, but still occasionally occurs in refrigeration plant using ammonia refrigerant.

6 Hydrogen Attack/Damage (High Temperature Hydrogen Attack)
Steels are also damaged by hydrogen blistering at high temperatures. There are three categories of hydrogen damage (Figure 8-93):

(a) High temperature hydrogen attack (hydrogen damage)
(b) Hydrogen blistering
(c) Hydrogen embrittlement

For the HIC Figure 8-94, to occur, the following conditions must occur:

(a) The presence of water phase.
(b) The presence of atomic hydrogen.
(c) An agent that retards the formation of molecular hydrogen at the surface.
(d) Presence of grain boundaries or inclusions.
(e) Maintenance of an active surface.
(f) Discontinuity in metal, such as slag, inclusion and/or void.

7 Sulfide Stress Corrosion Cracking
Mechanisms sulfide stress corrosion cracking is shown in Figure 8-95.

8.7.8.5 Storage Tank Cathodic Protection

Tanks are subject to corrosion if not protected (Figure 8-96).

Corrosion can be initiated and propagated throughout the tankage system including the following locations

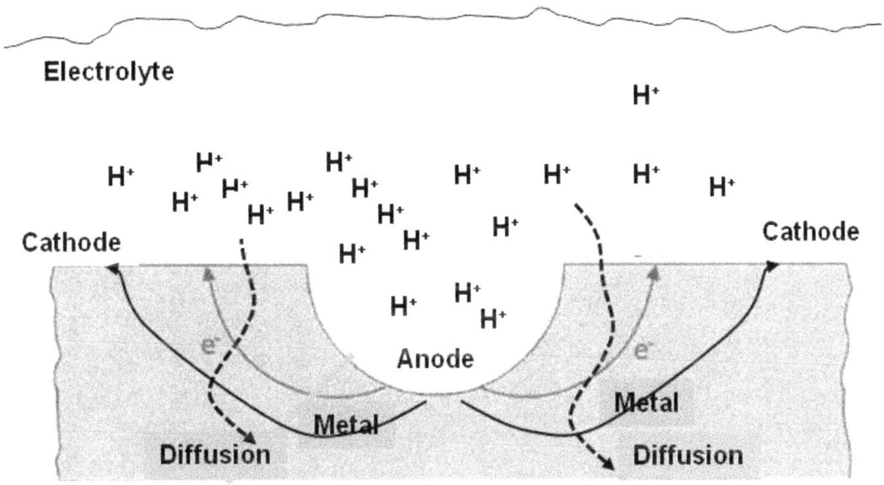

Figure 8-93. Hydrogen diffusion and attack (reproduced from [41])

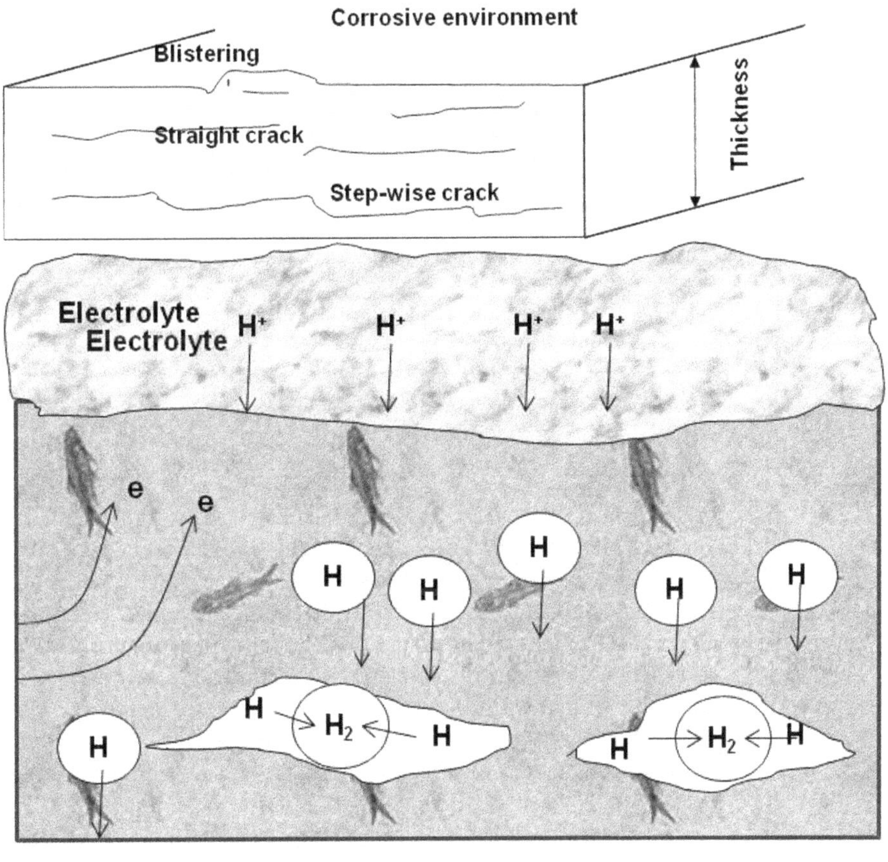

Figure 8-94. Hydrogen induced cracking (HIC) (reproduced from [41])

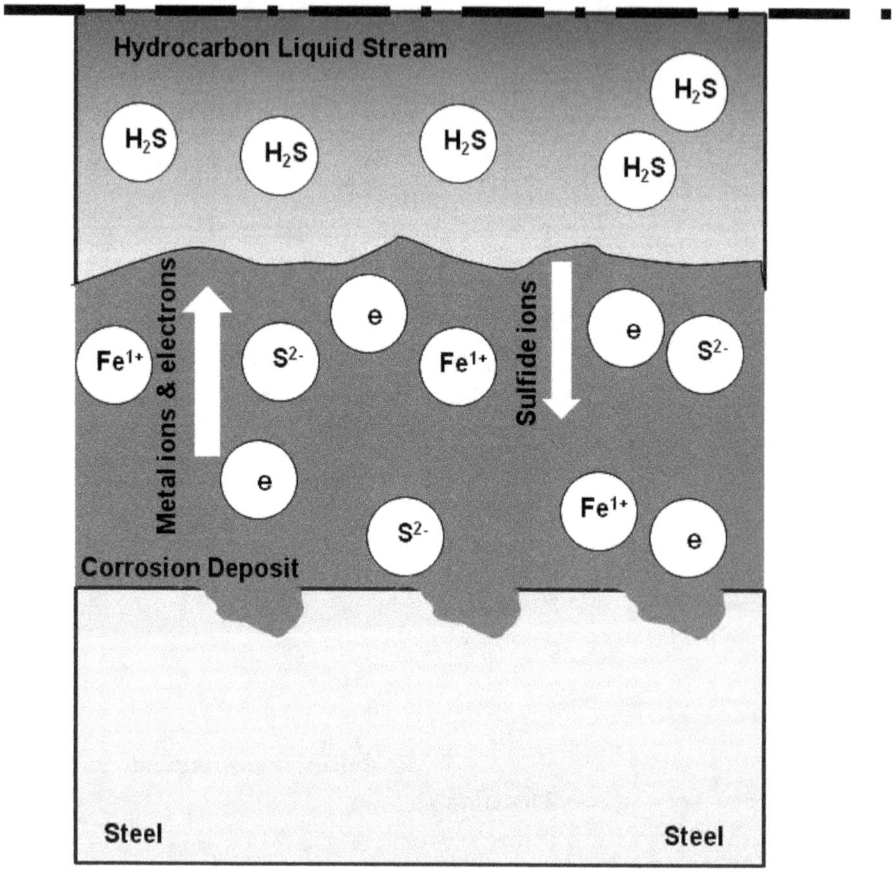

Figure 8-95. Sulfide stress corrosion cracking mechanism (reproduced from [41])

- Tank underside [43] (Figure 8-97A)
- Atmospheric corrosion
- Presence of corrosive sludge
- Contaminants causing corrosion
- Water seepage

As indicated previously corrosion is an electrochemical process as it involves transport of electron(s) from a source (anode, in this case defect in steel) to a receptor (cathode) through an electrical path created by a conductive medium (electrolyte) (Figure 8-97B). To avoid corrosion, the aim would be to supply additional electrons to a metallic structure. Such additional electrons would cause an increase in the rate of cathodic reaction thus reducing the rate of anodic reaction. This would eventually minimize or eliminate corrosion (Figure 8-98).

The anode would become more negative and the cathode more positive. Cathodic protection is, therefore, achieved by supplying an external negative current to the corroding metal to make the surface acquire the same potential to eliminate the anodic areas. The anodic areas are eliminated by transfer of electrons.

After a sufficient current flow, the potential of anodic areas would become negative enough for corrosion to stop.

Figure 8-96. Underside corrosion in hydrocarbon storage tanks

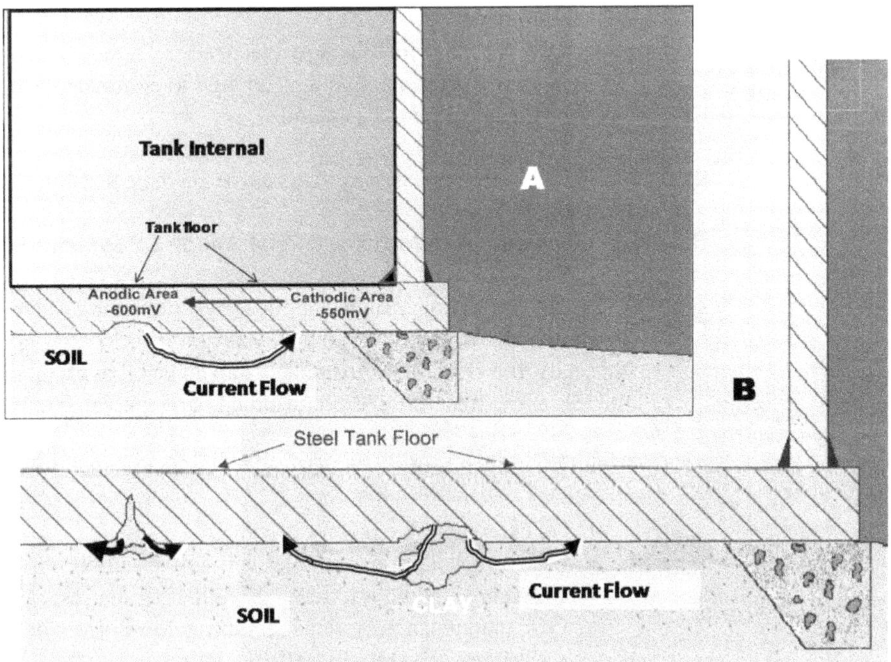

Figure 8-97. A—External corrosion of tank bottom and B—corrosion cell [44,45]

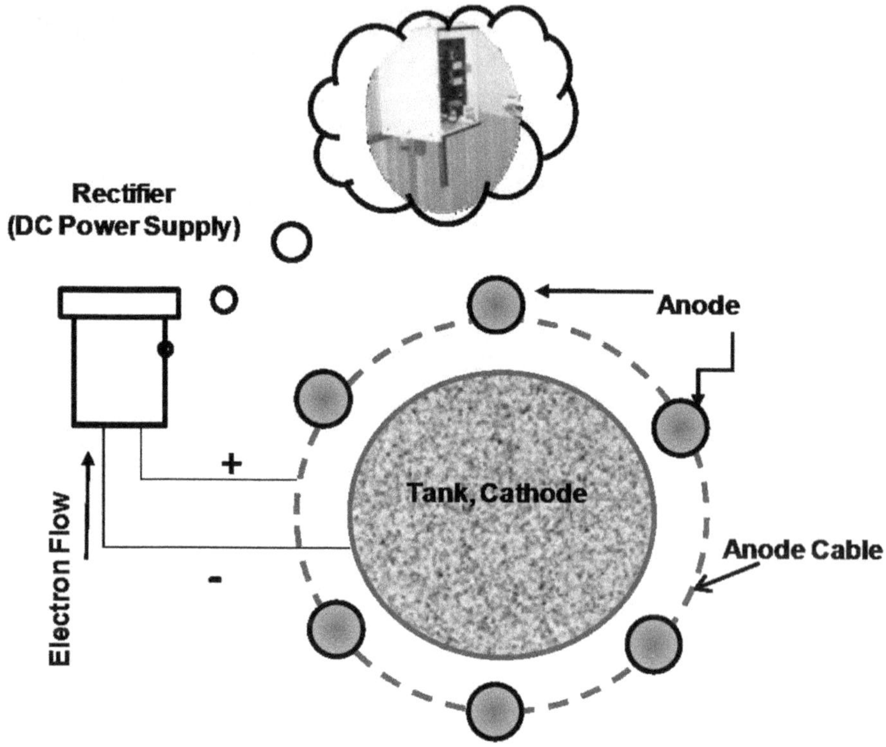

Figure 8-98. Cathodic protection of hydrocarbon storage tank

(a) There must be an anode, a cathode, an electrolyte and a metallic path for the transfer of electrons.
(b) A source of DC current to supply electrons.
(c) Sufficient direct current should be applied to eliminate the potential difference between the anode and the cathode.

Two types of cathodic protection systems exist: Gavanic and Impressed current.

8.7.8.5.1 Galvanic Anode or Sacrificial Anode CP System (Figure 8-99)

- Cathodic protection can be applied by connecting sacrificial anodes to a structure.
- Basically, the principle is to create a galvanic cell, with the anode representing the less noble material that is consumed in the galvanic interaction.

  The following advantages are associated with sacrificial anode CP systems:

  - No external power sources required.
  - Ease of installation (and relatively low installation costs).
  - Unlikely cathodic interference in other structures.
  - Low-maintenance systems (assuming low current demand).
    - System is essentially self-regulating.
    - Relatively low risk of overprotection.
    - Relatively uniform potential distributions

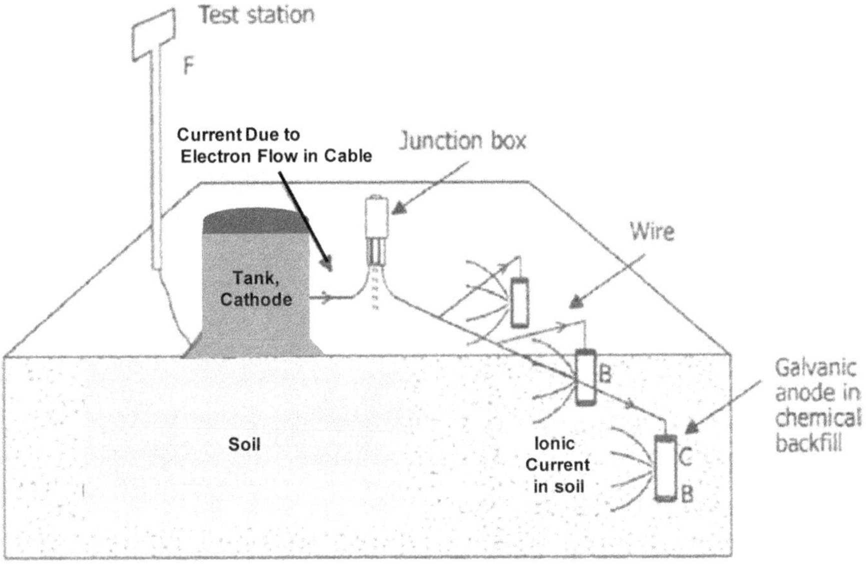

Figure 8-99. Storage tank sacrificial anode CP system [41]

However, these relatively simple systems also have some limitations such as:

- Limited current and power output.
- High-resistivity environments or large structures may require excessive number of electrodes. Maximum resistivity of 6000 to 10,000 ohm-cm is generally regarded as the limit, depending on coating quality.
- Anodes may have to be replaced frequently under high current demand.
- Anodes can increase structural weight if directly attached to a structure.

Anode types:
For land-based CP applications of structural steel, anodes based on zinc or magnesium are the most important.

1 Zinc Anode (Figure 8-100)

- Zinc anodes employed underground are high-purity Zn alloys, as specified in ASTM B418-95a.
- Only the Type II anodes in this standard are applicable to buried soil applications.
- For zinc anodes, the mass-based theoretical capacity is relatively low at 780 Ah/kg, but efficiencies are high at around 90%.

2 Magnesium Anode (Figure 8-101)

- Magnesium anodes generally have a low efficiency at 50% or even lower.
- The theoretical capacity is around 2200 Ah/kg (Amper.hour/kg)
- The magnesium alloys are also high-purity grades and have the advantage of a higher driving voltage.

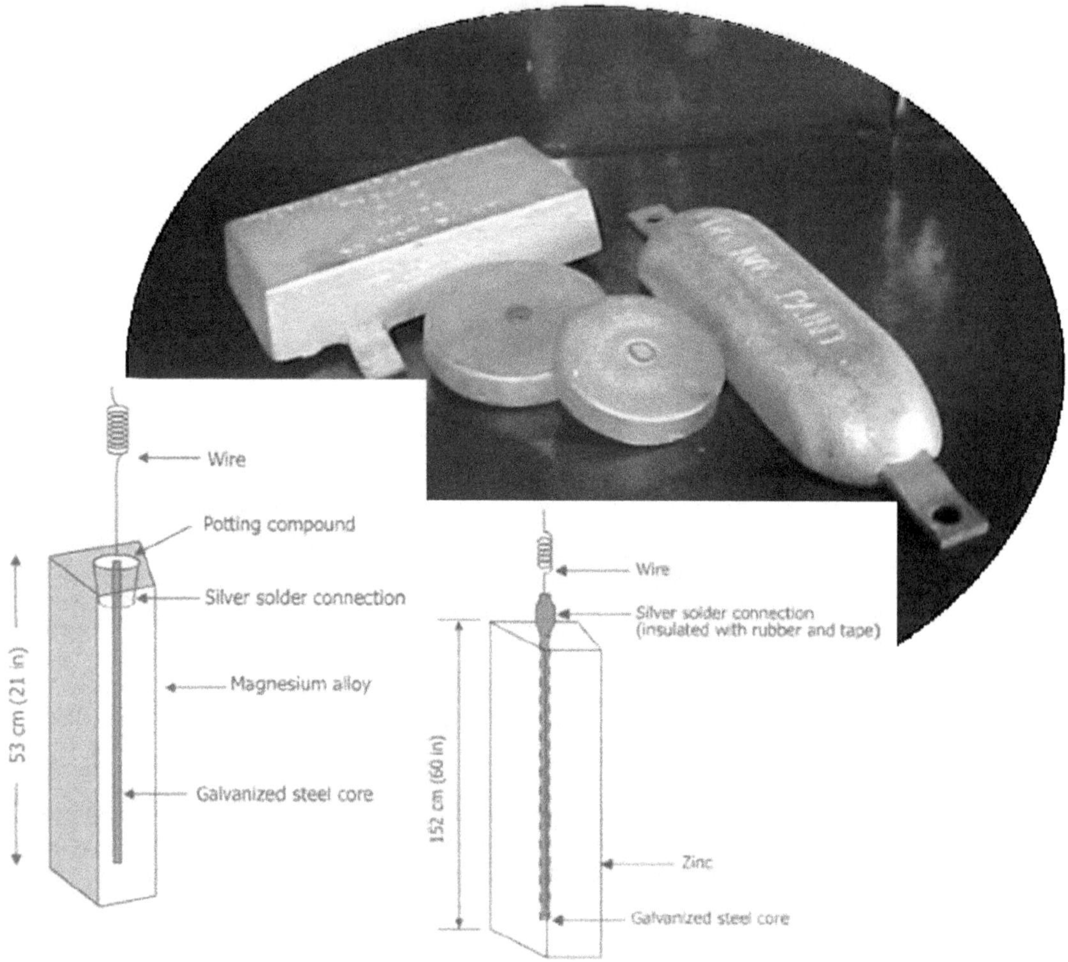

Figure 8-100. Typical zinc and magnesium anodes

3 Aluminum-Indium Anode

- These are mostly employed for seawater applications.
- The base metal contains 98-99% of aluminum.
- The rate of consumption varies between 7 and 9 lb/A-year. The efficiency varies between 87 and 95%.

8.7.8.5.2 Impressed Current CP System In impressed current systems cathodic protection is applied by means of an external power current source (Figure 8-102).

In contrast to the sacrificial anode systems, the anode consumption rate is usually much lower. Unless a consumable "scrap" anode is used, a negligible anode consumption rate is actually a key requirement for long system life.

Impressed current systems typically are favored under high-current requirements and/or high-resistance electrolyte.

Advantages:

- High current and power output range.
- Ability to adjust ("tune") the protection levels.

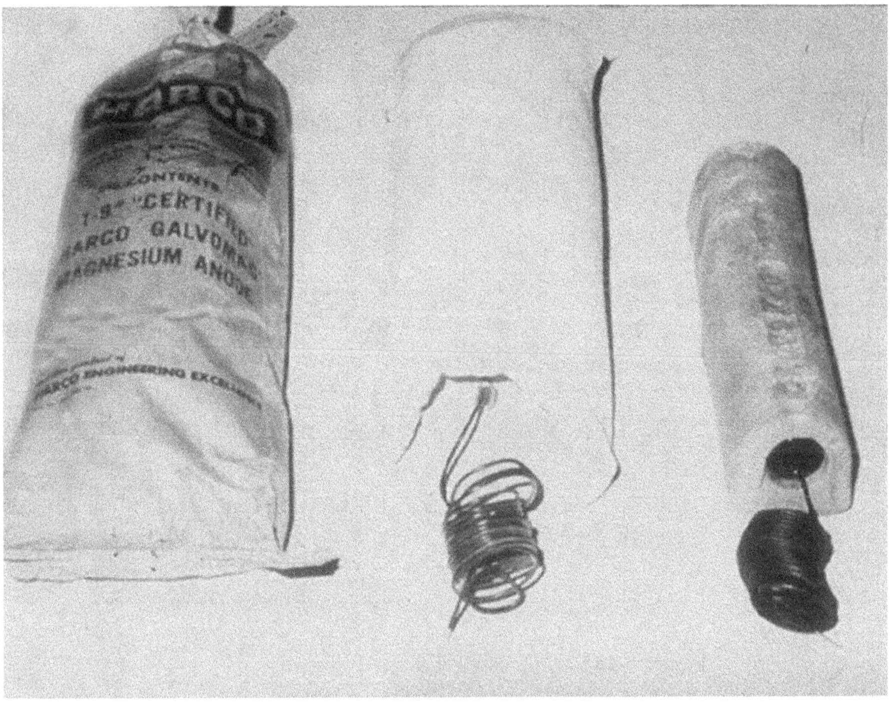

Figure 8-101. Magnesium anodes (courtesy of Corrpro Canada)

- Large areas of protection.
- Low number of anodes, even in high-resistivity environments.
- May even protect poorly coated structures.

Disadvantages:

- Relatively high risk of causing interference effects.
- Lower reliability and higher maintenance requirements.
- External power has to be supplied.
- Higher risk of overprotection damage.
- Running cost of external power consumption.
- More complex and less robust than sacrificial anode systems in certain applications.

Typical in-ground anode installation is depicted in Figure 8-103.

8.7.8.6 Above Ground Storage Tank CP System

Storage Tanks are applied two different and independent cathodic protection systems. These are;

1 -External cathodic protection system of the tank.
2 -Internal cathodic protection system of the tank.

8.7.8.6.1 External Cathodic Protection System of the Tank

- External part of the tank is embedded to the soil over the tank base (Figure 8-104).

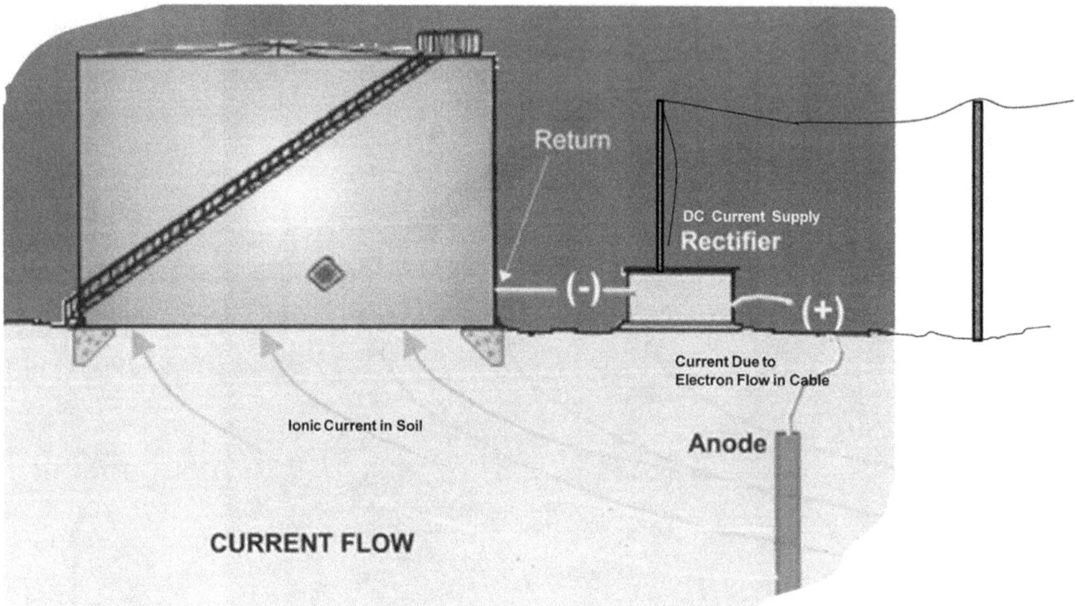

Figure 8-102. Impressed current CP system

- Soil is very resistive and corrosive media.
- The tank base is protected with impressed current cathodic protection system.
- The reason of applying the impressed current system is requirement lots of galvanic anode. So, galvanic CP system is more expensive than Impressed current CP system.

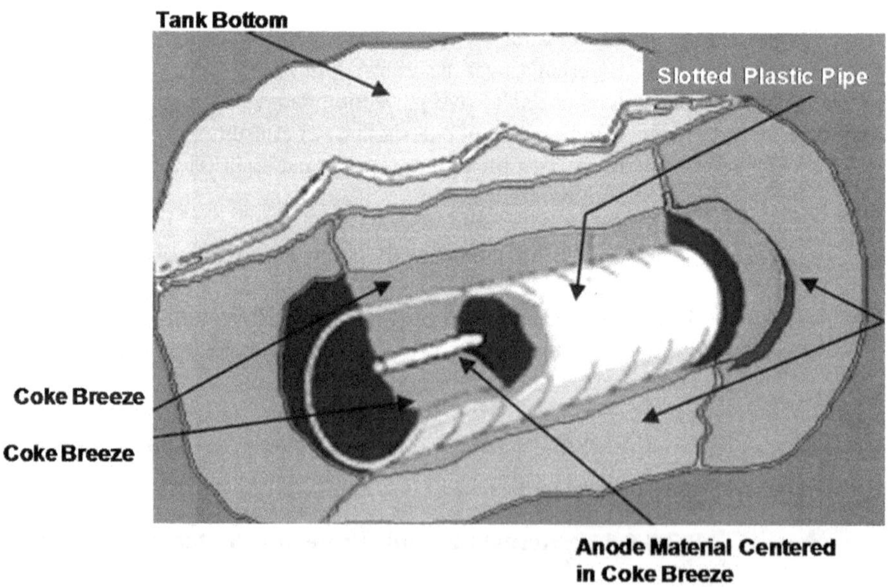

Figure 8-103. Typical in ground anode design and installation

8.7.8.6.2 Internal Cathodic Protection System of the Tank Some hydrocarbon storage tanks may require protection from inside because of ingress of rain water into the stored hydrocarbon from roof drain system, rim seals etc.

- Such water typically accumulates on the tank bottom. Due to lighter density of the stored hydrocarbon liquid.
- The hydrocarbon entrained with water is a corrosive media. So, that area has to be protected against the corrosion.
- Other example is Internal surface of fire water protection tanks.
- The Internal CP system of the tank is galvanic anode cathodic protection.
- Al-In anodes are typically used in these CP systems.
- The anodes are installed to the tank base.

The other alternative is to build a new floor with a liner on the old floor (Figure 8-105) and thus protecting the tank from both internal adverse/corrosive conditions and external seepage of water, that may occur on the underside of the tank lower bottom.

8.7.8.7 Typical CP Installation for Above Ground Storage Tanks

Figures 8-106 through 8-110 highlight the industry's typical design and installation patterns of anodes installation and the cathodic protection system in above ground storage tanks (ASTs).

8.7.8.8 Applicable CP Standards

The standards that the industry uses for the design, installation, operation and monitoring of CP systems are:

- Recommended Practices API-651 - Cathodic Protection of Aboveground Petroleum Storage Tanks: "Galvanic anodes method is not practical for protection of large bare structures."

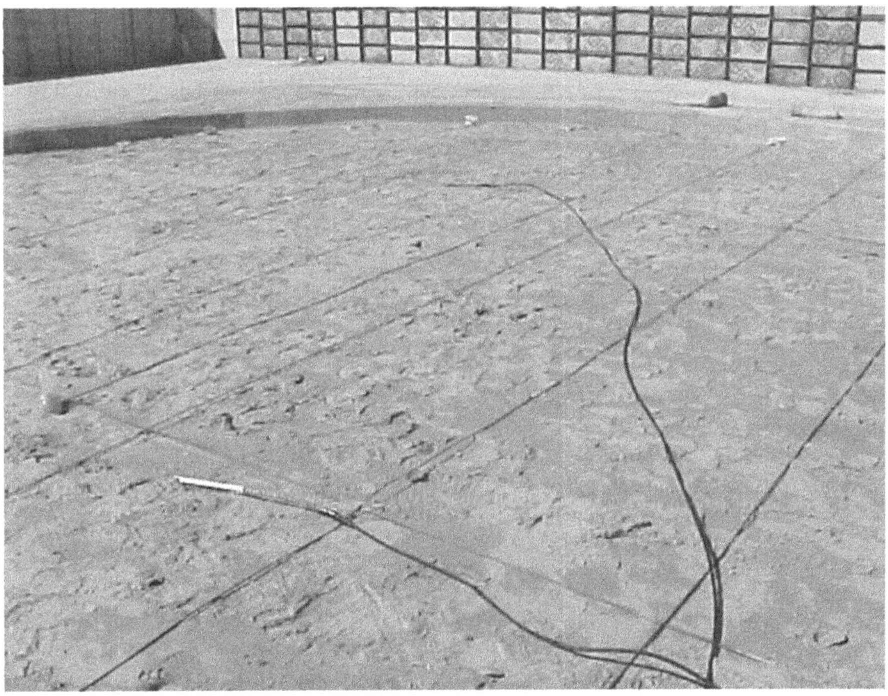

Figure 8-104. Anode and reference cell placement in high resistance sand under tank bottom

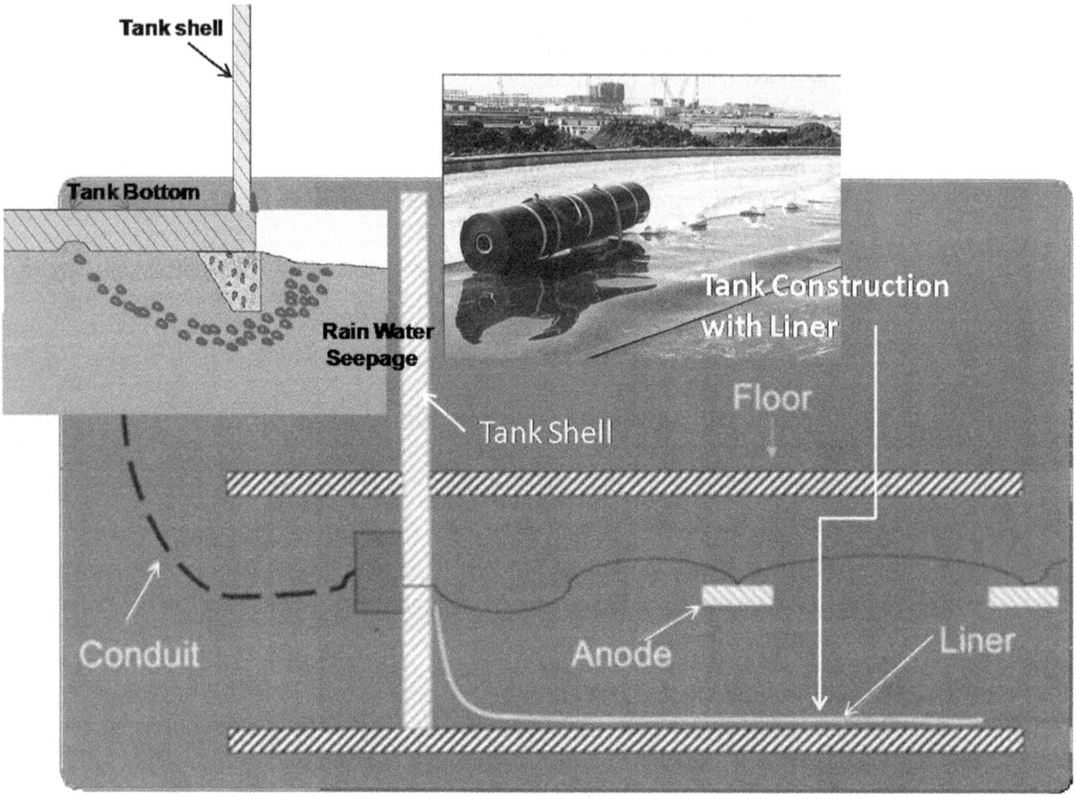

Figure 8-105. CP installation on double tank bottom with liner [44, 45]. Inset: rain water seepage under tank bottom

- NACE RP0193-01 - External Cathodic Protection of On-Grade Metallic Storage Tanks: "Galvanic protection systems can be applied to tank bottoms where the metallic surface area exposed to the electrolyte can be minimized through the application of a dielectric coating or the area is small due to the tank size or configuration."

8.8 TANK FAILURES AND EMERGENCY RESPONSE

8.8.1 Tank Failures

Catastrophic failures of aboveground atmospheric storage tanks (AST) can occur when flammable vapors in the tank explode and break either the shell-to-bottom or side seam. Flaws can also cause failures.

These failures have caused the tanks to rip open and, in some cases, hurled the tanks through the air. A properly designed and maintained storage tank will break along the shell-to-top seam. For example tanks up to ~16 m in diameter can be designed to fail at the shell-to-roof weld. This is called a frangible joint and is designed to limit damages to the tank and minimize the extent of a resulting fire/spill. This sacrificial joint is primary designed to ensure integrity of the AST shell-to-bottom joint in the event of an over-pressurization of a tank to assure containment of the stored

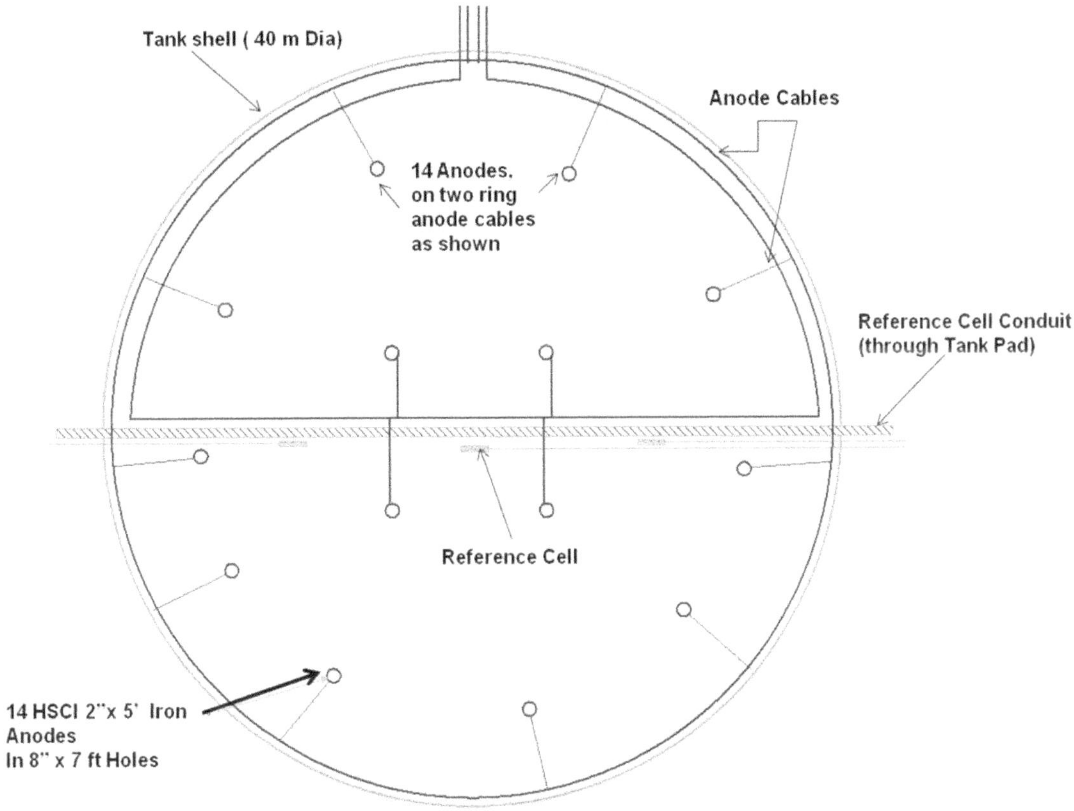

Figure 8-106. Anode installation and CP cable conduit under an Above Ground Storage Tank (AST)

Figure 8-107. CP installation on re-bottomed tank

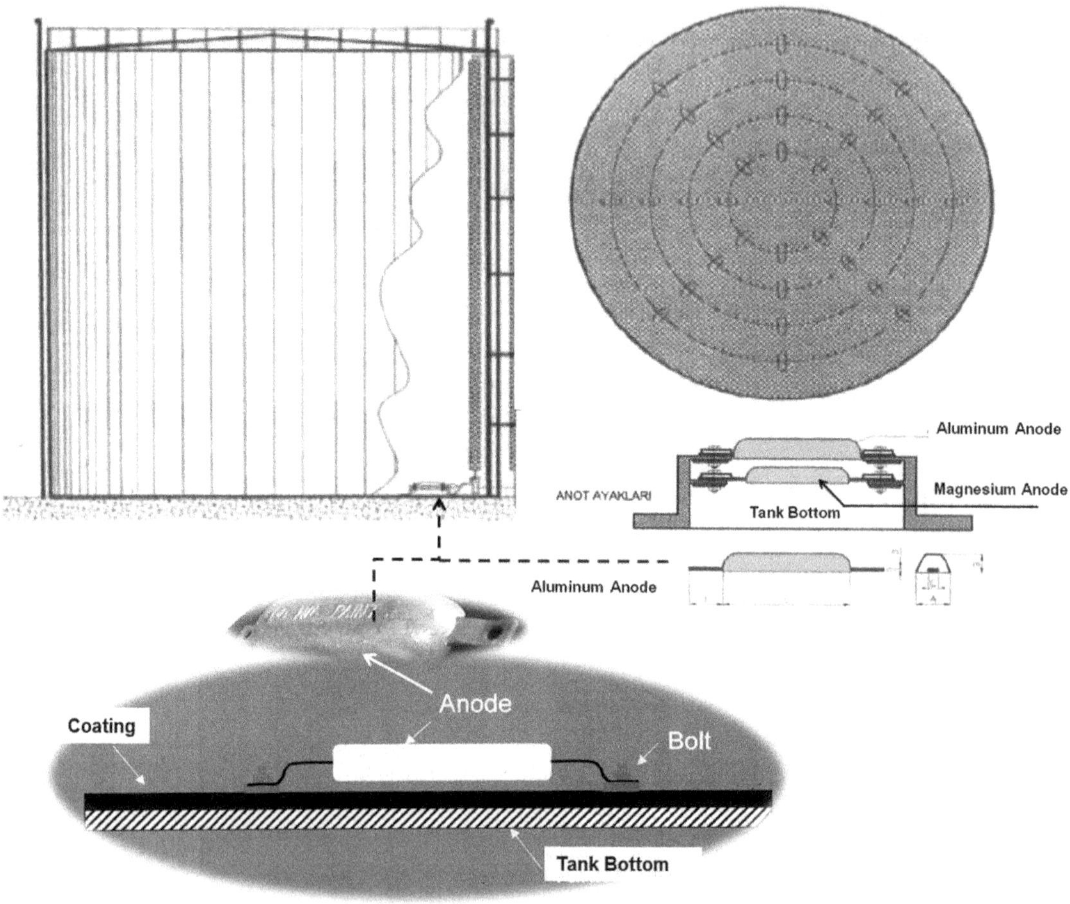

Figure 8-108. Typical anode installation design in storage tank bottom

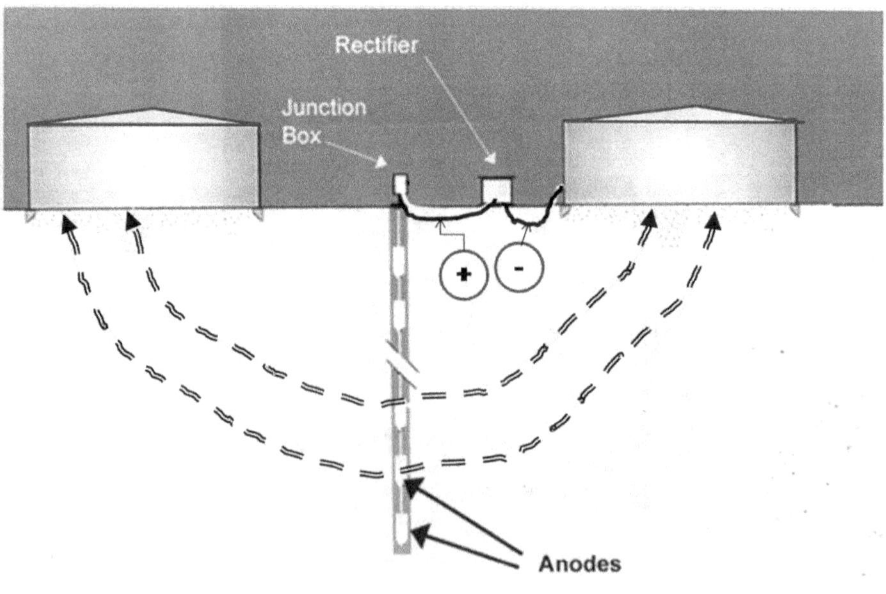

Figure 8-109. Cathodic protection of multiple ASTs with deep anode

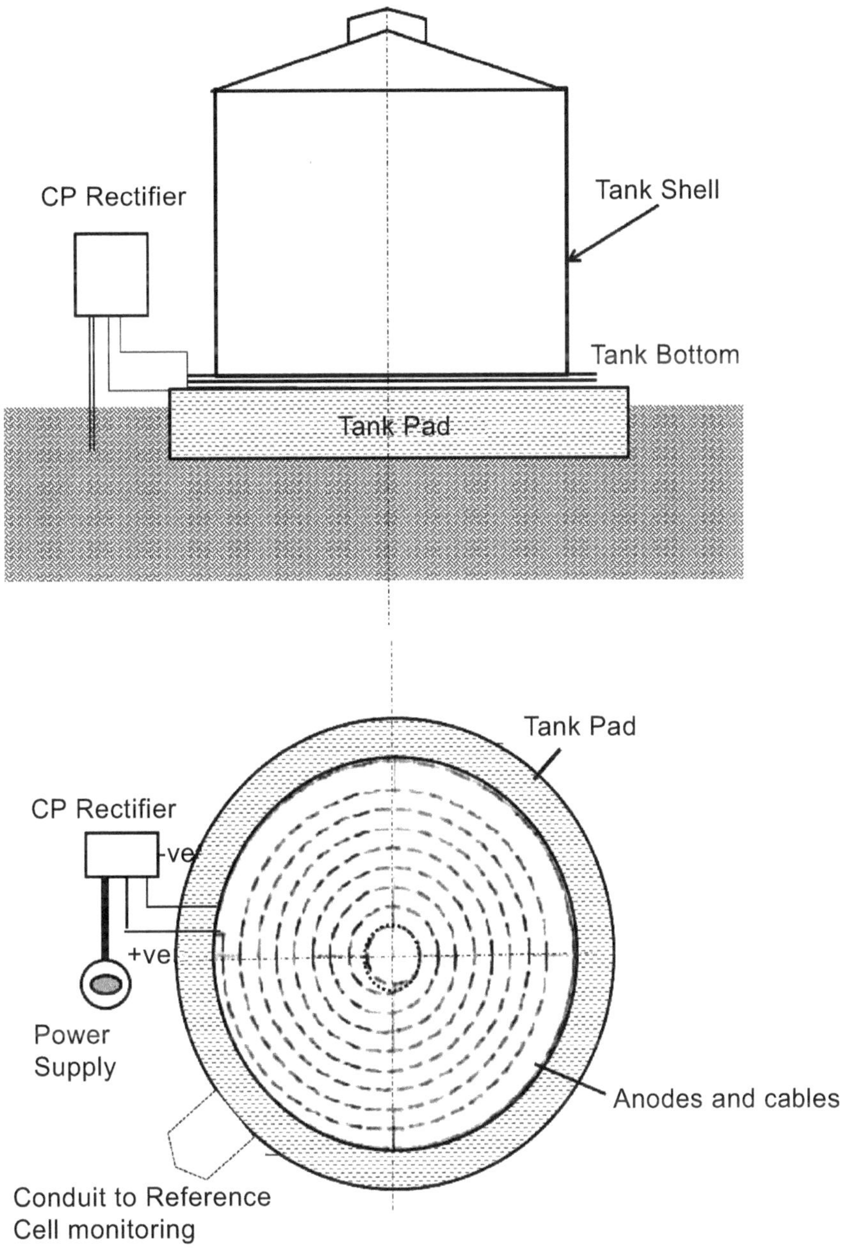

Figure 8-110. Typical impressed current CP system in above ground storage tank

liquid. In such a design, the fire would more likely be limited to the damaged tank and the contents would not be spilled. This section describes the types of tanks that may be prone to catastrophic failure and maintenance practices that can help prevent the accidents [46].

8.8.1.1 Past Accidents

Several accidents have occurred in which storage tanks have failed catastrophically when the flammable vapors inside an atmospheric tank exploded (Table 8-8). Table

8-8 indicates some of the incidences due to catastrophic hydrocarbon storage tank failures.

The specific incidents demonstrate the potential dangers posed to personnel, the public, and the environment when these storage tanks fail catastrophically. Often maintenance personnel were performing tank maintenance or other activities that introduced an ignition source. The vapors were ignited either inside the tank or outside and then flashed back into the tank.

From Table 8-8, it is clear that catastrophic failures of tanks have occurred in the past, resulting in either complete removal of the tank wall when the tank rockets due to an explosion in the vapor space, or an "unzipping" due to rapid brittle fracture initiated at a defect. Failures have also occurred in earthquakes, although it is not clear how rapid the failures were.

8.8.1.2 Causes of Tank Failures Hazards

Tank design and inspection/maintenance practices are factors directly related to catastrophic tank failure.

Tank design: Historically, accidents where the shell-to-bottom seam fails are more common among older storage tanks. Steel storage tanks built before 1950 generally do not conform to current industry standards for explosion and fire venting. Atmospheric tanks used for storage of flammable and combustible liquids should be designed to fail along the shell-to-roof seam when an explosion occurs in the tank. This prevents the tank from propelling upward or splitting along the side. Several organizations have developed standards and specifications for storage tank design. Published standards relevant to this design feature include API-650" Welded Steel Tanks for Oil Storage" issued by the American Petroleum Institute (API). Additional codes and standards, published by API and other organizations, address tank design, construction, venting, and safe welding and are listed at the end of this alert.

Inspection, maintenance, and repair practices: Tanks that are poorly maintained, rarely inspected, or repaired without attention to design, risk catastrophic failure in the event of a vapor explosion. Either weakening of the shell-to-bottom seam through corrosion or strengthening the shell-to-roof seam relative to the shell-to-bottom seam will increase the vulnerability of the tank to failure along the shell-to-bottom seam. The practice of placing gravel and spill absorbants around the base of the tank, may increase the likelihood of bottom corrosion. Given years of this practice, the bottom of some tanks, especially older ones, may be below ground level, thereby trapping moisture along the tank bottom. This

TABLE 8-8. Historical records of some hydrocarbon storage tank failures

Incident year	Brief description and comment
1919	**USA**: a United States Industrial Alcohol Company's distilling tank (2.5 million gallons) which recently had received a shipment of molasses in from Puerto Rico, exploded
1924	**Ponca City**: Failure of oil tank due to a dramatic drop in temperature, with bund overtopped due to momentum from release. No fire or fatalities listed.
1957	**Meraux**: Petrol tank ruptured and fell across bund. Fire occurred, but no fatalities listed. Records show that an operator closing a valve and also the presence of cast iron fittings (hence brittle failure is the likely cause) and also a wave spreading over the bund.
1970	**USA**: Failure of a shell to floor seam due to lightning igniting vapor in slop oil tank.
1976	**Addyston**: Methanol tank struck by lightning. Tank rocketed and burning contents overflowed surrounding dykes
1978	**USA**: Three tanks failed catastrophically in an earthquake.
1987	**South Dakota**: Above ground Tank Leakage from bottom shell – caused school closure.
1988	**Floreffe, PA**: Catastrophic rupture of 48 year old diesel tank on initial fill, after it had been relocated and reconstructed. Testing included only partial x-ray of welds and hydrotest to 5 feet (i.e. about 10% of tank height - 100% is now normal practice). According to a report "The investigation found that the rupture occurred due to low temperature embrittlement initiated at a flaw in the tank shell base metal, about 20 cm up from the bottom". No fire or fatalities are listed. Note: this is probably the most famous bund overtopping incident, also referred to as the Ashland or Monongahela tank collapse after the company and the river.
1989	**Richmond**: Earthquake ruptured a gasoline storage tank. The spill was contained in the bund and was not ignited
1992	**USA**: incident, while workers were welding the outside of a tank empty of liquid, the residual vapor in the storage tank exploded and propelled the tank upward and into an adjacent river. Three workers were killed and one was injured
1994	**USA:** incident, during a grinding operation on a tank holding petroleum based sludge, the tank was propelled upward, injuring 17 workers and spilling its contents over a containment berm into a nearby river
1995	**USA**: incident, during a welding operation on the outside of a tank, the combustible vapor inside two large, 30-ft. diameter by 30-ft. high, storage tanks exploded and propelled the tanks upward — one landing more than 50 feet away. The flammable liquid inside was instantly released and ignited, resulting in a massive fire that caused five deaths and serious injuries
1997	**USA**: a fixed roof gas-oil tank located in the tank farm of Ashdod Oil Refinery exploded, causing the death of one Sample Man, and a fire in two adjacent tanks located in the same dike.
2000	**USA**: catastrophic tank failure (Capacity one million-gallon bulk storage tank), Southside River Rail Co
2001	**USA**: failure of a Crude Oil AST (Capacity :100,000 gallons)
2003	**Japan Tomakomai, Hokkaido**: A fire occurred at a 33,000,000 l (33,000 BBLS) crude oil floating roof tank and attached piping at a refinery. There was an earthquake with a seismic intensity of a little under the 6th grade and a magnitude of eight. The fire was extinguished after about seven hours. Two days after the earthquake, a fire occurred at a 33,000 kL floating roof storage tank containing naphtha, which was damaged by the earthquake. Due to the earthquake, the floating roof sank and naphtha floated above the roof and ignited. This was the first fire of its type at a floating roof tank in Japan. It took 44 hours to extinguish the fire. There was insufficient foam available, and it had to be collected from the whole country
2003	**USA, Glenpool, Oklahoma**: Tankfire due to a static charge caused by using flowrates that were too high for the transfer operation
2009 (Figure 8-111)	**Puerto Rico Refinery**: Tank Farm Explosion, multiple tanks exploded or caught fire due to 2.8 Richter earthquake. Flames were as high as 30m and a black toxic smoke stack could be seen from satellite images. 2 people (from a nearby US Army base) were injured

can weaken the bottom and the shell-to-bottom seam. Alternatively, changes to the roof seam such as modifications to or replacement of the roof, or attachments to the roof, could make the roof-to-shell seam stronger relative to the shell-to-bottom seam.

Other hazards that can contribute to a tank explosion and possible consequences are:

- Emission of Combustible vapor
- Presence of ignition source
- Proximity

Figure 8-111. Puerto Rico refinery/tank farm explosion (Oct 2009, www.ffti.com.au)

Combustible vapors: Generation of combustible vapors is a hazard not only for the storage of pure flammable liquids but also for the storage of any sludge or mixture where a combustible component is present or can be produced by reaction. Sludge (slop tanks) and mixture (e.g., oil/water) tanks may be particularly vulnerable because they are sometimes open to the air; explosive atmospheres may form inside and outside the tank. Facilities may not always recognize this hazard. In addition, even tanks appearing to be empty may pose a hazard if they still contain combustible vapors.

In the cited cases, the potential for combustible vapors was not clearly recognized and materials were stored in tanks that were not equipped with flame arresters to prevent external fire from reaching the vapor space inside the tank or with vapor control devices to limit vapor emissions from the tank.

Ignition sources: When combustible vapors escape from their containment and mix with air in the presence of an ignition source, combustion may occur. To minimize this hazard, all possible ignition sources must be isolated from potential combustible vapors, e.g., welding equipment or other maintenance equipment that can spark or arc, sources of static electricity, lightning, "hot work" in adjacent areas, and any electrical equipment in the vicinity of tanks that does not conform to National Fire Protection Association (NFPA)-70, "National Electric Code."

Proximity to personnel and environment: The danger posed by hydrocarbon liquid storage tanks is often increased when the location of the tank does not conform with current minimum spacing requirements. Sections 2.3.2 to 2.3.3 of NFPA-30 discuss minimum spacing. This subject is previously discussed in Section 7.1 (Figure 7-45). For mitigating consequences to personnel/workers, the environment, and other tanks, proper secondary containment (diking/bunding) should be considered for containment.

Hazard Identification:

Facilities should evaluate their storage tanks for potential to catastrophically fail and identify factors that could cause storage tank explosion. Some of the factors to look for include, but are not limited to, the following:

- Atmospheric storage tanks that do not meet API-650 or other applicable code(s) and contain flammable liquids or liquids that may produce combustible vapor.
- Tanks with corrosion around the base and/or steel tanks whose base is in direct contact with ground and exposed to moisture.
- Tanks or associated structures (e.g., pipes) with weakened or defective welds.
- Tanks used to store mixtures containing water and flammables where the water phase is at the tank bottom and may contribute to internal bottom corrosion.
- Tanks containing combustible vapor and not equipped with flame arrestors or vapor control devices to limit emissions.
- Possible ignition sources near tanks containing combustible vapor.

Safety Areas for Hazard Reduction

Storage tanks should comply with all regulations, industry codes and standards, including inspection and maintenance requirements to keep tanks in proper condition. Facilities with storage tanks that can contain flammable vapors should review their equipment and operations. Areas to review should include, but not be limited to, the following:

1) **Design of atmospheric storage tanks:** API and other organizations have standards and codes that address recommended practices for tank design and construction. It is imperative to evaluate whether the liquids or certain components of liquid mixtures may generate combustible vapors. Design measures include fire protection, flame arrestors, emergency venting (such as part of the API-650), prevention of flash back (for tanks containing flammable liquids), and proper berming or diking.
2) **Inspection and maintenance of storage tanks:** API-653 has tank inspection guidelines and procedures for periodic inspections and testing, especially for older tanks. These procedures call for written documentation of inspections by API Certified Tank Inspectors. Measures to review include procedures for pressure testing, welding inspections, and checks for corrosion or metal fatigue. API-650 specifies welding procedures and welding qualifications as well as joint inspection (e.g., radiograph and magnetic particle examination). Programs for tank inspection and maintenance should be developed in accordance with these standards.
3) **Hot-work safety:** Both the Occupational Safety and Health Administration's (OSHA) regulations concerning hot work and NFPA's standards on welding should be reviewed for compliance. Hazard reduction measures include proper hot-work procedures such as obtaining a hot work permit, having a fire watch and fire extinguishing equipment present, and proper testing of atmosphere for explosivity; covering and sealing all drains, vents, manways, and open flanges; sealing all sewers (to prevent gas or vapor migration); and training workers and providing them with appropriate protective equipment.
4) **Ignition source reduction:** Both OSHA regulations and NFPA standards should be reviewed for compliance. Hazard reduction measures may include: having all electrical equipment in a hazardous environment conform with the requirements of the National Electric Code (NFPA-70), grounding tanks to dissipate static charge, using only "non-spark producing" tools and equipment in flammable atmospheres, and taking care to not create sufficient heat or sparks to cause ignition of flammable vapors.

8.8.2 Designing Tankage Systems to Minimize Hazards

Storage and handling of flammable and combustible liquids can be conducted without fire losses to storage tanks with proper awareness of the hazards associates with operation of storage tanks and hydrocarbon fuel transfer. Proper precautions begin with the design and installation of storage tanks taking into account the accepted safety measures for the use, storage and handling of flammable and combustible liquids [46, 47].

Application of effective fire prevention to the protection of storage tanks and facilities related to handling hydrocarbon fuels at a storage site can be achieved through an evaluation of fire hazards and installation of protective measures, operating practices and proper emergency response plans. The facility designer has available a number of tools to provide a fire safe facility including good engineering practices as outlined in NFPA Standards.

8.8.2.1 Effective Steps

Effective steps to tank fire safety can be assessed through examination of following:

- Fluid in the tank
- Type of tank
- Tank Location
- Spill control technique
- Tanks venting arrangement
- Control of ignition sources
- Fire prevention measures
- Fire protection in place for the tank
- Emergency Response planning and facilities

Some of the above steps not covered elsewhere in this document are described below.

8.8.2.1.1 Fluid in the Tank The information on type and characteristics of the stored liquid hydrocarbon is essential to understanding the inherent hazards of the liquid. This should include specific data and whether or not the liquid is a flammable or combustible liquid.

Such a data is essential for developing an effective site fire protection plan. Data related to specific hazards of fuel stored should include data from the Material Safety Data Sheet (MSDS) on flash and boiling points, molecular weight, vapor density, and flammable range. The key terms that define flammable and combustible liquids fire hazards are Flashpoint and Vapor Pressure. These information provide the vital information used for emergency planning and fire prevention.

> Flashpoint — This is the liquid temperature at which a liquid releases sufficient vapor to form an ignitable mixture with air, either near the liquid surface or within a storage tank. Flashpoint can be determined by a flashpoint tester and is the basis for hazard classification.
>
> There is a direct connection between volatility and flashpoint. A liquid with a flashpoint near normal temperatures, without being heated, will produce vapor that can be ignited by a small ignition source; such as a spark or pilot flame. Flashpoint and boiling point temperatures are reduced as altitude increases since liquid volatility increases with reduced atmospheric pressure. Hydrocarbon liquids that are combustible at sea level may be more hazardous as atmospheric pressure is reduced.
>
> Vapor Pressure — This is the pressure, measured in kPa (Psia) is exerted by vapor against the atmosphere. Just as the atmosphere exerts pressure on the hydrocarbon

liquid surface, the hydrocarbon liquid pushes back. Vapor pressure is normally less than atmospheric pressure and is a measure of the evaporation or change in state from liquid to the gaseous state. This characteristic is often termed volatility and liquids that easily evaporate are termed as "volatile" liquids. The safety concern is that the higher the vapor pressure, the more the hydrocarbon liquid evaporates and the lower the boiling point, resulting in more vapors within an increased risk.

Other important information of the stored hydrocarbon liquid is its classification. Such a classification will define flammability for application of fire and safety code requirements. Flammable hydrocarbon liquids are Classified as Class I liquids and combustible hydrocarbon liquids are classified as Class II or III liquids.

Flammable Liquids — Any liquid with a closed-cup (type of flashpoint test) flashpoint below 37.8°C (100°F) and a Reid vapor pressure not exceeding 2068 mm Hg (6 kPa, 40 psia) is a flammable liquid. These liquids are referred to as Class I liquids. Liquid classifications are further divided into three classifications as follows:

- o Class IA liquids have flashpoints below 22.8°C (73°F) and boiling points below 37.8°C (100°F).
- o Class IB liquids have flashpoints below 22.8°C (73°F) and boiling points at or above 37.8°C (100°F).
- o Class IC liquids have flashpoints at or above 22.8°C (73°F), but below 37.8°C (100°F).

Gasoline, Jet B, and JP 4 are typical examples of a flammable liquid.

Combustible Liquid — A liquid with a closed-cup flashpoint at or above 37.8°C (100°F) is a combustible liquid. These liquids are referred to as either a Class II or Class III liquids based on the following:

- o Class II Liquids have flashpoints at or above 37.8°C (100°F) and below 60°C (140°F).
- o Class IIIA Liquids have flashpoints at or above 60°C (140°F), but below 93°C (200°F).
- o Class IIIB Liquids have flashpoints at or above 93°C (200°F).

Class IIIB liquids are generally not included in the scope of codes and standards with the exception of the International Fire Code which does not differentiate between Class IIIA and IIIB liquids requiring the same safety measures despite the lower hazard. Fuel Oil No. 2, kerosene, and some Jet A and A-1 fuels are typical examples of a combustible liquid.

8.8.2.1.2 Type of Tank and Fire Risk The type of hydrocarbon liquid storage tanks can influence the assessment of safety issues depending on their design, configuration, and operating pressure, some of which are indicated below [48]:

Tank Type	Overfill Fire	Vent Fire	Rim-Seal Fire	Obstructed Full Surface Fire	Unobstructed Full Surface Fire
Floating roof					
Internal	Yes	Yes	Yes	Yes	No
External	Yes	No	Yes	Yes	Yes
Domed	Yes	Yes	Yes	Yes	No
Fixed roof	Yes	Yes	No	Yes	Yes

The following only considers vertical configuration tanks that are normally designed to operate at atmospheric pressure. While some of the information were described previously, they are however detailed herein to address subsequent safety issues.

Atmospheric Above Ground Storage Tanks: Atmospheric storage tanks operate at pressures ranging from atmospheric up to and including 0.7 kPag (1.0 psig). Larger atmospheric storage tanks may have a maximum operating pressure below 0.7 kPag (1.0 psig) in order to eliminate roof plate weld stress from continuous exposure to higher operating pressure.

Some locations may have low-pressure tanks designed to operate at pressures greater than 0.7 kPag (1.0 psig) but not more than 100 kPag (15 psig). The primary differences between an atmospheric and low-pressure tank is that protection is provided to the low-pressure tanks to prevent explosive tank failure when exposed to fire.

Atmospheric Tank Design and Safety Standards: The most commonly used standards for the design and construction of atmospheric storage tanks are listed in Section 5. The definitive standard that is commonly used for fire safety in the handling, storage, and use of flammable and combustible liquids is the National Fire Protection Association, Standard 30, *The Flammable and Combustible Liquids Code.* Most building and fire codes and the standards developed by other organizations use NFPA 30 as the definitive standard for flammable and combustible liquids fire safety.

Vertical Storage Tanks: The configuration of a vertical aboveground tank design can be either an **open top** with the roof floating on the stored liquid or a **fixed roof**. The safe design of floating roof tank offers a considerable level of fire safety over other vertical tank designs. As a result, fire codes allow closer spacing between floating roof tanks and for separation from adjacent properties or operations providing a cost advantage in tank farm layout and arrangement.

Open Top Tanks – An open top tank has a floating roof exposed to the environment. Fire scenarios in Open Roof Tanks can be summarized in Figure 8-112.

Open top tanks have a low fire potential with the most likely fire exposure being a limited amount of liquid in the seal area between the shell and the roof. The fire potential is very low this area, which is the only space on the tank roof where an ignitable mixture normally exists. Fires in the seal area can be readily extinguished by a hose line or a portable extinguisher on small to medium sized tanks; while a fixed foam system is normally installed on larger tanks.

Lightning is the most prevalent ignition source for a seal area fire and is preventable by installing a bond between the roof and the shell. Routine inspection of lightning protection bonding straps is required to assure that protection measures are in place and not damaged by roof movement, maintenance activities or by weather effects. Another ignition source is hot work on the roof or near the seal area without proper precautions to prevent ignition. Fire in a dike area near or impinging on the tank can ignite seal area vapors. Fire activities should minimize tank shell fire impingement to prevent fire spread to the seal area.

> ***Covered Floating Roof Tanks*** — This tank design is a floating roof which is protected by another fixed roof for weather or environmental control. During filling, an ignitable mixture can be present in the vapor space between roofs. Gas tests should be conducted before personnel entry or during hot work in the vapor space above the floating roof. Personnel entries into the vapor space should be conducted as a confined space entry. It is possible for the floating roof to be less substantial or stable than a standard open top floating roof tank.

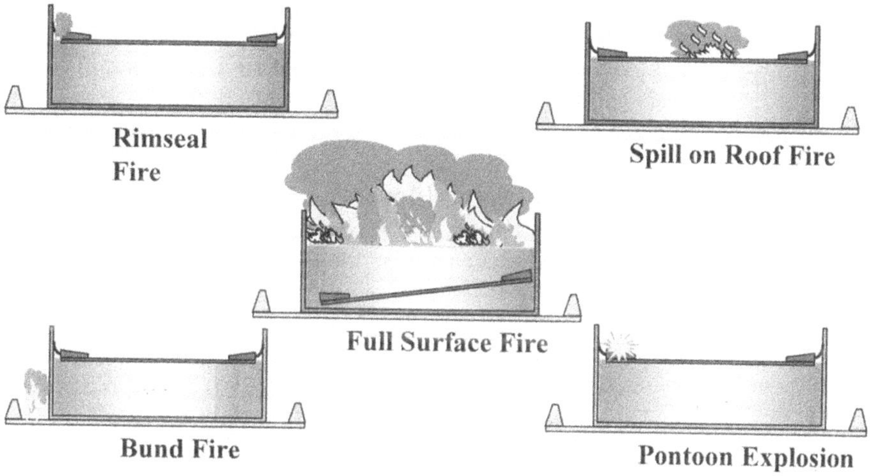

Figure 8-112. Fire scenario in above ground open floating roof tank

Where the roof is a pan configuration, see discussion below on Fixed Roof Tanks.

Fixed (Cone) Roof Tanks — Fire exposure to a tank increases internal pressure from boiling liquids or, very rarely, from an internal explosion. Fixed roof tanks are designed to vent during an emergency by opening weak roof-to-shell joint seams at the point of panel attachment to the tank shell. The weak seam allows the roof to tear free from the tank shell to prevent failure of other joints.

An internal pan roof floats on the liquid. These roofs are typically constructed of honeycombed aluminum panels, metal on plastic/foam floats or similar style roof design. This type of roof lacks substantial support and stability inherent in an open or covered floating roof tank and is not fire resistant. A common failure mode for this style of roof is tilting or sinking resulting in fire exposure to the tank. This style of roof is commonly used for emission control purposes, but is unlikely to have a substantial effect on tank fire safety. Installation of a pan roof can also result in blocking application of foam during a fire, especially where the tank is protected by sub-surface foam and/or the roof sinks blocked foam inlets.

8.8.2.1.3 Tank Location/Accessibility The location of a tank facility has a direct impact on fire safety. The tank location should be arranged to prevent exposing fires from spreading into the tank farm; and conversely adjacent buildings should be located so that a fire in the tank farm will not spread off of the property or spread to adjacent tanks.

NFPA 30 contains requirements for spacing between tanks and to or from property lines and adjacent structures and facilities. These distances are minimal and increased spacing may be beneficial when constructing a new facility since increased separation will reduce risk of fire exposure. Reference should be made to NFPA 30 minimum spacing requirements during emergency planning to prepare for exposure protection during fire fighting operations. See also previous Figure 8-45 Storage Tank Spacing for Chemical/Oil Plants.

Accessibility is a key factor in effective storage tank fire control. Fire fighting operations require separation from the tanks in order to avoid fire exposure to fire equipment and personnel. Where subsurface inlets are installed for tank fire control, access to foam inlets should be outside the dike area or in an easily accessible location where tank shell connections are utilized.

Where foam trucks and other portable equipment will be used for firefighting operations, the tank farm should be accessible from at least two directions. The ideal arrangement provides complete access around the tank dike areas with roads and access points sufficiently wide for fire trucks. Vehicles should be able to move by fire trucks parked at fire department connections to foam systems and at hydrants.

8.8.2.1.4 Spill Control Technique Hydrocarbon liquid spills occur when tanks are overfilled, as pump seals or pipeline flanges leak or corrosion holes occur in tank shells. The ground area around the tanks should be arranged to direct spilled liquids away from the tanks to lower sections of dike areas, into diked impoundment areas or to a basin, remote impounding. If space is available in a tank farm the most desirable and safest arrangement is a containment area that is remote from equipment, tanks, and exposed structures.

Dike/bund area drains should be a fire sealed and extend through the dike/bund wall. Each drain valve should be clearly labeled and located outside the dike discharging to the site drainage system for treatment of small spills in an environmentally correct manner.

8.8.2.1.5 Tanks Venting Arrangement Tanks exposed to a fire in contact with or near the tank shell will have an increase in pressure as tank contents are heated. Tank vents are arranged to facilitate pressure changes as liquid is added or withdrawn or changes resulting from atmospheric temperature and pressure. Fire exposure to an aboveground storage tank heats the hydrocarbon liquid contents to their boiling point which substantially increases storage tank pressure. Fire exposure also can damage tank supports and un-wetted portions of the tank shell. Increased pressure can be controlled either by tank design or by tank vent devices. Emergency pressure relief through tank design includes

- lifter or floating roofs,
- a weak roof to shell tank roof seam or
- emergency vent devices.

Without pressure relief the tank is likely to rupture randomly, or even rocket as a result of a failure of the shell to bottom seam.

Hydrocarbon Liquids, such as LPG in pressure tanks having boiling points below atmospheric temperature usually are fitted with relief valves which are set to maintain the tank at a higher pressure setting. Fire exposure to the portions of the vessel shell not in contact with the liquid can cause a loss of vessel shell strength as the fire heats unprotected metal to failure. The resulting metal tear is likely to completely encircle the tank and to release tank contents; often with pieces of the tank rocketing explosively away from the vessel. This action, known as a Boiling Liquid Expanding Vapor Explosion (or BLEVE), is the catastrophic failure of a container into at least two major pieces and simultaneously releasing the contained liquid which is at a temperature well above its boiling point at normal atmospheric pressure. BLEVES have occurred when LPG filled railroad cars or storage tanks were fire exposed. Pressurized tanks are therefore used for storage of LPG.

8.8.2.1.6 Control of Ignition Sources Precautions need to be taken to prevent ignition of flammable vapors through the control of:

- Open flames
- Over heat or failure of mixers
- Lightning, floating roof shunt design

- Floating roof contact with tank shell
- Hot surfaces
- Radiant heat
- Smoking
- Cutting and welding
- Static electricity
- Electrical sparks and stray currents
- Heating equipment and other ignition sources
- Lightning strikes/Floating roof shunt
- Hot work on live tanks
- Flare stack fall-out

Welding, cutting and similar spark producing operations should only be conducted under a strict hot work control system. Adequate controls for hot work should include comprehensive testing of the area to assure that flammable vapors are not present in the work area. All combustibles that may be ignited by hot sparks from hot work should be removed or properly protected to prevent against ignition.

Static electricity controls start with proper operating practices and controls to minimize the potential for generation of static electricity. Metallic equipment should be bonded or grounded when used in areas where an ignitable mixture could be present. All non-metallic equipment and piping, where an ignitable mixture could be present, should be carefully reviewed to assure that the generation of static is not a possibility.

NFPA 70, *The National Electrical Code addresses electrical equipment design and installation*. In areas where there is likely to be the presence of an ignitable vapor mixture, the equipment would be required to meet requirements for electrical area classification in accordance with NFPA 30 and 70.

NFPA 30 requires an engineering evaluation of the installation and operation to determine the extent of fire prevention and control measures followed by the application of sound fire protection and process engineering principles. The analysis is required to include:

- Analysis of fire and explosion hazards of the facility;
- Analysis of local conditions, such as exposure to and from adjacent properties, flood potential or earthquake potential; and
- Fire department or mutual aid response.

8.8.2.1.7 Fire Prevention Measures Fire Protection measures for storage tanks would be best to consider the relative hazard of the stored materials instead of on every tank regardless of hazard. Consideration would include the relative hazard of the stored materials, the hazards inherent in the type of tanks in the storage facility, and the availability of public and private fire fighting support and equipment.

NFPA 30 requires that a fire-extinguishing system be provided for vertical atmospheric fixed-roof storage tanks storing Class I flammable liquids larger than 1.2×10^6 I (1200 bbls) capacity. Fire Protection is especially desirable where tanks are located in a congested area with an unusual exposure hazard between the tank and adjacent property. Fixed roof tanks storing Class II or III liquids at temperatures below their flashpoints and floating roof tanks storing any liquid, generally do not require protection when the tanks are installed in compliance with spacing and installation requirements of NFPA 30.

Relative Hazard of Stored Materials: Class IIB liquids (flashpoints above 93°C (200°F)) can be safely stored without fire protection systems unless the oils are being heated to or above their flashpoint. Tank spacing, containment measures and fire extinguishing systems are not required by most codes for combustible liquids with a flashpoint of 93°C (200°F) or above. A low curb or other spill control means should be provided and spacing for maintenance activities should be provided between tanks.

Class III liquids (flashpoints over 60°C (140°F) and below 93°C (200°F)) are not normally fire protected, except when heated to or above their flashpoints. Unheated tanks in this category will usually not require foam protection unless there are unusual conditions such as:

- There is a potential that the tanks will be filled with materials with a flashpoint below 60°C (140°F).
- Overheating of the tanks can occur resulting in storage temperatures exceeding the flashpoint of the stored materials.
- The tank is vulnerable to fires involving adjacent equipment or facilities.

8.8.2.1.8 Fire Protection Tanks containing flammable liquids, those with a flashpoint below 60°C (140°F), may require fire protection. Industry experience with gasoline and other flammable liquids stored in open top or covered floating roofs show minimal fire experience; and with good spacing between tanks and minimal fire exposure, these types of tanks storing flammable liquids have a low fire hazard. NFPA 30 requirements for protection for Class I liquids was discussed previously.

Basis for Foam Protection Requirements: NFPA 30 contains requirements for foam protection for Class I liquids. Extinguishment of the largest cone roof tank fire is typically the design basis for a tank foam system. Cone roof tanks are typically protected by sub-surface foam systems, unless the tanks containing polar solvents and alcohol resistant foams are not available or pan roofs are installed in the tank. Open-top floating roof tanks require protection from rim fires.

Foam Application: Foam application rates are detailed in NFPA Standard 11 as *Low, Medium and High Expansion Foam* for various hydrocarbon storage tanks. Foam application, once started, is required to be applied continuously at the minimum recommended rate and duration in order to extinguish a tank fire. Intermittent or shortened foam application without fire extinguishment results in foam breakdown and destruction of any established foam blanket. Additional foam concentrate will be required to re-establish the lost foam blanket and blanket the entire liquid surface. The NFPA II recommended quantities of foam-producing material are sufficient for fire extinguishment, however, application of additional foam-producing concentrate can decrease the time for fire extinguishment.

Cone Roof Tanks — Foam application on cone roof tank fires can be achieved by various methods: subsurface foam injection; portable foam towers; fixed foam chambers mounted on the tank shell or portable hose streams or monitors. Application by subsurface injection, towers or fixed foam chambers is usually applied at a rate of 4 l/minute/m^2 (0.1 gallons per minute per square foot) and application rates for portable hose streams or monitors is 6 liter/minute/m^2 (0.16 gallons per minute per square foot). Fires in more volatile materials such as gasoline may require higher rates of up to 8 liter/minute/m^2 (0.2 gallons per minute per square foot), depending on pre-burn time.

Cone Roof Tanks with Internal Floating Covers (Pans) — Foam protection using an over-the-top system designed for coverage of the entire tank liquid surface area of the tank should be provided when internal covers are a pan-type design, constructed of combustible materials, or are thin aluminum skin and floats. Subsurface applications are not recommended since the inlets may be blocked by a sunken roof and thus prevent foam application into the tank.

Open Top Floating Roof Tanks — Open-top floating roof tanks are protected for seal fires. For tanks up to 45 m (150 ft) in diameter, foam can be applied from portable hose lines supplied from fire trucks for tanks up to 36 m (120 ft) diameter and from a foam solution standpipe on tanks 36 to 45 m (120 to 150 ft) diameter. Tanks over 45 m (150 ft) diameter are more suited for a fixed foam application system for rim fire protection. The rim fire protection system (catenary design) can be supplied foam from a fire truck or from a piped foam system. The rim seal protection system should meet NFPA 11 requirements for application rate dependant upon the type of seal and foam outlet arrangement.

Portable hose streams should be a minimum of two 3.15 to 3.6 l/s (50 to 60 gpm) foam nozzles supplied from foam fire trucks or a foam-proportioning system. A foam dam is required when hose streams are used for protection when the seal is a tube seal or includes metal weather shields or a noncombustible secondary seal.

Covered Floating Roof Tanks — When the floating roof in a covered floating roof tank is a steel floating roof of the single or double-deck design, a foam system is normally not required due to the low risk of a fire inherent with this tank design. However, other styles and designs for covered floating roof tanks should be provided with foam systems for control of seal fires meeting requirements for open top floating tanks when the tank diameter exceeds 45 m (150 feet) or when the tank has significant value or risk to the site.

Portable Monitors and Hose Nozzles/handlines: Portable foam monitors and hose streams have limited ability to extinguish tank fires. Protection from monitors is generally ineffective on cylindrical or cone roof tanks over 18 m (60 ft) in diameter and hose lines are not effective for tanks over 10 m (30 ft) in diameter and over 6 m (20 ft high). While foam monitors have been successful in extinguishing fires that involved tanks up to 40 m (130 feet) in diameter and 13 m (42 feet) high, the use of large monitors should not be depended upon as a primary means for extinguishment of large cone roof tank fires.

There are substantial limitations when using portable monitors or hose streams for tank fire extinguishment. Low tank product levels, wind effects and intense fire updrafts can prevent foam from reaching the product surface and forming a foam blanket. Foam requires continuous and even application to establish a foam blanket, since foam streams should be directed against the inner tank shell so that the foam flows gently onto the burning liquid surface without undue submergence. Due to the difficulties in use of portable foam devices – hose and monitor nozzles – a foam solution rate of at least 0.16 gallon per minute per square foot of tank liquid surface areas is required. Due to the difficulty in directing the foam into the narrow annular space of the roof seal, and the potential to sink the roof, ground level monitors are not recommended for seal fires. Often poorly executed monitor application for foam fires has resulted in roofs sinking or tilting, which results in a larger fuel surface fire in the tank. Boom mounted nozzles on fire trucks may be useful in application of foam providing there is foam truck accessibility and the tanks are relatively small in diameter

Water Supply: Subsurface injection or topside foam application rates are determined by the tank size, the application rate and the amount of water required for cooling the tank and exposures. For flammable and combustible liquids the water application rate for subsurface or topside application is typically 4 liter/minute/m^2 (0.1 gallon per minute per square foot). Additional water will be required for foam hose streams and water cooling streams.

Water Requirements - The following example illustrates the additive quantities of water required to meet the full water demand for tank protection:

***Example 8.1 – 12 m* (40 ft) diameter Tank**

- Required Foam Solution - 473 Liter/minute (125 gpm)
- Foam Hose Streams 189 Liter/minute (50 gpm)
- Water Cooling Streams 1890 Liter/minute (500 gpm)
- Total Water Requirement 2552 Liter/minute (675 gpm)
- Foam Solution Rate (@4 liter/minute/m^2, 0.1 gpm/sq ft) = 665 Liter/minute (176 gpm)

***Example 8.2 – 45 m* (150 ft) diameter Tank**

- Required Foam Solution - 6680 liter/m^2 (1,767 gpm)
- Foam Hose Streams 567 liter/m^2 (150 gpm)
- Water Cooling Streams 3780 liter/m^2 (1,000 gpm)
- Total Water Requirement 10962 liter/m^2 (2,900 gpm)
- Foam Solution Rate (@4 liter/minute/m^2, 0.1 gpm/sq ft) = 7246 liter/m^2 (1,917 gpm)

Water Pressure – Water cooling streams can be taken directly from fire hydrants, however, hydrant residual pressure less than 550 kPag (80 psig) will require fire truck booster pumps to supply hose streams. As an example, a 945 liter/minute (250 gpm) cooling stream will reach over the curb angle of about 22 m (75 foot) high tank from a horizontal distance of 20 m (65 ft) with 345 kPag (50 psig) nozzle pressure using a solid stream nozzle of 28 mm (1 1/8 inches). Minimum residual hydrant outlet pressure should be 550 kPag (80 psig) in order to supply 189 liter/minute (50 gpm) through 75 m (250 ft) of 38 mm (1 ½ in.) fire hose.

Fire Mains and Hydrants – The storage tank farm fire water demand is based on the quantity of water required to protect the largest cone roof tank, plus the water needed for cooling hose streams to protect exposed tanks or adjacent facilities. As a general rule of thumb sufficient water will be required to extinguish a fire in the largest tank, plus the amount of water required for protection of unshielded neighboring tanks. Water pressure in the hydrants and the water system should be adequate to provide 700 kPag (100 psig) residual pressure when flowing sufficient water to supply cooling streams directly from hydrants.

A fire water system looped around the facility provides flexibility during use. Isolation of damaged sections without impairment of the entire fire protection system can be aided by provision of division valves arranged so that any section of the fire main grid or loop may be removed from service while fire water continues to be supplied for tank fire protection. System hydraulic calculations can assist in determining that there is an adequate water flow rate. The number of valves needed for system reliability should be based on no more than six hydrants out of service as a result of a single main break. Hydrants should be provided in sufficient number to permit fire fighting operations regardless of wind direction.

Hydrants should be located outside of tank dike/bund walls, adjacent to roadways and a minimum of 1 tank diameter; but no less than 15 m (50 feet) from the tank shell to avoid heat exposure to fire trucks when connected to the hydrant. Hydrant outlets should be no more than about 2 m (6 feet) from the approach of a fire truck to the hydrant.

***Tank Protection Using Passive and Active Systems*:** Fire exposure protection can be provided by passive systems including fire proof insulation, concrete encasement or similar insulation materials applied on the tank shell. Protection could be provided by methods tested by the Underwriters Laboratories (UL) to a 2 to 4 hour hydrocarbon fire exposure in order to protect the tank from direct hydrocarbon fire exposure. Insulation systems are difficult to apply to vertical tanks and are more suited for horizontal vessels. Careful consideration should be given to the type of fire exposure and potential fire duration when applying hydrocarbon fire exposure insulation methods.

Active tank protection systems include water spray systems for application of water onto the vessel shell. This type of system, designed to requirements of NFPA 15, *Water Spray Systems*, typically applies water through a system of specially designed nozzles at minimum application rates of 6 to 16 l/minute/m^2 (0.15 to 0.40 gpm/square foot) of vessel surface area. Water spray systems require a strong fire water supply and a means to automatically apply the water in order to avoid delay which will heat tank contents and possible result in tank rupture.

8.8.3 Design of a Foam System for Fire Protection of Storage Tanks

8.8.3.1 Identifying Flammable Liquid

Foam and foam system selection depends on the liquid protection classification/requirement. There are two basic classifications of flammable and combustible liquids:

- Hydrocarbon (non water miscible) and
- Polar Solvent (water miscible)

The Hydrocarbon family typically consists of standard petroleum products such as - Gasoline,

- Kerosene,
- Diesel,
- Jet Fuel,
- Heptane,
- Crude Oil

The above products do not mix with water.

The Polar Solvent group typically consists of:

- Ethanol,
- Methanol,
- Ketone,
- Acetone

or products that will mix readily with water. It may be noted that pure MTBE is only slightly water miscible (approx. 5% - 7%.)

8.8.3.2 Types of Foam Discharge Outlets

Under Writer Laboratory (U.L.) has established two different types of foam discharge outlets:

- ***Type II Discharge Outlet*** - A fixed device that delivers foam onto the burning liquid and partially submerges the foam and produces minimal agitation of the surface.

  Examples of this type of device are Foam Chambers and Foam Makers.

- ***Type III Discharge Outlet*** - A fixed or portable device that delivers foam in a manner that causes the foam to fall directly onto the surface of the burning liquid in such a manner that causes severe agitation.

  Examples of this type of device are Hose Stream Nozzles/hand-lines and Monitors.

8.8.3.3 Foam System for Fire Protection of Storage Tanks

There are two basic methods of fire protection systems for storage tanks:

1. Sub-surface Base Injection
2. Over the Top - (Subdivided as follows)
   - Foam Chambers
   - Foam Makers
   - Portable Foam Monitor
   - Foam Tower

These are described below:

8.8.3.3.1 Sub Surface Base Injection: The sub-surface method of fire protection produces foam with a "High Back Pressure Foam Maker" usually located outside the storage tank. This system delivers the expanded foam mass through piping into the base of the tank. The pipe may be an existing product line or can be a dedicated fire protection foam line. The expanded foam entering the tank through a discharge outlet is injected into the flammable liquid. The discharge outlet must be a minimum of 30 cm (1 ft) above any water that may be present at the base of the tank. The foam will be destroyed if injected into the water layer. When injected into the fuel, the foam will rise through the fuel and form a vapor tight foam blanket on the fuel surface.

Advantages of Sub-surface

- The rising foam can cause the fuel in the tank to circulate which can assist in cooling the fuel at the surface. If there is an explosion and fire that may damage the top of the tank, the sub-surface injection system is not likely to suffer damage.
- The discharging foam is more efficiently directed to the fuel surface without any interruption from the thermal updraft of the fire.

Disadvantages of Sub-surface

- This technique **cannot** be used in storage tanks containing polar solvent type fuels (i.e. ethanol, Methanol, Ketone, Acetone.)

- Not Recommended for use in either Open Top Floating Roof or Internal Floating Roof type tanks.
- Caution must be used so that the maximum foam inlet velocity is not exceeded; otherwise, there will be an excessive fuel pickup by the foam as it enters the tank.
- Not to be used for protection of Class 1A hydrocarbon liquids.
- Sub-surface injection of foam is generally not recommended for fuels that have a viscosity greater than 2,000 ssu (440 centistokes) at their minimum anticipated storage temperature.

High Back Pressure Foam Maker (HBPFM): The HBPFM is mounted in the foam line used to aspirate the foam solution before it is discharged into the storage tank. It will typically give an expansion ratio of between 2 -1 and 4 - 1. The device is capable of discharging against considerable back pressure which can be as high as 40% of the operating pressure. The back pressure is an accumulation of the head pressure of the liquid hydrocarbon inside the storage tank and any friction loss between the foam maker and the tank. A minimum of 700 kPa (100 psi) inlet pressure into the HBPFM is normally required to ensure correct operation.

The foam velocity through the piping to the hydrocarbon storage tank from the HBPFM is very critical. With flammable liquids, the foam velocity entering the tank should NOT exceed 3 m/s (10 ft. per sec) and with combustible liquids the foam velocity should NOT exceed 6 m/s (20 ft. per second).

The following Table 8.9 shows the minimum discharge times and application rates for **Sub-surface Base Injection** application:

8.8.3.3.2 Over the Top — Foam Chambers

8.8.3.3.2.1 Foam Chambers

Type II Discharge Device: The Foam Chamber is normally used on cone roof storage tanks. The chamber is bolted or welded on the outside of the tank shell near the roof joint. A deflector is mounted on the inside of the tank so that the discharging foam from the foam chamber will be diverted back against the inside of the tank wall.

The foam chamber is mounted on the cone roof storage tank vertical shell wall in a position just below the roof joint, or approximately 20 to 30 cm (8″ to 12″) down from the roof joint to the center point of the foam chamber outlet. When the foam chamber is mounted correctly, the internal glass vapor seal of the foam chamber will be just slightly higher in elevation than the roof joint on the storage tank.

Each foam chamber mounted on a cone roof storage tank should have its own individually valved riser supplying the foam solution from outside the dike area. For

TABLE 8-9. Sub-surface base foam injection system – design rates

Hydrocarbon Type Fuel	Minimum Discharge Time	Minimum Application Rate
Flash point between 100 °F and 140 °F (37.8 °C and 93.3 °C)	30 min.	0.10 gpm/ft. 4.1 L/min./m2
Flash point below 100 °F (37.8 °C) liquids heated above their flash points.	55 min.	0.10 gpm/ft. 4.1 L/min./m2
Crude Petroleum	55 min.	0.10 gpm/ft. 4.1 L/min./m2

NOTE: The maximum application rate shall be 0.20 gpm/ft. (8.1 L/min./m.)

correct operation, a minimum of 280 kPa (40 psi) is required at the inlet to the foam chamber.

8.8.3.3.2.2 Foam Makers **Type II Discharge Device**: The foam maker is normally used to aspirate foam solution before being discharged inside a dike (bund) area or when used with external floating roof tanks to supply foam to the rim seal area.

The discharge pipe downstream of the foam maker is sized to slow the velocity of the expanded foam and shaped to deflect the foam back against the inside of the dike wall or onto a splash board or the tank shell wall when used for floating roof seal protection. The splash board is to be mounted above the top of the floating roof tank. The foam discharging pipe must be correctly size for dike/bund protection system.

When mounted on a storage tank or used for a dike/bund protection system, the foam maker can be mounted in either a horizontal or vertical position without any detrimental effect on foam performance. It is recommended that a minimum 30 cm (12″) length of straight pipe be installed upstream from the foam maker during the installation.

When using a 38 mm (1 1/2″) Foam Maker for a dike fire protection system, a 10 mm (3″) diameter pipe with minimum length of 0.6 m (28″) and a maximum of 2.25 m (100″) are usually connected to the foam maker outlet (downstream side). This length of discharge pipe allows for the correct foam expansion to take place and slows the discharge velocity.

A 63 mm (2 1/2″) Foam Maker requires a length of 100 mm (4″) pipe to be connected to the discharge side of the maker. This length of pipe should also be a minimum of 0.6 m (28″) but can have a maximum length of 2.7 m (120″). The discharge pipe in both instances should be directed back against the inside wall of the dike. This installation allows a more gentle application to the flammable liquid within the dike and lessen the submergence of the foam.

8.8.3.3.3 Criteria for Sizing a Foam System for a Cone Roof Storage Tank

- Identify the fuel inside the tank.
- Type of foam concentrate to be used.
- Calculate the fuel surface area (r^2).
- Application rate.
- Type of discharge device required and quantity (based on fuel flash point and tank diameter).
- Calculate discharge duration.
- Supplementary hose lines required and discharge duration.
- Quantity of foam concentrate required*.
- Establish bill of materials.

* **NOTE:** To determine the quantity of foam concentrate in a given quantity of foam solution, use the following formula: Multiply the foam solution by:

- × 0.01 if using a 1% type of concentrate
- × 0.03 if using a 3% type of concentrate
- × 0.06 if using a 6% type of concentrate

Calculation Example 8.1:
Cone roof tank - 150 ft (45 m) diameter
Fuel — Gasoline
Foam Concentrate — 3% **Aqueous Film Forming Foams (AFFF)**

- Surface area — 75' × 75' × 3.1417 = 17,672 sq. ft. (3.1417 × 22.5^2 = 163 m^2)
- App. Rate at 0.10 gpm per sq. ft. (Per NFPA 11) 0.10 × 17,672 sq. ft. = 1767.2 gpm of foam solution required.

- Discharge device × Foam Chamber, Qty. 4 required.
- Discharge Duration — 55 min. 1767.2 × 55 = 97,196 gallons of foam solution × 0.03 = 2916 gallons of 3% **Aqueous Film Forming Foams (AFFF)** concentrate required.
- Supplementary Hose Lines required (Per NFPA 11) — Qty. 3 required (each minimum 50 gpm) (Tank dia. over 120 ft.).
- Hose Line discharge duration (Per NFPA 11) — 30 min.
- (Tank dia. over 95 ft.) 3 × 50 = 150 × 30 = 4500 gallons of foam solution × 0.03 = 135 gallons of 3% **Aqueous Film Forming Foams (AFFF)**. 3,051 (135 + 2,916) gallons of foam concentrate required.

The appropriate bill of materials of major components for the above system using a bladder tank could be.

- 1 × 3,200 Gallon horizontal style bladder tank.
- 1 × 6" Between flange style ratio controller.
- 4 × 6" foam chambers each with a flow rate of 442 gpm at appropriate incoming pressure that exceeds
- 40 psi.
- 1 × 2 1/2" Threaded type ratio controller (For supplementary system.)
- 3 × 50 gpm hand-line nozzles.
- foam concentrate.

NOTE: When protecting multiple storage tanks the foam system is to be sized to protect the single largest hazard. Table 8-10 identifies the number of Foam Chambers required for the protection of a flammable liquid contained in a vertical cone roof atmospheric storage tank where the discharge device is attached to the tank. Where two or more outlets are required, the outlets are to be equally spaced around the tank periphery and each outlet is to be sized to deliver foam at approximately the same rate.

It is suggested that for tanks above 200 ft. (60 m) in diameter at least one additional discharge outlet be added for each additional 5000 sq. ft. (465 sq. m.) of liquid surface or fractional part.

Figure 8-113 depicts a foam pump skid system illustrating piping, valves, discharge devices, ratio controller, foam pump and foam storage tank for the above system.

Figure 8-114 shows a Cone Roof Storage Tank with a Semi-Fixed Foam System.

Calculation Example 8.2: The following example shows the foam system requirements for seal protection of a 150 ft. diameter open top floating roof tank.

- Type of Tank — Open Top Floating Roof Tank
- Diameter of Tank — 150 ft.
- Type of Fuel — Gasoline

TABLE 8-10. Foam chambers for storage tanks

Tank diameter, ft (or equivalent of discharge area)	Tank diameter, m	Minimum number outlets
Up to 80 ft.	24	1
Over 80 to 120	24 to 36	2
Over 120 to 140	36 to 42	3
Over 140 to 160	42 to 48	4
Over 160 to 180	48 to 54	5
Over 180 to 200	54 to 60	6

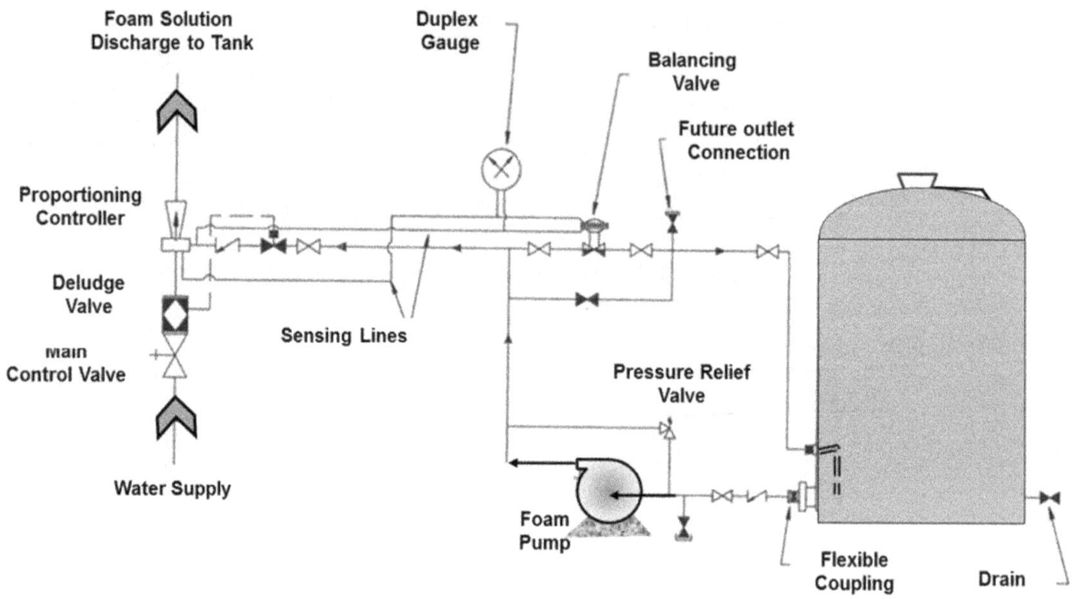

Figure 8-113. Typical foam system: balance pressure pump skid with proportioner (from [49] www.Buckeyef.com, [11])

- Foam Dam installed on roof — Yes — 2 ft. from tank wall and 2 ft. in height
- Sq. ft. area of annular ring — 930 sq. ft.
- Application Rate — 0.30 gpm per sq. ft. (Per NFPA 11),0.30 × 930 sq. ft. = 279 gpm of foam solution required.
- Type of Discharge Device — Foam Makers
- Discharge Duration — 20 min. 279 × 20 = 5,580 gallons of foam solution — × 0.03 (3% AFFF) = 167.4 gallons of foam concentrated
- Quantity of Foam Makers Required (Per NFPA 11) 6 required.
- Supplementary hose lines may be added as per example for Cone Roof.

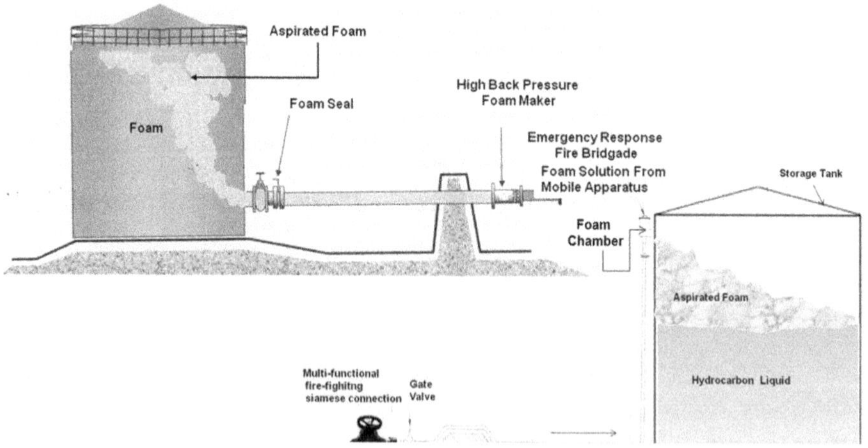

Figure 8-114. Typical semi fixed foam system (with mobile apparatus supplying the foam solution), Buckeye — www.buckeyef.com, [49], Inset: Chemguards [48]

NOTE: The number of fixed foam discharge points on an open top floating roof tank is determined by the circumference of the tank.

The maximum spacing between discharge points is 40 ft.(12.2 m) of tank circumference when using a 12″ (305 mm) high foam dam and every 80 ft. (24.4 m) of tank circumference when using a 24″ (610 mm) high foam dam.

8.8.3.4 Foam Dam Design for Tanks

In accordance with NFPA 11 the following will apply in foam dam design for foam tanks and monitors/handlines.

Foam Dam Design — The foam dam should be circular and constructed of at least No. 10 US Standard Gage Thickness (0.134 in.) (3.4 mm) steel plate. The dam is to be welded or otherwise securely fastened to the floating roof. The foam dam is designed to retain foam at the seal area at a sufficient depth to cover the seal area while causing the foam to flow laterally to the point of seal rupture.

Dam height is to be at least 12″ (305 mm) and should extend at least 2″ (51 mm) above any metal secondary seal or a combustible secondary seal using a plastic foam log. It is to be at least 2″ (51 mm) higher than any burnout panels in metal secondary seals.

Foam dams are to be at least 1 ft. (0.3 m) but no more than 2 ft. (0.6 m) from the edge of the floating roof.

Foam solution & rain water is to be drained by, slotting the bottom of the dam on the basis of 0.04 sq. in. of slot area per sq. ft. (278 mm sq./sq. m) of diked area while restricting the slots to 3/8 in. (9.5 mm) in height. Excessive dam openings for drainage should be avoided to prevent loss of foam through the drainage slots.

Foam Monitors and Hand-lines: NFPA II states that monitors are not to be considered as the primary means of protection for fixed roof tanks over 20 m (60 ft) in diameter. Foam hand-lines are not to be considered as the primary means of protection for tanks over 10 m (30 ft) in diameter or those over 6 m (20 ft) in height.

Application Rates Using Monitors and Hand-lines

The minimum foam solution application rate is based on the assumption that all discharging foam will reach the area being protected. In considering actual solution flow requirements, consideration should be given to potential foam losses from such as climatic conditions and thermal updraft of the fire.

The following Table 8.11 shows application density and duration for monitors and hand-lines on tanks containing liquid hydrocarbons.

Included in the above table are gasohols and unleaded gasolines containing no more than 10% of an oxygenated additive by volume.

On tanks containing water miscible/polar solvent flammable liquids the recommended foam application duration is 65 minutes.

Flammable liquids having a boiling point less than 100 °F and products that have been burning for some time can develop a heat layer which might require foam solution application rates as high as 0.2 or 0.25 gpm per sq. ft.

Where monitors or handlines are used to protect storage tanks containing polar solvent or water miscible liquids the discharge duration shall be a minimum of 65 minutes at the recommended application rate.

8.9 EMERGENCY RESPONSE PLANNING AND FACILITIES

Emergency planning can never be successful if it is started when the facility is on fire. Emergency planning requires a great deal of effort including time to develop an emergency plan, arrange mutual aid agreements, identify sources of foam and other

TABLE 8-11. Foam application rates using monitors and hand-lines

Hydrocarbon Type	Minimum Application Rate gpm/ft^2	(L/min.)/m^2 (min.)	Minimum Discharge Time
Flash point between (37.8 °C and 93.3 °C) (100 °F and 140 °F)	0.16	6.5	50
Flash point below 37.8 °C (100 °F) or liquids heated above their flash points	0.16	6.5	65
Crude Petroleum	0.16	6.5	65

fire fighting chemicals, develop procedures for all operations and plans for controlling emergencies, and last but not least, conducting drills to practice emergency plans and procedures.

8.9.1 Planning for the Emergency

A comprehensive plan should be developed with participation of all parties involved in emergency operations for storage, handling and use of bulk quantities of hydrocarbon liquids. Guidance on development of emergency plans for fighting fires in atmospheric storage tanks can be found in API Recommended Practice 2021, *Management of Atmospheric Storage Tank Fires, some of the requirement are highlighted in Section 8.1.10 of API.*

8.9.2 Responding to Oil Spill Emergencies

Emergency Operations Center (EOC): In the event of a working incident inside a Tank Farm complex, the industry recommends setting up and announcing an ***Emergency Operations Center (EOC***).

At such an EOC industry expects that representatives of all operating groups within the complex usually assemble at this location and be available if any action involving their facilities is necessary and to provide information or advice.

The responsible party for the facility involved in the incident will report to the Forward Command Post.

Command will assign a Command Officer and at least one supervisor to the Emergency Operations Center. The Emergency Operations Center will provide support as directed by Command. Complete sets of site plans, and photographs must be maintained at the Complex.

It is usual that all public contact, and liaison functions be conducted at the Emergency Operations Center. However is usual for the Command Post be located in the most appropriate position to direct tactical operations.

The first unit arriving at the scene of an incident at the Tank Farm Complex is expected to provide the following information in the initial report:

- Specific location - name of shipper involved
- Type of incident - leak, spill, fire or no fire
- Extent of spill, leak or fire

The first unit arriving at the scene of an incident at the Tank Farm Complex also is expected to provide the following information in the initial report:

- Operation of any automatic fire protection, liquid level control or pipeline product delivery systems
- Tank number(s) and location(s)

Command Post will normally direct alarm to notify the 24-hour duty Operator at Pipelines Operation Centre of any working incident. If a major leak or spill is involved, the direction must be given to shut down the incoming pipeline flow.

It is also necessary to notify the responsible party for the involved property to respond. The Tank Farm Complex has usually a call up system to notify all key personnel to respond to the designated emergency Operations Center.

Initial actions should be directed toward the tactical priorities listed below.

- Action should proceed cautiously
- High level of safety
- Avoid committing personnel to dangerous situations

8.9.3 Tactical Priorities

Major incidents at the Tank Farm Complex will involve either a leak or a spill of a petroleum product.

The situation may or may not involve a fire. The tactical priorities are:

1. Ensure that all personnel are not within a hazardous atmosphere or have the potential to be exposed.
2. Cover the spill with a foam blanket to control fire and/or prevent ignition.
3. Control potential sources of ignition.
4. Have a HAZ MAT unit monitor the foam blanket to determine its effectiveness.
5. Contain the spill or run-off.
6. Identify and control the source of the spill or leak.
7. Maintain foam blanket until product can be picked up.
8. Keep all personnel and vehicles out of the spill area.
9. Maintain an adequate volume of foam solution on scene for the duration of the incident.

It may be noted that a large spill can create an extremely large vapor problem and may flash back from ignition sources at significant distances. While covering the spill to suppress vapors, the direction and extent of vapor travel must be determined.

8.9.4 Foam Application

When attempting to control a large flammable liquid fire, the strategy is generally to wait until enough foam concentrate to control the fire is on the scene before beginning the attack the fire. If the fire attack runs out of foam before the fire is controlled, all of the foam will have been wasted. The minimum foam solution supply and the total amount of foam water solution required for each storage tank must be calculated in advance and be available in the Storage Tank Tactical Guidelines for the complex.

Fires which are controllable with the foam supply on hand are generally attacked without delay. This applies to most spill fires and tank vehicle incidents. If the fire is too large to be controlled by the initial attack capability, Command should consider a holding action to protect exposures and prevent spread until additional foam supplies can be assembled and prepared for use.

Subsurface Injection: Most Tank complexes have a system of connections available to pump foam directly into the main piping manifold at the Tank Farm Complex. The connections must be clearly indicated in the complex site map. By opening valves and directing the flow, it is possible to direct this flow to designated tank in the complex and accomplish subsurface injection. The details of the required connections and pre- calculated flow rates for each tank must be carried out in advance for the required type of foam and available at a strategic location with the Tank Farm Complex.

Before beginning subsurface injection, it may be necessary to transfer hydrocarbon product out of the involved tank to make room for product in the lines that will be pushed ahead of the foam.

If subsurface foam application is being contemplated the following items are generally considered

- As the polar solvents will absorb the water in the foam, subsurface foam injection cannot be used on ethanol or other polar solvents.
- The volume of liquid in the pipeline must be displaced by the foam water solution. Pipeline volumes can be calculated and depend on their diameter and distance from the subsurface injection manifold.
- The inlet valve of the storage tank must be open.
- It is industry's experience that if the velocity of the foam water solution exceeds 3 m/s (10 feet/second) in the pipeline, the water will separate from the foam, making it useless.
- Depending on the length of the line and the tank height, it may take 30-60 minutes of time from the time that foam water solution is pumped into the injection manifold until it reaches the burning surface of the tank.

Fixed Systems: Fixed piping is provided for some tanks to provide direct delivery of **Aqueous Film Forming Foams** (**AFFF**), from Foam supply into the affected hydrocarbon storage tank. These systems will deliver foam onto the product via topside application at the tank. In such cases a Siamese connection (Figure 8-115) is provided to allow the foam lines to be connected.

The use of these systems may require complicated operations to manipulate valves, drain lines or remove product. These operations require liaison and cooperation with the responsible parties.

Hose Streams: **Aqueous Film Forming Foams** (**AFFF**) may also be applied through 1 1/2 inch hand-lines from the company or third party fire fighting facilities/departments which also may have the capability to supply hand- lines and/or master streams, including elevated streams on ladder trucks.

When using hose streams, caution must be taken to use a large enough line to penetrate the heat of the fire from a safe distance.

8.9.4.1 Foam Supply

If the foam supply on hand at the incident scene is not adequate for the incident, Command usually direct Dispatch to notify Resource Management to begin to assemble a larger supply. This shall be obtained from:

- Truck and Trailer
- Fire Department Warehouse
- Mutual Aid Fire Departments
- Emergency Purchase from vendors

Because of the large foam demands required for ground spill fires, Command should consider assigning at least one group to ensure the availability of required volume and type of foam. This group is responsible for ensuring that a sufficient volume of foam is available to control the fire and to assist in maintaining a constant supply during foam water solution application. The group is also responsible for ensuring that enough foam remains on-hand for continuous application if, after extinguishment, fire firefighters

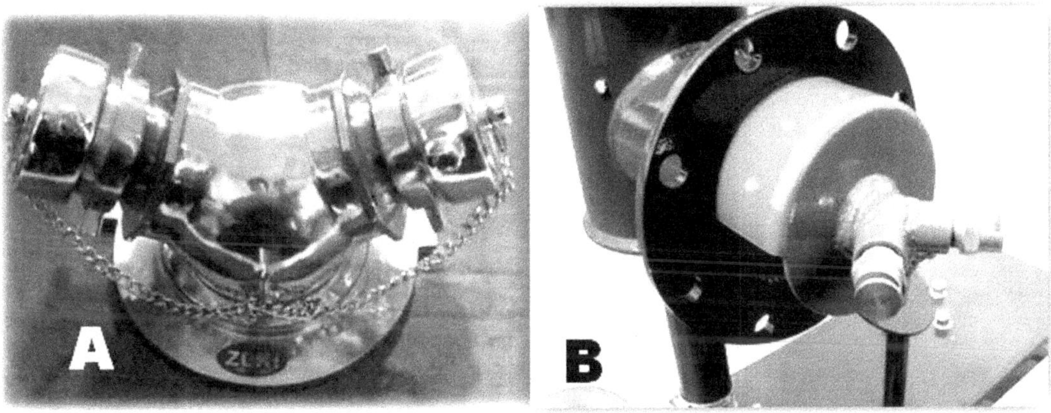

Figure 8-115. Typical connection for firefighting facilities, A Siamese Twin and B: dual fitting foam connection

must enter into the spill area. It is usual that at least 50% of the amount of foam required for extinguishment be available, especially if tank overhaul is required.

8.9.4.2 Water Supply

The Complex's Tank Tactical Guidelines must contain water supply data and available sources for the tank farm.

8.9.4.3 Exposure Protection

When exposure protection is required, large volume water streams should be used for reach and cooling capacity. Water application must be managed to avoid breaking-up foam blankets or increasing the problems of fuel spills.

Steam production should be used as a guide to protecting exposures: If steam is created when water strikes the surface of the tank, the need for protection is indicated. Tanks generally require little protection on vertical surfaces below the liquid level.

Some of the tank farm facilities have fixed monitor nozzles that can be used to apply cooling water onto tanks that require exposure protection. Such facilities are usually identified in the Storage Tank Tactical Guidelines book for the Tank farm complex.

a- Valve Protection

Product control valves on the storage tanks are beneficial because they can be used to route liquid from a fire-involved tank to an empty or partially full tank. Failing to protect these valves in the event of a ground spill fire may prevent tank farm personnel from routing liquid from fire exposed or damaged tanks. In the event of a ground spill fire that does not submerge the valve in liquid, a fire stream should be applied to each valve that may be subjected to heat damage. The use of protective streams will protect the operating components of the valves so that they are not damaged.

b- Utility Control

If disconnecting the electric power is considered as part of the incident tactics, it is necessary to discuss the requirement with the tank farm personnel of the implications of this disconnection. Disconnecting power can shut down transfer pumps used to remove spilled products at loading racks, cause motor operated valves on storage tanks to close which prevents subsurface foam injection, and disable controls for all the storage tanks.

c- Traffic Control

One of the more likely events at the tank farms is a spill resulting from an accidentally overfilled tank. Given the volatility of some fuels, a large area may be covered with vapors that can be within their flammable range. Depending on the location of the Tank Farm complex, traffic control should be established early in the incident to limit the potential for vehicles becoming ignition sources.

Command should recognize that all of the trucking companies that transport fuel would have access cards that may control access gates. These vehicles can access the tank farm area unless the major intersections and roadways surrounding the tank farm complex are blocked.

REFERENCES

[1] Wikipedia, 2010, "Oil Depot," http://en.wikipedia.org/wiki/Oil_depot.

[2] Industrial Fire Journal, 2006, "Buncefield – Anatomy of a Disaster," January.

[3] Mann, A. L., 2009, "Some Petroleum Pioneers of Pittsburgh," *Western Pennsylvania History Magazine*. Summer Issue. http://www.heinzhistorycenter.org/uploads/Media/4_MicrosoftWord-Mann_Petroleum_web.pdf.

[4] Saadat Nouri, M., 2008, "First Iranian City Where Largest Oil Field of Middle East Was Explored," *Persian Mirror* http://www.persianmirror.com/Article_det.cfm?id=2096&getArticleCategory=58&getArticleSubCategory=32.

[5] Lugoff, and Camden, 2010, Plate Arrangement & Rivet Pattern, http://www.etraxx.com/wordpress_train/?page_id=304.

[6] Big Inch Petroleum, 2007, "Hydrocarbon Vapour Pressure," July 17. http://www.eng-tips.com/viewthread.cfm?qid=191613&page=10.

[7] OTEC (Overseas Technical Engineering & Construction Pte Ltd.), 2007, http://www.otec.com.sg/omj.pdf.

[8] Godoy, L. A., and Mendez-Degro, J. C., 1989, "Introduction to Above Ground Steel Tanks," http://www.efn.uncor.edu/investigacion/e-learning/tanques/documentos/pub/typestanks.pdf.

[9] ARFF (Airport Rescue Fire Fighting Professional Services LLC), 1993 (1st published) "Tactical Preparation for Major Incidents Involving Fuels," https://docs.google.com/viewer?url=http://www.apssafety.net/sitebuildercontent/sitebuilderfiles/lviatankfarms.pdf&chrome=true.

[10] Tempcore Rollwell, 2009, "Temcor Aluminum Domes have Revolutionized Tank Covers" http://www.temcorrollwell.com.

[11] SCT (System Creative Technology), 2006, "Tankage in Refinery," October 27. http://www.scteng.co.kr/customer2.php.

[12] European Commision (EC), 2001, "Best Available Techniques on Emissions from Storage," Institute for Prospective Technological Studies http://eippcb.jrc.es.

[13] US EPA (US Environmental Protection Agency), 1994, "Review of Guide Pole Fittings Analyses Conducted in Support of 5-72 An Addendum to API Publication 2517 for External Floating Roof Tanks," May 25, http://www.epa.gov/ttnchie1/ap42/ch07/bgdocs/b07s01.pdf.

[14] US EPA (US Environmental Protection Agency), 2006, "Emission Factor Documentation for AP-42, Section 7.1,Organic Liquid Storage Tanks- Final Report," Sept. http://www.epa.gov/ttnchie1/ap42/ch07/bgdocs/b07s01.pdf.

[15] Ferry, R., 2002, "Emissions from Aboveground Storage Tanks-The Basics," TGB Partnership http://www.tgbpartnership.com/linked_data_files/tanks_3_0.pdf.

[16] Clement, F., Riethmuller, M., and Chauveau, D., 2005, "Esempi applicativi Guida AFIAP. LPG Tanks," Milan Workshop « Inspection of Underground Gas Tanks by AE, 13th Oct. http://www.afiap.org/pdf/MILAN/MilanChauveau.pdf.

[17] Irving, B., and Hart, L., 1994, "A Pictorial History of Welding as Seen Through the Pages of the Welding Journal," *The Welding Journal*, June, (75th anniversary of AWS).

[18] Kamyab, H., and Palmer, S. C., 1989, "Analysis of displacements and stresses in oil storage tanks caused by differential settlement," *Proc. Inst. Mechanical Engineers*, UK, Vol. 203(C1), pp. 61-70.

[19] API RP575 (American Petroleum Institute), 2008, "Inspection Practices for of Existing Atmospheric and Low-Pressure Pressure Storage Tanks," http://ballots.api.org/cre/sci/ballots/docs/RP575SecondBallotDraft.pdf.

[20] Hiner, L. C., 2006, "Secondary Seals & Selected Floating Roof Issues," *CBI Storage Tank Conference*, September 27.

[21] Patrol Limited, 2007, "Tank Farm LHDC Fire Detection - Product Guide," D1138 Issue 2 - 15 September.

[22] Pennsylvania DEP(Department of Environmental Protection), 2008, "Pennsylvania Code Title 25, Chapter 245- Administration of the Storage Tank and Spill Prevention Program," March, http://www.elibrary.dep.state.pa.us/dsweb/Get/Document-70528/2570-BK-DEP1790.pdf.

[23] API (American Petroleum Institute), 2008, "List of Frequently Utilized Storage Tank Standards and Practices," http://standard.digibooks.cn/standard/industry_156_1.html.

[24] API (American Petroleum Institute), 2010, "Storage Tank Publications," http://www.api.org/Publications/2010-catalog-pages.cfm.

[25] SOD (State of Delaware), 2004 & 2005, *Regulations Governing Aboveground Storage Tanks*, Dept Nat Res. & Env. Control, Tank Management Branch. http://www.dnrec.state.de.us/dnrec2000/Divisions/AWM/ast/.

[26] Shaw-Shong, L., Deng-Ing, T., and Yew-Hup L., 2010, "Piling Foundation Design & Construction Problems of Tank Farm in Reclaimed Land over Untreated Soft Marine Clay in Malaysia," *The 17th Southeast Asian Geotechnical Conference*, Taipei, Taiwan, May 10–13.

[27] AIChE (Aamerican Institute of Chemical Engineers) Center for Chemical Process Safety, 2004, "Guidelines for Engineering Design for Process Safety," http://www.knovel.com/web/portal/browse/display?_EXT_KNOVEL_DISPLAY_bookid=848.

[28] Rivers, K., 2007, "Safety and Environmental Standards for Fuel Storage Sites-Buncefield Standards Task Group (BSTG) Final Report," July 24. http://www.hse.gov.uk/comah/buncefield/bstgfinalreport.pdf.

[29] HSE (Health and Safety Executive), 2009, "Safety and Environmental Standards for Fuel Storage Sites-Final Report," http://www.hse.gov.uk/comah/buncefield/fuel-storage-sites.pdf.

[30] GPSA (Gas Processors Association), 1994, "Storage", Engineering Data Book, Volume 1, Section 6, 10th Ed.

[31] Kresmer, A., 1930, *National Petroleum News*, **22**(21): pp. 43–49. 67.

[32] API (American Petroleum Institute), 1982, *Evaporation Loss From Internal Floating Roof Tanks*, 3rd Edition, Bulletin No. 2519, API, Washington, DC.

[33] Akhavan-Zanjani, A., 2009, "Settlement Criteria for Steel Oil Storage Tanks," *Edge (The Electronic Journal of Geotechnical Engineering)*, Bundle A. ed., Vol. 13, http://www.ejge.com/2009/Ppr0904X/Ppr0904.pdf.

[34] Kobelco (Kobe Steel, Ltd.), 2006, *Kobelco Welding Today*, **9**(1), Jan. http://www.kobelco.co.jp/english/welding/files/kwt2006-01.pdf.

[35] Rogantea, M., Battistella, P., and Cesari, F., 2006, "Hydrogen Interaction and Stress-Corrosion in Hydrocarbon Storage Vessel and Pipeline Weldings," *International Journal of Hydrogen Energy*, **31**, pp. 597–601.

[36] Ikawa, H., Godai, T., 1978, "Welding of Heat-Resistant Steel and Heat Resistant Materials,"The Complete Book of Welding— Series 4, Sanpo Publications Inc.

[37] M+F Systems Technology, 2011 "Fuel Management in Tank Farms and Terminals" http://www.mfx-systems.de/doku_1/1816/01%20Additivation/Sales%20Literature/1739-0001-0044.pdf

[38] EPA (US Environmental Protection Agency), 2006, *AP 42*, 5th Edition, Volume I, Chapter 7: Liquid Storage Tanks, http://www.epa.gov/ttn/chief/ap42/ch07/index.html.

[39] ARB(Air Resources Board) 2005, "Draft Approach to Estimating Aboveground Storage Tank Emission Factors Using the AP-42 Method," http://www.arb.ca.gov/vapor/ast/astemissionfactorapproach.pdf.

[40] Sung, H. M., Sue, 2001, "Testing VOC Evaporation Losses from Floating Roof Tanks Turnover," *Oil and Gas Journal*, Dec 10.

[41] Özçelik, E., 2003, "Cathodic Protection of Aboveground Petroleum Storage Tanks," http://tankstorageinternational.com/pdf/09/ERGIL2pdf.PDF.

[42] Özçelik, E., 2005, "Cathodic Protection of Aboveground Petroleum Storage Tanks" http://www.tankstorageinternational.com/pdf/09/ERGIL2pdf.PDF.

[43] Rim–Rukeh, and Okokoyo, A., 2005, "Underside Corrosion of above Ground Storage Tanks (ASTs)," *J. Appl. Sci. Environ. Mgt.* **9**(1), pp. 161–163.

[44] Fitzgerald, J., 2004, "Cathodic Protection for On-Grade Storage Tanks and Buried Piping," *Freshwater Spills Symposium*, April, http://www.epa.gov/oem/docs/oil/fss/fss04/fitzgerald_04.pdf.

[45] API (American Petroleum Institute), 2007, *Cathodic Protection of Above ground Petroleum Storage Tanks*, 3rd Edition, Jan.

[46] Cornell, J. R., and Baker, M. A., 2002, "Catastrophic Tank Failures: Highlights of Past Failures Along With Proactive Tanks Designs," *The US EPA Fourth Biennial Freshwater Spills Symposium*, Sheraton Cleveland City Centre Hotel, Cleveland, Ohio, USA March 19–21.

[47] Slye Orville, M., Bud, 2007, "Fire Safe Fuel Farm Operations," *Presented at the 30th Annual Airport Conference*, Hershey, PA, http://www.faa.gov/airports/eastern/airports_news_events/hershey/media_30/slyde.doc.

[48] Chemguards, 2005, "Fixed or Semi-Fixed Fire Protection Systems for Storage Tanks," Rev 9. http://www.chemguard.com/pdf/design-manuals/D10D03192.pdf.

[49] Buckeye, 2012 download, "Foam storage Storage tanks_- Fixed or Semi-Fixed Systems" "http://www.buckeyetest.com/foampdfs/storage/fixed.pdf, (see: http://www.buckeyef.com/, refer to Products, Foam Concentrate and Hardware, Foam Storage Tank Systems)

[50] CAPP (Candain association of Petroleum Producers), 2007, "A Recommended Approach to Completing the National Pollutant Release Inventory (NPRI) for the Upstream Oil and Gas Industry," Report #: 2007-0009, www.capp.ca.

[51] Gossaman, D., 2006, *Tank Farm Design and Operation – The Early Years*, Gossman Consulting, Inc. Publication Volume 11, Number 3 March.

[52] NSTB, 2003, "Storage Tank Explosion and Fire in Glenpool," OklahomaPipeline Accident Report.

[53] NTSB/PAR-04/02, April 7, www.ntsb.gov/doclib/reports/2004/PAR0402.pdf.

[54] Ritchie, R., 2009, "Preventing Storage Tank Fores, "Hydrocarbon Processing, November. http;//www.us.sgs.com/sgs-preventing_storage-tank_fires-hydrocarbon_processing-nov09-en-09.pdf.

[55] Okada M., and Suzuki, H 1970i, "Metallurgy of Welding", K. K. Sanpo, Tokyo, 112 , https://www.jstage.jst.go.jp/article/isijinternational1966/21/4/21_4_260/_pdf

[56] Toja, M., Okuzumi, T. I., 2005, "Welding of Crude Oil Storage Tanks," *KOBELCO Bulletin*, http://www.kobelcowelding.com/Kobelco%20Welding%20Today/Vol-9(No1).pdf.

Chapter 9

LIQUID PIPELINE OPERATION

The operations directly related to hydraulics are covered in Section 5.1 and batch operations in Section 5.2. This chapter discusses various techniques and tools necessary to improve safety and efficiency in operations; SCADA, leak detection, DRA injection, tank farm operation and volume measurement, and power cost control.

9.1 SUPERVISORY CONTROL AND DATA ACQUISITION (SCADA)

9.1.1 Introduction

Pipeline systems are automated to provide the capabilities of operating pipeline systems reliably, efficiently and thus economically. Pipeline operation involves monitoring and controlling of a pipeline system, and monitoring is required for checking the pipeline states and controlling facilities such as pump and valve stations. Modern pipeline system operation is centralized because a centralized operation of the pipeline systems benefits the stakeholders including the pipeline company, producer of the product, and the shipper of the product. A centralized system provides the capability to monitor and control the complete pipeline system in a safe and efficient manner. It allows the stakeholders to meet the changing demands for the product being shipped expediently and to move the product from source to market safely and quickly in the most economical way possible. A SCADA system provides the pipeline companies with centralized monitoring and controlling capabilities [1].

SCADA is an acronym for Supervisory Control and Data Acquisition; supervisory because human operators always issue control commands, not providing a closed-loop control function. A SCADA system is a computer-based data acquisition system designed to gather operating data from an array of geographically remote field locations, and to transmit this data via communication links to one or more control center location(s) for monitoring, controlling, and reporting.

A SCADA system is designed to assist pipeline operators in the operation of the pipeline system using real-time and historical information. Pipeline operators typically regulate pipeline pressure and flow, start and stop pumps at stations, and monitor the status of pumps and valves through the SCADA system. Local equipment control systems monitor and control the detailed process for the pump and its associated driver. They may then issue commands of a supervisory nature to the remote locations in response to the incoming data. Additionally, software programs implemented within the SCADA host can provide for specific responses to changes in field conditions, by reporting such changes or automatically sending commands to remote field locations.

Pipeline system control is accomplished by setting a controlling variable at the desired level and the control system responds to reach the set point. Depending on the controlling functions, the controlling variable can be pressure, flow, and sometimes temperature. The controllers monitor and change the controlling variables through the SCADA system, which transmits the control signals to remote stations such as pump,

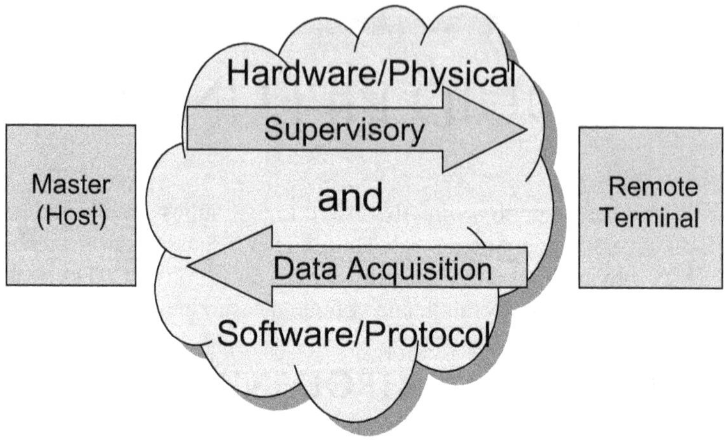

Figure 9-1. Supervisory control and data acquisition system

lifting, delivery or valve station. Figure 9-1 illustrates the relationship between the master and remote terminal through the computer and communication system.

Traditional SCADA users include the pipeline system dispatchers or controllers, operation engineers, system engineers, maintenance and measurement staff. System dispatchers use the SCADA for safe and efficient pipeline system operation while meeting transportation requirements. Operations engineers analyze pipeline operational problems to increase operation reliability, efficiency, and throughput as well as troubleshooting, while system engineers configure and maintain the SCADA system including instruments and remote terminals. Maintenance staff analyzes equipment performance based on historical data, and measurement staff validates volume measurements.

Current business environment requires fast access to operational information. As a result, other groups use the SCADA data to improve the pipeline business. These groups include accounting personnel who account liquid volumes and issue invoices, liquid marketers who use estimated batch data to schedule and market liquids movements, and management who make management decisions regarding normal and abnormal conditions including emergency situations.

In order to accommodate a rapidly changing business condition or environment, corporate-wide information access has become critical to the efficient operation and management of a pipeline system. Not only is it important to provide accurate information to operation and management staff, but timely access to this information is of vital importance to the successful operation of the pipeline company's business. Companies that are able to acquire, process, and analyze information more efficiently than their competitors have a distinct market advantage. Such expansion of the scope, functionality, and capabilities is made possible by continuing improvements in computer and telecommunication technologies.

A properly designed, installed, and operating SCADA system is a keystone in the operation and management of a pipeline in today's competitive deregulated pipeline market. The SCADA system has become the hub for corporate information systems. Refer to Figure 9-2 for an overview of an integrated corporate and SCADA system. Looking at the information requirements of a pipeline company and considering both operational and business aspects, the key requirements can be broadly grouped into the following categories [2]:

- **Measurement information** — Measurement information is used for the safe and efficient operation of the pipeline system. It includes pipeline data

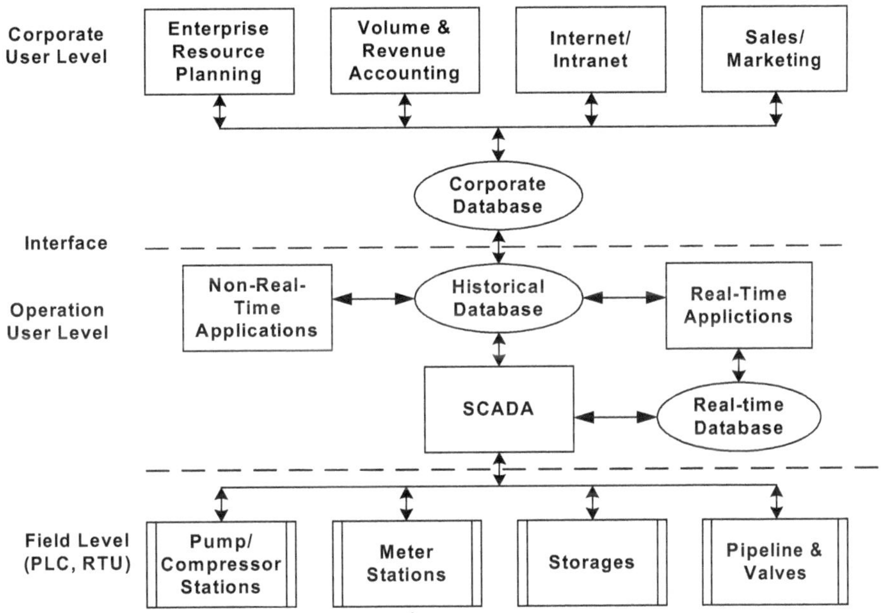

Figure 9-2. Integrated corporate and SCADA system

acquired from field telemetry equipment such as volumes, flows, pressures, temperatures, product quality, and equipment status. It would also include any calculated data originating from the SCADA host.

- **Simulation information** — Simulation information incorporates measurement data and simulated data to diagnose current pipeline states and predict future behavior of the pipeline. The simulation information can be used for system optimization, line pack and capacity management, storage management, product scheduling, and training-related applications on the pipeline system. This data would originate from a modeling application that may use measurement information.
- **Business information** — Business information combines measurement data and possibly simulated data along with business and economic data. The information is used in business applications related to custody transfer, preventative maintenance, cost tracking, contracting, marketing, inventory, scheduling and accounting. This is where SCADA and simulation data are aggregated with other business data to feed into business processes.
- **Decision support information** — Decision support information is a summary of the key measurement, simulation, and business data required for executive level decision support. Extracting this key data is generally the function of a Management Information System (MIS). Such a system has the ability to gather and aggregate data from numerous corporate and operational databases to supply key performance data.

It is becoming more and more common for pipeline applications to be tightly integrated with SCADA systems and to be part of a higher level Management Information System (MIS). The advances in computer and communication technologies have made it much easier to connect SCADA systems to business systems. This allows for both physical integration of SCADA and business systems as well as business process integration. Process integration means that SCADA systems are becoming a key part of business processes. This provides for both proactive business processes as well as the ability to provide better information and thus better service for customers.

9.1.2 Pipeline System Monitoring and Control

The pipeline system has to be monitored for control. The control of a pipeline system is achieved mainly at pump stations located in a tank farm and along the pipeline, meter stations in receipt and delivery locations, and valves along the pipeline and in these stations. In addition, flow direction and tank levels in a tank farm have to be monitored and controlled. Therefore, their status and values have to be monitored first and then controlled if required.

- **Pump station monitoring and control** — Since a pump is a primary pressure control device, the pipeline pressure is controlled by starting or stopping pump units or adjusting the pressure set point at the operating pump station. Most pump stations are equipped with multiple pump units. Depending on the pressure or flow requirements, the operating pumps can be arranged in series, parallel or a combination of series and parallel through open/close operations of valves. The pump station pressure generated by fixed speed pumps are controlled by a control valve at the station, while the pressure generated by variable speed pumps is controlled by the pump rotation speed. Refer to Chapter 4 for a detailed discussion of pump station control.
- **Pipeline monitoring and control** — The main function of a pipeline control system is to monitor flows, pressures, and sometimes temperatures along the pipeline. Pipeline companies receive nominations, schedule volumes, and lift and deliver them by controlling flow rates while maintaining the pipeline pressure within the operating pressure limits and the flow rates within the capacity. Valves along the pipeline are used to change the flow direction, isolate some sections of the pipeline, or open other sections. The quality of the products has to be monitored and batch movements have to be tracked for proper delivery.
- **Meter station monitoring and control** — An active pressure control does not take place at a meter station. Instead, the flow rate and volumes are monitored to control the flow directions to an appropriate location such as tank or pump. The most important function of a meter station is custody transfer by providing metering information.
- **Tank farm monitoring and control** — Petroleum products are either lifted at or delivered to a tank farm. A tank farm is composed of multiple tanks, booster pumps, meters, valves and piping. The correct product should be lifted from the correct tank at the injection station and delivered to the correct tank at the correct tank farm. The product movement, flow direction, and tank level have to be monitored and controlled by opening or closing various valves along the flow path to make this happen.

9.1.3 Control Center and SCADA System

Since the SCADA system plays a critical role in the success of the pipeline business, it must satisfy the following requirements:

- Capability to operate the pipeline system safely and economically,
- Provide timely and accurate data for monitoring and controlling the pipeline system in real-time,

- Be reliable with high availability,
- Provide security, protecting valuable corporate information from inside and outside intruders.

Most modern SCADA systems can provide the functionality to meet these requirements. However, the combination of the SCADA system together with its control center should be configured to fulfill them.

There are three basic tiers in a SCADA system as shown in Figure 9-2, namely, field device, control room, and corporate. The field to SCADA connection is some form of a telecommunications network, and the connection between SCADA host and the corporate or enterprise environment is made with a communication network. A backup control center located at an offsite may be connected to the main control system.

In US, PHMSA incorporates American Petroleum Institute (API) recommended practices 1165, 1167 and 1168, which are the recommended practices for Pipeline SCADA Displays, Pipeline SCADA Alarm Management, and Pipeline Control Room Management, respectively. Each document describes the following:

- API RP 1165 — Pipeline SCADA Displays [3] focuses on the design and implementation of displays used for the monitoring and control of information on SCADA.
- API RP 1167 — Pipeline SCADA Alarm Management [4] provides guidance on industry practices that include alarm definition and determination, alarm philosophy, alarm functionality and design, alarm handling, alarm documentation, alarm audit and performance monitoring, roles and responsibilities, management of change, etc.
- API RP 1168 — Pipeline Control Room Management [5] addresses pipeline control room personnel roles, guidelines for shift turnover, pipeline control room fatigue management, and pipeline control room management of change.

The operational nerve center of today's pipelines is the pipeline control center. It is from this central location that a geographically diverse pipeline is monitored and operated. It is also the center for gathering information in real time that is used for real-time operation, for making business decisions and for operational planning. Figure 9-3

Figure 9-3. Control Console (Cerda J., 2008, "Oil Pipeline Logistics" Instituto de Desarrollo Tecnológico para la Industrial, August 11–21, Mar del Plata, Argentina, http://cepac.cheme.cmu.edu/pasi2008/slides/cerda/library/slides/jcerda-pasi-2008-1page.pdf)

shows a modern control console that the pipeline operator uses for minutes by minutes system operations. Usually, several large screens are made available to monitor the entire pipeline system.

Since the control center provides real-time information, it may also include an emergency situation-room adjacent to the control room. This room may be dedicated to addressing dispatching issues and particularly to resolving emergency or upset conditions. Several stakeholders, including technical support and management, may be assembled to address emergencies.

A backup control center may be required in order to operate the pipeline system continuously in the event that the main control center is severely disrupted. This backup is normally in a physically separate location from the main control room. The backup center is equipped with the same equipment and devices as the main control center. One option is that the backup system receives the real-time data directly from the field devices each cycle, so that it is the exact replica of the primary system. The other option is that the entire backup system is refreshed with the required data received from the primary system at a regular interval.

The division of control between a central location and the local pump station varies widely. A large complex pipeline system may be divided into multiple control sections defined in terms of size of the pipeline network, complexity of the network, or number of shippers. This division allows the operators, assigned to each section, to efficiently monitor and safely control the pipeline system.

A control center houses most of the equipment used by the operators on a daily basis. The equipment required includes the SCADA system computers and terminals, printers, communication devices, and network equipment used to implement Local Area Networks (LAN) and/or Wide Area Networks (WAN). In addition, pipeline system maps and schematics may be displayed, and operator manuals and other information required for performing dispatching functions can be made available.

A SCADA system consists of three main components; host or master, communication system, and remote terminals. A SCADA host or "master" is a collection of computer equipment and software located at the control center and used to centrally monitor and control the activity of the SCADA network, receive and store data from field devices and send commands to the field. A SCADA system gathers the data from a variety of field instrumentation, typically connected to remote terminals. See Figure 9-4

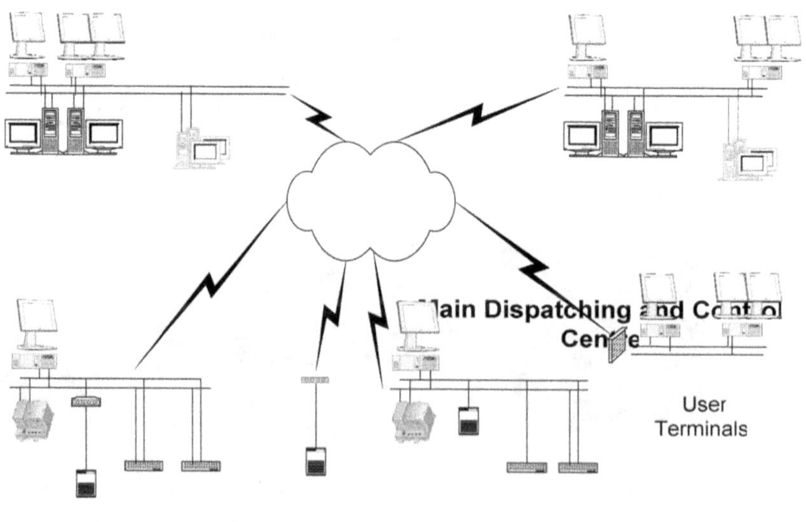

Figure 9-4. Typical SCADA system

for a modern SCADA architecture of both the main control center and backup control center.

The architecture of SCADA systems can vary from a relatively simple configuration of a computer and modems to a complicated network of equipment. In whatever form it takes, however, SCADA architecture will incorporate the following key hardware and software capabilities:

- Ability to interface with field devices and facilities for control and/or monitoring, usually through remote terminals.
- Provision of a communication network capable of two-way communication between the remote terminals and the control center. This network might also provide communication between the control center and a backup control center.
- Ability to process all incoming data and enable outgoing commands through a collection of equipment and software called the SCADA host.

Modern SCADA systems provide additional capability:

- Business applications such as meter ticketing, volume accounting, nomination management, etc.
- Application software such as leak detection, inventory management, and training
- Interface to corporate systems

The network is normally an internal private network. However, there are now SCADA systems that utilize secure connections to the Internet that replaces the private network. Web-based SCADA systems are ideal for remote unattended applications, assuming that an RTU or flow computer is available. In other words, they are suitable to pipeline systems or remote locations where centralized computing or control requirements are not intense and the primary function is remote data gathering. A web-based SCADA system offers several benefits. The main advantages are as follows:

- It provides an economical solution with wireless technology using the Internet infrastructure.
- It allows data access from anywhere without extra investment in communication and software.

Here, it needs to be mentioned that a distributed control system (DCS), instead of a SCADA, can be used for controlling pipeline systems. The goals of DCS and SCADA are quite different. A DCS is process oriented. It looks at the controlled process (the gas processing plant or chemical plant) as the center of the universe, and it presents data to the operators as part of its job. SCADA is data-gathering oriented; the control center and operators are the center of its universe and the remote equipment is merely there to collect the data — though it may also do some very complex process control.

DCS systems were developed to automate process control systems. These systems are characterized by having many closed loop control elements controlling an analogue process in real time. The key differences and characteristics of DCS and SCADA are as follows:

- A DCS normally does not have remotely (i.e., off-site) located components and is always connected to its data source. Redundancy is usually handled by parallel equipment.
- SCADA needs to have secure data and control over a potentially unreliable and slow communication medium, and needs to maintain a database of 'last

known good values' for prompt operator display. It frequently needs to do event processing and data quality validation. Redundancy is usually handled in a distributed manner.

- A DCS does not poll data but rather needs to be able to process a high number of transactions at a high speed in order to implement multiple real-time closed loop control.
- The majority of operations, such as start/stop commands and alarm detection of a SCADA system are digital. They also gather/poll analogue readings but do not implement closed loop control; humans determine if set points need to be adjusted. A DCS is process control oriented and therefore is designed to be able to implement many control loops as well as standard operator initiated start/stop commands.
- When the DCS operator wants to see information, he usually makes a request directly to the field I/O and gets a response. Field events can directly interrupt the system and the operator is advised automatically of their occurrence.

The remote terminals are located where the process values are monitored and interfaced with the host SCADA. They can be a remote terminal unit (RTU), programmable logic controller (PLC), or flow computer. The remote terminals collect data from the process devices, transmit data to the host SCADA, receive supervisory commands from the host SCADA, and issue these commands to the process devices. Supervisory commands may include pump/compressor station or unit start and stop commands, valve opening and closing commands, and set point settings.

An RTU acquires process values independent of the host by scanning hardware and software points, and communicates with the host, field I/O points, and other computer systems. It can detect and report alarm conditions, which include I/O error, bad measurement, high/low limit violations, rate-of-change alarm, and other deviations from set-points. An RTU provides limited control functions at field devices. The functions range from simple on-off or open-close control to logical control sequences such as ESD. It supports diagnostic checks with diagnostic software running in the remote watching for a number of possible problems. Some RTUs provide electronic flow measurement capability, by performing calculations of AGA, API and other standards, storing the measurement data, and allowing instant access of the measurement data.

A PLC provides extensive control, communication and operator interface capabilities. PLCs are used as remote terminals on a SCADA system, the heart of station control for field equipment (pumps, drivers, lube oil systems), communicating with the host. At a pump station, it can perform all the monitoring and control functions of pump unit and driver, station valve, station suction and discharge, station electrical and auxiliary equipment. It may have its own memory for the data to be transferred, or logic control for the gathering of data and error-checking with the host. PLCs can also be networked to provide a complete control system for a complex station.

It has to be noted that DCSs are not only economic for large installations but can be a solution choice for larger pump stations. They would certainly be considered for installations where there is a station and an associated processing facility or a refinery that would utilize a DCS for its control. The traditional boundaries between various control system solution options have become blurred due to the flexibility of today's control equipment. For small systems, the control system will generally be implemented using a PLC. As the facility gets larger and more complex, several options are now available of choosing between installing a control system using networked PLCs or a DCS system, requiring a careful consideration to ensure the operating requirements are met while at the same time the design dovetails with corporate business information gathering and processing.

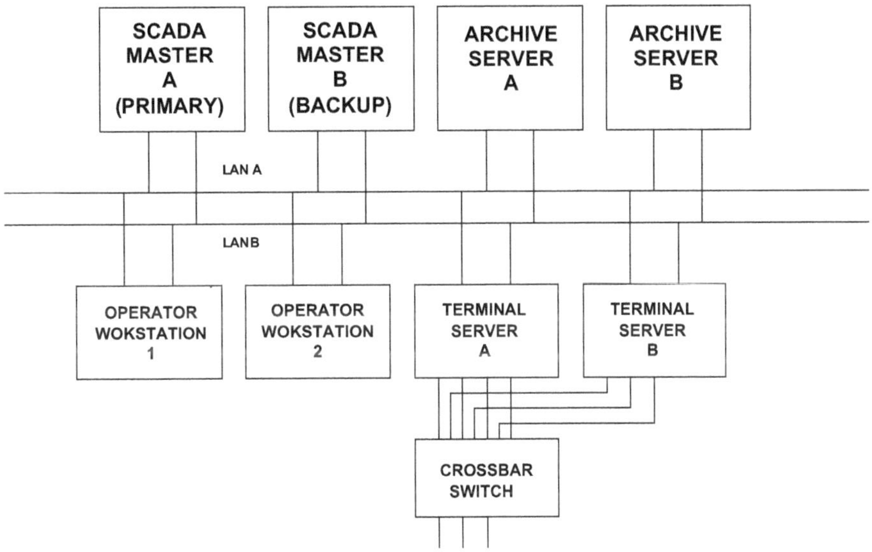

Figure 9-5. Redundant SCADA system

Reliability and availability requirements particular to individual installations will determine the configuration of redundant SCADA and database computers and redundant networks. Reliability provides an indication of how frequently a system or device will fail, while availability is the amount of time a system is fully functional divided by the sum of the time a system is fully functional plus the time to repair failures. Figure 9-5 illustrates a fully redundant SCADA architecture, in which both computer and communication systems including the associated equipment are duplicated.

SCADA host software architecture is different for every product. However, they all have the following key components:

- Operating system such as Unix, Windows or Linux
- Relational database for historical data management, interfacing with corporate databases
- Real-time database for processing real-time data quickly
- Real-time event manager, which is the core of the SCADA
- HMI manager for user interfaces

In addition, various utilities and development software are important for system development, configuration, and maintenance.

The SCADA will manage the polling of data, processing of that data and the passing of it to the real-time database. It will make data available to the presentation layer consisting of the HMI Manager and interfaces to other applications, as well as process control and data requests.

9.1.4 Data Communications

Data communications for a SCADA system require various components; modem, protocols, network, transmission media, and polling. A modem is an electronic device that encodes digital data on to an analog carrier signal (a process referred to as *modulation*), and also decodes modulated signals (*demodulation*). This enables computers' digital data to be carried over analog networks, such as the conventional telephone network. In general, modems are used for the connection between an RTU and the SCADA network or where it is not feasible to have a high-speed network, connection directly to the RTU.

In the context of data communication, a network protocol is a formal set of rules, conventions, and data structure that governs how computers and other network devices exchange information over a network. In other words, a protocol is a standard procedure and format that two data communication devices must understand, accept, and use to be able to exchange data with each other. A wide variety of network protocols exist, which are defined by worldwide standards organizations and technology vendors over years of technology evolution and development. For a retrofit or upgrade project, it is important to ensure that the SCADA system can support all of the protocols that exist in the legacy equipment that will be connected to the SCADA system. In some cases where there may be proprietary protocols, converters may need to be implemented. On a new SCADA system, there is no need to be concerned about existing equipment and protocols. However, it is important to ensure that the SCADA system utilizes industry standard protocols and not just proprietary ones. This will make expansion and addition of new equipment easier. It will also provide more flexibility in being able to choose equipment from a wide range of vendors and not be tied to a specific vendor's equipment.

A SCADA system will usually incorporate a local area network (LAN) within a control center and one or more wide-area networks (WANs). The major improvement in current generation SCADA systems comes from the use of WAN protocols. Not only does this facilitate the use of standard third party equipment but more importantly it allows for the possibility to distribute SCADA functionality across a WAN and not just a LAN. In some WAN distributed systems, pipeline controls are not assigned to a single central location. Instead, control operations can be switched or shared between numerous control centers. Responsibilities can be divided vertically according to a control hierarchy or horizontally according to geographic criteria. In both cases co-ordination and integration of control commands issued from various centers are maintained. In the event of the loss of one or more control centers, the operation can be switched to another center.

The SCADA network requires some form of communication media to implement the WAN connection between the SCADA host and remote locations. Ultimately the choice of which media to use to implement a connection to a remote site will be based on cost, availability of a particular medium and technical factors such as reliability, data transfer rate, geography, etc. A second choice to be made is whether the communication should be leased from a 3rd party or owned and operated by the pipeline company. This decision needs to be consistent with the corporate IT and operating guidelines. Commonly adopted communication media include:

- Metallic line is a hardwired physical connection between the SCADA host and the remote location. This is a good practical choice in SCADA applications where the distances between the SCADA host and the remote locations are not significant and there may be a limited choice of other media. An equivalent is usually leasing "lines" from a telephone company. The connection will utilize the internal network of the telephone company and may be any combination of wire, fibre optic cable, and radio. Another alternative is to utilize mobile telephone networks which provide good coverage in populated areas.
- Application of radio transmission on a pipeline SCADA usually takes two forms. The simple case is where a radio link is used as the last communication link between the SCADA and a remote site. The main communication backbone of the SCADA system is some other media other than simple radio. A long distance pipeline that may be geographically located in remote areas as well as

near occupied areas may well incorporate a mix of radio links and fixed links (leased lines, fibre optic, etc.)

- A fibre optic cable uses coherent laser light sent along a "cable." The cables are not lossless and repeater equipment is required at spacing of up to 100 km. The growth of the need for data transfer capability for the internet and private networks has spurred advances in fibre optic equipment. Because a fibre optic cable uses light and not electricity to transmit data it has the benefit of being unaffected by electromagnetic interference. On new pipeline projects, some pipeline companies have installed fibre optic cable in the same right of way as the pipeline. This can be a cost effective way of providing a transmission medium to implement the SCADA WAN.
- A satellite can provide a cost effective communication solution for pipelines under certain conditions. This solution is usually considered when the RTU is in a very remote location where the ability to utilize other media is not practical or very expensive. The capital cost is typically more than alternative techniques but when operating costs are factored in, this option can be a cost effective solution. However, poor weather conditions can adversely affect the reliability of communications.

Polling is the term used to describe the process of the SCADA host communicating with a number of RTUs connected on a network and exchanging data with each RTU. The arrangement between the SCADA host and the remote RTU is sometimes referred to as 'master-slave' implying that the SCADA host is in charge of each communication session with an RTU. The types of polling schemes are as follows:

- Polled Only or multi-drop scheme: The SCADA host will sequentially initiate communication with each RTU in sequence on a fixed schedule. There will be a fixed number of attempts to establish communication with an RTU before reporting that communications with the RTU are faulty. One can imagine that for a system with a large number of points to be updated at the SCADA host, this may take some time and therefore there will be some time lag between the sample time for the first data point and the last.
- Freeze scheme: One variation of multi-drop scheme is the ability of the master to issue a freeze command to all RTUs. The RTUs then store their data samples and the master begins polling and retrieves the data. This results in a database update at the master where all data was taken more or less at the same time. One way of mitigating this is to have all the RTUs take and store data samples at the same time. The major disadvantage of the above two schemes is that the status and value of all data base points are transmitted every polling cycle, which can be costly.
- Polled Report by Exception (RBE): In this scheme, a local history of each data point is saved and the RTU will only send back those points that have changed since the last poll. In the case of an analogue value, these will have a dead band that the value must exceed before a new value is sent back to the SCADA host. This reduces the amount of data traffic on the network. The user must be careful in choosing dead bands for analogue values for example to ensure that information is not lost.
- Unsolicited RBE: In this case, the host does not poll on a regular basis, but each RTU "pushes" data back to the host when it has updated data to send. This can reduce data traffic even more than the polled RBE. However, it has the disadvantage of the host not knowing if data points have not changed or failed. A variation can be to have a system that incorporates a guaranteed polling time. For example, all RTUs may be scanned at least once every 15 minutes.

9.1.5 Data Management

Typical data required for the safe and efficient operation of pipeline systems from various locations include the following:

- Pump station — values and quality of suction, casing and discharge pressures. Sometimes, temperature value and its quality are made available, especially if a heater or chiller is installed. Status of various valves is also required. For variable speed pumps, unit speeds are made available for the operator to review the performance of the unit. Data to allow monitoring of unit operating point is also useful for determining operating efficiency.
- Meter station — values and quality of flow, pressure, and temperature. In addition, a densitometer reading may be available for liquid pipelines, particularly batch pipelines. The status of various valves is also required.
- Control or pressure reducing valve station — values and quality of pressure or flow.
- Pipeline — values and quality of pressures along the pipeline. Sometimes, values and quality of temperature are available. These may be retrieved from automated block valve sites to take advantage of the need for an RTU for valve control. The incremental cost of pressure and temperature measurements in this situation is minimal.

Alarm messages are generated to signal the potential or real interruption of normal operation at any monitored location on the pipeline.

There are four basic data types in a SCADA system, namely, "discrete," "analogue," "internal," and "parameter":

- The term "discrete" reflects the fact that these points can only be in one of two (or more) predefined states. Discrete points are generally binary in nature, i.e., they only have distinct states. This can represent open/closed, on/off, normal/alarm, etc. They are referred to as digital, status or binary points. Some systems will implement three or four state points, such as a valve status, to indicate that the valve is "open," "in transit," or "closed," or a pump is running, in start-up, in shutdown or off.
- "Analogue" or "Analog" refers to points that have a numeric value rather than two or more discrete states. Analogue inputs are field data points with a value that represents a process variable at any given remote location such as pipeline pressure, oil temperature or pressure set point on a control valve. Analogue output points can also be sent as commands from the SCADA host, such as set points for controllers.
- A third type of data point is determined internally by the SCADA host as opposed to being sent by an RTU. The internal data type is also called derived data. This can range from a simple calculation to change the engineering units of a field value to more complicated calculations such as the corrected volume measurement in a tank based on tank level, temperature, and product density of a flow calculation that uses API corrections.
- Parameters or factors are generally used to calculate derived values. Examples include orifice plate sizes, AGA calculation parameters, and performance curves.

All SCADA systems work in a real-time environment, consequently they have a real-time database to process real-time data. The real-time database must be able to process large amounts of real-time data quickly. A typical corporate relational database cannot meet such requirements. Conventional database systems are typically not used in real-time applications due to their poor performance and lack of predictability.

The SCADA host must be able to meet the requirements of a real-time environment and easily interface to standard external databases for the purposes of making key data available to other business processes. One method used is to utilize some form of a data repository or data historian to store SCADA data for access by other applications. This reduces the transactions in the real-time database and improves response performance.

Creating the SCADA database consists of populating the database with each of the individual data sources in the SCADA network. Each point will require a number of information fields to be entered to complete a record in the database. This effort is a time-consuming task and must be done accurately. Typically, the SCADA host provides a high-level software utility for interactive creation and modification of the system database. A key feature of a SCADA system is the ability to download RTU configuration information from the database thus eliminating the need to re-enter data at each RTU. This also eliminates another source of possible error. Database changes (e.g., addition, deletion, or modification of points) can generally be performed on-line and should not require recompiling the system software

All data points will be stored with a time stamp indicating when they were sampled by the RTU. A "quality" flag may also be stored indicating the quality of the value. Some examples of quality indicators are "Good" meaning that the data has been scanned recently and is within range, "Stale" indicating that the point has not been refreshed for some configurable period, "Bad" meaning that the point's value cannot be relied upon, etc. Analogue values are processed by the SCADA host and stored in the real-time database, usually along with the original or raw value received from the RTU. Typical processing of analogue points could include conversion to engineering units, alarm checking against pre-set values for each reading, rate of change alarm, instrument failure alarm, averaging, and totalizing such as volume going into a tank.

SCADA data security and integrity features must be consistent with the corporate IT standards and should be outlined during the development of the SCADA requirements. SCADA manuals should include detailed procedures for generating accurate and complete copies of records, while the system should allow for each user's account to limit the access and function the user can execute. All SCADA historical records should use secure, computer-generated, time-stamped audit trails to independently record the date and time of operator entries and actions that create, modify, or delete electronic records.

A historical database provides for internal analysis and reference as well as meeting the requirements of regulating agencies to review pipeline system operation. For example, operation engineers use the historical data for operational analysis for performance enhancement. The regulator may require emergency scan data to track events leading to and following an emergency condition and eventually to determine the cause/effect relationship. SCADA historical data includes time-stamped analogue values and other control-related analogue values. It can also include digital points and host generated points including alarm and event logs. Operator task logs are also typically included.

Since a large amount of data can be accumulated, the historical data needs to be archived periodically. Archived data refers to data that has been stored on archival media (CD, digital tape, etc.) and is stored in a separate location from the SCADA host system as required by corporate policy. The period of time after which data should be archived is determined by corporate policy. The data archive should include all analogue and digital data, alarms, events and operator actions. The SCADA system must be able to retrieve archived data without interrupting ongoing process operations.

To facilitate analyzing system upsets and events, the SCADA system can have a feature known as "playback." This functions much like a rewind on a VCR and allows a user to replay historical data through an off-line operator terminal in order to more easily analyze and determine the root cause of an upset. It can also be used to do "post-mortems" with operators to provide feedback on actions that were taken and to determine if remedial action taken was done correctly and in a timely fashion. The SCADA data base manager needs to store and time tag all operator actions (alarm acknowledgment, commands, etc.) as well as all incoming and outgoing data to get the most benefit from this feature.

9.1.6 Alarms

For safe pipeline operation, potential alarm situations should be addressed by annunciating alarm messages. High-priority alarms may require audible alarming. Alarm conditions are expected during the course of pipeline system operation. The alarm processing function can help to identify potential alarm conditions before actual alarm conditions occur. Examples of potential alarms include high-pressure violation, high-temperature violation at a compressor discharge, leak detection, etc.

The alarm processing function should be able to limit the number of alarms to those that are important. If the number of alarms is too large, the operator's attention is consumed reviewing and acknowledging alarms instead of monitoring and controlling the pipeline system. An overabundance of alarms also desensitizes the operators and can result in them ignoring critical alarm conditions.

In general, alarms are prioritized according to their critical nature in order to give the operator an indication of which alarms need to be attended to first. Emergency alarms require the operators to take immediate action to correct the condition, while communication alarms may require them to contact support staff immediately. Warning alarms are not usually critical, requiring preventive measure without immediate action. The severity of alarms should be configured to be one of multiple levels of severity (for example, high, medium, or low) for all alarm generating points. Alarms are usually color coded, requiring a different color for each level of alarm. In addition, an audible signal should be generated for high-level alarms.

Analogue alarms are generated when a current value for an analogue point reaches a limit pre-defined in the data base attribute for that point. Figure 9-6 is a typical alarm summary display, which in this case shows the conditions both "in alarm" and "not in alarm" as well as both the "unacknowledged" and "acknowledged" statuses. The first two alarm messages are in an alarm condition because the tank is in "Low-Low" level.

Alarm levels will typically include the following:

- High-High (or Alarm) means that the point has reached its maximum allowable value. This will generally mean that it is close to or has reached a point where local automatic protection systems may be initiating action.
- High (or High Warning) means that the point has reached a warning level. If remedial action is not taken, the point may reach High-High. The trending system will allow an operator to display such a point to see how long it has taken the point to get to the warning level.
- Low-Low (or Alarm) similar to High-High but for a lower limit
- Low (or Low Warning) similar to High but for a lower limit
- Rate of Change: The slope of a trend line has exceeded a pre-defined limit. This means the process value is changing more rapidly than would be expected.

Figure 9-6. Alarm summary (courtesy of Telvent)

Discrete alarms are generated upon a change of state of the data base point. These can represent:

- Change from normal to abnormal such as a high-temperature alarm on a compressor station outlet.
- A change of status that was not the result of an operator control action. For example, a valve closes or a pump shuts down with no initiation from the operator.

All such alarms will be reported and logged, as will any change of status of a point. This will provide not only a record of all abnormal events but will also show when equipment was acted upon by an operator.

A basic alarm management scheme consists of detecting the alarm and reporting the alarm to the operator. An alarm management system will also log and provide an audit trail of each alarm. This will include the time that the alarm was reported, when it was acknowledged by the operator and when the alarming point returned to normal. This information along with the database log will provide key information for post-event analysis.

In any system upset, there will be an initiating event followed by secondary indications or alarms. For example, a control valve may fail causing pressure to rise, which may then cause pressure relief valves to operate and flow rates to exceed expected values. Some SCADA systems may incorporate some form of artificial intelligence to process alarms automatically to advise the operator of what the potential root cause may be.

The SCADA database will have the ability to assign various levels of alarm severity to individual points to provide an easy means of reporting high-priority alarms to an operator. In an emergency condition, it is important to not overload an operator and allow concentration on priority items.

The alarm message includes the date and time of the alarm, the point that caused the alarm, the severity of the alarm denoted by color and an audible signal, and the state of the point. The message is displayed in the alarm window and in the tabular summary of alarms. The alarm window lists all unacknowledged alarms, which should be made available on the screen at all times. Alarms are always logged in an event summary, including not only all the information in the alarm message but also the time when the alarm was acknowledged and by whom.

The operators should be able to easily monitor alarm messages and quickly respond to the messages. Therefore, messages should be made readily available to the operator. The current alarm summary is mainly used for monitoring and acknowledging the messages, while the alarm history summary is mainly used for reviewing the alarm status and pipeline system operation.

9.1.7 Human Machine Interface (HMI) and Reporting

Key features of displays and reports are discussed in this section. Typical data *x*/*y* included in displays and reports are as follows:

- Telemetered data, including analogue, digital, and derived values and quality
- Parameter data, such as orifice plate size
- Schematic information, including station yard piping, facility locations on the pipeline system, and other pertinent information

The displays need to be designed to meet the needs of individual operators, because they are the prime users of SCADA displays. Displays need to:

- provide a fixed area on the screen for alarm and emergency annunciation
- refresh the displays dynamically and within a short time (at most a few seconds) after a command is issued
- allow the operators to be able to navigate the displays easily and quickly
- maintain a consistent "look and feel" and use intuitive design industry-accepted display design methodologies and standards.

All SCADA vendors will have a comprehensive HMI system, which will include tools for creating and modifying displays and reports. In fact, the capabilities of most systems can be bewildering and intimidating. Since a typical SCADA host will have a large real-time database, the challenge is to design an HMI that presents relevant information to the operator in an easy to understand set of displays.

It is suggested that a fixed area be reserved on the screen for alarm and event messages, system performance monitoring, and annunciation of emergency scan. In other words, this information remains on the screen always until it has been acknowledged. It is important to develop some guiding principles for each system before the displays are created. These guidelines should include some variation of the following:

1. Have a hierarchical approach: Top-level displays will show key summary information but also have the ability to "zoom" in quickly for more detail. Typically, the top level display is a pipeline system overview or a pipeline system schematic. The system overview display allows the operator not only to view the current pipeline states including set points and alarms of the system but also to access a particular station for viewing control points and/or modifying their values. It not only displays all pump/compressor stations and current alarm messages but also flow, pressure and temperature including set points. In addition,

this display may show the link to pump/compressor, meter, or valve station control panels through which the operator can send a control command.
2. Screen navigation should follow the current expected features found in most window-type navigation software to reduce operator-learning time and to make the system as intuitive as possible.
3. Ensure a consistent "look and feel" of displays to minimize training and the chance of operator error. These will include the use of color and a consistent and logical approach to the use of buttons, menus and toolbars. A judicious choice of colors is important as certain types of color blindness can result in some colors appearing the same to some people.
4. Keep screens as uncluttered as possible while still supplying the required information. The possibility of confusion should be minimized and care taken to reduce the possibility of information being lost or "buried" on the screen.

The following displays are considered to be key display requirements for effective operation of the pipeline system:

- Display or schematics of an entire pipeline system
- Pump station overview including measurements on piping, unit and driver
- Meter station information including flow rate and accumulated volume, total station flows, etc.
- Pipeline elevation and pressure profiles with MAOP
- Batch tracking information along the pipeline system
- Tank and storage information such as tank inventory
- Alarm and event annunciation and summary
- Communication summary
- Measurement and equipment status summary
- Security-related information including system status and police contacts

The displays are either in tabular or graphical format. In some cases, it may be useful to have both tabular and graphical formats for displaying data. The selection of format depends on how the data is used. For example, it is more useful to display pressure drop along the pipeline in graphical format. Most modern SCADA systems use several display mechanisms, which include textual and graphical images augmented with real-time information. Color and shape can be used to relate discrete information in an intuitive manner, plot and trend display types can be used to display graphs of analog data in an *x*/*y* format or a horizontal or vertical bar graph.

There are other display types such as pushbutton for selecting a button to perform a specific function, meter/gauge for showing a meter/gauge device with values, and region for marking a location on a display. Some SCADA display systems support display format control. The format control functions include popup and pan/zoom. For example, the functions such as set-point control and communication control can be supported by pop-ups. A large display area can be easily navigated by means of a panning/zooming feature of the display system.

Figure 9-7 is a display of a pipeline system. It shows the operating statuses and parameters of the entire pipeline system; pump stations and the operating pump units, current station pressures and flow rate, density, and list of alarm messages at the bottom of the display. From this display, a desired pump station or alarm message is selected to review the detailed data for the station.

Figure 9-8 shows a typical pump station diagram. To monitor or control a pump station, it can be directly selected from the display of the pipeline system. Then, the operator can monitor the measured variables and controlled parameters of the selected

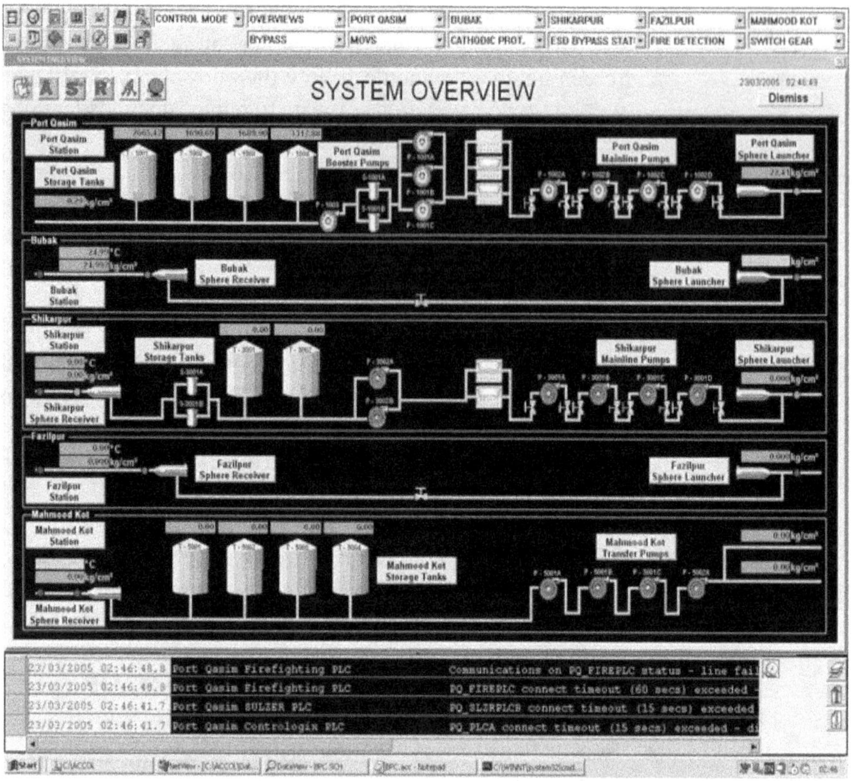

Figure 9-7. Screen display of a pipeline system (courtesy of Telvent)

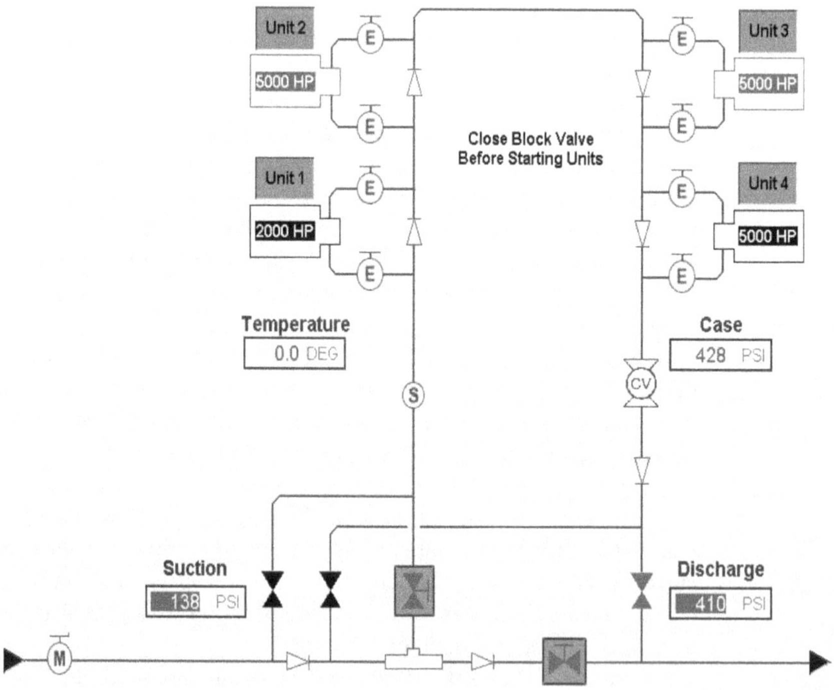

Figure 9-8. Pump station diagram (courtesy of Telvent)

station as well as its station operation statuses including pump unit and valve statuses. From the pump station display, the pump units and valves can be controlled and the control point can be set. Normally, the suction or discharge pressure is controlled from this screen.

Figure 9-9 displays a meter station in a tank farm, installed with a meter prover. The operator can view the current meter station statuses and meter data including flow meter and valve positions as well as control the meter station from this display. The operator can acknowledge any alarm messages related to the meter station operation. These alarms are listed in the alarm summary at the bottom of the screen. In addition, the flow meter can be proved by means of a meter prover.

Figure 9-10 displays the elevation profile, and pressure profile with MAOP. The pressures can be presented in terms of head so that all three units are the same. This allows the operator to visually detect trouble spots such as slack flow conditions along the pipeline.

Data trending capability is one of the most important functions of any SCADA system because it helps the dispatchers and operations staff to identify potential problems before they arise and to diagnose alarm conditions. Data trending is used to display any analogue values which are stored in the historical database over time at a specific location or locations. Data trending displays are in graphical format due to the

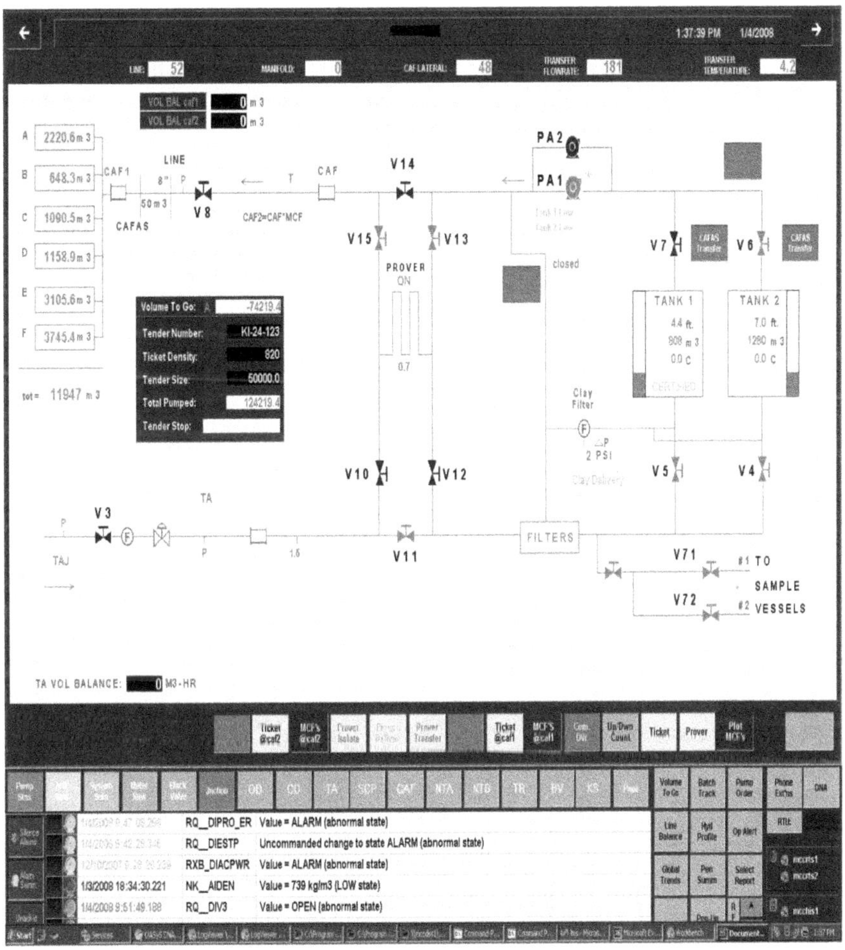

Figure 9-9. Meter station with a prover (courtesy of Telvent)

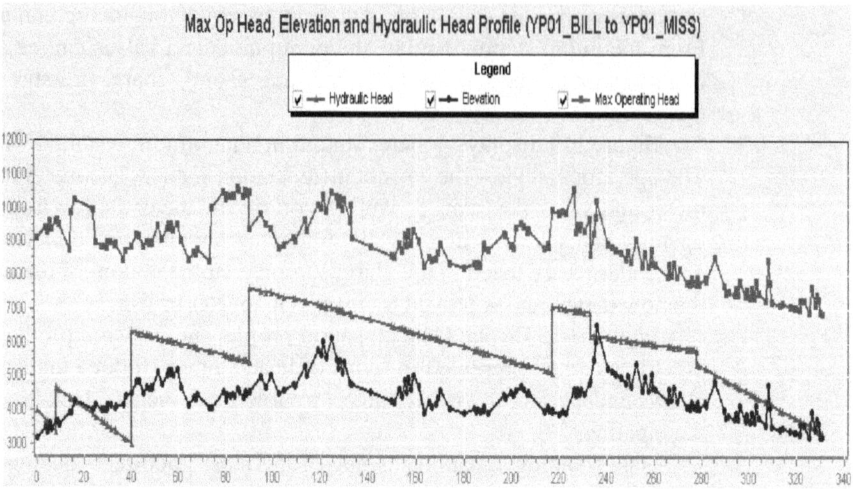

Figure 9-10. MAOP, pressure and elevation profiles (courtesy of Telvent)

large amount of data. Figure 9-11 shows a typical trend plot of flow rate, pressure, temperature, and API gravity at any specific point of a pipeline. Any data can be trended, and the trended data can be analyzed to detect any anomaly at the point.

All SCADA systems have some type of reporting capability. This will typically consist of both standard reports generated automatically by the system and user-defined reports. These reports are generated from the SCADA databases containing real-time, historical and calculated data. The standard reports are of a predefined structure, while the user-defined reports meet the user's specific needs. Examples of standard reports include operating summary reports and billing reports, and those of user-defined reports include such things as command/alarm log sorted by station.

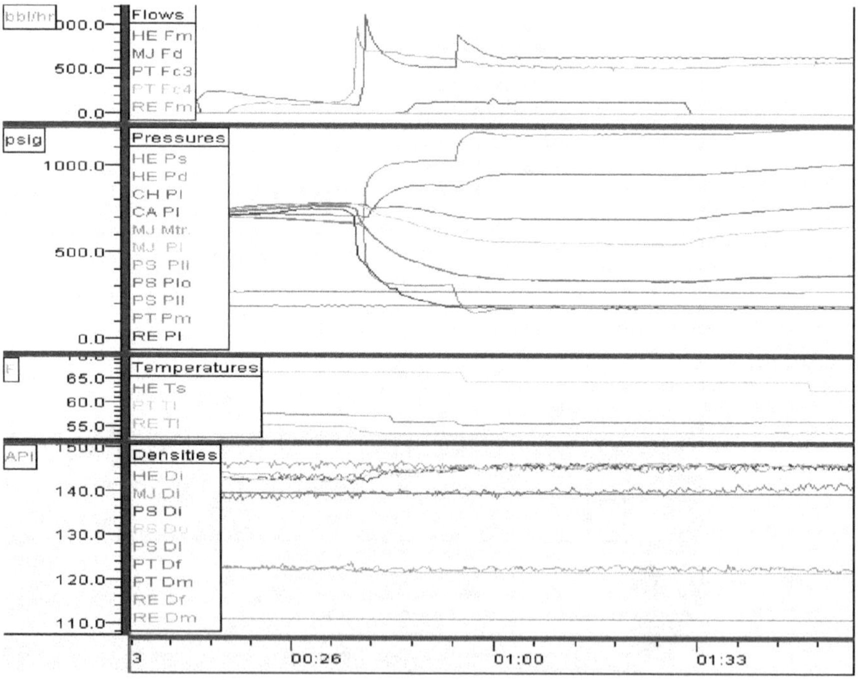

Figure 9-11. Data trending

The types of reports usually found on a pipeline SCADA system would include some of the following:

- Operating reports
- Shift or daily operating summary reports
- Product movement report
- Alarm summary report
- System availability, communication and reliability report
- Emergency scan report, containing operating data during emergency conditions

Government regulators may require pipeline companies to submit regulatory reports. Normal operation reports may need to be submitted regularly, but emergency reports are mandatory in the event of emergency conditions.

The SCADA system provides system administration tools to configure and maintain the system, and allows the SCADA users to access various logs.

- Command log, containing a record of all commands issued by the operator
- Alarm log, containing all generated and acknowledged alarm messages for tracking operational problems
- Database maintenance log for recording commands used to change any SCADA database values
- System log for recording the SCADA system performance including error data such as the start/stop time, abnormal running time, etc.
- Communication log for recording the statistics of the communications with the RTUs such as the number of attempts, the number and types of error, etc.

The number, content, and style of reports will vary widely depending on the pipeline type, the business requirements, and the regulatory environment. It is important that the SCADA system provides an easy to use, flexible reporting package that does not require programming changes to create and implement reports.

9.1.8 Security

A SCADA system will provide for user password access and the ability to configure specific levels of access for each user. For example, there may be users who may access the SCADA system but are allowed only the ability to read some pre-configured reports. For example, only those who are directly responsible for the database are allowed to make changes to the database and this is done with password protection.

SCADA systems have long been thought to operate in a secure environment because of their closed networks, which are not exposed to external entities. In addition, the communication protocols employed were primarily proprietary and not commonly published. Recent advances, such as Web-based reporting and remote operator access, have driven the requirement to interface with the Internet. This opens up physical access over the public network and subjects SCADA systems to the same potential malicious threats as those that corporate networks face on a regular basis.

Typically, compliance with industry standards and technologies is regarded as a good thing. However, in the case of newer SCADA systems, recent adoption of commonly used operating systems and standards makes for a more vulnerable target. Newer SCADA systems have begun to use operating systems such as Windows that are commonplace in corporate networks. While this move offers benefits, it also makes SCADA systems susceptible to numerous attacks related to these operating systems.

RTU to host protocols are now utilizing industry standard protocols, which may compromise their security.

Due to cyber terrorism, the security associated with the SCADA network needs to be designed and assessed by the same policies utilized in other areas of the company. If there are no such clear network security policies in place, then they need to be established before taking specific actions on the SCADA network. For detailed information on SCADA security, refer to API Standard 1164 — Pipeline SCADA Security [6]. This standard provides guidance to the operators of pipeline systems for managing SCADA system integrity and security.

9.2 OVERVIEW OF PIPELINE LEAK DETECTION SYSTEM

9.2.1 Introduction

This section discusses various aspects of pipeline leak detection without an emphasis on any particular techniques. Anyone, who is interested in the detailed discussion of the leak detection techniques and their implementation considerations, are referred to other volumes [1]. This section introduces the selection criteria of a leak detection system and various leak detection techniques.

Pipeline leak detection is only one aspect of a pipeline leak management program; it encompasses leak prevention, detection and mitigation procedures. In order to minimize the consequences of a leak, pipeline companies require a comprehensive leak management program. A leak detection system by itself does not improve on a pipeline's integrity nor reduce potential failures of a pipeline system. However, such a program will not only help prevent and monitor the degradation of a pipeline that may eventually lead to failure, but will also minimize the consequences of pipeline leaks if they occur.

Pipeline companies minimize leaks through a leak prevention program. The main causes of leaks are: outside or third party damage such as excavation equipment hitting the pipeline, geophysical forces such as floods and landslides, improper control of the pipeline system, and pipe corrosion. Figure 9-12 shows leak statistics in US, Canada and Europe.

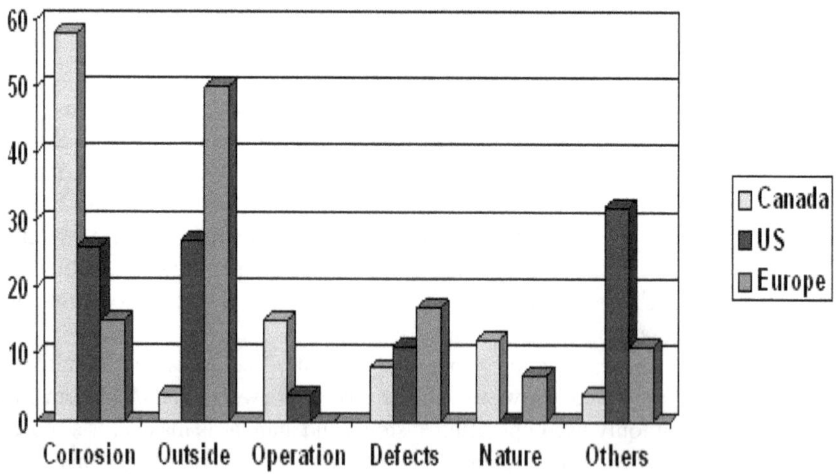

Figure 9-12. Leak statistics

Even though the statistics are about ten years old, they can be relevant to address key issues on leaks. Incidents resulting from damage by a third party are significantly higher in Europe than those in Canada, mainly because the population density in Europe is much higher. Proper control of third-party damage is achieved through: marking of the right of way; education of employees, contractors, and the public; and effective use of systems such as "One-Call." Geophysical forces cannot be controlled but can be monitored and their effects can be mitigated. Pipeline integrity management is a significant subject by itself and discussed in separate volumes [7].

Leak mitigation is the attempt to reduce the consequences of a leak when it occurs. If a leak can be detected quickly and isolated quickly, the spillage can be minimized. This requires that the leak alarm and its associated information are reliable and accurate. Having effective procedures in place and the proper resources and tools to enact them are critical in addressing the leak mitigation issues efficiently. The leak confirmation and isolation issues should be part of leak detection. The scope of leak detection does not normally include spillage management issues such as cleanup procedures and manpower mobilization.

Historical data indicates that leaks were predominantly detected by local operation staff and third parties. Successful detection by means of a single leak detection system was random. This was because no single leak detection system could detect leaks quickly and accurately or provide reliable leak detection continuously and cost-effectively. Therefore, more systematic approaches to leak detection are required, such as a combination of line patrol, sensing devices and/or SCADA-based systems with automated leak detection capability.

Since SCADA systems have become an integral part of pipeline operations, a particular consideration has to be given to leak detection methods that can be easily implemented on the SCADA system. API Publication 1130 [8] addresses various Computational Pipeline Monitoring (CPM) methodologies, integrated with a host SCADA system. In association with the CPM, API Publication 1149 [9] and API Publication 1155 [10] are briefly discussed with respect to how they are used for specifying and evaluating leak detection performance.

Pipeline Leaks

This chapter uses the definition of leaks as defined in "Petroleum Pipeline Leak Detection Study [11]." There are two types of leaks: an incipient leak and an actual leak. "Incipient leaks" are defined as those that are just about to occur. Certain incipient leaks can be discovered by inspecting the pipeline and dealt with before they become actual leaks. Here, an actual leak is called a pipeline leak when fluid is leaking out of a pipeline system.

All pipeline leaks are associated with certain external and internal phenomena. External phenomena include the following:

- Spilled product around the pipeline
- Noise generated from leakage at the hole in the pipe
- Temperature changes around the hole

Internal phenomena include:

- Pressure drops and flow changes
- Noise around the hole
- Temperature drop for gas pipeline

All leak detection systems take advantage of the presence of one or more leak phenomena.

Standards on Leak Detection

In North America, a leak detection system is normally required on new liquid pipelines, but not on existing pipelines unless mandated otherwise by the appropriate regulatory agency. In general, there is no leak detection requirement on gas pipelines other than a few new gas pipelines. The same is true of multi-phase gathering pipelines.

Pipeline companies are using various leak detection methods with varying degrees of success. At present, no single method truly stands out as an ideal system able to detect the wide ranges of leaks with accuracy and reliability, and having low installation and operating costs. Some are accurate and reliable but too expensive, and some are economical but unreliable.

Different countries have developed different leak detection regulations and practices. A few references and standards are introduced below. However, in general, the codes and standards on pipeline leak detection are not well defined.

American Petroleum Institute (API) has published several standards on pipeline leak detection. They are listed below:

- API 1130 "Computational Pipeline Monitoring" addresses the design, implementation, testing and operation of Computational Pipeline Monitoring (CPM) systems. It is intended as a reference for pipeline operating companies and other service companies. The publication is used as a standard by regulatory agencies in many parts of the world.
- API 1149 "Pipeline Variable Uncertainties and Their Effects on Leak Detectability" discusses the effects of variable uncertainties and leak detectability evaluation procedures for a computational pipeline monitoring methodology. This publication describes a method of analyzing detectable leak sizes theoretically using physical parameters of the target pipeline. It can be used for assessing leak detectability for new and existing pipelines.
- API 1155 "Evaluation Methodology for Software Based Leak Detection Systems" describes the procedures for determining CPM's leak detection performance. Unlike API 1149, this publication addresses the performance evaluation procedures based on physical pipeline characteristics and actual operating data collected from pipeline operations.

The Canadian standards applicable to oil and gas pipelines are specified in Z662, "Oil and Gas Pipeline Systems." Section 10.2.6 of Z662 specifies leak detection for liquid hydrocarbon pipeline systems. The specifications in Section 10.2.6 for liquid pipeline systems states:

> "Operating companies shall make periodic line balance measurements for system integrity. Operating companies shall periodically review their leak detection methods to confirm their adequacy and effectiveness. Installed devices or operating practices, or both, shall be capable of early detection of leaks. Measuring equipment shall be calibrated regularly to facilitate proper measurement."

The title of Annex E is "Recommended Practice for Liquid Hydrocarbon Pipeline System Leak Detection." The annex describes a practice for leak detection based on computational methods, particularly material balance techniques. It does not exclude other leak detection methods that are equally effective. The annex emphasizes that operating companies shall comply as thoroughly as practicable with the record retention, maintenance, auditing, testing, and training requirements, regardless of the method of leak detection used.

Leak Detection System Selection Criteria

It is essential that the objectives and requirements for employing the leak detection system are defined. The objectives of the leak detection system are to assist the pipeline operators with:

- Reducing spillage of product and thus reducing the consequences of leaks,
- Reducing operator's burden by detecting leaks quickly and consistently without relying heavily on operator experience,
- Satisfying regulatory requirements.

A leak would be initially detected and located by the leak detection system and then confirmed by some means such as visual inspection. After, or even before the leak is confirmed (depending on the company's leak response procedures), the leak must be isolated by closing block valves adjacent to the leak. After the leak is isolated, a significant volume of product can be lost depending on the leak location and terrain of the leaked pipeline section. The spillage during the detection phase is often relatively small compared to potential total spillage. Therefore, the importance of rapid detection time as a valuable feature of a detection system cannot be over-emphasized.

It is important to define a set of selection criteria for use in assessing the performance and selection of various leak detection systems. Typical performance criteria are listed in Table 9-1 [12, 13]:

TABLE 9-1. Leak detection system performance criteria

Criteria	Description
Detectability	Detectability of leaks is measured in terms of leak detection time and range of leak size.
Sensitivity	Sensitivity is defined as the minimum leak size that the leak detection system can detect.
Reliability	Reliability of a leak detection system is defined in terms of false alarm rate. If the frequency of false alarms is high, the operators may not trust the leak detection system, increasing the confirmation time and thus spillage volume.
Robustness	Robustness is defined as a measure of the leak detection system's ability to continue to operate and provide useful information in all pipeline operating conditions.
Operability	The leak detection system needs to operate not only continuously but also in all operating conditions (shut-in, steady state and transient state). In addition, the system should not interfere with normal operations.
Accuracy	Accuracy is defined as a measure of the leak detection system's ability to estimate how close the estimated leak location and size is to the actual leak location and size.
Cost	The cost includes the installation and operating costs of a leak detection system, including instrumentation or sensing devices.

An effective leak detection system helps pipeline operators mitigate the risks and consequences of any leak. It can shorten leak detection time, increase reliability (not miss actual leaks and at the same time not produce false alarms), and reduce leak confirmation and isolation time with accurate leak location estimates. Simply put, overall cost of a leak can be reduced using an effective leak detection system. However, there are costs to implement and operate a leak detection system.

Therefore, the decision-making process of implementing and operating a leak detection system can be made by balancing the risk and consequences of possible leaks against the cost of a leak detection system and mitigation program. The following process may help in analyzing potential risks of leaks in terms of cost and the cost savings resulting from the implementation of a leak detection system:

- Estimate likely probabilities of various leaks and thus the potential number of leaks.
- Estimate the direct and indirect costs of leaks over a period of time without a leak detection system by using historical data for the consequences of the leaks.
- Assess attainable leak detection performances of various leak detection systems by applying the above criteria.
- Determine the costs of implementing and operating these leak detection systems over the same period of time used in the cost calculation above.
- Estimate potential cost savings from the use of a leak detection system.

9.2.2 Overview of Leak Detection Techniques

Broadly, there are three different types of leak detection methods: Inspection Methods, Sensing Devices, and Computational Pipeline Monitoring Methods.

9.2.2.1 Inspection Methods

In general, inspection methods provide very accurate, sensitive and reliable results. Particularly, ultrasonic and magnetic inspection techniques can detect both actual and incipient leaks by determining the pipe wall thickness. However, internal inspection methods are very expensive requiring specialized tools and expertise, and a pipeline cannot be inspected continuously. Due to the nature of intermittent operation, only leaks that occurred prior to the inspection will be detected and any occurring after will remain undetected until the next inspection. Inspection techniques include visual inspection, magnetic flux technique, ultrasonic technique, hydrostatic test, and others:

Visual Inspection — Current visual inspection methods rely on detecting hydrocarbons along the pipeline right of way either visually or by using an instrument. Spillage evidence includes spilled hydrocarbons, vegetation changes caused by hydrocarbons, odor released from the pipeline, or noise generated by product escaping from a pipeline hole. For inspecting transmission lines, pipeline companies often use an inspection airplane equipped with hydrocarbon detection sensors and cameras.

Magnetic Flux Technique — Strong magnets are mounted on a magnetic inspection pig. When a strong magnetic field is applied to steel pipe, magnetic flux is formed in the pipe. If the pipe is uniform, so is the resulting magnetic flux. If the magnetic flux is distorted, the magnetized pipe may contain defects. Since changes in magnetic flux induce electric current, transducers measure the induced current. A magnetic inspection pig can detect pipe defects reliably and locate them accurately. It can run without interrupting normal pipeline operations. In general, it can produce a wealth of information for detailed defect assessment. However, a magnetic inspection pig tends to miss longitudinal defects and cracks, and is expensive to purchase or operate.

Ultrasonic Technique — Pigs mounted with high-frequency ultrasonic equipment are used to inspect internal and external defects and pipe welds on manufactured pipes and operational pipelines. An ultrasonic inspection tool can detect small defects accurately under clean conditions assuming that it is well coupled with the pipe surface. This technique does not interfere with normal pipeline operations nor adversely affect the pipeline system safety. However, it is sometimes difficult to maintain good coupling between the transducer and pipe wall. Recent advancements with these tools have resulted in the ability to detect stress corrosion cracking (SCC).

Hydrostatic Test — Hydrostatic testing must be performed on new pipelines, as specified in ASME B31.4 and other standards, prior to in-service use. The main advantage is that it can detect not only incipient failure but also existing small pin hole size leaks. However, non-critical cracks may not be found and crack growth can accelerate due to pressurizing at the time of testing.

Hydrostatic testing was also used on operating pipelines to assess their structural integrity. It is another method of identifying SCC problems. When an operating pipeline is tested at a pressure above normal operating pressure using the fluid normally transported in the pipeline, it is called a dynamic pressure test. The purpose of this test was not to accommodate the increase in operating pressure level, but to confirm the pressure capability of the pipeline system. However, a pressure test can be destructive if a line break occurs during the test. As a result, pressure testing of operating pipelines with hydrocarbon liquids is not allowed by code and is not practiced anymore. Procedures for hydrostatic testing and ILI are outlined in API Standard 1160 — Managing System Integrity for Hazardous Liquid Pipelines.

9.2.2.2 Sensor Methods

Sensing Devices can be used to continuously sense particular characteristics of leaks such as sudden pressure drop, noise, electrical impedance, or other signals caused by a leak or interference around a pipe. Some sensing devices can detect not only leaks but also third party interference around the pipeline system. Traditionally, these techniques have been relatively unreliable and impractical. There are a few emerging technologies in sensing devices such as fiber optics that are showing increasing promise. Certain techniques such as specialized fiber optic cables can be expensive for existing pipelines, as the pipeline has to be retrofitted with the cable or sensing devices.

Acoustic Sensing Device — The figure above shows the schematics of this technique. The principle of this technique relies on the fact that when a fluid passes through a hole under high pressure, the resulting turbulence creates acoustic pressure waves that travel through the fluid and pipe. Acoustic sensors are placed on the pipe, regularly spaced along the pipeline, to detect these acoustic waves. An acoustic leak detection system continuously monitors the pipeline for the sound characteristic of a leak. The signals, after the background noise including operation characteristics are filtered, are compared to the appropriate thresholds to confirm or reject a leak. The acoustic leak detection system can also determine the leak location by correlating the sensor spacing, velocity of sound, and propagation time difference.

The advantages, if it is installed properly, include the detectability of small leaks in a short time, accurate leak location and continuous operation. However, it tends to generate frequent false alarms particularly for small leaks in the presence of large background noise in the pipeline and can be expensive for a long transmission line, because of the need for many acoustic sensors (Figure 9-13).

Optical Fiber Sensor System — This is an emerging technology that uses an optical fiber sensor to detect leaks and/or impending pipe damage [14]. It requires the installation of an optical fiber cable along the entire length of the pipeline. It operates by detecting optical properties, temperature change, and/or micro bends of the pipe. The latter capability allows the detection of activities outside the pipeline which can be picked up by micro-strain sensors.

The advantages of this system, if it is installed properly, include the capability of detecting and locating outside third party damage and fluid theft as well as continuous operation. However, the installation cost on an existing pipeline can be high. Even though several successful trials were reported [15], its performance has not yet been fully proven for long pipelines.

Vapor Monitoring System — A vapor monitoring leak detection system [16] detects leaks by placing a sensor tube next to the pipeline. In the event of a leak, the hydrocarbon vapors will diffuse into the sensor tube. Its operation is shown in the figure below (Figure 9-14).

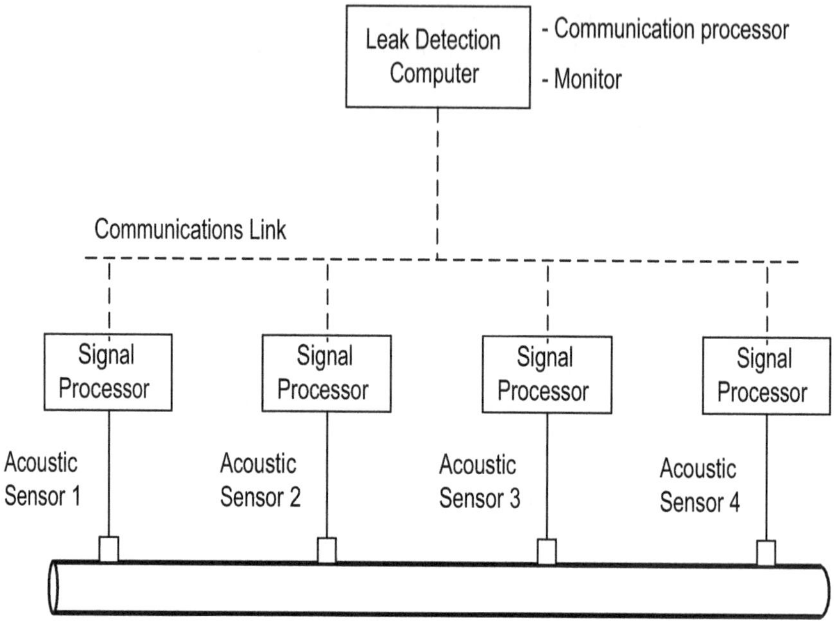

Figure 9-13. Schematics of acoustic pressure wave technique

The system consists of a suction pump, gas detector and a plastic cable or tube that is installed adjacent to the pipeline. When a leak occurs, some hydrocarbon molecules diffuse into the tube as a result of the hydrocarbon concentration difference between the inside and outside of the tube around the leaking section. In due course, the affected area of the tube will have a higher hydrocarbon concentration than the rest of the tube. When the pump pulls the air, the affected air is also pulled toward the detection unit, which analyzes the hydrocarbon concentration. Because the air is pulled at a constant speed, the system can determine the leak location. Leak size can be estimated from the concentration of hydrocarbons.

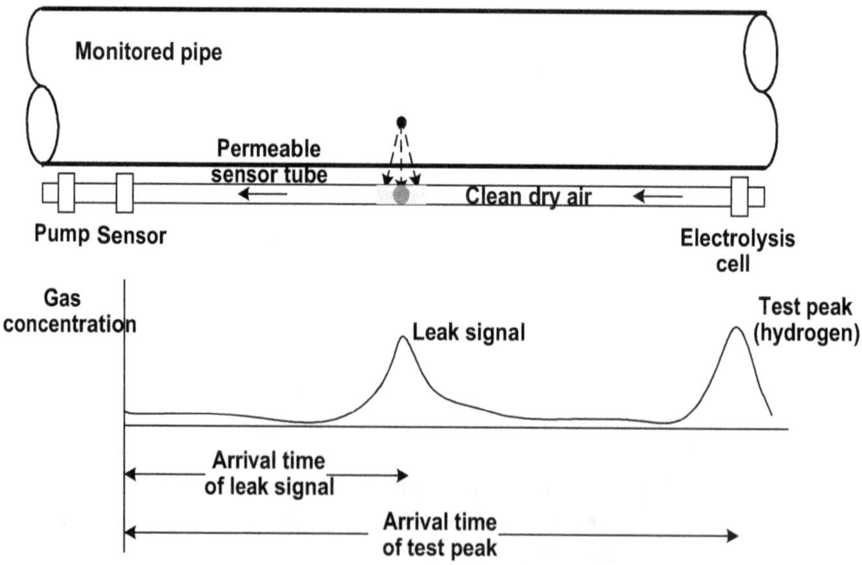

Figure 9-14. Vapor monitoring method

This method of leak detection and location can detect a very small leak and locate it accurately. It can be used for both onshore and offshore pipelines as well as multiphase leak detection. In addition, this methodology can be used to detect many different substances. A system based on this technology has been used in an Arctic pipeline (Northstar Development). However, this method may be too slow to react to large leaks, and the installation and operation costs can be very high. This system should be used in conjunction with other leak detection systems in environmentally sensitive areas.

9.2.2.3 Computational Pipeline Monitoring (CPM) Methods

The CPM methods are based on mathematical or statistical computations of certain quantities using commonly available measured values such as flows and pressures obtained through the host SCADA system. Each scan, a CPM system receives an updated set of SCADA data and sends a set of the modelled results back to SCADA through the SCADA interface software. In general, the cost is relatively reasonable but the sensitivity is lower than other methods.

Any pipeline monitoring system that continuously checks for leaks can be considered a real-time leak detection system. All CPM methodologies are classified as real-time leak monitoring systems. Real-time leak detection as discussed in this section includes only the methods based on leak detection software operating in conjunction with a host SCADA system.

Any CPM system consists of the following components:

- Field instrumentation and RTU which sends the field data to the host SCADA
- SCADA system, which collects the field data, sends them to the real-time leak detection system, and annunciates event and alarm messages. The SCADA system requirements for leak detection are discussed in ref. [1].
- Hardware and software interfaces which integrate the functions of the host SCADA and real-time leak detection system
- Real-time leak detection computer and software

The key advantage of the CPM methods is that they seldom need additional instruments and equipment to those that already exist for normal pipeline operations. As a result, the implementation and operating costs are typically lower than the costs for inspection and sensor methods.

API Publication 1130 defines the following eight CPM methodologies:

- Line balance technique
- Volume balance technique
- Modified volume balance technique
- Compensated mass balance technique
- Real-Time transient model (RTTM) method
- Flow/pressure monitoring method
- Acoustic/Negative pressure wave method
- Statistical techniques

The first five methodologies are based on mass balance principle and will be discussed in that context. The mass balance principle applied to a pipeline means that the difference between the amount of fluid that enters and leaves the pipe over a given time must be the same as the change in fluid inside the pipe over the same period of time. This principle is expressed mathematically as follows:

$$\text{Imb} = V_{\text{in}} - V_{\text{out}} - \Delta\text{LP} \qquad (9-1)$$

where

V_{in} = corrected volume entering the pipeline over a fixed time interval
V_{out} = corrected volume leaving the pipeline over the same time interval
ΔLP = change in line pack over the same time interval
Imb = imbalance

The non-zero imbalance can be attributed to a number of factors including measurement errors and line pack calculation errors. It is interpreted as a leak if the imbalance is positive beyond a predefined limit, and as an unaccounted flow if it is negative, assuming that the measured flows are accurate. The flows that go into and out of the pipe are measured quantities and line pack changes are calculated quantities. Depending on how line pack changes are treated, the mass balance method has several forms.

Line Balance — Line balance methodology uses only the measured flows into and out of the pipeline system, ignoring line pack changes. It assumes that a leak may have occurred if more fluid enters the pipeline than leaves the pipeline over a certain time period. Since the inlet and outlet flow rates are measured at different pressures and temperatures, they need to be corrected to base conditions and then compared. If pressure and temperature measurements are not available, leak detection is less accurate.

The line balance methodology is very simple to implement and the computational requirement is not extensive. It is a suitable method for pipeline systems with very small pipe size and short pipe length, assuming that reliable flow measurements are available at both injection and delivery ends. This technique is less popular now, because it does not offer much advantage over the other techniques.

Volume Balance — This technique uses both the flow difference and line pack change terms in the mass balance equation, roughly compensating for the difference between the volumes into and out of the pipeline with line pack changes over a certain time period. Imbalance beyond a set limit is interpreted as a potential leak. Line pack change depends on fluid properties such as compressibility and thermal expansion, pressure, and temperature, and pipe data such as size and length. Since line pack change is a dynamic quantity, the fluid compressibility and pressures are the most important parameters in calculating line pack change. Even though pressure changes in a transient condition, it is assumed that the pressures are in a steady state at the time of the pressure measurements. Because of the steady state assumption, the calculated line packs and its changes under transient operations are not very accurate.

At a minimum, this method requires flow and pressure measurements at all fluid injection and delivery points. Short-pressure measurement spacing helps to reduce error in line pack calculation, but no additional instrumentation is required. The volume balance technique is better suited to heavier hydrocarbon liquid pipelines with relatively small line pack and short-pressure measurement spacing, because the line pack calculation error can be minimized. Its implementation is simple and its operation is reliable. However, it has limited leak detection capability for long pipelines with highly compressible fluid in which the line pack changes significantly. Figure 9-15 illustrates a trend of volume balance, which is composed of the volume in and out difference and the estimated line pack change.

Modified Volume Balance — The modified volume balance method is a modified version of the volume balance method; it differs in that it uses a more accurate accounting of product properties. The method tracks batches along the pipeline and calculates the average bulk modulus dynamically in each pipe segment. The segment bulk modulus is applied to calculate the segment line pack and its changes, and the whole line pack and its changes and then all changes are added together.

Most comments made for the volume balance method are valid for this method, except that its implementation is not so simple but it generally calculates the line pack change more accurately than the volume balance particularly for batch pipelines.

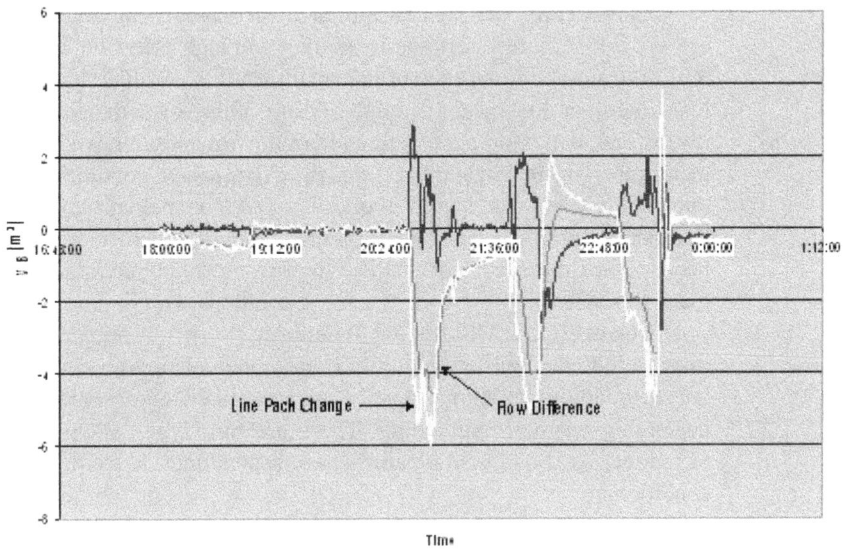

Figure 9-15. Trend of volume balance

Compensated Volume Balance — The compensated volume balance method is an enhanced version of the modified volume balance. It calculates temperature profiles along the pipeline by solving an energy equation with the temperatures at the injection as a boundary condition. The method takes into account the fluids movements including batch, fluid blending, and product characteristics. To reduce line pack calculation error during transient operations, a filtering technique is applied to line pack changes.

Most comments made for the volume balance method are valid for this method, except that its implementation is more complex whereas it generally calculates the line pack change more accurately than the volume balance particularly for batch pipelines and light hydrocarbon liquids such as propane and ethane. Also, this method may be able to estimate a leak location if the pipeline state after the leak reaches a steady state condition.

RTTM — The Real-Time Transient Model (RTTM)-based leak detection methodology performs the functions of determining the pipeline state in terms of flow, pressure, temperature and density profiles based on real-time data and then detecting anomalies of pipeline state including leaks. API Publication 1130 defines the Real-Time Transient Model-based leak detection methodology as follows:

"The fundamental difference that a RTTM provides over the CMB method is that it compares the model directly against measured data, i.e., primarily pressure and flow) rather than use the calculated values as inputs to volume balance. Extensive configuration of physical pipeline parameters (length, diameter, thickness, pipe composition, route topology, internal roughness, pumps, valves, equipment location, etc.), commodity characteristics (accurate bulk modulus value, viscosity, etc.), and local station logic (e.g., pressure/flow controllers) are required to design a pipeline specific RTTM. The application software generates a real-time transient hydraulic model by this configuration with field inputs from meters, pressures, temperatures, densities at strategic receipt and delivery locations, referred to as software boundary conditions. Fluid dynamic characteristic values will be modelled throughout the pipeline, even during system transients. The RTTM software compares the measured data for a segment of pipeline with its corresponding modelled conditions."

Theoretically, the RTTM approach of real-time modelling and leak detection can provide the most accurate modelling and leak detection sensitivity results. Most RTTMs can provide a wealth of information on the pipeline state. In practice, however, real-time data quality and availability are often not sufficient for reliable operation of this leak detection approach, and certain values such as viscosity are not readily measurable on-line. In addition, modeling in transient conditions sometimes increases uncertainty when data quality is questionable. The main disadvantages include longer installation time, requirements for a high level of expertise to operate, and strong dependency of measurement quality. As a result, many companies have attempted to make this methodology work in actual operations with limited success.

Pressure/Flow Monitoring Technique — This technique is used on liquid pipelines to indicate unusual conditions and potential rupture conditions. This monitoring methodology monitors rapid or unexpected changes in pressure and/or flow rate, depending on their availability. There are four types of pressure/flow monitoring techniques used on liquid pipelines to indicated unusual conditions and potential leak conditions:

- Pressure/Flow Limit Monitoring — ensures that measurements stay within predefined operating conditions and emergency limits.
- Pressure/Flow Deviation Monitoring — ensures that measurements stay within a predefined tolerance of an expected operating value. Often, separate deviation limits are established for active and inactive conditions and for positive and negative deviations.
- Pressure/Flow rate of change (ROC) Monitoring — ensures that any rapid measurement change, above a predefined value per defined time period, is made known. Often, separate ROC limits are established for the positive and negative directions.
- Pressure/Flow ROC deviation — modified version of the Pressure/Flow ROC Monitoring, that projects expected ROC values during transient conditions. Often, separate ROC deviation limits are established for positive and negative directions.

Mathematically, a projected value is expressed in terms of a linear regression to predict the next pressure or flow rate using a specified number of pressures or flow rates collected over a specified period. In principle, if the current measurement drops outside a predefined threshold from the predicted value, an alarm condition is satisfied. Normally, a second violation check is performed with the next value in order to avoid generating frequent alarms. If a second consecutive violation is detected, pressure and/or flow rate violation alarms are generated.

This method is simple and easily implemented on the host SCADA system. The main difficulties with this method are as follows:

- Normal operations can produce rapid changes in pressure and flow rate that do not necessarily indicate a leak.
- Pipeline pressure increases can mask a leak.

This method may be useful for detecting unusual events or ruptures. For leak detection purposes, it is normally used in conjunction with other leak detection methods.

Acoustic/Negative Pressure Wave Technique — This technique works similarly to the acoustic sensing technique, except that pressure sensors are used instead of acoustic sensors. API Publication 1130 defines this method as follows:

"The acoustic/negative pressure wave technique takes advantage of the rarefaction waves produced when the commodity breaches the pipe wall. The leak produces a sudden drop in pressure in the pipe at the leak site which generates two negative pressure or rarefaction waves, travelling upstream and downstream. High response rate/moderate accuracy pressure transmitters at select locations on the pipeline continuously measure the fluctuation of the line pressure. A rapid pressure drop and recovery will be reported to the central facility. At the central facility, the data from all monitored sites will be used to determine whether to initiate a CPM alarm."

This technique provides accurate leak location and rapid detection of relatively large leaks, assuming that the leaks occur rapidly, the sensor spacing is close and operating pressure is high. However, it tends to generate frequent false alarms.

Statistical Technique — This technique is not a leak detection technique in itself, but a decision tool based on the data generated by some of the above methods. Two statistical techniques specifically applied to leak detection are described below. Even though it is time-consuming to determine all possible "no leak" conditions for all possible operating scenarios, several successful implementations have been reported.

Sequential Probability Ratio Test (SPRT) Technique — A statistical leak detection method applied to this problem is a sequential probability ratio test (SPRT) technique to determine an alarm status. It provides a means of making a leak alarm decision by analyzing time series data statistically. For pipeline leak detection, the SPRT is applied to the time series data of the volume imbalances or flow differences.

To achieve reliable and sensitive leak detection performance, test values such as imbalance data, should be reliable and the statistical parameters properly set during a tuning process. A sufficient amount of normal operational data must be analyzed in order to obtain the correct statistical tuning parameters. The tuning parameters include the number of time series data points, probabilities that determine the thresholds, leak sizes to be detected with minimum standard deviation, and mean value correction.

The SPRT offers good fault detection capability including pipeline leak detection. The sequential probability ratio test expression includes the standard deviation and mean value terms that indicate variability of the incoming data and inherent measurement bias. Therefore, the equation automatically takes into account the pipeline operations in terms of changes in test values and bias correction. This technique responds to changes quickly, and if properly tuned, it can provide sensitive and reliable leak detection capability.

However, successful operation of the SPRT technique requires that the smooth time series data to be tested be reliable. Since it relies on other calculation methods for its test values such as volume imbalance, the selection of a proper imbalance calculation method is an important factor in achieving good leak detection performance. In general, the SPRT tends to use a lot of test data for proper trending analysis, and thus it may respond too slowly to respond to pipeline ruptures that require immediate leak detection and confirmation.

Bayesian Inference Technique — Another statistical approach to leak detection uses a Bayesian inference technique in order to make a leak/no leak alarm decision. In other words, assuming known prior probabilities of no leak for a set of no leak patterns, the Bayesian inference technique applies the Bayes' rule to determine the probability of a no leak alarm condition. The same Bayes' rule is applied to a leak condition to determine the probability of a leak occurring. This technique has been successfully implemented in a simple pipeline, because prior probabilities can be determined. However, it is not simple to apply it to complex pipelines with several pump stations because it is much more complex to determine prior probabilities.

Initially, a large number of operation scenarios including leaks are simulated or alternatively past operational data is used off-line to obtain the leak and non-leak patterns. This data becomes the basis for patterns in the pipeline system. These patterns are then refined with actual operation data obtained while the system is operating in real-time. As a result, reliability and detection sensitivity can be improved as more operating data is accumulated over time and used to refine the pipeline state and probabilities.

A Bayesian inference method has been successfully applied for fault detection, and this pattern recognition technique, at least in theory, can be applied to any pipeline operation. For pipelines with a simple configuration, with no more than one intermediate pump station, the probability of a leak and no leak condition may be easily determined. For complex pipeline systems, however, it is time-consuming to determine prior probabilities and establish pipeline operation patterns. Since the Bayesian inference technique needs to build an accurate probability database for almost all possible operations, extensive field and maintenance tuning efforts are required for reliable operation; this may take a long time to acquire for a complex pipeline system.

9.2.3 Implementation and Operation

Whatever the leak detection method or methods used, the implementation and operation issues are critical to get the full benefits of the leak detection system. For example, installing adequate instrumentation or sensing devices is critical to the implementation of an effective CPM or sensor-based system. For optimum performance, it is important that the installed instrumentation or sensing devices be consistent with leak detection requirements.

The real-time leak detection systems such as CPM and sensing methods are closely integrated with the host SCADA. Therefore, CPM and sensing methods require an interface with a host SCADA system. The interface allows all field data used and data generated by these systems to be exchanged with the SCADA system, so that the pipeline operators can respond to an emergency expediently according to the company's emergency response procedure.

As part of the implementation phase, the commissioning and tuning tasks have to be performed after these systems are installed on site. The following tasks are usually performed (the tasks listed below are not necessarily required for all these systems):

- Check the SCADA functions and interface
- Check the instrumentation or sensing devices for their availability, accuracy and other behaviors
- Check and tune the performance of the installed system or systems during normal pipeline system operations
- Check if other operation-related problems exist and correct them before the system is put into service
- Perform several levels of system acceptance tests to check if the system satisfies the leak detection requirements

The pipeline operator identifies and analyzes pipeline operation problems via the user interface. The user interface should be easy to use and provide appropriate information in order to make correct decisions. It is critical to have accurate and timely information in an easy-to-interpret format. The decision may include leak confirmation and location, so that the operator can shut down the pipeline in accordance with the pipeline's operating and alarm conditions. A proper response must be made quickly when an emergency such as a leak occurs. In addition, the user interface should be

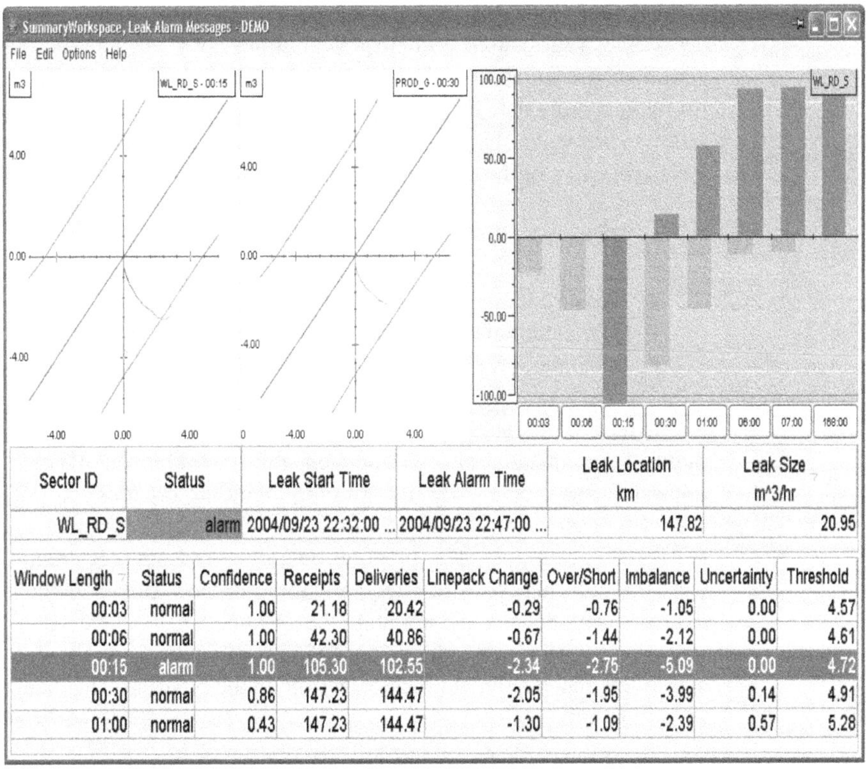

Figure 9-16. Example display of leak alarm messages

consistent with the SCADA system, because the operator gets the information through the SCADA. Alarm messages are the most critical information that the operators must pay attention to, and a pipeline map is required to efficiently execute emergency response procedures. The map gives detailed information about the names and contact numbers of responsible parties, pipeline route and terrain, population close to the pipeline route, responsible officials including police, critical environmental concerns, etc.

An example display of leak alarm messages is shown in Figure 9-16. It shows the leak alarm status, estimated leak location and size, and other information that helps the operator to quickly identify the potential problem.

Alarm messages are critical information that the operations staff must pay attention to. It is strongly recommended to display alarm messages including leak detection alarms on the SCADA alarm display screens. The following features and qualities should be part of the alarm displays:

- Consistent with SCADA system alarms and have an appropriate priority.
- Have different colors for each category of alarm.
- Acknowledged and unacknowledged alarms should be accessible to the pipeline operator in one step. Acknowledged alarms still in the alarm state should remain readily available to the pipeline operator.
- Have a time stamp as part of the displayed alarm.
- Should have both audible and visual cues. Each alarm should have a unique audible tone. Visual cues for any given alarm should persist for a long enough period of time so as not to be overwritten irrevocably by newer alarms.
- Not easily defeated, or inhibited without just cause. The use of screen savers or any other screen blanking is strongly discouraged.

Trending measured and calculated values of the SCADA and CPM system help determine what caused an alarm. Trending may be in graphical and tabular forms: Graphical presentation makes it easier to identify anomalies. The tabular form is useful for analyzing data in detail. API 1130 suggests that a trend cover a long enough duration to see values before a CPM alarm occurred and continue right through to when the alarm ends, or the current time. The following values need to be trended:

- Measured pressures and temperatures
- Measured densities, particularly for batch pipelines
- Measured flow rates and their differences between inlet and outlet flows
- Calculated line pack changes if they are made available
- Imbalances for the CPM methodologies using mass balance principle

Effective operation of a leak detection system requires a thorough understanding of the system installation, operation, capabilities and maintenance. The pipeline operations staff must have extensive training including practical on-line operating experience. Emphasis is placed on how to operate the system effectively and how to analyze the results accurately. The pipeline operators and other operation engineers should learn the skills necessary to effectively monitor the system and diagnose anomalies and to effectively manage and maintain the system.

A leak detection system manual should be readily available for reference by those employees responsible for leak detection on the pipeline. The manual may contain the following information:

- A system map, profile, and detailed physical description of each pipeline segment
- A summary of the characteristics of each product transported
- A tabulation of the measurement devices used in the leak detection procedure for each pipeline segment and a description of how the data is gathered
- A list of special considerations or step-by-step procedures to be used in evaluating leak detection results
- Details of the expected performance of the leak detection system under normal and line upset conditions
- The effects of system degradation on the leak detection results

API 1130 recommends on-going testing — establishing the policy of periodic testing, test frequency, and test methods. Testing methods may include removal of fluid from the pipeline, if permitted. The main purposes of the testing are to check if the installed system is effectively operating and to test whether operators follow the company's emergency response procedure. Effective emergency response is one of the key tasks for mitigating the consequences of the leak when a leak is detected. Emergency response procedures must not only be clearly written but understood and practiced by pipeline operating staff.

API 1130 recommends keeping design records, software changes and test records, and specifies the record retention length. Records of tests should include the following:

- Date, time and duration of the test
- Method, location, and description of the commodity withdrawal
- Operating conditions at the time of the test
- Analysis of the performance of the CPM system and, for tests, the effectiveness of the response by operating personnel
- Documentation of corrective measures taken or mitigated as a result of the test
- SCADA data generated during the test

It also recommends that records detailing the initial or retest results should be retained until the next test.

To maximize performance of the implemented CPM system, operating companies need to establish a procedure and schedule for maintaining all instruments, communication tools, and hardware and software that affect the leak detection system. Refer to API 1130 which describes several aspects of a system maintenance and support program.

9.2.4 Leakage Response

A pipeline leak, particularly a large leak, is treated as an emergency. At the event of a leak, line pressure should be reduced as fast as possible in the leaking section to minimize spillage caused by the leak. The leak detection system should be able to identify at least the leak section. To reduce the line pressure, the originating station should be taken off quickly, and then other stations should be dropped.

If possible, the operator should shut down the closest station upstream of the leak section and keep the downstream station on as long as possible to pull pressure down at the leak site. Practically, the operator is able to shut down the station downstream of the leak after reducing the suction set point as low as possible. This allows the operator to monitor line pressure between the stations. When the line pressure drops significantly, the leaking section should be isolated by closing the upstream sectionalizing or isolating valves first and then the downstream valves.

Whenever a line is shut down due to either a confirmed or suspected leak, clearance must be obtained prior to starting the line back up. All relevant data must be retained along with any other information that would help analyze the causes of the leak and facilitate the leak report preparation. This report is also required for the management and for the next shift operator in case line start-up is delayed.

9.2.5 Summary

A leak detection system is a tool for mitigating the consequences associated with a leak by fast but reliable detection and accurate location. The operator should be well trained in using the leak detection system so that any emergency due to a leak can be effectively managed. If a leak detection system is effective, it can be good insurance for reducing risks. An appropriate leak detection system should help pipeline companies operate their pipeline systems safely by reducing the consequences related to a leak.

A SCADA system is an integral part of daily pipeline operations. The CPM and sensing methods of leak detection take advantage of real-time capability and the effectiveness of the SCADA system as a monitoring and controlling tool. As the historical data indicates, the current CPM technologies are far from satisfactory in their performance. They need further improvement in their reliability and leak detection sensitivity. Also, a single CPM system may not satisfy all the criteria of an effective leak detection system. Combining a few CPM and sensing methodologies, however, may be able to satisfy not only most regulatory requirements but also effectiveness criteria.

9.3 DRAG REDUCING AGENT (DRA)

9.3.1 Introduction

A drag reducing agent (DRA) is primarily used to increase the capacity of petroleum liquid pipelines. It is also known as drag reducer, flow improvers, or just as DRA. It is reported [17] that the Trans-Alaska Pipe Line System (TAPS) started to use a commercially available DRA since 1979, after its effectiveness in increasing throughput was proven. Nowadays, many North American pipeline companies use DRA to increase throughput well beyond their design capacities. In addition to oil transmission pipelines, DRA has been applied to district heating and cooling systems.

Different types of DRA are used in these systems. Most DRA products consist of a long chain and high molecular weight polymer which is injected into the petroleum liquid in small amount (in the range of 10 to 30 ppm of the petroleum liquid) for reducing the frictional pressure drop. It is injected into the petroleum liquid flow stream to raise line throughput or lower line pressure to maintain the line within operating parameters while increasing flow for a given power input. Therefore, it can save power costs and also relieve the pipeline company of capacity restriction.

Main classes of additives and chemicals used for transport are summarized in Table 9-2 [18].

TABLE 9-2. Main classes of additives and chemicals

Additive	Remark
Drag reducer	For both liquid and gas transport
Corrosion inhibitors	Both chemical inhibitors and biocides
Paraffin inhibitors	Prevent paraffin depositions
Pour point depressants	Lower pour point of waxy crudes
Gas hydrate inhibitors	For gas and multi-phase transport to prevent hydrate deposits
Surfactants	For multi-phase transport of heavy crudes
Odorizing additives	For safety

9.3.1.1 Drag Reduction Mechanism

Drag reduction is a phenomenon in which the friction of a liquid flowing in a pipeline in turbulent flow is decreased by using a small amount of an additive. The DRA is believed to damp flow turbulence of petroleum products near the pipeline wall, so that friction and thus frictional pressure drop can be reduced. This dampening effect reduces frictional pressure drop, reducing energy consumption or increasing flow rate. Its concentration in the liquid affects the turbulent characteristics of the liquid in the pipeline. DRAs do not coat the pipe wall or change fluid properties such as density and compressibility.

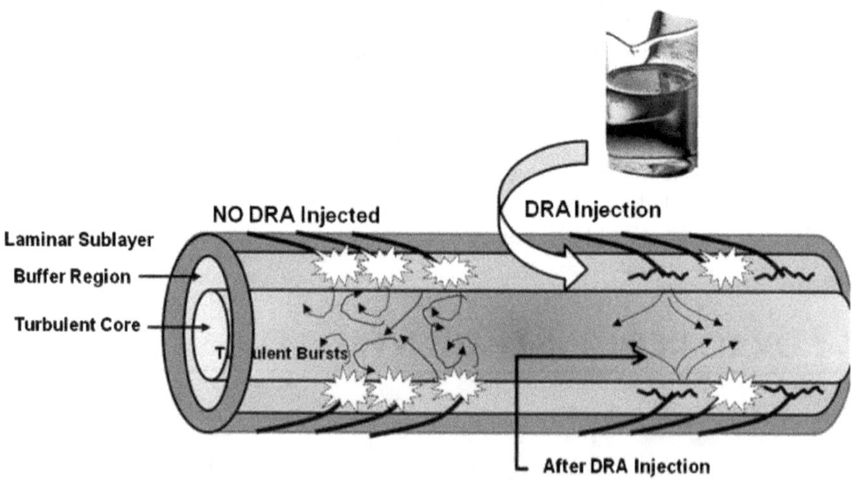

Figure 9-17. Mechanism of drag reduction

In most petroleum pipelines, the liquid flows through the pipeline in a turbulent flow regime. In this flow regime, the molecules move in a random motion. As shown in Figure 9-17 [19], a turbulent flow regime has three flow zones; laminar sub layer, buffer region, and turbulent core. Working with the molecules in the second and third zones, a DRA can reduce the energy waste caused by the random motion. In other words, drag reduction takes place through an interaction between DRA and the turbulence of flowing fluid.

The DRA is effective only for reducing friction and thus friction pressure loss. Since the total pressure drop is caused by both friction and elevation gain, it is beneficial only in the sections where the frictional pressure loss is significant but not useful where the primary pressure drop is caused by elevation gain.

Figure 9-18 [20] shows that lowering these internal fluid pressure losses increases the bulk throughput of the pipeline for a given pumping energy, hence operating costs are reduced.

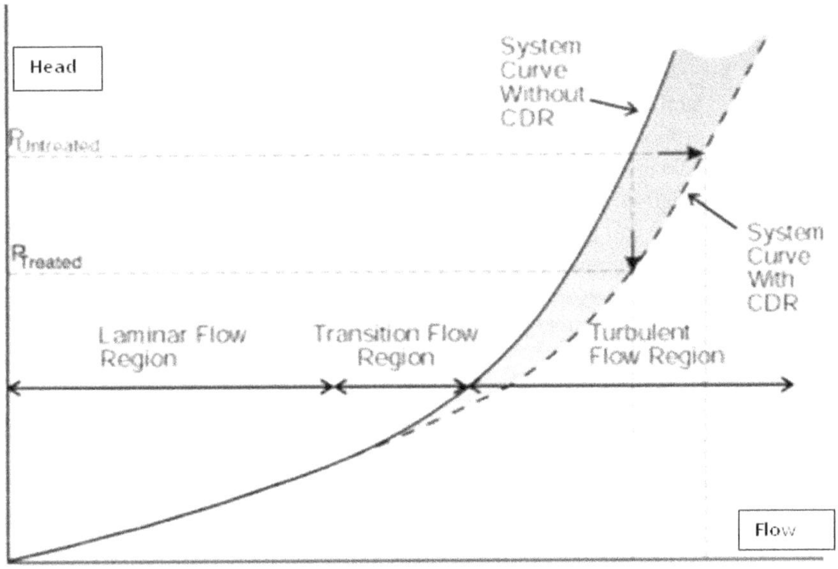

Figure 9-18. Effect of chemical drag reducers on pipeline pressure and flow

9.3.1.2 Benefits of Using a DRA

There are several options to increase the pipeline capacity; installing additional pump stations, adding parallel loops, and increasing the pipe diameter. However, these options are capital intensive and time-consuming. The key benefits of using a DRA are operating cost saving and flexibility in dealing with throughput with relatively small investment in facility. Specifically, the following benefits can be realized:

- The construction of new pump stations is not required to satisfy throughput increase, especially if the increased throughput requirements over the design capacity do not occur frequently.
- Significant increase in pipeline throughput can be handled with relatively small investment in the construction of DRA facilities. In some instances, it is reported that the throughput increase is expected to be more than 30% [21].
- Significant throughput can be maintained even for a de-rated line during main pump maintenance.
- Energy cost savings, particularly if the power cost is high.

- The DRA facility construction time is shorter and the cost is significantly lower compared to the time and cost for the construction of pump stations.

However, the cost of DRA is relatively high, so a break-even analysis has to be performed to justify the DRA facility installation and operation.

9.3.2 DRA Characteristics and Performance

To be effective, the DRA should have the following characteristics [22]:

- Solubility: The DRA must be soluble within the fluid being pumped.
- Shear stability: The DRA must be stable enough not to break down in turbulent pipe line flow.
- Downstream effects: The DRA should have no downstream effects on refinery equipment when used in crude pipe lines and, on product systems, no effects on engine performance.

A DRA can be degraded due to shearing while flowing through the pipeline and facilities such as pump and valve stations. The degradation through pipes is roughly proportional to travelling distance. The degradation of DRA through the facilities may be caused by the operating pumps as well as the station piping and fittings. The degradation due to the station piping and fittings may be small if the station is bypassed, but it can be larger if the DRA flows through the pump station. The degradation due to the pumps may be different for each pump, and the degradation for each pump in series is larger than that for a single operating unit. Field data indicates that the DRA passing through a running pump degrades roughly by one third of its value. Therefore, the DRA degrades roughly by two thirds if a pump station is operated with two pumps in series.

Several properties of the fluid being pumped can also affect DRA performance as follows:

- Turbulence: The pipeline must be operated at turbulent flow conditions for the DRA to be effective. Most DRAs are not effective for heavy crude transportation flowing in laminar flow conditions [23].
- Viscosity: Decreasing the viscosity of the fluid increases the effectiveness of the DRA.
- Temperature: Increase in the temperature of the fluid decreases the fluid viscosity and increases the solubility of the DRA, thus improving performance.
- Wax or water contents: It is indicated that high wax or water contents in the fluid reduces the effectiveness of the DRA.

9.3.3 DRA Operations

9.3.3.1 DRA Facilities

Figure 9-19 below shows an example of a test installation for application of the new ExtremePower™ DRA (warm climate). Some additional equipment would be required for cold climate installation.

A DRA injection facility is constructed of injection and feed pumps, flow meter, storage tank, pressure sensors, control equipment, and equipment for safety such as pressure relief valves. Figure 9-20 shows a DRA injection system. Multiple tanks are normally used for DRA storage, and gear pumps driven by variable speed motors for injecting it. The DRA injection rate is measured by a positive displacement flow meter.

Figure 9-19. Example of field test facilities (courtesy of ConocoPhillips Specialty Products, www.liquidpower.com)

9.3.3.2 DRA Injection

The effectiveness of DRA is measured in terms of the reduction in frictional losses in the pipeline. It varies with the DRA concentration, viscosity of the solvent fluid, pipeline temperature, fluid velocity, and pipeline diameter. Since a DRA is composed of long polymer strands, the DRA can be sheared when it passes through the pipeline and equipment such as pumps and control valves. This results in degradation of its effectiveness. The DRA effectiveness depends on the length of the pipeline and the amount of shearing. Therefore, DRA must be injected downstream of all pumps, meters and valves to prevent shear degradation, requiring the DRA pump pressure be higher than the pump station discharge pressure.

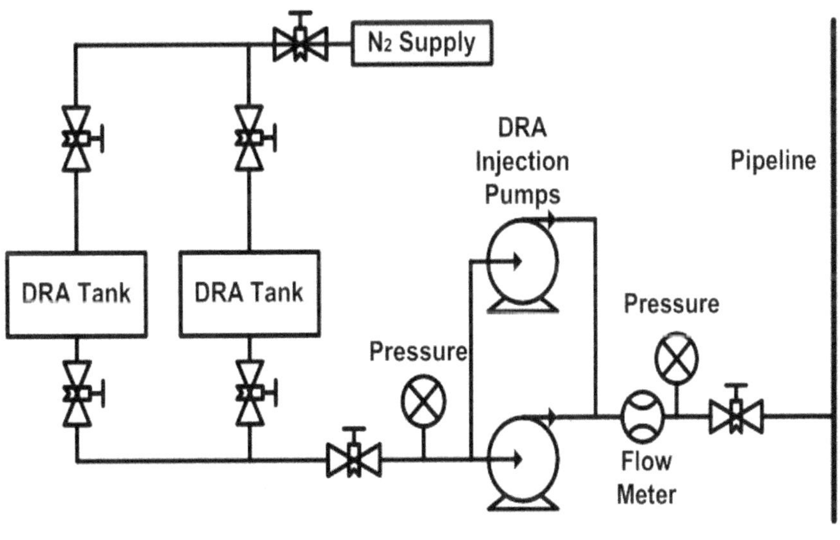

Figure 9-20. DRA injection system

Startup of DRA Injection: The DRA injection system is usually automated and controlled remotely by an operator from the control center. Refer to Figure 9-20, which illustrates a DRA injection system at a pump station. In anticipation that the remote control system may not work, the facility needs to provide local manual control capability. Described below is a normal DRA injection starting procedure:

- Select stations where DRA is to be injected, while checking to ensure that incompatible product (such as jet fuel) will not be affected by wrongly injecting a DRA into the passing batches.
- Determine the DRA flow rate set point based on the target flow rate in order to obtain an optimum drag reduction. The required DRA flow rate is calculated automatically if the line flow rate is known. Normally, DRA injection initially begins at a high flow rate, and then lowers to the required flow rate.
- Select and start the DRA pump if there are several DRA pump units.
- Check if the DRA flow rate agrees with the DRA flow rate set point.

DRA injection operation is shown in Figure 9-21. The DRA flow rate is controlled to reach the DRA set point. The pump station is equipped with two variable speed pumps (two VFD drivers) in series, and the DRA injection system is installed downstream of the station. A booster pump is installed to boost the suction pressure of the mainline pump.

In order to make sure that any jet fuel batch is not contaminated with DRA, the following steps for the startup and shut-down of DRA injection are taken:

- DRA injection should not start about one hour after a jet fuel batch has passed the DRA injection station.
- DRA injection should be shut-down about an hour before the jet fuel batch arrives at the injection station.
- Where a batch tracking application is employed, DRA lockout can be triggered by the approach of a jet fuel batch to the DRA injection station with a status returned to SCADA that can be used to lockout the DRA pump.

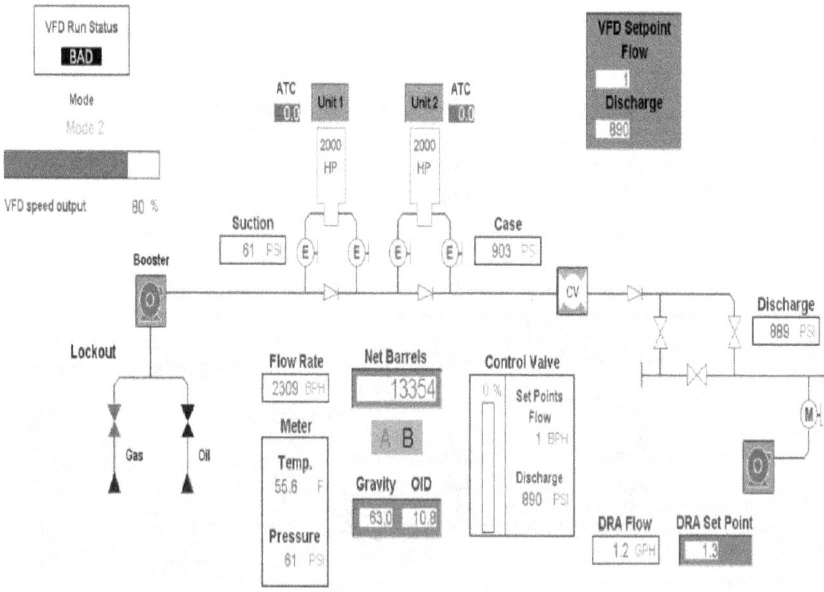

Figure 9-21. DRA injection at a pump station (courtesy of Telvent)

Shut-down of DRA Injection: When a pipeline operates near the pressure operating limit, shut-down of DRA injection could cause line operating pressure to be exceeded if the flow rate remains constant. Described below is a normal DRA injection shut-down procedure:

- Select the station where DRA injection is to be shut down.
- Check if DRA is no longer required to obtain line flow rate for pending batches or to maintain the line pressure within the operating limits.
- Select and stop the DRA pump at the station where DRA injection is shut down.
- Check if the DRA flow rate decreases to zero.

The above figure shows a DRA injection and shut-down control through the SCADA screen of the pump station, where the DRA injection takes place.

9.3.3.3 DRA Concentration Tracking

The DRA concentration is measured in parts-per-million in the flowing product. The DRA concentration is tracked as it moves down the pipeline, and the concentration in the subsequent section includes the degradation due to moving along the pipeline and passing through running pumps. The inclusion of the DRA will create a new 'batch' blended with the DRA when the DRA is being injected.

Both sheared and non-sheared DRA concentrations need to be tracked to properly operate DRA injection. A DRA injection rate is used with measured or calculated product flow rate to calculate the DRA concentration. When a DRA passes through a pump, it is sheared and no longer active. The DRA tracking function tracks the sheared and active DRA concentrations and checks the concentration against the maximum DRA concentration allowable in the product. For example, DRA is not allowed in jet fuel and thus its concentration should be checked against zero concentration level. A graphic view of the DRA contents within a pipeline can show active, sheared and total concentration of DRA in the product as well as the positions relative to DRA injectors or pump stations.

9.3.3.4 DRA Limitations on Operation and Design

If the throughput is restricted by the pipeline capacity, it is generally cost-effective to install DRA facilities at pump stations. However, if the desired flow rate is higher than the pump capacity, the pumping capacity must be increased to accommodate the increased throughput requirements.

Figure 9-22 illustrates the pump operating point change due to capacity increase. Since the throughput increases in the presence of a DRA, the existing pumps may not be able to accommodate the flow rate increase without modifying the pump characteristics. Note that the pump does not operate at the best efficiency point (BEP) when DRA is injected into the fluid.

DRA can be used in the transportation of crude oil and refined petroleum products (except jet fuel) in order to increase pipeline throughput. The DRA can accumulate on turbine blades and may damage the turbine. Therefore, it cannot be used for jet fuel transportation, not because of its effectiveness but because of its potential safety concern. The original DRA did not work with heavy crude, but ConocoPhillips has recently developed a DRA [23] that has proven effective for heavy crude.

Since much higher flow rate can be achieved with DRA, the flow velocity can be fast. It has to be noted that the higher velocity can also increase the surge pressure. Therefore, a check must be made to see if the existing pipeline can meet the new transient pressure requirements.

As the DRA is injected into the pipeline section, the throughput increase takes place slowly because the increased rate is linearly dependent on the flow velocity.

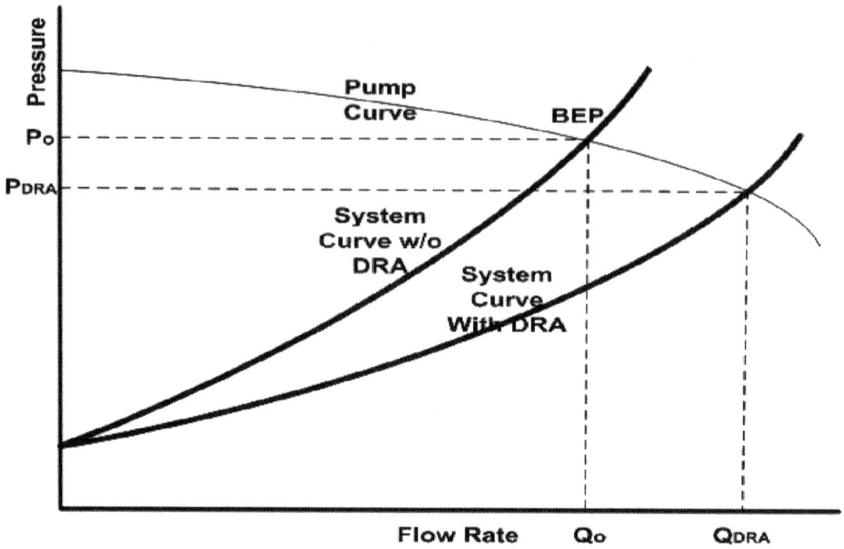

Figure 9-22. DRA injection and changes in pump operating point

The desired throughput can be achieved only after all the liquid in the pipeline section contains the required DRA concentration. For example, it would take about 10 hours to reach the intended throughput in a 100 km section if the flow velocity is in the order of 10 km an hour.

As discussed in the previous section, DRA is sheared as it moves through pipes and particularly pumps. In other words, it has to be injected at every operating pump station. Therefore, it can be very costly for a long pipeline with short pump station spacing.

9.3.4 DRA Correlations

When undertaking pipeline hydraulics simulations, generally, information required is the type of DRA used plus the injection rate in ppm (parts per million). The injection rate depends on the following:

- Type of DRA,
- Supplier,
- Pressure,
- Temperature,
- Pipeline product flow rate,
- Distance travelled.

The DRA injection rate needs to be determined to achieve the desired pressure or flow rate. DRA manufacturers provide their own DRA correlations. When studying a DRA operation, one of the following correlations can be used: AESOP [24], Burger [25], Conoco or Simplified Conoco correlations [26]. They relate the effective friction factor with the DRA concentration. The AESOP correlation has been recently developed by a joint consortium of academic and industrial bodies under European Union funding.

Aesop: This is a correlation that requires data which is specific to the fluid in a line.

Burger: This is a widely-used correlation that depends on the DRA concentration, product viscosity, fluid viscosity, and the pipe diameter.

Conoco and **Simplified Conoco**: This option is a correlation that requires data which is specific to the fluid in a line.

The drag reduction factor is a direct multiplier to the calculated friction factor.

$$f = fm \times (1 - F)$$

where

f = effective friction factor
F = drag reduction factor
fm = model-calculated Moody friction factor

Aesop drag reduction factor

The Aesop correlation is of the form

$$F = B\left(\frac{AC_e}{1+AC_e}\right)\left(1+C\cdot 10^{-5} N_{Re}\right)$$

$$C_e = e^{-\lambda d}\, ppm$$

where A, B, and C are AESOP coefficients

F = drag reduction factor (%)
C_e = Effective concentration of additive (ppm)
N_{Re} = Reynolds number
d = Distance travelled (km)
γ = Degradation coefficient

Burger drag reduction factor

$$F = k_1 * \ln\left(\frac{v * \left(\frac{ppm}{C_s}\right)^{0.5}}{d^{0.2}}\right) + k_2$$

where

k_1 and k_2 = Burger equation constants
v = local fluid velocity, ft/sec
ppm = DRA concentration, parts per million
C_s = fluid viscosity, centistokes
d = pipe diameter, ft

Conoco drag reduction factor

The Conoco (CDR) drag reduction factor is of the form

$$F = \frac{ppm}{(a * ppm + b)}$$

where a and b are product-specific constants and ppm is DRA concentration in parts per million. For full details of the Conoco CDR Correlation contact Conoco Speciality Products, Inc. This DRA correlation is only active for velocities in excess of 0.6m/s and Reynolds Number in excess of 7500.

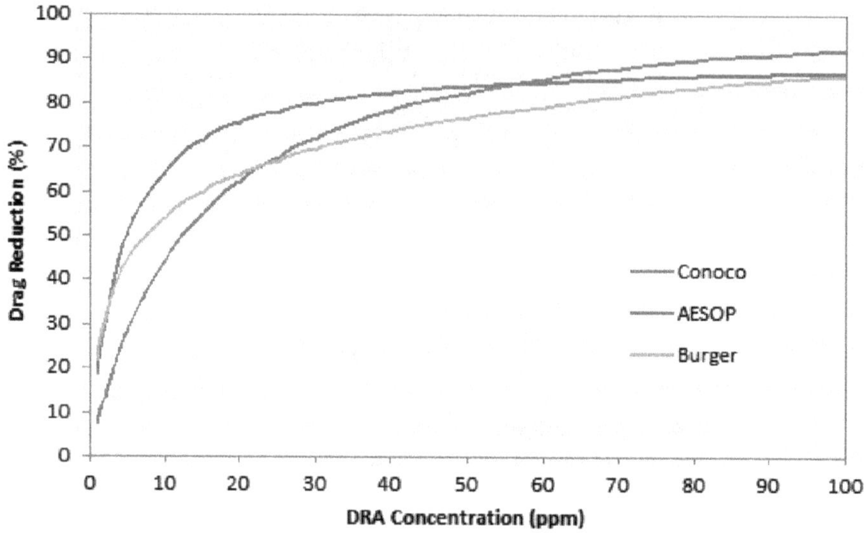

Figure 9-23. DRA concentrations and their effects (courtesy of Energy Solutions International Inc.)

Simplified Conoco drag reduction factor

The Simplified Conoco drag reduction factor is of the form

$$F = \frac{ppm}{\left(A * ppm + B\right)}$$

where A and B are product-specific constants. The equation above is used directly with the product-specific constants entered by the user.

Figure 9-23 shows a set of typical DRA effect curves for the Conoco, Burger, and AESOP correlations. These curves can only be regarded as typical as the drag reduction effects depend on the pipeline dimensions, fluid properties and velocity as well as the DRA properties themselves. Note that the DRA takes effect even when the DRA concentration level is very low, less than 10 ppm, and that drag is not reduced significantly even when the concentration keeps increasing beyond 30 ppm.

9.4 TANK FARM OPERATION AND VOLUME MEASUREMENT

Crude oil and petroleum products, including light hydrocarbons, are likely to deliver their products to/from tankage. These products are often stored in tanks in various locations such as producing areas, refineries, petrochemical plants, and/or distribution centers. Petroleum liquids are stored underground or in aboveground storage tanks. Storage allows for flexible pipeline transportation and efficient transportation management through the existing pipeline system and minimizes supply/delivery disruptions. The stored liquids need to be measured and accounted for accurately in order to keep track of all volume movements including custody transfer when required. Refer to Chapter 8 for the detailed tank and tank farm design.

Oil and petroleum product pipelines lift their products from and deliver to tank farms. A tank farm refers to a collection of tanks located at a refinery, shipping terminal or pipeline terminal. A tank farm at a refining operation is used to store feedstock and various products produced by the refinery and to hold them until they are scheduled for

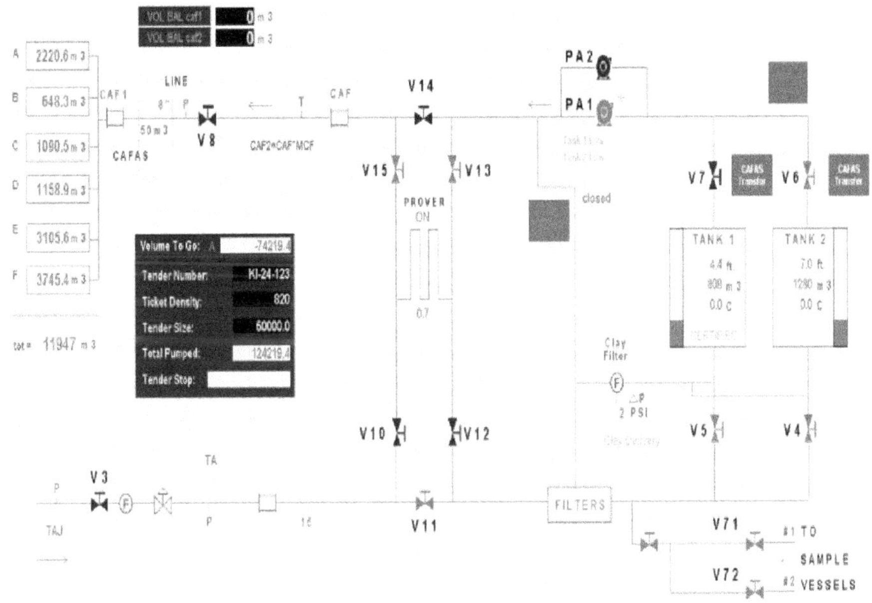

Figure 9-24. Simple tank farm (courtesy of Telvent)

injection into a pipeline for transportation. Similarly, tank farms at shipping terminals hold products until a shipping route is scheduled. The shipping route may be via tanker ship, truck, railcar, or another pipeline. Tank farm operation covers tank control, volume measurement and inventory.

A schematic for a typical tank farm is shown in Figure 9-24. On the left hand side, distillate and gasoline tanks are separately connected to each set of valves, which in turn are connected to a booster pump. The distillate or gasoline is lifted through the dedicated booster pump into the mainline pumps. The lifted fluid passes through the flow meter and in this case the meter prover before it reaches the mainline pumps. Any alarm or warning messages related to tank or tank farm operation are listed at the bottom of the SCADA screen, to which the operator can respond remotely. The operator responses may include simply acknowledging the message or taking corrective action.

9.4.1 Tank Farm Operation

Tank farms are located at the flow lifting or receiving and delivery terminals as well as side-stream injection and side-stream delivery points in the pipeline system. Full stream injection or delivery takes place at the lifting and delivery terminals. Either partial or full stream injection can take place at the side-stream injection points, while strip or full stream delivery at the side-stream delivery point. All flows in and out of the tank farms in the pipeline system must be measured and recorded for custody transfer.

9.4.1.1 Normal Batch Lifting Sequence at a Product Lifting Tank Farm

The first step of lifting a product is to select the tank or header from which the batch will be supplied to the first main line pump station, and then the appropriate booster pumps to be used for the batch.

The next step involves the valve manipulation in sequence:

- Open the valve on the tank or header from which the batch is to be supplied.
- Open the suction valves on the selected booster pumps.

- Close the valve on the booster pumps bypass lines.
- Close the valve on the pump discharge lines.
- Close the valve downstream of the minimum flow bypass control valves.

Once the conditions to start are satisfied, the start sequence is initiated by opening the bypass valve on the selected booster pump(s). After the flow through the bypass line has stabilized, the valve in the discharge line is opened to direct the flow to the main line pump suction through the meter designated for the batch.

- The flow rate for the batch is metered and accumulated volumes are recorded.
- Pumping of the leading batch would have been just completed prior to the introduction of the new batch.
- Close the booster pump discharge header valve to stop lifting the leading batch.
- Stop the booster pumps that were being used for pumping the leading batch. Close the booster pump suction valves and tank valves.
- Set the flow accumulator to zero at the start of the new batch.

9.4.1.2 Operation at the Delivery Terminal

The pressure levels of tanks are low unless the tanks are for high vapor pressure liquids, while the pressure level of the incoming fluid may be high. Therefore, a pressure control valve is installed at the entrance of the delivery terminal to keep the pressure below the maximum tank pressure.

The combination of estimated time of arrival of batches, the interface detection by the densitometer located upstream of the delivery terminal, and the detection of batch arrival by the densitometer is used to identify the arrival of various batches. Depending on the tank management procedures established for the delivery terminal, the batch will be directed into designated tanks by opening and closing of tank valves.

One or more densitometers or dye detectors are provided near the delivery terminal for batch interface detection. One detector will be installed a few kilometers from the terminal in order to provide lead time for taking actions to direct batches into appropriate tanks, and the other should be installed near the tank manifold.

The batch interfaces or transmix may be directed into a designated tank at the delivery terminal. The start and finish of batch interfaces will be indicated by the densitometer. It is normal practice to collect samples of the batch at frequent intervals to confirm the start and completion of batch interface. This will ensure the purity of the arriving batch. Any off-spec interfaces must be sent to a slop tank for separate treatment such as re-processing.

9.4.1.3 Side-Stream Injection

The side-stream injection facility, including densitometers, is normally installed on the upstream or suction side of the intermediate pump station. The upstream densitometer provides lead time to initiate or terminate the incoming batch injection. The injection of the batch starts or stops when batch interface arrival at the intermediate station is confirmed by a change in densitometer reading on the suction side of the pump station.

For partial injection, the injected product is the same as that of the incoming batch, resulting in the reduction of the upstream flow rate while maintaining the previous flow rate downstream of the injection point. For full stream injection, the line upstream of the injection point is shut in, but the downstream flow rate can be maintained unless the delivery rate is changed.

For partial injection, it is a normal practice to start injection by opening the injection valve after the batch interface has passed in order to avoid any unnecessary mixing. The flow rate at the lifting point has to be cut to compensate for the injection, because the flow rate upstream of the injection point has to be reduced if the downstream flow is maintained. The pressures in the upstream section have to be adjusted to maintain the selected injection rate. It may be necessary to take off some pump units and/or stations right after the partial injection starts. For a full stream injection, the mainline valve just upstream of the injection point is closed and all pump stations are shut down in the sections upstream of the injection point.

Figure 9-25 illustrates a side stream injection into the mainline at a pump station. A booster pump takes a batch or fluid from a tank and pumps the fluid into the mainline pump. The flow rate is measured before it is injected into the pump.

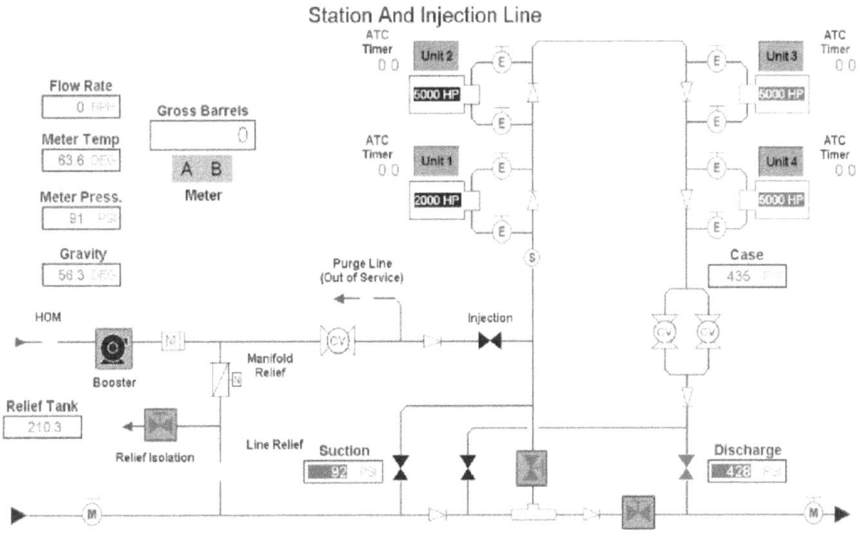

Figure 9-25. Side stream injection (courtesy of Telvent)

9.4.1.4 Side-Stream Delivery

Normally, two densitometers are installed at the delivery location to detect batch interface during batch operation. The first densitometer is installed a few kilometers upstream of the take-off location on the main line and the second densitometer installed at the delivery station. The flow into the take-off is shutdown by closing the appropriate valve on the take-off when the head of trailing batch arrives near the take-off as indicated by the upstream densitometer. Downstream of the take-off point, a pressure control valve is installed to maintain the required delivery pressure.

Normally, the take-off valve is opened when part of the stream is passing. Care must be taken to maintain the pre-determined downstream rates to prevent stretching out interfaces between batches. Therefore, the operator has to calculate the downstream flow rate before the strip delivery starts. If the delivery rate is large relative to the mainline flow, it is essential to take off some pump units and/or stations right after the strip delivery starts.

For a full stream delivery, the operator has to open the take-off valve and close the mainline block valve immediately downstream of the take-off connection. If the delivery is to be full stream, the downstream section should shut down with reasonable pressures in the section.

9.4.1.5 Break-Out Operation

The figure below shows a simple tank farm that allows a break-out operation (Figure 9-26).

At a certain location, a fluid breaks out full stream into tankage and other fluid is simultaneously pumped out of another tank, called a break-out operation. The break-out

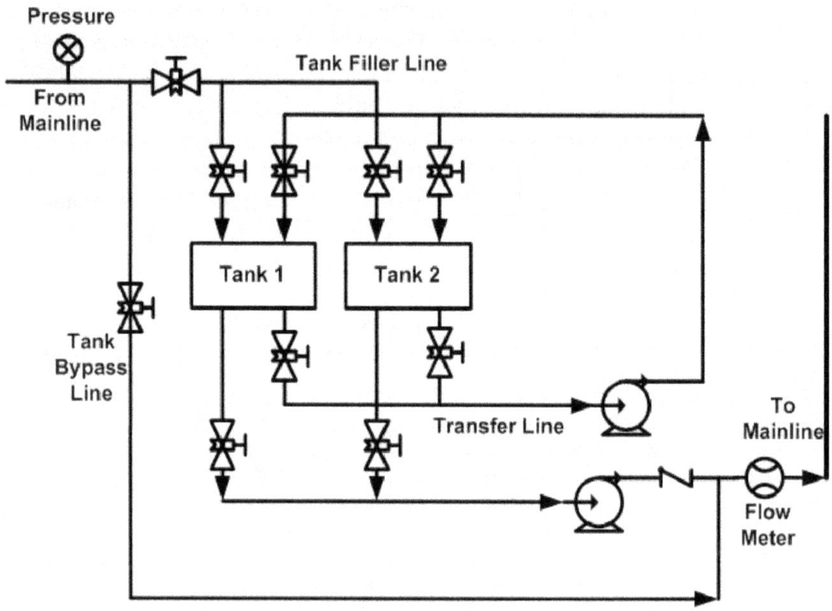

Figure 9-26. Simple break-out tank farm

operation tankage is usually located in the middle of a pipeline, not at the lifting or delivery terminal unless the terminal is connected to another pipeline system. The batch volume has to be determined by providing flow accumulators on the flow meters.

In the above tank farm diagram, the incoming fluid can bypass the tank farm or be stored in a tank. For example, a batch received from the mainline is stored in Tank 2 and the other batch in Tank 1 is injected into the mainline on the downstream side of the tank farm.

9.4.1.6 Sump System

A sump system collects drainage from various sources such as pumps and pig traps, and pumps them to other facilities when a sump tank is full. The system may include a tank that collects any slop or off-spec liquids such as used lubricating oil of a pump. Sump tanks are installed at pump stations and receiving/delivery points. When a sump tank is getting full, the liquid in the tank is transported to a refinery or if acceptable can be blended with other liquids being transported at the location.

9.4.2 Tank Control

A tank farm operation includes two levels; tank farm operation and tank operation. From the perspective of the control center, tank and tank farm operations are fully automated and controlled by a tank farm control system.

The purpose of a tank farm control system is to assist the operator in moving products and maintaining the inventory of the products. Terminals that handle multiple products (i.e., a batched pipeline) with a large number of tanks and interconnecting pipelines can have quite a complicated routing within the terminal. There will be a significant number of motor-operated valve controls and tank level monitoring systems. The system generates and stores product delivery and shipment feed information in business applications such as inventory tracking, billing for product receipts and deliveries, as well as feeding the same information into a pipeline scheduling system.

A tank farm control system can assist the operator by verifying that proposed valve line-ups represent a valid path before he initiates the sequencing and starts the pumps to move the product. This ability helps avoid an error of lifting a product from or delivering to a wrong tank. This is important, as an error such as the injection of crude oil into a refined product tank would be costly.

The tank control requires the tank information and several control variables. The tank information for a tank farm includes product name, tank type (net or gross), tank volume correction factor, maximum and minimum tank volumes, maximum and minimum working level, and others. The tank volumes are normally measured in tank level, which is converted into tank volume (see Chapter 7). The tank control variables are as follows:

- Flow rate: Flow rate into and out of a tank is calculated by dividing the net/gross volume change by the elapsed time.
- Volume-to-fill: The net/gross volume in the tank is subtracted from the maximum safe tank volume to calculate the volume-to-fill.
- Volume-to-pump: The tank bottom volume is subtracted from the net/gross volume to calculate the volume of liquid that can be pumped from the tank.
- Time-to-fill/empty: The time to fill or empty the tank is calculated from the current flow rate into or out of the tank, and the volume-to-fill or volume-to-pump.

A tank control requires alarms and events to be generated in response to various conditions. These may include alarms when maximum or minimum tank levels have been violated and alarms for abnormal rates of change. Typically, the information contained in a tank report includes such data as tank level and water level, measured and corrected gravity, temperature, gross and net wet volume, S&W volume, net dry volume, and flow rate (Figure 9-27).

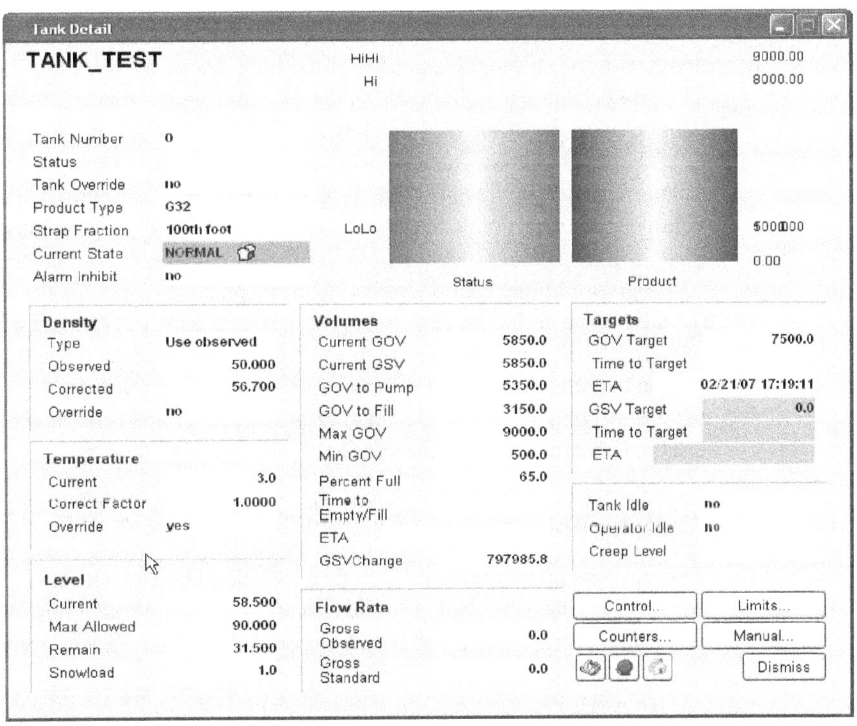

Figure 9-27. Tank control (courtesy of Telvent)

The above figure displays the information on the tank status, product in the tank, tank volume and level, and other tank control data. The tank level should be between the maximum and minimum levels, and the level gauge is used to control the flow moved into or out of a tank. Tank level conversion to tank volume is discussed below.

9.4.3 Tank Volume Measurement

One way to measure the volume of a stored liquid is to determine the level of the liquid in the tank and then calculate the volume from a capacity or strapping table that relates the level to the corresponding gross volume of liquid in the tank. The strapping table is established during the tank proving process, using a tank prover which has thermometers mounted in the measuring section to accurately measure temperature. The API Standard 2550, Measurement and Calibration of Upright Cylindrical Tanks, describes the strapping procedures, and API MPMS describes the strapping procedures for cylindrical as well as other types of tanks. See Chapter 7 for detailed description of the calibration procedures for tankage.

Tank level to volume conversion requires that the parameters and strapping table or equation associated with the tank be defined. In addition to the level measurement, the gravity and suspended sediment and water (S&W) content and the temperature of the liquid and ambient temperature near the tank need to be measured to determine the net volume and liquid head stress caused by high hydrostatic pressure on a large tank. The accurate calculation of the volume in the tank requires parameters such as tank roof types (fixed or floating) and the level of free water. The volume conversion can be performed by a field automation system device such as a PLC.

Once the tank level has been measured, whether manually or automatically, the level data is converted to a gross volume using a volume conversion process. The process uses either the strapping table data for each individual tank or an incremental table that defines incremental volumes per number of level increments for the tank. The conversion equation associated with the tank can be used for the volume conversion. The gross volume should be corrected for tanks with a floating roof by taking into account the weight of the roof and any snow load.

The level of free water is also required to determine the gross volume of the petroleum product in the tank. This value is converted to its equivalent volume using the volume conversion table and then subtracted from the gross volume to determine the gross volume of the product by assuming that the water is on the bottom of the tank.

A gross volume is converted to a net volume using the density and temperature of the fluid in the tank. The density or API gravity is used to calculate the temperature correction factor, which is detailed in the API Standard 2550. Once the temperature correction factor is determined, it is multiplied by the gross volume to obtain its equivalent net wet volume. If the sediment and water (S&W) contents are present, their values are used to determine the net dry volume.

9.4.4 Tank Inventory

The tank inventory functions include tank calculations such as flow rates and volume conversions, volume validation and correction, floating roof adjustments and tank tickets, limit violation alarming, and inventory data collection and storage.

- API 2550 standard is used to measure and calibrate tank volumes.
- The tank inventory and ticket data are used for daily scheduling, volume balancing, and gain/loss analysis.

- The tank data includes the tank ID, inventory date and time, shipper, product name and gravity, temperature, tank gauge level, and roof loading value.
- The host collects the tank data from each tank and stores them in the tank inventory database.

The tank inventory volume is determined from the measured gauge level through a multi-step process (API 2550 procedure):

- The tank level gauge is converted to a gross volume using a volume conversion table, which may be built by means of either increment or strapping table.
- An increment table defines incremental volumes per number of level increments, while a strapping table defines levels with corresponding tank volume. Linear interpolation is performed if the level is between two defined increments or strapping table entries.
- The volume of a tank with a floating roof has to be corrected by applying a correction factor to compensate for the effect of the floating roof weight.
- Assuming that sediment and water (S&W) are on the bottom of the tank for strapping purposes, a free water level is subtracted from the measured tank level to obtain a true gross volume of product.
- Gross volumes are converted to a net volume using the temperature and density of the liquid in the tank.
- First, the current API gravity is measured and corrected to base condition of 15°C.
- Second, the temperature correction factor is calculated using the API equation or API tables.
- The temperature correction factor is multiplied by the gross volume to obtain a net volume.

In addition to individual tank inventory, a tank farm inventory needs to be taken. Tank farm inventory is a balancing process, typically performed on a regular hourly or daily time period. All transactions at a tank farm are analyzed, receipts to tanks, tank transfer to pipeline, pipeline transfer to tanks, tank to tank transfers, etc. to ensure that the accumulated transaction volumes from all inputs and outputs match the actual inventory in the tanks.

As a result of individual tank inventory, leak detection can be performed on each tank by monitoring "dead" tanks for changes in level. A tank is determined to be dead if all valves to the tank have been closed for a certain period of time measured in minutes. The time delay allows the tank level to stabilize. The tank readings are captured in regular intervals over a specified period. If the tank readings show a downward trend over the period, a leakage from the tank can be suspected and should be investigated for its confirmation.

9.5 POWER COST CONTROL

In general, the energy and payroll costs are the two highest in pipeline system operation, depending on the locations of pipelines. As the energy cost increases, the energy cost control is even more important and pipeline companies should address the cost issue. The purpose of the energy cost control is to minimize the energy cost and thus pipeline system operating costs. Assuming that pumps are driven by electrical drivers, power is mostly consumed by pumps and thus this section discusses the power cost control in pump station operation.

Power cost can be reduced by controlling power demand at each pump station run by electrical drivers. Power demand can be controlled by scheduling products or batches properly and by minimizing the usage of power or energy. A proper schedule includes such factors as product or batch lifting and delivery schedule, throughputs, pumping order, etc.

The power cost can be minimized in four different ways; controlling power demand based on the power contracts, monitoring and collecting pump unit operating statistics, operating pump station and units at or near the best efficient point, and selecting an optimum set of pump stations and their control pressure in the entire pipeline system.

9.5.1 Power Demand Control

Power demand charge is the first area that has to be addressed to control power cost [27]. Power demand is the maximum rate of electrical energy used for a given period of time. Normally, power is measured in kilowatts (kW).

Typically, the power company is responsible to determine the quantity of output that will be supplied to each station, and to provide real-time data at the request of the pipeline company. The power contract specifies electricity rates, time of the day when and the locations where power is delivered. Also, the contract includes penalty clauses that are applied to either side when a clause is breached.

Power contracts have power demand charges with on-peak and off-peak hour charges, penalty clauses for high-power demands, and charges for unused power. On-peak hours are the power contract time period when power usage charges are at the highest rate, while off-peak hours are the time period when power usage charges are at a reduced rate.

The demand charge is most likely based on the maximum power demand during the month, but the billing is based on the power company's peak power used at one point in time during the billing period. Therefore, if the power usage is limited to a contracted level or even zero during the peak periods, the penalties can be reduced to zero. In order to reduce the power cost, power companies may encourage the customers to use power during off-peak hours when power is readily available and charges are low. Since the on-peak and off-peak hours of the day are defined in the contracts, the operator can control the power usage during the on-peak periods.

For a long pipeline, it is likely that several power companies provide power to pump stations and the time zones of the stations may be different. This implies that the on/off-peak hours can be different. Therefore, a summary table of power contracts can be made available to the operator in order to check the different on/off-peak hours of all the pump stations in the pipeline system and select the least cost power company to each pump station.

9.5.2 Pump Unit Operating Statistics

The unit operating statistics are supplied through the station PLC. The pump unit operating statistics are useful to run pumps efficiently and safely and to determine the pump and driver maintenance schedule. The unit operating statistics may include the following data:

- A count of limit violations such as power.
- A count of unit starts and unit total operating hours to check against the allocated number of annual starts for a pump unit. The count of unit starts is segregated into the number of attempted and successful starts.
- Measured input power, calculated output power and station efficiency
- Accumulated driver operating hours for maintenance purposes

These statistics are made available to the operator through the SCADA system. They are determined at regular intervals and displayed on the SCADA screen along with the efficiency at all pump stations. As a minimum, the following data may be required for each pump unit:

- On-peak run time, products and volume moved
- Off-peak run time, products and volume moved
- Total run time
- Total number of on-peak starts
- Total number of starts
- Date and time the unit was last running and started
- Limit violations and their counts
- Measured input power
- Calculated output power and station efficiency

Station efficiency for each station is calculated by dividing the calculated output power by the measured input power. The output power can be obtained by multiplying the flow rate with the differential pressure, which is the difference between the case pressure and suction pressure.

When a pump unit start is initiated, the operator has to check if starting the unit will violate the following constraints:

- Exceed the on-peak or off-peak maximum station power specified in the contract. This check is intended to reduce power cost.
- Exceed the maximum number of times the unit is allowed to be started and the minimum time required between starts. This check is intended to protect the unit from overuse.

The ultimate purpose of collecting unit operating statistics and validation is to control the overall operating cost.

9.5.3 Pump Station Monitoring

The pump station monitoring functions monitor and display the pump unit and driver performances in order to operate pumps efficiently. Operation efficiency of a pump station can be improved by monitoring unit performance and taking corrective action if required. A pump performance monitoring function calculates pump station performance, monitors the trends of each unit's performance, and displays the performance of pump units including alarms for deviation in performance. Based on the monitoring result, the operator tries to operate pumps at or near the best efficient points (BEP).

The pump station monitoring function is concerned mostly with the pump unit efficiency at the operating point on the pump performance curve. If the driver is a variable speed, then the pump performance curves are bounded by the minimum and maximum speeds, with the efficiency related to the flow, head, and speed. On the other hand, the fixed speed pump has a single performance curve with the efficiency related to the flow and the head controlled by a control valve. The operating efficiency is determined from the flow and pressure at the operating points, control logic, fixed or variable speed pump performance curves, and different combinations of pump units.

The operating point of a pump is plotted on the pump performance curves. Plots of the current and historical operating points are superimposed onto static performance curves including the minimum and maximum operating ranges. The operators

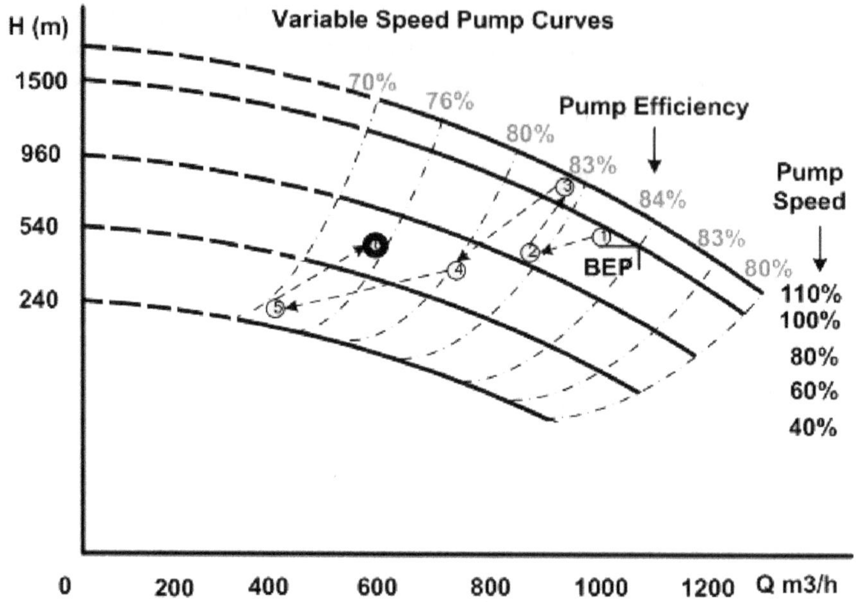

Figure 9-28. Operating point trajectory on performance curve

use these plots to operate pumps close to the BEP of the pumps. In addition, the efficiency can be trended to identify improper throttling operations or degradation of pump unit efficiency. Such information can be used to determine the operator training and equipment maintenance requirements and to re-rate the pump curves.

Figure 9-28 illustrates current operating point and historical trajectory of operating points superimposed on performance curves of a variable speed pump. It shows how efficiently this pump is and has been operating for varying flow rates. Based on this operating data, more efficient pump operating strategies can be developed. A similar trajectory can be plotted for fixed speed pumps to exhibit how efficiently throttling actions have been taken.

9.5.4 Power Optimization

Power optimization refers to short-term power minimization for current pipeline operations and off-line optimization for future operation planning. It is mainly concerned with system-wide optimum operations of facilities such as pump stations and pressure reducing stations. The results of a short-term power optimization are typically treated as recommendations and are not used for a closed-loop control.

For a large pipeline system, a mathematical model is used to obtain system-wide optimum solutions. The power optimization model deals with power consumption and DRA usage for liquid pipelines. It determines an optimum pump station selection and unit line-up as well as pressure set points at the on-line stations so as to minimize power/DRA cost. It may compare the DRA cost against the power cost for the given flow rates. The model may adjust flow rates to take advantage of lower energy costs during off-peak hours.

An optimization model can provide information regarding the following:

- Pump stations and units to be brought on-line
- Optimum pump station suction or discharge pressure set points, pump unit on/off switching schedules, and minimum power cost for a specified time period.

- Pump unit line-up and operating point, considering that a station may consist of different pump units and that the units can be combined in various modes. The operating points, overlaid on the pump performance curve, can be displayed on the host SCADA screen.
- Calculation of the overall pumping costs. When drag reducing agent (DRA) is injected for a liquid pipeline operation, the cost without DRA is compared against the cost with DRA.

In addition, some optimization systems may provide the following information for analysis to help improve pipeline operation efficiency performed by operation staff:

- Key optimization results and historical records
- Flow rate vs. suction/discharge pressure trends with set point change records
- Flow rate vs. number of pump units brought on-line and power consumption
- Cumulative pump operating records
- Pump efficiency trends

The model employs the following data in addition to the pipeline configuration and facility data:

- Pipeline hydraulics and equipment such as pumps
- Pipeline and facility availability data
- Power contract data
- DRA cost, if the DRA injection systems are installed
- Line fill and batch schedule data and injection and delivery flow rates for batched liquid pipelines

The primary criterion for an optimization model is to minimize power costs. A secondary criterion is to balance pump unit operating hours, avoiding frequent unit start-ups and shut-downs. The solution from the optimization model should not violate any pipeline and facility constraints. These constraints may include maximum and minimum pressures and flows at certain points in the pipeline network such as minimum delivery flow, maximum and minimum pump flows and compression ratio, maximum power, etc.

Optimization models are difficult to apply on complex network configuration and pump stations. However, it was reported that certain mathematical techniques were successfully implemented for liquid pipeline energy optimization [28, 29].

A power optimization system can be implemented as a part of the host SCADA system, and connected via an interface with the SCADA system. Through the interface, the SCADA system sends the current pipeline states required for an optimization run, controls its execution with the data, and receives the optimization results along with alarm and event messages such as new batch lifting and station startup or shutdown. The current states may include the following data:

- Lifting and delivery flow rates
- Pump stations and units which are on-line and off-line
- Batch and DRA tracking data for liquid pipelines
- Pipe roughness or efficiency to improve hydraulic calculation accuracy
- Unit utilization data and maintenance schedule

If it is implemented on the SCADA system, the accuracy of the batch/DRA tracking data and friction factor need to be improved in order to calculate the hydraulic profiles accurately. In order to calculate the pipeline hydraulics accurately, accurate pipe roughness or pipe efficiency along the pipeline may be required. A real-time batch tracking capability can provide a more accurate hydraulic calculation. Some optimization models can re-rate pump performance curves by analyzing recent data automatically to determine actual pump performance and efficiency.

A power optimization system is typically configured to run at regular intervals as well as on demand by the operator. Running the system at regular intervals ensures that the system will notify the operator of any system changes required due to changes in the pipeline line fill (e.g., batched operation, etc.). When there is a need for flow rate change, the operator will enter the new parameters and obtain new system changes.

REFERENCES

[1] Yoon, M., Warren, B., and Adam, S., 2007, *Pipeline System Automation and Control*, ASME Press, New York, N.Y.

[2] Chudiak, G. J., and Yoon, M., 1996, "Charting a course in the 90s — From field measurement to management information systems," *Proc. of International Pipeline Conference.*

[3] American Petroleum Institute, 2007, *API RP 1165 — Pipeline SCADA Displays*, API Publication, 1st Edition.

[4] American Petroleum Institute, 2010, *API RP 1167 — Pipeline SCADA Alarm Management*, API Publication, 1st Edition.

[5] American Petroleum Institute, 2008, *API RP 1168 — Pipeline Control Room Management*, API Publication, 1st Edition.

[6] American Petroleum Institute, 2009, *API Standard 1164 — Pipeline SCADA Security*, API Publication, 2nd Edition.

[7] Mohitpour, M., Murray, A., McManus, M., and Colquhoun, I., 2010, *Pipeline Integrity Assurance — A Practical Approach*, ASME Press, New York, N.Y.

[8] "Computational Pipeline Monitoring," API Publication 1130, 3rd Edition, American Petroleum Institute, 2007.

[9] American Petroleum Institute, 1993, *Pipeline Variable Uncertainties and Their Effects on Leak Detectability*, API Publication 1149, 1st Edition.

[10] American Petroleum Institute, 1995, *Evaluation Methodology for Software Based Leak Detection Systems*, API Publication 1155, 1st Edition.

[11] Yoon, M. S., and Yurcevich, J., 1985, "A Study of the Pipeline Leak Detection Technology," Government of Canada, Contract No. 05583-00106.

[12] Yoon, M. S., Mensik, M. and Luk, W. Y., 1988, "Spillage Minimization Through Real-Time Leak Detection," *Proceedings of OMAE Conference*, ASME.

[13] Yoon, M. S., Jacobs, G. B., and Young, B. E., 1991, "Leak Detection Performance Specification," *Proceedings of ETCE Conference*, ASME.

[14] Jeffrey, D., et al., 2002, "An effective and Proven Technique for Continuous Detection and Location of Third Party Interference Along Pipelines" *Proceedings of IPC*, ASME.

[15] Strong, A., et al., 2008, "A Comprehensive Distributed Pipeline Condition Monitoring and its Field Trial," *Proceedings of IPC*, ASME.

[16] "Northstar Development Project Buried Leak Detection System," Intec Project No. H-0660.03, 1999.

[17] Small, S. R., 1983, *Increase Throughput With Drag Reducing Additives*, Pipe Line Industry, June.

[18] Fink, J., 1998, "Additives and Chemicals Used for Transport," ULS, NWP 0905.

[19] Johnston, R., Lauzon, P., and Pierce, P., 2008. "New Heavy Crude Oil DRA Enhances Dilution for Flow Increase," Hydrocarbon Engineering, March.

[20] Seto, S. P., 2005, "Investigation of Pipeline Drag Reducers in Aviation Turbine Fuels" CRC Report#642, Project CA-68-97. http://www.crcao.com/reports/recentstudies2005/CRC%20642.pdf.

[21] http://hercules.us.es/aesop/aesop_presentation.pdf for the Assessment of Energy Saving in Oil Pipelines (AESOP) project.

[22] Dreher, W. R., et al, 2008, "New Heavy Crude Oil Flow Improver Increases Production — Application Scenarios," *Proceedings of IPC*, ASME.

[23] ConnocoPhillips' ExtremePower Flow Improvers website for heavy oil applications, www.extremepowerflowimprovers.com.

[24] ESI uses these correlations to model DRA and then try to use original references for the correlations themselves. The AESOP correlation was developed under EU Project ENK6-CT2000-00096 and provided by ESI.

[25] Burger, E. D., Munk, W. R., and Wahl, H. A., 1982, "Flow Increase in the Trans-Alaska Pipeline through use of a Polymeric Drag reducing Additive," *J. Petroleum Technology*, **34**(2), pp. 377–386.

[26] These and other DRA correlations and parameters are available from ConocoPhillips and other DRA vendors.

[27] Neaderhouser, D. L., and Wray, B. C., 2000, *Monitoring Electric Pump Costs in Real Time*, PSIG.

[28] Short, M., and Meller, S., 1996, "Elements of Comprehensive Pipeline Optimization," *Proceedings of International Pipeline Conference*, ASME, New York, N.Y.

[29] Jefferson, J. T., 1998, "Procedure allows calculation of ideal DRA levels in products line," *OGJ*.

Appendix 1

GLOSSARY OF TERMS AND ACRONYMS

The following is a glossary of some of the terms and abbreviations used in this book and acronyms that provide description of significant terms frequently used or referred to in the hydrocarbon liquids and oil industry including specifications, shipping, pipeline transportation and processing, **CAPP 2011**:

Absolute humidity	The measure of the amount of water vapor actually in the air. Often stated as pounds of water vapor per kg of dry air. At any given temperature, relative humidity equals 100 times the absolute humidity divided by the greatest amount of water vapor that can be present in the air at that temperature.
Absolute pressure	Gauge pressure plus barometric pressure.
AGA	American Gas Association.
AGO	Atmospheric gas oil.
Alarm	A warning given by a control system of a limit violation, abnormal change of state, or a failure.
Alkylate	A gasoline blending component composed of isobutane and propylene or butylene.
Ammonia	Organic amines and inorganic compounds that are liquids at standard temperature and pressure that, when discharged, release free amonia (NH_3), or ammonium ion (NH_4^+).
Aniline point	Reports the aromatics content of a mixture.
ANS	Alaska North Slope.
American National Standards Institute (ANSI)	A standards setting organization concerned with transmission.
API	American Petroleum Institute.
API gravity	Specific gravity scale for petroleum liquids at 60 °F developed by API with reference to the specific gravity of water being equivalent to 10 °API. The relationship between the API gravity and specific gravity is given below: API gravity = (141.5/SG at 60 °F) – 131.5.
Aromatics	Group of petrochemicals characterized by a ring structure, they are produced in refinery reformers and petrochemical plants. The most commonly traded are benzene, toluene, and xylenes. They are used for chemical production or as high-octane components for gasoline blending.
ASPH	Asphaltene
Asphalt	A mixture of bitumen and mineral aggregate as prepared for the construction of roads or in other paving uses. In the US it refers to the product.

AST Aboveground storage tank(s).

ASTM American Society for Testing and Materials.

Audit trail Log that documents changes that were made or the occurrence of an event in computer records or databases. The log should include the change, the date and time, the person that made each change, and the reason of the change.

Avails Availabilities.

Avgas High octane aviation gasoline used in piston type aircraft engines.

Backfilling Backfilling is the process of filling the trench where a newly constructed or recently unearthed pipeline is installed. Adequate fill material is provided and compacted around the pipe to completely fill the excavation and to ensure that the pipe is properly supported and not subjected to added stresses due to soil subsidence or movement.

Backwardation Market situation where prices are highest in the nearest date periods and are lower in forward delivery dates.

Barge A vessel carrying oil usually on rivers - containing between 8000 and 50,000 bbl (weighing 1,000 to 10,000 mt).

Barrel A volumetric unit of measure for crude oil and petroleum products. 1 barrel equals 42 US gallons, 35 imperial gallons or 159 liters.

Base assessment plan A baseline assessment plan (BAP) is the plan a pipeline operator must develop to assess the integrity of all of the lines included in its integrity management program. The BAP must, as a minimum: 1) identify all segments of a pipeline system that could impact a High Consequence Area (HCA); 2) identify the specific integrity assessment method(s) to be conducted on those segments; 3) specify the schedule by which those integrity assessments will be performed; and 4) provide the technical justification for the selection of the integrity assessment method(s) and the risk basis for establishing the assessment schedule. (reference 49CFR 195.452 [3]).

Base conditions Pressure and temperature reference conditions used in determining a fluid quantity for custody transfer.

Batch A batch refers to a contiguous product entity that remains whole throughout its journey through the pipeline system. A batch has the attributes of product type, volume, identification or name, and lifting and delivery locations and times with flow path.

Batching cycle Batching cycle is a specific period during which a predefined set of products is transported. Multiple cycles are repeated during the nomination period, usually a month.

Batch interface Interface volume of two adjacent batches. Refer to transmix

Batching meter A flow meter normally used for batch movement.

Battery Small field plants which remove natural gas, NGLs, water, and sand from crude oil delivered by gathering systems from individual wells.

Bell Hole A bell hole is an excavation made to permit a survey, inspection, maintenance, repair, or replacement of pipe sections. It is so called because of its upside-down bell shape, wide at the top and narrowing to a smaller diameter around the pipeline to be examined. The walls of the hole are angled according to OSHA (Occupational Safety and Health Administration) rules to prevent cave-ins and risks to worker safety.

Bend Radius Bend radius is the measurement of the radius of a bend in a pipeline. Bends can be included in the manufacture of the pipe or created in the field during construction and installation of a pipeline.

Bias error The difference between the average and true values or measurements. It is directional, but it is difficult to determine a true bias error in practice.

Bill of Lading Documentation associated with a specific cargo of oil and is signed by the captain of the ship and the contract supplier.

Bitumen A thick, sticky form of crude that is generallyheated or diluted to transport through a pipeline. Petroleum that exists in the semi-solid or solid phase in natural deposits.

Bleeder A valve or pipe through which bleeding is done.

Blending Mixing of two or more products by injecting one product stream into another.

Blendstock A component combined with other materials to produce a finished refined product.

Block valves Gate valves found in the pipeline on either side of a pumping station or terminal, river crossing, or other points where the line may have to be blocked. Block fluid flow in both directions when closed. Block valves are installed along the pipeline, called isolation or sectionalizing valve, and in a pump station (suction valve and discharge valve).

Blow-down valve Valve used to exhaust gas from a section of pipe when necessitated by repairs, emergency, or other conditions.

Breakout point An intermediate location or tank farm on a pipeline system that joins two or more pipeline sections, where batches can be simultaneously injected into and delivered out of the pipeline or a batch can be tight-lined.

Breakout tank A tank used to relieve surges in a liquid pipeline system, or to receive and store liquid for re-injection and continued transportation by pipeline.

BS&W Bottom sediment and water (expressed as a %by weight).

BTU British Thermal Unit.

BTX Benzene, toluene and xylene.

BTX extraction The process for removing benzene, toluene, and xylene from reformate or pyrolysis asoline.

Bubble point The temperature at a given pressure, as opposed to vapor pressure, at which vapor starts forming above a liquid.

Buckle A buckle is a partial collapse of the pipe wall due to excessive bending associated with soil instability, landslides, washouts, frost heaves, earthquakes, etc. Buckles can also occur in pipeline construction during a field bending operation using a side boom. Buckles cause localized stress concentrations and must not be installed in new construction or, if found, must be removed from existing systems.

Buffer A temporary product injected between two batches to reduce mixing of the two batches.

Bulkhead A bulkhead is a wall built or installed along a coastline or waterway to protect an adjacent pipeline from washout or soil erosion.

Bunker C A residual fuel used as ship's fuel (usually has a high sulfur content and is highly viscous).

Bypass valve Valve allowing flow around a metering system or equipment.

Calibration Adjustment of a measuring instrument against a known quantity to improve its performance or to conform to an applicable standard.

Caliper tool A caliper tool is an inline inspection device that measures deviations in the geometry of a pipeline's internal surface. A caliper tool uses a set of mechanical fingers that ride against the inside diameter of the pipe and record deviations caused by the relative movement of these fingers as the tool passes through the pipe.

Capacity The volume of fluid that a pipeline system can transport for given design condition.

CAPP Canadian Association of Petroleum Producers [1].

Carbon dioxide (CO_2) Carbon dioxide is a naturally occurring gas consisting of molecules formed of one carbon and two oxygen atoms. Carbon dioxide is a heavy gas that is transported by pipeline as a compressed fluid consisting of more than 90% carbon dioxide molecules. If released into the atmosphere from a pipeline leak carbon dioxide is considered hazardous due to its ability to displace breathing air (reference 49CFR 195.2 [3]).

Carbon steel All steel is formed by alloying iron with carbon. Steel is commonly considered to be carbon steel when:

(1) no minimum content is specified or required for aluminum, boron, chromium, cobalt, columbium, molybdenum, nickel, titanium, tungsten, vanadium, zirconium, or any other element added to obtain a desired alloying effect;

(2) the specified minimum content does not exceed 1.65% for manganese or 0.60% for copper. All carbon steels may contain small quantities of unspecified residual elements unavoidably retained from raw materials. These elements (copper, nickel, molybdenum, chromium, etc.) are considered incidental and are not normally determined or reported.

Cast iron Cast iron applies to gray cast iron, which is a cast ferrous material in which a major part of the carbon content occurs as free carbon in the form of flakes interspersed through the metal. Because the carbon flakes do not bond with the ferrous material on the molecular level, the metal is brittle and susceptible to stress cracking under high-pressure situations. Many older, low-pressure gas systems were constructed with cast iron pipe.

Catfeed Feedstock to a catalytic cracker, usually vacuum gasoil.

Cat gasoline (cat naphtha) A gasoline blending component made in a cat cracker.

Cathode A cathode is a component of an electrochemical corrosion cell. It is the metal that attracts ions and gains mass through the corrosion process. In a cathodic protection system, the pipeline acts as the cathode and is protected from corrosion by the sacrificial activity of the anode.

Cathodic protection A method of preventing corrosion of a metal surface by making that surface the cathode of an electrochemical cell through the use of devices such as galvanic anodes or impressed current.

Cathodic protection monitoring or survey Monitoring cathodic protection systems is an activity performed by pipeline operators to assess the adequacy of cathodic protection on the pipeline.

Cathodic Protection Tester A person who can demonstrate an understanding of the principles and measurements of all common types of cathodic protection systems as applied to buried or submerged metal piping and tank systems.

Cavitation Bubbles form when the pump pressure, most likely suction pressure, is equal to or below the liquid vapor pressure at flowing temperature.

Centering The process of approximating the location of a pipeline leak. Centering can be done manually using gas detection equipment, such as acoustical equipment, or through a more sophisticated SCADA-based leak detection system using mathematical modeling of critical pressures, temperatures, flow-rates, etc. to predict volume loss and location estimates.

Centrifugal pump A pump consisting of one or more impellers fixed on a rotating shaft. The liquid enters the impeller at the shaft and is impelled outward from the center by centrifugal force at high velocity into the volute of the pump casing. This pump has the advantage of constant pressure.

CEPA The Canadian Energy Pipeline Association.

Cetane A paraffinic hydrocarbon used as an additive in diesel fuel.

Cetane number Percentage by volume of cetane added to basic diesel fuel to achieve specific ignition performance characteristics.

Check valve A valve used to prevent the back flow. It remains open by flowing fluid but closed automatically when fluid flow is reversed or even stops.

Class location Class location is a criterion for gas pipeline design set by the U.S. Code of Federal Regulations, 49CFR 192.5. A class location is based on the number and type of buildings intended for human occupancy that are situated in an area that extends 220 yd (200 m) on either side of the centerline of any continuous 1.0-mile (1.6-km) length of a gas pipeline. Class locations are specified as Class 1, 2, 3, or 4. Class 1 indicates the least heavily populated of the class locations, representing an offshore area or an area with 10 or fewer buildings intended for human occupancy. In comparison, Class 4 indicates the most heavily populated of the class locations, representing an area where buildings with four or more stories above ground are prevalent.

Chlorine Organic and inorganic compounds that are liquids at standard temperature and pressure that, when discharged, may release free chlorine (Cl_2) or chlorides (Cl^-).

CIF Cost, insurance, and freight charges for shipping products. CIF prices include these charges.

Cleaning Pig A cleaning pig is a device that is placed inside a pipeline to remove unwanted debris from the inside of the pipeline. The pig can be drawn or pushed through a pipeline but often is moved through the line as a result of the flow of the product in the line. A type of utility pig, a cleaning pig uses cups, scrapers, or brushes to remove dirt, rust, mill scale, and other debris from the internal surface of a pipeline. Cleaning pigs are used periodically or as necessary to increase the operating efficiency of a pipeline or to facilitate inspection of the pipeline.

Cloud Point The temperature at which a fuel, when cooled, begins to congeal and take on a cloudy appearance due to bonding of paraffins.

Codes Documents that regulate design and construction to protect the safety of people. Codes establish minimum standards but do not guarantee efficiency or quality.

Coker The processing unit in which bitumen is cracked into lighter fractions and withdrawn to start the conversion of bitumen into upgraded crude oil.

Common carrier Pipeline that provides transportation service to all parties equally. It also applies to any transportation system available for use by the public for transporting cargo.

Common Ground Alliance The Common Ground Alliance (CGA) is a nonprofit organization dedicated to shared responsibility in damage prevention and promotion of the damage prevention Best Practices identified in the Common Ground Study Report. The purpose of the CGA is to ensure public safety, environmental protection, and the integrity of services by promoting effective damage prevention practices. The CGA can be reached at 703-818-3217 or for more information see www.commongroundalliance.com.

Compatible substance The ability of two or more substances to maintain their respective physical and chemical properties upon contact with one another for the design life of the storage tank system under conditions likely to be encountered in the storage tank system.

Component A component in a pipeline system is any part of the system that is subject to the internal system pressure. System components must be designed to not leak; thereby assuring that the oil or gas is not released from the system and that the system operating pressure is maintained. Examples of components include the pipe, valves, flanges, and other fittings.

Computational Pipeline Monitoring (CPM) Computational Pipeline Monitoring is a method of monitoring the operation of a pipeline system, as specified in API 1130. It makes use of a software-based monitoring tool that alerts the pipeline dispatcher of possible pipeline operating abnormalities that could indicate a commodity release.

Condensate The natural gas product, mostly pentanes and heavier, is separated and recovered as liquids at field facilities or gas processing plants.

Confirmation digs Confirmation digs are excavations performed at selected pipeline locations to expose the pipe to allow confirmation of the existence and characteristics of potential anomalies in the pipe wall that were identified by inline inspection techniques.

Consensus standards Consensus standards are standards for performance that are established in open and voluntary forums by consensus of parties affected by the standards. These standards provide the best available professional guidance related to specific subjects. National consensus standards reflect the professional knowledge and judgment of people having appropriate experience, training, and education in the subjects to which the standards apply. Some organizations that produce consensus standards include: American Petroleum Institute (API), American National Standards Institute (ANSI), ASTM International (ASTM), American Society of Mechanical Engineers (ASME).

Consent order Once PHMSA has issued a Corrective Action Order or given notice to an operator of other proposed enforcement action, PHMSA and the operator may occasionally agree to the issuance of a Consent Order in lieu of a Compliance Order or the imposition of civil penalties. In a Consent Order, the operator agrees to the jurisdictional facts of the case and agrees to implement the specific actions in the Consent Order by the required deadlines. Should the operator not comply with the Consent Order terms, PHMSA can enforce its terms as the parties have agreed. Consent Orders are described in 49 CFR 190.219.

Consequence A consequence is often noted as the cumulative, undesirable result of an accident. Consequences are usually measured in health and safety effects, environmental impacts, loss of property, and/or business costs. Consequence descriptions may be qualitative or quantitative estimates of the effects of an accident.

Consequence analysis A consequence analysis is the evaluation or analyses of a postulated action or condition or series of events to determine the logical result. A consequence analysis may be performed to determine the expected effects of events leading to a pipeline accident, independent of the likelihood of such events occurring.

Contamination or contaminated The presence of substances in surface water, groundwater, soil, sediment, or on the land, in quantities that result in exceedances of applicable cleanup target or usage levels.

Continual evaluation of pipeline integrity Under pipeline integrity management regulations, pipeline operators must develop programs for conducting periodic pipeline integrity assessments and evaluating the results of those assessments to understand current pipeline conditions and identify integrity issues. This periodic assessment and evaluation cycle is recognized as a continual evaluation of pipeline integrity.

Contract An agreement between the pipeline company and a shipper which specifies the type of service and minimum/maximum volumes.

Corrosion Corrosion is the deterioration of a material, usually a metal, which results from a reaction with its environment. Common rust is an example of corrosion of iron. Steel pipe is subject to corrosion damage.

Corrosion professional A person who, by reason of knowledge of the physical sciences and the principles of engineering and mathematics acquired by a professional education and related practical experience, is qualified to engage in the practice of corrosion control on buried or submerged metal components of a storage tank system.

Cost/benefit analysis A cost benefit analysis is a determination and comparison of the expected costs to implement a proposed activity and the expected safety, environmental, or other benefits resulting from the activity. Cost/benefit analysis can be used to establish priorities among various activities, to compare and select among alternate ways of accomplishing an objective, or to decide whether a proposed activity should be implemented. Cost/benefit analysis is used by regulators to justify new regulations, and by operators to define and allocate resources to operational and maintenance activities.

Cost/Benefit ratio The monetary costs required to complete a proposed activity (for example, replacing a section of pipe) divided by the monetized value of the benefits expected to be derived from that activity (for example, decreased pipe leak rate or likelihood of rupture). Cost/benefit ratios are usually used as an input to, but not sole determinate of, a decision to proceed with the proposed activity.

Cracked Molecules broken by some refining processes. the opposite of straight-run.

Cracked fuel Residue remaining after processing a straight run fuel by enhanced refining methods such as catalytic cracking

Cracks Cracks in line pipe are separations in the molecular structure of the base metal and form as a result of improper manufacturing or operational stresses. Cracks are detrimental to the pipe's pressure restraining capabilities and can propagate into complete failure or rupture zones.

Critical bond Buried pipelines in close proximity or crossing over/under one another will often have different levels of cathodic protection and thus a strong possibility of different electrical potentials (voltage differences). If the soil resistance path between the pipelines is low, electrical currents will flow from one pipeline to the other. The pipeline with the current leaving will experience metal loss or corrosion. To prevent this occurrence, pipeline companies electrically connect or bond their pipes to one another using a copper wire. The wire provides an electrical path for the current to flow through rather than allowing current to flow off the pipe and into the soil, thus eliminating corrosion. The connecting wire is referred to as an electrical bond. Critical bonds are those that if not attached, would allow corrosion to occur and jeopardize the safe operation of one of the pipelines in question.

Critical defect A critical defect is an identified pipeline defect for which analysis indicates immediate attention is required.

Critical point The meeting point of critical pressure and critical temperature at which the properties of the liquid and gas phases become identical.

Crude oil A naturally occurring liquid mixture of hydrocarbons; liquid petroleum.

Crude oil (Conventional) A mixture of pentanes and heavier hydrocarbons that is recovered or is recoverable at a well from an underground reservoir. It is liquid at the conditions under which its volumes is measured or estimated and includes all other hydrocarbon mixtures so recovered or recoverable except raw gas, condensate, or bitumen.

Crude oil (heavy) Crude oil that has an API of 27° or less. No differentiation is made between sweet and sour crude oil that falls in the heavy category because heavy crude oil is generally sour.

Crude oil (medium) Crude oil is has an API greater than 27° but less than 30°. Generally, no differentiation is made between sweet and sour crude oil that falls in the medium category because medium crude oil is generally sour.

Crude oil (synthetic) A mixture of hydrocarbons, similar to crude oil, derived by upgrading bitumen from the oil sands.

CST Centistokes — a measure of viscosity similar to seconds.

Custody transfer Custody transfer is the change of ownership of petroleum products at a given transfer point, most likely at a meter station.

Customer Entity such as a local distribution company or marketer that generates a net outflow of gas or liquid from the pipeline company.

Cut and cover tank A tank that is constructed with steel that is surrounded by soil above the natural surface of the ground.

DCS Distributed Control System. This is a type of automated control system that is used to monitor and control a process facility.

Damage prevention initiatives The Office of Pipeline Safety has undertaken a variety of other initiatives to promote pipeline damage prevention. These initiatives, some of which involve collaboration with various stakeholder industry groups, promote public awareness of pipelines and other underground facilities; education of the public and emergency officials, excavators, emergency responders, legislators and other stakeholders on how to recognize and respond to pipeline leaks; and other aspects of pipeline safety. Some more notable recent efforts include: Common Ground Alliance (CGA) and the One-Call Systems Study (OCSS).

Damage prevention regulations Damage prevention regulations for pipelines are found in the Code of Federal Regulations. They require pipeline operators to implement programs to prevent damage to buried pipelines from excavation activities. The regulations specify requirements for operators to interact with the public near pipelines and with companies/persons likely to perform excavation near pipelines. Pipeline operators may comply with some of the regulatory requirements for damage prevention by participating in public service programs such as a one-call system. However, such participation does not relieve the operator of the responsibility of compliance to the regulation.

Degrees API Gravity = (141.5/Specific Gravity at 60 °F) – 131.5

Defect A defect is an imperfection of sufficient magnitude in a pipeline that it should be analyzed using a recognized and approved engineering procedure, such as ASME B31G. Analysis determines if the defect is severe enough to require removal or repair, depending on prescribed operating requirements. Examples of defects include dents, corrosion pits, and cracks.

Delivery The transfer of a quantity of fluid out of a pipeline system, typically into a tank, either at the end or at an intermediate location. This is the point of custody transfer for fluid moving out of the pipeline company's system.

Delivery station Also known as terminal, this is where the product delivery takes place. It is equipped with tanks and distribution facilities. It is a place where one product is transferred from one party to another.

Demurrage The detention of a vessel in loading or unloading beyond the time agreed upon.

Dense phase A dense phase is defined as the phase whose pressure exceeds the pressure of the critical point of the fluid. The dense phase fluid exhibits high density and high compressibility (low viscosity).

Density Mass or weight per unit volume.

Dent A dent is a depression in the pipeline surface caused by pressure or a blow. A dent does not reduce the thickness of the pipe wall. Dents are usually local depressions caused by outside mechanical force, such as the pipe resting on a rock or being hit by a backhoe. Pipeline repairs or replacement may be needed depending on the severity of the deformation.

Derating Process of reducing the level less than its design capacity.

Design factor Ratio of the hoop stress developed in the pipeline by the design pressure and the Specified Minimum Yield Stress (SMYS) of the pipeline material.

Destructive testing Destructive testing refers to testing of a specimen in which the specimen is rendered unusable (i.e., destroyed) for its intended or practical use. Destructive tests are typically conducted on representative specimens to prove the strength or chemical characteristics of the sample piece.

Dew point The temperature at a given pressure or the pressure at a given temperature, at which liquid starts condensing from a vapor.

Diesel Index A measure of the ignition quality of a diesel fuel calculated from a formula involving the gravity of the fuel and its aniline point.

Dike field area The area around the tank or tanks that extends from the circumference of the base of an AST to the top of the berm, dike, or retaining wall surrounding the tank.

DilBit Bitumen that has been reduced in viscosity through addition of a diluent (or solvent) such as condensate or naphtha.

Diluent Lighter viscosity petroleum products that are used to dilute bitumen for transportation in pipelines.

Direct assessment Direct assessment refers to a method of evaluating the integrity of a pipeline. Various indirect measurement tools are used to determine locations on the pipeline that may require direct examination to verify pipeline integrity. These locations are then excavated and examined to verify that the pipe is in good condition or to make necessary repairs.

Disbonded coating Disbonded coating refers to any loss of bond or adhesion between the protective coating applied to the outside of a steel pipe and the pipe itself. Disbondment can result from adhesive failure, chemical attack, mechanical damage, hydrogen concentrations, or other causes.

Discharge Any spilling, leaking, seeping, pouring, misapplying, emitting, emptying, or dumping of any substance which occurs and which affects lands and the surface and ground waters.

Discharge pressure The pressure of the pipeline as the fluid exits a pump station.

Distribution line A distribution line is a line used to supply natural gas to the consumer. A distribution line is located in a network of piping located downstream of a natural gas transmission line. As defined in natural gas pipeline safety regulations, a distribution line is a pipeline other than a gathering or transmission line.

DOT U.S. Department of Transportation.[2].

Double-bottomed An AST that has secondary containment in the form of an outer- tank bottom having a closed interstitial space between the primary tank bottom and the secondary outer tank bottom.

Double-walled A storage tank that has an outer tank wall, or integral piping that has an outer wall that provides secondary containment of the primary tank or piping.

DRA Drag reducing agent, a long chain polymer. This is a fluid injected into a pipeline to reduce friction along the pipeline and thus increase the throughput.

Durability Durability is the ability of a material to resist fatigue, which is a weakening of the material resulting from the repeated application and removal of stress. Pipelines experience fatigue as a result of periodic increases (application of stress) and decreases (removal of stress) in operating pressures. Durability refers to the pipe's ability to resist fatigue and continue to perform its required function.

Easement A right-of-way agreement between a pipeline company and the property owner.

E-4 C.I.S. (Chemical Information system) high sulfur straight-run feedstock. Formerly called F-10.

Electric resistance weld pipe Electric resistance welding (ERW) is a method used for making pipe in which the two edges of a curved plate of metal are heated by passing an electric current through them and the edges are then forced together to create a solid longitudinal weld, thus turning the plate into a section of pipe.

Emergency flow restricting device An emergency flow restricting device (EFRD) is a device used to restrict or limit the amount of oil or gas that can release out of a leak or break in a pipeline. Check valves and remote control valves are types of EFRDs.

Empty All substances have been removed so that no more than 1 in. in depth or 0.3% by weight of total system remains in the storage tank system.

Encroachment Encroachment refers to the unauthorized use of a right-of-way in violation of the terms by which the right-of-way was established (e.g., easement).

Environmental impact statement An environmental impact statement is preceded and determined by an environmental assessment. An EIS must include a description of: (1) the environmental impact of the proposed action, (2) any adverse environmental effects which cannot be avoided should the proposal be implemented, (3) alternatives to the proposed action, (4) the relationship between local, short-term uses of the environment and the maintenance and enhancement of long-term productivity, and (5) any irreversible and irretrievable commitments of resources which would be involved in the proposed action should it be implemented.

EPA US Environmental Protection Agency

Environmentally sensitive area An environmentally sensitive area (ESA) is an area of environmental importance that is sensitive to environmental damage from a liquid pipeline release. (Reference 49CFR 195.2).

ESD Emergency Shutdown Detection. This is a controller that is independent of the station control system that detects conditions requiring an immediate shut down of the pump/compressor station.

ETA Estimated time of arrival. This is the ETA of a tracked object such as a batch front or scraper at a specified location or facility.

Excavation Excavation refers to any operation involving movement of earth, rock or other materials below existing grade or surface level. Excavation can include digging, blasting, boring, tunneling, and backfilling.

External floating roof A storage vessel cover in an open top tank consisting of a double deck or pontoon single deck which rests upon and is supported by the petroleum liquid being contained and is equipped with a closure seal or seals to close the space between the roof edge and tank shell.

Extraction A process generally unique to the oil sands industry, in which bitumen is separated from their source (oil sands).

Eye (of an impeller) Internal pressure point at the inlet of the first impeller of the compressor. A "suction-to-eye" differential pressure is typically representative of flow.

Failure While a pipeline that is actually leaking product is the most obvious indication of failure, failure is often also defined as the point at which the pipe material is stressed beyond its elastic or yield point. At that point, the material is deformed and does not return to its original shape. A pipeline rupture is an example of this type of failure.

Failure frequency The number of failure events that occur divided by the total elapsed calendar time during which those events occur or by the total number of demands, as applicable.

Failure probability Failure probability is the probability that a structure, device, equipment, system, etc. will fail on demand or will fail in a given time interval, expressed as a value from 0 to 1.

Failure rate Failure rate is the number of failure events that occur divided by the total elapsed operating time during which those events occur or by the total number of demands, as applicable.

Fatigue Pipelines experience fatigue as a result of periodic increases (application of stress) and decreases (removal of stress) in operating pressures. Because fatigue can cause a failure to occur at stress levels well below those that a material can withstand in a single, non-repetitive loading, materials that must resist repeated stress cycles must be specially designed for this service. Durability is the ability of the material to resist fatigue.

Feedstock Material used in or supplied to a processing plant, refinery or oil sands upgrader.

Field-erected storage tank An AST that is constructed by assembling it on-site at the facility.

FERC The Federal Energy Regulatory Commission is an independent regulatory agency within the Department of Energy that regulates the transmission and sale for resale of natural gas in interstate commerce; regulates the transmission of oil by pipeline in interstate commerce; regulates the transmission and wholesale sales of electricity in interstate commerce; licenses and inspects private, municipal, and state hydroelectric projects; oversees related environmental matters; administers accounting and financial reporting regulations and conducts of jurisdictional companies; and approves sitting and abandonment of interstate pipeline facilities. FERC can be contacted at 866-208-FERC or at www.ferc.gov.

Fittings The small pipes and valves that are used to make up a system of piping. This may include fabricated assemblies (such as valves, pumps, and pipe fittings).

Flammable fluid A fluid having a flash point lower than 100°C.

Flange A projecting rim, edge, lip, or rib to provide strength for guiding or for attachment to another object, as a plate to close a pipe opening or other orifice.

Flash point The lowest temperature under very specific conditions at which a combustible liquid will give off sufficient vapor to form a flammable mixture with air in a standardized vessel. Assesses the volatility of the product.

Flow computer (FC) Field device for collecting measurement data in real time, performing certain calculations such as AGA, storing historical measurement and calculation data, and uploading to/downloading from the host SCADA.

Flow line A pipeline (including valves and fittings) for transporting untreated hydrocarbons and other reservoir.

Flow profile An expected yearly flow pattern for the liquid supply or demand over the project period that would be required to design the pipeline system.

Flow-through process tank A tank that forms an integral part of a production process through which there is a steady, variable, recurring, or intermittent flow of materials during the operation of the process. Flow-through process tanks include tanks associated with vapor recovery units and oil-water separators. Flow-through process tanks do not include the storage of finished products or by-products from the production process.

FOB Free on board. FOB prices exclude all, insurance and freight charge.

Fracture toughness Fracture toughness is the ability of a material to deform under increasing tensile stress in the presence of a defect or crack without exhibiting rapid and extensive fracture propagation. Materials that have high fracture toughness can absorb larger amounts of energy (i.e., can withstand higher pressures or levels of stress) before an existing crack spreads.

Full or tight line flow Flow that completely fills a pipeline, opposite to slack flow.

Fungible batch A batch that can be combined with other batches of the same product.

Gathering Lines The piping networks installed in oil fields to transport petroleum to a processing plant or bulk shipping point.

Gathering system A system of pipelines laid to bring gas or oil from wells to a central point.

Gasoil An intermediate distillate product used for diesel fuel (interchangeably with No. 2 heating oil).

GIS Geographic Information Systems, a computerized information system for storing, manipulating and analyzing spatially referenced information.

Geometry tools Geometry tools are inline inspection tools that are used to identify changes in the internal shape/diameter of a pipeline. Changes in the internal shape or diameter of pipe can be caused by dents, wrinkles, ovality, bend radius and angle, and occasionally indications of significant internal corrosion.

Gross volume Raw volume of fluid at the measured conditions of pressure and temperature before it is corrected and before water and sediment (for liquid measurement) are accounted for.

Hazard and Operability Analysis A hazard and operability analysis (HAZOP) is a systematic method for evaluating hazards. It often involves the review of detailed system drawings, specifications, and operating procedures. Process hazards and potential operating problems are identified through a qualitative investigation of deviations from normal process conditions.

Hazardous site A hazardous site is the location of a hazard. This includes locations where hazardous materials are used, stored, transported or produced. It could also include a location where a hazardous liquid has been accidentally released.

Hazardous Substances Substances that are hazardous to human and environmental well being.

HDA Hydrodealkylation, a process used for making benzene from toluene.

Heating oil Any petroleum based fuel used in the operation of heating equipment, boilers, or furnaces.

Heavy crude oil Has API gravity lower than 28°. The lower the API gravity, the heavier the oil.

Heavy oil Oil having an API gravity less than 22.3°. Includes some oil that will flow, however slowly, but most heavy oil requires heat or dilution to flow to a well or through a pipeline.

High consequence area A high consequence area (HCA) is a location that is specially defined in pipeline safety regulations as an area where pipeline releases could have greater consequences to health and safety or the environment. For oil pipelines, HCAs include high population areas, other population areas, commercially navigable waterways and areas unusually sensitive to environmental damage. Regulations require a pipeline operator to take specific steps to ensure the integrity of a pipeline for which a release could affect an HCA and, thereby, the protection of the HCA.

High viscosity A pollutant with a viscosity of 30 centistokes (cSt) and higher at 40 °C, such as American Society for Testing and Materials (ASTM) grades 5 and 6 residual oils, intermediate fuel oils, or Bunker C fuel.

HMI Human Machine Interface. This refers to the interface between a user and a SCADA or computer system.

Host The centrally located collection of hardware and software of a SCADA system.

HSFO High sulfur fuel oil.

Highly volatile liquid	A highly volatile liquid (HVL), as defined in pipeline safety regulations, is a hazardous liquid that will form a vapor cloud when released to the atmosphere and has a vapor pressure exceeding 276 kPa (40 psia) at 37.8 °C (100 °F).
HVP products	The acronym for high vapor pressure products. These are light hydrocarbons whose vapor pressure exceeds 110 kPa at 38 °C.
Hoop stress	Hoop stress is stress that acts around the circumference of a pipe (i.e., perpendicular to the pipe length) and is caused by the internal pressure pushing outward against the pipe wall. As pressure within the pipe increases, the stress in the pipe wall must be capable of acting against that pressure to contain it. Hoop stress is a critical factor in determining a pipe's pressure holding capacity and thus in determining its appropriate application.
Hydrocarbons	A large class of liquid, solid, or gaseous organic compounds, consisting of only carbon and hydrogen, which are the basis of almost all petroleum products.
Hydrogen induced damange	Examples of hydrogen induced damage are: (1) formation of internal cracks, blisters, or voids in steels; (2) embrittlement (i.e., loss of ductility); and (3) high temperature hydrogen attack (i.e., surface decarbonation and chemical reaction with hydrogen).
Hydrant piping	The pressurized integral piping system, including hydrant pits, associated with petroleum storage tank systems.
Hydrocarbon bubble point	The pressure and temperature point of a hydrocarbon liquid, at which vaporization is about to occur.
Hydrostatic pressure testing	Hydrostatic pressure testing (or 'hydro testing') is a method of testing pipeline integrity in which the line is filled with a liquid, usually water, and then the pressure of the liquid is raised to a specified pressure that is maintained for a specified period of time. Any ruptures or leaks revealed during the test must be repaired and the test repeated until no problems are noted.
Immediate repair condition	An immediate repair condition is a defect or anomaly in the condition of the pipe for which pipeline safety regulations require immediate action to repair the anomaly (reference 49CFR 195.452 [3]).
Impeller	The rotating element of a centrifugal pump that imparts the kinetic energy to the liquid.
In-service	A storage tank system that is being actively maintained and operated.
In-service pipeline	An in-service pipeline is a pipeline that is being used to transport natural gas or liquid.
Inactive pipeline	An inactive pipeline is a pipeline that is not presently being used to transport oil or natural gas but that is being maintained in accordance with regulatory requirements so that it can be put back in to active use if needed.
In Situ recovery	The process of recovering crude bitumen from oil sands by drilling.
Incident	A condition or situation indicating that a discharge may have occurred from a storage tank system.

Incident command system In widespread or major emergencies that could affect large populations or have significant environmental impacts an Incident Command System (ICS) is usually established to coordinate the combined actions of various emergency response personnel.

Incipient leak A leak that is just about to occur.

Increment Strapping Table A strapping table of level increments vs. tank volumes, which is used to convert level increments into gross volumes.

Industrial occupancy building An enclosed structure that contains an AST system that is used in association with an industrial or manufacturing process, or for electric power generating utilities, provided that the building was constructed and is used primarily for industrial, manufacturing, or electric power generating purposes, and not solely for the purpose of storing approved petroleum products.

Industry cost of incident The industry cost of an incident is defined as all costs to the operator and its contractors.

Initial injection station Supply or inlet station, located at the beginning of the pipeline system, where the product is lifted into the pipeline. Storage facilities, pumps and flow meters are installed at the station.

Injection The process whereby a fluid is moved from a tank into a pipeline at the head or intermediate location within the pipeline company's system.

Inline inspection (ILI) Inline inspection is a method of inspecting a pipeline using an inline inspection tool or smart pig. ILI is also known as Internal Inspection or Smart Pigging. Different ILI techniques and tools are designed to detect defects in the pipe wall and on the internal and external surfaces of the pipe. Defects can include areas of corrosion, dents, metal loss, and the presence of cracks.

ILI tool An inline inspection tool is a device used to perform inline inspection. An ILI tool is inserted into a pipeline and, usually, is pushed through the line by the pressure of the fluid being transported. As the ILI tool travels through the pipeline it uses nondestructive testing techniques and technology to identify and record potential pipe defects or abnormalities. An ILI tool is also known as an Intelligent or Smart Pig.

Integral piping On-site piping, originating, or terminating at storage tank or tanks, that conveys petroleum products. Vapor recovery lines, pipeline facilities, and vent lines are not considered integral piping. Integral piping is not considered on-site if the piping crosses two boundaries. Integral piping includes all valves, elbows, joints, flanges, pumps, and flexible connectors, up to the forwarding pump.

Integration of data As used in pipeline integrity management regulations, integration of data (or data integration) is the process of bringing together all available risk and integrity-related data and information. Data integration is necessary and useful in evaluating the combined impact of diverse factors on pipeline risk.

Integrity Integrity is being of sound and unimpaired condition. Pipeline integrity assures that the pipeline is in sound and unimpaired condition and can safely carry out its function under the conditions and parameters for which it was designed.

Integrity assessment An integrity assessment is an evaluation to determine pipeline integrity. Acceptable assessment methods for pipelines include the use of internal inspection tools, hydrostatic pressure testing, or other technology that the operator demonstrates can provide an equivalent understanding of the pipe condition (reference 49CFR 195.452 [3]).

Integrity management system In accordance with pipeline safety regulations for oil pipelines, an operator's integrity management program must include, at a minimum, the following elements: a process for determining which pipeline segments could affect a High Consequence Area (HCA), a Baseline Assessment Plan, a process for continual integrity assessment and evaluation, an analytical process that integrates all available information about pipeline integrity and the consequences of a failure, repair criteria to address issues identified by the integrity assessment method and data analysis (the rule provides minimum repair criteria for certain, higher risk, features identified through internal inspection), a process to identify and evaluate preventive and mitigative measures to protect HCAs, methods to measure the integrity management program's effectiveness, and a process for review of integrity assessment results and data analysis by a qualified individual (reference 49CFR 195.452 [3]).

Internal floating roof A cover or roof in a fixed roof tank which rests upon or is floated upon the petroleum liquid being contained, and is equipped with a closure seal or seals to close the space between the roof edge and tank shell.

Internal lining A material that is applied internally on AST bottoms to protect the tank from internal corrosion.

Interruptible service A service that can be interrupted if the pipeline capacity is not sufficient to serve a higher priority transportation service. Interruptible service is less expensive than firm service

Interstate pipeline An interstate pipeline is a pipeline or that part of a pipeline that is used in transportation of hazardous liquids or natural gas in interstate or foreign commerce.

IPE International Petroleum Exchange.

ISO International Standards Organization.

Isomerate A gasoline blendstock made in an isomerization unit.

ITT In-tank transfer.

Kero Kerosene.

Landed Cost (of oil) The cost of oil off-loaded at a port. Similar to CIF, but would also include any duties fees or taxes.

Lateral A lateral is a segment of a pipeline that branches off of the main or transmission line to transport the product to a termination point, such as a tank farm or a metering station.

Launcher A launcher is a pipeline component that is used for inserting an inline inspection tool, cleaning pig, or other device into a pressurized pipeline. After performing its task, the tool or pig is removed via receiver.

Leak A leak is a small opening, crack, or hole in a pipeline allowing a release of oil or gas. Pipeline operators periodically perform leak surveys as leaks may not be readily or immediately detected.

Leak survey A leak survey is a systematic inspection for the purpose of finding leaks on a pipeline. The frequency and methods of performing leak surveys are regulated and may vary depending on several factors.

Lifting Injection of a batch at the head of a pipeline system, also tankers and barges loading petroleum at a terminal or transfer point.

Light crude oil Crude oil with API gravity higher than 33°.

Light products, light ends The group of petroleum products with lower boiling temperatures including gasolines and distillate fuels.

Line fill The volume of petroleum products within a pipeline or a pipeline segment during transportation.

Line markers Line markers are signs or other markers installed above the pipeline to indicate the approximate pipeline route and provide emergency phone numbers. Requirements for line markers are specified in federal pipeline safety regulations (reference 49CFR 192.707 [4] or 49CFR 195.410) [3].

Line pack The volume of fluid in a pipeline segment or entire pipeline system. Line pack can increase or decrease depending on whether the fluid volumes received are larger than the volumes delivered or vice versa.

Line packing/unpacking Increasing/decreasing process of line pack.

Line pressure Pressure generated by station pumps and transmitted to the line for purpose of moving product.

Liner An impervious material that is used externally as a method of secondary containment.

Line section A line section is a designated section of a continuous run of pipeline. Line sections may designate sections that run, for example, between adjacent compressor stations or pump stations, between a compressor/pump station and a storage facility, between a compressor/pump station and a block valve, or between adjacent block valves. A line section can also be designated for testing purposes. For example, a line section may be a pipeline segment designated for hydrostatic testing that runs between two mainline valves.

Liquid-mounted seal A primary seal mounted so the bottom of the seal covers the liquid surface between the tank shell and the floating roof.

Liquid trap Sumps, well cellars, and other traps used in association with oil and gas production, gathering and extraction operations.

LLS(USA) Light Louisiana Sweet.

LNG Liquefied natural gas.Supercooled natural gas that is maintained as a liquid at or below –160 °C; LNG occupies 1/600th of its original volume and is therefore easier to transport if pipelines cannot be used.

Load factor The ratio of average demand to maximum demand (peak load) during a specific period such as yearly or seasonal period.

Load sharing Sharing of load among pump units at a multiple unit station to achieve increased efficiency.

L/P Low pour.

LPG The acronym for liquefied petroleum gas, mostly being composed of normal butane, iso-butane, and propane.

LS Low sulfur.

LSFO Low sulfur fuel oil.

LSWR Low sulfur waxy residual fuel oil.

Lt Sweet Synthetic Light sweet synthetic crude.

MAOP Maximum allowable operating pressure, at which a pipeline or a section of the pipeline is allowed to operate in normal conditions.

Magnetic flux leakage Magnetic flux leakage (MFL) refers to an inline inspection technology in which a magnetic field is induced along a pipe wall through the use of a smart pig. As the smart pig travels through the pipeline, measurements are taken of the magnetic flux density at the internal surface of the wall. Changes in measured flux density indicate the presence of potential defects. A number of different MFL tool designs are used for pipeline inspection. These tools are often categorized as standard resolution or high resolution devices. High resolution MFL tools contain a greater number of sensors and thus provide more accurate measurements of potential defects and anomalies.

Mainline A mainline consists of one or more pipeline sections that directly connect an origin point to a final delivery or break out point. All mainline sections are hydraulically coupled to one another.

Mainline valves A mainline valve is a valve positioned at a location along the pipeline system that can be closed down to isolate a line section in an emergency. Mainline valves are especially important to minimize hazards and damage or pollution from an accidental release of oil or natural gas. Mainline valves may also be referred to as sectionalizing block valves for natural gas pipelines. (Reference 49CFR 192.179 [4] and 49CFR 195.260 [3])

Maintenance Maintenance is the process of maintaining property or equipment, including pipelines and tanks, to preserve it and prevent it from failure and ensure that it will continue to perform its intended function.

Manifold A collection of pipe in which tanks are connected to other equipment or a flow meter is mounted.

Manometer A U-shaped tube containing a liquid (usually mercury or water) to measure the fluid pressure. The liquid level on one side of the tube changes with respect to the level on the other side with changes in pressure.

Mass flow meter	Mass flow meter is a flow meter measuring the flow rate directly in mass.
Master	Another term used for a SCADA host.
Merchant upgrader	Processing facilities designed to accept raw bitumen on a contract basis from producers.
Metal loss	Metal loss refers to any of a number of types of anomalies in pipe in which metal has been removed from the pipe surface, usually due to corrosion or gouging.
Meter factor	Correction factor applied to a meter's reading to obtain a corrected reading, typically on a custody flow meter.
Meter run	Flow measurement unit consisting of the primary and secondary metering elements in the metering manifold.
Meter prover	A meter prover determines the meter factor of a turbine or positive displacement meter, i.e., the relationship between the number of counts or revolutions of the meter and the volume flowing through the meter. The number of counts on the meter being 'proved' is related to the volume passing the detectors on the 'prover'.
Metering stations	Measure the quantity of natural gas or liquids flowing through a pipeline.
Metric Ton	A standard measurement: averages 7.33 barrels of crude oil.
Mid-continent crude	Oil produced mainly in Kansas, Oklahoma, and North Texas.
Midgrade unleaded	Un-leaded gasoline with a 89 R+M/2 octane rating.
MMCF	**Million cubic feet.** The cubic foot is a standard unit of measure for quantities of gas at atmospheric pressure and temperature.
Modem	A device that converts computer generated data streams into analog form so that they can be transmitted over a transmission line.
Mobile tank	An AST that is moved to a different location at least once every 180 days.
MOGAS	Abbreviation for motor gasoline.
MON	Motor Octane Number.
MT	Metric ton.
MTBE	Methyl tertiary butyl ether — an octane booster and oxygenate used for gasoline blending.
Naphtha	Straight-run gasoline fractions (for feedstock for reforming and as a petrochemical feedstock).
N+A	Naphthenes and aromatics.
Naphthenic	A high naphthenes content.
Naphthenic naphtha	Usually favored as reformer feedstock.
NACE	The National Association of Corrosion Engineers (NACE) is a membership association focused on the study of corrosion and development of corrosion prevention and control standards. NACE International can be contacted at 281-228-6200 or for more information see www.nace.org.

NPMS — The National Pipeline Mapping System (NPMS) is a geographic information system (GIS) database that contains the locations and selected attributes of natural gas transmission lines, hazardous liquid trunklines, and liquefied natural gas (LNG) facilities operating in onshore and offshore territories of the United States. The NPMS is managed, operated, and maintained by the Office of Pipeline Safety (OPS) and is being developed under a joint government-industry effort involving OPS, other federal and state agencies, and the pipeline industry. OPS can be contacted at 202-366-6267 or for more information see www.npms.phmsa.dot.gov.

NTSB — The National Transportation Safety Board (NTSB) is an independent US federal agency charged by Congress with investigating significant accidents in all modes of transportation — pipeline, aviation, railroad, highway, and marine. NTSB issues safety recommendations aimed at preventing future accidents. NTSB can be contacted at 202-314-6000 or for more information see www.ntsb.gov.

Nationally recognized laboratory — An organization that can perform quantitative and qualitative tests on storage tank system equipment, evaluate the test data and equipment performance, and make determinations of the equipment's capability of meeting the technical standards.

Natural gas — Gaseous petroleum consisting of methane with some amounts of ethane, propane, butane and pentane, and heavier hydrocarbons. It can include carbon dioxide, hydrogen sulfide, and water.

Natural gas transmission pipeline — A natural gas transmission pipeline is a pipeline, other than a gathering line, that 1) transports gas from a gathering line or storage facility to a distribution center, storage facility, or large-volume customer that is not downstream from a distribution center; 2) operates at a hoop stress of 20% or more of specified minimum yield strength (SMYS); or, 3) transports gas within a storage field.

Net volume — Measured volume corrected to base pressure and temperature, used in accordance with an accepted standard such as API 11.2.2, ASTM tables or ISO 9770.

NFPA — National Fire Protection Association.

NGL — The acronym for natural gas liquid, being composed of propane, butane, pentane, hexane, heptane, and possibly trace amount of ethane. These components are produced from natural gas stream and recovered in liquids from gas processing plants.

NIST — National Institute of Science and Technology, a U.S. government organization that develops standards and applies technology and measurements.

Nomination — Nomination is a request for transportation service including the quantity of petroleum fluid that a shipper requests a pipeline company to transport for the nomination period.

Nomination allocation — A process by which capacity available in a pipeline is distributed to parties in the event that nominations are in excess of the available supply or pipeline capacity. Typically, the allocation is based on service type, contract type, and a company's tariff provisions.

Non-active corrosion Non-active corrosion is corrosion that has been discovered and treated (coated, chemically inhibited, or otherwise controlled) to stop the metal particle loss. Non-active corrosion areas are often monitored over time to verify the corrosion process is in fact halted.

NDT Pipeline non-destructive testing typically consists of: 1. Radiography (X-rays): identifies laminations and weld discontinuities. 2. Ultrasonics: locates lamination in the walls of pipe; determines wall thicknesses. 3. Magnetic particle inspection: tests for surface cracks in welds and component bodies. 4. Dye penetrant: locates surface cracks in welds or component bodies. 5. Ammonium persulfate: identifies hard spots in welds due to arc burns.

Noteworthy industry practices In pipeline integrity management, noteworthy industry practices are unique methods or procedures found to be in use by one or more pipeline operators and deemed to be notable such that their adoption by other operators could significantly enhance the implementation of pipeline integrity management requirements by those other operators.

NOR Notice of readiness.

NYMEX New York Mercantile Exchange.

Octane number A measure of the detonation quality of gasoline. The higher the octane number, the higher the resistance to engine knock.

Office of Pipeline Safety OPS is the agency within the U.S. Department of Transportation, Pipeline and Hazardous Materials Safety Administration (PHMSA), that is responsible for regulating the safety of design, construction, testing, operation, maintenance, and emergency response of U.S. oil and natural gas pipeline facilities. OPS can be contacted at 202-366-4595 or for more information see <http://phmsa.dot.gov/pipeline.

Off-peak period The period of low power demand, as opposed to on-peak demand.

Offshore pipeline An offshore pipeline is a pipeline located beyond the line of ordinary low water along that portion of the coast that is in direct contact with the open seas or beyond the line marking the seaward limit of inland waters.

Oil Condensate, crude oil, or a constituent of raw gas, condensate, or crude oil that is recovered in processing and is liquid at the conditions under which its volume is measured or estimated.

Oil pool An underground reservoir containing oil. An oil field may contain one or more pools, each of which has its own pressure system.

Oil rig A drilling rig for drilling oil and gas wells.

Oil run 1 — The production of oil during a specified period of time. 2 — a tank of oil gauged, tested, and put on a pipeline.

Oil sands A mixture of sand and other rock materials containing crude bitumen or the crude bitumen contained in those sands.

Oil Sands Deposit A natural reservoir containing or appearing to contain an accumulation of oil sands separated or appearing to be separated from any other such accumulation.

Oil Sands Heavy Upgraded heavy sour crude oil, and bitumen to which light oil fractions (i.e. diluent or upgraded crude oil) have been added in order to reduce its viscosity and density to meet pipeline specifications.

Oil shale A fine-grained, sedimentary rock containing kerogen, a partially formed oil.

Oilfield services Service companies that do work (e.g. cementing, perforating, trucking, logging, etc.) in and for the oilfield.

Olefins A group of petrochemicals characterized by their straight or branched structure (such as ethylene, propylene and butadiene).

One-Call system A one-call system is a system that allows excavators (individuals, professional contractors, and governmental organizations) to make one telephone call to provide notification of their intent to dig to underground facility operators. The one call center will then notify all underground facility operator members of the intended excavation along with the date and location of the excavation. The facility operators or, in some cases, the one-call center can then locate the facilities before the excavation begins so that extra care can be taken to avoid damaging the facilities.

On the pump A well that no longer flows from natural reservoir energy but is produced by means of a pump.

On-peak period The period of relatively high power demand.

On-site Facilities located on the same or geographically contiguous property. Piping connecting ASTs with pipeline facilities are considered on-site up to the point where it crosses through the dike wall surrounding the AST.

OPEC Organization of Petroleum Exporting Countries — an international oil cartel formed in 1960.

Operating stress Operating stress is the stress imposed on a pipe or structural member under operating conditions. This term normally refers to stress resulting from the internal forces due to the pressure of the gas or liquid in the pipeline; however, other forces such as thermal growth, expansion, or contraction may impose stress as well.

Operational life The period from the start of installation of the storage tank system to the completion of the closure of the storage tank system

O&M manual Operations and maintenance (O&M) manual is the common term for the set of comprehensive procedures that federal pipeline safety regulations require pipeline operators to develop and follow. Procedures in the O&M manual should cover pipeline operations, maintenance, and response to abnormal events and emergencies.

Operator The individual or person controlling a pipeline from a central control room using a SCADA system. Some pipeline companies also call this person a "dispatcher" or "controller." Also the company responsible for the drilling, completion, production, or pipeline transportation operations.

OQ Operator Qualification, the act or method for a pipeline operator to become qualified.

Organization costs Direct costs incurred in the creation of a new business organization.

OSHA (Administration) Occupational Safety and Health Administration.

Out-of-service A storage tank system that:
(a) is designated as an out-of-service system by owner or operator notification by the relevant regulatory authorities;
(b) is empty; and
(c) Does not have substances transferred into or withdrawn from.

Out-turn Quantity of oil unloaded from a vessel at discharge point.

Ovality Ovality is a condition in which the internal diameter of the pipe is not consistent around the entire circumference of the pipe. This can be thought of as the internal diameter of the pipe being, to varying degrees, egg-shaped. Ovality can be measured by inline inspection tools known as caliper pigs or geometry pigs.

Overfill A release or discharge that occurs when a tank is filled beyond its capacity.

Owner Any person/company as owning a facility.

Oxygenate Oxygen containing blend stocks favored for their octane and their clean burning quality. Includes MTBE and ethanol.

PADD Petroleum Administration for Defense District that defines a market area for crude oil in the U.S.

Paraffinic High in paraffin content.

Paraffinic naphtha A favored quality of naphtha for ethylene plant feedstock.

P/C Part-cargo.

Pentanes Plus A mixture mainly of pentanes and heavier hydrocarbons that ordinarily may contain some butanes and is obtained from the processing of raw gas, condensate, or crude oil.

Permafrost Soil that has remained frozen continuously for more than two years.

Performance curves Performance characteristics for a pump produced by the pump manufacturer.

Petrochemicals Chemicals derived from petroleum; feedstocks for the manufacture of plastics and synthetic rubber. Petrochemicals include benzene, toluene, xylene, styrene, and methanol.

Petroleum Oil, including crude petroleum oil, and other hydrocarbons, regardless of gravity, which are not the result of condensation of gas after it leaves the reservoir.

Petroleum gas Petroleum gas means propane, propylene, butane (normal butane or isobutanes), and butyline (including isomers) or mixtures composed predominantly of these gases, having a vapor pressure not exceeding 208 psi (1434 kPa) gage at 100 °F (38 °C).

Petroleum liquids Crude oil, condensate, and any finished or intermediate products manufactured or extracted in a petroleum refinery.

Petroleum Pipeline Facilities through which, crude oil, natural gas or other petroleum products are moved in transportation, including pipe, valves, and other equipment attached to the pipe, compressor units, metering stations, regulator stations, delivery stations, holders, and fabricated assemblies.

Petroleum Products The end result of the refining process (include fuels such as gasoline, aviation fuel, diesel and heating oil; products such as solvents and lubricants; and raw materials for manufacturing petrochemicals.

Petrochemicals Chemicals derived from petroleum; feedstock for the manufacture of plastics and synthetic rubber. Petrochemicals include benzene, toluene, xylene, styrene, and methanol.

Pig Pig is a generic term signifying any independent, self-contained device, tool, or vehicle that is inserted into and moves through the interior of a pipeline for inspecting, dimensioning, or cleaning. These tools are commonly referred to as 'pigs' because of the occasional squealing noises that can be heard as they travel through the pipe.

Pigging The act of pushing a pig through a pipeline to clean the inner surface of the pipeline, to remove debris from the pipeline, to conduct internal inspections of the pipeline, or to separate batch interfaces.

Pigging facilities Facilities for launching and receiving pigs.

Pipe or piping Any hollow cylindrical or tubular conveyance through which hydrocarbon liquids flow.

Pipe locator Device used for locating underground pipe.

Pipeline All parts of the physical facility through which, crude oil, natural gas or other petroleum derivatives are moved in transportation, including pipe, valves, and other equipment attached to the pipe, compressor units, metering stations, regulator stations, delivery stations, holders, and fabricated assemblies.

Pipeline capacity Pipeline capacity refers to the maximum flow rate that can be transported through the pipeline system in a given period of time under the conditions that prevail in the available facilities.

Pipeline components A pipeline component is some feature or part of a pipeline, such as pipe, valves, fittings, flanges, closures, tees, or cathodic protection.

Pipeline corridor A pipeline corridor is a linear area where two or more pipelines (either part of the same or different pipeline systems) are closely grouped in a single right-of-way.

Pipeline crossing A pipeline crossing is a point where two or more pipelines cross without a physical connection existing between the pipelines.

Pipeline facilities Are pipe systems, rights-of-way and any associated equipment, gathering lines, pumping facilities, buildings.

Pipeline integrity State of a pipeline that demonstrates the ability to withstand the stresses imposed during operations.

Pipeline intersection A pipeline intersection is a point where a physical connection between two pipelines occurs.

PLC Programmable Logic Controller is a field device that performs real-time data gathering, calculating, storing and controlling functions including close loop control based on current operating conditions. It can upload to and download data from the host SCADA. A distinguishing function of a PLC, compared to a flow computer or an RTU, is its ability to control valves, regulators and even pump/compressor stations.

Pressure relief valve A valve designed to open automatically to relieve pressure in a segment of a pipeline in order to keep the pressure below the maximum allowable pressure level.

Polymers Made from monomers. The most common include widely traded plastics like polyethylene, polypropylene and polystyrene.

PONA Paraffin, olefins, naphthenes and aromatics content.

Posted Price A statement of the price requested by a seller of crude oil or products. The "list price."

Pour Point Lowest temperature which oil will readily flow without disturbance when chilled.

PPM Parts per million.

Pressure relief valve A pressure relief valve is a mechanical safety device that provides protection to a pressurized container, such as a pipeline, by reducing the internal pressure by releasing it outside the container.

Pressure surge A sudden increase of pressure moving through the pipeline at acoustic speed, generated by a sudden change in flow rate or velocity.

Pressure test A test to determine the integrity of integral piping.

Pressure transient Pressure which occurs in a pipeline with limited frequency and within a limited period of time, such as surge pressures and thermal expansions, if not occurring most of the time.

Pressurized piping Piping through which hydrocarbon liquid flow due to a pump that is not located at the delivery terminal.

Primary device Primary device is the device directly or indirectly in contact with the fluid and generating a signal according to a known physical principle when applied to the fluid. For example, the primary devices for an orifice metering system include the orifice plate, meter tube, fitting and pressure taps.

Primary energy Energy (e.g., heat or motive power) that is produced first (from the combustion of coal, crude oil and natural gas; and electricity produced from nuclear, wind or hydropower), versus secondary energy (e.g., thermal electricity and motive power from gasoline) which is produced by consuming primary energy.

Processing Plant A facility designed to separate substances or make new substances through chemical reactions, procedures or physical actions.

Product Any commodity made from oil or gas and includes refined crude oil, crude tops, topped crude, processed crude petroleum, residue from crude petroleum, cracking stock, uncracked fuel oil, fuel oil, treated crude oil, residuum, gas oil, casinghead gasoline, natural gas gasoline, naphtha, distillate, condensate, gasoline, used oil, kerosene, benzene, wash oil, blended gasoline, lubricating oil, blends or mixtures of oil with one or more liquid products or byproducts derived from oil or gas, and blends or mixtures of two or more liquid products or byproducts derived from oil or gas, etc.

Products line A pipeline used for the shipment of refined products.

Products Pipeline System A pipeline with pumping stations, delivery pipeline terminals, tankage, and other apparatus used for the transportation of various liquid hydrocarbon products as opposed to crude oil.

Protective coating A protective coating is a substance or material applied to the exterior of the pipe or a pipeline component to prevent contact with the outside environment. Protective coating is generally applied to prevent corrosion. Many types of protective coatings are used on pipelines, including coal tar, tape wraps, and fusion bonded epoxy compounds.

Pump casing pressure Pressure of the fluid inside of a centrifugal pump just before the fluid leaves the pump.

Pump efficiency Ratio of the pump hydraulic power output to pump power input.

Pump stations Facilities placed along the route of a pipeline to keep oil moving along with pressure or suction.

Pygas Pyrolysis gasoline. A naphtha-range product with a high aromatics content used either for gasoline blending or as a feedstock for a BTX extraction unit. Pygas is produced in an ethylene plant that processes butane, naphtha or gasoil.

Rack pricing Selling to petroleum jobbers or other resellers FOB at the refinery, with the customer picking up transportation charges. The price of petroleum products at the refinery loading rack; cash and carry at the refinery's loading dock.

Random error A random error, also called precision error, is determined by calculating standard deviations of measured values.

Rangeability The range of linear flow rate over which the meter can retain its accuracy. The ratio of the maximum to the minimum linear flow rate is called turndown ratio.

Rated point Pump manufacturers test the performance at the rated point, where the pump head, capacity, efficiency, and power are determined.

Real time Real time is the actual time that a physical process is taking place.

Receipt A receipt occurs when fluid is moved from a shipper or feeder into the pipeline system either at the end or at an intermediate location in the pipeline. This location is the point of custody transfer into the pipeline company's system.

Receiver A pipeline component used for removing an inline inspection tool, cleaning pig, or other device from a pressurized pipeline. The device is inserted into the pipeline via a launcher.

Recirculation Recirculation of discharge liquid back to suction in order to keep the flow rate through a pump above the minimum capacity of the pump.

Reciprocating pump A pump that relies upon the action of a piston or plunger moving within a cylinder for positive displacement of a given volume of liquid on each discharge stroke.

Reformate A high-aromatics, high-octane product made in a reformer and used to blend motor gasoline or aviation gasoline.

Refined Petroleum Products End products in the refining process (e.g., gasoline).

Reg Unl Regular unleaded gasoline.

Regulator station Facilities along a pipeline where the flow rate can be adjusted. Also, known as pressure reducing station (PRS), a regulator station is a special type of valve station, where some of the pressure is relieved from the line. A regulator station is usually located at the downhill side of a peak point.

Release A discharge; or a loss of hydrocarbon liquid from a storage tank system into the system's secondary containment.

Release detection A method of determining a discharge or detecting the presence of hydrocarbon liquid within a storage tank.

Remote control valve A remote control valve is a type of valve that can be operated from a location other than at the valve site.

Repeatability The variation in measurements of an item taken by an instrument under the same conditions.

Repair To restore or replace any defective or damaged parts of a storage tank system. Replacement of a non-defective part is not a repair.

Repair conditions Repair conditions are identified pipeline conditions (defects and anomalies) that require repair. Under pipeline safety regulations for integrity management programs, characteristics are defined to reflect the severity of a condition and determine the timeframe in which repairs must be made (reference 49CFR 195.452 [3]).

Repair criteria Repair criteria are standards and requirements that govern the methods and timeframes to make pipeline repairs, based on the type, extent or configuration of identified anomalies, defects, or damages.

Resids Residual fuel oils.

Right-of-way 1-right to pass over someone else's property (plural: rights-of-way) 2- the actual property subject to a right-of-way (plural: right-of-ways).

Risk Risk is a measure of both the likelihood that an adverse event could occur and the magnitude of the expected consequences should it occur.

Risk assessment A risk assessment is a comprehensive evaluation of the pipeline system to identify hazardslikelihood and consequences of potential pipeline incidents or accidents. A pipeline risk assessment should provide a significant understanding of pipeline risks and significant risk contributors. (sometimes referred to as risk analysis or risk evaluation) and estimate the risk factors, which are contributing factors that can influence the likelihood and/or consequence of a pipeline accident.

Risk management Risk management is the process by which an organization understands and makes decisions about its risks.

Risk reduction Risk reduction is the act of making changes to controllable factors (e.g., operating pressure, patrol frequency) to reduce risk. Risk reduction can also refer to a quantitative measurement of the magnitude of a reduction in risk.

RON Research octane number.

Route The path taken by a batch as it moves through the pipeline, including the lifting and delivery locations of the batch.

R+M/2 Research Octane Number plus Motor Octane Number divided by 2. A gasoline octane designation commonly used in the US.

RSFO Regular sulfur fuel oil.

RTTM Real Time Transient Model.

RTU Remote Terminal Unit is a field device for collecting real-time data, calculating process variables, storing historical data, and performing uploading to and downloading from the host SCADA.

Rupture A rupture is the propagation or growth of a defect to such an extent that the pipe becomes completely unserviceable.

RVP Reid vapor pressure, a measure of a gasoline's volatility.

SCADA Supervisory Control and Data Acquisition.

Scan Process of obtaining and updating real time data by an RTU.

Scan rate The time required to update all real time data and derived values. The required rate depends on the fluid and required response time of the pipeline system. For example, liquid pipelines require a faster scan rate than gas pipelines because of the fast transient times in liquid pipelines.

Secondary containment A release detection and prevention system that includes dispenser liners, piping sumps, double-walled tanks and piping systems, or single-walled tanks or piping systems that are contained within a liner or an impervious containment area. A Release Prevention Barrier, as specified in API Standard 650, Appendix I, is considered secondary containment for field-erected aboveground storage tank bottoms.

Secondary device The device that responds to the signal generated by the primary device and converts it to an output signal that can be displayed. For example, the secondary devices for an orifice metering system include the electronic system, flow recorder or computer, and timer.

Segregated batch Batch that must be lifted and delivered without having other batches added to, blended with, or taken away from it.

Serious pipeline incident A serious pipeline safety incident is an event involving a fatality or injury requiring in-patient hospitalization.

Set point control The supervisory control system that sets the target value that a process controller should achieve.

Shipper A legal entity that contracts with a pipeline company to transport petroleum fluid over its pipeline system.

SG Specific gravity.

Shop-fabricated storage tank An AST that is constructed at the tank manufacturer's plant and transported to the facility for installation.

Shrinkage The reduction in volume of a blended liquid. The volume of a blended liquid is smaller than the separate volumes of individual liquids.

Shut-off The point on the pump curve where flow is zero.

Side stream injection Injecting volume of a batch at an intermediate injection location into the main pipeline.

Significant loss or gain The sum of losses and gains of a hydrocarbon fluid over a 30-day or monthly period.

Siting The process of determining the location where a facility or structure, such as a pipeline, will be constructed.

Slack flow A slack flow condition occurs when the pipeline pressure drops below the fluid vapor pressure. In practice, it can arise where a large elevation drop occurs with low back pressure at the downstream of the high elevation point.

Sleeve A sleeve is a full-encirclement of a pipeline with a reinforcing material at the location of a damage or defect to restore the strength of the pipe. Sleeves may be fabricated from steel or composite material. Sleeves may be for reinforcement only (Type A) or for reinforcement and pressure retention (Type B).

SMYS Specified minimum yield stress is the level of stress which produces 0.5% total strain (API definition). This is specified by the Company and shall be guaranteed by the Manufacturers/Suppliers/Vendors.

Sour/Sweet Crude The degree of a given crude's sulfur content. Sour refers to high sulfur and sweet to low-sulfur.

Specific gravity The ratio of the density of a liquid to the density of water at a given temperature for liquid, or the ratio of the molecular weight of a gas to the molecular weight of air at a given temperature for gas.

Specification Defined properties of a crude oil or refined petroleum product.

Spill response plan A plan required of hazardous liquid pipeline operators for responding, to the maximum extent practicable, to a worst case discharge (spill) of oil, or the substantial threat of such a discharge.

SPR (US) Strategic Petroleum Reserve.

Station valves Valves at a station such as suction, discharge, by-pass, or block valve.

Steam cracker (Ethylene plant) A petrochemical plant that produces olefins, particularly ethylene, and in some cases, aromatics.

Storage facilities Storage facilities are facilities designed and used for storage of oil or natural gas. Storage facilities vary greatly in size and design based on the product stored and the location and need within the pipeline system. Natural gas storage facilities typically operate so that large volumes of natural gas can be readily available for delivery to customers upon demand. The largest storage facilities are created in underground caverns, such as salt domes, or in porous rock formations.

Storage tank system	A tank used to contain hyrdocarbon liquid its integral piping, and all its components, including dispensing systems, spill containment devices, overfill protection devices, secondary containment systems, and any associated release detection equipment.
Straight-run	Material which has come straight from an atmospheric distillation unit has not been cracked or reformed, and which is usually used as a feedstock or as a utility fuel.
Strapping table	A table to convert liquid level in a storage tank to gross volume.
Stress	Stress is the resultant internal forces within a material that resist change in the size or shape of the material when the material is acted on by external forces.
Stress corrosion cracking (SCC)	SCC is environmentally-assisted cracking that can result when the combined action of stress, an electrochemical cracking environment, and temperature causes cracks to initiate and grow in susceptible line-pipe steel.
Strip (side stream) delivery	The delivery of one or more batches at an intermediate delivery location out of the main pipeline.
Suction pressure	Pressure of the fluid as it enters a pump station.
Supervisory control	Method in which information about a process is sent to a remote control location but the controlling action is taken by an operator.
Supplier	A petroleum fluid producer or another pipeline company that supplies fluid into the pipeline system.
Surge pressure	Surge pressure is internal pressure produced in a pipeline by a change in velocity of the product stream flowing through the pipeline. Surge pressure can result from shutting down a pump, closure of a valve, or any other blockage of the flow.
Survey	Survey refers to measurements, inspections, or observations performed to discover and identify events or conditions that indicate a departure from normal operation of the pipeline. One example is a leak survey.
SynBit	A blend of bitumen and synthetic crude oil that has similar properties to medium sour crude oil.
Synthetic crude oil	A mixture of hydrocarbons (similar to crude oil), derived by upgrading bitumen from oil sands.
Tank	An enclosed stationary container or structure that is designed or used to store hydrocarbon liquids.
Tank farm	A tank farm is a location of a group of large tanks maintained by a pipeline company and used to store crude oil and refined products. Tank farms normally store many different types of petroleum products, which are transported to and from the tank farm through pipelines.
Temporary out-of-service	A designation of a service status for a field-erected storage tank system that is emptied solely for the purpose of cleaning, routine maintenance, or change of product for a time period exceeding 30 days, but less than six months.
Terminal	A delivery point, usually the final delivery point as opposed to an intermediate terminal.

Thermal expansion Change in fluid volume due to a change in fluid temperature.

Third party damage prevention Third-party damage prevention includes all efforts and programs designed to prevent outside force damage to underground facilities (e.g., pipelines) that can occur during excavation activities. Responsibility for preventing underground facility damage is shared by all stakeholders. Advanced planning, effective use of one-call systems, accurate locating, and marking of underground facilities, and the use of safe-digging practices can all be very effective in reducing underground facility damage.

Throttling Controlling the flow rate, resulting in the reduction of the pump station discharge pressure, by partially closing a valve located downstream of a pump.

Throughput The amount of fluid transported through a pipeline from the lifting point to the delivery terminal.

Ticket A record of metered batch receipt/delivery volume according to the daily batch schedule. For single product operation, a ticket of metered volume is issued daily or at a specified interval.

Tight line receipt Fluid moved directly from a shipper or feeder into the pipeline without going through an intermediate tank.

Tight line delivery Fluid moving directly from a pipeline to a delivery facility outside the pipeline company's system, without going through an intermediate tank.

Transient The dynamic response of a pipeline system to changes in one or more variables such as flow and pressure.

Transmission or trunk line A major, long-distance pipeline, usually of large diameter.

Transmix Commingled product whose volume is from two interfacing products. It is also referred to as a batch interface or contamination batch.

Trap A trap is a pipeline component used to launch or receive inline inspection tools and pigs.

Trench A trench is a long narrow ditch dug into the ground and embanked with its own soil and used for concealment and protection of line pipe. Trenches are usually dug by a backhoe or by a specialized digging machine.

Trends Changes in system variables over time at a fixed location.

Unmanned station A station which is totally controlled by a central control center without station personnel's intervention.

Upgrading The process that converts bitumen or heavy crude oil into a product with a lower density and viscosity.

UST An underground storage tank.

Unmaintained storage A storage tank system that was not closed in regulatory requirements, an out-of-service storage tank system that is not returned to in-service status within:

- five years of its being out-of-service for ASTs; or
- ten years of its being out-of-service for storage tank systems with secondary containment.

Upgrade The addition or retrofit of cathodic protection, internal lining, spill prevention, overfill protection, or secondary containment, to a storage tank system.

Utility pig A utility pig is a pig that performs relatively simple mechanical functions, such as cleaning the pipeline.

Valve A valve is a mechanical device installed in a pipeline and used to control the flow of gas or liquid.

Vanadium Metal present in certain types of fuel.

Vapor-mounted seal A primary seal mounted so there is an annular vapor space underneath the seal. The annular vapor space is bounded by the bottom of the primary seal, the tank shell, the liquid surface, and the floating roof.

Vapor pressure The pressure at a given temperature at which vapor starts forming above a liquid.

VGO Vacuum gasoil, also known as cat feed. Feedstock for fluid catalytic cracker used to make gasoline, No.2 oil and other byproducts.

VIS or VISC Viscosity. A measure of a liquid's resistance to flow. Several viscosity scales are used depending on the type of oil being measured.

Volume correction factor for temperature Factor used to convert gross volume to net volume by taking into account the temperature difference between the operating temperature and base temperature.

Wall thickness The thickness of the pipe wall material. It is the dimensional measurement between a point on the inside surface of the pipe and the closest point on the outside surface of the pipe.

Wrinkles Anomalies that are usually caused by excessive bending or curvature being applied to the pipe. Depending on the severity of the wrinkles, they may represent an undesirable anomaly that can cause localized stress concentrations and must be repaired.

WTI West Texas Intermediate. WTI is a light sweet crude oil, produced in the United States, which is the benchmark grade of crude oil for North American price quotations.

WTS West Texas Sour.

WYO SWT Wyoming Sweet (US).

Yield strength The stress level above which the pipe will yield or bend.

APPENDIX A REFERENCES

[1] CAPP (Canadian Association of Petroleum Producers), 2011 "Crude Oil—Forecast, Markets & Pipelines" http://www.capp.ca/getdoc.aspx?DocId=190838

[2] US DOT PHMSA (U.S. Department of Transportation Pipeline and Hazardous Materials Safety Administration) 2010 " Glossary of Terms", http://208.109.252.161/residents-businesses/glossary/

[3] U.S. Department of Transportation 2012 "Code of Federal Regulation (USDOT CFR) Title 49, Part 195— Transportation of Hazardous Liquids by Pipelines"

[4] U.S. Department of Transportation 2012 "Code of Federal Regulation (USDOT CFR) Title 49, Part 192— Transportation of Natural and Other Gas By Pipelines: Minimum Federal Safety Standards"

INDEX

Page numbers followed by f and t indicate figures and tables, respectively.

www.ingramcontent.com/pod-product-compliance
Lightning Source LLC
LaVergne TN
LVHW080356011026
849161LV00009B/2320
* 9 7 8 0 7 9 1 8 6 0 0 0 7 *